Biomarkers of Neurodegenerative Disorders

Rawan Tarawneh
Editor

Biomarkers of Neurodegenerative Disorders

Current Progress and Future Directions

Springer

Editor
Rawan Tarawneh
Department of Neurology
University of New Mexico
Albuquerque, NM, USA

ISBN 978-3-031-80848-7 ISBN 978-3-031-80849-4 (eBook)
https://doi.org/10.1007/978-3-031-80849-4

This Springer imprint is published by the registered company Springer Nature Switzerland AG
The registered company address is: Gewerbestrasse 11, 6330 Cham, Switzerland

To my family for their love and unwavering support,

To my patients and their families who inspire me every day with their strength and resilience,

Rawan Tarawneh, MD

Foreword

Alzheimer disease (AD) is a global and rapidly growing health crisis, estimated to affect more than 140 million people worldwide over the next 25 years. In the absence of effective treatment or prevention strategies, AD and other related dementias (ADRD) are considered a major cause of morbidity and mortality among older adults and a significant source of caregiver distress.

Biomarker research has emerged as an important and exciting field of ADRD research. Significant advances in ADRD biomarker research over the last decade, including the advent of highly innovative technologies in biofluid-based biomarker analysis and advanced imaging techniques, have led to major breakthroughs in ADRD biomarker and drug target discovery. By providing measurable in vivo surrogates of ADRD pathology, biomarkers allow early detection, improve diagnostic accuracy, predict future cognitive outcomes, help track disease progression, and guide therapeutic interventions. Further, by providing molecular characterization of clinical trial cohorts, biomarkers are becoming increasingly important tools in participant selection and stratification in clinical trials of new therapies that target various ADRD pathologies and can provide valuable dynamic indicators of disease progression or response to potential disease-modifying treatments.

Importantly, fluid and imaging biomarkers that can reliably measure various pathological substrates of ADRD have revolutionized our understanding of the molecular underpinnings of these disorders, and helped identify the earliest molecular and pathway alterations contributing to disease onset and progression. This knowledge will ultimately pave the way towards newer and more effective therapies. By examining changes at the molecular and cellular levels, biomarkers provide valuable insights into the complex interplay of age, pathology, genetic, and environmental factors that contribute to cognitive impairment, and are an important step towards a more personalized approach to patient care. As several ADRD biomarkers can now be reliably measured in the blood, advances in cerebrospinal fluid (CSF) and imaging biomarker research now have strong potential to be more easily translated into the clinic, marking the beginning of an exciting era of ADRD research and patient care.

This book consists of 21 chapters and an epilogue, which explore the current progress and future directions in the use of blood, CSF, and imaging biomarkers in ADRD. The first section consists of chapters that describe biomarkers associated with specific ADRD pathologies, many of which are shared across different neurodegenerative disorders. The second section includes chapters that discuss biomarkers associated with specific neurodegenerative disorders. Proteomics, genomics, and multi-omics of ADRD, the role of biomarkers in clinical and translational research, and promising emerging areas in biomarker research are each discussed in dedicated sections of this book. Each chapter, authored by leading experts in the field, provides an in-depth analysis of the latest advancements in biomarker research, as well as current challenges and future directions.

As we stand on the cusp of a new era in ADRD research, this book provides a comprehensive overview of where we are now and where we are headed. It is our hope that the knowledge and insights contained within these pages will inspire further research and innovation, ultimately bringing us closer to a cure for AD and other devastating neurodegenerative diseases.

Henrik Zetterberg
Professor of Neurochemistry, Chief Physician
Institute of Neuroscience and Physiology,
University of Gothenburg, Sweden and University College,
London, UK

Rawan Tarawneh
Associate Professor of Neurology
University of New Mexico,
Albuquerque, NM, USA

Contents

About the Editor

Rawan Tarawneh, MD is Associate Professor of Neurology and Director of the Cognitive Neurology Division at the University of New Mexico. She is a board-certified neurologist with UCNS- certification in Behavioral Neurology and Neuropsychiatry. Dr. Tarawneh completed her neurology residency at Washington University in St Louis and a combined clinical and post-doctoral research fellowship at the Knight-Washington University in St Louis Alzheimer Disease Research Center. Her NIH-funded research utilizes state-of-the-art proteomic and genomic methods to identify novel biomarkers and drug targets for Alzheimer disease (AD) and related dementias (ADRD). Dr. Tarawneh's pioneering work has been fundamental in identifying visinin-like protein-1, neurogranin, and vascular endothelial-cadherin as novel markers of neuronal, synaptic, and endothelial injury, respectively, in AD. As a national leader in AD and ADRD biomarker research, Dr. Tarawneh has led several clinical trials in aging and dementia. Her highly cited research in ADRD biomarkers has been published in prestigious journals and featured in national and international conferences.

Part I
Fluid and Imaging Markers of Pathologies Associated with Alzheimer Disease or Other Related Dementias

Fluid and Imaging Markers of Amyloid Pathology

Tim West, Matthew R. Meyer, Kristopher M. Kirmess, Kevin E. Yarasheski, Joel B. Braunstein, and Philip B. Verghese

Abstract Amyloid brain pathology consists of amyloid plaques and is temporally the first brain pathology present in patients with Alzheimer's disease. The presence of amyloid plaques in the brain can only be definitively established by neuropathology performed at autopsy. However, the development of positron emission tomography ligands that specifically bind to amyloid plaques has allowed for spatial resolution of the presence of amyloid plaques in the brain with high, but not perfect, accuracy. Similarly, biomarkers present in both cerebrospinal fluid and blood can be used to identify the presence of amyloid plaques in the brain. In this chapter, we describe the current state of the art for both imaging and fluid biomarkers developed to identify the presence or absence of amyloid brain pathology.

Keywords Blood biomarker · Alzheimer's disease · Amyloid beta · Tau

1 Introduction

Alzheimer's disease (AD) brain pathology consists of the presence of amyloid plaques and tau tangles. Amyloid brain pathology is caused by the aggregation of amyloid beta peptides and other proteins into amyloid plaques in the brain. Amyloid plaques and their precise location within the brain can be visualized in postmortem brain tissue using histopathology. Radiolabeled positron emission tomography (PET) ligands allow for low-resolution spatial visualization and quantification of amyloid plaques in living tissue. As described originally by Braak et al., AD brain pathology follows a very consistent pattern of spread of amyloid and tau pathologies from one brain region to another, allowing for the staging of disease progression by observing which brain regions are affected by which proteinopathies [1].

T. West (✉) · M. R. Meyer · K. M. Kirmess · K. E. Yarasheski · J. B. Braunstein · P. B. Verghese
C2N Diagnostics, St. Louis, MO, USA
e-mail: twest@c2n.com

R. Tarawneh (ed.), *Biomarkers of Neurodegenerative Disorders*,
https://doi.org/10.1007/978-3-031-80849-4_1

While blood and cerebrospinal fluid (CSF) biomarkers do not provide spatial information about the location of amyloid pathology in different regions of the brain, they can be used to follow various stages of amyloid accumulation [2, 3]. Some fluid biomarkers show a binary change from one state to another during amyloid accumulation; for example, the Aβ42/40 ratio declines during the early stages of amyloid accumulation but does not continue to decline as amyloid continues to accumulate. Other biomarkers show a correlative progression with amyloid accumulation. Tau phosphorylation, for example, starts to increase at some phosphorylation sites once amyloid accumulation is initiated, and at other phosphorylation sites only once tau pathology starts to accumulate. Thus, quantitation of phosphorylation at various tau phosphosites can correlate with both amyloid plaque and tau tangle burden (Fig. 1).

In this chapter, we outline some of the advantages and disadvantages of imaging, blood, and CSF biomarkers of amyloid pathology as the field stands in the first quarter of 2024. Especially, the field of blood biomarkers is rapidly evolving with multiple recent exciting developments.

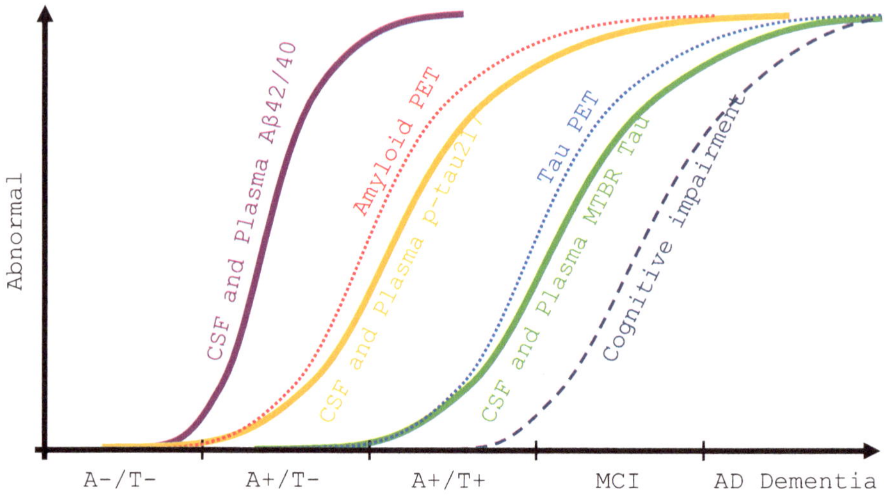

Fig. 1 Changes in fluid and imaging biomarkers as well as cognition in the various stages of AD (A and T on the x-axis refer to the presence or absence of amyloid and tau brain pathology). CSF and plasma Aβ42/40 ratios start to become abnormal right before the first signs of amyloid plaques in the brain. Once amyloid plaques start accumulating, or perhaps even slightly before, there is a steady increase in tau phosphorylation at specific amino acid residues (e.g., p-tau217) which mirrors increases in standardized uptake value ratios (SUVR) for amyloid PET tracers. Amyloid PET SUVR and p-tau217 levels continue to increase as tau tangles form and SUVR for tau PET tracers increase. By the time patients start to show signs of cognitive impairment (i.e., symptomatic AD), both amyloid and tau tangle pathology are present in the brain. Abbreviations: CSF cerebrospinal fluid, MCI mild cognitive impairment, MTBR microtubule-binding region, PET positron emission tomography

2 Amyloid Imaging

Over the past decades, the landscape of neurodegenerative disorders has been transformed by significant advancements in neuroimaging techniques allowing visualization of aggregated misfolded proteins in the living brain. Among these groundbreaking innovations, amyloid imaging stands as a pivotal tool that has revolutionized AD clinical research and diagnostics.

In the early 2000s, researchers developed Pittsburgh compound B (PiB), a first-generation amyloid imaging agent labeled with ^{11}C; a positron emitting isotope of carbon [4]. PiB binds to amyloid plaques with high specificity, enabling noninvasive PET scans to visualize the spatial topography of amyloid deposition in the brain of living humans. This groundbreaking development marked the inception of in vivo amyloid imaging, adding a crucial tool for both researchers and neurologists. However, due to the complexities of implementing ^{11}C-labeled imaging compounds (i.e., short half-life and need for a cyclotron for production), there was a need for amyloid imaging compounds that could gain wider adoption. In 2008, several ^{18}F-amyloid imaging compounds were developed, which would go on to receive approval from the US Food and Drug Administration (FDA) for use in clinical practice for the evaluation of AD as a possible cause of cognitive impairment [5–8]. However, the high cost and restricted availability of amyloid PET imaging have prevented wide adoption of amyloid imaging in routine clinical practice, and amyloid imaging is still mostly used as a research tool and for enrollment and therapeutic monitoring in clinical trials [9, 10].

With the development of amyloid imaging, it became possible to follow amyloid accumulation over time in living individuals. This has given rise to a better understanding of how fluid (CSF and blood) biomarkers correlate with brain amyloid and clinical symptoms and the insight that the first signs of amyloid accumulation can be observed as early as 10–15 years before clinical symptoms [2, 11–13]. Due to the slow and continuous process of amyloid deposition in the brain, care must be exercised when assessing if clinical symptoms attributed to a possible diagnosis of AD are due to the level of amyloid burden as measured by amyloid PET, especially if a patient's level of amyloid burden is close to the SUVR cut point for amyloid PET positivity.

2.1 Visual Interpretation of Amyloid Imaging

Visual interpretation of amyloid PET imaging is a common method used to identify the presence of amyloid plaques in the brain and is currently the only interpretation of amyloid imaging that is recognized by the FDA. While visual interpretation can provide valuable insights, there are several potential issues and challenges associated with this approach.

First, visual interpretation is inherently subjective and is influenced by the experience and biases of the interpreter or reader of the image. Different experts may have varying thresholds for what they consider as "positive" or "negative" scans, leading to inconsistencies in the classification of the images. While ensuring consistent training of image readers and adherence to standardized protocols can minimize interpretation bias, ultimately visual interpretation is subjective, and even among trained experts, there is variability in how amyloid scans are visually interpreted. This variability can impact the reliability and reproducibility of PET image interpretation and results, affecting clinical decision-making, clinical trials, and research studies.

Second, visual interpretation does not provide a quantitative measurement of amyloid burden, which can be crucial for tracking disease progression, evaluating treatment efficacy, and conducting longitudinal studies. As such, visual interpretation is also less accurate in detecting subtle levels of amyloid burden found in patients in the early stages of amyloid deposition. The lack of quantitation is especially critical for amyloid PET scans that may exhibit borderline or ambiguous findings. In these scans, it can be challenging to confidently assign a dichotomized classification of the presence or absence of amyloid pathology.

Finally, while the visual read is the FDA-accepted gold standard for determining amyloid positivity, studies comparing antemortem visual read results to postmortem neuropathology have shown that the overall agreement between the two methods for assessing plaque presence is good, but not perfect, with a sensitivity in the 88%–98% range, and specificity ranging from 80% to 95% [10, 14, 15].

2.2 *Quantitative Amyloid Imaging*

Amyloid imaging provides information about both the location and overall amount of amyloid plaques in the brain. In order to transform this three-dimensional data into a single measure that can be used in research, processing pipelines for quantitative amyloid imaging have been developed which transform the raw PET data into a quantitative output. This analysis aims to provide an accurate and objective measurement of the overall amyloid burden by quantifying the tracer uptake in regions of the brain that are known to have amyloid-specific binding of the tracer. Quantitation of amyloid burden is valuable for diagnostic purposes, monitoring disease progression, and assessing the effectiveness of potential therapies.

Standardized Uptake Value Ratios (SUVR) was the first method developed to quantify brain amyloid abundance and location and is one of the most widely used methods for amyloid image analysis [16]. It involves quantifying the tracer's retention in specific brain regions relative to a reference region known to have minimal amyloid accumulation (e.g., the cerebellum). The ratio of the standardized uptake value (SUV) in the target region to that in the reference region is calculated, providing a simple and semi-quantitative measure of amyloid burden [17]. Since the various PET tracers have different regions of nonspecific binding, the SUVR scale is tracer-specific.

In order to facilitate comparison of data between tracers, the Centiloid scale was developed [18]. The Centiloid scale relies on large datasets of PET scans acquired using different tracers and scanners to create reference curves representing the amyloid load for each tracer. Using these reference curves, SUVR values obtained from individual scans can be converted into Centiloid units, providing a standardized and comparable measure of amyloid burden across different tracers, studies, and centers [19, 20]. However, since the SUVR to Centiloid conversion relies on anchoring the SUVR measures at the two extremes (0 and 100), differences in the linearity of the SUVR scales for different tracers can lead to Centiloid values in the mid-range not being completely translatable between tracers.

While quantitative amyloid imaging is the pinnacle of in vivo amyloid imaging and generally considered to be the most accurate way to assess the result of an amyloid PET scan, this method, like all quantitative methods, has inherent variability. Multiple studies have looked at test-retest variability in both SUVR and Centiloid outcomes. These studies show that the variation in the same patient imaged twice in a short period of time (2–12 weeks) is around 4–10%, with a Pearson R correlation of around 0.9 between the two measures [21–25]. In addition, the specifics of the image processing pipeline can contribute around 2% variability when processing the same image [19]. Thus, while patients who are far away from the SUVR or Centiloid cut point chosen for a study are unlikely to change classification upon re-imaging, patients who fall within +/− 5–10% of the prespecified Centiloid cut point are at risk of having their classification being in flux due to variability in the quantitative measure.

3 Cerebrospinal Fluid Biomarkers

The use of CSF in AD diagnostics has evolved significantly over the years, playing a crucial role in improving diagnostic accuracy and understanding disease progression. Early work showed that amyloid beta (Aβ) 1-42 concentrations were lower, and that tau concentrations were elevated, in patients with AD [26–29]. Subsequent studies throughout the 2000s solidified the utility of CSF biomarkers in identifying AD brain pathology and distinguishing AD from other dementias, thus establishing their significance in diagnostic criteria and clinical guidelines [30–33]. Technological advancements in the 2010s propelled the field of CSF biomarkers forward, defining sample collection and handling procedures, refining assay techniques, and uncovering biomarkers beyond Aβ and tau proteins [34–40]. Today, while CSF sampling is common practice in some parts of the world, the lack of easy access to lumbar punctures and the risks associated with the procedure have limited the integration of CSF biomarkers into routine clinical practice in most of the world.

3.1 Amyloid Beta

The measurement of CSF Aβ concentrations has emerged as a valuable tool in the detection and diagnosis of AD. Aβ peptides, particularly Aβ42, are central to the amyloid cascade hypothesis of AD pathogenesis, wherein the aggregation and deposition of Aβ in the brain are considered early events in disease development [41]. CSF Aβ42 levels are lower in individuals with AD compared to healthy patients, due to sequestration of the Aβ42 peptide into amyloid plaques, making Aβ42 a promising core biomarker for AD pathology [42]. Numerous studies have demonstrated the utility of CSF Aβ42 measurement in distinguishing AD from healthy controls and other dementias, with lower levels showing high sensitivity and specificity for a diagnosis of AD [43–47]. Inter- and intraperson variability in concentrations of Aβ42 can be accounted for by calculation of the Aβ42/40 ratio, normalizing the concentration of the pathological Aβ42 species to the less pathologically affected Aβ40 (both Aβ species are produced by processing of the same amyloid precursor protein) [48–50]. While the difficulties of CSF collection have limited the widespread clinical use of the CSF Aβ42/40 ratio, the CSF Aβ42/40 ratio has seen extensive use in both research and clinical trials, with increases in the CSF Aβ42/40 ratio being generally accepted as associated with a positive outcome for participants who are exposed to experimental anti-amyloid therapies. The CSF Aβ42/40 ratio is a valuable biomarker for early detection, differential diagnosis, and monitoring of AD progression, with applications in clinical trials and personalized patient management strategies.

3.2 Tau Biomarkers

Studies in the 1990s demonstrated elevated levels of total tau (t-tau) and phosphorylated tau (p-tau) in the CSF of AD patients compared to healthy controls. These findings provided evidence of tau protein release into the CSF as a result of neuronal injury and neurofibrillary tangle formation, both hallmark features of late-stage AD neuropathology. Over the following decades, research efforts focused on validating the utility of CSF tau biomarkers in AD diagnosis and disease monitoring. The inclusion of CSF tau biomarkers in diagnostic criteria and clinical guidelines underscores their importance in clinical practice [30–33]. Technological advancements in assay techniques and biomarker discovery further refined the measurement and interpretation of CSF tau biomarkers, enhancing their clinical utility. Today, CSF tau biomarkers remain integral components of AD diagnostics, providing valuable insights into disease pathology, progression, and treatment response.

While increases in total tau concentrations in CSF appear to happen as part of the death of neurons with tau tangles, phosphorylation of tau at various residues can occur earlier in the AD neuropathological process, even prior to significant tau tangle formation [51–53]. Early proteomic work sought to map the various phosphorylation sites on tau that are found in patients with AD and from this work, several promising phosphorylated tau sites and peptide candidates for AD diagnosis were uncovered [54, 55]. Phosphorylation of tau at threonine-181, −217, and −231 are indicators of early-stage AD pathology (amyloid plaque pathology) in the brain [53, 56–59], whereas phosphorylation of tau at threonine-205 and -235 are indicators of late-stage AD neuropathology (tau tangles) [60–63]. This data suggests that following the progression of tau phosphorylation in CSF can be used as a way to assess the stage of the disease process.

In addition to following the phosphorylation of tau, recent work has shown that tau cleavage products that include the microtubule-binding region (MTBR) of tau can be used as indicators of AD brain pathology. Specifically, the MTBR-tau243 peptide has been quantified in CSF samples using liquid chromatography–tandem mass spectrometry (LC-MS/MS) methods and has been shown to correlate highly with tau tangle pathology [64]. This tau peptide is likely to become an important tool for use in clinical trials to aid in participant recruitment and to monitor treatment response, and in the neurology clinic for AD diagnosis and disease staging.

3.3 *FDA-Approved CSF Tests to Aid in the Diagnosis of AD*

There are three FDA-approved and cleared CSF assays used for evaluating AD as a possible cause of cognitive impairment: the Lumipulse® G Amyloid β42/40 Ratio CSF Assay, Elecsys® β-Amyloid (1-42) CSF II with Elecsys® Phospho-Tau181P CSF Assay (used to derive a pTau181/Abeta42 ratio), and Elecsys® β-Amyloid (1-42) CSF II with Elecsys® Total-Tau CSF Assay (used to derive a tTau/Abeta42 ratio). These tests aid in the evaluation of cognitive impairment in patients aged 55 years and older and are used to complement clinical assessments. However, it is important to note that a positive or negative result, alone, does not establish a diagnosis of AD or other cognitive disorders.

3.3.1 Lumipulse® G Amyloid β42/40 Ratio CSF Assay

The Lumipulse® G Amyloid β42/40 Ratio CSF Assay combines the results of amyloid Aβ42 and amyloid Aβ40 into a ratio using the LUMIPULSE G 1200 System. A ratio greater than or equal to 0.073 indicates a negative result consistent with a negative amyloid PET scan, while a ratio less than or equal to 0.058 indicates a positive result consistent with a positive amyloid PET scan. A ratio between 0.059 and 0.072 is considered likely positive.

The pivotal clinical study included 292 patients aged 55 to 90 years from the Alzheimer's Disease Neuroimaging Initiative (ADNI) sample bank [65]. The study participants were diagnosed with subjective cognitive decline (SCD, 18/292, 6.2%), early mild cognitive impairment (EMCI, 111/292, 38.0%), late mild cognitive impairment (LMCI, 59/292, 20.2%), or AD dementia (104/292, 35.6%). The study's prevalence of amyloid positivity was 68.2%. Diagnostic performance metrics for this assay are as follows: positive percent agreement (PPA) = 92.5%, negative percent agreement (NPA) = 83.9%, positive predictive value (PPV) = 92.5%, negative predictive value (NPV) = 83.9%, and overall percent agreement (OPA) = 89.7%.

3.3.2 Elecsys® β-Amyloid (1-42) CSF II and Elecsys® Phospho-Tau181P CSF Assays

These assays measure β-Amyloid (1-42) (Abeta42) and Phospho-Tau181P (pTau181) protein concentrations in CSF to generate a pTau181/Abeta42 ratio. A ratio >0.023 indicates a positive result consistent with a positive amyloid PET scan.

The pivotal clinical study included 646 patients from ADNI [66]. The analysis population was diagnosed with significant memory concerns (SMC) (14.5%), EMCI (42%), LMCI (23.5%), and AD dementia (19.8%). The study involving these assays had a prevalence of amyloid positivity of 53.7%. Diagnostic performance metrics are as follows: PPA = 88.2%, NPA = 92.6%, PPV = 93.3%, NPV = 87.1%, and OPA = 90.3%. The predicate device for this test was Lumipulse® G Amyloid β42/40 Ratio CSF test, and this ratio assay was found to be substantially equivalent to the predicate.

3.3.3 Elecsys® β-Amyloid (1-42) CSF II and Elecsys® Total-Tau CSF Assays

These assays measure β-Amyloid (1-42) (Abeta42) and Total-Tau (tTau) protein concentrations in CSF to generate a tTau/Abeta42 ratio. A ratio >0.33 indicates a positive result consistent with a positive amyloid PET scan.

The tTau/Abeta42 ratio cutoff was validated using retrospectively collected CSF samples in ADNI [67]. The analysis population was diagnosed with SMC (14.5%), EMCI (42%), LMCI (23.5%), and AD dementia (19.8%). The study involving these assays had a prevalence of amyloid positivity of 53.7%. Diagnostic performance metrics for this assay are as follows: PPA = 85.0%, NPA = 94.0%, PPV = 94.2%, NPV = 84.4%, and OPA = 89.2%.

4 Blood Biomarkers

While amyloid imaging and CSF biomarkers both are proximal to the brain and thus in theory have the highest chance of diagnostic success, recent developments have afforded accurate measurement of biomarkers in the blood and shown that these can be highly accurate for detecting the presence of brain amyloid pathology [68, 69]. Whether by active transport across the blood–brain barrier (BBB), or through the BBB becoming leaky with age, several protein biomarkers from the central nervous system can be detected in the blood. Due to blood sampling being both cheaper and less invasive than amyloid imaging and lumbar punctures, blood biomarkers for amyloid pathology are poised to revolutionize the field of AD diagnostics. Also, access to blood sampling facilities is widely available, opening up access to biomarker-informed AD diagnoses to a much larger population than those who can be served by either PET imaging or CSF biomarkers [70, 71].

Since the discovery that there is a drop in CSF Aβ42 concentrations in the presence of amyloid pathology, there has been intense development towards accurate measurement of Aβ peptides in the blood. Measuring brain-derived proteins in the blood presents a unique set of challenges when compared to CSF: (1) Blood is a much more complex biological matrix with high concentrations of proteins, lipids, and complex metabolites; and (2) the concentration of brain derived proteins is around 100-fold lower in blood compared to CSF. This combination means that there is a significant drop in signal while the noise from interfering proteins increases—leading to a 100–10,000-fold reduction in signal-to-noise ratio in blood samples. As such, initial efforts to convert CSF-based assays to blood samples were met with mixed success with some studies suggesting that Aβ could be used as a diagnostic in plasma while others showed no difference in blood Aβ by amyloid status [45, 72]. From this early work, it became clear that standardization of blood sampling workflows would be critical for the precise and accurate measurement of brain-derived proteins in blood samples [73, 74].

Another issue with blood biomarkers is the effect that comorbidities and factors such as the fed/fasting state of the patient at the time of blood sampling can have on overall protein concentrations in blood [75–77]. When measuring a single biomarker against a prespecified cut point, the concentration of that biomarker can be above or below the cut point either due to the presence of brain amyloidosis or due to interpatient differences in overall protein concentrations in blood. The Aβ42/40 ratio addresses this issue by normalizing the amount of Aβ42 to the amount of Aβ40. Thus, a patient who may have twofold higher overall Aβ concentrations (both Aβ42 and 40) in blood compared to another patient, would have their Aβ42 concentrations normalized to their Aβ40 concentration, normalizing this individual's variability in Aβ concentrations. This same phenomenon has been observed for the p-tau blood biomarkers. For these biomarkers, it has been shown that normalizing the amount of phosphorylated tau to the amount of nonphosphorylated tau or the total amount of tau in a blood sample can mitigate some of the changes in blood biomarker concentrations caused by comorbidities [78].

4.1 Amyloid Beta

With the advent of highly sensitive mass spectrometers as well as immunoassays that use amplification or single molecule detection methods, it has become possible to measure the concentration of amyloid beta peptides in blood samples with high precision and accuracy [79–84]. These studies have confirmed that the Aβ42/40 ratio in blood behaves similarly to the Aβ42/40 ratio in CSF with the ratio dropping early during amyloid accumulation in the brain [85, 86]. Also, similar to what is observed in CSF, once amyloid accumulation has begun and the Aβ ratio has decreased in the blood, there is no further decrease in the Aβ ratio as more amyloid accumulates in the brain. Thus, the Aβ ratio does not correlate with the amount of amyloid accumulated in the brain and serves more as a binary/dichotomized readout of the presence or absence of brain amyloid.

The diagnostic accuracy of the Aβ42/40 ratio in the blood is somewhat limited by the narrow dynamic range of the ratio—with patients with brain amyloidosis measuring on average 10–14% lower than patients without brain amyloidosis [86–88]. Thus, it is of extreme importance to have high accuracy and precision of the Aβ42/40 measurements in blood, and mass spectrometry-based assays have afforded this level of precision [89, 90]. Diagnostic performance of the Aβ ratio as measured by the area under the receiver operating characteristic curve (AUC-ROC) ranges between 0.6 (poor) and 0.86 (excellent) depending on the assay, the population tested, and the way that amyloid positivity is defined (CSF, qualitative or quantitative amyloid PET) [91].

In addition to being used as a single cross-sectional biomarker, the Aβ42/40 ratio can be used as a longitudinal biomarker to investigate both natural disease progression and pharmacodynamic changes in the setting of AD disease-modifying therapies [92, 93]. In the ClarityAD phase 3 clinical trial of Lecanemab, the Aβ42/40 ratio in blood samples was measured as a pharmacodynamic marker and the study calculated change from baseline in this ratio at 6, 12, and 18 months after the start of dosing [94]. In this analysis, for patients who were on the drug, there was a positive change in the Aβ42/40 ratio over time compared to patients on a placebo, and this change in the blood biomarker mirrored changes in the Centiloid observed in the treatment arm. In the long-term extension of the Lecanemab trial, the treatment benefit observed with the blood Aβ42/40 ratio, regressed during a gap in treatment faster than it did for amyloid imaging changes, showing that the blood Aβ42/40 ratio has the potential to be a dynamic marker of both target engagement and response to treatment [95, 96].

At the time of writing this section at the start of Q3 2023, there was only one commercial diagnostic test available under Clinical Laboratory Improvement Amendments (CLIA) that measures just the Aβ42/40 ratio in blood samples: AD-Detect® offered by Quest Diagnostics. Highlighting the rapid development of new blood tests for brain amyloidosis, at the time of editing this section in Q1, 2024, an additional blood test had been launched by LabCorp that also measures Aβ42/40 in blood and is based on the immunoassay that was developed by Sysmex [97]. AD-Detect uses LC-MS/MS to quantify the two Aβ peptides from a single sample and the ratio between the two can then be calculated. A presentation abstract from 2022 lists that this test had an AUC-ROC of 0.86 with a sensitivity of 89% and a specificity of 71% at a single cut point when assessed in a cohort of 209 clinically assessed individuals with PET data for comparison [98]. More recent publications outline the analytical validation of the test and both clinical validation and utility of the test [99, 100]. In this latest study of 250 patients from an Alzheimer's Disease Research Center, the AUC-ROC of the test was 0.84, and the single cut point resulted in an overall accuracy of 82% with a sensitivity of 91% and a specificity of 76%.

The test initially required a physician order, but in 2023 the test was introduced as a direct-to-consumer test, leading to pushback from key opinion leaders in the AD diagnostic field who were concerned that the consumer lacks the sophistication to interpret the result in the context of their clinical presentation [101]. This pushback led to Quest no longer marketing the test as a direct-to-consumer test in the fourth quarter of 2023 [102]. The concern from the AD physician community seems well based as the clinical utility publication for the test showed that multiple patients aged less than 25 years had ordered the test [99].

4.2 Amyloid Beta-Based Algorithms

Due to the narrow dynamic range of the Aβ42/40 ratio in blood and the significant amount of overlap of Aβ42/40 ratios between patients with and without amyloid pathology, the Aβ42/40 ratio performs best as a diagnostic tool when combined with other risk factors for AD brain pathology. The two best-established risk factors for AD are age and the apolipoprotein E (apoE) genotype. ApoE exists in three different genetic variants (E2/E3/E4), with the E4 variant being associated with a marked increase in the likelihood of developing AD [103–105]. One copy of apoE4 is associated with an approximate odds ratio for AD of 3 and having two copies of apoE4 has an odds ratio of around 12. Conversely, one copy of apoE2, which is protective against AD, is associated with an odds ratio for AD of around 0.6. With the significant differences in odds ratios associated with these risk factors, come significant differences in prevalence and pre-test probabilities of disease. Incorporating these differences in prevalence can significantly increase the overall accuracy of a blood test's ability to identify patients with brain amyloid among individuals with cognitive impairment.

Logistic regression models can be used to predict the likelihood of a dichotomous outcome based on one or more binary or continuous variables. In 2020, C_2N Diagnostics launched the PrecivityAD® CLIA blood test for amyloid brain pathology. The PrecivityAD test was rigorously validated both analytically and clinically, and C_2N's laboratory received College of American Pathologists (CAP) accreditation in January 2022 [87, 90]. The result of the PrecivityAD test is the amyloid probability score (APS), a 0–100 scaled likelihood that a patient will have a Centiloid value greater than 25, signifying the presence of brain amyloid pathology. The test measures the Aβ42/40 ratio and the presence of peptides from apoE (apoE proteotype) in blood samples using LC-MS/MS, and the PrecivityAD logistic regression algorithm takes these blood results as well as the patient's age as input. By incorporating the apoE and age risk factors, the AUC-ROC improved from 0.84 for the Aβ42/40 ratio alone to 0.88 for the APS [87]. The PrecivityAD algorithm was developed and cross-validated using blood samples from two clinical trials: PARIS, an IDEAS sub-study [106], and MissionAD, a phase 3 study testing the safety and efficacy of Elenbecestat [107]. Performance of the PrecivityAD test was subsequently confirmed in a completely independent cohort (AIBL) and in a more diverse population of patients from the Knight-Washington University Alzheimer Disease Research Center [108, 109]. The PrecivityAD test is offered as a CLIA test, and a study of the clinical utility of the test showed that the results of the test informed clinical management through both changes in the reported probability of disease and changes in AD drug therapy, demonstrating – for the first time – the clinical utility of blood biomarkers in AD diagnosis and treatment [110].

4.3 Tau and Phosphorylated Tau

Several assays exist that can accurately measure tau protein in blood samples, but unlike CSF, total tau in blood samples does not show great correlation with amyloid pathology [52, 111, 112]. This is likely due to tau in blood originating partially from organs other than the brain [52, 113]. However, the phosphorylated tau species that are specific to AD brain pathology appear to be unique to brain-derived tau, and therefore assays measuring the AD-specific forms of phosphorylated tau still have high diagnostic value when measured in blood samples [58, 114]. While the tau protein is an integral part of the tau tangle brain pathology, changes in tau biomarkers do not always happen at the same time in AD disease progression as tau tangle formation. Specifically, some tau phosphorylation sites may start getting occupied during the early stages of AD brain pathology where amyloid plaques are forming, while other tau phosphorylation sites will become sequentially phosphorylated during later stages when tau tangles are also starting to form. In this section, we will review some of the main phosphorylation sites for tau that have been measured in blood samples and correlate with amyloid plaque pathology.

4.3.1 P-tau217

Tau phosphorylation at threonine-217 (p-tau217) is one of the best corollaries of amyloid pathology that has been observed in blood samples, with the AUC-ROC in the range of 0.85–0.96 depending on the study cohort and the assay used for measuring p-tau217 in blood [52, 114, 115]. The utility of p-tau217 as a diagnostic biomarker for amyloidosis is due to the large dynamic range of p-tau217 concentrations and the fact that, in the absence of amyloid plaques, the concentration of p-tau217 is very low and sometimes undetectable. Not only does phosphorylation at threonine-217 increase as amyloid accumulates in the brain, but there is a strong correlation between p-tau217 and the amount of amyloid present in the brain as measured by quantitative amyloid PET, suggesting that tau phosphorylation on threonine-217 is a highly specific marker of amyloid accumulation [116, 117].

The correlation between p-tau217 concentrations and brain amyloid burden allows p-tau217 to be used to detect various levels of amyloid accumulation and longitudinal measures of p-tau217 have the potential to be used to follow disease progression. The concentration of p-tau217 in blood was used as a pharmacodynamic marker in Eli Lilly's Donanemab Phase 3 clinical trial [118]. This study showed that there was a decrease from baseline in blood p-tau217 in patients who received the study drug compared to patients in the placebo arm, and this decline in p-tau217 mirrored declines in brain amyloid measured by amyloid PET. This shows that the p-tau217 blood biomarker has future potential for use as a dynamic marker of changes in brain amyloid.

While immunoassays rely on measuring the concentration of p-tau217 in blood samples, mass spectrometry-based assays can measure the concentration of both the phosphorylated and nonphosphorylated tau peptides that contain the phosphorylated threonine-217. This makes it possible to calculate the ratio of phosphorylated to nonphosphorylated tau217 (commonly expressed as a percentage, %p-tau217), normalizing the phosphorylation to the total amount of epitope available for phosphorylation [119]. Similar to observations with the Aβ42/40 ratio, %p-tau217 has improved diagnostic performance over the p-tau217 concentration, since it performs better in the presence of comorbidities that may alter general tau levels by decreasing protein clearance from blood [78]. In addition to improved diagnostic performance of the %p-tau217, measurement of p-tau217 by mass spectrometry has been shown to have an improved diagnostic performance when compared to the same biomarker measured by immunoassays, showing that the best diagnostic accuracy for the p-tau217 marker is achieved when measured using mass spectrometry [115].

4.3.2 Combining p-tau217 and Amyloid Beta

Since both Aβ and p-tau217 are indicators of the presence of amyloid pathology, diagnostic accuracy can be improved by combining the two analytes. Such a strategy was applied for enrolling individuals into the AHEAD study [120]. Here, the authors developed an algorithm based on the Aβ ratio, p-tau217, apoE, and age of the individuals and used it to identify individuals qualified to continue in the screening funnel for the study. The model that combined data from multiple biomarkers as well as risk factors performed better than any of the biomarkers alone. Of interest, the authors created multiple models designed to predict various Centiloid levels, and then investigated the relative importance of Aβ and p-tau217 in these models. This analysis showed that the Aβ ratio performs best when the model is designed to identify early stages of amyloid accumulation (low Centiloid values). Conversely, p-tau217 had a significant contribution to the models designed to identify both low and high levels of amyloid accumulation. This result is consistent with the observation that the Aβ42/40 ratio has a binary change early in amyloid accumulation but does not continue to get worse as amyloid accumulates, while p-tau217 continues to increase as amyloid burden increases.

In the summer of 2023, C_2N Diagnostics launched a second generation of their CLIA blood test for amyloid brain pathology: the PrecivityAD2™ test. This test measures both the Aβ42/40 ratio and the %p-tau217 in blood samples and uses a proprietary algorithm to produce an Amyloid Probability Score 2 (APS2), a likelihood score that a patient will be amyloid positive by the criteria of a Centiloid score greater than 25 [119]. The PrecivityAD2 test uses a binary cut point for the APS2 to assign either a positive or negative category to the patient. However, as for most diagnostic tests, the numerical value of the APS2 provides information about the likelihood that a patient will be amyloid-positive, with values further from the cut point having greater diagnostic certainty. The AUC-ROC for the APS2 was 0.94 (outstanding) [91], and at the chosen cut point, the overall agreement with amyloid PET was 88% resulting from a sensitivity of 88% and a specificity of 89%. With this, the PrecivityAD2 test has an overall accuracy that is comparable with that of visual reads for detecting amyloid plaques by neuropathology [10, 119].

4.3.3 Other Tau Species

In addition to p-tau217, there are two phospho-tau species in the blood that correlate well with amyloid pathology, p-tau181 and p-tau231 [58, 117, 121]. The utility of p-tau181 in blood is limited by the relatively narrow dynamic range of p-tau181 concentrations measured in blood samples. Compared to p-tau217, which is almost undetectable in blood samples from healthy individuals, p-tau181 concentrations are relatively high in healthy individuals, suggesting that phosphorylation of tau on threonine-181 is not specific to AD brain pathology [61, 92, 121, 122]. While p-tau181 may have utility for longitudinal monitoring in clinical trials or for therapeutic monitoring, the comparatively lower accuracy (vs. p-tau217) for predicting the presence of brain amyloid pathology significantly reduces its usefulness for diagnostic applications.

Total tau in blood has shown little correlation with early stages of AD brain pathology. It is likely that total tau in plasma does not increase until the late stages of AD when neurodegeneration becomes prevalent and thus total tau in blood is not a good marker of amyloid pathology. Similar to CSF, phosphorylation of tau in blood at threonine-205 and -235 are indicators of tau tangle pathology and these biomarkers are actively under investigation to assess their specificity for tau brain pathology [60, 121]. At the Clinical Trials on Alzheimer's Disease meeting in Boston in November 2023, Dr. Horie presented results showing that the MTBR243-tau peptide from CSF can be measured in blood samples, and similar to CSF, the MTBR243-tau peptide concentration correlated with tau tangle pathology as measured by tau PET imaging [64].

4.4 Currently Available Blood Tests

Table 1 describes the currently commercially available blood biomarkers for AD as of the first quarter of 2024. Please note that this is a mix of research use-only assays and clinical diagnostic assays, but this is a rapidly evolving field, and as such this

Table 1 Blood biomarker tests for AD as of Q1 2024

Developer	Blood biomarker assay(s)	Method	Platform/instrument
Alzpath	p-tau217	USIA	Simoa
C_2N diagnostics	Aβ42/40, %p-tau217	LC-MS/MS	High-resolution MS
C_2N diagnostics	Aβ42/40 and ApoE Proteotype	LC-MS/MS	High-resolution MS
C_2N diagnostics	MTBR-tau	LC-MS/MS	High-resolution MS
Fujirebio	p-tau217	CLEIA	Lumipulse
Janssen assay	p-tau217	USIA	Simoa
Lilly	p-tau217	USIA	Simoa
Quanterix	Aβ42	USIA	Simoa
Quanterix	Aβ42, 4-plex (Aβ40,Aβ42, GFAP, NFL)	USIA	Simoa
Quanterix	p-tau181	USIA	Simoa
Quanterix	GFAP	USIA	Simoa
Quanterix	NFL	USIA	Simoa
Quanterix	p-tau231	USIA	Simoa
Quest	Aβ42/40	LC-MS/MS	TQMS
Quest	p-tau181	ECLIA	Elecsys
Roche	p-tau181	ECLIA	Elecsys
Roche	NFL	ECLIA	Elecsys
Sysmex	Aβ42/40	CLEIA	Sysmex

Abbreviations: *GFAP* glial fibrillary acidic protein, *NFL* neurofilament light chain, *LC-MS/MS* liquid chromatography–tandem mass spectrometry, *TQMS* triple quadrupole mass spectrometry, *USIA* ultrasensitive immunoassay, *CLEIA* chemiluminescence enzyme immunoassay, *ECLIA* electrochemiluminescence immunoassay, *Simoa* single-molecule arrays

table is likely to be dated and we encourage the reader to seek up-to-date information online. The inclusion of this table is meant primarily to demonstrate the diversity of types of platforms and the large number of companies developing blood biomarker assays.

5 Conclusions

It has been known for decades that amyloid pathology in the brain exists on a continuum [1, 123]. The development of imaging and biofluid biomarkers of amyloid pathology has allowed researchers and clinicians to assess the level of amyloid pathology in living individuals. This has led to immense improvements in research, clinical trials, and clinical diagnostics for AD. High-performing blood tests hold great promise for bringing highly accurate AD diagnostics to the general population. Development of blood tests that can accurately measure the core proteins indicating the presence of brain amyloid pathology (Aβ42/40 ratio and p-tau217) is a very active field with changes to the competitive landscape occurring daily.

5.1 Considerations for Implementation of Biomarkers in Clinical Practice

Implementing AD biomarkers into clinical practice requires careful consideration due to the complexity involved. Here are some key considerations:

1. Clinical Use Case: Biomarkers should be used in clinical care according to their intended use, as specified on respective test labels. For instance, no recommendations currently exist to support the use of biomarker testing in healthy individuals without symptoms or signs of memory or cognitive impairment. Instead, testing currently has a role to aid in diagnosis among individuals with cognitive concerns undergoing evaluation for a possible diagnosis of AD.
2. Accuracy and Reliability: Biomarkers must be accurate and reliable in detecting the presence of AD-related brain pathology. This includes sensitivity to early-stage AD and specificity to differentiate AD from other neurodegenerative diseases.
3. Cost-Effectiveness and Equitable Access: Biomarker tests should be cost-effective and easy to implement for widespread adoption in clinical practice. Tests that require specialized equipment (PET imaging) and access to highly skilled doctors (CSF sampling through lumbar punctures) may be less likely to be available in non-urban areas.
4. Clinical Utility: Biomarkers should provide clinically actionable information to guide diagnosis, prognosis, and treatment decisions. This requires evidence demonstrating the utility of the biomarkers in improving patient outcomes and informing personalized treatment strategies.
5. Integration with Clinical Guidelines: Biomarkers should be integrated into existing clinical guidelines for the diagnosis and management of AD. This involves collaboration between researchers, clinicians, and guideline developers to ensure alignment with current best practices.
6. Education and Training: Healthcare professionals need education and training to effectively interpret biomarker results and incorporate them into clinical decision-making. Such education should include laying the groundwork for understanding the strengths and limitations of the different biomarkers and their implications for patient care.
7. Patient Education and Counseling: Patients and their families should receive education and counseling about the purpose, significance, and limitations of biomarker testing. The potential implications of test results for diagnosis, prognosis, and treatment planning should be discussed. For tests that involve the disclosure of genetic disposition to AD, such as apoE, and tests to detect mutations that cause AD, test results should be followed up with genetic counseling.
8. Longitudinal Monitoring: While not all biomarkers are suitable for longitudinal monitoring of disease progression and treatment response, some biomarkers may become suitable-once properly validated and a clear understanding of how to interpret longitudinal changes in any particular biomarker measurement has

been developed. Identification of biomarkers that can be used to follow disease progression and treatment response is currently a very actively researched topic. Longitudinal testing requires considering factors such as test reliability over time and the feasibility of repeated measurements in clinical practice both in terms of cost and exposure of the patient to the procedure of testing.

By addressing these considerations, healthcare providers can effectively incorporate biomarkers of AD into clinical practice to improve early detection, diagnosis, and management of AD. Multiple recent publications speak to the appropriate use of PET, CSF, and blood biomarkers in clinical care [124–126]. This is an area of clinical medicine that is expected to rapidly evolve as disease-modifying agents continue to shape a new standard of care for patients with early AD.

5.2 Future Directions for Biomarkers of AD Pathology

Future directions for biomarkers of AD pathology are likely to focus on enhancing early detection and monitoring of the disease, as well as improving specificity and precision in diagnosis. Advances in technology, such as the development of novel imaging techniques and CSF and blood-based biomarkers, hold promise for accelerating the detection of AD pathology when interventions may be more effective. Additionally, there is growing interest in identifying biomarkers that reflect the specific molecular pathways implicated in AD pathogenesis, such as neuroinflammation, synaptic dysfunction, and protein aggregation. Integration of multimodal biomarker approaches, including neuroimaging, fluid biomarkers, and genetic markers, may provide a more comprehensive assessment of AD pathology and improve diagnostic accuracy.

Disease staging and prediction of future conversion using biomarkers for AD represent critical endeavors in the quest for early intervention and personalized treatment strategies. Biomarkers offer invaluable insights into the underlying pathological processes and trajectory of AD, enabling more accurate staging of disease severity and prediction of progression and future conversion. By integrating biomarkers from various modalities, including neuroimaging, blood and CSF, and genetic markers, researchers and clinicians can delineate distinct disease stages and identify individuals at heightened risk of conversion from preclinical stages to symptomatic AD. Advances in machine learning algorithms can further enhance the predictive power of biomarker-based models, enabling the identification of subtle patterns and biomarker combinations associated with future disease conversion. Ultimately, the integration of biomarker-based disease staging and prediction models into clinical practice holds promise for facilitating early diagnosis, monitoring disease progression, identifying patients that may not respond to, or who may experience an adverse effect of, an approved drug, and guiding targeted interventions to mitigate the devastating impact of AD.

Introduction of novel biofluid biomarkers that can monitor proteins and biological processes involved in other neurodegenerative diseases can lead to the development of a brain health panel of biomarkers. Such a panel can aid in the differential diagnosis of patients who may suffer from a combination of neurodegenerative disorders (e.g., Lewy Body disease, Parkinson's disease, progressive supranuclear palsy, frontotemporal dementia, corticobasal degeneration, and prion diseases). By integrating information from a brain health biomarker panel, neurologists can achieve a more accurate and comprehensive assessment of AD and other neurodegenerative pathologies, allowing an earlier and more reliable diagnosis. This holistic approach may enable neurologists to identify individuals at risk of conversion to symptomatic AD in preclinical stages and track disease progression in symptomatic individuals. Overall, a brain health biomarker panel equips neurologists with valuable tools to enhance the accuracy and effectiveness of AD diagnosis and management, ultimately improving patient outcomes.

References

1. Braak H, Braak E. Neuropathological stageing of Alzheimer-related changes. Acta Neuropathol [Internet]. 1991 [cited 2014 Jan 21];82(4):239–59. Available from: http://www.ncbi.nlm.nih.gov/pubmed/1759558
2. Jack CR, Knopman DS, Jagust WJ, Shaw LM, Aisen PS, Weiner MW, et al. Hypothetical model of dynamic biomarkers of the Alzheimer's pathological cascade. Lancet Neurol [Internet]. 2010 [cited 2014 Jul 16];9(1):119–28. Available from: http://www.pubmedcentral.nih.gov/articlerender.fcgi?artid=2819840&tool=pmcentrez&rendertype=abstract
3. Jack CR, Knopman DS, Jagust WJ, Petersen RC, Weiner MW, Aisen PS, et al. Tracking pathophysiological processes in Alzheimer's disease: an updated hypothetical model of dynamic biomarkers. Lancet Neurol. 2013;12(2):207–16.
4. Klunk WE, Engler H, Nordberg A, Wang Y, Blomqvist G, Holt DP, et al. Imaging brain amyloid in Alzheimer's disease with Pittsburgh Compound-B. Ann Neurol [Internet]. 2004;55(3):306–19. Available from: http://www.ncbi.nlm.nih.gov/pubmed/14991808
5. Heurling K, Leuzy A, Zimmer ER, Lubberink M, Nordberg A. Imaging β-amyloid using [18F]flutemetamol positron emission tomography: from dosimetry to clinical diagnosis. Eur J Nucl Med Mol Imaging. 2016;43(2):362–73.
6. Rowe CC, Ackerman U, Browne W, Mulligan R, Pike KL, O'Keefe G, et al. Imaging of amyloid beta in Alzheimer's disease with 18F-BAY94-9172, a novel PET tracer: proof of mechanism. Lancet Neurol. 2008;7(2):129–35.
7. Carpenter AP, Pontecorvo MJ, Hefti FF, Skovronsky DM. The use of the exploratory IND in the evaluation and development of 18F-PET radiopharmaceuticals for amyloid imaging in the brain: a review of one company's experience. Q J Nucl Med Mol Imaging. 2009;53(4):387–93.
8. Wong DF, Rosenberg PB, Zhou Y, Kumar A, Raymont V, Ravert HT, et al. In vivo imaging of amyloid deposition in Alzheimer disease using the radioligand 18F-AV-45 (florbetapir [corrected] F 18). J Nucl Med. 2010;51(6):913–20.
9. Gil D Rabinovici. IDEAS Press Release. 2017. Clinical impact of brain amyloid PET scans—Interim results from the IDEAS study. Available from: http://www.alz.org/aaic/releases_2017/AAIC17-Wed-briefing-Developing-Topics.asp
10. Chapleau M, Iaccarino L, Soleimani-Meigooni D, Rabinovici GD. The role of Amyloid PET in imaging neurodegenerative disorders: a review. J Nucl Med [Internet]. 2022;63(Suppl 1):13S–9S. Available from: http://www.ncbi.nlm.nih.gov/pubmed/35649652

11. Jack CR, Bennett DA, Blennow K, Carrillo MC, Feldman HH, Frisoni GB, et al. A/T/N: an unbiased descriptive classification scheme for Alzheimer disease biomarkers. Neurology. 2016;87(5):539–47.
12. Sutphen CL, Jasielec MS, Shah AR, Macy EM, Xiong C, Vlassenko AG, et al. Longitudinal cerebrospinal fluid biomarker changes in preclinical Alzheimer disease during middle age. JAMA Neurol [Internet]. 2015;72(9):1029–42. Available from:. https://doi.org/10.1038/nm.3810.
13. Jack CR, Wiste HJ, Algeciras-Schimnich A, Figdore DJ, Schwarz CG, Lowe VJ, et al. Predicting amyloid PET and tau PET stages with plasma biomarkers. Brain. 2023;146:2029.
14. Clark CM, Pontecorvo MJ, Beach TG, Bedell BJ, Coleman RE, Doraiswamy PM, et al. Cerebral PET with florbetapir compared with neuropathology at autopsy for detection of neuritic amyloid-β plaques: a prospective cohort study. Lancet Neurol. 2012;11(8):669–78.
15. Sabri O, Sabbagh MN, Seibyl J, Barthel H, Akatsu H, Ouchi Y, et al. Florbetaben PET imaging to detect amyloid beta plaques in Alzheimer's disease: phase 3 study. Alzheimers Dement. 2015;11(8):964–74.
16. Thie JA. Understanding the standardized uptake value, its methods, and implications for usage. J Nucl Med. 2004;45(9):1431–4.
17. Zhou Y, Resnick SM, Ye W, Fan H, Holt DP, Klunk WE, et al. Using a reference tissue model with spatial constraint to quantify [11C]Pittsburgh compound B PET for early diagnosis of Alzheimer's disease. NeuroImage. 2007;36(2):298–312.
18. Klunk WE, Koeppe RA, Price JC, Benzinger TL, Devous MD, Jagust WJ, et al. The Centiloid project: standardizing quantitative amyloid plaque estimation by PET. Alzheimers Dement [Internet]. 2015;11(1):1–15.e4. Available from:. https://doi.org/10.1016/j.jalz.2014.07.003.
19. Battle MR, Pillay LC, Lowe VJ, Knopman D, Kemp B, Rowe CC, et al. Centiloid scaling for quantification of brain amyloid with [18 F]flutemetamol using multiple processing methods. EJNMMI Res. 2018;8
20. Coath W, Modat M, Cardoso MJ, Markiewicz PJ, Lane CA, Parker TD, et al. Operationalizing the centiloid scale for [18F]florbetapir PET studies on PET/MRI. Alzheimers Dement (Amst) [Internet]. 2023;15(2):e12434. Available from: http://www.ncbi.nlm.nih.gov/pubmed/37201176
21. Battle MR, Pillay LC, Lowe VJ, Knopman D, Kemp B, Rowe CC, et al. Centiloid scaling for quantification of brain amyloid with [18F]flutemetamol using multiple processing methods. EJNMMI Res. 2018;8(1):107.
22. Golla SS, Verfaillie SC, Boellaard R, Adriaanse SM, Zwan MD, Schuit RC, et al. Quantification of [18F]florbetapir: a test-retest tracer kinetic modelling study. J Cereb Blood Flow Metab [Internet]. 2019;39(11):2172–80. Available from: http://www.ncbi.nlm.nih.gov/pubmed/29897009
23. Villemagne VL, Pike KE, Chételat G, Ellis KA, Mulligan RS, Bourgeat P, et al. Longitudinal assessment of Aβ and cognition in aging and Alzheimer disease. Ann Neurol [Internet]. 2011;69(1):181–92. Available from: http://www.ncbi.nlm.nih.gov/pubmed/21280088
24. Chen CD, McCullough A, Gordon B, Joseph-Mathurin N, Flores S, McKay NS, et al. Longitudinal head-to-head comparison of 11C-PiB and 18F-florbetapir PET in a Phase 2/3 clinical trial of anti-amyloid-β monoclonal antibodies in dominantly inherited Alzheimer's disease. Eur J Nucl Med Mol Imaging. 2023;50(9):2669–82.
25. Lowe SL, Duggan Evans C, Shcherbinin S, Cheng YJ, Willis BA, Gueorguieva I, et al. Donanemab (LY3002813) phase 1b study in Alzheimer's disease: rapid and sustained reduction of brain amyloid measured by florbetapir F18 imaging. J Prev Alzheimers Dis. 2021:1–11.
26. Hulstaert F, Blennow K, Ivanoiu A, Schoonderwaldt HC, Riemenschneider M, De DPP, et al. Improved discrimination of AD patients using β-amyloid (1-42) and tau levels in CSF. Neurology. 1999;52(8):1555.

27. Motter R, Vigo-Pelfrey C, Kholodenko D, Barbour R, Johnson-Wood K, Galasko D, et al. Reduction of β-amyloid peptide 42 in the cerebrospinal fluid of patients with Alzheimer's disease. Ann Neurol. 1995;38(4):643–8.
28. Nitsch RM, Rebeck GW, Deng M, Richardson UI, Tennis M, Schenk DB, et al. Cerebrospinal fluid levels of amyloid β-protein in alzheimer's disease: inverse correlation with severity of dementia and effect of apolipoprotein e genotype. Ann Neurol. 1995;37(4):512–8.
29. Tato RE, Frank A, Hernanz A. Tau protein concentrations in cerebrospinal fluid of patients with dementia of the Alzheimer type. J Neurol Neurosurg Psychiatry. 1995;59(3):280–3.
30. Hort J, O'Brien JT, Gainotti G, Pirttila T, Popescu BO, Rektorova I, et al. EFNS guidelines for the diagnosis and management of Alzheimer's disease. Eur J Neurol. 2010;17(10):1236–48.
31. Albert MS, DeKosky ST, Dickson D, Dubois B, Feldman HH, Fox NC, et al. The diagnosis of mild cognitive impairment due to Alzheimer's disease: recommendations from the National Institute on Aging-Alzheimer's Association workgroups on diagnostic guidelines for Alzheimer's disease. Alzheimers Dement. 2011;7(3):270–9.
32. Dubois B, Feldman HH, Jacova C, Hampel H, Molinuevo JL, Blennow K, et al. Advancing research diagnostic criteria for Alzheimer's disease: the IWG-2 criteria. Lancet Neurol. 2014;13(6):614–29.
33. Jack CR, Albert MS, Knopman DS, McKhann GM, Sperling RA, Carrillo MC, et al. Introduction to the recommendations from the National Institute on Aging-Alzheimer's Association workgroups on diagnostic guidelines for Alzheimer's disease. Alzheimers Dement. 2011;7(3):257–62.
34. Pannee J, Portelius E, Minthon L, Gobom J, Andreasson U, Zetterberg H, et al. Reference measurement procedure for CSF Aβ1-42 and the CSF Aβ1-42 /Aβ1-40 ratio – a cross-validation study against Amyloid PET. J Neurochem [Internet]. 2016. Available from: http://doi.wiley.com/10.1111/jnc .13838;139:651.
35. Tarawneh R, Lee JM, Ladenson JH, Morris JC, Holtzman DM. CSF VILIP-1 predicts rates of cognitive decline in early Alzheimer disease. Neurology [Internet]. 2012 [cited 2012 Mar 8];78(10):709–19. Available from: http://www.ncbi.nlm.nih.gov/pubmed/22357717
36. Van Hulle C, Jonaitis EM, Betthauser TJ, Batrla R, Wild N, Kollmorgen G, et al. An examination of a novel multipanel of CSF biomarkers in the Alzheimer's disease clinical and pathological continuum. Alzheimers Dement [Internet]. 2020. Available from: https://onlinelibrary.wiley.com/doi/10.1002/alz.12204;17:431.
37. Pannee J, Gobom J, Shaw LM, Korecka M, Chambers EE, Lame M, et al. Round robin test on quantification of amyloid-β 1-42 in cerebrospinal fluid by mass spectrometry. Alzheimers Dement [Internet]. 2015:1–5. Available from: http://linkinghub.elsevier.com/retrieve/pii/S1552526015021275
38. Fourier A, Portelius E, Zetterberg H, Blennow K, Quadrio I, Perret-Liaudet A. Pre-analytical and analytical factors influencing Alzheimer's disease cerebrospinal fluid biomarker variability. Clin Chim Acta. 2015;449:9–15.
39. Hansson O, Rutz S, Zetterberg H, Bauer E, Hähl T, Manuilova E, et al. Pre-analytical protocol for measuring Alzheimer's disease biomarkers in fresh CSF. Alzheimers Dement. 2020;12(1)
40. Kester MI, Teunissen CE, Sutphen C, Herries EM, Ladenson JH, Xiong C, et al. Cerebrospinal fluid VILIP-1 and YKL-40, candidate biomarkers to diagnose, predict and monitor Alzheimer's disease in a memory clinic cohort. Alzheimers Res Ther [Internet]. 2015;7(1):59. Available from: http://alzres.com/content/7/1/59
41. Jack CR, Bennett DA, Blennow K, Carrillo MC, Feldman HH, Frisoni GB, et al. A/T/N: an unbiased descriptive classification scheme for Alzheimer disease biomarkers. Neurol Int. 2016:1–10. Available from: http://www.ncbi.nlm.nih.gov/pubmed/27371494
42. Glenner GG, Wong CW. Alzheimer's disease: initial report of the purification and characterization of a novel cerebrovascular amyloid protein. Biochem Biophys Res Commun. 1984;120(3):885–90.
43. Fagan AM, Xiong C, Jasielec MS, Bateman RJ, Goate AM, Benzinger TLS, et al. Longitudinal change in CSF biomarkers in autosomal-dominant Alzheimer's disease. Sci

Transl Med [Internet]. 2014 [cited 2014 Mar 4];6(226):226ra30–226ra30. Available from: http://stm.sciencemag.org/cgi/doi/10.1126/scitranslmed.3007901
44. Potter WZ. Mining the secrets of the CSF: developing biomarkers of neurodegeneration. J Clin Invest [Internet]. 2012 [cited 2012 Aug 30];364(26):1–3. Available from: http://www.ncbi.nlm.nih.gov/pubmed/22922253
45. Olsson B, Lautner R, Andreasson U, Öhrfelt A, Portelius E, Bjerke M, et al. CSF and blood biomarkers for the diagnosis of Alzheimer's disease: a systematic review and meta-analysis. Lancet Neurol [Internet]. 2016;15(7):673–84. Available from: http://www.ncbi.nlm.nih.gov/pubmed/27068280
46. Hertze J, Minthon L, Zetterberg H, Vanmechelen E, Blennow K, Hansson O. Evaluation of CSF biomarkers as predictors of Alzheimer's disease: a clinical follow-up study of 4.7 years. J Alzheimers Dis [Internet]. 2010 [cited 2010 Sep 17]; Available from: http://www.ncbi.nlm.nih.gov/pubmed/20693640
47. Hansson O, Zetterberg H, Buchhave P, Londos E, Blennow K, Minthon L. Association between CSF biomarkers and incipient Alzheimer's disease in patients with mild cognitive impairment: a follow-up study. Lancet Neurol [Internet]. 2006;5(3):228–34. Available from: http://linkinghub.elsevier.com/retrieve/pii/S1474442206703556
48. Lucey BP, Bateman RJ. Amyloid-β diurnal pattern: possible role of sleep in Alzheimer's disease pathogenesis. Neurobiol Aging [Internet]. 2014;35 Suppl 2:S29–34. Available from: http://linkinghub.elsevier.com/retrieve/pii/S0197458014003509
49. Huang Y, Potter R, Sigurdson W, Kasten T, Connors R, Morris JC, et al. β-amyloid dynamics in human plasma. Arch Neurol [Internet]. 2012 [cited 2012 Sep 11];1–7. Available from: http://www.ncbi.nlm.nih.gov/pubmed/22964799.
50. Bateman RJ, Wen G, Morris JC, Holtzman DM. Fluctuations of CSF amyloid-beta levels: implications for a diagnostic and therapeutic biomarker. Neurol Int. 2007;68(9):666–9. Available from: http://www.ncbi.nlm.nih.gov/pubmed/17325273
51. Barthelemy NR, Li Y, Sato C, Horie K, McDade E. Longitudinal changes of multiple CSF phosphorylated tau sites in dominantly inherited Alzheimer disease. Alzheimers Dement. 2021;17(S5)
52. Barthélemy NR, Horie K, Sato C, Bateman RJ. Blood plasma phosphorylated-tau isoforms track CNS change in Alzheimer's disease. J Exp Med. 2020;217(11)
53. Barthélemy NR, Bateman RJ, Marin P, Becher F, Sato C. Tau hyperphosphorylation on T217 in cerebrospinal fluid is specifically associated to amyloid- β pathology. bioRxiv. 2017;065667(2).
54. Barthélemy NR, Fenaille F, Hirtz C, Sergeant N, Schraen-Maschke S, Vialaret J, et al. Tau protein quantification in human cerebrospinal fluid by targeted mass spectrometry at high sequence coverage provides insights into its primary structure heterogeneity. J Proteome Res [Internet]. 2016;acs.jproteome.5b01001. Available from: http://pubs.acs.org/doi/10.1021/acs.jproteome.5b01001
55. Bros P, Vialaret J, Barthelemy N, Delatour V, Gabelle A, Lehmann S, et al. Antibody-free quantification of seven tau peptides in human CSF using targeted mass spectrometry. Front Neurosci [Internet]. 2015;9(September):1–8. Available from: http://journal.frontiersin.org/article/10.3389/fnins.2015.00302
56. Ashton NJ, Brum WS, Di Molfetta G, Benedet AL, Arslan B, Jonaitis E, et al. Diagnostic accuracy of a plasma phosphorylated tau 217 immunoassay for Alzheimer disease pathology. JAMA Neurol. 2024;81:255.
57. Fagan AM, Shaw LM, Xiong C, Vanderstichele H, Mintun MA, Trojanowski JQ, et al. Comparison of analytical platforms for cerebrospinal fluid measures of {beta}-amyloid 1-42, total tau, and P-tau181 for identifying Alzheimer disease amyloid plaque pathology. Arch Neurol [Internet]. 2011 [cited 2011 Jun 15];1–8. Available from: http://www.ncbi.nlm.nih.gov/pubmed/21555603.

58. Ashton NJ, Pascoal TA, Karikari TK, Benedet AL, Lantero-Rodriguez J, Brinkmalm G, et al. Plasma p-tau231: a new biomarker for incipient Alzheimer's disease pathology. Acta Neuropathol. 2021;141(5):709–24.
59. Ashton NJ, Benedet AL, Pascoal TA, Karikari TK, Lantero-Rodriguez J, Brum WS, et al. Cerebrospinal fluid p-tau231 as an early indicator of emerging pathology in Alzheimer's disease. EBioMedicine. 2022;76:103836.
60. Lantero-Rodriguez J, Montoliu-Gaya L, Benedet AL, Vrillon A, Dumurgier J, Cognat E, et al. CSF p-tau205: a biomarker of tau pathology in Alzheimer's disease. Acta Neuropathol. 2024;147(1):12.
61. Barthélemy NR, Saef B, Li Y, Gordon BA, He Y, Horie K, et al. CSF tau phosphorylation occupancies at T217 and T205 represent improved biomarkers of amyloid and tau pathology in Alzheimer's disease. Nat Aging. 2023;3(4):391–401.
62. Strain JF, Barthelemy N, Horie K, Gordon BA, Kilgore C, Aschenbrenner A, et al. CSF tau phosphorylation at Thr205 is associated with loss of white matter integrity in autosomal dominant Alzheimer disease. Neurobiol Dis. 2022;168:105714.
63. Lantero-Rodriguez J, Snellman A, Benedet AL, Milà-Alomà M, Camporesi E, Montoliu-Gaya L, et al. P-tau235: a novel biomarker for staging preclinical Alzheimer's disease. EMBO Mol Med. 2021;13(12):e15098.
64. Horie K, Salvadó G, Barthélemy NR, Janelidze S, Li Y, He Y, et al. CSF MTBR-tau243 is a specific biomarker of tau tangle pathology in Alzheimer's disease. Nat Med [Internet]. 2023; Available from: https://www.nature.com/articles/s41591-023-02443-z;29:1954.
65. FDA. De Novo Number: DEN200072. [cited 2024 Jan 6]. Evaluation of automatic class Ill designation for Lumipulse G 13-Amyloid Ratio (1–42/1–40): Decision summary Available from: https://www.accessdata.fda.gov/cdrh_docs/reviews/DEN200072.pdf
66. FDA. K221842 - 510(k) Substantial equivalence determination decision summary [Internet]. 2022 [cited 2024 Apr 27]. Available from: https://www.accessdata.fda.gov/cdrh_docs/reviews/K221842.pdf
67. FDA. K231348 - 510(k) Substantial equivalence determination decision summary [Internet]. 2022 [cited 2024 Apr 27]. Available from: https://www.accessdata.fda.gov/cdrh_docs/reviews/K231348.pdf
68. Teunissen CE, Verberk IMW, Thijssen EH, Vermunt L, Hansson O, Zetterberg H, et al. Blood-based biomarkers for Alzheimer's disease: towards clinical implementation. Lancet Neurol. 2022;21(1):66–77.
69. Leuzy A, Mattsson-Carlgren N, Palmqvist S, Janelidze S, Dage JL, Hansson O. Blood-based biomarkers for Alzheimer's disease. EMBO Mol Med. 2022;14(1):e14408.
70. Hampel H, Hu Y, Cummings J, Mattke S, Iwatsubo T, Nakamura A, et al. Blood-based biomarkers for Alzheimer's disease: current state and future use in a transformed global healthcare landscape. Neuron [Internet]. 2023; Available from: https://linkinghub.elsevier.com/retrieve/pii/S0896627323003902
71. Mattke S, Jun H, Chen E, Liu Y, Becker A, Wallick C. Expected and diagnosed rates of mild cognitive impairment and dementia in the U.S. Medicare population: observational analysis. Alzheimers Res Ther. 2023;15(1):128.
72. Henriksen K, O'Bryant SE, Hampel H, Trojanowski JQ, Montine TJ, Jeromin A, et al. The future of blood-based biomarkers for Alzheimer's disease. Alzheimers Dement. 2014;10(1):115–31.
73. Verberk IMW, Misdorp EO, Koelewijn J, Ball AJ, Blennow K, Dage JL, et al. Characterization of pre-analytical sample handling effects on a panel of Alzheimer's disease–related blood-based biomarkers: results from the standardization of Alzheimer's blood biomarkers (SABB) working group. Alzheimers Dement. 2022;18(8):1484–97.
74. Snyder HM, Carrillo MC, Grodstein F, Henriksen K, Jeromin A, Lovestone S, et al. Developing novel blood-based biomarkers for Alzheimer's disease. Alzheimers Dement [Internet]. 2014 [cited 2014 Mar 25];10(1):109–14. Available from: http://www.ncbi.nlm.nih.gov/pubmed/24365657

75. O'Bryant SE, Petersen M, Hall J, Johnson LA, HABS-HD Study Team. Medical comorbidities and ethnicity impact plasma Alzheimer's disease biomarkers: Important considerations for clinical trials and practice. Alzheimers Dement [Internet]. 2022; Available from: http://www.ncbi.nlm.nih.gov/pubmed/35235702
76. Pichet Binette A, Janelidze S, Cullen N, Dage JL, Bateman RJ, Zetterberg H, et al. Confounding factors of Alzheimer's disease plasma biomarkers and their impact on clinical performance. Alzheimers Dement [Internet]. 2022; Available from: https://onlinelibrary.wiley.com/doi/10.1002/alz.12787;19:1403.
77. Mielke MM, Frank RD, Dage JL, Jeromin A, Ashton NJ, Blennow K, et al. Comparison of plasma phosphorylated tau species with Amyloid and tau positron emission tomography, neurodegeneration, vascular pathology, and cognitive outcomes. JAMA Neurol. 2021;78(9):1108.
78. Janelidze S, Barthélemy NR, He Y, Bateman RJ, Hansson O. Mitigating the associations of kidney dysfunction with blood biomarkers of Alzheimer disease by using phosphorylated tau to total tau ratios. JAMA Neurol. 2023;80(5):516.
79. Ovod V, Ramsey KN, Mawuenyega KG, Bollinger JG, Hicks T, Schneider T, et al. Amyloid β concentrations and stable isotope labeling kinetics of human plasma specific to central nervous system amyloidosis. Alzheimers Dement [Internet]. 2017. Available from:;13:841. https://doi.org/10.1016/j.jalz.2017.06.2266.
80. Kaneko N, Nakamura A, Washimi Y, Kato T, Sakurai T, Arahata Y, et al. Novel plasma biomarker surrogating cerebral amyloid deposition. Proc Jpn Acad B [Internet]. 2014;90(9):353–64. Available from: http://jlc.jst.go.jp/DN/JST.JSTAGE/pjab/90.353?lang=en&from=CrossRef&type=abstract
81. Nakamura A, Kaneko N, Villemagne VL, Kato T, Doecke J, Doré V, et al. High performance plasma amyloid-β biomarkers for Alzheimer's disease. Nature [Internet]. 2018; Available from: http://www.nature.com/doifinder/10.1038/nature25456;554:249.
82. Thijssen EH, Verberk IMW, Vanbrabant J, Koelewijn A, Heijst H, Scheltens P, et al. Highly specific and ultrasensitive plasma test detects Abeta(1–42) and Abeta(1–40) in Alzheimer's disease. Sci Rep. 2021;11(1):9736.
83. Vergallo A, Mégret L, Lista S, Cavedo E, Zetterberg H, Blennow K, et al. Plasma amyloid β 40/42 ratio predicts cerebral amyloidosis in cognitively normal individuals at risk for Alzheimer's disease. Alzheimers Dement. 2019;15(6):764–75.
84. Palmqvist S, Janelidze S, Stomrud E, Zetterberg H, Karl J, Zink K, et al. Performance of fully automated plasma assays as screening tests for Alzheimer disease–Related β-amyloid status. JAMA Neurol. 2019;76(9):1060.
85. Brand AL, Lawler PE, Bollinger JG, Li Y, Schindler SE, Li M, et al. The performance of plasma amyloid beta measurements in identifying amyloid plaques in Alzheimer's disease: a literature review. Alzheimers Res Ther. 2022;14(1):195.
86. Schindler SE, Bollinger JG, Ovod V, Mawuenyega KG, Li Y, Gordon BA, et al. High-precision plasma β-amyloid 42/40 predicts current and future brain amyloidosis. Neurol Int. 2019;93(17):e1647–59. Available from: http://www.ncbi.nlm.nih.gov/pubmed/31371569
87. Hu Y, Kirmess KM, Meyer MR, Rabinovici GD, Gatsonis C, Siegel BA, et al. Assessment of a plasma amyloid probability score to estimate amyloid positron emission tomography findings among adults with cognitive impairment. JAMA Netw Open [Internet]. 2022;5(4):e228392. Available from: https://jamanetwork.com/journals/jamanetworkopen/fullarticle/2791438
88. West T, Kirmess KM, Meyer MR, Holubasch MS, Knapik SS, Hu Y, et al. A blood-based diagnostic test incorporating plasma Aβ42/40 ratio, ApoE proteotype, and age accurately identifies brain amyloid status: findings from a multi cohort validity analysis. Mol Neurodegener. 2021;16(1):30.
89. Janelidze S, Teunissen CE, Zetterberg H, Allué JA, Sarasa L, Eichenlaub U, et al. Head-to-head comparison of 8 plasma amyloid-β 42/40 assays in Alzheimer disease. JAMA Neurol. 2021;78(11):1375.

90. Kirmess KM, Meyer MR, Holubasch MS, Knapik SS, Hu Y, Jackson EN, et al. The precivityAD™ test: accurate and reliable LC-MS/MS assays for quantifying plasma amyloid beta 40 and 42 and apolipoprotein E proteotype for the assessment of brain amyloidosis. Clin Chim Acta. 2021;519:267–75.
91. Mandrekar JN. Receiver operating characteristic curve in diagnostic test assessment. J Thorac Oncol. 2010;5(9):1315–6.
92. Bilgel M, An Y, Walker KA, Moghekar AR, Ashton NJ, Kac PR, et al. Longitudinal changes in Alzheimer's-related plasma biomarkers and brain amyloid. Alzheimers Dement. 2023;19:4335.
93. Yakoub Y, Ashton NJ, Strikwerda-Brown C, Montoliu-Gaya L, Karikari TK, Kac PR, et al. Longitudinal blood biomarker trajectories in preclinical Alzheimer's disease. Alzheimers Dement. 2023;19:5620.
94. van Dyck CH, Swanson CJ, Aisen P, Bateman RJ, Chen C, Gee M, et al. Lecanemab in early Alzheimer's disease. N Engl J Med [Internet]. 2022; Available from: http://www.ncbi.nlm.nih.gov/pubmed/36449413
95. McDade E, Cummings JL, Dhadda S, Swanson CJ, Reyderman L, Kanekiyo M, et al. Lecanemab in patients with early Alzheimer's disease: detailed results on biomarker, cognitive, and clinical effects from the randomized and open-label extension of the phase 2 proof-of-concept study. Alzheimers Res Ther. 2022;14(1):191.
96. Hayato S, Takenaka O, Sreerama Reddy SH, Landry I, Reyderman L, Koyama A, et al. Population pharmacokinetic-pharmacodynamic analyses of amyloid positron emission tomography and plasma biomarkers for lecanemab in subjects with early Alzheimer's disease. CPT Pharmacometrics Syst Pharmacol. 2022;11:1578.
97. Yamashita K, Miura M, Watanabe S, Ishiki K, Arimatsu Y, Kawahira J, et al. Fully automated and highly specific plasma β-amyloid immunoassays predict β-amyloid status defined by amyloid positron emission tomography with high accuracy. Alzheimers Res Ther. 2022;14(1):86.
98. Weber DM, Kim JC, Goldman S, Racke MK, Clarke NJ. A new LC-MS/MS assay for the quantification of Aβ40 and Aβ42 in plasma: validation and clinical performance. Alzheimers Dement. 2022;18(S6)
99. Weber DM, Taylor SW, Lagier RJ, Kim JC, Goldman SM, Clarke NJ, et al. Clinical utility of plasma Aβ42/40 ratio by LC-MS/MS in Alzheimer's disease assessment. Front Neurol. 2024;15:1364658.
100. Weber DM, Kim JC, Goldman SM, Clarke NJ, Racke MK. EXPRESS: new plasma LC-MS/MS assays for the quantitation of beta-amyloid peptides and identification of apolipoprotein E proteoforms for Alzheimer's disease risk assessment. J Investig Med. 2024:10815589241246536.
101. Rubin R. Consumers can now buy a blood test to evaluate their Alzheimer Disease risk, but should they? JAMA. 2023;330(13):1214.
102. George J. Quest Alzheimer's blood test no longer marketed directly to patients. 2024 [cited 2024 Apr 27]; Available from: https://www.medpagetoday.com/neurology/alzheimersdisease/109142
103. Therriault J, Pascoal TA, Benedet AL, Tissot C, Savard M, Chamoun M, et al. Frequency of biologically defined Alzheimer disease in relation to age, sex, APOE ϵ4, and cognitive impairment. Neurology. 2021;96(7):E975–85.
104. Farrer LA, Cupples LA, Haines JL, Hyman B, Kukull WA, Mayeux R, et al. Effects of age, sex, and ethnicity on the association between apolipoprotein E genotype and Alzheimer disease: a meta-analysis. JAMA J Am Med Assoc. 1997;278(16):1349–56.
105. Mahley RW, Weisgraber KH, Huang Y. Apolipoprotein E4: a causative factor and therapeutic target in neuropathology, including Alzheimer's disease. Proc Natl Acad Sci U S A [Internet]. 2006[cited 2011 Jun 15];103(15):5644–51. Available from: http://www.pubmedcentral.nih.gov/articlerender.fcgi?artid=1414631&tool=pmcentrez&rendertype=abstract

106. Rabinovici GD, Gatsonis C, Apgar C, Chaudhary K, Gareen I, Hanna L, et al. Association of amyloid positron emission tomography with subsequent change in clinical management among medicare beneficiaries with mild cognitive impairment or dementia. JAMA [Internet]. 2019;321(13):1286–94. Available from: http://jama.jamanetwork.com/article.aspx?doi=10.1001/jama.2019.2000
107. Roberts C, Kaplow J, Giroux M. Amyloid and APOE status of screened subjects in the elenbecestat missionAD phase 3 program. J Prev Alzheimers Dis. 2021;8:1–6.
108. Fogelman I, West T, Braunstein JB, Verghese PB, Kirmess KM, Meyer MR, et al. Independent study demonstrates amyloid probability score accurately indicates amyloid pathology. Ann Clin Transl Neurol. 2023;10(5):765–78.
109. Schindler SE, Karikari TK, Ashton NJ, Henson RL, Yarasheski KE, West T, et al. Effect of race on prediction of brain amyloidosis by plasma Aβ42/Aβ40, phosphorylated tau, and neurofilament light. Neurology. 2022;99(3):e245–57.
110. Monane M, Johnson KG, Snider BJ, Turner RS, Drake JD, Maraganore DM, et al. A blood biomarker test for brain amyloid impacts the clinical evaluation of cognitive impairment. Ann Clin Transl Neurol. 2023;10:1738.
111. Müller S, Preische O, Göpfert JC, Yañez VAC, Joos TO, Boecker H, et al. Tau plasma levels in subjective cognitive decline: results from the DELCODE study. Sci Rep. 2017;7(1):9529.
112. Zetterberg H, Wilson D, Andreasson U, Minthon L, Blennow K, Randall J, et al. Plasma tau levels in Alzheimer's disease. Alzheimers Res Ther. 2013;5(2):9.
113. Dugger BN, Whiteside CM, Maarouf CL, Walker DG, Beach TG, Sue LI, et al. The presence of select tau species in human peripheral tissues and their relation to Alzheimer's disease. J Alzheimers Dis. 2016;51(2):345–56.
114. Palmqvist S, Janelidze S, Quiroz YT, Zetterberg H, Lopera F, Stomrud E, et al. Discriminative accuracy of plasma phospho-tau217 for Alzheimer disease vs other neurodegenerative disorders. JAMA [Internet]. 2020 [cited 2023 Jan 11];324(8):772. Available from: https://jamanetwork.com/journals/jama/fullarticle/2768841
115. Janelidze S, Bali D, Ashton NJ, Barthélemy NR, Vanbrabant J, Stoops E, et al. Head-to-head comparison of 10 plasma phospho-tau assays in prodromal Alzheimer's disease. Brain. 2023;146(4):1592–601.
116. Janelidze S, Christian BT, Price J, Laymon C, Schupf N, Klunk WE, et al. Detection of brain tau pathology in down syndrome using plasma biomarkers. JAMA Neurol [Internet]. 2022; Available from: https://jamanetwork.com/journals/jamaneurology/fullarticle/2793702;79:797.
117. Ashton NJ, Janelidze S, Mattsson-Carlgren N, Binette AP, Strandberg O, Brum WS, et al. Differential roles of Aβ42/40, p-tau231 and p-tau217 for Alzheimer's trial selection and disease monitoring. Nat Med. 2022;28:2555.
118. Sims JR, Zimmer JA, Evans CD, Lu M, Ardayfio P, Sparks J, et al. Donanemab in early symptomatic Alzheimer disease. JAMA. 2023;330(6):512.
119. Meyer MR, Kirmess KM, Eastwood S, Wente-Roth TL, Irvin F, Holubasch MS, et al. Clinical validation of the PrecivityAD2 blood test: a mass spectrometry-based test with algorithm combining %p-tau217 and Aβ42/40 ratio to identify presence of brain amyloid. Alzheimers Dement [Internet]. 2024; Available from: http://www.ncbi.nlm.nih.gov/pubmed/38491912
120. Rissman RA, Langford O, Raman R, Donohue MC, Abdel-Latif S, Meyer MR, et al. Plasma Aβ42/Aβ40 and phospho-tau217 concentration ratios increase the accuracy of amyloid PET classification in preclinical Alzheimer's disease. Alzheimers Dement. 2023;20:1214.
121. Montoliu-Gaya L, Benedet AL, Tissot C, Vrillon A, Ashton NJ, Brum WS, et al. Mass spectrometric simultaneous quantification of tau species in plasma shows differential associations with amyloid and tau pathologies. Nat Aging. 2023;3:661.
122. Chatterjee P, Pedrini S, Doecke JD, Thota R, Villemagne VL, Doré V, et al. Plasma Aβ42/40 ratio, p-tau181, GFAP, and NfL across the Alzheimer's disease continuum: a cross-sectional and longitudinal study in the AIBL cohort. Alzheimers Dement. 2023;19(4):1117–34.

123. Therriault J, Pascoal TA, Lussier FZ, Tissot C, Chamoun M, Bezgin G, et al. Biomarker modeling of Alzheimer's disease using PET-based Braak staging. Nat Aging. 2022;2(6):526–35.
124. Hansson O, Edelmayer RM, Boxer AL, Carrillo MC, Mielke MM, Rabinovici GD, et al. The Alzheimer's association appropriate use recommendations for blood biomarkers in Alzheimer's disease. Alzheimers Dement. 2022;18(12):2669–86.
125. Dubois B, von Arnim CAF, Burnie N, Bozeat S, Cummings J. Biomarkers in Alzheimer's disease: role in early and differential diagnosis and recognition of atypical variants. Alzheimers Res Ther. 2023;15(1):175.
126. Angioni D, Hansson O, Bateman RJ, Rabe C, Toloue M, Braunstein JB, et al. Can we use blood biomarkers as entry criteria and for monitoring drug treatment effects in clinical trials? A report from the EU/US CTAD task force. J Prev Alzheimers Dis. 2023;10(3):418–25.

Fluid Markers of Tau Pathology

Naazneen Khan and Jeffrey L. Dage

Abstract Alzheimer's disease (AD) is a neurodegenerative disorder characterized by the progressive deterioration of cognitive functions, primarily memory, with evidence of the accumulation of amyloid-beta (Aβ) plaques and tau neurofibrillary tangles (NFTs) in the brain. Related dementias include conditions such as Lewy body dementia (LBD), frontotemporal dementia (FTD), corticobasal syndrome (CBS), and vascular dementia, which exhibit similar cognitive impairments but are associated with different underlying pathologies at autopsy. The diagnosis of AD and related dementias (ADRD) involves comprehensive evaluations that include patient history, physical examination, neurological assessments, and cognitive testing. In recent years, commercial diagnostics using cerebrospinal fluid (CSF) and positron emission tomography (PET) scans have been increasingly used to detect the presence of Aβ plaques and tau proteins, providing a more accurate diagnosis. However, a definitive diagnosis is still largely dependent on clinical history followed by postmortem assessments. Research studies from the Dage lab at Eli Lilly were the first to demonstrate robust and reliable measurements, as well as clinical utility, of measuring phosphorylated tau (P-tau) forms in plasma. Now, several P-tau immunoassay tests are rapidly advancing into clinical trials and practice. Biomarkers of tau pathology accessible in living subjects are of great importance to understand the development and role of tau pathology in the onset and progression of neurodegenerative diseases. In addition, ongoing treatment trials targeting tau pathology now call for reliable and quantifiable tau biomarkers for the early identification of eligible study participants and the in vivo quantification of potential treatment effects. In this chapter, we summarize the current state of biofluid biomarkers of tau

N. Khan
Department of Neurology, Indiana University School of Medicine, Indianapolis, IN, USA

J. L. Dage (✉)
Department of Neurology, Indiana University School of Medicine, Indianapolis, IN, USA

Department of Medical and Molecular Genetics, Indiana University School of Medicine, Indianapolis, IN, USA
e-mail: jdage@iu.edu

R. Tarawneh (ed.), *Biomarkers of Neurodegenerative Disorders*,
https://doi.org/10.1007/978-3-031-80849-4_2

pathology and explain the association of plasma P-tau217 levels with Alzheimer's disease neuropathologic change (ADNC).

Keywords CSF · Plasma · Tau · Phosphorylated tau (P-tau) · P-tau217

1 Introduction

After Alois Alzheimer's description of neurofibrillary tangles (NFTs) in 1906, it took another 80 years until hyperphosphorylated tau was identified as a component of AD-related paired helical filaments (PHF), the main components of NFTs [1, 2]. Another decade later, in vivo assessment of tau became possible through the development of assays for the detection of total tau (T-tau) and phosphorylated tau (P-tau) in the cerebrospinal fluid (CSF) [3].

Since 2004, positron emission tomography (PET) ligands binding to tau with high affinity have been introduced, enabling the visualization, mapping, and quantification of tau in the living brain [4]. Most recently, T-tau (year 2013) and P-tau (year 2018) were reliably measured in human plasma demonstrating an association with AD, and – for the first time – the association of P-tau with tau pathology as measured by tau PET [5, 6]. Biomarkers of tau pathology accessible in living subjects are of great importance to understand the development and role of tau pathology in the onset and progression of diseases. In addition, ongoing treatment trials targeting tau pathology call for reliable and quantifiable tau biomarkers for the early identification of eligible study participants and the in vivo quantification of potential treatment effects [7].

Although the neuropathologic changes in the brain, i.e., accumulation of amyloid-β (Aβ) plaques and NFTs containing hyperphosphorylated tau (P-tau), can be detected in living people using PET scanning, there is a need for affordable and accessible options to track neuropathologic features over time. Fluid biomarkers involving quantification of Aβ peptides and P-tau protein isoform levels in CSF and, recently plasma, make this possible (for a review with more details, see Hansson et al., 2023) [8]. However, tau in CSF and plasma does not exist as an intact single entity but is a diverse population of fragments and post-translationally modified forms of tau [9]. This suggests that what is measured is dependent highly on the epitopes of the reagents or assay design and has led to the development of many fluid-based tau assays with varied performance as biomarkers. However, the existence of these multiple forms has led to opportunities to discover forms that best report on tau pathology and develop high-quality and robust assays for CSF, plasma, serum, or other fluids.

Biofluid biomarkers of tau pathology have been developed over recent decades and their understanding guided by classical methods of postmortem tissue characterization as well as novel PET imaging tools. How and why these biofluid tests can inform on risk, diagnosis, or prognosis of Alzheimer's disease (AD) and related dementias (ADRD) are aided by the biomarker staging model first proposed by Cliff Jack Jr. of the Mayo Clinic [10]. There have been several unique discoveries made possible by these biofluid biomarker tests that also have extended our understanding of AD through the careful application of these tests in health, disease, and treatment trials. Through all these advancements, specific contexts of use of the plasma biomarkers of tau pathology are forming and guiding widespread implementation in clinical practice.

2 The Complexities Associated with Measuring Tau

The microtubule-associated protein tau gene (*MAPT*, OMIM: 157140) is located on chromosome 17q21 and encodes six tau isoforms that range in size from 352 to 441 amino acids. This gene is highly conserved, with 97–100% homology among primates [11]. Tau is expressed in the brain and peripheral tissues (Fig. 1). The isoforms result from complex, regulated alternate splicing differing in the number of amino terminal inserts (0/1/2 N) and the number of microtubule-binding domain repeats, either three (aka 3-repeat or 3R tau) or four (aka 4-repeat or 4R tau) [12]. In the normal adult brain, both 3R and 4R tau isoforms are present but their relative expression changes across the lifespan. There are also a high number of mutations in the *MAPT* gene that disrupt exon 10 splicing thereby altering the 3R/4R ratio. For

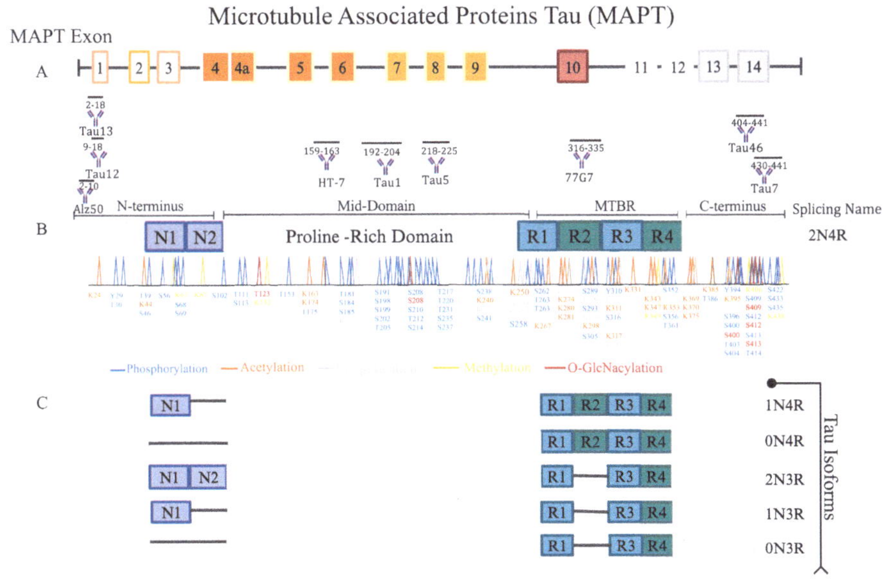

Fig. 1 Tau gene, isoforms, and post-translational modifications. (**a**) Schematic representation of the *MAPT* gene. (**b**) The longest form of tau encompasses an N-terminal domain, proline-rich domain, and microtubule-binding domain. The post-translational modifications are labeled with different colors and epitopes for major tau antibodies. (**c**) Tau isoforms based on the presence and absence of N terminal and R regions

example, the mutations altering *MAPT* splicing via the inclusion of exon 10 resulted in 3R < 4R tau isoforms (e.g., frontotemporal dementia with parkinsonism linked to chromosome 17 [FTDP-17]), whereas the mutations associated with exclusion of exon 10 resulted in 3R > 4R tau isoforms (e.g., Pick's disease [PiD]) [13]. In terms of tissue-specific splicing, *MAPT* expression and splicing are differentially regulated within the central nervous system and peripheral tissues. There is also significant regional variation in mRNA expression and splicing of *MAPT* within the human brain. The regional distribution of mRNA expression and total tau protein expression levels seem highly correlated. The genetic risk factors for neurodegenerative diseases at the *MAPT* locus are likely to operate by changing mRNA splicing in different brain regions, as opposed to the overall expression of the *MAPT* gene. However, two haplotypes, H1 and H2, arise from an inversion in the *MAPT* location that generates a ~1.3 megabase region of linkage disequilibrium (LD). The association of the *MAPT* haplotype H2 with progressive supranuclear palsy (PSP) was observed in many recent publications and PSP genome-wide association studies (GWAS) [14, 15].

There has been evidence that other post-translational modifications (PTMs), such as tau acetylation and ubiquitination, contribute directly or indirectly to neurodegeneration. There are several implications of tau-mediated tauopathies resulting from the crosstalk between tau acetylation and ubiquitination or phosphorylation [16]. Other PTMs such as methylation and glycosylation are also involved in tau aggregation and oligomerization [17, 18]. Under normal conditions, PTMs regulate the interaction of tau with microtubules and its role in axonal transport or localization. However, in ADRD, abnormal hyperphosphorylation and other modifications of tau are associated with its aggregation into NFTs, a hallmark of tauopathies. Thus, the intricate interplay of alternative splicing and PTMs contributes to the diverse roles of tau in health and disease. The research area of tau PTMs is certainly likely to continue to grow over the coming years.

Creating targeted assays for the tau forms found in biofluids that are also closely associated with specific pathological variants enhances individual biomarker utility. This strategy has led to the development of P-tau-specific isoform assays in both CSF and blood. In both biofluids, threonine (T)181, T217, and T231 have been explored and found to correlate between the two fluids and have a similar rank ordering in terms of performance in each context of use. However, their performance describing the pathological forms of tau is limited by the qualitative nature of neuropathology.

3 Definitive Neuropathological Diagnosis of AD

The neuropathology of different tauopathies exhibits varying isoform compositions, distinct neuroanatomical distributions, and relative amounts of different intracellular tau inclusions. Tauopathies include: variants of frontotemporal lobar degeneration (FTLD) such as FTLD-tau or autosomal-dominant FTLD-17; PiD, characterized by so-called Pick bodies, round inclusions formed almost exclusively by 3R tau;

PSP, characterized by predominant 4R tau inclusions; corticobasal degeneration (CBD) with 4R-dominant tau pathology grossly overlapping with PSP; and AD, which accounts for the vast majority of tauopathy cases and is characterized by mixed 3R/4R tau pathology accumulating in distinct spatial and temporal stages (Braak stages). In early studies, aggregated tau isolated from AD brains had threefold to fourfold higher phosphorylation levels which were detected with specific phosphorylated tau antibodies such as the monoclonal antibody AT8 [19].

The 2018 National Institute on Aging-Alzheimer's Association guidelines for the neuropathologic assessment of AD provide an approach to assess the likelihood of AD neuropathologic change (ADNC) [10]. The guidelines use an "ABC" score for ADNC that incorporates histopathologic assessments of amyloid β deposits (A), staging of NFTs (B), and scoring of neuritic plaques (C) [20]. After assessing each category, ADNC is indicated with not, low, intermediate, or high likelihood. An intermediate or high likelihood score is sufficient to explain clinical symptoms to be associated with AD pathology. This equates to the presence of at least a moderate degree of Aβ plaques or neuritic plaques containing Aβ in a low-power microscopic section of some region of the neocortex and the extent of the regional distribution of NFT pathology corresponding to Braak and Braak stage III or higher. These changes may occur even in the apparent absence of cognitive impairment. The guidelines also provide more detailed approaches for assessing commonly comorbid conditions such as Lewy body disease (LBD), vascular brain injury, hippocampal sclerosis, and TAR DNA-binding protein (TDP)-43 immunoreactive inclusions. These conditions often coexist with AD and can also contribute to cognitive impairment. Ultimately, the definitive neuropathological diagnosis of AD is usually present with other comorbid pathologies that are all qualitative and difficult to assign to be the cause of the clinical syndrome during life. Having biofluid biomarkers to report on the development and progression of specific pathological forms of tau during life will help improve our understanding of the development and progression of ADRD.

Before the development of PET or CSF biomarkers for AD pathologies, the diagnosis of AD was largely based on clinical symptoms, cognitive assessments, and exclusion of other possible causes for the observed symptoms using the Diagnostic and Statistical Manual of Mental Disorders (DSM)-IV or the National Institute of Neurological and Communicative Disorders and Stroke- The Alzheimer's Disease and Related Disorders Association (NINCDS-ADRDA) AD Criteria. Updates to these criteria were proposed by workgroups commissioned by the Alzheimer's Association and the National Institute on Aging in 2011 with the proposed integration of biomarkers [21]. The clinical diagnosis is not definitive until postmortem examination of the brain tissue reveals the presence of ADNC. In fact, there is a high (15–20%) error rate in clinical diagnosis without biomarker evidence. When cases go to autopsy, over half of identified AD cases will have mixed pathologies making the etiology of dementia difficult to predict with great certainty. The high diagnostic error rate makes the identification of a diagnostic biomarker very challenging.

The "ABC" score for ADNC incorporates histopathologic assessments of amyloid β deposits (A), staging of NFTs (B), and scoring of neuritic plaques (C) [22]. It is important to note that these pathological changes are often associated with other age-related processes such as cerebrovascular disease and LBD. Additionally, more recently described pathologies including limbic-predominant age-related TDP-43 encephalopathy (LATE), chronic traumatic encephalopathy (CTE), and aging-related tau astrogliopathy (ARTAG) may also be present. Although neuropathologic assessments are considered the "gold standard" for a definitive diagnosis, they are generally qualitative. As quantitative neuropathology techniques are developed and used more broadly, they will have a dramatic effect on our understanding of biofluid biomarkers.

The biofluid biomarkers of tau pathology have undergone many studies with postmortem outcomes, and despite the qualitative nature of the pathology data, there appears to be a consistent association across Braak stages. Unfortunately, more advanced analysis has not been regularly possible; however, as digital pathology techniques expand and resolution of specific tau pathologies at the cellular level increases, opportunities to discover and develop more tools for specific tau pathologies will expand dramatically.

4 Measuring Insoluble Tau Pathology in Living Patients

With the advent of new technologies such as tau PET imaging, we can now visualize insoluble tau in the brain of living patients. Longitudinal studies have suggested that spreading to connected brain regions happens and a tau transmission hypothesis has led to several novel approaches to block the spread. However, it is unclear whether the tau PET tracers bind all the forms of tau pathology or if there are distinct forms that are characteristic of specific tauopathies. The first tau PET tracer to gain US Food and Drug Administration (FDA) approval was F18-flortaucipir which is a first-generation tau PET tracer. However, that was limited to determining the density and distribution of aggregated tau NFTs during the workup for an AD diagnosis. F18-Flortaucipir has not shown sufficient specific binding to be generally useful for other primary tauopathies such as FTLD-tau, PSP and CBD [23]. However, imaging can be used to detect the presence of tau NFTs in the AD brain with a high degree of precision, but low resolution and analytical sensitivity [24]. In a study by Fleisher A.S. et al., F18-flortaucipir PET images were obtained and compared to immunohistology examination of postmortem brain tissues for AD tau and amyloid-β plaque accumulation patterns. It was demonstrated that tau PET was reliably able to identify tau in Braak stages V-VI with the presence of some Aβ plaques. There was very limited sensitivity for Braak stages III and IV regardless of the presence or absence of Aβ plaques and this is reflected in the label. Braak III and IV can be identified easily with immunohistochemistry (IHC) and contribute to tauopathy diagnoses including intermediate likelihood ADNC which may result in many false negatives when compared to ADNC in a diverse cohort study. In the study of Fleisher A.S. et al., where the autopsy diagnosis only considered Braak V and VI as positive, it was revealed that F18-flortaucipir PET scan delineated tau pathology with higher sensitivity (92–100%) but lower specificity (51–92%) [24]. This disconnect between

ADNC (intermediate/high likelihood) diagnosis and the positive categorization of data supporting the F18-flortaucipir label is important to consider when comparing biomarkers associated with tau pathology. Additional studies have begun to use the brain regions defined by neuropathological Braak staging to calculate regional PET standardized uptake value ratios (SUVR) in order to monitor progression of tau deposition [25].

The development of tau PET tracers aided the development of fluid biomarkers of tau pathology, but there are key differences between the two measures that are critical to understanding their relationship. First, in autopsy outcome studies, tau PET has been shown to reliably detect Braak stage V-VI pathology, while stages III-IV require unique localization analysis to obtain some signal above normal. Plasma measures are not much different but do appear slightly elevated earlier than tau PET such that – on average – levels in antemortem fluids are increased at Braak stages III-IV. This key factor is often obscured by PET-only studies that use the brain regions defined by Braak and Braak to quantify the PET signal and then report it similarly to neuropathology. Unfortunately, this terminology has confused some scientists who are not spending significant time in the field. Another key difference is the types of tau measured by tau PET vs. biofluid biomarkers. Tau PET measures insoluble forms present within cells, while the biofluid biomarkers measure soluble forms that have made their way to CSF or blood. Although these have shown similar associations with qualitative neuropathology, they are measuring inherently different forms of tau. There have been many scientists working to identify alternate forms of tau in biofluids that have stronger associations with the tau PET signal. However, it is a moving target, and as PET tracers also improve, there will be a new understanding of the relationship with the fluid tau markers. For this reason, it is likely best to anchor on the neuropathological assessment of tau pathology when exploring the relationship with novel measures.

5 Dynamic Model of AD

The dynamic model of AD is a conceptual framework used to understand the complex interactions between various biological, genetic, environmental, and lifestyle factors that contribute to the development and progression of AD. This is a very important concept to apply in the understanding and interpretation of biofluid biomarkers of tau pathology. The timing of each factor in this multifactorial complex disease as well as the effect size and prevalence could explain statistically significant associations. For example, the association of plasma P-tau levels with amyloid (by PET or CSF) could simply be a result of the relationship between amyloid and tau in AD. However, amyloid removal therapies result in a pharmacodynamic change in plasma P-tau levels more so than the insoluble tau measured by tau PET, indicating a specific and unique component of P-tau being attributed more directly to amyloid effects on production of soluble P-tau.

Genetic studies have identified a few key gene mutations causative of amyloid accumulation and early-onset AD such as in the amyloid precursor protein (*APP*), presenilin-1 (*PSEN1*), and presenilin-2 (*PSEN2*) genes. Although mutations present on the *MAPT* gene can lead to some primary tauopathies, they have not been associated with AD. Studying carriers of the dominantly inherited AD mutations has been extremely useful in the discovery and characterization of both imaging and fluid biomarkers. There have been many key discoveries within the Dominantly Inherited Alzheimer's Network (DIAN) study on the timing of biomarker changes and their associations with one another as well as symptom onset [26].

In 2012, Dr. Clifford Jack from the Mayo Clinic proposed a dynamic model that integrated five biomarkers into a timeline of AD progression that has gone through several revisions and updates (Fig. 2) [27]. The model suggests that changes in amyloid-beta (Aβ) occur first in plasma, CSF, and PET, possibly as early as 20 years before symptoms appear. Shortly after the amyloid changes, P-tau changes in plasma and CSF are observable. However, these changes are well ahead of those observed by tau PET. This occurs before changes in fluid and imaging markers of neurodegeneration such as neurofilament light chain (NfL) or volumetric magnetic resonance imaging (vMRI). This model has proven to be a very useful framework for understanding the development and progression of AD [28]. There have been several studies that have validated the general ordering first proposed in 2012, yet more recent evidence measuring pathologies with biofluid markers, MRI, and PET suggests the pathologies are closer together and overlapping to a greater extent than the original model suggested [28]. This dynamic model provides a useful framework for the understanding of the relationship and interactions between the biomarkers of specific pathologies involved in AD. Recent work has extended these relationships to understand the time before symptom onset when these biomarkers become abnormal [29]. Since biofluid biomarkers change many years before symptom onset, they are very likely to be good risk markers and may have opportunities

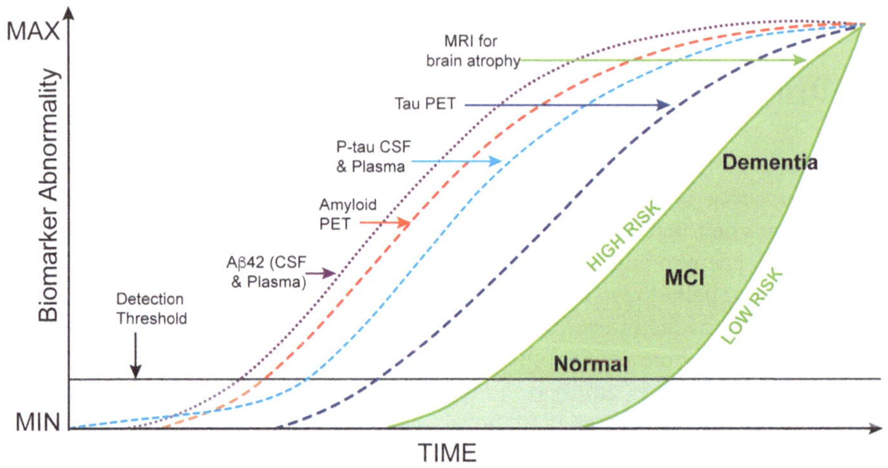

Fig. 2 AD biomarker trajectories with the disease course. (The figure was adapted with permission from Jack et al. 2010 and Palmqvist et al. 2019 [27, 28]. Open access through a Creative Commons license. https://creativecommons.org/licenses/by/4.0/). Abbreviations: CSF cerebrospinal fluid, MCI mild cognitive impairment, MRI magnetic resonance imaging, PET positron emission tomography

to help identify methods to detect earlier cognitive changes than are currently possible.

6 Measuring Tau Forms in CSF

The discovery of P-tau as a primary component of NFTs in CSF [30] highlighted tau proteins as a pivotal target for quantitative immunoassays. Today, both CSF T-tau and P-tau assessments (in addition to low Aβ42 levels) play an essential role in the research criteria for AD established by the International Working Group (IWG) [31] and the NIA-AA [10]. In AD, neurons release elevated amounts of tau proteins into the extracellular space, leading to increased concentrations of both CSF T-tau and P-tau [32]. These alterations in biomarkers are interconnected with strong correlations, especially across stages of AD. They have demonstrated clinical utility in AD and have been shown to identify signs of ADNC co-pathology in subjects diagnosed with Parkinson's disease, LBD, FTLD, and vascular dementia [33]. CSF T-tau and P-tau levels are linked to AD-related neurodegeneration and tangle pathology in postmortem analyses connecting the fluid markers with insoluble pathology in the brain [34].

CSF T-tau levels are thought to indicate the severity of acute injury or ongoing neurodegeneration [35]. In individuals with AD, higher T-tau values are linked to faster cognitive decline [36], more rapid hippocampal atrophy, and hypometabolism in Aβ+ individuals [37], and more specifically linked to Aβ pathology across diagnostic categories. CSF T-tau levels were found to be elevated in cognitively normal (CN) Aβ+, mild cognitive impairment (MCI) Aβ+, and AD dementia Aβ+ individuals compared to CN Aβ- participants suggesting a connection between T-tau levels and the intensity of neurodegeneration and thus, CSF T-tau serves as a marker for AD severity. In a longitudinal study, AD patients with higher CSF T-tau levels (>800 ng/L) showed lower Mini-Mental State Examination (MMSE) scores initially and experienced a more rapid decline over 3 years compared to those with lower CSF T-tau levels (<800 ng/L) [38]. Higher CSF T-tau levels are also observed in non-amnestic AD patients, suggesting more extensive cortical involvement presenting with early non-amnestic symptoms, including aphasia, in early symptomatic stages of atypical AD subtypes [39]. The interpretation of the changing CSF tau levels in disease is difficult as the levels of tau proteoforms are strongly correlated. Many studies look at these markers one at a time and do not consider the influence of co-linearity on the findings or interpretation. In the future, using multiple measures and panels of biomarkers may provide better resolution on the finer details of these interpretations of CSF markers.

Although there have been many proteoforms of tau measured, P-tau (P-tau181, P-tau217, P-tau231) has been the most extensively studied. In a cross-sectional study of 194 participants from the Swedish BioFINDER cohort, assessment of CSF from cognitively unimpaired (CU) individuals (n = 65), and those with MCI (n = 29), and AD dementia (n = 43) using the Meso Scale Discovery (MSD) platform revealed that tau phosphorylated at threonine 217 (P-tau217) shows stronger correlations with the tau PET tracer F18-flortaucipir, more accurately identifies individuals with abnormally increased F18-flortaucipir retention, correlates better than P-tau181 with CSF and PET measures of neocortical amyloid-β burden, and more accurately distinguishes AD dementia from non-AD neurodegenerative disorders [40]. Similarly, another study in the Swedish BioFINDER-2 cohort with a total of 629 participants compared the diagnostic accuracy of P-tau181, P-tau217, and P-tau231 in CSF for AD and their association with Aβ and tau PET. P-tau217$_{\text{Lilly}}$ exhibited higher accuracy than other P-tau variants for separating AD dementia from non-AD (area under the curve [AUC], 0.98 vs. 0.88 [$p < 0.0001$] – 0.96 [$p < 0.05$]) and for identifying Aβ-PET- (AUC, 0.86 vs. 0.74 [$p < 0.0001$] and 0.83 [$p < 0.001$]) and tau PET- (AUC, 0.94 vs. 0.80–0.92, $p < 0.0001$) positivity [41]. The performance of P-tau217 was further compared in a large sample size and this study found a large difference in CSF P-tau 217 levels compared to P-tau181 when distinguishing amyloid-positive (Aβ+) and amyloid-negative (Aβ-) participants with MCI [42]. Another head-to-head comparison of the CSF and plasma immunoassays to distinguish AD (prodromal or dementia stage) from non-AD in symptomatic individuals in a clinical setting revealed interesting findings. Participants with an AD CSF profile (n = 127) were significantly older and had lower cognitive scores (i.e., MMSE) as compared with participants with non-AD CSF (n = 70) [43]. Additionally, the effect sizes of all CSF biomarkers in the AD CSF profile group, compared to the non-AD CSF profile group, were large ($r > 0.5$); ADx P-tau181 ($r = 0.75$), ADx P-tau231 ($r = 0.7$), Janssen CSF P-tau217 ($r = 0.72$), Lilly CSF P-tau181 ($r = 0.74$), CSF P-tau217 ($r = 0.79$), CSF T-tau ($r = 0.57$), Quanterix CSF P-tau181 ($r = 0.76$) and Ugot CSF P-tau181 ($r = 0.73$) [43].

Along with the immunoassay methods for measuring tau proteoforms, there have also been advancements in using immunoprecipitation mass spectrometry (IP/MS) to perform precise measures of multiple tau proteoforms in a single run [44–46]. In a longitudinal study focusing on patients with dominantly inherited AD (DIAD), the sequential emergence of different P-tau species was observed over four decades of disease progression [47]. The study found that levels of P-tau217 in CSF were elevated 21 years before the estimated age of symptom onset (based on the age of onset in others with the same genetic mutation). This increase was followed by elevated levels of P-tau181 (19 years prior), T-tau (17 years prior), and then tau phosphorylated at threonine 205 (P-tau205) (13 years prior). As symptom onset approached, concentrations of P-tau217 and P-tau181 decreased significantly, the increases of P-tau205 reached a plateau, while T-tau levels continued to rise. These changes were associated with other markers of disease progression, such as structural, metabolic, or clinical indicators. These findings suggest that tau phosphorylation at specific epitopes in DIAD, and possibly sporadic AD, follows a dynamic and predictable trajectory over time. However, the plateau and reduction in the later stages have been reported in sporadic AD and are suggested to result from brain atrophy and a significant increase in ventricular volume. In an early-onset dementia study using immunoassays, 165 participants across diagnostic groups (37 CN, 96 early-onset

AD [EOAD], and 32 early-onset non-AD [EOnonAD]) showed medium to high correlations of tau proteoform markers with neurodegeneration, synaptic, and glial cell markers [48]. Most of the CSF biomarkers (Aβ42/40, P-tau181, synaptosomal-associated protein-25 [SNAP-25], neurogranin [Ng] and T-tau) differed significantly in an EOAD group compared with the CN and EOnonAD groups ($p < 0.001$) [48]. These results highlight that due to the correlations across many of tau proteoforms as well as the other CSF biomarkers, it takes unique study designs to observe subtle differences of interpretations.

Comparing three CSF tau biomarkers, P-tau231 has shown robust specificity in some studies, indicating its potential as an early AD biomarker, and P-tau217 has recently emerged as a more promising diagnostic biomarker than P-tau181, showing greater accuracy in distinguishing AD from other neurodegenerative disorders and in predicting tau PET abnormalities and Aβ burden [41]. These findings highlight the evolving landscape of CSF P-tau as a diagnostic and prognostic tool in AD and related disorders. Understanding the role and detailed pace of pathological tau changes is crucial for developing tau-targeted therapies for AD and other tau-related conditions [41].

7 Measuring Soluble Tau Forms in Plasma

Prior to 2018, levels of P-tau in plasma were thought to be very low and likely to have limited utility in associations with pathology in the brain. This was first suggested to be incorrect in work by Tatebe et al., in 2017 which showed levels in Down syndrome and AD to be elevated compared with controls but were largely undetectable or below the quantifiable limit of the assay [49]. However, work from the Dage lab has designed a robust and reliable measurement of plasma P-tau181 on the MSD platform. This assay incorporated an antibody with specificity for the six brain isoforms of the tau protein. In their first collaborative study, they demonstrated an association of plasma measures of T-Tau and P-tau181 with tau pathology as measured by PET [6]. This work was conducted in the Mayo Clinic Study of Aging (MCSA), a community-based cohort based in Rochester, MN. With the robustness and reliability of this assay, the results were rapidly replicated across cohorts including the Swedish BioFINDER Study, University of California San Francisco (UCSF), and the Washington Heights-Hamilton Heights-Inwood Community Aging Project (WHICAP) (Table 1) [6, 50–52].

WHICAP is a community-based longitudinal study of aging and dementia among elderly, urban-dwelling residents. Together with MCSA, WHICAP demonstrates the broad applicability of the assay across diverse populations and in both urban and rural settings. Despite reported parameters being slightly different between studies, there is a clear replication of significant fold change between amyloid-negative (Aβ−) CU and amyloid-positive (Aβ+) clinically diagnosed AD, diagnostic accuracies, and correlation with tau PET. The fold differences vary between clinical

Table 1 Initial studies demonstrating reliable measurements of P-tau181 with replication of clinical utility for association with AD pathology (Aβ and tau) [6, 40, 51, 52]

Study	CU A-	AD A+	Fold change	AUROC for Aβ (95% CL)	Correlation with tau PET
MCSA, Mean (SD)	6.0 (7.6)	12.0 (3.8)	2	0.80 (0.75, 0.86)	0.19
BioFINDER, Median (IQR)	1.3 (0.9–2.4)	4.4 (3.3–6.4)	3.4	0.78 (0.69–0.88)	0.68
UCSF, Mean (SD)	2.4 (3)	8.4 (4)	3.5	0.91 (0.87–0.96)	0.73
WHICAP, Mean (SD)	1.06 (0.81)	1.93 (1.14)	1.8	0.77 (0.67 to 0.87)	NA

1. MCSA - Mayo Clinic Study of Aging, community based cohort in Rochester MN. Tau PET used region in entorhinal cortex for SUVR
2. BioFINDER - Swedish BioFINDER study, clinical cohort based in Lund, Sweden. FC was calculated from median levels. Tau PET used region in inferior temporal cortex
3. UCSF - Clinical cohort from two sources: University of California San Francisco (UCSF) Memory and Aging Center and Advancing Research and Treatment for Frontotemporal Lobar Degeneration (ARTFL) consortium. Tau PET used Cortex for SUVR
4. WHICAP - Diverse community-based cohort in New York City. Participants in the biomarker study were selected cases from the entire cohort. FC and AUROC are from pathologically confirmed cases. There are differences in assay design that incorporated the 4G10E2 antibody in place of the LRL antibody for the P-tau181 assay. Abbreviations: *AUROC* area under the receiver operating characteristic curve, *CU* cognitively unimpaired, *IQR* interquartile range, *FC* fold change, *SD* standard deviation

cohorts and community cohorts which is expected as populations become more diverse and proportions across AD disease stages are different.

The diagnostic accuracies reported as areas under the receiver operating characteristic curve (AUROC) for the determination of amyloid positivity are similar, with the greatest difference being in the MCSA study where the dynamic range in tau PET SUVR and number of tau-positive subjects was the lowest. Three of these studies included a sub-cohort that was followed to autopsy and neuropathology data were available. These data show a reliable increase across Braak stages with a pattern similar to that shown with tau PET [24, 50–52]. The correlations with tau PET are expected to be highly variable as the regions and methods for tau PET were not standardized at the time of these studies.

As these studies were completed, an assay specific for P-tau217 was developed and measured using a novel assay in the same cohorts. Results of studies carefully comparing P-tau181 and P-tau217 consistently show a high correlation, with P-tau217 also consistently having slightly better diagnostic accuracies and correlations with amyloid and tau PET. This is shown in the WHICAP study but also clearly seen in a follow-up UCSF study [53]. In both cases, the anti-tau antibody was changed from LRL to 4G10E2; both are specific for the six brain isoforms, but 4G10E2 was designed to improve the assay's sensitivity and enable detection of P-tau217 in most clinically unaffected controls. Despite consistent results of these initial studies with P-tau181 and P-tau217, showing P-tau217 to be a superior biomarker, many additional assays followed using IP/MS or ultra-sensitive enzyme-linked immunosorbent assay (ELISA) platforms such as the Quanterix Simoa-HDx [43, 54]. There have been several studies comparing the available plasma P-tau immunoassays. P-tau217, regardless of how it is measured, has the best performance. There are multiple methods for measuring P-tau217, at least four immunoassays developed by AlzPATH Inc., Fujirebio, Lilly Research Laboratories, and Janssen Research. These immunoassays have all been shown to accurately predict abnormal CSF or PET amyloid status, and several studies have been evaluating these to predict the risk of future conversion to AD dementia (ADD) [55]. In contrast, a certain degree of variability has been found in the performance of different P-tau181 or P-tau231 immunoassays [56]. Interestingly, differences in the performance between plasma P-tau217 and P-tau181 appear much smaller when both biomarkers are measured with similarly designed immunoassays that only differ in phospho-specific capture antibodies. Understanding how variations in analysis platforms, assay design, and even performing laboratories, can influence the diagnostic accuracy and clinical utility of biomarkers is an important factor to consider.

There are now many assays for multiple phosphorylation sites, and the field continues to expand as these are moved into clinical trials and clinical practice uses. With the successes seen for P-tau217, several additional assays have been developed and evaluated for unique uses. Unfortunately, many of the published findings highlight only differences due to positive reporting bias in scientific publications where only statistically significant novel findings are highlighted. This discounts the fact that many of these studies are showing strong replication of results if analyzed accordingly. There have also been many new assays for total tau, N-terminal, mid-domain, brain-derived, and microtubule-associated tau. The exact utility of these assays and head-to-head comparisons are necessary to explore the similarities and differences in clinical utility and help understand how the complex tau PTMs influence its role in disease. The next frontier is to identify tau forms that are specifically useful for non-AD primary tauopathies. There remains a current need to continue to advance this area of research and to explore novel tau measures.

8 Contexts of Use (COU)

The "context of use" (COU) of a biomarker refers to the specific way the result will be used in drug development. It must be a concise description and limited to a single use; however, there can be multiple COUs for a single biomarker. Each requires its own set of data to support each specific use. A defined COU can help focus experimental design during biomarker development. Some common categories are as follows: *Susceptibility/Risk Biomarker*, *Diagnostic Biomarker*, *Monitoring Biomarker*, *Prognostic Biomarker*, *Predictive Biomarker*, *Pharmacodynamic/Response Biomarker*, and *Safety Biomarker* [57].

Outside of drug development uses for biomarkers, there are clinical uses either as a laboratory-developed test or in vitro diagnostic (IVD) tests approved by the governing regulatory authority such as the FDA in the United States. Laboratory-developed tests can be highly varied in their development and implementation and should be limited in use by a single laboratory. An FDA-cleared IVD test will have an approved label that contains information about the IVD and its safe and effective use. This will include describing the test, how it should be used and interpreted for the specific purpose defined in the label. There will also be warnings of risks of use of the test, which can include the risks associated with false-positive or false-negative results. In this section, we provide brief descriptions of how biofluid measures of tau might be used within each COU category as these tests have been employed in several aspects of drug development.

8.1 *Susceptibility/Risk Biomarker*

Screening potential participants for inclusion in clinical trials based on their disease susceptibility or risk of developing cognitive impairment due to AD involves both tests of cognitive performance and tests to identify the presence of ADNC as defined earlier in the chapter. Current practices can include amyloid PET, tau PET, CSF Aβ42/Aβ40, P-tau/Aβ42, or plasma P-tau217. Each of these tests identifies similar participants with only slight differences in their groupwise risk or susceptibility of developing cognitive symptoms due to AD. These differences depend on the selection of cut-off points when any two tests are compared. The positive participants are at an increased risk of developing cognitive symptoms due to AD within the next 2–4 years and are good candidates for trials aimed at preventing the onset of cognitive decline. Patients identified as negative might still develop cognitive symptoms in 2–4 years but these symptoms would be less likely due to AD. Plasma P-tau217 is regarded as an easily accessible, economical, and non-invasive strategy for risk screening of clinically normal individuals [58–61]. Plasma P-tau217 can be used in the clinically normal population to evaluate for Aβ as well as tau pathology [58–61].

8.2 *Diagnostic Biomarker*

A diagnosis of symptomatic AD requires both the clinical symptoms and the associated neuropathological changes to be aligned. Most AD clinical trials prior to 2012 identified patients solely based on clinical symptoms. This has changed, and now most studies use CSF or amyloid PET imaging to identify symptomatic patients with signs of ADNC. Recently, plasma measures of P-tau217 and P-tau181 have been similarly used for this purpose citing lower costs, higher trial screening efficiency, and more options for global trial site selection. The differences between risk and diagnostic use are largely in the measurement and staging of cognitive symptoms. The clinical test most widely used is the Clinical Dementia Rating (CDR®) scale. This and other tests are good for the general detection of dementia; however, biomarker tests can identify ADNC as many as 10 years before the ability of clinical tests such as the CDR to detect symptoms.

8.3 Monitoring or Pharmacodynamic/Response Biomarker

The use of tau biomarkers as monitoring tools or pharmacodynamic measures in disease-modifying clinical trials has largely relied upon tau PET imaging as the anatomical localization can be assessed. However, CSF and plasma markers have begun to be used more often and plasma P-tau217 has shown easy implementation, more frequent measurements, and larger dynamic effects compared with other tests [62, 63]. The sample size and frequency of testing can depend on the expected effect size attributable to the hypothesized mechanism of action of the drug being tested.

8.4 Prognostic Biomarker

Biofluid measures of tau pathology have been shown to be prognostic of cognitive decline in observational research studies; however, there are very few examples of this specific use in clinical trials [61, 64, 65]. The cut-off point that predicts more rapid disease progression (i.e., for prognosis) is slightly higher than that for the identification of ADNC. As more data are obtained, future incorporation of plasma P-tau217 in this context will likely improve the design of clinical trials.

8.5 Predictive Biomarker

Predictive biomarkers identify individuals who are more likely to respond to a particular treatment. They play a crucial role in personalized medicine, aiding in the selection of patients for clinical trials to improve overall effect size. This particular use for plasma tau biomarkers is being explored in the recent clinical trial datasets from the lecanemab and donanemab studies. Future studies investigating this potential use of plasma P-tau217 will be performed as more tau-directed therapies show efficacy.

9 Future Directions

In a relatively short amount of time, fluid markers of tau pathology have made significant progress in transforming what we know about AD. They have helped build our understanding of the relationship between pathology, onset of measurable symptoms, time course of disease and prevalence of mixed dementia cases. The utility of plasma P-tau in AD research and clinical practice has been demonstrated and replicated in many studies around the world. The use of these measures in clinical trials and novel drug discovery will continue to lead researchers in new directions and help them appreciate the details behind the amyloid effects on neurons, leading to increased tau expression, phosphorylation, and secretion of soluble tau forms. The complexities of the tau protein will continue to provide opportunities to explore novel measures that inform on primary tauopathies and will one day lead to trials and therapies for these unmet medical needs.

References

1. Grundke-Iqbal IIK, Tung YC, Quinlan M, Wisniewski HM, Binder LI. Abnormal phosphorylation of the microtubule-associated protein tau (tau) in Alzheimer cytoskeletal pathology. Proc Natl Acad Sci U S A. 1986;83:4913.
2. Grundke-Iqbal IIK, Quinlan M, Tung YC, Zaidi MS, Wisniewski HM, Microtubule-associated protein tau. A component of Alzheimer paired helical filaments. J Biol Chem. 1986;261:6084.
3. Blennow KWA, Agren H, Spenger C, Siegfried J, Vanmechelen E. Tau protein in cerebrospinal fluid: a biochemical marker for axonal degeneration in Alzheimer disease? Mol Chem Neuropathol. 1995;26:231.
4. Klunk WEEH, Nordberg A, Wang Y, Blomqvist G, Holt DP, Bergström M, Savitcheva I, Huang GF, Estrada S, Ausén B, Debnath ML, Barletta J, Price JC, Sandell J, Lopresti BJ, Wall A, Koivisto P, Antoni G, Mathis CA, Långström B. Imaging brain amyloid in Alzheimer's disease with Pittsburgh Compound-B. Ann Neurol. 2004;55:306.
5. Zetterberg H, Wilson D, Andreasson U, Minthon L, Blennow K, Randall J, et al. Plasma tau levels in Alzheimer's disease. Alzheimers Res Ther. 2013;5(2):9.
6. Mielke MM, Hagen CE, Xu J, Chai X, Vemuri P, Lowe VJ, et al. Plasma phospho-tau181 increases with Alzheimer's disease clinical severity and is associated with tau- and amyloid-positron emission tomography. Alzheimers Dement. 2018;14(8):989–97.
7. Cummings JLG, Ritter A, Zhong K. Alzheimer's disease drug development pipeline. Alzheimers Dement (N Y). 2018;
8. Hansson O, Blennow K, Zetterberg H, Dage J. Blood biomarkers for Alzheimer's disease in clinical practice and trials. Nat Aging. 2023;3(5):506–19.
9. Meredith JE Jr, Sankaranarayanan S, Guss V, Lanzetti AJ, Berisha F, Neely RJ, et al. Characterization of novel CSF Tau and ptau biomarkers for Alzheimer's disease. PLoS One. 2013;8(10):e76523.
10. Jack CR Jr, Bennett DA, Blennow K, Carrillo MC, Dunn B, Haeberlein SB, et al. NIA-AA Research Framework: toward a biological definition of Alzheimer's disease. Alzheimers Dement. 2018;14(4):535–62.
11. Caillet-Boudin MLBL, Sergeant N, Lefebvre B. Regulation of human MAPT gene expression. Mol Neurodegener. 2015;10:28.

12. Goedert MSM, Jakes R, Rutherford D, Crowther RA. Multiple isoforms of human microtubule-associated protein tau: sequences and localization in neurofibrillary tangles of Alzheimer's disease. Neuron. 1989;3:519.
13. Minaya MAMS, Iyer AK, Eteleeb AM, Martinez R, Huang G, Budde J, Temple S, Nana AL, Seeley WW, Spina S, Grinberg LT, Harari O, Karch CM. Conserved gene signatures shared among MAPT mutations reveal defects in calcium signaling. Front Mol Biosci. 2023;
14. Höglinger GU, Melhem NM, Dickson DW, Sleiman PM, Wang LS, Klei L, et al. Identification of common variants influencing risk of the tauopathy progressive supranuclear palsy. Nat Genet. 2011;43(7):699–705.
15. Melquist S, Craig DW, Huentelman MJ, Crook R, Pearson JV, Baker M, et al. Identification of a novel risk locus for progressive supranuclear palsy by a pooled genomewide scan of 500,288 single-nucleotide polymorphisms. Am J Hum Genet. 2007;80(4):769–78.
16. Min SW, Cho SH, Zhou Y, Schroeder S, Haroutunian V, Seeley WW, et al. Acetylation of tau inhibits its degradation and contributes to tauopathy. Neuron. 2010;67(6):953–66.
17. Alonso A, Zaidi T, Novak M, Grundke-Iqbal I, Iqbal K. Hyperphosphorylation induces self-assembly of tau into tangles of paired helical filaments/straight filaments. Proc Natl Acad Sci USA. 2001;98(12):6923–8.
18. Arnold CS, Johnson GV, Cole RN, Dong DL, Lee M, Hart GW. The microtubule-associated protein tau is extensively modified with O-linked N-acetylglucosamine. J Biol Chem. 1996;271(46):28741–4.
19. Köpke E, Tung YC, Shaikh S, Alonso AC, Iqbal K, Grundke-Iqbal I, Microtubule-associated protein tau. Abnormal phosphorylation of a non-paired helical filament pool in Alzheimer disease. J Biol Chem. 1993;268(32):24374–84.
20. Montine TJPC, Beach TG, Bigio EH, Cairns NJ, Dickson DW, Duyckaerts C, Frosch MP, Masliah E, Mirra SS, Nelson PT, Schneider JA, Thal DR, Trojanowski JQ, Vinters HV, Hyman BT, National Institute on Aging, Alzheimer's Association. National Institute on Aging-Alzheimer's Association guidelines for the neuropathologic assessment of Alzheimer's disease: a practical approach. Acta Neuropathol. 2012;123:1.
21. Jack CR Jr, Albert MS, Knopman DS, McKhann GM, Sperling RA, Carrillo MC, et al. Introduction to the recommendations from the National Institute on Aging-Alzheimer's Association workgroups on diagnostic guidelines for Alzheimer's disease. Alzheimers Dement. 2011;7(3):257–62.
22. Montine TJ, Phelps CH, Beach TG, Bigio EH, Cairns NJ, Dickson DW, et al. National Institute on Aging-Alzheimer's Association guidelines for the neuropathologic assessment of Alzheimer's disease: a practical approach. Acta Neuropathol. 2012;123(1):1–11.
23. Schöll M, Maass A, Mattsson N, Ashton NJ, Blennow K, Zetterberg H, et al. Biomarkers for tau pathology. Mol Cell Neurosci. 2019;97:18–33.
24. Fleisher AS, Pontecorvo MJ, Devous MD Sr, Lu M, Arora AK, Truocchio SP, et al. Positron emission tomography imaging with [18F]flortaucipir and postmortem assessment of Alzheimer disease neuropathologic changes. JAMA Neurol. 2020;77(7):829–39.
25. Scholl M, Lockhart SN, Schonhaut DR, O'Neil JP, Janabi M, Ossenkoppele R, et al. PET imaging of tau deposition in the aging human brain. Neuron. 2016;89(5):971–82.
26. Llibre-Guerra JJ, Iaccarino L, Coble D, Edwards L, Li Y, McDade E, et al. Longitudinal clinical, cognitive and biomarker profiles in dominantly inherited versus sporadic early-onset Alzheimer's disease. Brain Commun. 2023;5(6):fcad280.
27. Jack CR et al. Hypothetical model of dynamic biomarkers of the Alzheimer's pathological cascade. Lancet Neurol. 2010;9(1):119–28
28. Palmqvist S, Insel PS, Stomrud E, Janelidze S, Zetterberg H, Brix B, et al. Cerebrospinal fluid and plasma biomarker trajectories with increasing amyloid deposition in Alzheimer's disease. EMBO Mol Med. 2019;11(12):e11170.
29. Jia J, Ning Y, Chen M, Wang S, Yang H, Li F, et al. Biomarker changes during 20 years preceding alzheimer's disease. N Engl J Med. 2024;390(8):712–22.

30. Grundke-Iqbal I, Iqbal K, Tung YC, Quinlan M, Wisniewski HM, Binder LI. Abnormal phosphorylation of the microtubule-associated protein tau (tau) in Alzheimer cytoskeletal pathology. Proc Natl Acad Sci USA. 1986;83(13):4913–7.
31. Dubois B, Feldman HH, Jacova C, Hampel H, Molinuevo JL, Blennow K, et al. Advancing research diagnostic criteria for Alzheimer's disease: the IWG-2 criteria. Lancet Neurol. 2014;13(6):614–29.
32. Olsson B, Lautner R, Andreasson U, Öhrfelt A, Portelius E, Bjerke M, et al. CSF and blood biomarkers for the diagnosis of Alzheimer's disease: a systematic review and meta-analysis. Lancet Neurol. 2016;15(7):673–84.
33. Skillbäck T, Farahmand BY, Rosén C, Mattsson N, Nägga K, Kilander L, et al. Cerebrospinal fluid tau and amyloid-β1-42 in patients with dementia. Brain. 2015;138(Pt 9):2716–31.
34. Tapiola T, Alafuzoff I, Herukka SK, Parkkinen L, Hartikainen P, Soininen H, et al. Cerebrospinal fluid {beta}-amyloid 42 and tau proteins as biomarkers of Alzheimer-type pathologic changes in the brain. Arch Neurol. 2009;66(3):382–9.
35. Blennow K, Hampel H. CSF markers for incipient Alzheimer's disease. Lancet Neurol. 2003;2(10):605–13.
36. Buchhave P, Minthon L, Zetterberg H, Wallin AK, Blennow K, Hansson O. Cerebrospinal fluid levels of β-amyloid 1-42, but not of tau, are fully changed already 5 to 10 years before the onset of Alzheimer dementia. Arch Gen Psychiatry. 2012;69(1):98–106.
37. Mattsson N, Insel PS, Palmqvist S, Portelius E, Zetterberg H, Weiner M, et al. Cerebrospinal fluid tau, neurogranin, and neurofilament light in Alzheimer's disease. EMBO Mol Med. 2016;8(10):1184–96.
38. Sämgård K, Zetterberg H, Blennow K, Hansson O, Minthon L, Londos E. Cerebrospinal fluid total tau as a marker of Alzheimer's disease intensity. Int J Geriatr Psychiatry. 2010;25(4):403–10.
39. Pillai JA, Bonner-Jackson A, Bekris LM, Safar J, Bena J, Leverenz JB. Highly elevated cerebrospinal fluid total tau level reflects higher likelihood of non-amnestic subtype of Alzheimer's disease. J Alzheimers Dis. 2019;70(4):1051–8.
40. Janelidze S, Stomrud E, Smith R, Palmqvist S, Mattsson N, Airey DC, et al. Cerebrospinal fluid p-tau217 performs better than p-tau181 as a biomarker of Alzheimer's disease. Nat Commun. 2020;11(1):1683.
41. Leuzy A, Janelidze S, Mattsson-Carlgren N, Palmqvist S, Jacobs D, Cicognola C, et al. Comparing the clinical utility and diagnostic performance of CSF P-Tau181, P-Tau217, and P-Tau231 assays. Neurology. 2021;97(17):e1681–e94.
42. Mielke MM, Aakre JA, Algeciras-Schimnich A, Proctor NK, Machulda MM, Eichenlaub U, et al. Comparison of CSF phosphorylated tau 181 and 217 for cognitive decline. Alzheimers Dement. 2022;18(4):602–11.
43. Ashton NJ, Puig-Pijoan A, Milà-Alomà M, Fernández-Lebrero A, García-Escobar G, González-Ortiz F, et al. Plasma and CSF biomarkers in a memory clinic: head-to-head comparison of phosphorylated tau immunoassays. Alzheimers Dement. 2023;19(5):1913–24.
44. Barthelemy NR, Gabelle A, Hirtz C, Fenaille F, Sergeant N, Schraen-Maschke S, et al. Differential mass spectrometry profiles of tau protein in the cerebrospinal fluid of patients with Alzheimer's disease, progressive supranuclear palsy, and dementia with lewy bodies. J Alzheimers Dis. 2016;51(4):1033–43.
45. Barthelemy NR, Fenaille F, Hirtz C, Sergeant N, Schraen-Maschke S, Vialaret J, et al. Tau protein quantification in human cerebrospinal fluid by targeted mass spectrometry at high sequence coverage provides insights into its primary structure heterogeneity. J Proteome Res. 2016;15(2):667–76.
46. Barthelemy NR, Li Y, Joseph-Mathurin N, Gordon BA, Hassenstab J, Benzinger TLS, et al. A soluble phosphorylated tau signature links tau, amyloid and the evolution of stages of dominantly inherited Alzheimer's disease. Nat Med. 2020;26(3):398–407.

47. Barthélemy NR, Li Y, Joseph-Mathurin N, Gordon BA, Hassenstab J, Benzinger TLS, et al. A soluble phosphorylated tau signature links tau, amyloid and the evolution of stages of dominantly inherited Alzheimer's disease. Nat Med. 2020;26(3):398–407.
48. Dage JL, Eloyan A, Thangarajah M, Hammers DB, Fagan AM, Gray JD, et al. Cerebrospinal fluid biomarkers in the Longitudinal Early-onset Alzheimer's Disease Study. Alzheimers Dement. 2023;19 Suppl 9(Suppl 9):S115–s25.
49. Tatebe H, Kasai T, Ohmichi T, Kishi Y, Kakeya T, Waragai M, et al. Quantification of plasma phosphorylated tau to use as a biomarker for brain Alzheimer pathology: pilot case-control studies including patients with Alzheimer's disease and down syndrome. Mol Neurodegener. 2017;12(1):63.
50. Janelidze S, Mattsson N, Palmqvist S, Smith R, Beach TG, Serrano GE, et al. Plasma P-tau181 in Alzheimer's disease: relationship to other biomarkers, differential diagnosis, neuropathology and longitudinal progression to Alzheimer's dementia. Nat Med. 2020;26(3):379–86.
51. Thijssen EH, La Joie R, Wolf A, Strom A, Wang P, Iaccarino L, et al. Diagnostic value of plasma phosphorylated tau181 in Alzheimer's disease and frontotemporal lobar degeneration. Nat Med. 2020;26(3):387–97.
52. Brickman AM, Manly JJ, Honig LS, Sanchez D, Reyes-Dumeyer D, Lantigua RA, et al. Plasma p-tau181, p-tau217, and other blood-based Alzheimer's disease biomarkers in a multi-ethnic, community study. Alzheimers Dement. 2021;17(8):1353–64.
53. Thijssen EH, La Joie R, Strom A, Fonseca C, Iaccarino L, Wolf A, et al. Plasma phosphorylated tau 217 and phosphorylated tau 181 as biomarkers in Alzheimer's disease and frontotemporal lobar degeneration: a retrospective diagnostic performance study. Lancet Neurol. 2021;20(9):739–52.
54. Mielke MM, Frank RD, Dage JL, Jeromin A, Ashton NJ, Blennow K, et al. Comparison of plasma phosphorylated tau species with amyloid and tau positron emission tomography, neurodegeneration, vascular pathology, and cognitive outcomes. JAMA Neurol. 2021;78(9):1108–17.
55. Groot C, Villeneuve S, Smith R. Tau PET imaging in neurodegenerative disorders. J Nucl Med. 2022;
56. Bayoumy SVI, den Dulk B, Hussainali Z, Zwan M, van der Flier WM, Ashton NJ, Zetterberg H, Blennow K, Vanbrabant J, Stoops E, Vanmechelen E, Dage JL, Teunissen CE. Clinical and analytical comparison of six Simoa assays for plasma P-tau isoforms P-tau181, P-tau217, and P-tau231. Alzheimers Res Ther. 2021;13:198.
57. https://www.fda.gov/drugs/biomarker-qualification-program/about-biomarkers-and-qualification
58. Gonzalez-Ortiz F, Kac PR, Brum WS, Zetterberg H, Blennow K, Karikari TK. Plasma phospho-tau in Alzheimer's disease: towards diagnostic and therapeutic trial applications. Mol Neurodegener. 2023;18(1):18.
59. Janelidze S, Palmqvist S, Leuzy A, Stomrud E, Verberk IMW, Zetterberg H, et al. Detecting amyloid positivity in early Alzheimer's disease using combinations of plasma Abeta42/Abeta40 and p-tau. Alzheimers Dement. 2021;17
60. Janelidze S, Berron D, Smith R, Strandberg O, Proctor NK, Dage JL, et al. Associations of plasma phospho-tau217 levels with tau positron emission tomography in early Alzheimer disease. JAMA Neurol. 2021;78(2):149–56.
61. Cullen NC, Leuzy A, Janelidze S, Palmqvist S, Svenningsson AL, Stomrud E, et al. Plasma biomarkers of Alzheimer's disease improve prediction of cognitive decline in cognitively unimpaired elderly populations. Nat Commun. 2021;12(1):3555.
62. Pontecorvo MJ, Lu M, Burnham SC, Schade AE, Dage JL, Shcherbinin S, et al. Association of donanemab treatment with exploratory plasma biomarkers in early symptomatic alzheimer disease: a secondary analysis of the TRAILBLAZER-ALZ randomized clinical trial. JAMA Neurol. 2022;79:1250.

63. McDade E, Cummings JL, Dhadda S, Swanson CJ, Reyderman L, Kanekiyo M, et al. Lecanemab in patients with early Alzheimer's disease: detailed results on biomarker, cognitive, and clinical effects from the randomized and open-label extension of the phase 2 proof-of-concept study. Alzheimers Res Ther. 2022;14(1):191.
64. Palmqvist S, Janelidze S, Quiroz YT, Zetterberg H, Lopera F, Stomrud E, et al. Discriminative accuracy of plasma phospho-tau217 for alzheimer disease vs other neurodegenerative disorders. JAMA. 2020;324(8):772–81.
65. Palmqvist S, Tideman P, Cullen N, Zetterberg H, Blennow K, Alzheimer's Disease Neuroimaging I, et al. Prediction of future Alzheimer's disease dementia using plasma phospho-tau combined with other accessible measures. Nat Med. 2021;27(6):1034–42.

Fluid and Imaging Markers of Synuclein Pathology

Federico Paolini Paoletti, Lorenzo Gaetani, Giovanni Bellomo, and Lucilla Parnetti

Abstract Synucleinopathies are a group of clinically and neuropathologically heterogenous disorders characterized by the abnormal aggregation of α-synuclein in the form of Lewy bodies, Lewy neurites, and glial cytoplasmic inclusions. The antemortem diagnosis of synucleinopathies relies mostly on clinical features. However, clinical manifestations appear several years after the onset of neuropathological changes, with a considerable degree of overlap in clinical phenotypes across these disorders. Biomarkers reflecting α-synuclein pathology could allow for accurate and early diagnosis, also serving prognostic purposes, i.e., identifying those patients with the highest risk of clinical progression. With respect to clinical trials with disease-modifying treatments, biomarkers reflecting α-synuclein pathology could be valuable to include biologically homogeneous patients in which the molecular target of the therapy is actually represented, also providing outcome measures to verify the biological impact of treatment. In this chapter, current evidence on fluid and imaging biomarkers reflecting α-synuclein pathology will be discussed. Among fluid biomarkers, seeding amplification assays, which are able to identify α-synuclein prone to aggregation in cerebrospinal fluid and other peripheral tissues, so far, appear to offer the most promising techniques for in vivo identification of synucleinopathies. Clinical application of α-synuclein imaging of the brain is limited by several technical issues. To date, the combination of different biological matrices and the association of α-synuclein with other biomarkers could pave the way for an early and accurate diagnosis of synucleinopathies and better prognostication of associated clinical outcomes.

Keywords α-synuclein · Fluid biomarkers · Cerebrospinal fluid · Imaging biomarkers

Federico Paolini Paoletti and Lorenzo Gaetani contributed equally to this Chapter.

F. Paolini Paoletti · L. Gaetani · G. Bellomo · L. Parnetti (✉)
Section of Neurology, Laboratory of Clinical Neurochemistry, Department of Medicine and Surgery, University of Perugia, Perugia, Italy
e-mail: lucilla.parnetti@unipg.it

R. Tarawneh (ed.), *Biomarkers of Neurodegenerative Disorders*,
https://doi.org/10.1007/978-3-031-80849-4_3

1 The Need for Early and Accurate Diagnosis of Synucleinopathies

Synucleinopathies include a group of pathologically and clinically heterogeneous neurodegenerative disorders which are characterized by intracellular aggregates of α-synuclein (α-syn) as the core common neuropathological hallmark.

α-syn is a 140-amino acid protein which exerts several physiological functions in cells. It is mainly localized in the pre-synaptic compartment where it works as a key regulator of vesicular trafficking and neurotransmitter exocytosis. In dopaminergic neurons, it regulates dopamine metabolism by inhibiting dopamine synthesis and inducing both storage in pre-synaptic vesicles and extracellular re-uptake of the neurotransmitter. Additionally, α-syn participates in mitochondrial functioning and motility, and interacts with several cytoskeletal components. At the nuclear level, it is likely able to bind directly to more than 300 different regions of DNA and may downregulate the transcription of a variety of genes, mainly involved in cell cycle control and DNA repair systems. It is an intrinsically disorder protein, with aggregation-prone properties. Misfolding and aggregation of α-syn is a multistep process progressing from accumulation of monomeric α-syn to nucleation of small soluble oligomers and formation of large insoluble fibrils, leading to the deposition of neuronal inclusions, i.e., Lewy bodies and Lewy neurites in the so-called Lewy body disorders (LBD; which include Parkinson's disease [PD], PD with dementia [PDD] and Dementia with Lewy bodies [DLB]), and oligodendrocytic inclusions in multiple system atrophy (MSA) [1].

From a clinical point of view, synucleinopathies include PD, which is the most frequent disorder among this group, followed by DLB and MSA. The definitive diagnosis of synucleinopathies is based on post-mortem pathological confirmation of abnormal α-syn aggregation.

The diagnostic criteria of various synucleinopathies rely mostly on clinical manifestations as core features, with brain imaging characteristics as supportive aspects [2–4] (Fig. 1).

Since clinical symptoms/signs can be mild and overlap among different synucleinopathies, the differential diagnosis can be challenging during the early disease stages. Even in more advanced phases, when the clinical picture is fully manifest, the accuracy of a clinical diagnosis is unsatisfactory in most cases. Diagnostic misclassification of PD is common in clinical practice, with error rates ranging from 15% to 24% in different series [5, 6]. Several clinico-pathological studies show that a considerable proportion of individuals who are clinically diagnosed as PD have, in fact, an alternative post-mortem pathological diagnosis, with diagnostic accuracy of clinical assessments alone being lower than 80% [7].

Similarly to PD, diagnostic criteria for DLB and MSA rely mostly on clinical features. DLB shares overlapping clinical features with "PD with dementia" (PDD) and other neurodegenerative diseases associated with dementia including Alzheimer's disease (AD) [8]. Clinical and imaging characteristics of MSA can be present in other neurodegenerative diseases including hereditary ataxias [9, 10].

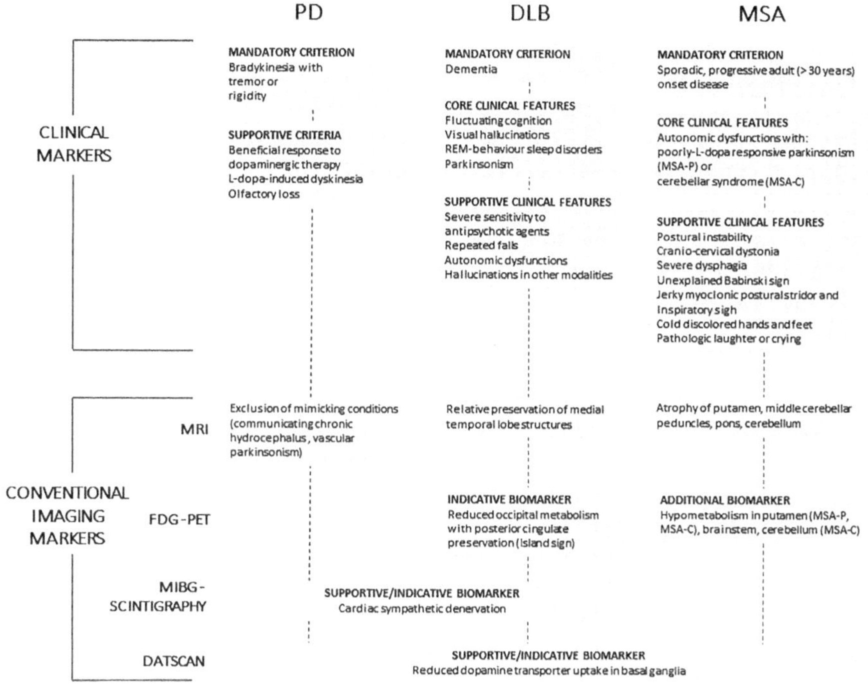

Fig. 1 Summary of the diagnostic criteria for Parkinson's disease (PD), dementia with Lewy bodies (DLB), and multiple system atrophy (MSA)

Only half of patients with a clinical diagnosis of MSA have pathologically confirmed disease, with most causes of misdiagnosis being represented by DLB, progressive supranuclear palsy (PSP), and PD [11]. The most updated diagnostic criteria for DLB [2] and MSA [4] need to be validated, and their sensitivity and specificity need to be examined in clinico-pathological cohorts.

The pathophysiological mechanisms underlying synucleinopathies start several years before the appearance of clinical manifestations. The "clinically silent" phase of neurodegeneration is of utmost relevance for the development of disease-modifying therapies or neuroprotective strategies as these have the greatest chance to be effective in the earliest disease stages. A three-phase disease has been defined for PD, with preclinical and prodromal disease preceding the motor stage [12]. In preclinical PD, the disease-related pathology is assumed to be present, without any evidence of clinical symptoms/signs, while in prodromal PD subtle motor signs and non-motor manifestations are evident.

Converging evidence indicates that idiopathic REM-sleep behavior disorder (iRBD) is the most specific premotor symptom of synucleinopathies. A considerable proportion (~80%) of patients initially diagnosed as iRBD eventually develop a parkinsonism/dementia syndrome, mainly LBD (PD and DLB), and less frequently, MSA. The risk of appearance of a clinically fully-manifest synucleinopathy increases over follow-up, with a rate of conversion ranging from 33% after a 5-year follow-up to 91% after a 14-year follow-up of iRBD. The time frame between iRBD and subsequent development of a neurodegenerative disorder can range up to 50 years, suggesting that α-syn pathology starts several decades before the appearance of the first subtle symptoms of PD, MSA, and DLB. Additionally, iRBD patients, even when remaining clinically free from motor signs, show imaging findings suggestive of dopaminergic nigro-striatal degeneration, i.e., reduced dopamine transporter (DAT)-uptake on ^{123}I-FP-CIT single-photon emission computed tomography (SPECT) and substantia nigra hyperechogenicity on transcranial sonography.

Diagnostic criteria for prodromal disorders have been recently proposed in the field of synucleinopathies [13–15]. The criteria for prodromal PD are based on environmental and genetic risk factors, combined with clinical and imaging markers. After their application in different, independent cohorts including subjects with iRBD, these criteria showed good specificity (>90% in most studies), but unsatisfactory sensitivity (values ranging from 30% to 80%) [16–18]. The best sensitivity was obtained in the identification of patients converting from RBD to DLB, rather than to PD [19]. The issue of reaching an early and accurate diagnosis remains unsolved even after applying criteria for prodromal disease.

2 The Importance of Predicting Clinical Outcomes in Synucleinopathies

Clinical progression of synucleinopathies is variable, with DLB and MSA showing a more aggressive course compared to PD. Clinical outcomes also vary within the same disorder.

In PD, an advanced age at disease onset, rigid-akinetic subtypes and early cognitive decline predict worse motor outcomes [20, 21]. The annual motor progression—as measured by the Unified Parkinson's Disease Rating Scale (UPDRS) score change—was found to range from 0.62% to 3.67%, following three distinct phases: improvement, stability, and steady progression [22]. To date, several studies have described the progression of motor symptoms and disability in PD. However, available data are not conclusive, since the natural history of PD has been often described in treated individuals, with an obvious confounding symptomatic effect due to concomitant dopaminergic medications [23–26].

Cognitive changes can occur at the time of diagnosis or even decades later, in the form of either subjective cognitive decline, mild cognitive impairment (MCI), or—when severe enough to affect daily functions—dementia (referred to as PDD), with a significant variability in the rate and percentage of progression to PDD among different individuals [27]. Individuals with PD-MCI can either progress to dementia or revert back to normal cognition [28]. Although findings vary among studies, the cumulative prevalence of PDD, in individuals with a mean age of 54-70 years at the time of diagnosis, is higher than 80%, after 20 years of their PD diagnosis [29]. Meanwhile, it has also been reported that about 30% of individuals with PD-MCI revert to normal cognition over a follow-up of 3-5 years [28, 30]. Male gender, older age at disease onset, lower educational level, longer disease duration, more advanced Hoehn and Yahr (H&Y) stage, more severe motor symptoms, the motor subtype of postural instability/gait difficulties, and mood or behavioral disturbances have been independently associated with cognitive decline in PD [31, 32].

Demographic and clinical features are not sufficient to provide a comprehensive prognostic characterization of the disease. The clinical presentation and rate of progression of synucleinopathies strongly rely on the underlying molecular and neuropathological substrates. Deposition of misfolded α-syn aggregates—Lewy bodies and Lewy neurites in PD and DLB, and glial cytoplasmic inclusions (GCI) in MSA—is the common neuropathological feature of synucleinopathies. However, the amount and the topographical distribution of these aggregates in the brain vary among the synucleinopathies and within each of these disorders, contributing to the presence of different clinical phenotypes and variable clinical progression [33]. Previous clinico-pathological investigations demonstrated that the regional distribution of α-syn pathology influences clinical features and outcomes of LBD, including the occurrence of dementia, the presence of hallucinations, and survival rates [33, 34]. The stereotypical progression of α-syn pathology through the brain over time suggests a cell-to-cell transmission of misfolded α-syn in a prion-like fashion. An initial seed of misfolded α-syn in one neuron may be released and taken up by another vulnerable neuron, thereby initiating the misfolding of α-syn in the latter [35]. Several data from experimental models and clinical studies suggest that abnormal accumulation of α-syn in different cell types and brain areas might be explained by the existence of distinct misfolded α-syn strains. Distinct fibrillary assemblies of α-syn may have a causative role in the different disease phenotypes and explain, at least to some extent, differences in clinical progression [36]. Therefore, the identification of different α-syn strains could serve as a novel and promising diagnostic tool to differentiate synucleinopathies, also representing a potential prognostic biomarker for characterizing clinical progression.

In addition to the α-syn-related pathology, the presence of other pathologies can influence the clinical spectrum and the progression profile of synucleinopathies. Aggregates composed of one protein, i.e., α-syn, usually seed the aggregation of the same protein. However, an aggregate may also seed the misfolding and the aggregation of a different protein, in a process termed "heterologous seeding" or "cross-seeding." Cross-seeding events may explain the frequent finding of mixed pathologies in the brain [37]. The co-occurrence of AD-related pathology was found in up to 50% of LBD cases [38]. Accordingly, a cerebrospinal fluid (CSF) AD profile, suggestive of the presence of brain amyloid and tau pathologies (i.e., A+/T+/N±) [39] has been reported in PDD, ranging from 9% to 28% in different observations, with an even higher prevalence in DLB [40, 41], and is predictive of worse cognitive outcomes when found in PD [42–44]. The detection of the CSF biomarker signature of AD in individuals with DLB correlates with memory impairment and disease severity [45], also predicting a high risk of institutionalization and death [46].

Overall, data from neuropathological and clinical studies suggest that different α-syn strains, and the presence of other proteins interacting with α-syn in its prion-like spread, drive the clinical heterogeneity and influence differences in disease progression rates across different synucleinopathies. Molecular characterization is mandatory to define the prognostic profile in each patient, and will also be important in applying emerging therapies which may potentially mitigate the underlying pathophysiological mechanisms responsible for disease progression.

3 Detection of α-Syn Species in CSF and Other Matrices

α-syn misfolding plays a key role in the development of PD and other α-syn aggregation disorders [47]. Post-translational modifications of the α-syn protein, such as phosphorylation, ubiquitination, nitration, truncation, and oxidization, may play a role in enhancing protein misfolding [48]. Several investigations into the identification and validation of biomarkers for synucleinopathies have therefore focused on the measurement of total α-syn (t-α-syn) and other specific α-syn species including oligomeric α-syn (o-α-syn) and α-syn phosphorylated at residue Ser129 (p-α-syn) in different biofluids. The CSF, which is in direct contact with the brain parenchyma, is an ideal biospecimen to measure biomarkers as it closely reflects the molecular changes occurring in the central nervous system (CNS). α-syn is secreted into extracellular spaces and can be found in fluids other than CSF, such as blood [49, 50] and saliva [51, 52]. Furthermore, typical non-motor symptoms of PD, including olfactory and autonomic symptoms (e.g., constipation and abnormal sweating patterns), are thought to occur because of the abnormal accumulation of α-syn aggregates in peripheral sites including olfactory mucosa (OM), gut mucosa, and skin nerve fibers, respectively, which inspired studies to investigate the presence of α-syn as a diagnostic biomarker in peripheral tissues [53–55] (Table 1 and Fig. 2).

3.1 Cerebrospinal Fluid (CSF)

3.1.1 Diagnostic and Prognostic Value of Total α-Syn

Total α-syn (t-α-syn) can be measured in the CSF using different immunoassays such as enzyme-linked immunosorbent assays (ELISA), electrochemiluminescence, and xMAP technology. Different immunoassays for CSF t-α-syn have been compared in different laboratories, showing high analytical precision, excellent inter-laboratory correlations, and good inter-assay correlations. Although absolute values of total α-syn measurements varied between the different assays, measurements were in line with those of immunoprecipitation-mass spectrometry [56].

Table 1 Diagnostic and prognostic value of biomarkers reflecting α-syn pathology in different biofluids and tissues

Biomarker	Diagnostic value	Prognostic value
	Cerebrospinal fluid	
t-α-syn	↓ in PD, DLB, MSA vs. HC	↑ baseline levels predict motor progression and cognitive decline in PD
o-α-syn	↑ in PD vs. HC	↑ levels correlate with motor progression in PD
p-α-syn	↑ in PD vs. HC	NA
	Blood	
t-α-syn	↑/=/↓ in PD vs. HC (serum, plasma) ↑ in PD vs. DLB (serum)	↑ levels predict worse cognitive outcomes in prodromal PD (plasma)
o-α-syn	↑ in PD vs. HC (serum, RBCs)	NA
p-α-syn	↑ in PD vs. HC (plasma)	NA
	Olfactory mucosa	
t-α-syn	= in PD vs. HC	NA
p-α-syn	↑ in LBD vs. HC	NA
	Saliva	
t-α-syn	↓ in PD vs. HC ↓ in MSA-P vs. PD (salivary EV-derived α-syn)	NA
o-α-syn	↑ in PD vs. HC	↑ in advanced PD stages
	Salivary glands	
t-α-syn	↑ in PD and incidental LBD vs. HC and MSA ↑ in iRBD vs. HC	NA
p-α-syn	↑ in PD vs. HC ↑ in iRBD vs. HC	NA
	Gut mucosa	
t-α-syn	↑/= in PD vs. HC ↑ in prodromal PD	% of positivity is related to L-dopa unresponsive features
p-α-syn	↑ in PD vs. HC	NA
	Skin	
t-α-syn	↑/= in PD vs. HC	NA
p-α-syn	↑ in PD vs. VaP and tauopathies ↑ in iRBD vs. HC	NA
α-syn amyloid fibrils	Somatic fibers at distal sites in MSA vs. autonomic fibers at proximal sites in PD, DLB, and PAF	NA

Abbreviations. *DLB* dementia with Lewy bodies, *EV* extracellular vesicle, *HC* healthy controls, *iRBD* isolated rapid eye movement (REM) behavior disorder, *LBD* Lewy body disease, *MSA* multiple system atrophy, *o-α-syn* oligomeric α-synuclein, *p-α-syn* phosphorylated α-synuclein, *PAF* pure autonomic failure, *PD* Parkinson's disease, *t-α-syn* total α-synuclein, *VaP* vascular parkinsonism

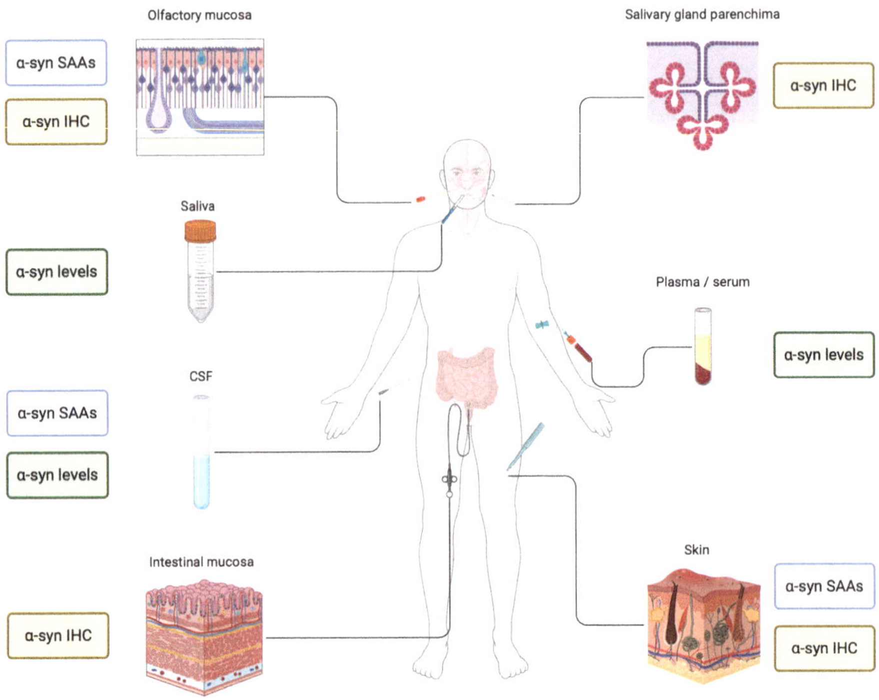

Fig. 2 Main body sources of different matrices for measuring α-synuclein (Abbreviations: IHC immunohistochemistry, SAAs seed amplification assays)

These results support the analytical validity of these various immunoassay techniques for measuring CSF t-α-syn.

Four meta-analyses examining thousands of individuals with LBD/MSA, or other causes of parkinsonism not related to α-syn pathology and healthy controls, verified the potential diagnostic value of CSF t-α-syn in PD [57–60]. In all of these meta-analyses, CSF t-α-syn was consistently found to be lower in individuals with PD compared to healthy controls and other neurological controls. However, when considering CSF t-α-syn as a potential biomarker of PD, its diagnostic accuracy was poor, with low sensitivity and specificity in discriminating PD from healthy controls. In the studies considered, there was a high variability of CSF t-α-syn values, probably due to the enrollment of PD patients with different disease characteristics in terms of both disease severity and disease duration, as well as different control populations, ranging from healthy controls to individuals with cognitive impairment due to other neurological diseases [61]. Additionally, pre-analytical and analytical factors represent a relevant source of heterogeneity in measured values, hampering the clinical use of CSF t-α-syn in clinical practice. Among the former, blood contamination of CSF samples deserves consideration as red blood cells (RBCs) are one of the major sources of α-syn and even a small amount of blood contamination can alter α-syn measurements in the CSF [62]. Other differences in sample processing and the use of different immunoassays could potentially limit the ability to directly compare results across different studies [56]. Furthermore, the great majority of immunoassays targeting α-syn have been developed using full-length α-syn as a standard protein, being unable to capture all species and forms of the protein present in the CSF [63].

With respect to MSA and DLB, a trend towards lower levels of CSF t-α-syn compared to healthy controls was also found, although with unsatisfactory diagnostic accuracy [59]. Recent meta-analyses found no difference in CSF t-α-syn between PD and DLB [57–60], and only one meta-analysis concluded for a difference between PD and MSA, with MSA showing lower CSF t-α-syn compared to PD [60]. Low CSF t-α-syn has also been found in other neurodegenerative diseases that do not involve α-syn pathology, such as corticobasal syndrome (CBS), and frontotemporal dementia (FTD), with CSF t-α-syn values largely overlapping with those of individuals with PD [64–66]. Meanwhile, a trend towards higher CSF t-α-syn values in PSP compared to PD has been described ([59, 60]). A significant increase in CSF t-α-syn has also been reported in AD and in Creutzfeldt-Jakob disease (CJD) compared to PD, and to a larger extent than what is observed in PSP [67, 68].

The increase in CSF t-α-syn levels found across several neurodegenerative disorders reflects the dual nature of the biomarker: although CSF t-α-syn can offer value as a marker of brain α-syn aggregation as reflected by its low levels, CSF t-α-syn levels may also increase as a non-specific marker of synaptic damage due to the synaptic localization of the protein which is released in case of neuronal damage/death. These observations suggest the presence of variable confounding factors in the interpretation of CSF t-α-syn concentrations. It is likely, for instance, that individuals with PD who have an aggressive disease course could have CSF t-α-syn levels that are comparable to healthy controls, due to a combination of factors: decreased values (due to α-syn sequestration in the setting of aggregation), which is counterbalanced by increased release (due to neurodegeneration) [1]. Accordingly, longitudinal measurements of CSF t-α-syn suggest increased levels in individuals with PD with long-disease duration (> 5 years) after a 2-year follow-up, indicating that, as PD progresses, t-α-syn increases as a result of worsening neuronal damage [69]. Individuals with PD with increased CSF t-α-syn (i.e., with more evident ongoing neurodegeneration) showed faster progression in motor and/or cognitive impairment [70]. Hall and colleagues found that CSF t-α-syn is positively associated with the Hoehn and Yahr (H&Y) stage, Unified Parkinson's Disease Rating Scale Part III (UPDRS-III), and prolonged Timed Up and Go test over 2 years [69].

Findings from DATATOP (Deprenyl and Tocopherol Antioxidative Therapy of Parkinsonism) and BioFINDER (Biomarkers For Identifying Neurodegenerative Disorders Early and Reliably) cohorts showed that baseline high levels of CSF t-α-syn predict rapid cognitive decline [70, 71]. Particularly, Stewart et al. found that high levels of CSF t-α-syn correlated with faster decline in memory performance and visual memory as measured by the Selective Reminding Test-Total and Delayed and the New Dot Test, respectively [71]. Similarly, Hall et al. [70] associated high levels of CSF t-α-syn with lower cognitive processing speed.

In conclusion, CSF t-α-syn might be a predictive marker for motor progression and cognitive decline in PD, but it does not seem to be a reliable diagnostic marker for synucleinopathies, due to several confounding factors that influence its CSF levels. The study of other more disease-specific species of α-syn, as discussed below, may provide promising results.

3.1.2 Diagnostic and Prognostic Value of Other α-Syn Species

The oligomerization of α-syn precedes its aggregation into mature amyloid fibrils within Lewy bodies. In vitro and in vivo studies provide evidence that oligomeric α-syn (o-α-syn) may play a primary role in PD pathophysiology [72]. CSF o-α-syn has consistently been found to be higher in individuals with PD compared to controls [57, 60, 73], with a pooled sensitivity of 71% and specificity of 64% [57].

Data from the DeNoPa cohort have recently shown that only individuals with PD who are in an advanced disease stage (H&Y stage II), but not those in less severe disease stages, have higher CSF o-α-syn levels compared to healthy controls. Additionally, in the same study, CSF o-α-syn levels longitudinally increased over the 4-year follow-up in all individuals with PD and such increase was associated with motor progression [74].

The CSF o-α-syn/t-α-syn ratio improves diagnostic performance of o-α-syn alone, with an area under the curve (AUC) up to 0.78 (sensitivity 82%, specificity 64%) in differentiating PD from healthy controls [43]. Similarly, combining CSF t-α-syn and o-α-syn has been shown to be of high diagnostic value in discriminating DLB from AD [75]. Additionally, data from the BioFINDER cohort showed an association between the longitudinal change of o-α-syn/t-α-syn ratio over 2 years and worsening of motor signs, particularly in individuals with PD who had predominant symptoms of postural instability and gait difficulty [76].

One of the major limitations related to o-α-syn lies in the immunoassay methods used for its measurement, based on the same monoclonal antibody for both the capture and the detection steps. Recognition can occur only if the epitope used by the capture antibody is also displayed (i.e., in an oligomer-specific situation). However, processes leading to α-syn aggregation and the subsequent formation of oligomers can mask the epitopes to be recognized, thus precluding the possibility of accurate assessments of the levels of o-α-syn [48].

Phosphorylated α-syn (p-α-syn) is one of the main disease-associated forms of α-syn, and this modification accounts for more than 90% of α-syn found in LBs [77]. Only a few studies have focused on CSF p-α-syn as a diagnostic marker, finding increased levels in PD compared to controls and PSP [57, 78]. Similar to o-α-syn, the diagnostic accuracy of p-α-syn increases when examined in combination with other α-syn species and neurodegenerative biomarkers (e.g., o-α-syn/t-α-syn ratio together with p-α-syn and phosphorylated tau protein, p-tau, provide a diagnostic AUC of 0.86, sensitivity 79%, specificity 67%) in differentiating PD from controls [76]. These studies on p-α-syn require validation in other cohorts and laboratories. In addition, p-α-syn has very low concentrations in CSF so that many assays cannot reliably measure its levels [48].

Studies focusing on oligomeric and phosphorylated species of α-syn in MSA and DLB are sparse. In one of these investigations, post-mortem ventricular CSF samples were obtained from patients affected by pathologically confirmed synucleinopathies (PD, n=13; PDD, n=26; DLB, n=17; MSA n=8), other parkinsonisms (PSP, n=12), and controls (n=20) to measure concentrations of different α-syn species. Mean CSF levels of t-α-syn, o-α-syn, and p-α-syn did not significantly vary among the diagnostic groups, while mean CSF concentrations of phosphorylated oligomeric α-syn (p-o-α-syn) were significantly higher in MSA patients compared to all other groups, suggesting the utility of this biomarker in differentiating MSA from other degenerative diseases including PD and DLB [79]. Histopathological confirmation of clinically diagnosed degenerative disorders is extremely helpful to make the results of CSF biomarkers more reliable. Nevertheless, findings from postmortem samples should be considered cautiously since both the breakdown of the blood-brain barrier (BBB) with subsequent contamination of blood α-syn and the autolysis of brain cells could alter the measurement of CSF biomarkers, making the results difficult to interpret [80].

3.2 Blood

Measurement of t-α-syn levels in the plasma and serum can be achieved through various techniques, including ELISAs and other ultrasensitive assays such as electrochemiluminescence and single-molecule counting arrays (e.g., SIMOA from Quanterix, USA). Unfortunately, t-α-syn levels in the blood may be significantly influenced by factors such as RBC contamination and hemolysis [62].

Available studies showed conflicting results, with different works suggesting that plasma and serum t-α-syn levels are increased [81–85], unchanged [23, 50, 86, 87], or even reduced [88–90] in PD compared to healthy controls. Serum levels of t-α-syn have recently been reported to be higher in PD compared to DLB [90]. Overall, these findings, together with the issue concerning RBC contamination, limit the diagnostic utility of plasma and serum t-α-syn measurements in PD and other synucleinopathies. Discrepancies in findings across different studies could be attributed to hemoglobin contamination as well as to differences in the immunoassay protocols or pre-analytical factors across different studies, including sample handling, storage, processing, and other confounding variables. Together, these limitations make it difficult to draw definite conclusions from these data.

Studies on blood o-α-syn have provided more concordant results, showing increased levels in PD in serum [91] and RBCs [92, 93] from individuals with PD compared to controls with serum values showing good diagnostic accuracy (sensitivity 75%, specificity 100%) [91]. Similarly, plasma p-α-syn was found to be higher in PD compared to healthy controls (AUC 0.71) [50]. Notably, RBC measurements of a panel of post-translationally modified forms of α-syn including Tyr125-p-α-syn, nitrated α-syn, glycated α-syn, and SUMOylated α-syn showed high discriminatory power in distinguishing individuals with PD from healthy controls (AUC 0.84) [94]. The cohorts in these studies were relatively small and, therefore, such results need to be replicated in larger cohorts. However, it appears that levels of t-α-syn alone are not sufficient to distinguish individuals with PD from healthy controls, while the combination of different α-syn species may hold greater promise as biomarker-based assessments in the blood.

Available findings referring to the potential predictive value of plasma t-α-syn are sparse. Data from the PARS (Parkinson's Associated Risk Syndrome) study revealed that baseline and longitudinal increases in plasma t-α-syn predict the progression of cognitive decline in hyposmic individuals with dopamine transporter binding reduction, who have not yet been diagnosed with PD. In these participants, longitudinal decreases in plasma exosomal-derived α-syn predicted worse cognitive scores [95]. Overall, these findings suggest that plasma t-α-syn and exosomal-derived α-syn are likely associated with PD progression, including cognitive impairment, even in the early stages that precede the onset of motor symptoms.

More recently, increasing attention has been paid to blood α-syn in extracellular vesicles (EVs) derived from the CNS. EVs are small membrane-bound structures released by cells, playing crucial roles in intercellular communication and regulation of various physiological processes. It seems that, in pathological conditions, they are involved in the cell-to-cell transmission of aggregates of misfolded proteins characterizing neurodegenerative disorders. These vesicles encompass a wide range of biomolecules, including proteins, lipids, and nucleic acids, which reflect the state of the parent cell. Due to their ability to cross the BBB, EVs offer a unique window into the brain's biochemistry, allowing for study of CNS processes and potentially providing a valuable source of biomarkers for neurodegenerative disorders [96].

In the last 10 years, many studies have employed the approach of measuring α-syn in neuronal and oligodendroglial EVs (nEVs and oEVs) to differentiate Parkinsonian disorders from one another or from healthy controls. With respect to nEV α-syn, results from independent studies are contradictory, showing either increased or decreased levels in PD compared to healthy controls [97].

Similarly, the measurement of α-syn in oEVs is not reliable, as the available data are inconsistent. Two studies from one research group reported higher oEVs α-syn levels in patients with MSA compared to healthy controls [98, 99], while one study from another group found decreased oEVs α-syn concentrations in MSA [100]. Noteworthy, in a very recent meta-analysis, oEV α-syn was not found to be different in individuals with MSA compared to PD and healthy controls, although oligodendrocyte α-syn pathophysiology is believed to be exclusive to MSA [97]. All these inconclusive findings, taken together, highlight the need to overcome pre-analytical and analytical variability to have standardized and rigorous methodologies to measure α-syn in EVs.

3.3 Olfactory Mucosa

Olfactory mucosa (OM) is a window to the olfactory bulb, which is one of the earliest sites affected by α-syn pathology in PD and other synucleinopathies. Samples from the OM can be collected by a non-invasive procedure, referred to as nasal brushing.

Using immunohistochemistry, it was demonstrated that the main proteins involved in neurodegenerative diseases (α-synuclein, β-amyloid, tau, and TAR DNA-binding Protein 43 [TDP-43]) are expressed in olfactory neurons of healthy individuals [101].

Immunohistochemical analysis of olfactory epithelium from individuals with PD, individuals with hyposmia and anosmia due to other causes, and healthy individuals, found similar levels of t-α-syn expression across these different groups [102]. In contrast, in a small study, p-α-syn staining positivity was reported in the olfactory epithelium of six out of eight individuals with PD but not in healthy controls [103]. Consistent with the latter finding, a postmortem immunohistochemical analysis of the olfactory bulb from individuals with PD/DLB in another study showed that almost all patients (55 out of 58) tested positive for p-α-syn staining, with 95% sensitivity and 91% specificity in discriminating patients from age-matched healthy controls [104], which suggests that detection of p-α-syn in the OM might be a more useful biomarker for LBD than t-α-syn. There are limited data on OM assessments of α-syn pathology in MSA.

3.4 Saliva

The source of salivary α-syn is not definitively known. α-syn can be actively secreted by neurons [105] as it is likely that nerve fibers innervating salivary glands release α-syn into the saliva. However, it is also possible that α-syn in the saliva may originate from CSF and/or plasma through mechanisms that are yet-to-be defined or from salivary cellular components [51].

Compared to CSF and blood, saliva is free of hemolysis. Salivary levels of α-syn have been measured by ELISAs, Luminex assays, and the combination of gel filtration chromatography and Western blots.

Salivary t-α-syn showed a trend towards decreased levels in PD compared to healthy controls. However, such differences did not reach statistical significance in any of the investigations [51, 52, 106, 107], suggesting that salivary t-α-syn, alone, may not be useful as a diagnostic biomarker.

Conversely, salivary o-α-syn may have greater potential as a diagnostic biomarker in PD; different studies reported significantly higher levels in PD compared to healthy controls [52, 106, 107]. Salivary o-α-syn and o-α-syn/t-α-syn were also found to be higher in more advanced disease according to the H&Y stage [106], suggesting that these biomarkers may also be useful in monitoring disease progression.

More recently, α-syn species (t-α-syn, o-α-syn, p-α-syn) derived from EVs have been measured in the saliva from individuals with PD and MSA-parkinsonism predominant (MSA-P) using an electrochemiluminescence assay [108]. T-α-syn levels in EVs were lower in MSA-P compared to PD, and discriminated the two diagnostic groups with good diagnostic accuracy (AUC of 0.804), while no significant differences in EV-derived o-α-syn and p-α-syn levels were found between the two groups.

3.5 *Salivary Glands*

Core needle biopsy of salivary glands, particularly of the submandibular gland, a major salivary gland, has been shown to be a reliable and safe minimally invasive procedure to obtain sufficient glandular parenchymal and neural structures to evaluate α-syn pathology [109, 110].

Over a decade ago, a retrospective autopsy-based study of the submandibular gland was conducted in a neuropathologically confirmed series including: 9 individuals with PD, 3 individuals with incidental LBD (asymptomatic individuals with evidence of brain LB pathology), 2 individuals with MSA, and 19 healthy controls [111]. Positive t-α-syn staining was found in all individuals with PD and in 2 out of 3 subjects with incidental LBD, but not in MSA or healthy controls. These findings suggest that t-α-syn immunostaining in the submandibular gland may serve as a potential biomarker for LBD, being able to discriminate PD/DLB from MSA. Similarly, immunohistochemical analysis of p-α-syn in the submandibular gland was conducted in a large cohort including neuropathologically confirmed PD, incidental LBD, MSA, PSP, and AD and age-matched healthy controls, showing positive staining in all PD patients [112].

More recently, a pilot study analyzed p-α-syn immunoreactivity in nerve fibers obtained from biopsies of minor salivary glands in individuals with PD and healthy controls, demonstrating positive staining in 5 out of 7 individuals with PD and none of the controls [113].

Different investigations revealed positive immunochemistry for α-syn (t-α-syn and p-α-syn) in the submandibular [114] and labial salivary glands [109, 115] in individuals with iRBD, suggesting that α-syn in the salivary glands might be a useful biomarker of synucleinopathies even in the premotor stages of disease.

3.6 *Gut Mucosa*

Clinical and experimental data indicate that the gut may play a role in PD susceptibility. Gastrointestinal symptoms such as constipation often precede the cardinal motor manifestations of PD. In rodent models, exogenous α-syn fibrils introduced into the gut were demonstrated to spread to the brain via the vagal nerve [116]. The vagal route of α-syn transport has also been documented following exposure to the environmental toxin, rotenone, causing α-syn misfolding [117], as well as direct injections of adeno-associated viral vectors overexpressing human α-syn in vagal neurons [118]. It was also demonstrated that vagotomy prevented the formation of brain aggregates when α-syn fibrils were injected into the gut of susceptible mice [119].

According to the "dual-hit hypothesis", an origin of PD-pathology outside the brain has been proposed, placing the initial source of the disease in the gastrointestinal tract after infection by an unknown pathogen that could promote α-syn misfolding and aggregation in the gut with the subsequent spreading to the brain via the autonomic plexuses [120].

Abnormal α-syn aggregates have been histopathologically identified in the enteric nervous system prior to the development of PD. Evidence supporting a role for the enteric nervous system involvement in α-syn pathology includes the observation that α-syn immunoreactive inclusions localize to neurons of the submucosal plexus, whose axons project to the gut mucosa and myenteric plexus, and parallel input of the vagus nerve [121].

To date, the colon has been the most investigated part of the gastrointestinal tract in synucleinopathies. Several studies revealed positive staining for p-α-syn in colon biopsies from PD patients differently from healthy controls [122–128], suggesting that colon biopsies are useful for the detection of α-syn pathology. The presence of α-syn-positive Lewy neurites has been reported in colon biopsies of individuals with PD, specifically in the submucosal plexus [122, 125].

In one study, immunohistochemical analysis of colon biopsies provided positive results for α-syn in individuals with PD 2–5 years prior to the onset of motor symptoms [126], suggesting that α-syn in colon biopsies might be a biomarker for prodromal PD. The amount of α-syn-positive lymph nodes in colon biopsies was also shown to be related to L-dopa-unresponsive features, suggesting its potential role as a prognostic biomarker [125].

However, these findings have not been supported by other studies which showed no differences in α-syn staining of gastric and colonic biopsies between individuals with PD and healthy controls [129, 130]. Inconsistencies among different studies might be due to the use of different quantification/detection methods, pre-analytical confounding factors, or differences in disease stage or severity across the study cohorts.

3.7 Skin

In one of the first studies aimed at investigating α-syn pathology in the skin, Michell et al [131] assessed t-α-syn by using immunoblotting in skin biopsies collected from a small cohort of individuals with PD and healthy controls. In this study, α-syn was not detected in the skin using this technique in most individuals with PD. However, subsequent larger studies using immunochemistry consistently revealed increased t-α-syn staining in PD compared to healthy controls [132, 133].

Many studies on skin biopsies focused on p-α-syn which can be detected in skin nerve fibers by immunohistochemistry and/or immunofluorescence. P-α-syn in proximal peripheral skin nerves was suggested to be a sensitive biomarker for PD, helping to differentiate it from other parkinsonian disorders which are not associated with α-syn pathology (e.g., vascular parkinsonism, tauopathies, and *Parkin* mutation-related parkinsonism) [134]. Consistently, Donadio and colleagues [135] investigated α-syn deposits in multiple skin nerves obtained from 44 individuals with different synucleinopathies including PD, DLB, MSA, and pure autonomic failure (PAF, which is considered a synucleinopathy characterized by predominantly peripheral deposition of α-syn in autonomic ganglia and nerves, with no evidence of CNS dysfunction but with the possibility to progress to other synucleinopathies with CNS involvement, i.e., PD and more frequently MSA) using antibodies that target native α-syn, p-α-syn, and α-syn mature amyloid fibrils. The two main findings from this study were: (i) p-α-syn provided the best sensitivity and specificity for detecting skin α-syn deposits compared to other α-syn species investigated, and (ii) abnormal deposits in MSA were only found in somatic fibers mainly at distal sites differently from PAF, PD, and DLB which demonstrate α-syn deposits in autonomic fibers mainly at proximal sites. These observations suggest that the amount and regional distribution of α-syn aggregates in different cutaneous nerves may provide helpful clues to differentiate various types of synucleinopathies.

More recently, a 30-site, cross-sectional study of academic and community-based neurology practices was performed, and included patients aged 40–99 years with a clinical diagnosis of PD, DLB, MSA, or PAF based on clinical consensus criteria and confirmed by an expert review panel and control participants, in order to evaluate the positivity rate of cutaneous α-syn deposition in these diagnostic groups. The proportions of individuals with cutaneous p-α-syn detected by skin biopsy were 92.7% with PD, 98.2% with MSA, 96.0% with DLB, and 100% (22 of 22) with PAF, while only 3.3% of the controls had cutaneous p-α-syn detected [136].

Notably, p-α-syn deposition in skin nerves has recently been found in a large majority (up to 75%) of individuals with iRBD, suggesting that skin α-syn might be a reliable biomarker reflecting α-syn pathology even in the prodromal stages of synucleinopathies [137].

4 How to Detect Prone-to-Aggregation α-Syn: Seed Amplification Assays (SAAs)

4.1 General Overview of α-Synuclein SAAs

Seed amplification assays (SAAs), including real-time quaking-induced conversion (RT-QuIC) and protein misfolding cyclic amplification (PMCA), are ultrasensitive tools which are capable of amplifying and detecting minute amounts (down to femtograms) of prone-to-aggregation proteins in biofluids and tissues [138–140]. These assays are currently being applied for the diagnosis of CJD, holding promise for the identification of synucleinopathies. A trace amount of α-syn fibrils in tissue extracts can be amplified using these techniques with their conformations conserved [141].

In SAAs, a biological matrix (fluid or tissue) is incubated in a reaction mix containing monomers of the prone-to-aggregation protein of interest, usually in the form of recombinant protein. If present, preformed fibrils and protofibrils (also referred to as "seeds") elongate, using free monomers from the reaction solution. The application of cycles of vigorous shaking or sonication promotes fragmentation of the fibrils thus increasing the number of seeds. This leads to an exponential amplification of the aggregated protein in the reaction mix [142]. In most SAAs, the amplification process can be monitored in real-time by recording the fluorescence of thioflavin-T, a fluorophore with a high affinity for cross-β-sheet motifs of fibrillary aggregates [143].

The first PMCA protocol was developed by Soto's group in 2001 to detect the misfolded prion protein [140]. The multiplication of the template units was performed by sonication followed by an incubation phase to allow the aggregates to grow. These steps were repeated several times in a cyclic process to allow the detection of the misfolded proteins in the samples, i.e., brain homogenates, urine, blood, CSF, and saliva; at the end of the process, proteinase K digestion and Western blot analysis were used to characterize and recognize the presence of pathological aggregates. The PMCA technique was tested in the subsequent years on biological samples coming from animals and patients affected by transmissible spongiform encephalopathy. Atarashi et al., taking advantage of the PMCA method, developed the QuIC assay by introducing some variants in the protocol. In the QuIC, the cellular prion protein (PrPC) substrate coming from hamsters' brain homogenates was replaced by recombinant PrPC and sonication was replaced with a vigorous intermittent shaking which promoted seeded aggregation of the monomeric substrate [144]. Moreover, the Western blot analysis was substituted by a real-time monitoring (hence the name RT-QuIC) of the fluorescence emitted by the amyloid-sensitive thioflavin-T dye during the aggregation process [138].

Among SAAs, prion RT-QuIC applied in CSF and OM is now considered a reliable tool to support the clinical diagnosis of sporadic CJD [145, 146], while classical prion PMCA is being used to detect variant CJD [147]. Although the main clinical applications of SAA are for prion diseases, these assays have also been used to amplify and detect pathological aggregates of α-syn and Aβ [148].

4.2 *Role of α-Syn SAAs in Synucleinopathies and Other Neurodegenerative Disorders*

Since 2016, SAAs have shown great promise as diagnostic tools for synucleinopathies [142, 149–151] (Table 2). Indeed, α-syn SAAs can reliably detect the presence of α-syn aggregates in various synucleinopathies including prodromal phases such as iRBD [152].

Different investigations revealed excellent performance of CSF α-syn SAAs in discriminating PD and DLB from controls [153, 154]; however, CSF α-syn SAAs

Table 2 Diagnostic performance of α-syn SAA in different matrices in discriminating between synucleinopathies and control individuals, between synucleinopathies and other neurodegenerative disorders, and among different synucleinopathies

Cohort	Matrix	Sensitivity	Specificity
PD vs. HC	CSF	86–100%	83–100%
	OM	46–69%	90–100%
	Skin	75–94%	83–98%
	Sliva	76%	94%
DLB vs. HC	CSF	92–100%	91–100%
	OM	81%	92%
	Skin	100%	95%
MSA vs. HC	CSF	6–85%	96–100%
	OM	90%	91–100%
	Saliva	61%	94%
RBD vs. HC	CSF	64–100%	90–98%
	OM	44%	90%
PD vs. PSP/CBS	CSF	84–94%	89–100%
	OM	56%	84%
DLB vs. PSP/CBS	CSF	97%	100%
MSA vs. PSP/CBS	CSF	4–35%	89–100%
	OM	82%	84%
PD vs. MSA	CSF	91–94%	96–97%
PD vs. MSA	OM	65–90%	84–100%
PD vs. MSA	Saliva	61%	84%

Abbreviations. *CBS* corticobasal syndrome, *CSF* cerebrospinal fluid, *DLB* dementia with Lewy bodies, *HC* healthy controls, *RBD* REM behavior disorders, *MSA* multiple system atrophy, *OM* olfactory mucosa, *PSP* progressive supranuclear palsy, *PD* Parkinson's disease

seem to have lower diagnostic utility for differentiating MSA from controls [155], which may further vary by the MSA phenotype (MSA-C or MSA-P) [156].

Some reports have highlighted different kinetic responses (shape of thioflavin-T fluorescence tracers) for MSA and LBD. This is the case, for instance, of the findings from Shahnawaz et al. who showed that lower seeding activities and lower fluorescence were observed in CSF samples from MSA individuals compared to PD [157]. Additionally, the reaction end-products (amplified aggregates) showed structural differences between MSA and LBD [35, 158, 159]. The biochemical, morphological, and structural properties of the final SAA reaction products might be influenced by α-syn strains, which are thought to be different between PD, DLB, and MSA, underlying their neuropathological and phenotypic heterogeneity [157]. The identification of different α-syn strains requires highly specialized technologies, including transmission electron microscopy and nuclear magnetic resonance spectroscopy which can be expensive and time-consuming, thus limiting their applicability in clinical settings. However, these findings indicate that, through specialized analyses of SAAs aggregates, it might be possible to distinguish between different synucleinopathies.

In addition to CSF, α-syn SAA has been examined in other biological fluids. In particular, two easily accessible biological matrices, i.e., the OM and skin biopsy sections, have now gained attention in this field. In skin biopsies, α-syn SAAs showed comparable performance to that of CSF α-syn SAAs in differentiating individuals with PD and DLB from healthy controls and those with other neurological disorders [55, 135, 160, 161].

SAA analyses of the OM have allowed for the identification of α-syn which is prone to aggregation in samples collected from individuals with PD, MSA, and DLB [158, 162, 163]. Notably, MSA has been detected with high sensitivity in OM samples, with the best results in MSA-P (sensitivity 90%) [158, 162]. Conversely, the sensitivity in detecting PD and DLB in this biological matrix ranged between 46% and 81%, albeit with high specificity (90–100%).

Overall, the different outcomes produced by SAAs in amplifying α-syn aggregates from MSA and PD/DLB in different biospecimens, i.e., CSF and OM, may also be explained by considering that the elongation of a particular strain may be favored or hindered in different SAA reaction mixes. Because the chemical environment has been shown to influence the formation and growth of aggregates belonging to different strains [164], it is thus likely that a protocol optimized for the amplification of a PD strain may not efficiently work for the detection of MSA and vice versa. In addition to the strain effect, the distribution of misfolded α-syn in peripheral tissues/biofluids may also depend on the type of synucleinopathy, making a biomatrix able to reflect ongoing synucleinopathy better than another.

To our knowledge, only one published study has evaluated the diagnostic accuracy of α-syn SAAs in saliva samples collected from patients with PD and MSA, showing satisfactory results in discriminating PD and MSA from controls [165].

α-syn SAAs can also differentiate synucleinopathies from other disorders with similar clinical features. CSF samples collected from α-syn-unrelated parkinsonism (e.g., PSP, CBS, and vascular parkinsonism) showed markedly reduced seeding activity in the latter conditions, reliably differentiating them from LBD [154, 155]. While CSF α-syn SAA did not provide good discrimination between MSA and PSP/CBS (sensitivity 4-35%) in one study, likely due to the weak seeding activities of CSF samples from individuals with MSA, α-syn SAA using OM samples distinguished MSA from PSP/CBS with good sensitivity and specificity (82% and 84%, respectively) in another investigation [159].

Genetic analyses can also assist in interpreting results of α-syn SAAs. In one study by Garrido et al. [153], fewer α-syn SAA-positive samples were found in CSF collected from individuals with PD who had mutations in the *LRRK2* gene (46%) compared to individuals with idiopathic PD (90%). This finding is consistent with neuropathological evidence that Lewy bodies are not typically present in the brains of individuals with PD who harbor *LRRK2* mutations [166]. Brockmann et al. [167] investigated the α-syn seeding activity in a wide range of PD-associated genotypes and found that CSF collected from individuals with PD and DLB who harbor *GBA* mutations had the fastest seeding activity and highest prevalence of α-syn SAA-positive samples. In contrast, individuals with PD who carry bi-allelic mutations in recessively inherited genes (e.g., *PARK2* or *PINK1*) did not show CSF α-syn seeding activity, while those carrying heterozygous mutations in these genes had a reduced positivity rate compared to idiopathic PD.

In a recent investigation by Quadalti et al. [168], it was demonstrated that measuring CSF levels of neurofilament-light chain (NfL), which is a marker of axonal damage, could significantly improve the ability of α-syn SAA to differentiate PD from MSA, PSP, and CBS. This suggests that the detection of α-syn by SAA in the CSF, when combined with high levels of neuronal or axonal injury markers (e.g., NfL), may indicate a synucleinopathy characterized by significant neurodegeneration, particularly MSA rather than PD.

SAAs can also be employed to detect α-syn-related pathology in other neurodegenerative disorders in which concomitant α-syn pathology may be present (e.g., AD). Since aggregates of misfolded α-syn can incorporate different aggregation-prone proteins, other proteins such as Aβ, tau, and TDP-43 deposition may occur concurrently in synucleinopathies. AD may also exhibit synuclein aggregation as a secondary proteinopathy [169].

DLB arises from a combination of synucleinopathy and AD-related pathology. The co-occurrence of CSF α-syn SAA positivity and a CSF AD-like profile may indicate a condition in which both synucleinopathy and AD-related pathology coexist. In one study, the prevalence of positive α-syn SAA is reported to be slightly higher in AD compared to controls [154]. This finding is consistent with previous neuropathological studies that have shown Lewy bodies to be present in approximately 50% of clinically diagnosed AD cases [170]. Nonetheless, further adhoc studies with larger sample sizes are needed to confirm the utility of SAA measures of α-syn in AD. It has been suggested that adding fluid or imaging biomarkers of brain α-syn pathology, which is present in 20–50% of participants with AD, will improve the current diagnostic classification for AD which emphasizes the presence of amyloid (A), tau (T), and neurodegeneration (N) and is referred to as the ATN framework [171, 172]. In summary, to improve the diagnosis and prognosis of synucleinopathies, a comprehensive biological profiling approach similar to the A/T/(N) model currently used in AD could be employed [39]. This approach would include the analysis of α-syn SAAs, AD-related biomarkers, NfL, and genetic data, and could potentially allow the identification of biologically homogeneous cohorts for clinical trials of disease-modifying therapies, and thus increase the likelihood of treatment efficacy, as the molecular targets would be more precisely defined.

5 Is It Possible to Image α-Syn Pathology in the Brain?

5.1 α-Syn PET

SPECT and positron emission tomography (PET) imaging are used in routine clinical practice when PD or other degenerative parkinsonisms are suspected. Specifically, SPECT using ioflupane I-123 ([123I]FP-CIT SPECT) reflecting the integrity of the nigro-striatal dopamine transporter shows decreased uptake of the tracer at the level of putamen and caudate nucleus in PD, DLB, and MSA. Abnormal cardiac sympathetic imaging ([123I]-MIBG-scintigraphy) is found in PD and DLB, being able to discriminate LBD from MSA in which [123I]-MIBG-scintigraphy is normal. With respect to DLB, there is normal glucose metabolism as reflected by normal 18F-fluorodeoxyglucose (18F-FDG) uptake in the posterior cingulate region, surrounded by reduced 18F-FDG uptake in the adjacent occipital cortex, representing the so-called "cingulate island" sign, which is able to differentiate DLB from AD [173].

Presynaptic dopaminergic imaging, i.e., [123I]FP-CIT SPECT and [18F]F-DOPA PET, has been used to distinguish parkinsonism due to neurodegenerative disorders from other similar conditions in terms of clinical presentation including essential tremor, vascular, and drug-induced parkinsonism with high sensitivity (98%) and specificity (98%) [174]. Presynaptic dopaminergic imaging is also useful to distinguish DLB from AD [175]. Although dopaminergic deficits detected with SPECT and PET radiotracers can support the clinical diagnosis, two main issues should be considered. First, dopaminergic imaging cannot detect the underlying pathology associated with nigral cell loss, and therefore cannot reliably discriminate synucleinopathies from tauopathies. Further, more than 50% of nigral dopaminergic neurons are already lost at motor symptom onset; therefore, these imaging modalities are not sensitive in early, including prodromal, stages. Therefore, the identification of novel imaging tracers that can discriminate synucleinopathies from α-syn-unrelated parkinsonism, also detecting prodromal disease, would be of utmost relevance in clinical practice. The Michael J. Fox Foundation established the α-Synuclein Imaging Consortium in 2011, launched a $2 million prize for the development of a selective α-syn PET radiotracer in 2016, and in 2019 announced the $10 million Ken Griffin Alpha-synuclein Imaging Competition. Despite these efforts, to date, sensitive and reliable radiotracers that can be used to image α-syn pathology have not yet been established [176]. So far, attempts to target α-syn pathology with neuroimaging biomarkers have been met with limited success. Small-molecule radiotracers and antibody-based radiotracers represent the main strategies to image α-syn in vivo (Table 3).

The main issues for the development of these radiotracers refer to the specificity and selectivity for α-syn aggregates. Specificity of agent binding means that the tracer should only recognize the aggregated form of the protein which is in β-sheet conformation. Selectivity means the agent should ideally distinguish α-syn inclusions from other β-sheet deposits, such as those of tau and amyloid-β proteins [177].

So far, small molecules as ligands mostly bind to the β-sheet structure of α-syn, which is similar to that of neurofibrillary tangles and amyloid-β plaques [178]. Inclusions of α-syn co-localize with numerous other proteins inside the inclusion bodies including tau [179], and Lewy body pathology in PD and DLB often co-occurs with β-amyloid and tau pathologies [180]. To overcome this challenge, more efforts are needed with the current and novel in silico approaches to develop small-molecule ligands that are specific for β-sheet aggregates of α-syn as binding sites [181, 182].

Antibody-based radiotracers offer higher binding affinity and specificity than small-molecule radiotracers [176]. However, as antibodies are large molecules by

Table 3 Summary of the characteristics of radiotracers for α-syn brain imaging

Ligand	Affinity for α-syn fibrils	Binding to α-syn-positive brain homogenates	Selectivity
Small-molecule radiotracers			
[^{11}C]BIP	$K_d = 4$ nM	No binding to LB/LN in DLB and PD brain homogenates	Poor binding selectivity for α-syn vs. Aβ42
[^{11}C]BF227	$K_d = 9.6$ nM	No binding to LB/LN in DLB brain homogenates–High uptake to GCI-rich brain region in MSA	Poor binding selectivity for α-syn vs. Aβ42
[^{125}I]SIL23	$K_d = 148$ nM	Binding to α-syn aggregates in PDD brain	Poor binding selectivity for α-syn vs. Aβ42 and tau
[^{3}H]C05–01	$K_d = 24$ nM	Binding to α-syn aggregates in PD and MSA	Poor binding selectivity for α-syn vs. Aβ42 and tau
[^{3}H] MODAG-001	$K_d = 0.6 \pm 0.1$ nM	No binding to α-syn aggregates in DLB brain	Good binding selectivity for α-syn vs. Aβ42 and tau
[^{18}F]46a	$K_d = 8.9$ nM	NA	Good binding selectivity for α-syn vs. Aβ42 and tau
[^{125}I]61	$K_d = 1.06$ nM	Binding to α-syn-rich regions in PD mouse model	Good binding selectivity for α-syn vs. Aβ42
[^{18}F]2FBox	$K_d = 3.3 \pm 2.8$ nM	No binding to LB in PD and MSA brains	Good binding selectivity for α-syn vs. Aβ42
Antibody-based radiotracers			
MEDI1341	$K_d = 74$ pM	Binding to extracellular α-syn aggregates	NA
BIIB054	$K_d \sim 120$ pM	NA	NA

Abbreviations. *DLB* dementia with Lewy bodies, *GCI* glial cytoplasmic inclusions, *LB* Lewy bodies, *LN* Lewy neurites, *MSA* multiple system atrophy, *PD* Parkinson's disease, *PDD* Parkinson's disease with dementia

nature, the main limitation for radiolabeled antibodies pertains to their ability to effectively cross the BBB [183, 184]. Potential strategies to overcome this limitation include receptor-mediated transcytosis [185] and microbubble-mediated focused ultrasound, a non-invasive imaging technique which is also being investigated as a potential tool to enhance drug delivery to the brain by reversibly increasing the permeability of the BBB [186]. Another relevant limitation lies in the slow pharmacokinetics of antibodies, showing slow target accumulation and clearance, which requires the use of long-lived radionuclides for PET imaging, thus resulting in a high risk for radiation exposure [187]. Further research in this area is warranted.

5.2 Magnetic Resonance Imaging

Magnetic resonance imaging (MRI) is more widely available compared to PET/SPECT and is used to exclude secondary causes of parkinsonism such as vascular lesions, non-obstructive chronic hydrocephalus, and, less frequently, brain tumors. Therefore, MRI is commonly used in routine clinical settings as a screening diagnostic tool to differentiate degenerative parkinsonian syndromes from other causes of parkinsonism.

More sophisticated sequences are mostly used in advanced research center settings. This is the case, for instance, of the iron-sensitive MRI sequences (i.e., susceptibility-weighted imaging, SWI) for the quantitative assessment of iron in the basal ganglia [188], and diffusion tensor imaging (DTI) for the evaluation of potential abnormalities in the olfactory bulb and olfactory tract [189].

Apart from its wide availability, MRI has the advantage of high spatial resolution, allowing resolution of anatomical areas affected by α-syn inclusions. Additionally, the development and increasing availability of high-field (7T) imaging is likely to be an important factor. High-field MR imaging allows significantly improved resolution of even deep sub-cortical nuclei and substantia nigra involved in PD and other synucleinopathies. Therefore, the combination of novel MRI contrast agents with this level of resolution could be beneficial to detect α-syn inclusions by MRI. The imaging of molecular pathology by MRI has also been previously demonstrated by the use of non-contrast-enhanced MRI to visualize amyloid plaques in the brains of transgenic mouse models of AD. However, amyloid plaques contain iron, which renders them detectable without exogenous contrast agents using T2-weighted spin echo pulse sequences. The substantia nigra has a high iron content, which is further elevated in PD. While iron may play a role in α-syn aggregation, Lewy bodies themselves do not contain iron. Therefore, it seems unlikely to detect α-syn aggregates without using non-contrast-enhanced MRI, particularly in the iron-rich substantia nigra, limiting the application of MRI [177].

MRI contrast agents have several disadvantages including: (i) poor circulation time, (ii) low capability of readily crossing the BBB, and (iii) cell toxicity. Several expedients can be potentially used to overcome such issues. Poor circulation time may be overcome by reducing contrast agent size or equipping them with stealth peptides, allowing these agents to stay longer in the circulation and enhancing the possibility to cross the BBB. To facilitate the penetration across the CNS, actively transported peptides or proteins can be used. Targeted fluorinated MRI contrast agents might overcome the issue of toxicity, although they suffer from a low signal-to-noise ratio [177].

Overall, the current application of MRI to detect α-syn pathology is very limited, not only in clinical practice but also in research settings.

6 Final Remarks and Future Directions

While a definite diagnosis of PD and other synucleinopathies can only be made post-mortem, there is a great need to identify and validate fluid or imaging biomarkers that can improve in vivo clinical diagnostic and prognostic assessments. The development of novel—potentially disease-modifying—therapies has been limited by the lack of early diagnostic tools and prognostic markers that can monitor disease progression over the course of treatment. Thus, there is an urgent need to identify in vivo biomarkers that can reliably reflect α-syn-related pathology, with the aim of reaching an early and accurate diagnosis and prognostic characterization of these neurodegenerative disorders.

CSF is directly in contact with the brain tissue, thus representing a valuable source of biomarkers able to reflect the early pathophysiological mechanisms taking place in the CNS. However, the utility of CSF t-α-syn is limited by confounding factors. α-syn aggregation in brain tissue is associated with lower CSF t-α-syn levels; however, synaptic injury due to the ongoing degeneration is associated with higher CSF t-α-syn levels, which limits the accurate interpretation of abnormal CSF t-α-syn levels. Further, α-syn species (i.e., o-α-syn and p-α-syn) measured in the CSF offer limited sensitivity and specificity as diagnostic biomarkers.

T-α-syn and other α-syn species have also been investigated in other biofluids, including blood and saliva. However, the measurements of α-syn species in these peripheral biofluids resulted in inconclusive findings, mostly due to variability in several pre-analytical and analytical factors across studies.

Immunohistochemical analysis of α-syn in peripheral tissues suggests that OM, gut mucosa, and salivary glands might be useful for detecting α-syn pathology even in the early stages. Further studies in larger cohorts with longitudinal follow-up are required to assess the usefulness of such matrices, also taking into account the need to implement standardized methods and to minimize pre-analytical and analytical variability.

At present, α-syn SAAs represent the most promising techniques for the identification of aggregation-prone α-syn species, even for prodromal disease (i.e., in patients with iRBD). The possibility to apply SAAs in easily accessible peripheral tissues such as skin biopsies and OM samples makes them of utmost interest both for diagnostic and monitoring purposes. The combination of different biospecimens (e.g., CSF and OM), as well as the combination of biomarkers reflecting different pathophysiological mechanisms, represents the way forward for an accurate early diagnosis, for better prognostic predictions, and for a personalized therapeutic approach. Overall, SAAs, which are currently the most reliable indicators of synucleinopathies, need to be validated before they can be translated into effective diagnostic and prognostic tools in the clinic.

References

1. Barba L, Paolini Paoletti F, Bellomo G, Gaetani L, Halbgebauer S, Oeckl P, Otto M, Parnetti L. Alpha and beta synucleins: from pathophysiology to clinical application as biomarkers. Mov Disord. 2022;37(4):669–83. https://doi.org/10.1002/mds.28941.
2. McKeith IG, Boeve BF, Dickson DW, Halliday G, Taylor J-P, Weintraub D, Aarsland D, Galvin J, Attems J, Ballard CG, Bayston A, Beach TG, Blanc F, Bohnen N, Bonanni L, Bras J, Brundin P, Burn D, Chen-Plotkin A, et al. Diagnosis and management of dementia with Lewy bodies: fourth consensus report of the DLB consortium. Neurology. 2017;89(1):88–100. https://doi.org/10.1212/WNL.0000000000004058.
3. Postuma RB, Berg D, Stern M, Poewe W, Olanow CW, Oertel W, Obeso J, Marek K, Litvan I, Lang AE, Halliday G, Goetz CG, Gasser T, Dubois B, Chan P, Bloem BR, Adler CH, Deuschl G. MDS clinical diagnostic criteria for Parkinson's disease. Mov Disord. 2015;30(12):1591–601. https://doi.org/10.1002/mds.26424.
4. Wenning GK, Stankovic I, Vignatelli L, Fanciulli A, Calandra-Buonaura G, Seppi K, Palma J-A, Meissner WG, Krismer F, Berg D, Cortelli P, Freeman R, Halliday G, Höglinger G, Lang A, Ling H, Litvan I, Low P, Miki Y, et al. The movement disorder society criteria for the diagnosis of multiple system atrophy. Mov Disord. 2022;37(6):1131–48. https://doi.org/10.1002/mds.29005.
5. Jellinger KA. How valid is the clinical diagnosis of Parkinson's disease in the community? J Neurol Neurosurg Psychiatry. 2003;74(7):1005–6. https://doi.org/10.1136/jnnp.74.7.1005.
6. Rajput AH, Rajput A. Accuracy of Parkinson disease diagnosis unchanged in 2 decades. Neurology. 2014;83(5):386–7. https://doi.org/10.1212/WNL.0000000000000653.
7. Rizzo G, Copetti M, Arcuti S, Martino D, Fontana A, Logroscino G. Accuracy of clinical diagnosis of Parkinson disease: a systematic review and meta-analysis. Neurology. 2016;86(6):566–76. https://doi.org/10.1212/WNL.0000000000002350.
8. McCann H, Stevens CH, Cartwright H, Halliday GM. α-Synucleinopathy phenotypes. Parkinsonism Relat Disord. 2014;20 Suppl 1:S62–7. https://doi.org/10.1016/S1353-8020(13)70017-8.
9. Kim H-J, Jeon BS, Shin J, Lee W-W, Park H, Jung YJ, Ehm G. Should genetic testing for SCAs be included in the diagnostic workup for MSA? Neurology. 2014;83(19):1733–8. https://doi.org/10.1212/WNL.0000000000000965.
10. Paolini Paoletti F, Prontera P, Nigro P, Simoni S, Cappelletti G, Filidei M, Calabresi P, Parnetti L, Tambasco N. Small-expanded allele spinocerebellar ataxia 17: imaging and phenotypic variability. Neurol Sci. 2021;42(10):4309–15. https://doi.org/10.1007/s10072-021-05313-z.
11. Koga S, Aoki N, Uitti RJ, van Gerpen JA, Cheshire WP, Josephs KA, Wszolek ZK, Langston JW, Dickson DW. When DLB, PD, and PSP masquerade as MSA: an autopsy study of 134 patients. Neurology. 2015;85(5):404–12. https://doi.org/10.1212/WNL.0000000000001807.
12. Stern M, Lang A, Poewe W. Toward a redefinition of Parkinson's disease. Mov Disord. 2012;27(1):54–60. https://doi.org/10.1002/mds.24051.
13. Berg D, Postuma RB, Adler CH, Bloem BR, Chan P, Dubois B, Gasser T, Goetz CG, Halliday G, Joseph L, Lang AE, Liepelt-Scarfone I, Litvan I, Marek K, Obeso J, Oertel W, Olanow CW, Poewe W, Stern M, Deuschl G. MDS research criteria for prodromal Parkinson's disease. Mov Disord. 2015;30(12):1600–11. https://doi.org/10.1002/mds.26431.
14. McKeith IG, Ferman TJ, Thomas AJ, Blanc F, Boeve BF, Fujishiro H, Kantarci K, Muscio C, O'Brien JT, Postuma RB, Aarsland D, Ballard C, Bonanni L, Donaghy P, Emre M, Galvin JE, Galasko D, Goldman JG, Gomperts SN, et al. Research criteria for the diagnosis of prodromal dementia with Lewy bodies. Neurology. 2020;94(17):743–55. https://doi.org/10.1212/WNL.0000000000009323.
15. Xia C, Postuma RB. Diagnosing multiple system atrophy at the prodromal stage. Clin Auton Res. 2020;30(3):197–205. https://doi.org/10.1007/s10286-020-00682-5.
16. Fereshtehnejad S-M, Montplaisir JY, Pelletier A, Gagnon J-F, Berg D, Postuma RB. Validation of the MDS research criteria for prodromal Parkinson's disease: longitudinal assessment in

a REM sleep behavior disorder (RBD) cohort. Mov Disord. 2017;32(6):865–73. https://doi.org/10.1002/mds.26989.

17. Giagkou N, Maraki MI, Yannakoulia M, Kosmidis MH, Dardiotis E, Hadjigeorgiou GM, Sakka P, Ntanasi E, Anastasiou CA, Xiromerisiou G, Stefanis L, Scarmeas N, Stamelou M. A prospective validation of the updated movement disorders society research criteria for prodromal Parkinson's disease. Mov Disord. 2020;35(10):1802–9. https://doi.org/10.1002/mds.28145.
18. Mirelman A, Saunders-Pullman R, Alcalay RN, Shustak S, Thaler A, Gurevich T, Raymond D, Mejia-Santana H, Orbe Reilly M, Ozelius L, Clark L, Gana-Weisz M, Bar-Shira A, Orr-Utreger A, Bressman SB, Marder K, Giladi N, AJ LRRK2 Consortium. Application of the Movement Disorder Society prodromal criteria in healthy G2019S-LRRK2 carriers. Mov Disord. 2018;33(6):966–73. https://doi.org/10.1002/mds.27342.
19. Phillips JR, Matar E, Ehgoetz Martens KA, Moustafa AA, Halliday GM, Lewis SJG. Exploring the sensitivity of prodromal dementia with Lewy bodies research criteria. Brain Sci. 2022;12(12):1594. https://doi.org/10.3390/brainsci12121594.
20. Alves G, Wentzel-Larsen T, Aarsland D, Larsen JP. Progression of motor impairment and disability in Parkinson disease: a population-based study. Neurology. 2005;65(9):1436–41. https://doi.org/10.1212/01.wnl.0000183359.50822.f2.
21. Jankovic J, Kapadia AS. Functional decline in Parkinson disease. Arch Neurol. 2001;58(10):1611–5. https://doi.org/10.1001/archneur.58.10.1611.
22. Reinoso G, Allen JC, Au W-L, Seah S-H, Tay K-Y, Tan LCS. Clinical evolution of Parkinson's disease and prognostic factors affecting motor progression: 9-year follow-up study. Eur J Neurol. 2015;22(3):457–63. https://doi.org/10.1111/ene.12476.
23. Ding J, Zhang J, Wang X, Zhang L, Jiang S, Yuan Y, Li J, Zhu L, Zhang K. Relationship between the plasma levels of neurodegenerative proteins and motor subtypes of Parkinson's disease. J Neural Transm (Vienna). 2017;124(3):353–60. https://doi.org/10.1007/s00702-016-1650-2.
24. Evans JR, Mason SL, Williams-Gray CH, Foltynie T, Brayne C, Robbins TW, Barker RA. The natural history of treated Parkinson's disease in an incident, community based cohort. J Neurol Neurosurg Psychiatry. 2011;82(10):1112–8. https://doi.org/10.1136/jnnp.2011.240366.
25. Marras C, Rochon P, Lang AE. Predicting motor decline and disability in Parkinson disease: a systematic review. Arch Neurol. 2002;59(11):1724–8. https://doi.org/10.1001/archneur.59.11.1724.
26. Simuni T, Siderowf A, Lasch S, Coffey CS, Caspell-Garcia C, Jennings D, Tanner CM, Trojanowski JQ, Shaw LM, Seibyl J, Schuff N, Singleton A, Kieburtz K, Toga AW, Mollenhauer B, Galasko D, Chahine LM, Weintraub D, Foroud T, et al. Longitudinal change of clinical and biological measures in early Parkinson's disease: Parkinson's Progression Markers Initiative cohort. Mov Disord. 2018;33(5):771–82. https://doi.org/10.1002/mds.27361.
27. Aarsland D, Batzu L, Halliday GM, Geurtsen GJ, Ballard C, Ray Chaudhuri K, Weintraub D. Parkinson disease-associated cognitive impairment. Nat Rev Dis Primers. 2021;7(1):47. https://doi.org/10.1038/s41572-021-00280-3.
28. Pedersen KF, Larsen JP, Tysnes O-B, Alves G. Natural course of mild cognitive impairment in Parkinson disease: a 5-year population-based study. Neurology. 2017;88(8):767–74. https://doi.org/10.1212/WNL.0000000000003634.
29. Hely MA, Reid WGJ, Adena MA, Halliday GM, Morris JGL. The Sydney multicenter study of Parkinson's disease: the inevitability of dementia at 20 years. Mov Disord. 2008;23(6):837–44. https://doi.org/10.1002/mds.21956.
30. Pedersen KF, Larsen JP, Tysnes O-B, Alves G. Prognosis of mild cognitive impairment in early Parkinson disease: the Norwegian ParkWest study. JAMA Neurol. 2013;70(5):580–6. https://doi.org/10.1001/jamaneurol.2013.2110.
31. Baiano C, Barone P, Trojano L, Santangelo G. Prevalence and clinical aspects of mild cognitive impairment in Parkinson's disease: a meta-analysis. Mov Disord. 2020;35(1):45–54. https://doi.org/10.1002/mds.27902. Epub 2019 Nov 19

32. Marinus J, Zhu K, Marras C, Aarsland D, van Hilten JJ. Risk factors for non-motor symptoms in Parkinson's disease. Lancet Neurol. 2018;17(6):559–68. https://doi.org/10.1016/S1474-4422(18)30127-3.
33. Halliday GM, Holton JL, Revesz T, Dickson DW. Neuropathology underlying clinical variability in patients with synucleinopathies. Acta Neuropathol. 2011;122(2):187–204. https://doi.org/10.1007/s00401-011-0852-9.
34. Irwin DJ, White MT, Toledo JB, Xie SX, Robinson JL, Van Deerlin V, Lee VM-Y, Leverenz JB, Montine TJ, Duda JE, Hurtig HI, Trojanowski JQ. Neuropathologic substrates of Parkinson disease dementia. Ann Neurol. 2012;72(4):587–98. https://doi.org/10.1002/ana.23659.
35. Koga S, Sekiya H, Kondru N, Ross OA, Dickson DW. Neuropathology and molecular diagnosis of Synucleinopathies. Mol Neurodegener. 2021;16(1):83. https://doi.org/10.1186/s13024-021-00501-z.
36. Malfertheiner K, Stefanova N, Heras-Garvin A. The concept of α-synuclein strains and how different conformations may explain distinct neurodegenerative disorders. Front Neurol. 2021;12:737195. https://doi.org/10.3389/fneur.2021.737195.
37. Soto C, Pritzkow S. Protein misfolding, aggregation, and conformational strains in neurodegenerative diseases. Nat Neurosci. 2018;21:1332–40. https://doi.org/10.1038/s41593-018-0235-9.
38. Irwin DJ, Grossman M, Weintraub D, Hurtig HI, Duda JE, Xie SX, Lee EB, Van Deerlin VM, Lopez OL, Kofler JK, Nelson PT, Jicha GA, Woltjer R, Quinn JF, Kaye J, Leverenz JB, Tsuang D, Longfellow K, Yearout D, et al. Neuropathological and genetic correlates of survival and dementia onset in synucleinopathies: a retrospective analysis. Lancet Neurol. 2017;16(1):55–65. https://doi.org/10.1016/S1474-4422(16)30291-5.
39. Jack CR, Bennett DA, Blennow K, Carrillo MC, Dunn B, Haeberlein SB, Holtzman DM, Jagust W, Jessen F, Karlawish J, Liu E, Molinuevo JL, Montine T, Phelps C, Rankin KP, Rowe CC, Scheltens P, Siemers E, Snyder HM, et al. NIA-AA research framework: toward a biological definition of Alzheimer's disease. Alzheimers Dement. 2018;14(4):535–62. https://doi.org/10.1016/j.jalz.2018.02.018.
40. Bellomo G, Paolini Paoletti F, Chipi E, Petricciuolo M, Simoni S, Tambasco N, Parnetti L. A/T/(N) profile in cerebrospinal fluid of Parkinson's disease with/without cognitive impairment and dementia with Lewy bodies. Diagnostics (Basel, Switzerland). 2020;10(12):1015. https://doi.org/10.3390/diagnostics10121015.
41. van Steenoven I, Aarsland D, Weintraub D, Londos E, Blanc F, van der Flier WM, Teunissen CE, Mollenhauer B, Fladby T, Kramberger MG, Bonanni L, Lemstra AW, European DLB, consortium. Cerebrospinal fluid Alzheimer's disease biomarkers across the spectrum of Lewy body diseases: results from a large multicenter cohort. J Alzheimers Dis. 2016;54(1):287–95. https://doi.org/10.3233/JAD-160322.
42. Fereshtehnejad SM, Romenets SR, Anang JB, Latreille V, Gagnon JF, Postuma RB. New clinical subtypes of Parkinson disease and their longitudinal progression: a prospective cohort comparison with other phenotypes. JAMA Neurol. 2015;72(8):863–73. https://doi.org/10.1001/jamaneurol.2015.0703.
43. Parnetti L, Farotti L, Eusebi P, Chiasserini D, De Carlo C, Giannandrea D, Salvadori N, Lisetti V, Tambasco N, Rossi A, Majbour NK, El-Agnaf O, Calabresi P. Differential role of CSF alpha-synuclein species, tau, and Aβ42 in Parkinson's disease. Front Aging Neurosci. 2014;6:53. https://doi.org/10.3389/fnagi.2014.00053.
44. Siderowf A, Xie SX, Hurtig H, Weintraub D, Duda J, Chen-Plotkin A, Shaw LM, Van Deerlin V, Trojanowski JQ, Clark C. CSF amyloid {beta} 1–42 predicts cognitive decline in Parkinson disease. Neurology. 2010;75(12):1055–61. https://doi.org/10.1212/WNL.0b013e3181f39a78.
45. Parnetti L, Tiraboschi P, Lanari A, Peducci M, Padiglioni C, D'Amore C, Pierguidi L, Tambasco N, Rossi A, Calabresi P. Cerebrospinal fluid biomarkers in Parkinson's disease with dementia and dementia with Lewy bodies. Biol Psychiatry. 2008;64(10):850–5. https://doi.org/10.1016/j.biopsych.2008.02.016.

46. Lemstra AW, de Beer MH, Teunissen CE, Schreuder C, Scheltens P, van der Flier WM, Sikkes S, a. M. Concomitant AD pathology affects clinical manifestation and survival in dementia with Lewy bodies. J Neurol Neurosurg Psychiatry. 2017;88(2):113–8. https://doi.org/10.1136/jnnp-2016-313775.
47. Poewe W, Seppi K, Tanner CM, Halliday GM, Brundin P, Volkmann J, Schrag A-E, Lang AE. Parkinson disease. Nat Rev Dis Primers. 2017;3:17013. https://doi.org/10.1038/nrdp.2017.13.
48. Schmid AW, Fauvet B, Moniatte M, Lashuel HA. Alpha-synuclein post-translational modifications as potential biomarkers for Parkinson disease and other synucleinopathies. Mol Cell Proteomics. 2013;12(12):3543–58. https://doi.org/10.1074/mcp.R113.032730.
49. Cariulo C, Martufi P, Verani M, Azzollini L, Bruni G, Weiss A, Deguire SM, Lashuel HA, Scaricamazza E, Sancesario GM, Schirinzi T, Mercuri NB, Sancesario G, Caricasole A, Petricca L. Phospho-S129 alpha-synuclein is present in human plasma but not in cerebrospinal fluid as determined by an ultrasensitive immunoassay. Front Neurosci. 2019;13:889. https://doi.org/10.3389/fnins.2019.00889.
50. Foulds PG, Diggle P, Mitchell JD, Parker A, Hasegawa M, Masuda-Suzukake M, Mann DMA, Allsop D. A longitudinal study on α-synuclein in blood plasma as a biomarker for Parkinson's disease. Sci Rep. 2013;3:2540. https://doi.org/10.1038/srep02540.
51. Devic I, Hwang H, Edgar JS, Izutsu K, Presland R, Pan C, Goodlett DR, Wang Y, Armaly J, Tumas V, Zabetian CP, Leverenz JB, Shi M, Zhang J. Salivary α-synuclein and DJ-1: potential biomarkers for Parkinson's disease. Brain. 2011;134(Pt 7):e178. https://doi.org/10.1093/brain/awr015.
52. Kang W, Chen W, Yang Q, Zhang L, Zhang L, Wang X, Dong F, Zhao Y, Chen S, Quinn TJ, Zhang J, Chen S, Liu J. Salivary total α-synuclein, oligomeric α-synuclein and SNCA variants in Parkinson's disease patients. Sci Rep. 2016;6:28143. https://doi.org/10.1038/srep28143.
53. Fenyi A, Leclair-Visonneau L, Clairembault T, Coron E, Neunlist M, Melki R, Derkinderen P, Bousset L. Detection of alpha-synuclein aggregates in gastrointestinal biopsies by protein misfolding cyclic amplification. Neurobiol Dis. 2019;129:38–43. https://doi.org/10.1016/j.nbd.2019.05.002.
54. Tanei Z-I, Saito Y, Ito S, Matsubara T, Motoda A, Yamazaki M, Sakashita Y, Kawakami I, Ikemura M, Tanaka S, Sengoku R, Arai T, Murayama S. Lewy pathology of the esophagus correlates with the progression of Lewy body disease: a Japanese cohort study of autopsy cases. Acta Neuropathol. 2021;141(1):25–37. https://doi.org/10.1007/s00401-020-02233-8.
55. Wang Z, Becker K, Donadio V, Siedlak S, Yuan J, Rezaee M, Incensi A, Kuzkina A, Orrú CD, Tatsuoka C, Liguori R, Gunzler SA, Caughey B, Jimenez-Capdeville ME, Zhu X, Doppler K, Cui L, Chen SG, Ma J, Zou W-Q. Skin α-synuclein aggregation seeding activity as a novel biomarker for Parkinson disease. JAMA Neurol. 2020;78:1. https://doi.org/10.1001/jamaneurol.2020.3311.
56. Mollenhauer B, Bowman FD, Drake D, Duong J, Blennow K, El-Agnaf O, Shaw LM, Masucci J, Taylor P, Umek RM, Dunty JM, Smith CL, Stoops E, Vanderstichele H, Schmid AW, Moniatte M, Zhang J, Kruse N, Lashuel HA, et al. Antibody-based methods for the measurement of α-synuclein concentration in human cerebrospinal fluid—method comparison and round robin study. J Neurochem. 2019;149(1):126–38. https://doi.org/10.1111/jnc.14569.
57. Eusebi P, Giannandrea D, Biscetti L, Abraha I, Chiasserini D, Orso M, Calabresi P, Parnetti L. Diagnostic utility of cerebrospinal fluid α-synuclein in Parkinson's disease: a systematic review and meta-analysis. Mov Disord. 2017;32(10):1389–400. https://doi.org/10.1002/mds.27110.
58. Gao L, Tang H, Nie K, Wang L, Zhao J, Gan R, Huang J, Zhu R, Feng S, Duan Z, Zhang Y, Wang L. Cerebrospinal fluid alpha-synuclein as a biomarker for Parkinson's disease diagnosis: a systematic review and meta-analysis. Int J Neurosci. 2015;125(9):645–54. https://doi.org/10.3109/00207454.2014.961454.

59. Sako W, Murakami N, Izumi Y, Kaji R. Reduced alpha-synuclein in cerebrospinal fluid in synucleinopathies: evidence from a meta-analysis. Mov Disord. 2014;29(13):1599–605. https://doi.org/10.1002/mds.26036.
60. Zhou B, Wen M, Yu W-F, Zhang C-L, Jiao L. The diagnostic and differential diagnosis utility of cerebrospinal fluid α -synuclein levels in Parkinson's disease: a meta-analysis. Parkinsons Dis. 2015;2015:567386. https://doi.org/10.1155/2015/567386.
61. Mollenhauer B, Parnetti L, Rektorova I, Kramberger MG, Pikkarainen M, Schulz-Schaeffer WJ, Aarsland D, Svenningsson P, Farotti L, Verbeek MM, Schlossmacher MG. Biological confounders for the values of cerebrospinal fluid proteins in Parkinson's disease and related disorders. J Neurochem. 2016;139(Suppl 1):290–317. https://doi.org/10.1111/jnc.13390.
62. Barbour R, Kling K, Anderson JP, Banducci K, Cole T, Diep L, Fox M, Goldstein JM, Soriano F, Seubert P, Chilcote TJ. Red blood cells are the major source of alpha-synuclein in blood. Neurodegener Dis. 2008;5(2):55–9. https://doi.org/10.1159/000112832.
63. Petricca L, Chiki N, Hanna-El-Daher L, Aeschbach L, Burai R, Stoops E, et al. Comparative analysis of total alpha-synuclein (αSYN) immunoassays reveals that they do not capture the diversity of modified αSYN proteoforms. J Parkinsons Dis. 2022;12:1449. https://doi.org/10.3233/JPD-223285.
64. Bäckström DC, Eriksson Domellöf M, Linder J, Olsson B, Öhrfelt A, Trupp M, Zetterberg H, Blennow K, Forsgren L. Cerebrospinal fluid patterns and the risk of future dementia in early, incident parkinson disease. JAMA Neurol. 2015;72(10):1175–82. https://doi.org/10.1001/jamaneurol.2015.1449.
65. Magdalinou NK, Paterson RW, Schott JM, Fox NC, Mummery C, Blennow K, Bhatia K, Morris HR, Giunti P, Warner TT, de Silva R, Lees AJ, Zetterberg H. A panel of nine cerebrospinal fluid biomarkers may identify patients with atypical parkinsonian syndromes. J Neurol Neurosurg Psychiatry. 2015;86(11):1240–7. https://doi.org/10.1136/jnnp-2014-309562.
66. Wennström M, Surova Y, Hall S, Nilsson C, Minthon L, Boström F, Hansson O, Nielsen HM. Low CSF levels of both α-synuclein and the α-synuclein cleaving enzyme neurosin in patients with synucleinopathy. PLoS One. 2013;8(1):e53250. https://doi.org/10.1371/journal.pone.0053250.
67. Barba L, Abu Rumeileh S, Bellomo G, Paolini Paoletti F, Halbgebauer S, Oeckl P, Steinacker P, Massa F, Gaetani L, Parnetti L, Otto M. Cerebrospinal fluid β-synuclein as a synaptic biomarker for preclinical Alzheimer's disease. J Neurol Neurosurg Psychiatry. 2023;94(1):83–6. https://doi.org/10.1136/jnnp-2022-329124.
68. Oeckl P, Metzger F, Nagl M, von Arnim CAF, Halbgebauer S, Steinacker P, Ludolph AC, Otto M. Alpha-, beta-, and gamma-synuclein quantification in cerebrospinal fluid by multiple reaction monitoring reveals increased concentrations in Alzheimer's and Creutzfeldt- Jakob disease but no alteration in synucleinopathies. Mol Cell Proteomics. 2016;15(10):3126–38. https://doi.org/10.1074/mcp.M116.059915.
69. Hall S, Surova Y, Öhrfelt A, Swedish BioFINDER Study, Blennow K, Zetterberg H, Hansson O. Longitudinal measurements of cerebrospinal fluid biomarkers in Parkinson's disease. Mov Disord. 2016;31(6):898–905. https://doi.org/10.1002/mds.26578.
70. Hall S, Surova Y, Öhrfelt A, Zetterberg H, Lindqvist D, Hansson O. CSF biomarkers and clinical progression of Parkinson disease. Neurology. 2015;84(1):57–63. https://doi.org/10.1212/WNL.0000000000001098.
71. Stewart T, Liu C, Ginghina C, et al. Cerebrospinal fluid α-synuclein predicts cognitive decline in Parkinson disease progression in the DATATOP cohort. Am J Pathol. 2014;184:966–75. https://doi.org/10.1016/j.ajpath.2013.12.007.
72. Conway KA, Lee SJ, Rochet JC, Ding TT, Harper JD, Williamson RE, Lansbury PT. Accelerated oligomerization by Parkinson's disease linked alpha-synuclein mutants. Ann N Y Acad Sci. 2000;920:42–5. https://doi.org/10.1111/j.1749-6632.2000.tb06903.x.
73. Xiang C, Cong S, Tan X, Ma S, Liu Y, Wang H, Cong S. A meta-analysis of the diagnostic utility of biomarkers in cerebrospinal fluid in Parkinson's disease. NPJ Parkinsons Dis. 2022;8(1):165. https://doi.org/10.1038/s41531-022-00431-7.

74. Majbour NK, Abdi IY, Dakna M, Wicke T, Lang E, Ali Moussa HY, Thomas MA, Trenkwalder C, Safieh-Garabedian B, Tokuda T, Mollenhauer B, El-Agnaf O. Cerebrospinal α-synuclein oligomers reflect disease motor severity in DeNoPa longitudinal cohort. Mov Disord. 2021;36(9):2048–56. https://doi.org/10.1002/mds.28611.
75. van Steenoven I, Majbour NK, Vaikath NN, Berendse HW, van der Flier WM, van de Berg WDJ, Teunissen CE, Lemstra AW, El-Agnaf OMA. α-Synuclein species as potential cerebrospinal fluid biomarkers for dementia with lewy bodies. Mov Disord. 2018;33(11):1724–33. https://doi.org/10.1002/mds.111.
76. Majbour NK, Vaikath NN, Eusebi P, Chiasserini D, Ardah M, Varghese S, Haque ME, Tokuda T, Auinger P, Calabresi P, Parnetti L, El-Agnaf OMA. Longitudinal changes in CSF alpha-synuclein species reflect Parkinson's disease progression. Mov Disord. 2016;31(10):1535–42. https://doi.org/10.1002/mds.26754.
77. Fujiwara H, Hasegawa M, Dohmae N, Kawashima A, Masliah E, Goldberg MS, Shen J, Takio K, Iwatsubo T. Alpha-Synuclein is phosphorylated in synucleinopathy lesions. Nat Cell Biol. 2002;4(2):160–4. https://doi.org/10.1038/ncb748.
78. Constantinides VC, Majbour NK, Paraskevas GP, Abdi I, Safieh-Garabedian B, Stefanis L, El-Agnaf O, Kapaki E. Cerebrospinal fluid α-synuclein species in cognitive and movements disorders. Brain Sci. 2021;11(1):119. https://doi.org/10.3390/brainsci11010119.
79. Foulds PG, Yokota O, Thurston A, Davidson Y, Ahmed Z, Holton J, et al. Post mortem cerebrospinal fluid α-synuclein levels are raised in multiple system atrophy and distinguish this from the other α-synucleinopathies, Parkinson's disease and Dementia with Lewy bodies. Neurobiol Dis. 2012;45:188. https://doi.org/10.1016/j.nbd.2011.08.003.
80. Laurens B, Constantinescu R, Freeman R, Gerhard A, Jellinger K, Jeromin A, et al. Fluid biomarkers in multiple system atrophy: A review of the MSA biomarker initiative. Neurobiol Dis. 2015;80:29. https://doi.org/10.1016/j.nbd.2015.05.004.
81. Bougea A, Stefanis L, Emmanouilidou E, Vekrelis K, Kapaki E. High discriminatory ability of peripheral and CFSF biomarkers in Lewy body diseases. J Neural Transm. 2020;127(3):311–22. https://doi.org/10.1007/s00702-019-02137-2.
82. Chang C-W, Yang S-Y, Yang C-C, Chang C-W, Wu Y-R. Plasma and serum alpha- synuclein as a biomarker of diagnosis in patients with Parkinson's disease. Front Neurol. 2019;10:1388. https://doi.org/10.3389/fneur.2019.01388.
83. Chen C-H, Lee B-C, Lin C-H. Integrated plasma and neuroimaging biomarkers associated with motor and cognition severity in Parkinson's disease. J Parkinsons Dis. 2020;10(1):77–88. https://doi.org/10.3233/JPD-191766.
84. Lee PH, Lee G, Park HJ, Bang OY, Joo IS, Huh K. The plasma alpha-synuclein levels in patients with Parkinson's disease and multiple system atrophy. J Neural Transm (Vienna). 2006;113(10):1435–9. https://doi.org/10.1007/s00702-005-0427-9.
85. Zubelzu M, Morera-Herreras T, Irastorza G, Gómez-Esteban JC, Murueta-Goyena A. Plasma and serum alpha-synuclein as a biomarker in Parkinson's disease: a meta-analysis. Parkinsonism Relat Disord. 2022;99:107–15. https://doi.org/10.1016/j.parkreldis.2022.06.001.
86. Chahine LM, Beach TG, Brumm MC, Adler CH, Coffey CS, Mosovsky S, Caspell-Garcia C, Serrano GE, Munoz DG, White CL, Crary JF, Jennings D, Taylor P, Foroud T, Arnedo V, Kopil CM, Riley L, Dave KD, Mollenhauer B, Systemic Synuclein Sampling Study. In vivo distribution of α-synuclein in multiple tissues and biofluids in Parkinson disease. Neurology. 2020;95(9):e1267–84. https://doi.org/10.1212/WNL.0000000000010404.
87. Oosterveld LP, Verberk IMW, Majbour NK, El-Agnaf OM, Weinstein HC, Berendse HW, Teunissen CE, van de Berg WDJ. CSF or serum neurofilament light added to α- synuclein panel discriminates Parkinson's from controls. Mov Disord. 2020;35(2):288–95. https://doi.org/10.1002/mds.27897.
88. Ishii R, Tokuda T, Tatebe H, Ohmichi T, Kasai T, Nakagawa M, Mizuno T, El-Agnaf OMA. Decrease in plasma levels of α-synuclein is evident in patients with Parkinson's disease after elimination of heterophilic antibody interference. PLoS One. 2015;10(4):e0123162. https://doi.org/10.1371/journal.pone.0123162.

89. Mollenhauer B, Locascio JJ, Schulz-Schaeffer W, Sixel-Döring F, Trenkwalder C, Schlossmacher MG. α-Synuclein and tau concentrations in cerebrospinal fluid of patients presenting with parkinsonism: a cohort study. Lancet Neurol. 2011;10(3):230–40. https://doi.org/10.1016/S1474-4422(11)70014-X.
90. Schulz I, Kruse N, Gera RG, Kremer T, Cedarbaum J, Barbour R, Zago W, Schade S, Otte B, Bartl M, Hutten SJ, Trenkwalder C, Mollenhauer B. Systematic assessment of 10 biomarker candidates focusing on α-synuclein-related disorders. Mov Disord. 2021;36(12):2874–87. https://doi.org/10.1002/mds.28738.
91. Williams SM, Schulz P, Sierks MR. Oligomeric α-synuclein and β-amyloid variants as potential biomarkers for Parkinson's and Alzheimer's diseases. Eur J Neurosci. 2016;43(1):3–16. https://doi.org/10.1111/ejn.13056.
92. Daniele S, Frosini D, Pietrobono D, Petrozzi L, Lo Gerfo A, Baldacci F, Fusi J, Giacomelli C, Siciliano G, Trincavelli ML, Franzoni F, Ceravolo R, Martini C, Bonuccelli U. α-synuclein heterocomplexes with β-amyloid are increased in red blood cells of Parkinson's disease patients and correlate with disease severity. Front Mol Neurosci. 2018;11:53. https://doi.org/10.3389/fnmol.2018.00053.
93. Wang X, Yu S, Li F, Feng T. Detection of α-synuclein oligomers in red blood cells as a potential biomarker of Parkinson's disease. Neurosci Lett. 2015;599:115–9. https://doi.org/10.1016/j.neulet.2015.05.030.
94. Vicente Miranda H, Cássio R, Correia-Guedes L, Gomes MA, Chegão A, Miranda E, Soares T, Coelho M, Rosa MM, Ferreira JJ, Outeiro TF. Posttranslational modifications of blood-derived alpha-synuclein as biochemical markers for Parkinson's disease. Sci Rep. 2017;7(1):13713. https://doi.org/10.1038/s41598-017-14175-5.
95. Wang H, Atik A, Stewart T, Ginghina C, Aro P, Kerr KF, Seibyl J, Jennings D, Investigators PARS, Jensen PH, Marek K, Shi M, Zhang J. Plasma α-synuclein and cognitive impairment in the Parkinson's Associated Risk Syndrome: a pilot study. Neurobiol Dis. 2018;116:53–9. https://doi.org/10.1016/j.nbd.2018.04.015.
96. Jan AT, Malik MA, Rahman S, Yeo HR, Lee EJ, Abdullah TS, Choi I. Perspective insights of exosomes in neurodegenerative diseases: a critical appraisal. Front Aging Neurosci. 2017;9 https://doi.org/10.3389/fnagi.2017.00317.
97. Taha HB, Ati SS. Evaluation of α-synuclein in CNS-originating extracellular vesicles for Parkinsonian disorders: a systematic review and meta-analysis. CNS Neurosci Ther. 2023;29:3741. https://doi.org/10.1111/cns.14341.
98. Dutta S, Hornung S, Kruayatidee A, Maina KN, del Rosario I, Paul KC, et al. α-Synuclein in blood exosomes immunoprecipitated using neuronal and oligodendroglial markers distinguishes Parkinson's disease from multiple system atrophy. Acta Neuropathol. 2021;142:495. https://doi.org/10.1007/s00401-021-02324-0.
99. Taha HB, Hornung S, Dutta S, Fenwick L, Lahgui O, Howe K, et al. Toward a biomarker panel measured in CNS-originating extracellular vesicles for improved differential diagnosis of Parkinson's disease and multiple system atrophy. Transl Neurodegener. 2023;12:14. https://doi.org/10.1186/s40035-023-00346-0.
100. Yu Z, Shi M, Stewart T, Fernagut PO, Huang Y, Tian C, et al. Reduced oligodendrocyte exosome secretion in multiple system atrophy involves SNARE dysfunction. Brain. 2021;143:1780. https://doi.org/10.1093/BRAIN/AWAA110.
101. Brozzetti L, Sacchetto L, Cecchini MP, Avesani A, Perra D, Bongianni M, Portioli C, Scupoli M, Ghetti B, Monaco S, Buffelli M, Zanusso G. Neurodegeneration-associated proteins in human olfactory neurons collected by nasal brushing. Front Neurosci. 2020;14:145. https://doi.org/10.3389/fnins.2020.00145.
102. Witt M, Bormann K, Gudziol V, Pehlke K, Barth K, Minovi A, Hähner A, Reichmann H, Hummel T. Biopsies of olfactory epithelium in patients with Parkinson's disease. Mov Disord. 2009;24(6):906–14. https://doi.org/10.1002/mds.22464.

103. Saito Y, Shioya A, Sano T, Sumikura H, Murata M, Murayama S. Lewy body pathology involves the olfactory cells in Parkinson's disease and related disorders. Mov Disord. 2016;31(1):135–8. https://doi.org/10.1002/mds.26463.
104. Beach TG, White CL, Hladik CL, Sabbagh MN, Connor DJ, Shill HA, Sue LI, Sasse J, Bachalakuri J, Henry-Watson J, Akiyama H, Adler CH, Arizona Parkinson's Disease Consortium. Olfactory bulb alpha-synucleinopathy has high specificity and sensitivity for Lewy body disorders. Acta Neuropathol. 2009;117(2):169–74. https://doi.org/10.1007/s00401-008-0450-7.
105. Lee H-J, Patel S, Lee S-J. Intravesicular localization and exocytosis of alpha-synuclein and its aggregates. J Neurosci Off J Soc Neurosci. 2005;25(25):6016–24. https://doi.org/10.1523/JNEUROSCI.0692-05.2005.
106. Vivacqua G, Latorre A, Suppa A, Nardi M, Pietracupa S, Mancinelli R, Fabbrini G, Colosimo C, Gaudio E, Berardelli A. Abnormal salivary total and oligomeric alpha-synuclein in Parkinson's disease. PLoS One. 2016;11(3):e0151156. https://doi.org/10.1371/journal.pone.0151156.
107. Vivacqua G, Suppa A, Mancinelli R, Belvisi D, Fabbrini A, Costanzo M, Formica A, Onori P, Fabbrini G, Berardelli A. Salivary alpha-synuclein in the diagnosis of Parkinson's disease and Progressive Supranuclear Palsy. Parkinsonism Relat Disord. 2019;63:143–8. https://doi.org/10.1016/j.parkreldis.2019.02.014.
108. Cao Z, Wu Y, Liu G, Jiang Y, Wang X, Wang Z, Feng T. Differential diagnosis of multiple system atrophy-parkinsonism and Parkinson's disease using α-synuclein and external anal sphincter electromyography. Front Neurol. 2020;11:1043. https://doi.org/10.3389/fneur.2020.01043.
109. Iranzo A, Borrego S, Vilaseca I, Martí C, Serradell M, Sánchez-Valle R, Kovacs GG, Valldeoriola F, Gaig C, Santamaria J, Tolosa E, Gelpi E. α-Synuclein aggregates in labial salivary glands of idiopathic rapid eye movement sleep behavior disorder. Sleep. 2018;41(8) https://doi.org/10.1093/sleep/zsy101.
110. Shin J, Park S-H, Shin C, Kim J-H, Yun TJ, Kim H-J, Jeon B. Submandibular gland is a suitable site for alpha synuclein pathology in Parkinson disease. Parkinsonism Relat Disord. 2019;58:35–9. https://doi.org/10.1016/j.parkreldis.2018.04.019.
111. Del Tredici K, Hawkes CH, Ghebremedhin E, Braak H. Lewy pathology in the submandibular gland of individuals with incidental Lewy body disease and sporadic Parkinson's disease. Acta Neuropathol. 2010;119(6):703–13. https://doi.org/10.1007/s00401-010-0665-2.
112. Beach TG, Adler CH, Dugger BN, Serrano G, Hidalgo J, Henry-Watson J, Shill HA, Sue LI, Sabbagh MN, Akiyama H, Arizona Parkinson's Disease Consortium. Submandibular gland biopsy for the diagnosis of Parkinson disease. J Neuropathol Exp Neurol. 2013;72(2):130–6. https://doi.org/10.1097/NEN.0b013e3182805c72.
113. Carletti R, Campo F, Fusconi M, Pellicano C, De Vincentiis M, Pontieri FE, Di Gioia CR. Phosphorylated α-synuclein immunoreactivity in nerve fibers from minor salivary glands in Parkinson's disease. Parkinsonism Relat Disord. 2017;38:99–101. https://doi.org/10.1016/j.parkreldis.2017.02.031.
114. Vilas D, Iranzo A, Tolosa E, Aldecoa I, Berenguer J, Vilaseca I, Martí C, Serradell M, Lomeña F, Alós L, Gaig C, Santamaria J, Gelpi E. Assessment of α-synuclein in submandibular glands of patients with idiopathic rapid-eye-movement sleep behaviour disorder: a case-control study. Lancet Neurol. 2016;15(7):708–18. https://doi.org/10.1016/S1474-4422(16)00080-6.
115. Mangone G, Houot M, Gaurav R, Boluda S, Pyatigorskaya N, Chalancon A, Scilhcan D, Prigent A, Lehéricy S, Arnulf I, Corvol J-C, Vidailhet M, Duyckaerts C, Degos B. Relationship between substantia nigra neuromelanin imaging and dual alpha-synuclein labeling of labial minor in salivary glands in isolated rapid eye movement sleep behavior disorder and Parkinson's disease. Genes. 2022;13(10):1715. https://doi.org/10.3390/genes13101715.
116. Kim S, Kwon SH, Kam TI, Panicker N, Karuppagounder SS, Lee S, et al. Transneuronal propagation of pathologic α-synuclein from the gut to the brain models Parkinson's disease. Neuron. 2019;103:627. https://doi.org/10.1016/j.neuron.2019.05.035.

117. Pan-Montojo F, Schwarz M, Winkler C, Arnhold M, O'Sullivan GA, Pal A, et al. Environmental toxins trigger PD-like progression via increased alpha-synuclein release from enteric neurons in mice. Sci Rep. 2012;2 https://doi.org/10.1038/srep00898.
118. Ulusoy A, Rusconi R, Pérez-Revuelta BI, Musgrove RE, Helwig M, Winzen-Reichert B, Di Monte DA. Caudo-rostral brain spreading of α-synuclein through vagal connections. EMBO Mol Med. 2013;5:1119. https://doi.org/10.1002/emmm.201302475.
119. Svensson E, Horváth-Puhó E, Thomsen RW, Djurhuus JC, Pedersen L, Borghammer P, Sørensen HT. Vagotomy and subsequent risk of Parkinson's disease. Ann Neurol. 2015;78:522. https://doi.org/10.1002/ana.24448.
120. Hawkes CH, Del Tredici K, Braak H. Parkinson's disease: a dual-hit hypothesis. Neuropathol Appl Neurobiol. 2007;33:599. https://doi.org/10.1111/j.1365-2990.2007.00874.x.
121. Braak H, Rüb U, Gai WP, Del Tredici K. Idiopathic Parkinson's disease: possible routes by which vulnerable neuronal types may be subject to neuroinvasion by an unknown pathogen. J Neural Transm. 2003;110:517. https://doi.org/10.1007/s00702-002-0808-2.
122. Clairembault T, Leclair-Visonneau L, Neunlist M, Derkinderen P. Enteric glial cells: new players in Parkinson's disease? Mov Disord. 2015;30(4):494–8. https://doi.org/10.1002/mds.25979.
123. Gold A, Turkalp ZT, Munoz DG. Enteric alpha-synuclein expression is increased in Parkinson's disease but not Alzheimer's disease. Mov Disord. 2013;28(2):237–40. https://doi.org/10.1002/mds.25298.
124. Lebouvier T, Chaumette T, Paillusson S, Duyckaerts C, Bruley des Varannes S, Neunlist M, Derkinderen P. The second brain and Parkinson's disease. Eur J Neurosci. 2009;30(5):735–41. https://doi.org/10.1111/j.1460-9568.2009.06873.x.
125. Lebouvier, T., Neunlist, M., Bruley des Varannes, S., Coron, E., Drouard, A., N'Guyen, J.-M., Chaumette, T., Tasselli, M., Paillusson, S., Flamand, M., Galmiche, J.-P., Damier, P., & Derkinderen, P. (2010). Colonic biopsies to assess the neuropathology of Parkinson's disease and its relationship with symptoms. PLoS One, 5(9), e12728. https://doi.org/10.1371/journal.pone.0012728.
126. Shannon KM, Keshavarzian A, Dodiya HB, Jakate S, Kordower JH. Is alpha-synuclein in the colon a biomarker for premotor Parkinson's disease? Evidence from 3 cases. Mov Disord. 2012a;27(6):716–9. https://doi.org/10.1002/mds.25020.
127. Shannon KM, Keshavarzian A, Mutlu E, Dodiya HB, Daian D, Jaglin JA, Kordower JH. Alpha-synuclein in colonic submucosa in early untreated Parkinson's disease. Mov Disord. 2012b;27(6):709–15. https://doi.org/10.1002/mds.23838.
128. Visanji NP, Marras C, Hazrati L-N, Liu LWC, Lang AE. Alimentary, my dear Watson? The challenges of enteric α-synuclein as a Parkinson's disease biomarker. Mov Disord. 2014;29(4):444–50. https://doi.org/10.1002/mds.25789.
129. Antunes L, Frasquilho S, Ostaszewski M, Weber J, Longhino L, Antony P, Baumuratov A, Buttini M, Shannon KM, Balling R, Diederich NJ. Similar α-Synuclein staining in the colon mucosa in patients with Parkinson's disease and controls. Mov Disord. 2016;31(10):1567–70. https://doi.org/10.1002/mds.26702.
130. Chung SJ, Kim J, Lee HJ, Ryu H-S, Kim K, Lee JH, Jung KW, Kim MJ, Kim M-J, Kim YJ, Yun S-C, Lee J-Y, Hong S-M, Myung S-J. Alpha-synuclein in gastric and colonic mucosa in Parkinson's disease: limited role as a biomarker. Mov Disord. 2016;31(2):241–9. https://doi.org/10.1002/mds.26473.
131. Michell AW, Luheshi LM, Barker RA. Skin and platelet alpha-synuclein as peripheral biomarkers of Parkinson's disease. Neurosci Lett. 2005;381(3):294–8. https://doi.org/10.1016/j.neulet.2005.02.030.
132. Gibbons CH, Wang N, Freeman R. Cutaneous alpha-Synuclein from paraffin embedded autopsy specimens in Parkinson's disease. J Parkinsons Dis. 2017;7(3):503–9. https://doi.org/10.3233/JPD-171088.
133. Rodriguez-Leyva I, Chi-Ahumada E, Mejía M, Castanedo-Cazares JP, Eng W, Saikaly SK, Carrizales J, Levine TD, Norman RA, Jimenez-Capdeville ME. The presence of alpha-

synuclein in skin from melanoma and patients with Parkinson's disease. Mov Disord Clin Pract. 2017;4(5):724–32. https://doi.org/10.1002/mdc3.12494.
134. Donadio V, Incensi A, Leta V, Giannoccaro MP, Scaglione C, Martinelli P, Capellari S, Avoni P, Baruzzi A, Liguori R. Skin nerve α-synuclein deposits: a biomarker for idiopathic Parkinson disease. Neurology. 2014;82(15):1362–9. https://doi.org/10.1212/WNL.0000000000000316.
135. Donadio V, Incensi A, El-Agnaf O, Rizzo G, Vaikath N, Del Sorbo F, Scaglione C, Capellari S, Elia A, Stanzani Maserati M, Pantieri R, Liguori R. Skin α-synuclein deposits differ in clinical variants of synucleinopathy: an in vivo study. Sci Rep. 2018;8(1):14246. https://doi.org/10.1038/s41598-018-32588-8.
136. Gibbons CH, Levine T, Adler C, Bellaire B, Wang N, Stohl J, et al. Skin biopsy detection of phosphorylated α-synuclein in patients with synucleinopathies. JAMA. 2024;331:1298.
137. Antelmi E, Donadio V, Incensi A, Plazzi G, Liguori R. Skin nerve phosphorylated α- synuclein deposits in idiopathic REM sleep behavior disorder. Neurology. 2017;88(22):2128–31. https://doi.org/10.1212/WNL.0000000000003989.
138. Atarashi R, Sano K, Satoh K, Nishida N. Real-time quaking-induced conversion: a highly sensitive assay for prion detection. Prion. 2011;5(3):150–3. https://doi.org/10.4161/pri.5.3.16893.
139. Ferreira N d C, Caughey B. Proteopathic seed amplification assays for neurodegenerative disorders. Clin Lab Med. 2020;40(3):257–70. https://doi.org/10.1016/j.cll.2020.04.002.
140. Saborio GP, Permanne B, Soto C. Sensitive detection of pathological prion protein by cyclic amplification of protein misfolding. Nature. 2001;411(6839):810–3. https://doi.org/10.1038/35081095.
141. Jung BC, Lim YJ, Bae EJ, Lee JS, Choi MS, Lee MK, et al. Amplification of distinct α-synuclein fibril conformers through protein misfolding cyclic amplification. Exp Mol Med. 2017;49:e314. https://doi.org/10.1038/emm.2017.1.
142. Bellomo G, De Luca CMG, Paoletti FP, Gaetani L, Moda F, Parnetti L. α-Synuclein seed amplification assays for diagnosing synucleinopathies: the way forward. Neurology. 2022;99(5):195–205. https://doi.org/10.1212/WNL.0000000000200878.
143. Paciotti S, Bellomo G, Gatticchi L, Parnetti L. Are we ready for detecting α-synuclein prone to aggregation in patients? The case of "protein-misfolding cyclic amplification" and "real-time quaking-induced conversion" as diagnostic tools. Front Neurol. 2018;9:415. https://doi.org/10.3389/fneur.2018.00415.
144. Atarashi R, Wilham JM, Christensen L, Hughson AG, Moore RA, Johnson LM, et al. Simplified ultrasensitive prion detection by recombinant PrP conversion with shaking. Nat Methods. 2008;5:211. https://doi.org/10.1038/nmeth0308-211.
145. Hermann P, Laux M, Glatzel M, Matschke J, Knipper T, Goebel S, Treig J, Schulz-Schaeffer W, Cramm M, Schmitz M, Zerr I. Validation and utilization of amended diagnostic criteria in Creutzfeldt-Jakob disease surveillance. Neurology. 2018;91(4):e331–8. https://doi.org/10.1212/WNL.0000000000005860.
146. Orrú CD, Groveman BR, Foutz A, Bongianni M, Cardone F, McKenzie N, Culeux A, Poleggi A, Grznarova K, Perra D, Fiorini M, Liu X, Ladogana A, Sbriccoli M, Hughson AG, Haïk S, Green AJ, Geschwind MD, Pocchiari M, et al. Ring trial of 2nd generation RT-QuIC diagnostic tests for sporadic CJD. Ann Clin Transl Neurol. 2020;7(11):2262–71. https://doi.org/10.1002/acn3.51219.
147. Ciaccone G, Moda F. PMCA applications for prion detection in peripheral tissues of patients with variant Creutzfeldt-Jakob disease. Biomol Ther. 2020;10(3):405. https://doi.org/10.3390/biom10030405.
148. Salvadores N, Shahnawaz M, Scarpini E, Tagliavini F, Soto C. Detection of misfolded Aβ oligomers for sensitive biochemical diagnosis of Alzheimer's disease. Cell Rep. 2014;7(1):261–8. https://doi.org/10.1016/j.celrep.2014.02.031.
149. Fairfoul G, McGuire LI, Pal S, Ironside JW, Neumann J, Christie S, Joachim C, Esiri M, Evetts SG, Rolinski M, Baig F, Ruffmann C, Wade-Martins R, Hu MTM, Parkkinen L, Green

AJE. Alpha-synuclein RT-QuIC in the CSF of patients with alpha- synucleinopathies. Ann Clin Transl Neurol. 2016;3(10):812–8. https://doi.org/10.1002/acn3.338.
150. Groveman BR, Orrù CD, Hughson AG, Raymond LD, Zanusso G, Ghetti B, Campbell KJ, Safar J, Galasko D, Caughey B. Rapid and ultra-sensitive quantitation of disease- associated α-synuclein seeds in brain and cerebrospinal fluid by αSyn RT-QuIC. Acta Neuropathol Commun. 2018;6(1):7. https://doi.org/10.1186/s40478-018-0508-2.
151. Shahnawaz M, Tokuda T, Waragai M, Mendez N, Ishii R, Trenkwalder C, Mollenhauer B, Soto C. Development of a biochemical diagnosis of Parkinson disease by detection of α-synuclein misfolded aggregates in cerebrospinal fluid. JAMA Neurol. 2017;74(2):163–72. https://doi.org/10.1001/jamaneurol.2016.4547.
152. Iranzo A, Fairfoul G, Ayudhaya ACN, Serradell M, Gelpi E, Vilaseca I, Sanchez-Valle R, Gaig C, Santamaria J, Tolosa E, Riha RL, Green AJE. Detection of α-synuclein in CSF by RT-QuIC in patients with isolated rapid-eye-movement sleep behaviour disorder: a longitudinal observational study. Lancet Neurol. 2021;20(3):203–12. https://doi.org/10.1016/S1474-4422(20)30449-X.
153. Garrido A, Fairfoul G, Tolosa ES, Martí MJ, Green A, Barcelona LRRK2 Study Group. α-Synuclein RT-QuIC in cerebrospinal fluid of LRRK2-linked Parkinson's disease. Ann Clin Transl Neurol. 2019;6(6):1024–32. https://doi.org/10.1002/acn3.772.
154. Rossi M, Candelise N, Baiardi S, Capellari S, Giannini G, Orrù CD, Antelmi E, Mammana A, Hughson AG, Calandra-Buonaura G, Ladogana A, Plazzi G, Cortelli P, Caughey B, Parchi P. Ultrasensitive RT-QuIC assay with high sensitivity and specificity for Lewy body- associated synucleinopathies. Acta Neuropathol. 2020;140(1):49–62. https://doi.org/10.1007/s00401-020-02160-8.
155. van Rumund A, Green AJE, Fairfoul G, Esselink RAJ, Bloem BR, Verbeek MM. α- Synuclein real-time quaking-induced conversion in the cerebrospinal fluid of uncertain cases of parkinsonism. Ann Neurol. 2019;85(5):777–81. https://doi.org/10.1002/ana.25447.
156. Poggiolini I, Gupta V, Lawton M, Lee S, El-Turabi A, Querejeta-Coma A, Trenkwalder C, Sixel- Döring F, Foubert-Samier A, Pavy-Le Traon A, Plazzi G, Biscarini F, Montplaisir J, Gagnon J-F, Postuma RB, Antelmi E, Meissner WG, Mollenhauer B, Ben-Shlomo Y, et al. Diagnostic value of cerebrospinal fluid alpha-synuclein seed quantification in synucleinopathies. Brain. 2022;145(2):584–95. https://doi.org/10.1093/brain/awab431.
157. Shahnawaz M, Mukherjee A, Pritzkow S, Mendez N, Rabadia P, Liu X, Hu B, Schmeichel A, Singer W, Wu G, Tsai A-L, Shirani H, Nilsson KPR, Low PA, Soto C. Discriminating α-synuclein strains in Parkinson's disease and multiple system atrophy. Nature. 2020;578(7794):273–7. https://doi.org/10.1038/s41586-020-1984-7.
158. Bargar C, De Luca CMG, Devigili G, Elia AE, Cilia R, Portaleone SM, Wang W, Tramacere I, Bistaffa E, Cazzaniga FA, Felisati G, Legname G, Di Fonzo A, Xu R, Gunzler SA, Giaccone G, Eleopra R, Chen SG, Moda F. Discrimination of MSA-P and MSA-C by RT-QuIC analysis of olfactory mucosa: the first assessment of assay reproducibility between two specialized laboratories. Mol Neurodegener. 2021;16(1):82. https://doi.org/10.1186/s13024-021-00491-y.
159. De Luca CMG, Elia AE, Portaleone SM, Cazzaniga FA, Rossi M, Bistaffa E, De Cecco E, Narkiewicz J, Salzano G, Carletta O, Romito L, Devigili G, Soliveri P, Tiraboschi P, Legname G, Tagliavini F, Eleopra R, Giaccone G, Moda F. Efficient RT-QuIC seeding activity for α-synuclein in olfactory mucosa samples of patients with Parkinson's disease and multiple system atrophy. Transl Neurodegener. 2019;8:24. https://doi.org/10.1186/s40035-019-0164-x.
160. Kuzkina A, Bargar C, Schmitt D, Rößle J, Wang W, Schubert A-L, Tatsuoka C, Gunzler SA, Zou W-Q, Volkmann J, Sommer C, Doppler K, Chen SG. Diagnostic value of skin RT-QuIC in Parkinson's disease: a two-laboratory study. NPJ Parkinsons Dis. 2021;7(1):99. https://doi.org/10.1038/s41531-021-00242-2.
161. Mammana A, Baiardi S, Quadalti C, Rossi M, Donadio V, Capellari S, Liguori R, Parchi P. RT-QuIC detection of pathological α-synuclein in skin punches of patients with Lewy body disease. Mov Disord. 2021;36(9):2173–7. https://doi.org/10.1002/mds.28651.

162. Perra D, Bongianni M, Novi G, Janes F, Bessi V, Capaldi S, Sacchetto L, Tagliapietra M, Schenone G, Morbelli S, Fiorini M, Cattaruzza T, Mazzon G, Orrù CD, Catalan M, Polverino P, Bernardini A, Pellitteri G, Valente M, et al. Alpha-synuclein seeds in olfactory mucosa and cerebrospinal fluid of patients with dementia with Lewy bodies. Brain Communications. 2021;3(2):fcab045. https://doi.org/10.1093/braincomms/fcab045.
163. Stefani A, Iranzo A, Holzknecht E, Perra D, Bongianni M, Gaig C, Heim B, Serradell M, Sacchetto L, Garrido A, Capaldi S, Sánchez-Gómez A, Cecchini MP, Mariotto S, Ferrari S, Fiorini M, Schmutzhard J, Cocchiara P, Vilaseca I, et al. Alpha-synuclein seeds in olfactory mucosa of patients with isolated REM sleep behaviour disorder. Brain. 2021;144(4):1118–26. https://doi.org/10.1093/brain/awab005.
164. Peng C, Gathagan RJ, Covell DJ, Medellin C, Stieber A, Robinson JL, et al. Cellular milieu imparts distinct pathological α-synuclein strains in α-synucleinopathies. Nature. 2018;557:558. https://doi.org/10.1038/s41586-018-0104-4.
165. Luan M, Sun Y, Chen J, Jiang Y, Li F, Wei L, Sun W, Ma J, Song L, Liu J, Liu B, Pei Y, Wang Z, Zhu L, Deng J. Diagnostic value of salivary real-time quaking-induced conversion in Parkinson's disease and multiple system atrophy. Mov Disord. 2022;37(5):1059–63. https://doi.org/10.1002/mds.28976.
166. Vilas D, Gelpi E, Aldecoa I, Grau O, Rodriguez-Diehl R, Jaumà S, et al. Lack of central and peripheral nervous system synuclein pathology in R1441G LRRK2 -associated Parkinson's disease. J Neurol Neurosurg Psychiatry. 2019;90:832. https://doi.org/10.1136/jnnp-2018-318473.
167. Brockmann K, Quadalti C, Lerche S, et al. Association between CSF alpha-synuclein seeding activity and genetic status in Parkinson's disease and dementia with Lewy bodies. Acta Neuropathol Commun. 2021;9:175. https://doi.org/10.1186/s40478-021-01276-6.
168. Quadalti C, Calandra-Buonaura G, Baiardi S, et al. Neurofilament light chain and α-synuclein RTQuIC as differential diagnostic biomarkers in parkinsonisms and related syndromes. npj Parkinsons Dis. 2021;7:1–8. https://doi.org/10.1038/s41531-021-00232-4.
169. Spires-Jones TL, Attems J, Thal DR. Interactions of pathological proteins in neurodegenerative diseases. Acta Neuropathol. 2017;134:187–205. https://doi.org/10.1007/s00401-017-1709-7.
170. Hamilton RL. Lewy bodies in Alzheimer's disease: a neuropathological review of 145 cases using alpha-synuclein immunohistochemistry. Brain Pathol. 2000;10(3):378–84. https://doi.org/10.1111/j.1750-3639.
171. Bellomo G, Toja A, Paolini Paoletti F, Ma Y, Farris CM, Gaetani L, et al. Investigating alpha-synuclein co-pathology in Alzheimer's disease by means of cerebrospinal fluid alpha-synuclein seed amplification assay. Alzheimers Dement. 2024;20:2444. https://doi.org/10.1002/alz.13658.
172. Cousins KAQ, Arezoumandan S, Shellikeri S, Ohm D, Shaw LM, Grossman M, et al. CSF biomarkers of Alzheimer disease in patients with concomitant α-synuclein pathology. Neurology. 2022;99:e2303. https://doi.org/10.1212/WNL.0000000000201202.
173. Brooks DJ, Tambasco N. Imaging synucleinopathies. Mov Disord. 2016;31:814. https://doi.org/10.1002/mds.26547.
174. Suwijn SR, van Boheemen CJ, de Haan RJ, Tissingh G, Booij J, de Bie RM. The diagnostic accuracy of dopamine transporter SPECT imaging to detect nigrostriatal cell loss in patients with Parkinson's disease or clinically uncertain parkinsonism: a systematic review. EJNMMI Res. 2015;5:12. https://doi.org/10.1186/s13550-015-0087-1.
175. Cousins O, Yousaf T, Wilson H, Pagano G, Politis M. Molecular imaging of dementia with Lewy bodies. Int Rev Neurobiol. 2019;144:59–93. https://doi.org/10.1016/bs.irn.2018.10.007.
176. Alzghool OM, van Dongen G, van de Giessen E, Schoonmade L, Beaino W. α-Synuclein radiotracer development and in vivo imaging: recent advancements and new perspectives. Mov Disord. 2022;37(5):936–48. https://doi.org/10.1002/mds.28984.
177. Vernon AC, Ballard C, Modo M. Neuroimaging for Lewy body disease: is the in vivo molecular imaging of?-synuclein neuropathology required and feasible? Brain Res Rev. 2010;65:28. https://doi.org/10.1016/j.brainresrev.2010.05.006.

178. Serpell LC, Berriman J, Jakes R, Goedert M, Crowther RA. Fiber diffraction of synthetic alpha-synuclein filaments shows amyloid-like cross-beta conformation. Proc Natl Acad Sci USA. 2000;97(9):4897–902. https://doi.org/10.1073/pnas.97.9.4897.
179. Wakabayashi K, Tanji K, Odagiri S, Miki Y, Mori F, Takahashi H. The Lewy body in Parkinson's disease and related neurodegenerative disorders. Mol Neurobiol. 2013;47(2):495–508. https://doi.org/10.1007/s12035-012-8280-y.
180. Toledo JB, Gopal P, Raible K, Irwin DJ, Brettschneider J, Sedor S, Waits K, Boluda S, Grossman M, Van Deerlin VM, Lee EB, Arnold SE, Duda JE, Hurtig H, Lee VM-Y, Adler CH, Beach TG, Trojanowski JQ. Pathological α-synuclein distribution in subjects with coincident Alzheimer's and Lewy body pathology. Acta Neuropathol. 2016;131(3):393–409. https://doi.org/10.1007/s00401-015-1526-9.
181. Cairns AG, Vazquez-Romero A, Mahdi Moein M, Ådén J, Elmore CS, Takano A, Arakawa R, Varrone A, Almqvist F, Schou M. Increased brain exposure of an alpha-synuclein fibrillization modulator by utilization of an activated Ester prodrug strategy. ACS Chem Neurosci. 2018;9(11):2542–7. https://doi.org/10.1021/acschemneuro.8b00236.
182. Hsieh C-J, Ferrie JJ, Xu K, Lee I, Graham TJA, Tu Z, Yu J, Dhavale D, Kotzbauer P, Petersson EJ, Mach RH. Alpha synuclein fibrils contain multiple binding sites for small molecules. ACS Chem Neurosci. 2018;9(11):2521–7. https://doi.org/10.1021/acschemneuro.8b00177.
183. Bard F, Cannon C, Barbour R, Burke RL, Games D, Grajeda H, Guido T, Hu K, Huang J, Johnson-Wood K, Khan K, Kholodenko D, Lee M, Lieberburg I, Motter R, Nguyen M, Soriano F, Vasquez N, Weiss K, et al. Peripherally administered antibodies against amyloid beta-peptide enter the central nervous system and reduce pathology in a mouse model of Alzheimer disease. Nat Med. 2000;6(8):916–9. https://doi.org/10.1038/78682.
184. Poduslo JF, Curran GL, Berg CT. Macromolecular permeability across the blood-nerve and blood-brain barriers. Proc Natl Acad Sci USA. 1994;91(12):5705–9. https://doi.org/10.1073/pnas.91.12.5705.
185. Sehlin D, Fang XT, Cato L, Antoni G, Lannfelt L, Syvänen S. Antibody-based PET imaging of amyloid beta in mouse models of Alzheimer's disease. Nat Commun. 2016;7:10759. https://doi.org/10.1038/ncomms10759.
186. Dasgupta A, Liu M, Ojha T, Storm G, Kiessling F, Lammers T. Ultrasound-mediated drug delivery to the brain: principles, progress and prospects. Drug Discov Today Technol. 2016;20:41–8. https://doi.org/10.1016/j.ddtec.2016.07.007.
187. Börjesson PKE, Jauw YWS, de Bree R, Roos JC, Castelijns JA, Leemans CR, van Dongen GAMS, Boellaard R. Radiation dosimetry of 89Zr-labeled chimeric monoclonal antibody U36 as used for immuno-PET in head and neck cancer patients. J Nucl Med. 2009;50(11):1828–36. https://doi.org/10.2967/jnumed.109.065862.
188. Tambasco N, Nigro P, Chiappiniello A, Paolini Paoletti F, Scialpi S, Simoni S, et al. An updated overview of the magnetic resonance imaging of brain iron in movement disorders. Behav Neurol. 2022;2022:1. https://doi.org/10.1155/2022/3972173.
189. Nigro P, Chiappiniello A, Simoni S, Paolini Paoletti F, Cappelletti G, Chiarini P, et al. Changes of olfactory tract in Parkinson's disease: a DTI tractography study. Neuroradiology. 2020;63:235. https://doi.org/10.1007/s00234-020-02551-4.

Fluid and Imaging Markers of Synaptic Injury

Rawan Tarawneh

Abstract Synaptic injury is an important pathological feature of Alzheimer disease (AD) which is present in even the earliest pre-symptomatic stages and continues with further disease progression. Synaptic loss in AD is thought to result from various pathological α-substrates including synaptic toxicity of abnormal protein aggregates, such as amyloid-β oligomers and various forms of hyperphosphorylated tau, abnormal α-synuclein and TDP-43 aggregation, and other mechanisms such as microglial and astrocytic dysregulation, oxidative stress, mitochondrial damage, and vascular or endothelial injury. Brain expression of several pre-synaptic and post-synaptic proteins is reduced in AD. Cortical and hippocampal synaptic density is reduced by 30% and 50%, respectively, in even the early symptomatic stages which is thought to result from both neuronal loss and reduced synaptic density per neuron. Recent evidence suggests that, while common mechanisms of synaptic injury are shared across different neurodegenerative disorders, other mechanisms of synaptic injury may be relatively specific to AD.

Clinicopathological studies converge on the notion that synaptic loss is the best surrogate for cognitive impairment and is more closely associated with cognitive outcomes than amyloid or tau pathologies. Therefore, there has been increasing interest in identifying novel fluid and imaging markers of synaptic injury as these may provide important prognostic information regarding clinical disease progression. Further, synaptic markers that reflect disease-specific mechanisms may improve early diagnosis and complement other diagnostic tools. Importantly, there has been an interest in integrating synaptic injury markers as biomarker surrogate endpoints to demonstrate disease modification in clinical trials of emerging therapies in AD and related dementias.

The last few years have witnessed considerable progress in identifying and validating new markers of synaptic injury in the cerebrospinal fluid (CSF). Some of these markers, such as neurogranin, offer great promise as dynamic measures that correlate with clinical progression in longitudinal studies. However, there remains a

R. Tarawneh (✉)
Department of Neurology, University of New Mexico, Albuquerque, NM, USA
e-mail: raltarawneh@salud.unm.edu

R. Tarawneh (ed.), *Biomarkers of Neurodegenerative Disorders*,
https://doi.org/10.1007/978-3-031-80849-4_4

need to identify new proteins that can be measured in the blood and serve as useful surrogates for synaptic injury in AD-vulnerable brain regions or identify new methods to reliably quantify synaptic markers, which have already been established in the CSF, in the periphery.

This chapter summarizes the current state of the most promising fluid and imaging markers of synaptic injury in AD and highlights important areas for future research.

Keywords Synaptic injury · Cognition · Fluid · Imaging · Biomarker

1 Introduction

Synaptic loss or injury is an early and primary pathological feature in Alzheimer disease (AD) brains [1]. Cortical and hippocampal synaptic density is reduced by approximately 30% and 50%, respectively, in even the earliest symptomatic stages and decreases further with disease progression [2, 3]. Synaptic loss in AD is due to both neuronal loss and reduced synaptic density per neuron, and reflects the cumulative effects of various AD pathologies, including abnormal protein aggregation (i.e., amyloid, tau, and possibly TAR DNA binding protein-43 [TDP-43] and α-synuclein), vascular disease, neuroinflammation, oxidative stress and immune dysregulation [4]. Clinicopathological studies of AD suggest that synaptic loss is more closely associated with cognitive outcomes than amyloid or tau pathologies and offers a useful biological surrogate for cognitive impairment [5]. Cognitive deficits attributed to synaptic dysfunction occur in the absence of, and prior to, neuronal loss in AD [6]. Neuropathological studies suggest that brain expression of several pre-synaptic and post-synaptic proteins is reduced in AD [7]; however, cerebrospinal fluid (CSF) levels of many synaptic proteins are increased in AD, likely reflecting the release of abundant synaptic constituents into the extracellular space in the setting of neurodegeneration and synaptic disintegration.

There has been great interest in the identification and validation of fluid or imaging markers that can reliably measure synaptic injury and monitor its progression across different disease stages [4, 8, 9]. As synaptic injury is considered a strong surrogate for cognitive impairment, markers of synaptic injury offer important tools to predict cognitive outcomes, including future cognitive decline in cognitively normal individuals, monitor or predict clinical progression in symptomatic individuals, and may serve as potential co-primary or secondary outcome measures in clinical trials of disease-modifying therapies. Further, several synaptic injury markers have been shown to improve the collective diagnostic utility of AD biomarkers of amyloid and tau and provide complementary information regarding the presence of brain amyloid pathology on amyloid positron emission tomography (PET) imaging [10]. Significant synaptic loss is a common feature across many neurodegenerative disorders. Therefore, biomarkers that reflect nonspecific mechanisms of synaptic injury may help differentiate individuals with AD or other neurodegenerative disorders from healthy controls. Conversely, a few synaptic markers have been shown to be more closely associated with disease-specific mechanisms of synaptic injury, and thus, may offer value in differentiating between neurodegenerative disorders [11]. Further, synaptic proteins whose levels may be influenced by the regional distribution of neurodegenerative pathology or disease severity may provide useful tools for disease staging.

2 Fluid Markers of Synaptic Injury

2.1 *Pre-synaptic Markers*

2.1.1 Synaptosomal-Associated Protein-25 (SNAP-25)

SNAP-25 is a pre-synaptic protein which is a component of the soluble *N*-ethylmaleimide-sensitive factor attachment protein receptor (SNARE) complex and is involved in vesicle docking, neurotransmitter release, and neurite outgrowth [12]. Through its cytoplasmic region, SNAP-25 interacts with other pre-synaptic proteins, such as vesicle-associated membrane protein-2 (VAMP2) and synaptotagmin-1 (Syt-1), to form a complex that facilitates calcium-mediated vesicle fusion and neurotransmitter release [13].

Several studies, using affinity purification and immunoprecipitation mass spectrometry (IP-MS), have demonstrated the diagnostic value of CSF SNAP-25 in AD with higher CSF levels of soluble SNAP-25 fragments offering a relatively high sensitivity (>90%) in differentiating AD from controls [14]. CSF SNAP-25 levels, measured using enzyme-linked immunosorbent assays (ELISAs), are increased in the earliest symptomatic stage of AD (mild cognitive impairment [MCI]; area under the curve [AUC] 0.95) and dementia due to AD (AUC 0.97) compared to healthy controls [15]. Similar observations have been reported by other studies which utilized the highly sensitive single molecule array technology (SIMOA, Quanterix) to measure CSF SNAP-25 levels. In these studies, CSF SNAP-25 levels provided high sensitivity in detecting brain amyloid status as they differentiated cognitively impaired individuals who were amyloid-positive (i.e., who had evidence of brain amyloid pathology) from cognitively normal or impaired individuals who were amyloid-negative [16]. The diagnostic value of CSF SNAP-25 was comparable when measured using IP-MS or SIMOA [16]. However, altered brain expression of SNAP-25 has also been observed in primary psychiatric and other neurodegenerative conditions, suggesting that this marker may not be specific to AD-related synaptic injury [12]. While a few studies suggest that CSF SNAP-25 levels may be higher in AD compared to several other neurodegenerative disorders, the overall utility of SNAP-25 levels in the differential diagnosis of AD appears to be limited [9, 12, 17]. Two splicing variants of SNAP-25, referred to as SNAP-25a and SNAP-25b, which differ in 9 amino acid residues, have been identified [18]. Further studies are needed to determine whether the diagnostic or prognostic value of CSF SNAP-25 may be different between these two isoforms.

CSF SNAP-25 levels correlate with CSF levels of other synaptic proteins, including post-synaptic proteins (i.e., post-synaptic density protein-95 [PSD-95] and neurogranin [Ng]), and brain atrophy in AD [19, 20], and predict future risk of cognitive impairment [20]. One study examined the associations of various synaptic markers including SNAP-25, Syt-1 (both measured by IP-MS), growth-associated protein-43 (i.e., GAP-43 or neuromodulin) and Ng (both measured by immunoassays) with amyloid and tau load on PET imaging and voxel-based morphometric magnetic resonance imaging (VBM-MRI) [21]. In this study, CSF levels of all synaptic markers were increased across different disease stages. All synaptic markers correlated with brain amyloid and tau burden and gray matter, but not white matter, volumes. CSF SNAP-25 demonstrated the strongest correlations with these imaging measures compared to other synaptic markers in this cohort. However, it will be important to validate these findings in larger studies. In another study which examined several synaptic markers in a cohort from a memory clinic, CSF levels of the SNAP25aa40 peptide were associated with the highest diagnostic accuracy, compared to other synaptic markers, in differentiating biomarker-confirmed AD from those with other neurological disorders [22].

Plasma SNAP-25 levels can be measured using a SIMOA assay and, in pilot studies, have been found to be elevated in the presence of brain amyloid and tau pathologies [23]. SNAP-25 levels have also been measured in neuron-derived exosomes obtained from serum samples and may offer potential value in differentiating cognitively impaired individuals clinically diagnosed with AD from healthy controls and correlate with cognitive measures [24].

2.1.2 Growth-Associated Protein-43 (GAP-43)

The pre-synaptic protein, GAP-43 (aka neuromodulin) plays an important role in axonal branching, synaptic plasticity and remodeling, and neuronal development and is expressed in brain regions that are vulnerable to early AD pathology, including the medial temporal lobe [25, 26]. Increased intracellular calcium levels result in GAP-43 phosphorylation which facilitates vesicular recycling through its interactions with synaptophysin and SNAP-25 [27]. Elevated CSF GAP-43 levels have been reported in early symptomatic AD compared to controls and several other neurodegenerative disorders across several cohorts [28–31]. Further, in a small cohort with neuropathologically-confirmed AD, CSF GAP-43 levels were found to correlate with the severity of amyloid plaques and neurofibrillary tangles in the hippocampus, amygdala, and neocortex, but not with the severity of other AD co-pathologies such as TDP-43 and α-synuclein [28]. While these preliminary studies suggest a relative specificity of GAP-43 to AD-related synaptic loss, further studies in larger cohorts are needed to better evaluate the diagnostic specificity of CSF GAP-43 for AD. Studies have also shown associations of CSF GAP-43 with lower cognitive performance and higher rates of clinical progression in AD [29, 32], further supporting its value as a surrogate of AD-related synaptic injury.

GAP-43 is abundantly expressed in the pre-synaptic compartment where it is involved in synaptic modeling and synaptogenesis (i.e., the formation of new synapses). Further, its neuronal expression appears to be regulated by neuronal activity with higher levels of neuronal GAP-43 expression reported in rodent models of epilepsy associated with increased excitatory neuronal activity [33]. Consistent with these findings, higher CSF GAP-43 levels have been linked to increased Aβ-mediated trans-synaptic spread of tau pathology in AD brains. In a longitudinal study from the Alzheimer's Disease Neuroimaging Initiative (ADNI) cohort, the investigators reported that higher baseline CSF GAP-43 levels were associated with faster Aβ-related tau accumulation across connected brain regions, particularly in regions that were the most closely connected to participant-specific epicenters of highest baseline tau pathology [34]. Further, in individuals who had both increased CSF GAP-43 and increased brain amyloid, functional connectivity patterns on functional MRI were more closely aligned with the progression of brain tau pathology on tau-PET [34]. While these interesting findings support an association between CSF GAP-43 and tau progression in individuals with brain amyloid pathology, further studies are needed to better understand the relationship between CSF GAP-43 levels and tau aggregation in AD. Future mechanistic studies will also be important to understand whether synaptic changes, measured by CSF GAP-43 levels in AD, are the cause or result of increased tau aggregation.

In another large study of 786 participants from the ADNI cohort, CSF GAP-43 levels were found to be increased in CSF Aβ-positive, compared to CSF Aβ-negative, participants regardless of cognitive status [32]. Further, Aβ-positive participants who had high baseline CSF GAP-43 levels had higher rates of cognitive decline, brain atrophy and progression of hypometabolism on 18F-fluorodeoxy-glucose (18F-FDG)-PET, and a higher risk of progression to AD dementia over time, compared to those with lower baseline CSF GAP-43 levels [32]. In this cohort, CSF GAP-43 levels were strongly associated with biomarkers of tau pathology (i.e., CSF tau phosphorylated at threonine 181 [p-tau181]), and to a lesser extent, amyloid pathology (i.e., amyloid PET) [32]. Together, these findings support the value of CSF GAP-43 as a diagnostic and prognostic biomarker of synaptic injury in AD. Recent studies suggest that CSF GAP-43 levels may be higher in AD than other neurodegenerative disorders and that it may offer relative specificity to AD-related synaptic injury [28]; however, further research in this area is warranted.

2.1.3 Synaptotagmin-1 (Syt-1)

Syt-1 is a pre-synaptic transmembrane protein which acts as a calcium sensor for fast synchronous release of neurotransmitters and facilitates calcium-dependent vesicle translocation [35, 36]. The quantification of CSF Syt-1 levels has mostly been performed using IP-MS assays [22, 37]. Higher CSF Syt-1 levels have been observed in MCI due to AD and AD dementia compared to healthy controls and other dementias [22, 37]; however, results of studies have been mixed. Other studies have shown elevated CSF Syt-1 levels in preclinical AD [38]. CSF levels of the Syt-1 peptide 215–223 (VPYSELGGK) and peptide 238–245 (HDIIGEFK) correlate with CSF tau and p-tau181 in individuals with AD dementia and healthy controls [37]. In a retrospective cross-sectional study, which examined the diagnostic utility of several CSF synaptic proteins (Syt-1, Ng, GAP-43, SNAP-25 and SNAP25aa40 peptide), all synaptic markers differentiated dementia due to AD from dementia due to other causes. With the exception of Syt-1, CSF synaptic markers also differentiated MCI due to AD or AD dementia from controls [22]. Several other studies have demonstrated the diagnostic value of CSF Syt-1 in differentiating AD from controls across various cohorts [9]; while a few others reported trends for higher CSF Syt-1 levels in AD that did not reach statistical significance [39]. Further, CSF Syt-1 and several other synaptic markers may differentiate atypical AD from frontotemporal dementia (FTD) as their levels were comparable between FTD and controls [39].

2.1.4 Neuronal Pentraxins (NPTX, NP)

Neuronal pentraxins (NPTX, NP) are predominantly pre-synaptic glycoproteins which are involved in the assembly of excitatory α-amino-3-hydroxy-5-methyl-4-isoxazoleproprionic acid (AMPA)-mediated synapses [40]. NP1 (NPTX1) is a pro-apoptotic protein which is detected within dystrophic neurites in human AD brains and has been shown to downregulate excitatory synaptic density [41, 42]. The pre-synaptic release of NP1 is promoted by Aβ oligomers and is likely one of the mechanisms that underly Aβ-mediated synaptic deficits observed in AD brains [42]. Conversely, knockdown of NP1 in animal models is associated with an increased number of cortical excitatory synapses, enhanced neuronal excitability, and facilitation of hippocampal long-term potentiation (LTP) [41]. Plasma NP1 levels are increased in MCI compared to controls and further increases are observed in MCI patients who progress to AD dementia [43]. There are also trends for higher plasma NP1 levels in individuals who carry at least one copy of the apolipoprotein-E4 (*APOE4*) allele (i.e., *APOE4* carriers) compared to *APOE4* non-carriers.

Conversely, studies which examined NP2 (NPTX2) have found that brain expression and CSF NP2 levels are reduced in AD, compared to controls, and correlate with cognitive outcomes and hippocampal volumes [44]. Lower CSF NP2 levels have also been observed in individuals with Down syndrome, who have a significantly higher risk for AD pathology given the presence of an extra copy of the amyloid precursor protein (*APP*) gene on chromosome 21 [45]. In another study, individuals who had higher baseline CSF concentrations of p-tau181 and total tau demonstrated a more rapid decline in CSF NP2 levels over time [46]. Findings from this study also suggest that CSF NP2 may offer predictive value for future cognitive decline in cognitively normal individuals [46].

A longitudinal study from the ADNI cohort, which used a targeted, stable isotope, quantitative mass-spectrometry method for CSF NP2 (i.e., NPTX2) quantification, demonstrated longitudinal reductions in CSF NP2 levels in individuals with MCI compared to controls, in participants who had CSF biomarker evidence of AD (i.e., a high CSF p-tau181/Aβ42 ratio) compared to those who did not, and in participants who progressed to dementia over follow-up compared to those who did not progress [47]. Longitudinal decreases in CSF NP2 levels were associated with faster rates of cognitive decline [47]. Other studies suggest that CSF NP2 (i.e., NPTX2) levels are associated with resting-state functional connectivity in non-demented older adults [48]. While these findings support the value of CSF NP2 as a prognostic biomarker in AD, further investigation is needed to better understand the relationship between NP2 and other CSF, plasma, or imaging markers of AD and evaluate the diagnostic or prognostic value of NP2 in AD.

The neuronal pentraxin receptor-1 (NPTXR1) has also been examined as a potential AD biomarker. Lower CSF NPTXR1 levels have been associated with increased dementia severity in early symptomatic AD, with lower levels observed in AD dementia compared to MCI or healthy controls [49].

2.2 *Post-synaptic Markers*

2.2.1 Neurogranin (Ng)

Ng is a calmodulin-binding neuronal protein [50] which is abundantly expressed in the post-synaptic membrane, including dendritic spines, and is involved in synaptic plasticity, LTP, and memory functions [10]. Ng was first proposed as a potential marker of synaptic injury in large-scale gene arrays [51] due to its preferential neuronal expression and widespread distribution in different brain regions [52]. Brain Ng expression is reduced in AD compared to controls [7, 53, 54], similarly to reduced brain expression of several other synaptic markers [7, 54, 55], and negatively correlates with dementia severity [1, 6, 53]. Conversely, CSF Ng levels are increased in AD, likely reflecting the release of abundant synaptic constituents into the extracellular space in the setting of synaptic disintegration [10].

Ng is among the most promising biomarkers of synaptic injury and has been studied by many groups across different centers [10, 11, 56, 57]. Previous studies across different cohorts and using different immunoassays have shown that CSF Ng is a useful diagnostic and prognostic biomarker in AD [10, 11, 56, 57]. In one of the largest studies of Ng, which examined a well-characterized longitudinal cohort of participants with AD and healthy controls who were enrolled at the Knight Washington University in St Louis- Alzheimer Disease Research Center (Knight ADRC), CSF Ng levels were found to be significantly higher in cognitively impaired individuals with early symptomatic AD, and in cognitively normal individuals with preclinical (i.e., pre-symptomatic) AD, compared to cognitively normal healthy controls [10]. CSF Ng also differentiated individuals who had brain amyloid pathology on PET (i.e., amyloid-positive) from those without amyloid (i.e., amyloid-negative) regardless of clinical status [10]. Further, in this cohort, baseline CSF Ng levels predicted future cognitive decline in cognitively normal individuals, defined as progression from Clinical Dementia Rating (CDR®) 0 to CDR $\geq$ 0.5 over follow-up, at least as well as CSF total tau and p-tau181 levels [10]. Importantly, the addition of CSF Ng to CSF total tau, p-tau181, and visinin-like protein-1 (VILIP-1) levels, when each of these markers were examined individually or in combination, improved the collective predictive ability of these markers for future cognitive decline in cognitively normal individuals [10]. This study is among the first and largest studies of CSF Ng to date and was the first to demonstrate the diagnostic and prognostic value of CSF Ng in early pre-symptomatic AD [10]. Findings from this large study and others have shown that baseline CSF Ng levels are closely associated with whole brain and regional atrophy of the hippocampus, entorhinal cortex, and parahippocampal gyrus in early symptomatic AD [10, 56], reflecting more severe synaptic loss accompanying brain volume loss in AD-vulnerable brain regions, and that higher baseline CSF Ng levels predict faster rates of cognitive decline and clinical progression in early symptomatic AD [10]. In a meta-analysis of CSF Ng studies, individuals with MCI who progressed to dementia had higher CSF Ng levels than those who did not progress [58]. Other studies have also shown that higher CSF levels of Ng fragments correlate with future cognitive decline, brain atrophy, and glucose hypometabolism, even in early disease stages [56, 59, 60]. CSF Ng levels also correlate with CSF markers of tau pathology (i.e., p-tau181) and neurodegeneration (i.e., total tau and VILIP-1) [10]. CSF Ng, VILIP-1, and total tau levels [10, 61, 62] correlate with cortical amyloid burden on PET scans in pre-symptomatic AD, which is consistent with the notion that increasing amyloid deposition [63, 64] is associated with ongoing neuronal or synaptic loss before symptom onset. Conversely, as cortical amyloid deposition likely plateaus near the time of symptom onset, these CSF biomarkers are not as closely associated with amyloid PET load in symptomatic AD [65, 66]. Together, these findings are consistent with the notion that synaptic injury is a close surrogate for cognitive decline, and further support the prognostic value of CSF Ng in pre-symptomatic and early symptomatic AD. Most studies suggest that high CSF Ng levels may offer relative diagnostic specificity for AD, and that CSF Ng may differentiate AD from most other neurodegenerative disorders [11, 67]. However, the mechanisms by which synaptic injury in AD, compared to synaptic injury observed in other neurodegenerative disorders, may preferentially influence Ng levels remain to be elucidated and more research in this area is warranted.

In addition to the full-length Ng protein, Ng fragments have also been measured in the CSF. Ng is cleaved by 2 intracellular enzymes, calpain-1 and prolyl endopeptidase, at the functional IQ domain and the C-terminal end, respectively [68]. CSF Ng consists mostly of truncated C-terminal fragments. Findings from comparative studies suggest that various ELISAs, which target different Ng segments, generally provide similar diagnostic value in AD [69]. Novel ultrasensitive SIMOA assays have been developed for measurement of full-length Ng (FL-Ng) and C-terminal Ng fragments (CT-Ng) in CSF. Although the median CSF levels for both were increased in MCI or dementia due to AD, the median of the CSF CT-Ng/FL-Ng ratio was decreased in AD dementia and MCI due to AD, compared to neurological controls [70].

Plasma Ng levels have limited utility as AD biomarkers [71] as they do not reliably differentiate AD from controls and have poor correlation with CSF Ng levels. Recent studies examining Ng in neuronally-derived extracellular vesicles (EVs) obtained from blood samples report lower levels, reflective of reduced brain expression, in AD compared to controls, and in individuals with MCI who progressed to dementia compared to those with MCI who did not progress [72, 73]. While this approach offers promise to measure brain expression levels of proteins in peripherally derived samples, its routine implementation may be limited due to the relative complexity and variability of EV extraction and enrichment protocols.

2.3 Other Synaptic Markers

Alpha (α)-synuclein (α-Syn) is a pre-synaptic protein which plays several important roles in lipid transport, neurotransmitter release, and synaptic plasticity [74]. Abnormal α-Syn aggregation is a characteristic pathologic feature of Parkinson's disease (PD), Dementia with Lewy bodies (DLB), and multiple system atrophy (i.e., collectively referred to as α-synucleinopathies). α-Syn aggregation is also a common comorbid pathology in AD as it is observed in 40–70% of AD brains [75]. α-Syn pathology in the neocortex, limbic system, and substantia nigra correlates with extrapyramidal symptoms, visual hallucinations, and more rapid cognitive decline [76, 77]. In AD brains, α-Syn pathology co-localizes with tau, and to a lesser extent, Aβ pathology [78, 79]. Interactions of α-Syn with tau and glycogen synthase kinase (GSK3β) appear to promote tau hyperphosphorylation in AD [75]. Several studies have examined CSF α-Syn levels in AD, although these appear to be considered mostly a surrogate of concomitant α-Syn pathology rather than a measure of synaptic injury. Biomarker measures of α-Syn are discussed in detail in other chapters.

Other potential synaptic markers have been proposed including synapsin, synaptophysin, syntaxin-1b, actin-associated protein Arc, neurofascin, members of the Rab family, SV2A, contactin-2, and neurexins [80–83]. Data regarding the utility of these proteins as CSF markers of synaptic injury is currently limited, and further research is warranted (Fig. 1).

3 Imaging Markers of Synaptic Density

3.1 Synaptic Vesicle Glycoprotein 2A (SV2A)

The synaptic vesicle glycoprotein 2 (SV2) protein family, including 3 isoforms (SV2A, SV2B, and SV2C), is a group of keratan sulfate proteoglycans which are expressed in the membranes of pre-synaptic vesicles and play an important role in

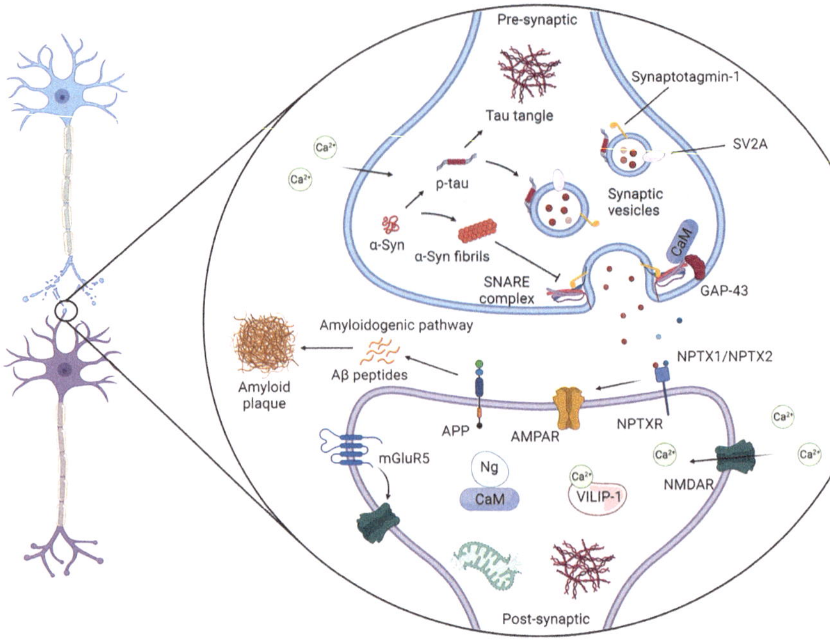

Fig. 1 Schematic Diagram of Synaptic Markers in AD. Synaptic loss in AD occurs in both the pre-synaptic and post-synaptic compartments. In the pre-synaptic compartment, misfolded tau and Aβ oligomers contribute to loss of synaptotagmin-1 and depletion of the synaptic vesicle pool. Alpha-synuclein (α-Syn) monomers and oligomers interfere with vesicular fusion by SNARE proteins, which include the pre-synaptic protein SNAP-25, and promote tau hyperphosphorylation. GAP-43 is an axonal protein which is phosphorylated following LTP. SV2A is a synaptic vesicle protein that regulates neurotransmitter release. In the post-synaptic compartment, Aβ binds to mGluR5, leading to increased tau phosphorylation and a decrease in NMDARs which contributes to increased LTD and diminished LTP, and promotes synaptic loss. Ng plays an important role in synaptic plasticity and LTP through the calcium and CaM signaling pathways. VILIP-1 is a neuronal calcium sensor protein that is mostly localized within the neuronal soma and proximal dendrites and is implicated in calcium-mediated excitotoxicity leading to neuronal loss. VILIP-1 is discussed in other chapters as a neuronal injury marker. Abbreviations: α-Syn alpha-synuclein, Aβ amyloid-beta, AD Alzheimer disease, AMPAR α-amino-3-hydroxy-5-methyl-4-isoxazolepropionic acid receptor, APP amyloid precursor protein, CaM calmodulin, GAP-43 growth-associated protein-43, LTD long-term depression, LTP long-term potentiation, mGluR5 metabotropic glutamate receptor 5, Ng neurogranin, NMDAR N-methyl-D-aspartate receptor, NPTX2 neuronal pentraxin 2, SNARE Soluble N-ethylmaleimide-Sensitive Factor Attachment Protein Receptor, SV2A synaptic vesicle glycoprotein 2A, VILIP-1 visinin-like protein 1. (Used with permission from Lista et al., 2024 [8] with no modifications; Copyright © 2024, The Author(s), under exclusive licence to Springer Nature Limited; permission conveyed through the Copyrights Clearance Center, Inc.)

regulating vesicular dynamics [84, 85]. Proposed functions of SV2A include regulating exocytosis and synaptic transmission and possible roles in promoting synaptogenesis [86]. SV2A dysfunction is associated with the accumulation of pre-synaptic Ca^{+2}, synaptic destabilization, and impaired synaptic signaling [87]. However, the exact function of SV2A and the pathways through which SV2A regulates neurotransmitter release and vesicle recycling are yet to be fully elucidated. Of the 3 SV2 isoforms, SV2A appears to be the predominant isoform in rodents with widespread and consistent expression in synapses across the brain gray matter [88–90]. Conversely, SV2B is found in some neocortical regions and the hippocampus, and SV2C is exclusively expressed in subcortical regions such as the striatum, pallidum, midbrain, brainstem, substantia nigra, and olfactory bulb [88–90]. SV2 mapping in human brains has not yet been completed.

Due to its ubiquitous expression in all brain synapses [88], several PET ligands that bind SV2A have been developed as potential tools for in vivo measurements of synaptic density in AD and other neurodegenerative disorders [91]. The potential of SV2A as an imaging marker of synaptic density is further supported by the observation that there is minimal variability in the number of SV2A copies per synaptic vesicle [92].

Among the first SV2A PET tracers was 11C-levetiracetam, based off the antiepileptic drug levetiracetam, which showed only poor binding affinity to SV2A [93, 94]. Other specific SV2A PET tracers which were also derived from levetiracetam, and had higher binding affinities to SV2A, were subsequently developed. These include 11C-UCB-A (4-(3,5-Difluorophenyl)-1-((1-methyl-1H-imidazol-5-yl)methyl)pyrrolidin-2-one), 18F-UCB-H (1-((3-Fluoropyridin-4-yl)methyl)-4-(3,4,5-trifluorophenyl) pyrrolidin-2-one), and 11C-UCB-J ((R)-1-((3-(11C-methyl-11C)pyridin-4-yl)methyl)-4-(3,4,5-trifluorophenyl)pyrrolidin-2-one). These SV2A PET ligands were first examined in non-human primates with differential binding characteristics observed between the 3 ligands. In primate studies, 11C-UCB-J appeared to have the most favorable imaging characteristics including high brain uptake, rapid and reversible binding kinetics, and high specificity to binding gray matter with a low rate of nonspecific binding to white matter, and rapid metabolism in the plasma [91, 95]. One study examined the distribution of 11C-UCB-J binding on PET and regional SV2A distribution in postmortem brains and found that these 2 measures were strongly correlated in all gray matter regions, and that the regional synaptic density of SV2A was also closely associated with that of synaptophysin, a commonly used measure of synaptic integrity [96]. Conversely, 18F-UCB-H demonstrated lower binding affinity to SV2A, and thus lower binding specificity [95], while the use of 11C-UCB-A was complicated by slow uptake [97].

Studies in humans yielded similar imaging characteristics of the 3 SV2A PET tracers compared to those in non-human primates, with 11C-UCB-J potentially demonstrating more favorable imaging properties as an SV2A ligand [96, 98]. A regional mapping study using 11C-UCB-J found specific binding of SV2A in all gray matter structures and consistent regional expression patterns between humans and non-human primates [89].

The first human PET studies of 11C-UCB-J were performed in individuals with epilepsy using the centrum semiovale as the reference region, which was later replaced by the cerebellum due to the presence of significant variability of SV2A binding in the centrum semiovale among healthy controls [96, 99, 100]. Despite its great promise, the production of 11C-UCB-J and its translation into clinical use has been significantly limited by the short half-life of the C-11 moiety (20 min) [91]. Therefore, 18F-fluorinated versions of UCB-J, which have longer half-lives (110 min), have been developed [101, 102]. These include 18F-SynVesT-1 (also known as 18F-MNI-1126 and 18F-SDM-8), 18F-SynVesT-2 (18F-SDM-2) and 18F-SDM-16. Compared to 11C-UCB-J, 18F-SynVesT-2 has faster pharmacokinetics, and slightly lower SV2A-binding potential, in non-human primate studies while 18F-SDM-16 showed the highest SV2A uptake affinity and metabolic stability compared to other ligands [103–105]. In human studies, 18F-SynVesT-1 had higher binding potential, yet similar pharmacokinetic properties, compared to 11C-UCB-J [101, 102]. Other 18F-labeled SV2A PET ligands are currently being developed and examined in humans [91].

3.2 SV2A PET in AD

While synaptic loss is an early and prominent pathological substrate in AD, results of recent studies utilizing the SV2A tracer, 3H-UCB-J, to measure synaptic density have been variable. Lower SV2A binding was found to be associated with higher AT8 immunoreactivity, reflective of hyperphosphorylated tau, in one study [106]; however, in that study, no differences in the frontal density of 3H-UCB-J binding were observed between cognitively impaired individuals with AD and non-demented controls. Conversely, other studies have demonstrated lower 3H-UCB-J binding, reflective of reduced synaptic density, in the middle frontal gyrus in AD compared to controls and that older age was associated with lower 3H-UCB-J binding [107]. Other studies which used 11C-UCB-J PET in biomarker-confirmed cohorts have found lower binding in several brain regions in AD compared to controls [108]. In an early study of 10 amyloid-PET positive individuals with MCI and mild AD dementia and 11 amyloid-PET negative healthy controls, hippocampal 11C-UCB-J binding was decreased by 41% in amyloid-PET positive individuals, which remained significant even after partial volume correction, and correlated with episodic memory scores [108]. Interestingly, hippocampal volume was only reduced by 22% in this group, which may suggest that low synaptic density, measured by 11C-UCB-J binding, results from both neuronal loss and reduced synaptic density per neuron [108]. Follow-up studies using the cerebellum, rather than the centrum semiovale, as a reference region showed reduced 11C-UCB-J binding in additional brain regions including the entorhinal cortex, amygdala, and neocortical regions in AD [109].

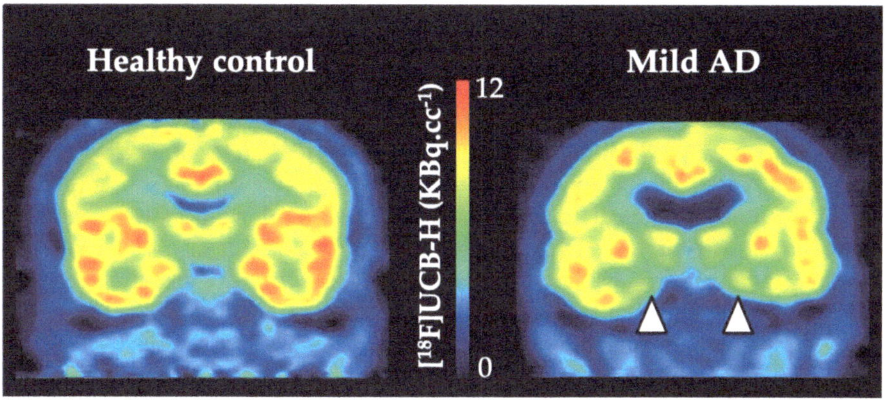

Fig. 2 18F-UCB-H binding in a healthy control and an individual with mild AD dementia. White arrows show reduced 18F-UCB-H uptake in the temporal lobe in AD. (Used with permission from Becker et al., 2020 [115]; Open Access through a Creative Commons license. http://creativecommons.org/licenses/by/4.0/)

Studies which have compared regional 11C-UCB-J binding with regional amyloid or tau PET burden have found associations between reduced regional synaptic density and regional tau, but not amyloid, burden on PET imaging. In one study which compared regional 11C-UCB-J binding and regional 18F-flortaucipir (tau) PET, inverse correlations between tau deposits in the entorhinal cortex and SV2A density in the hippocampus were observed in AD [110]. Higher cortical tau burden, measured by 18F-MK-6240 tau PET, was associated with lower 11C-UCB-J binding in the hippocampus and medial temporal regions in another study [111], while a third study found that 11C-UCB-J binding was negatively associated with tau burden only in individuals with medium to high tau signal [112]. Conversely, no correlations were observed between regional amyloid binding, measured by 11C-Pittsburgh Compound B (PiB) PET, and 11C-UCB-J binding in any of the examined regions in other studies [113].

Consistent with known associations of synaptic loss with cognitive outcomes, reduced synaptic density, measured by 11C-UCB-J binding, was found to be associated with worse global cognition and impairment in several cognitive domains, following partial volume correction, in one study. Interestingly, findings from this study suggest that 11C-UCB-J binding was more closely associated with global cognition than whole brain volumes [100]. Other studies using 18F-UCB-H PET by other groups have also shown associations between reduced SV2A binding on PET and cognitive outcomes, with the strongest synaptic density reductions being observed in the hippocampus and correlating with cognitive decline and unawareness of memory loss [114] (Fig. 2).

3.3 SV2A PET in Other Dementias

Despite the importance of synaptic loss as a pathological substrate of Lewy body dementias (LBD) [116–118], which include PD dementia (PDD) and DLB, similarly to what is observed in AD [119], only a few studies have examined SV2A PET imaging in LBD. In one study which performed immunostaining for SV2A and SV2C in the substantia nigra and striatum of individuals with LBD, progressive supranuclear palsy (PSP), AD, and healthy controls, no significant changes in SV2A immunostaining were observed between any of the disease groups and controls [91]. However, in individuals with PD, and to a lesser extent those with DLB, SV2C staining appeared as puncta rather than diffuse staining. Another analysis which used 3H-UCB-J autoradiography on substantia nigra and putamen in 15 individuals with moderate to severe LBD and 13 controls reported a 17% decrease in SV2A density in the substantia nigra in LBD compared to controls [91]. In another analysis, the authors applied 11C-UCB-J PET imaging to 12 individuals with PD and 12 controls and found significant reductions in 11C-UCB-J uptake in subcortical regions, including the substantia nigra (45%), red nucleus (31%), and locus coeruleus (17%) in PD [120]. In these analyses, PET changes were almost 3 times more than the autoradiography changes. Together, these observations suggest that, while 11C-UCB-J PET may have potential as an imaging marker of synaptic density in LBD, further studies are needed to better evaluate the in vivo specificity of this tracer.

Other studies support the potential of 11C-UCB-J PET in PD. Significant reductions in 11C-UCB-J volume of distribution were observed in the caudate (15.1%), thalamus (11.6%), putamen (9.7%), substantia nigra (6.9%) and several cortical regions compared to controls, and lower brain stem 11C-UCB-J binding negatively correlated with clinical measures of disease severity (Movement Disorder Society-Unified Parkinson's Disease Rating Scale part III [MDS-UPDRS part III] and Hoehn and Yahr [H&Y] scale) in one study [121]. However, other studies have shown significantly reduced 11C-UCB-J binding in the substantia nigra (15.2%), but not in the putamen, in PD compared to controls [122]. In a small study which included 2 participants with DLB and 10 controls, reduced binding potential in several areas, including the thalamus, parieto-occipital, middle and superior frontal cortices, but not the midbrain or putamen, was observed in DLB [123].

Based on these observations, further studies are needed to determine the utility of SV2A PET imaging in LBD, and whether other synaptic density markers such as SV2C may offer greater promise in measuring synaptic density in these cohorts. Future studies which perform fine mapping of SV2A density in LBD brains, a better characterization of the in vivo binding specificity of SV2A ligands such as 11C-UCB-J PET, and studies which include both cross-sectional and longitudinal PET imaging in LBD cohorts will be important to answer these questions.

Very few studies have examined SV2A PET tracers in FTD. In one study using 18F-UCB-H in a small cohort of AD, the behavioral variant of FTD (bvFTD), and healthy controls, there were trends for lower uptake in the right anterior parahippocampal gyrus in bvFTD but not other regions such as the medial prefrontal cortex even when atrophy of this region was evident [124]. Another study using 11C-UCB-J found reduced binding in regions known to be affected by frontotemporal lobar degeneration (FTLD) pathology, including the bilateral cingulate, insula, and frontal regions, a few of which also correlated with cognitive outcomes [125]. Results of these 2 small studies are conflicting; therefore, further studies in larger cohorts are needed to understand the potential value of SV2A PET imaging in FTD.

4 Conclusions and Future Directions

The last decade has witnessed significant advances in the identification and validation of several fluid and imaging markers that can reliably measure synaptic loss in AD. Most of these markers reflect injury to the pre-synaptic compartment, and, with the exception of Ng, there is a paucity of post-synaptic proteins in the current repertoire of AD biomarkers. The post-synaptic protein, CSF Ng, has been among the most widely investigated and validated markers in AD and may offer relative specificity to AD. Conversely, most synaptic injury markers likely reflect common mechanisms of synaptic degeneration that are shared across different neurodegenerative disorders. Therefore, there is a need to identify and validate new markers of post-synaptic injury, including those that are specific to AD, and to identify synaptic injury markers that can be reliably measured in the blood.

Biomarkers of synaptic injury offer important prognostic value for future onset or progression of cognitive decline and brain atrophy and complement diagnostic information provided by the core AD biomarkers of amyloid and tau. Importantly, as synaptic loss is a close surrogate of cognitive impairment, fluid or imaging markers that can reliably measure and track with clinical progression have strong potential to serve as surrogate biomarker endpoints in clinical trials of emerging disease-modifying AD therapies. CSF Ng has already been incorporated as a biomarker outcome measure in clinical trials of anti-amyloid and anti-inflammatory agents in AD.

While significant progress has been made, there are several areas that require further investigation. There is a need for large longitudinal studies that can demonstrate the ability of fluid or imaging markers of synaptic injury to change in response to treatment or with disease progression, and thus serve as dynamic measures of disease severity. It will also be important to understand the mechanisms by which synaptic marker alterations occur in AD or other neurodegenerative disorders, and whether these may be preferentially affected by certain pathologies or may reflect underlying differences in vulnerability of specific neuronal populations. Further, common challenges in the implementation of AD fluid biomarkers in research settings, as a step towards potential translation into the clinic, include the need to validate assays across different centers and cohorts, a better understanding of the influence of pre-analytical factors on biomarker measurements, and research dedicated to examining the impact of demographic factors, such as age, sex, race/ethnicity, and the *APOE4* genotype on biomarker levels.

Funding This work was supported by the National Institute of Health- National Institute of Aging Grant RF1AG083744 (RT).

References

1. DeKosky ST, Scheff SW. Synapse loss in frontal cortex biopsies in Alzheimer's disease: correlation with cognitive severity. Ann Neurol. 1990;27(5):457–64.
2. Terry RD, Masliah E, Salmon DP, Butters N, DeTeresa R, Hill R, et al. Physical basis of cognitive alterations in Alzheimer's disease: synapse loss is the major correlate of cognitive impairment. Ann Neurol. 1991;30(4):572–80.
3. Scheff SW, Price DA, Schmitt FA, Mufson EJ. Hippocampal synaptic loss in early Alzheimer's disease and mild cognitive impairment. Neurobiol Aging. 2006;27(10):1372–84.
4. Tarawneh R. Biomarkers: our path towards a cure for Alzheimer disease. Biomark Insights. 2020;15:1177271920976367.
5. Price JL, Ko AI, Wade MJ, Tsou SK, McKeel DW, Morris JC. Neuron number in the entorhinal cortex and CA1 in preclinical Alzheimer disease. Arch Neurol. 2001;58(9):1395–402.
6. Knobloch M, Mansuy IM. Dendritic spine loss and synaptic alterations in Alzheimer's disease. Mol Neurobiol. 2008;37(1):73–82.
7. Reddy PH, Mani G, Park BS, Jacques J, Murdoch G, Whetsell W Jr, et al. Differential loss of synaptic proteins in Alzheimer's disease: implications for synaptic dysfunction. J Alzheimers Dis. 2005;7(2):103–17; discussion 73–80
8. Lista S, Santos-Lozano A, Emanuele E, Mercuri NB, Gabelle A, López-Ortiz S, et al. Monitoring synaptic pathology in Alzheimer's disease through fluid and PET imaging biomarkers: a comprehensive review and future perspectives. Mol Psychiatry. 2024;29(3):847–57.
9. Camporesi E, Nilsson J, Brinkmalm A, Becker B, Ashton NJ, Blennow K, et al. Fluid biomarkers for synaptic dysfunction and loss. Biomark Insights. 2020;15:1177271920950319.
10. Tarawneh R, D'Angelo G, Crimmins D, Herries E, Griest T, Fagan AM, et al. Diagnostic and prognostic utility of the synaptic marker neurogranin in Alzheimer disease. JAMA Neurol. 2016;73(5):561–71.
11. Wellington H, Paterson RW, Portelius E, Törnqvist U, Magdalinou N, Fox NC, et al. Increased CSF neurogranin concentration is specific to Alzheimer disease. Neurology. 2016;86(9):829–35.
12. Antonucci F, Corradini I, Fossati G, Tomasoni R, Menna E, Matteoli M. SNAP-25, a known presynaptic protein with emerging postsynaptic functions. Front Synaptic Neurosci. 2016;8:7.
13. Toft-Bertelsen TL, Ziomkiewicz I, Houy S, Pinheiro PS, Sørensen JB. Regulation of Ca2+ channels by SNAP-25 via recruitment of syntaxin-1 from plasma membrane clusters. Mol Biol Cell. 2016;27(21):3329–41.
14. Brinkmalm A, Brinkmalm G, Honer WG, Frölich L, Hausner L, Minthon L, et al. SNAP-25 is a promising novel cerebrospinal fluid biomarker for synapse degeneration in Alzheimer's disease. Mol Neurodegener. 2014;9:53.
15. Öhrfelt A, Brinkmalm A, Dumurgier J, Zetterberg H, Bouaziz-Amar E, Hugon J, et al. A Novel ELISA for the measurement of cerebrospinal fluid SNAP-25 in patients with Alzheimer's disease. Neuroscience. 2019;420:136–44.
16. Nilsson J, Ashton NJ, Benedet AL, Montoliu-Gaya L, Gobom J, Pascoal TA, et al. Quantification of SNAP-25 with mass spectrometry and Simoa: a method comparison in Alzheimer's disease. Alzheimers Res Ther. 2022;14(1):78.

17. Wang C, Zhao D, Shah SZA, Yang W, Li C, Yang L. Proteome analysis of potential synaptic vesicle cycle biomarkers in the cerebrospinal fluid of patients with sporadic creutzfeldt-jakob disease. Mol Neurobiol. 2017;54(7):5177–91.
18. Bark IC, Wilson MC. Human cDNA clones encoding two different isoforms of the nerve terminal protein SNAP-25. Gene. 1994;139(2):291–2.
19. Kivisäkk P, Carlyle BC, Sweeney T, Quinn JP, Ramirez CE, Trombetta BA, et al. Increased levels of the synaptic proteins PSD-95, SNAP-25, and neurogranin in the cerebrospinal fluid of patients with Alzheimer's disease. Alzheimers Res Ther. 2022;14(1):58.
20. Zhang H, Therriault J, Kang MS, Ng KP, Pascoal TA, Rosa-Neto P, et al. Cerebrospinal fluid synaptosomal-associated protein 25 is a key player in synaptic degeneration in mild cognitive impairment and Alzheimer's disease. Alzheimers Res Ther. 2018;10(1):80.
21. Benedet AL, Ashton NJ, Pascoal TA, Brinkmalm A, Nilsson J, Kvartsberg H, et al. SNAP25 reflects amyloid- and tau-related synaptic damage: associations between PET, VBM and cerebrospinal fluid biomarkers of synaptic disfunction in the Alzheimer's disease spectrum. Alzheimers Dement. 2020;16(S5):e046358.
22. Tible M, Sandelius Å, Höglund K, Brinkmalm A, Cognat E, Dumurgier J, et al. Dissection of synaptic pathways through the CSF biomarkers for predicting Alzheimer disease. Neurology. 2020;95(8):e953–e61.
23. Ribaldi F, Mendes AJ, Zetterberg H, Blennow K, Grötschel L, Lathuiliere A, et al. Plasma SNAP-25 as a potential Alzheimer's disease biomarker. Alzheimers Dement. 2023;19(S15):e076863.
24. Agliardi C, Guerini FR, Zanzottera M, Bianchi A, Nemni R, Clerici M. SNAP-25 in serum is carried by exosomes of neuronal origin and is a potential biomarker of Alzheimer's disease. Mol Neurobiol. 2019;56(8):5792–8.
25. Denny JB. Molecular mechanisms, biological actions, and neuropharmacology of the growth-associated protein GAP-43. Curr Neuropharmacol. 2006;4(4):293–304.
26. Kanazir S, Ruzdijic S, Vukosavic S, Ivkovic S, Milosevic A, Zecevic N, et al. GAP-43 mRNA expression in early development of human nervous system. Brain Res Mol Brain Res. 1996;38(1):145–55.
27. Holahan MR. A shift from a pivotal to supporting role for the Growth-Associated Protein (GAP-43) in the coordination of axonal structural and functional plasticity. Front Cell Neurosci. 2017;11:266.
28. Sandelius Å, Portelius E, Källén Å, Zetterberg H, Rot U, Olsson B, et al. Elevated CSF GAP-43 is Alzheimer's disease specific and associated with tau and amyloid pathology. Alzheimers Dement. 2019;15(1):55–64.
29. Qiang Q, Skudder-Hill L, Toyota T, Wei W, Adachi H. CSF GAP-43 as a biomarker of synaptic dysfunction is associated with tau pathology in Alzheimer's disease. Sci Rep. 2022;12(1):17392.
30. Remnestål J, Just D, Mitsios N, Fredolini C, Mulder J, Schwenk JM, et al. CSF profiling of the human brain enriched proteome reveals associations of neuromodulin and neurogranin to Alzheimer's disease. Proteomics—Clin Appl. 2016;10(12):1242–53.
31. Sjögren M, Davidsson P, Gottfries J, Vanderstichele H, Edman A, Vanmechelen E, et al. The cerebrospinal fluid levels of tau, growth-associated protein-43 and soluble amyloid precursor protein correlate in Alzheimer's disease, reflecting a common pathophysiological process. Dement Geriatr Cogn Disord. 2001;12(4):257–64.
32. Öhrfelt A, Benedet AL, Ashton NJ, Kvartsberg H, Vandijck M, Weiner MW, et al. Association of CSF GAP-43 With the rate of cognitive decline and progression to dementia in amyloid-positive individuals. Neurology. 2023;100(3):e275–e85.
33. Nemes AD, Ayasoufi K, Ying Z, Zhou QG, Suh H, Najm IM. Growth Associated Protein 43 (GAP-43) as a novel target for the diagnosis, treatment and prevention of epileptogenesis. Sci Rep. 2017;7(1):17702.
34. Franzmeier N, Dehsarvi A, Steward A, Biel D, Dewenter A, Roemer SN, et al. Elevated CSF GAP-43 is associated with accelerated tau accumulation and spread in Alzheimer's disease. Nat Commun. 2024;15(1):202.

35. Courtney NA, Bao H, Briguglio JS, Chapman ER. Synaptotagmin 1 clamps synaptic vesicle fusion in mammalian neurons independent of complexin. Nat Commun. 2019;10(1):4076.
36. Chang S, Trimbuch T, Rosenmund C. Synaptotagmin-1 drives synchronous Ca(2+)-triggered fusion by C(2)B-domain-mediated synaptic-vesicle-membrane attachment. Nat Neurosci. 2018;21(1):33–40.
37. Öhrfelt A, Brinkmalm A, Dumurgier J, Brinkmalm G, Hansson O, Zetterberg H, et al. The pre-synaptic vesicle protein synaptotagmin is a novel biomarker for Alzheimer's disease. Alzheimers Res Ther. 2016;8(1):41.
38. Milà-Alomà M, Brinkmalm A, Ashton NJ, Kvartsberg H, Shekari M, Operto G, et al. CSF synaptic biomarkers in the preclinical stage of alzheimer disease and their association with MRI and PET. Neurology. 2021;97(21):e2065–e78.
39. Clarke MTM, Brinkmalm A, Foiani MS, Woollacott IOC, Heller C, Heslegrave A, et al. CSF synaptic protein concentrations are raised in those with atypical Alzheimer's disease but not frontotemporal dementia. Alzheimers Res Ther. 2019;11(1):105.
40. Xu D, Hopf C, Reddy R, Cho RW, Guo L, Lanahan A, et al. Narp and NP1 form hetero-complexes that function in developmental and activity-dependent synaptic plasticity. Neuron. 2003;39(3):513–28.
41. Figueiro-Silva J, Gruart A, Clayton KB, Podlesniy P, Abad MA, Gasull X, et al. Neuronal pentraxin 1 negatively regulates excitatory synapse density and synaptic plasticity. J Neurosci. 2015;35(14):5504–21.
42. Abad MA, Enguita M, DeGregorio-Rocasolano N, Ferrer I, Trullas R. Neuronal pentraxin 1 contributes to the neuronal damage evoked by amyloid-beta and is overexpressed in dystrophic neurites in Alzheimer's brain. J Neurosci. 2006;26(49):12735–47.
43. Ma QL, Teng E, Zuo X, Jones M, Teter B, Zhao EY, et al. Neuronal pentraxin 1: a synaptic-derived plasma biomarker in Alzheimer's disease. Neurobiol Dis. 2018;114:120–8.
44. Xiao MF, Xu D, Craig MT, Pelkey KA, Chien CC, Shi Y, et al. NPTX2 and cognitive dysfunction in Alzheimer's disease. elife. 2017;6:e23798.
45. Belbin O, Xiao MF, Xu D, Carmona-Iragui M, Pegueroles J, Benejam B, et al. Cerebrospinal fluid profile of NPTX2 supports role of Alzheimer's disease-related inhibitory circuit dysfunction in adults with Down syndrome. Mol Neurodegener. 2020;15(1):46.
46. Soldan A, Oh S, Ryu T, Pettigrew C, Zhu Y, Moghekar A, et al. NPTX2 in cerebrospinal fluid predicts the progression from normal cognition to mild cognitive impairment. Ann Neurol. 2023;94(4):620–31.
47. Libiger O, Shaw LM, Watson MH, Nairn AC, Umaña KL, Biarnes MC, et al. Longitudinal CSF proteomics identifies NPTX2 as a prognostic biomarker of Alzheimer's disease. Alzheimers Dement. 2021;17(12):1976–87.
48. Soldan A, Moghekar A, Walker KA, Pettigrew C, Hou X, Lu H, et al. Resting-state functional connectivity is associated with cerebrospinal fluid levels of the synaptic protein NPTX2 in non-demented older adults. Front Aging Neurosci. 2019;11:132.
49. Begcevic I, Tsolaki M, Brinc D, Brown M, Martinez-Morillo E, Lazarou I, et al. Neuronal pentraxin receptor-1 is a new cerebrospinal fluid biomarker of Alzheimer's disease progression. F1000Res. 2018;7:1012.
50. Díez-Guerra FJ. Neurogranin, a link between calcium/calmodulin and protein kinase C signaling in synaptic plasticity. IUBMB Life. 2010;62(8):597–606.
51. Laterza OF, Modur VR, Crimmins DL, Olander JV, Landt Y, Lee JM, et al. Identification of novel brain biomarkers. Clin Chem. 2006;52(9):1713–21.
52. Watson JB, Battenberg EF, Wong KK, Bloom FE, Sutcliffe JG. Subtractive cDNA cloning of RC3, a rodent cortex-enriched mRNA encoding a novel 78 residue protein. J Neurosci Res. 1990;26(4):397–408.
53. Masliah E, Mallory M, Alford M, DeTeresa R, Hansen LA, McKeel DW Jr, et al. Altered expression of synaptic proteins occurs early during progression of Alzheimer's disease. Neurology. 2001;56(1):127–9.
54. Davidsson P, Blennow K. Neurochemical dissection of synaptic pathology in Alzheimer's disease. Int Psychogeriatr. 1998;10(1):11–23.

55. Sze CI, Troncoso JC, Kawas C, Mouton P, Price DL, Martin LJ. Loss of the presynaptic vesicle protein synaptophysin in hippocampus correlates with cognitive decline in Alzheimer disease. J Neuropathol Exp Neurol. 1997;56(8):933–44.
56. Mattsson N, Insel PS, Palmqvist S, Portelius E, Zetterberg H, Weiner M, et al. Cerebrospinal fluid tau, neurogranin, and neurofilament light in Alzheimer's disease. EMBO Mol Med. 2016;8(10):1184–96.
57. Thorsell A, Bjerke M, Gobom J, Brunhage E, Vanmechelen E, Andreasen N, et al. Neurogranin in cerebrospinal fluid as a marker of synaptic degeneration in Alzheimer's disease. Brain Res. 2010;1362:13–22.
58. Mavroudis IA, Petridis F, Chatzikonstantinou S, Kazis D. A meta-analysis on CSF neurogranin levels for the diagnosis of Alzheimer's disease and mild cognitive impairment. Aging Clin Exp Res. 2020;32(9):1639–46.
59. Kvartsberg H, Duits FH, Ingelsson M, Andreasen N, Öhrfelt A, Andersson K, et al. Cerebrospinal fluid levels of the synaptic protein neurogranin correlates with cognitive decline in prodromal Alzheimer's disease. Alzheimers Dement. 2015;11(10):1180–90.
60. Portelius E, Zetterberg H, Skillbäck T, Törnqvist U, Andreasson U, Trojanowski JQ, et al. Cerebrospinal fluid neurogranin: relation to cognition and neurodegeneration in Alzheimer's disease. Brain. 2015;138(Pt 11):3373–85.
61. Tarawneh R, D'Angelo G, Macy E, Xiong C, Carter D, Cairns NJ, et al. Visinin-like protein-1: diagnostic and prognostic biomarker in Alzheimer disease. Ann Neurol. 2011;70(2):274–85.
62. Fagan AM, Mintun MA, Shah AR, Aldea P, Roe CM, Mach RH, et al. Cerebrospinal fluid tau and ptau(181) increase with cortical amyloid deposition in cognitively normal individuals: implications for future clinical trials of Alzheimer's disease. EMBO Mol Med. 2009;1(8–9):371–80.
63. Jack CR Jr, Knopman DS, Jagust WJ, Shaw LM, Aisen PS, Weiner MW, et al. Hypothetical model of dynamic biomarkers of the Alzheimer's pathological cascade. Lancet Neurol. 2010;9(1):119–28.
64. Perrin RJ, Fagan AM, Holtzman DM. Multimodal techniques for diagnosis and prognosis of Alzheimer's disease. Nature. 2009;461(7266):916–22.
65. Ingelsson M, Fukumoto H, Newell KL, Growdon JH, Hedley-Whyte ET, Frosch MP, et al. Early Abeta accumulation and progressive synaptic loss, gliosis, and tangle formation in AD brain. Neurology. 2004;62(6):925–31.
66. Rowe CC, Ng S, Ackermann U, Gong SJ, Pike K, Savage G, et al. Imaging beta-amyloid burden in aging and dementia. Neurology. 2007;68(20):1718–25.
67. Portelius E, Olsson B, Höglund K, Cullen NC, Kvartsberg H, Andreasson U, et al. Cerebrospinal fluid neurogranin concentration in neurodegeneration: relation to clinical phenotypes and neuropathology. Acta Neuropathol. 2018;136(3):363–76.
68. Becker B, Nazir FH, Brinkmalm G, Camporesi E, Kvartsberg H, Portelius E, et al. Alzheimer-associated cerebrospinal fluid fragments of neurogranin are generated by Calpain-1 and prolyl endopeptidase. Mol Neurodegener. 2018;13(1):47.
69. Willemse EAJ, De Vos A, Herries EM, Andreasson U, Engelborghs S, van der Flier WM, et al. Neurogranin as cerebrospinal fluid biomarker for Alzheimer disease: an assay comparison study. Clin Chem. 2018;64(6):927–37.
70. Öhrfelt A, Dumurgier J, Zetterberg H, Vrillon A, Ashton NJ, Kvartsberg H, et al. Full-length and C-terminal neurogranin in Alzheimer's disease cerebrospinal fluid analyzed by novel ultrasensitive immunoassays. Alzheimers Res Ther. 2020;12(1):168.
71. De Vos A, Jacobs D, Struyfs H, Fransen E, Andersson K, Portelius E, et al. C-terminal neurogranin is increased in cerebrospinal fluid but unchanged in plasma in Alzheimer's disease. Alzheimers Dement. 2015;11(12):1461–9.
72. Goetzl EJ, Kapogiannis D, Schwartz JB, Lobach IV, Goetzl L, Abner EL, et al. Decreased synaptic proteins in neuronal exosomes of frontotemporal dementia and Alzheimer's disease. FASEB J. 2016;30(12):4141–8.
73. Winston CN, Goetzl EJ, Akers JC, Carter BS, Rockenstein EM, Galasko D, et al. Prediction of conversion from mild cognitive impairment to dementia with neuronally derived blood exosome protein profile. Alzheimers Dement (Amst). 2016;3:63–72.

74. Burré J. The synaptic function of α-synuclein. J Parkinsons Dis. 2015;5:699–713.
75. Twohig D, Nielsen HM. α-synuclein in the pathophysiology of Alzheimer's disease. Mol Neurodegener. 2019;14(1):23.
76. Iseki E. Dementia with lewy bodies: reclassification of pathological subtypes and boundary with Parkinson's disease or Alzheimer's disease. Neuropathology. 2004;24(1):72–8.
77. Olichney JM, Galasko D, Salmon DP, Hofstetter CR, Hansen LA, Katzman R, et al. Cognitive decline is faster in Lewy body variant than in Alzheimer's disease. Neurology. 1998;51(2):351–7.
78. Ishizawa T, Mattila P, Davies P, Wang D, Dickson DW. Colocalization of tau and alpha-synuclein epitopes in Lewy bodies. J Neuropathol Exp Neurol. 2003;62(4):389–97.
79. Clinton LK, Blurton-Jones M, Myczek K, Trojanowski JQ, LaFerla FM. Synergistic Interactions between Abeta, tau, and alpha-synuclein: acceleration of neuropathology and cognitive decline. J Neurosci. 2010;30(21):7281–9.
80. Poirel O, Mella S, Videau C, Ramet L, Davoli MA, Herzog E, et al. Moderate decline in select synaptic markers in the prefrontal cortex (BA9) of patients with Alzheimer's disease at various cognitive stages. Sci Rep. 2018;8(1):938.
81. Wesenhagen KEJ, Teunissen CE, Visser PJ, Tijms BM. Cerebrospinal fluid proteomics and biological heterogeneity in Alzheimer's disease: a literature review. Crit Rev Clin Lab Sci. 2020;57(2):86–98.
82. Nikolaienko O, Patil S, Eriksen MS, Bramham CR. Arc protein: a flexible hub for synaptic plasticity and cognition. Semin Cell Dev Biol. 2018;77:33–42.
83. Lleó A, Núñez-Llaves R, Alcolea D, Chiva C, Balateu-Paños D, Colom-Cadena M, et al. Changes in synaptic proteins precede neurodegeneration markers in preclinical Alzheimer's disease cerebrospinal fluid. Mol Cell Proteomics. 2019;18(3):546–60.
84. Buckley K, Kelly RB. Identification of a transmembrane glycoprotein specific for secretory vesicles of neural and endocrine cells. J Cell Biol. 1985;100(4):1284–94.
85. Scranton TW, Iwata M, Carlson SS. The SV2 protein of synaptic vesicles is a keratan sulfate proteoglycan. J Neurochem. 1993;61(1):29–44.
86. Mendoza-Torreblanca JG, Vanoye-Carlo A, Phillips-Farfán BV, Carmona-Aparicio L, Gómez-Lira G. Synaptic vesicle protein 2A: basic facts and role in synaptic function. Eur J Neurosci. 2013;38(11):3529–39.
87. Janz R, Goda Y, Geppert M, Missler M, Südhof TC. SV2A and SV2B function as redundant Ca2+ regulators in neurotransmitter release. Neuron. 1999;24(4):1003–16.
88. Bajjalieh SM, Frantz GD, Weimann JM, McConnell SK, Scheller RH. Differential expression of synaptic vesicle protein 2 (SV2) isoforms. J Neurosci. 1994;14(9):5223–35.
89. Varnäs K, Stepanov V, Halldin C. Autoradiographic mapping of synaptic vesicle glycoprotein 2A in non-human primate and human brain. Synapse. 2020;74(10):e22157.
90. Pazarlar BA, Aripaka SS, Petukhov V, Pinborg L, Khodosevich K, Mikkelsen JD. Expression profile of synaptic vesicle glycoprotein 2A, B, and C paralogues in temporal neocortex tissue from patients with temporal lobe epilepsy (TLE). Mol Brain. 2022;15(1):45.
91. Bavarsad MS, Grinberg LT. SV2A PET imaging in human neurodegenerative diseases. Front Aging Neurosci. 2024;16:1380561.
92. Mutch SA, Kensel-Hammes P, Gadd JC, Fujimoto BS, Allen RW, Schiro PG, et al. Protein quantification at the single vesicle level reveals that a subset of synaptic vesicle proteins are trafficked with high precision. J Neurosci. 2011;31(4):1461–70.
93. Cai H, Mangner TJ, Muzik O, Wang MW, Chugani DC, Chugani HT. Radiosynthesis of (11) C-Levetiracetam: a potential marker for PET imaging of SV2A expression. ACS Med Chem Lett. 2014;5(10):1152–5.
94. Danish A, Namasivayam V, Schiedel AC, Müller CE. Interaction of approved drugs with synaptic vesicle protein 2A. Arch Pharm. 2017;350(3–4):1700003.
95. Nabulsi NB, Mercier J, Holden D, Carré S, Najafzadeh S, Vandergeten MC, et al. Synthesis and preclinical evaluation of 11C-UCB-J as a PET tracer for imaging the synaptic vesicle glycoprotein 2A in the brain. J Nucl Med. 2016;57(5):777–84.

96. Finnema SJ, Nabulsi NB, Eid T, Detyniecki K, Lin SF, Chen MK, et al. Imaging synaptic density in the living human brain. Sci Transl Med. 2016;8(348):348ra96.
97. Estrada S, Lubberink M, Thibblin A, Sprycha M, Buchanan T, Mestdagh N, et al. [(11)C]UCB-A, a novel PET tracer for synaptic vesicle protein 2A. Nucl Med Biol. 2016;43(6):325–32.
98. Finnema SJ, Nabulsi NB, Mercier J, Lin SF, Chen MK, Matuskey D, et al. Kinetic evaluation and test-retest reproducibility of [(11)C]UCB-J, a novel radioligand for positron emission tomography imaging of synaptic vesicle glycoprotein 2A in humans. J Cereb Blood Flow Metab. 2018;38(11):2041–52.
99. Mertens N, Maguire RP, Serdons K, Lacroix B, Mercier J, Sciberras D, et al. Validation of parametric methods for [(11)C]UCB-J PET imaging using subcortical white matter as reference tissue. Mol Imaging Biol. 2020;22(2):444–52.
100. Mecca AP, O'Dell RS, Sharp ES, Banks ER, Bartlett HH, Zhao W, et al. Synaptic density and cognitive performance in Alzheimer's disease: a PET imaging study with [(11) C]UCB-J. Alzheimers Dement. 2022;18(12):2527–36.
101. Li S, Naganawa M, Pracitto R, Najafzadeh S, Holden D, Henry S, et al. Assessment of test-retest reproducibility of [(18)F]SynVesT-1, a novel radiotracer for PET imaging of synaptic vesicle glycoprotein 2A. Eur J Nucl Med Mol Imaging. 2021;48(5):1327–38.
102. Naganawa M, Li S, Nabulsi N, Henry S, Zheng MQ, Pracitto R, et al. First-in-human evaluation of (18)F-SynVesT-1, a radioligand for PET imaging of synaptic vesicle glycoprotein 2A. J Nucl Med. 2021;62(4):561–7.
103. Cai Z, Li S, Finnema S, Lin S-f, Zhang W, Holden D, et al. Synthesis and preclinical evaluation of an ^{18}F-labeled synaptic vesicle glycoprotein 2A PET imaging probe: [^{18}F]SynVesT-2. J Nucl Med. 2017;58(supplement 1):547.
104. Constantinescu CC, Tresse C, Zheng M, Gouasmat A, Carroll VM, Mistico L, et al. Development and in vivo preclinical imaging of fluorine-18-labeled synaptic vesicle protein 2A (SV2A) PET tracers. Mol Imaging Biol. 2019;21(3):509–18.
105. Zheng C, Holden D, Zheng M-Q, Pracitto R, Wilcox KC, Lindemann M, et al. A metabolically stable PET tracer for imaging synaptic vesicle protein 2A: synthesis and preclinical characterization of [18F]SDM-16. Eur J Nucl Med Mol Imaging. 2022;49(5):1482–96.
106. Metaxas A, Thygesen C, Briting SRR, Landau AM, Darvesh S, Finsen B. Increased inflammation and unchanged density of synaptic vesicle glycoprotein 2A (SV2A) in the postmortem frontal cortex of Alzheimer's disease patients. Front Cell Neurosci. 2019;13:538.
107. Mikkelsen JD, Kaad S, Aripaka SS, Finsen B. Synaptic vesicle glycoprotein 2A (SV2A) levels in the cerebral cortex in patients with Alzheimer's disease: a radioligand binding study in postmortem brains. Neurobiol Aging. 2023;129:50–7.
108. Chen MK, Mecca AP, Naganawa M, Finnema SJ, Toyonaga T, Lin SF, et al. Assessing synaptic density in Alzheimer disease with synaptic vesicle glycoprotein 2A positron emission tomographic imaging. JAMA Neurol. 2018;75(10):1215–24.
109. Mecca AP, Chen MK, O'Dell RS, Naganawa M, Toyonaga T, Godek TA, et al. In vivo measurement of widespread synaptic loss in Alzheimer's disease with SV2A PET. Alzheimers Dement. 2020;16(7):974–82.
110. Mecca AP, Chen MK, O'Dell RS, Naganawa M, Toyonaga T, Godek TA, et al. Association of entorhinal cortical tau deposition and hippocampal synaptic density in older individuals with normal cognition and early Alzheimer's disease. Neurobiol Aging. 2022;111:44–53.
111. Vanhaute H, Ceccarini J, Michiels L, Koole M, Sunaert S, Lemmens R, et al. In vivo synaptic density loss is related to tau deposition in amnestic mild cognitive impairment. Neurology. 2020;95(5):e545–e53.
112. Coomans EM, Schoonhoven DN, Tuncel H, Verfaillie SCJ, Wolters EE, Boellaard R, et al. In vivo tau pathology is associated with synaptic loss and altered synaptic function. Alzheimers Res Ther. 2021;13(1):35.
113. O'Dell RS, Mecca AP, Chen MK, Naganawa M, Toyonaga T, Lu Y, et al. Association of Aβ deposition and regional synaptic density in early Alzheimer's disease: a PET imaging study with [(11)C]UCB-J. Alzheimers Res Ther. 2021;13(1):11.

114. Bastin C, Bahri MA, Meyer F, Manard M, Delhaye E, Plenevaux A, et al. In vivo imaging of synaptic loss in Alzheimer's disease with [18F]UCB-H positron emission tomography. Eur J Nucl Med Mol Imaging. 2020;47(2):390–402.
115. Becker G, Dammicco S, Bahri MA, Salmon E. The rise of synaptic density PET imaging. Molecules. 2020;25(10):2303.
116. Schulz-Schaeffer WJ. The synaptic pathology of alpha-synuclein aggregation in dementia with Lewy bodies, Parkinson's disease and Parkinson's disease dementia. Acta Neuropathol. 2010;120(2):131–43.
117. Villalba RM, Smith Y. Differential striatal spine pathology in Parkinson's disease and cocaine addiction: a key role of dopamine? Neuroscience. 2013;251:2–20.
118. Colom-Cadena M, Spires-Jones T, Zetterberg H, Blennow K, Caggiano A, DeKosky ST, et al. The clinical promise of biomarkers of synapse damage or loss in Alzheimer's disease. Alzheimers Res Ther. 2020;12(1):21.
119. Scheff SW, Price DA, Schmitt FA, DeKosky ST, Mufson EJ. Synaptic alterations in CA1 in mild Alzheimer disease and mild cognitive impairment. Neurology. 2007;68(18):1501–8.
120. Matuskey D, Tinaz S, Wilcox KC, Naganawa M, Toyonaga T, Dias M, et al. Synaptic changes in parkinson disease assessed with in vivo imaging. Ann Neurol. 2020;87(3):329–38.
121. Wilson H, Pagano G, de Natale ER, Mansur A, Caminiti SP, Polychronis S, et al. Mitochondrial complex 1, Sigma 1, and synaptic vesicle 2A in early drug-naive Parkinson's disease. Mov Disord. 2020;35(8):1416–27.
122. Delva A, Van Laere K, Vandenberghe W. Longitudinal imaging of regional brain volumes, SV2A, and glucose metabolism in huntington's disease. Mov Disord. 2023;38(8):1515–26.
123. Nicastro N, Holland N, Savulich G, Carter SF, Mak E, Hong YT, et al. (11)C-UCB-J synaptic PET and multimodal imaging in dementia with Lewy bodies. Eur J Hybrid Imaging. 2020;4(1):25.
124. Salmon E, Bahri MA, Pl'enevaux A, Becker G, Seret A, Delhaye E, et al. In vivo exploration of synaptic projections in frontotemporal dementia. Sci Rep. 2021;11(1):16092.
125. Malpetti M, Jones PS, Cope TE, Holland N, Naessens M, Rouse MA, et al. Synaptic loss in frontotemporal dementia revealed by [(11) C]UCB-J positron emission tomography. Ann Neurol. 2023;93(1):142–54.

Fluid Markers of Axonal and Neuronal Injury

Burak Arslan and Henrik Zetterberg

Abstract Over the past decade, there has been a surge in the interest in identifying, developing, and implementing biofluid-based biomarkers to enhance the diagnostic process for neurodegenerative diseases. This interest has been partly driven by the advent of highly sensitive detection techniques and disease-modifying therapies, underscoring the necessity for reliable, cost-effective, and easily obtainable biofluid-based biomarkers. These biomarkers would be essential for facilitating the development and monitoring of such therapies and for selecting appropriate participants for clinical trials. In this review, we will discuss the significance of various biofluid-based biomarkers, with the emphasis on neurofilament light, including those indicative of neuronal cell-body damage (neuron-specific enolase and ubiquitin C-terminal hydrolase-L1) and axonal damage (neurofilament proteins and myelin basic protein). Additionally, we will explore the integration of novel biomarkers into clinical practice and trials, highlighting their potential to revolutionize the management of neurodegenerative diseases. This chapter aims to provide a comprehensive overview of the current landscape and future prospects of biofluid biomarkers in neurodegenerative disorders.

B. Arslan
Department of Psychiatry and Neurochemistry, Institute of Neuroscience and Physiology, The Sahlgrenska Academy at the University of Gothenburg, Mölndal, Sweden

Clinical Neurochemistry Laboratory, Sahlgrenska University Hospital, Mölndal, Sweden

H. Zetterberg (✉)
Department of Psychiatry and Neurochemistry, Institute of Neuroscience and Physiology, The Sahlgrenska Academy at the University of Gothenburg, Mölndal, Sweden

Clinical Neurochemistry Laboratory, Sahlgrenska University Hospital, Mölndal, Sweden

Wisconsin Alzheimer's Disease Research Center, School of Medicine and Public Health, University of Wisconsin-Madison, Madison, WI, USA

Department of Neurodegenerative Disease, UCL Institute of Neurology, London, UK

UK Dementia Research Institute at UCL, London, UK

Hong Kong Center for Neurodegenerative Diseases, Clear Water Bay, Hong Kong, China
e-mail: henrik.zetterberg@clinchem.gu.se

R. Tarawneh (ed.), *Biomarkers of Neurodegenerative Disorders*,
https://doi.org/10.1007/978-3-031-80849-4_5

Keywords Alzheimer's disease · Neurodegeneration · Axonal injury · Neuronal cell injury · Neurofilament light

1 Introduction

In recent years, significant progress has been made in the accurate measurement of proteins derived from the central nervous system (CNS) in both cerebrospinal fluid (CSF) and blood, a development largely attributable to the advent of highly sensitive assays. Thanks to these advancements, it is now possible to detect axonal and neuronal damage by analyzing these biomarkers in both CSF and blood [1]. In this chapter, we initially explore the role of biofluid biomarkers in neurodegenerative diseases, highlighting their critical importance. Subsequently, we provide an overview of the biomarkers employed to identify neuronal and axonal damage through biofluid analysis, with a special focus on those biomarkers that can be detected in blood.

2 Why Biofluid Biomarkers Are Needed for Neurodegenerative Diseases

2.1 *Clinical Practice*

The urgent need for precise and early diagnosis of neurodegenerative diseases, including Alzheimer's disease (AD), frontotemporal dementia (FTD), dementia with Lewy bodies (DLB), vascular dementia, and Parkinson's disease (PD)—the latter being the second most prevalent neurodegenerative condition alongside related parkinsonian disorders—is paramount [2]. Research suggests that between 25% and 30% of patients clinically diagnosed with AD are incorrectly identified. This rate of misdiagnosis is even higher in the primary care setting [3]. This misdiagnosis rate occurs despite the management of these conditions often occurring in specialized centers. Autopsy remains the definitive standard for diagnosing these conditions, revealing the presence of core pathological hallmarks, such as amyloid beta plaques and neurofibrillary tau tangles in AD, which cannot be accurately identified through clinical evaluations alone [4]. Consequently, integrating biofluid-based biomarkers into clinical diagnostic algorithms could significantly bolster the accuracy of clinical diagnoses, in particular during early disease stages, and reduce the workload of clinicians.

2.2 *Clinical Trials and Research*

In clinical trials, relying solely on clinical diagnoses for participant selection can result in misleading outcomes in addressing the foundational disease pathology. Consequently, many trials, particularly those for AD, now mandate biomarker evidence for participant inclusion. Given that amyloid pathology can manifest in the brain approximately 20 years prior to the onset of symptoms, it is crucial to identify and include participants at the earliest possible stage, especially in trials focused on preventive therapies. In regard to clinical trials of non-AD neurodegenerative diseases, positive biofluid-based biomarkers for neurodegeneration are increasingly used as an inclusion criterion. Biomarkers for neuroaxonal injury and neurodegeneration have also become crucial to demonstrate disease-modifying effects of novel drugs.

3 The Selection of the Sample Matrix—CSF or Blood?

Although there has been significant interest in discovering and developing blood-based biomarkers for easier detection of brain pathology, CSF still offers advantages over blood in certain aspects. Given its proximity to the brain parenchyma, CSF can more accurately reflect brain pathology and its biomarkers. However, the collection of CSF samples is still considered invasive, leading some patients to avoid giving samples due to the perceived invasiveness of lumbar punctures. In contrast, blood draws are more appealing for several reasons: they are easier to perform, enable serial sampling, and are considered minimally invasive tests. Other alternative biofluids, such as urine, saliva, and tears, are also being explored. However, given the numerous compartments and barriers a marker must traverse, it is challenging to envision how a marker could yield reproducible results that accurately reflect changes in the brain. With that, our extensive analysis of a broad cohort of urine samples, which span the *AD continuum*, revealed that AD biomarkers can indeed be successfully detected in urine employing single-molecule array (Simoa) technology. However, it is important to note that there was no observed correlation between the concentrations of biomarkers in urine and those in CSF or blood (B.A., A.L.B., H.Z., and N.J.A., unpublished data).

4 Neuronal Cell-Body Injury Markers—NSE and UCH-L1

Primarily, two markers, neuron-specific enolase (NSE) and ubiquitin C-terminal hydrolase-L1 (UCH-L1), as well as visinin-like protein-1 (VILIP-1), are recognized for their ability to show damage to neuronal cell bodies within biofluids. These biomarkers, distinct from neurofilament-light (NfL), have a more confined

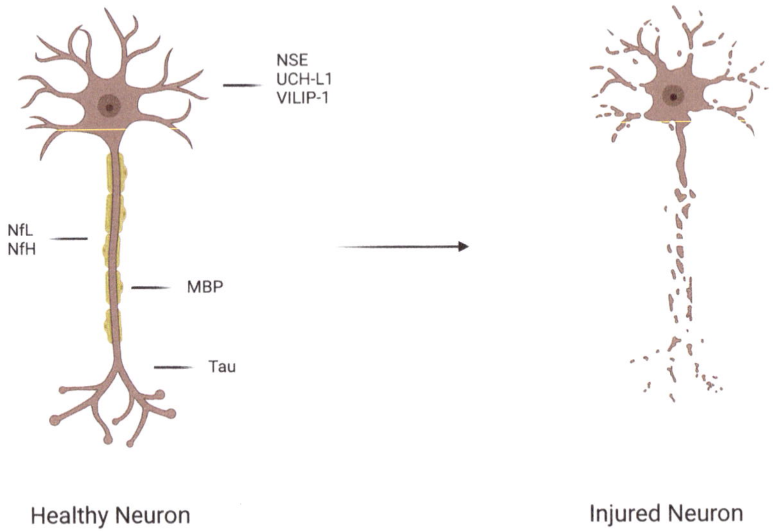

Fig. 1 Biofluid-based markers of neuronal body and axonal injury. This figure presents a schematic illustration of markers associated with neuronal body injury (NSE, UCH-L1, VILIP-1) and axonal injury (NfL, NfH, MBP, and tau). Abbreviations: NSE Neuron-specific enolase, UCH-L1 Ubiquitin C-terminal hydrolase-L1, VILIP-1 Visinin-like protein-1, NfL Neurofilament light, NfH Neurofilament heavy, MBP Myelin basic protein, Tau microtubule associated protein

application in the context of neurodegenerative diseases. Their significance is largely explored within the realms of traumatic injuries affecting the brain and spinal cord. When neuronal damage occurs, the concentrations of these markers typically increase, especially during the initial stages following the injury. Figure 1 shows a schematic illustration of these markers.

4.1 NSE

Neuron-specific enolase (NSE) is an enzyme, localized mainly in the neuronal cytosol, where it participates in the glycolytic cycle, and is not normally secreted by neurons. Under normal conditions, it is not present in extracellular body fluids. Nevertheless, its levels are increased during neuronal injury. To this date, its role in the neurodegenerative disorders has been investigated mostly in CSF given its association with neuronal damage. Its presence in CSF and serum correlates with the degree of neuronal damage observed in conditions like ischemic stroke, traumatic brain injury (TBI), and toxicity. However, there are mixed results regarding its diagnostic and prognostic role in AD and other dementias. Research has shown that in AD and PD/DLB, NSE mRNA levels in postmortem cortical brain tissue are notably reduced and, in AD, this reduction is particularly associated with AD progression [5]. Over the years, NSE concentrations in the CSF of individuals with AD have been variously reported as decreased [6], increased, or unchanged, compared to healthy controls or individuals with other forms of dementia [7, 8], leading to a debate on the diagnostic value of NSE levels in AD versus other dementias.

Certain studies [9–11] have identified higher NSE concentrations in AD compared to vascular dementia, FTD, mild cognitive impairment (MCI), and multi-infarct dementia. Meanwhile, other researchers propose that NSE levels in CSF could distinguish between various diseases, though this suggestion faces opposition from different studies, reflecting the variability in findings across the board. Interestingly, both increased and normal NSE levels have been linked to changes in key biomarkers like tau proteins and amyloid beta ($A\beta_{1\text{-}42}$) in AD patients, suggesting a complex relationship between NSE and AD pathology. In summary, when comparing AD patients to control subjects, the concentration of NSE in CSF shows a significant difference with a moderate effect size. However, significant differences are not reliably observed in NSE levels in serum or plasma [12]. Indeed, it is essential to recognize that as a cytosolic enzyme, NSE is greatly affected by hemolysis. The degree of hemolysis can lead to diverse alterations in its absolute concentrations in the CSF and blood. Therefore, in cases of hemolysis, interpreting the levels of this biomarker requires careful consideration [13].

4.2 UCH-L1

Ubiquitin C-terminal hydrolase-L1 (UCH-L1), primarily located in neurons, plays a vital role in the ubiquitin-proteasome system. This system is crucial for eliminating abnormal neuronal proteins in both healthy and diseased conditions. While UCH-L1 is predominantly associated with the CNS and constitutes 1–5% of total soluble proteins in the brain [14], it is also found in other tissues, including neurons of the peripheral nervous system, cells of the neuroendocrine system, endothelial and smooth muscle cells, and certain tumors [15]. Investigations into the CSF and serum levels of UCH-L1 have been extensive in scenarios such as head trauma and acute stroke. However, its role in neurodegenerative disorders, particularly in relation to cognitive functions, remains partially understood. There is evidence suggesting that diminished UCH-L1 activity might contribute to increased amyloid-beta production by impairing the degradation of BACE1. Decreased UCH-L1 levels have been observed in individuals with AD and AD animal models [16]. Furthermore, increased UCH-L1 levels in CSF and serum have been reported in amyotrophic lateral sclerosis (ALS) patients, though its prognostic significance warrants further exploration [17]. Recent studies have also identified reduced UCH-L1, alongside elevated NfL, levels in PD patients [18]. Given the multifunctional nature of UCH-L1 and the mixed findings in the literature, the significance of UCH-L1 in neurodegenerative diseases necessitates further investigation.

4.3 VILIP-1

Visinin-like protein 1 (VILIP-1), a neuron-specific calcium-sensing/binding protein, is liberated into the CSF in the event of neuronal damage. VILIP-1 is abundantly and widely expressed in the brain—specifically localized in neuronal cell bodies, dendrites, and certain axonal regions [19]—highlighting its potential as a biomarker for various conditions, including stroke, traumatic brain injury (TBI), AD, and epilepsy [20–23]. Within the AD research realm, VILIP-1 is considered indicative of the altered calcium homeostasis characteristic of AD pathology. Disturbed calcium homeostasis leads to the degeneration of susceptible neurons and the release of VILIP-1 into the extracellular space, positioning it as a marker of neuronal injury [24]. Despite its potential, there have been only a limited number of studies exploring the role of this biomarker in both CSF and blood. In one such study, CSF VILIP-1 levels were higher in AD dementia and MCI due to AD compared to non-neurodegenerative controls and MCI due to PD, and serum VILIP-1 levels were higher in AD dementia compared to controls [22]. However, these findings necessitate further validation in studies with larger sample sizes to clarify its role within the *AD continuum*. The utility of VILIP-1 as a biomarker in AD is discussed in more detail in other Chapters Fluid Markers of Alzheimer Disease and Biomarkers as Outcome Measures in Clinical Trials of Alzheimer Disease.

5 Axonal Injury Markers—Tau, NF Proteins, Myelin Basic Protein

Tau, NfL, and neurofilament heavy (NfH) proteins are well-established CSF markers of neuroaxonal injury. The microtubule-associated protein tau is abundant in axons, but can also be detected in the liver, kidney, and testis. NfL and NfH are key structural proteins found within axons, differing from tau primarily in their location and expression patterns. Unlike tau, which is mainly found in cortical thin unmyelinated axons, NfL and NfH are abundantly present in the thick, myelinated axons that reach below the cortex. Additionally, NfL and NfH are present in the peripheral nerves. Figure 1 shows a schematic illustration of these markers.

5.1 Tau

During axonal damage, tau protein can undergo proteolytic cleavage by activated caspases and calpains, producing fragments that are detectable in the bloodstream. Different tau species, total tau (t-tau) and tau phosphorylated at threonine 181 (p-tau181), can be measured in CSF. The levels of these tau biomarkers have been extensively investigated, with p-tau181 levels exclusively found to be increased in AD, but not in other neurodegenerative diseases. Considering tau can be phosphorylated at more than 40 positions, several p-tau assays (p-tau 205, 212, 217, 231) have been developed for both CSF and blood.

5.2 *NfL*

5.2.1 Structure, Function, and Overview

Neurofilament light (NfL) is a crucial scaffolding protein of the axonal cytoskeleton, alongside neurofilament medium (102.5 kDa) and NfH (112.5 kDa), classified as intermediate filaments (Ifs). As a smaller axonal protein (61.5 kDa) and the most soluble form of neurofilaments, NfL plays a vital role in the structural integrity of axons. Following an injury to the axon, NfL is released into the interstitial fluid and subsequently into the CSF. Under normal conditions, axons continuously release low levels of NfL, a process that increases with age. From there, it diffuses into the bloodstream, where its concentration is generally 50 to 100 times lower than in the CSF [25]. Technological advancements, particularly in methods such as Simoa, have revolutionized our ability to detect proteins in the blood that are present in very low quantities. This advancement has notably enhanced the measurement of NfL, a marker for axonal damage, directly from blood samples. The importance of this development stretches across the field of neurology, as it enables researchers and clinicians to study NfL levels in the context of various neurological conditions, including inflammatory (e.g., multiple sclerosis [MS]), neurodegenerative (e.g., FTD and AD), cerebrovascular (e.g., stroke), and traumatic diseases in a less invasive manner [26]. While NfL is not a specific marker, the elevation of its levels can facilitate the early detection of brain changes, assess the extent of active brain pathology, monitor the response to treatment, and predict clinical outcomes [27, 28].

5.2.2 Pre-analytical Considerations

NfL is noted for its robust stability under various pre-analytical conditions, leading to the observation that serum and plasma NfL levels remain stable even after multiple freeze-thaw cycles and extended exposure to room temperature. Regarding the choice of matrix, it has been observed that NfL concentrations in EDTA plasma are approximately 10% lower than in serum; nevertheless, there is a strong correlation between NfL levels in these two matrices [29–31]. Additionally, there is a significant correlation between NfL levels in CSF and blood. While there is no consensus on the need for fasting or specific blood sampling times, these factors are considered to have minimal impact on blood NfL concentrations [26]. In essence, NfL proves to be a resilient biomarker against pre-analytical variations, allowing for measurement in CSF, serum, and EDTA plasma despite the variations in absolute concentrations across these matrices. This underscores the need for matrix-specific reference limits. While NfL results are resistant to changes due to freeze-thaw cycles or storage temperatures, delayed centrifugation of samples could increase variability of NfL measurements [32–35].

5.2.3 Analytical Methods

Historically, the detection of NfL began with the use of immunoblotting techniques for its qualitative identification in CSF. This method was subsequently followed by the introduction of second-generation enzyme-linked immunosorbent assays (ELISAs), which allowed for the quantitative measurement of CSF NfL levels. Following this advancement, third-generation electrochemiluminescence (ECL) technology was developed, enhancing the detection sensitivity and specificity for NfL [36]. Progressing further, fourth-generation immunoassays, such as Simoa technology and microfluidic cartridge-based automated immunoassay platforms (e.g., ELLA from ProteinSimple), were introduced [37]. These latter technologies enabled the sensitive and precise detection of NfL in serum and plasma, marking a significant evolution in the methodologies used to assess NfL levels. Moreover, the development and implementation of NfL assays and fully automated platforms into clinical chemistry laboratories have rendered NfL a viable option for clinical applications [38, 39]. This advancement facilitates the widespread adoption of NfL measurement in routine diagnostic processes, supporting its use in monitoring neurological diseases, evaluating therapeutic responses, and potentially in the screening of neurodegenerative conditions.

5.2.4 Confounders and Clinical Relevance

Research has consistently demonstrated that age is the most significant factor that influences NfL levels in the blood, independent of disease. This finding underscores the importance of considering age when interpreting NfL concentrations, as higher levels may be observed in older individuals even in the absence of neurological conditions [40, 41]. In addition to aging, a higher body mass index (BMI) can lead to lower absolute values of NfL, due to the dilutional effect associated with the increased blood volume that accompanies a higher BMI [42]. This becomes particularly crucial when evaluating minor fluctuations in NfL levels, as opposed to more pronounced increases, indicating the need for careful consideration of BMI in the interpretation of NfL measurements. Finally, the absolute concentration of NfL in the blood may be affected by systemic conditions, including cirrhosis, diabetes, heart failure, hypertension, and kidney failure [43, 44]. This indicates that the presence of these diseases could potentially alter NfL levels, emphasizing the importance of considering these confounders when assessing NfL concentrations.

In numerous neurodegenerative disorders, including AD [45], FTD [46], ALS [47], and PD [48], elevated blood levels of NfL have been reported in comparison to healthy controls, with the highest increases observed in patients with FTD and

ALS. Moreover, NfL serves as an indicator of ongoing neurodegeneration; its levels are also associated with the severity of AD and brain atrophy [49]. Furthermore, NfL proves to be a valuable tool for predicting the progression of AD [50]. In clinical trials of drug candidates targeting neurodegenerative diseases, NfL has become a marker of choice to demonstrate disease modification.

In parallel with integrating NfL as a marker for neuroaxonal injury in clinical chemistry labs, akin to how C-reactive protein is used for inflammation, it is crucial to thoroughly understand the lifecycle of NfL from its production to its degradation. Additionally, it is essential to expand the currently suggested reference ranges to more accurately reflect the diversity found within broader populations. Given the limited research on NfL as a biomarker for psychiatric conditions, including schizophrenia, anorexia nervosa, and substance use disorders, there is a pressing need for further research to establish its significance in these areas.

5.3 AT(N) Classification

In 2018, a critical biological framework for AD, referred to as A/T/(N) classification, was established. This model is characterized by the biological markers of amyloid plaques (A)—indicated by lower CSF $A\beta_{1-42}$ levels or a decreased $A\beta_{1-42/A\beta 1-40}$ ratio and higher amyloid-positron emission tomography (PET) tracer uptake; tau pathology (T)—shown by elevated CSF phosphorylated tau levels or an increase in tau-PET tracer uptake; and neurodegeneration (N)—evidenced by higher CSF levels of total tau or NfL, hypometabolism on ^{18}F-fluorodeoxyglucose (FDG)-PET, or gray matter volume reduction on structural magnetic resonance imaging (MRI). It was designed exclusively for use within research settings as a research framework, rather than for clinical practice [51]. This methodology laid the groundwork for more accurate disease staging. As the scientific landscape evolves, this framework is under revision by an expert committee, taking into account recent breakthroughs in plasma biomarkers and the introduction of therapies that modify disease progression since 2018. The revised criteria differentiate AD biomarkers into two distinct groups: core 1 (A, Amyloid β proteinopathy and T1, phosphorylated and secreted AD-tau) and core 2 (T2, AD-tau proteinopathy [e.g., microtubule binding region (MTBR)-tau243 and p-tau205]), enhancing the precision of AD staging. This approach mandates the utilization of either fluid (CSF or plasma) biomarkers or imaging techniques (PET) for staging. Neurodegeneration (N), now categorized under nonspecific fluid biomarkers alongside inflammation markers (glial fibrillary acidic protein [GFAP]) and imaging modalities (anatomic MR and ^{18}F-FDG PET), reflects a nuanced understanding of AD complexity. Furthermore, the framework introduces classifications for non-AD pathologies, including V (vascular brain injury) and S (α-synuclein), expanding the scope of neurodegenerative disease characterization. While the presence of amyloid beta and tau in the brain can be identified by the concentrations of CSF Aβ_{1-42} or the A$\beta_{1-42/A\beta1-40}$ and CSF phosphorylated tau, extensive research has shown that the inclusion of blood biomarkers for tracking neuroaxonal injury (NfL) and neuroinflammation (GFAP) significantly increases the accuracy and precision of AD clinical diagnoses and the prediction of AD onset. However, universally recognized thresholds for these blood-based biomarkers have yet to be established. Current genome-wide association studies (GWAS) of sporadic late-onset AD have identified over 70 distinct loci associated with AD [52], suggesting the existence of multiple, interconnected molecular, and cellular pathways that could influence the development and progression of the disease. This implies that a deeper understanding of the diverse molecular pathways of AD will enhance our grasp of its mechanisms. Consequently, in the future, the present ATN classification system may undergo evolution and modifications to reflect these advanced insights [53]. Nevertheless, the histopathological examination of the brain at autopsy remains the definitive gold standard for diagnosing AD.

5.4 Clinical Assessment—Outcome Prediction in Preclinical AD

In the AT(N) framework, the "N" is enclosed in brackets to indicate the nonspecific nature of neuronal injury markers related to AD. The significance of NfL levels as an indicator of neurodegeneration has been explored in both CSF [54, 55] and blood [56, 57]. The severity of "N," or neurodegeneration, is associated with elevated NfL concentrations in both of these biological matrices. Additionally, studies have demonstrated that the diagnostic capabilities of NfL in both CSF and blood are similar, suggesting its potential as a noninvasive biomarker to enhance both the diagnosis and prognosis of AD [58]. Studies representing the *AD continuum* have indicated that plasma NfL levels are elevated in individuals with AD dementia compared to those without dementia. Furthermore, plasma NfL concentrations rise over time, a trend that aligns with the progression observed in core CSF biomarkers, MRI assessments of brain atrophy, brain hypometabolism, and declines in cognitive function [57, 59]. It is important to stress that increased NfL concentrations are not specific to AD; elevated levels can also be observed in various other neurodegenerative disorders and as part of the normal aging process [60]. Furthermore, monitoring and predicting the progression of AD can be achieved through longitudinal measurements of blood NfL levels [54, 57]. Collectively, the data demonstrate that NfL levels in plasma or serum can effectively track the progression of AD, and that elevated NfL levels during the preclinical phase of AD could be advantageous for predicting future deterioration [61, 62]. In this scenario, blood NfL measurements could prove to be particularly valuable for selecting participants for clinical trials involving presymptomatic individuals.

5.5 Outcome Measure in Clinical Trials of DMT

With the revised criteria for the biological definition of AD, there has been a nuanced understanding of the contributions made by these biomarkers. Amyloid beta (A) and tau (T) are more directly associated with the core pathological features of the disease. Amyloid beta is linked to the formation of amyloid plaques, a hallmark of AD pathology, while tau is associated with neurofibrillary tangles, another key pathological feature. These markers have shown considerable utility in diagnosing AD and in monitoring its progression.

Neurodegeneration (N), while still a critical aspect of AD pathology, has been somewhat deprioritized in the revised criteria compared to A and T. This shift reflects a nuanced understanding of disease progression and its underlying mechanisms. Despite this, NfL, as a marker of neurodegeneration, continues to play a vital role, particularly in the context of clinical trials for disease-modifying therapies.

NfL in plasma or serum has emerged as a valuable tool for indicating neurodegeneration across a spectrum of neurological diseases, including AD. Its utility lies in its ability to reflect neuronal damage and loss, making it an important biomarker for assessing the efficacy of disease-modifying therapies. In the setting of clinical trials, where the primary goal is often to evaluate the ability of a therapy to slow or reverse the progression of neurodegeneration, NfL serves as a crucial endpoint. It provides a quantifiable measure of neurodegeneration that can be monitored over time, offering insights into how a therapy may alter the trajectory of neuronal loss.

Therefore, despite the revised criteria and the less pronounced emphasis on neurodegeneration (N) compared to amyloid beta (A) and tau (T), NfL retains its relevance. It continues to be a useful biomarker for assessing changes associated with neurodegeneration during clinical trials. Its role in demonstrating the efficacy of disease-modifying therapies underscores its importance in AD research and treatment strategies. This utility of NfL in clinical trial settings highlights the ongoing need to consider a range of biomarkers for a comprehensive understanding of AD and the impact of therapeutic interventions.

The introduction of blood-based biomarkers has significantly facilitated longitudinal sampling in neurodegenerative disorders, making it easier to detect deterioration in patients than before.

For AD specifically, it is critically important to involve cognitively unimpaired individuals by assessing AD biomarker levels (i.e., in pre-symptomatic AD). This strategy is anticipated to enhance the design and management of clinical trials focused on disease-modifying treatments, potentially leading to more effective interventions. Consequently, blood NfL measurement, along with other promising blood-based biomarkers, can identify individuals experiencing neurodegeneration and more accurately predict disease progression. Furthermore, it allows for the monitoring of responses to therapy in trials of disease-modifying treatments, enhancing the ability to evaluate treatment efficacy.

Algorithms that integrate plasma NfL levels with other blood biomarkers (e.g., p-tau217 and $A\beta_{1-42/A\beta1-40}$) and demographic data have been demonstrated to predict cognitive decline, subsequent AD dementia in cognitively unimpaired older individuals, and brain Aβ deposition [63–65]. These models can more effectively identify participants for preclinical trials targeting AD, optimizing the selection process for these crucial studies.

In the recent trial involving lecanemab, a humanized IgG1 monoclonal antibody that targets Aβ-soluble protofibrils with high affinity, for early AD, the scope of biomarker evaluation was broadened to include CSF and plasma NfL, in addition to CSF $A\beta_{1-42}$, $A\beta_{1-40}$, total tau, p-tau181, neurogranin, and plasma $A\beta_{1-42/A\beta1-40}$, p-tau181, GFAP, and NfL. Observations from both the CSF substudy and plasma analysis of the wider participant group indicated that treatment with lecanemab led to more pronounced reductions in amyloid, tau, neurodegeneration, and neuroinflammation markers, in comparison to placebo. Notably, NfL was an exception, displaying less sensitivity to changes in neurodegeneration and a delayed rate of change than the other biomarkers evaluated [66].

5.6 Myelin Basic Protein

Myelin basic protein (MBP) is a primary constituent of the myelin sheath, produced by oligodendrocytes and Schwann cells. In conditions characterized by the degradation of the myelin sheath, such as MS, elevated levels of MBP have been detected in the CSF [67]. Furthermore, MBP is being explored as a potential biomarker for TBI, given that myelinated axons are often the target in cases of diffuse axonal injury. Notably, an increase in the serum levels of MBP has been observed following acute TBI [68]. Despite these findings, additional research is required to fully understand the potential of MBP as a biofluid marker in these contexts.

6 Future Directions

Biofluid markers hold significant potential for diagnosing, monitoring progression, predicting outcomes, evaluating treatment efficacy, and conducting screenings within populations for neurodegenerative diseases. With the use of highly sensitive detection methods, it is feasible to accurately measure the levels of biomarkers present in the blood, facilitating their integration into everyday clinical practice. Nevertheless, there is a pressing need to focus on enhancing the robustness of these assays to ensure that they reach a similar level of performance as the current gold standard methods.

References

1. Arslan B, Zetterberg H, Ashton NJ. Blood-based biomarkers in Alzheimer's disease–moving towards a new era of diagnostics. Clin Chem Lab Med. 2024;62:1063–9.
2. Hansson O. Biomarkers for neurodegenerative diseases. Nat Med. 2021;27(6):954–63.
3. Beach TG, Monsell SE, Phillips LE, Kukull W. Accuracy of the clinical diagnosis of Alzheimer disease at National Institute on Aging Alzheimer Disease Centers, 2005–2010. J Neuropathol Exp Neurol. 2012;71(4):266–73.
4. Zetterberg H, Blennow K. Moving fluid biomarkers for Alzheimer's disease from research tools to routine clinical diagnostics. Mol Neurodegener. 2021;16(1):10.
5. Coulson DT, Beyer N, Quinn JG, Brockbank S, Hellemans J, Irvine GB, et al. BACE1 mRNA expression in Alzheimer's disease postmortem brain tissue. J Alzheimers Dis. 2010;22(4):1111–22.
6. Cutler NR, Kay AD, Marangos PJ, Burg C. Cerebrospinal fluid neuron-specific enolase is reduced in Alzheimer's disease. Arch Neurol. 1986;43(2):153–4.
7. Blennow K, Wallin A, Ekman R. Neuron specific enolase in cerebrospinal fluid: a biochemical marker for neuronal degeneration in dementia disorders? J Neural Transm Park Dis Dement Sect. 1994;8:183–91.
8. Chaves ML, Camozzato AL, Ferreira ED, Piazenski I, Kochhann R, Dall'Igna O, et al. Serum levels of S100B and NSE proteins in Alzheimer's disease patients. J Neuroinflammation. 2010;7:1–7.
9. Palumbo B, Sabalich I, Tranfaglia C, Lucilla PM. Cerebrospinal fluid neuron-specific enolase: a further marker of Alzheimer's disease? Funct Neurol. 2008;23(2):93.
10. Sulkava R, Viinikka L, Erkinjuntti T, Roine R. Cerebrospinal fluid neuron-specific enolase is decreased in multi-infarct dementia, but unchanged in Alzheimer's disease. J Neurol Neurosurg Psychiatry. 1988;51(4):549–51.
11. Parnetti L, Palumbo B, Cardinali L, Loreti F, Chionne F, Cecchetti R, et al. Cerebrospinal fluid neuron-specific enolase in Alzheimer's disease and vascular dementia. Neurosci Lett. 1995;183(1–2):43–5.
12. Olsson B, Lautner R, Andreasson U, Öhrfelt A, Portelius E, Bjerke M, et al. CSF and blood biomarkers for the diagnosis of Alzheimer's disease: a systematic review and meta-analysis. Lancet Neurol. 2016;15(7):673–84.
13. Ramont L, Thoannes H, Volondat A, Chastang F, Millet M-C, Maquart F-X. Effects of hemolysis and storage condition on neuron-specific enolase (NSE) in cerebrospinal fluid and serum: implications in clinical practice. De Gruyter; 2005.
14. Wang KK, Yang Z, Sarkis G, Torres I, Raghavan V. Ubiquitin C-terminal hydrolase-L1 (UCH-L1) as a therapeutic and diagnostic target in neurodegeneration, neurotrauma and neuro-injuries. Expert Opin Ther Targets. 2017;21(6):627–38.
15. Zetterberg H, Blennow K. Fluid biomarkers for mild traumatic brain injury and related conditions. Nat Rev Neurol. 2016;12(10):563–74.
16. Guglielmotto M, Monteleone D, Vasciaveo V, Repetto IE, Manassero G, Tabaton M, et al. The decrease of Uch-L1 activity is a common mechanism responsible for Aβ 42 accumulation in Alzheimer's and vascular disease. Front Aging Neurosci. 2017;9:320.
17. Li R, Wang J, Xie W, Liu J, Wang C. UCHL1 from serum and CSF is a candidate biomarker for amyotrophic lateral sclerosis. Ann Clin Transl Neurol. 2020;7(8):1420–8.
18. Dong L, Chang Q, Ma J, Liu C, Guo D, Li X, et al. Associations of blood UCH-L1 and NfL levels with cognitive dysfunction in Parkinson's disease patients. Neurosci Lett. 2023;804:137219.
19. Bernstein H-G, Baumann B, Danos P, Diekmann S, Bogerts B, Gundelfinger ED, et al. Regional and cellular distribution of neural visinin-like protein immunoreactivities (VILIP-1 and VILIP-3) in human brain. J Neurocytol. 1999;28:655–62.
20. Saenger AK, Christenson RH. Stroke biomarkers: progress and challenges for diagnosis, prognosis, differentiation, and treatment. Clin Chem. 2010;56(1):21–33.

21. Groblewska M, Muszyński P, Wojtulewska-Supron A, Kulczyńska-Przybik A, Mroczko B. The role of visinin-like protein-1 in the pathophysiology of Alzheimer's disease. J Alzheimers Dis. 2015;47(1):17–32.
22. Halbgebauer S, Steinacker P, Riedel D, Oeckl P, Anderl-Straub S, Lombardi J, et al. Visinin-like protein 1 levels in blood and CSF as emerging markers for Alzheimer's and other neurodegenerative diseases. Alzheimers Res Ther. 2022;14(1):175.
23. Tan Z, Jiang J, Tian F, Peng J, Yang Z, Li S, et al. Serum visinin-like protein 1 is a better biomarker than neuron-specific enolase for seizure-induced neuronal injury: a prospective and observational study. Front Neurol. 2020;11:567587.
24. Dhiman K, Blennow K, Zetterberg H, Martins RN, Gupta VB. Cerebrospinal fluid biomarkers for understanding multiple aspects of Alzheimer's disease pathogenesis. Cell Mol Life Sci. 2019;76:1833–63.
25. Khalil M, Teunissen CE, Otto M, Piehl F, Sormani MP, Gattringer T, et al. Neurofilaments as biomarkers in neurological disorders. Nat Rev Neurol. 2018;14(10):577–89.
26. Arslan B, Zetterberg H. Neurofilament light chain as neuronal injury marker–what is needed to facilitate implementation in clinical laboratory practice? Clin Chem Lab Med. 2023;61(7):1140–9.
27. Khalil M, Pirpamer L, Hofer E, Voortman MM, Barro C, Leppert D, et al. Serum neurofilament light levels in normal aging and their association with morphologic brain changes. Nat Commun. 2020;11(1):812.
28. Shahim P, Tegner Y, Marklund N, Blennow K, Zetterberg H. Neurofilament light and tau as blood biomarkers for sports-related concussion. Neurology. 2018;90(20):e1780–e8.
29. Hviid CVB, Knudsen CS, Parkner T. Reference interval and preanalytical properties of serum neurofilament light chain in Scandinavian adults. Scand J Clin Lab Invest. 2020;80(4):291–5.
30. Ashton NJ, Suárez-Calvet M, Karikari TK, Lantero-Rodriguez J, Snellman A, Sauer M, et al. Effects of pre-analytical procedures on blood biomarkers for Alzheimer's pathophysiology, glial activation, and neurodegeneration. Alzheimers Demen. 2021;13(1):e12168.
31. van Lierop ZY, Verberk IM, van Uffelen KW, Koel-Simmelink MJ, in't Veld L, Killestein J, et al. Pre-analytical stability of serum biomarkers for neurological disease: neurofilament-light, glial fibrillary acidic protein and contactin-1. Clin Chem Lab Med. 2022;60(6):842–50.
32. Altmann P, Leutmezer F, Zach H, Wurm R, Stattmann M, Ponleitner M, et al. Serum neurofilament light chain withstands delayed freezing and repeated thawing. Sci Rep. 2020;10(1):19982.
33. Altmann P, Ponleitner M, Rommer PS, Haslacher H, Mucher P, Leutmezer F, et al. Seven day pre-analytical stability of serum and plasma neurofilament light chain. Sci Rep. 2021;11(1):11034.
34. Kuhle J, Kropshofer H, Haering DA, Kundu U, Meinert R, Barro C, et al. Blood neurofilament light chain as a biomarker of MS disease activity and treatment response. Neurology. 2019;92(10):e1007–e15.
35. Barro C, Benkert P, Disanto G, Tsagkas C, Amann M, Naegelin Y, et al. Serum neurofilament as a predictor of disease worsening and brain and spinal cord atrophy in multiple sclerosis. Brain. 2018;141(8):2382–91.
36. Kuhle J, Barro C, Andreasson U, Derfuss T, Lindberg R, Sandelius Å, et al. Comparison of three analytical platforms for quantification of the neurofilament light chain in blood samples: ELISA, electrochemiluminescence immunoassay and Simoa. Clin Chem Lab Med. 2016;54(10):1655–61.
37. Gauthier A, Viel S, Perret M, Brocard G, Casey R, Lombard C, et al. Comparison of SimoaTM and EllaTM to assess serum neurofilament-light chain in multiple sclerosis. Ann Clin Transl Neurol. 2021;8(5):1141–50.
38. Lee S, Plavina T, Singh CM, Rudick RA, Uzgiris AJ. Development of a highly sensitive neurofilament light chain assay on an automated immunoassay platform. Front Neurol. 2022;13:935382.

39. Wilson D, Chan D, Chang L, Mathis R, Verberk I, Montalban X, et al. Development and multicenter validation of a fully automated digital immunoassay for neurofilament light chain: toward a clinical blood test for neuronal injury. Clin Chem Lab Med. 2024;62(2):322–31.
40. Benkert P, Meier S, Schaedelin S, Manouchehrinia A, Yaldizli Ö, Maceski A, et al. Serum neurofilament light chain for individual prognostication of disease activity in people with multiple sclerosis: a retrospective modelling and validation study. Lancet Neurol. 2022;21(3):246–57.
41. Simrén J, Andreasson U, Gobom J, Suarez Calvet M, Borroni B, Gillberg C, et al. Establishment of reference values for plasma neurofilament light based on healthy individuals aged 5–90 years. Brain Commun. 2022;4(4):fcac174.
42. Syrjanen JA, Campbell MR, Algeciras-Schimnich A, Vemuri P, Graff-Radford J, Machulda MM, et al. Associations of amyloid and neurodegeneration plasma biomarkers with comorbidities. Alzheimers Dement. 2022;18(6):1128–40.
43. Berry K, Asken BM, Grab JD, Chan B, Lario Lago A, Wong R, et al. Hepatic and renal function impact concentrations of plasma biomarkers of neuropathology. Alzheimers Dem. 2022;14(1):e12321.
44. Fitzgerald KC, Sotirchos ES, Smith MD, Lord HN, DuVal A, Mowry EM, et al. Contributors to serum NfL levels in people without neurologic disease. Ann Neurol. 2022;92(4):688–98.
45. Leuzy A, Mattsson-Carlgren N, Palmqvist S, Janelidze S, Dage JL, Hansson O. Blood-based biomarkers for Alzheimer's disease. EMBO Mol Med. 2022;14(1):e14408.
46. Zetterberg H, van Swieten JC, Boxer AL, Rohrer JD. Fluid biomarkers for frontotemporal dementias. Neuropathol Appl Neurobiol. 2019;45(1):81–7.
47. Lu C-H, Macdonald-Wallis C, Gray E, Pearce N, Petzold A, Norgren N, et al. Neurofilament light chain: a prognostic biomarker in amyotrophic lateral sclerosis. Neurology. 2015;84(22):2247–57.
48. Hansson O, Janelidze S, Hall S, Magdalinou N, Lees AJ, Andreasson U, et al. Blood-based NfL: a biomarker for differential diagnosis of parkinsonian disorder. Neurology. 2017;88(10):930–7.
49. Smirnov DS, Ashton NJ, Blennow K, Zetterberg H, Simrén J, Lantero-Rodriguez J, et al. Plasma biomarkers for Alzheimer's disease in relation to neuropathology and cognitive change. Acta Neuropathol. 2022;143(4):487–503.
50. Sánchez-Valle R, Heslegrave A, Foiani MS, Bosch B, Antonell A, Balasa M, et al. Serum neurofilament light levels correlate with severity measures and neurodegeneration markers in autosomal dominant Alzheimer's disease. Alzheimers Res Ther. 2018;10:1–6.
51. Jack CR Jr, Bennett DA, Blennow K, Carrillo MC, Dunn B, Haeberlein SB, et al. NIA-AA research framework: toward a biological definition of Alzheimer's disease. Alzheimers Dement. 2018;14(4):535–62.
52. Wang P, Lynn A, Miskimen K, Song YE, Wisniewski T, Cohen M, et al. Genome-wide association studies identify novel loci in rapidly progressive Alzheimer's disease. Alzheimers Dement. 2024;
53. Young-Pearse TL, Lee H, Hsieh Y-C, Chou V, Selkoe DJ. Moving beyond amyloid and tau to capture the biological heterogeneity of Alzheimer's disease. Trends Neurosci. 2023;46(6):426–44.
54. Zetterberg H, Skillbäck T, Mattsson N, Trojanowski JQ, Portelius E, Shaw LM, et al. Association of cerebrospinal fluid neurofilament light concentration with Alzheimer disease progression. JAMA Neurol. 2016;73(1):60–7.
55. Olsson B, Portelius E, Cullen NC, Sandelius Å, Zetterberg H, Andreasson U, et al. Association of cerebrospinal fluid neurofilament light protein levels with cognition in patients with dementia, motor neuron disease, and movement disorders. JAMA Neurol. 2019;76(3):318–25.
56. Mattsson N, Andreasson U, Zetterberg H, Blennow K, AsDN I. Association of plasma neurofilament light with neurodegeneration in patients with Alzheimer disease. JAMA Neurol. 2017;74(5):557–66.
57. Mattsson N, Cullen NC, Andreasson U, Zetterberg H, Blennow K. Association between longitudinal plasma neurofilament light and neurodegeneration in patients with Alzheimer disease. JAMA Neurol. 2019;76(7):791–9.

58. Alagaratnam J, von Widekind S, De Francesco D, Underwood J, Edison P, Winston A, et al. Correlation between CSF and blood neurofilament light chain protein: a systematic review and meta-analysis. BMJ Neurol Open. 2021;3(1)
59. Lewczuk P, Ermann N, Andreasson U, Schultheis C, Podhorna J, Spitzer P, et al. Plasma neurofilament light as a potential biomarker of neurodegeneration in Alzheimer's disease. Alzheimers Res Ther. 2018;10:1–10.
60. Ashton NJ, Janelidze S, Al Khleifat A, Leuzy A, van der Ende EL, Karikari TK, et al. A multicentre validation study of the diagnostic value of plasma neurofilament light. Nat Commun. 2021;12(1):3400.
61. Benedet AL, Leuzy A, Pascoal TA, Ashton NJ, Mathotaarachchi S, Savard M, et al. Stage-specific links between plasma neurofilament light and imaging biomarkers of Alzheimer's disease. Brain. 2020;143(12):3793–804.
62. Mazzeo S, Ingannato A, Giacomucci G, Manganelli A, Moschini V, Balestrini J, et al. Plasma neurofilament light chain predicts Alzheimer's disease in patients with subjective cognitive decline and mild cognitive impairment: a cross-sectional and longitudinal study. Eur J Neurol. 2024;31(1):e16089.
63. Cullen NC, Leuzy A, Palmqvist S, Janelidze S, Stomrud E, Pesini P, et al. Individualized prognosis of cognitive decline and dementia in mild cognitive impairment based on plasma biomarker combinations. Nat Aging. 2021;1(1):114–23.
64. Cullen NC, Leuzy A, Janelidze S, Palmqvist S, Svenningsson AL, Stomrud E, et al. Plasma biomarkers of Alzheimer's disease improve prediction of cognitive decline in cognitively unimpaired elderly populations. Nat Commun. 2021;12(1):3555.
65. Palmqvist S, Stomrud E, Cullen N, Janelidze S, Manuilova E, Jethwa A, et al. An accurate fully automated panel of plasma biomarkers for Alzheimer's disease. Alzheimers Dement. 2023;19(4):1204–15.
66. Van Dyck CH, Swanson CJ, Aisen P, Bateman RJ, Chen C, Gee M, et al. Lecanemab in early Alzheimer's disease. N Engl J Med. 2023;388(1):9–21.
67. Cerri C, Silani V, Scarlato G. Oligoclonal immunoglobulins and immunoreactive myelin basic protein in the cerebrospinal fluid of patients with multiple sclerosis and other neurological diseases. Acta Neurol. 1985;7(3–4):311–4.
68. Yan EB, Satgunaseelan L, Paul E, Bye N, Nguyen P, Agyapomaa D, et al. Post-traumatic hypoxia is associated with prolonged cerebral cytokine production, higher serum biomarker levels, and poor outcome in patients with severe traumatic brain injury. J Neurotrauma. 2014;31(7):618–29.

Fluid and Imaging Markers of Glial Reactivity

Marc Suárez-Calvet

Abstract Biomarkers of glial reactivity in Alzheimer's disease and other neurodegenerative diseases have gained increased attention, highlighting the roles of inflammation, innate immunity, and glial reactivity in disease pathogenesis. Despite the historical focus on Amyloid-β and tau, current research in biomarkers also includes proteins secreted or released by microglia and astrocytes. This chapter focuses on key biomarkers of glial reactivity, specifically sTREM2, progranulin, and Galectin-3 (microglia-related) and GFAP and YKL-40 (astrocyte-related). It highlights the significance of these biomarkers in reflecting specific facets of the glial response, which is dynamic and varies with disease stage, age, and environmental factors. The discussion includes the need for a diverse array of biomarkers to comprehensively characterise glial responses, moving beyond binary classifications. This chapter also explores the potential of PET imaging for visualising glial reactivity. Future research directions emphasise the development of new biomarkers and their combinations, understanding the biological significance of glial markers, and integrating human and animal model data for drug development. The dynamic nature of glial responses, the potential beneficial or detrimental effects of glial reactivity, and the necessity of biomarkers in predicting and monitoring therapeutic responses are underscored. Although glial biomarkers are rarely useful for diagnostic purposes, they hold promise for prognostic and predictive applications in Alzheimer's disease and neurodegenerative disease treatments and can inform drug development and the identification of novel therapeutic targets.

Keywords Biomarker · Astrocyte · Microglia · Inflammation · Alzheimer's disease

M. Suárez-Calvet (✉)
Department of Neurology, Hospital del Mar, Barcelonaβeta Brain Research Center (BBRC), Barcelona, Spain
e-mail: msuarez@barcelonabeta.org

R. Tarawneh (ed.), *Biomarkers of Neurodegenerative Disorders*,
https://doi.org/10.1007/978-3-031-80849-4_6

1 Introduction

In the initial description of Alzheimer's disease (AD), Alois Alzheimer already noted the presence of glial cells (*Gliazelle*) surrounding the so-called senile plaques. While research in AD has been primarily focused on Amyloid-β (Aβ) and tau, the significance of inflammation, innate immunity, and glial reactivity has been increasingly recognised [1–3]. A clear indicator of this growing interest is the fact that there are currently 25 drugs related to inflammation/immunity in clinical trials [4]. Genetic studies have consistently implicated innate immunity and microglia in the pathogenesis of AD [5]. Furthermore, the Alzheimer's Association has introduced new criteria for diagnosing and staging AD, which include a new biomarker category labelled "I" (for inflammation or astrocyte reactivity), defined by the biomarker glial fibrillary acidic protein (GFAP).

Numerous biomarkers related to innate immunity and glial cells have been investigated in AD and other neurodegenerative diseases, including cytokines and chemokines, soluble immune receptors or shedded receptor ectodomains, complement proteins, and other innate immunity factors. In this chapter, our focus will be on proteins mainly secreted or released by microglia and astrocytes. Our aim here is not to comprehensively review all biomarkers associated with glial cells or innate immunity. Instead, we will examine the most significant ones. Our focus will be on soluble triggering receptor expressed on myeloid cells 2 (sTREM2), progranulin, and Galectin-3, primarily expressed by microglia, as well as GFAP and YKL-40, which are related to astrocytes. We will also provide a brief overview of positron emission tomography (PET) imaging to visualise glial reactivity, which has been recently examined in detail elsewhere [6, 7]. Given that the majority of evidence comes from AD, our main emphasis will be on this disease, with brief mentions of other neurodegenerative diseases.

When investigating glial biomarkers, several considerations arise. First, it is important to recognise that each biomarker indicates a partial aspect of the complex glial response that occurs in neurodegeneration. Consequently, relying on a single glial biomarker is insufficient to characterise the microglial or astrocytic responses comprehensively. Accordingly, sTREM2 does not encapsulate the entirety of the microglial response, nor does GFAP represent the entirety of the astrocytic response; instead, these biomarkers only capture specific facets of these responses. Second, microglia and astrocytes exhibit a wide range of status and functions that dynamically evolve in response to their environment. Hence, dynamic fluctuations in glial biomarkers are anticipated, influenced by disease stage, age, local milieu, and other variables. Glial cells in different states may exist simultaneously, and the measured levels of biomarkers might reflect the balance of the overall predominant response.

Considering the aforementioned factors, there are certain terminology aspects worthy of discussion. Herein, we will use the terms "biomarkers of microglial reactivity" and "astrocyte reactivity" or "reactive astrogliosis", while avoiding the ambiguous term "neuroinflammation". Additionally, we will discourage the use of the term "activation" when referring to a glial biomarker; microglia and astrocytes are always active, even under physiological conditions. As advocated in previous consensus publications [8, 9], we should move beyond the binary classification for microglial (e.g., M1 vs. M2, resting vs. activated) and astrocytic (A1 vs. A2) responses, also in the biomarkers field. Instead, we must recognise the diversity of microglial and astrocytic responses, acknowledging their diverse potential states, each characterised by a specific molecular profile, function, and impact on the disease. Arguably, there is still the need to identify new glial biomarkers, or combinations of them, that more accurately reflect the various patterns of the glial response in AD and neurodegeneration.

The heterogeneous glial response during the disease can have either a beneficial or a detrimental effect. Observational studies on glial biomarkers in humans may aid in elucidating this effect, but validation through experimental approaches is essential. It is important to emphasise once more the dynamic nature of the glial response. The same glial biomarker may be linked to either a beneficial or detrimental outcome at different stages of the disease.

2 Microglial Reactivity Biomarkers

2.1 sTREM2

Triggering receptor expressed on myeloid cells 2 (TREM2) is a type-1 transmembrane protein and an innate immune receptor belonging to the immunoglobulin family. It is predominantly expressed in myeloid lineage cells, including microglial cells in the central nervous system (CNS) [10, 11]. TREM2 is transported to the cell surface, where it can bind to various ligands, thereby triggering cell-autonomous signalling pathways through DAP12 and SYK phosphorylation. The microglial response triggered by TREM2 involves proliferation, chemotaxis, phagocytosis, increasing survival, and migration [12–21]. TREM2 activation is a key step in the microglial reactivity to Aβ deposition, neuronal death, demyelination, or other harmful stimuli affecting the brain. The termination of TREM2 signalling occurs when the full-length protein is cleaved from the cell surface by ADAM-10 and -17, specifically targeting the C-terminal region after histidine 157 [16, 22]. The resulting soluble N-terminal ectodomain, known as sTREM2, is then released into the extracellular space [16, 22]. In addition to being generated through proteolytic cleavage of the full-length form on the cell surface, sTREM2 can also arise from the translation of a transcript lacking the transmembrane domain due to alternative splicing [11, 23, 24]. It is still not known whether sTREM2, either the proteolytic shedding or alternative splicing form, has specific biological functions [25]. sTREM2 might act as a decoy receptor, preventing the binding of ligands to full-length TREM2 and ultimately preventing TREM2 signalling. Alternatively, sTREM2 might bind other unknown receptors in microglia or other cells. In cellular and animal model studies, sTREM2 has been shown to enhance microglial reactivity, reduce amyloid plaques, and rescue memory deficits [21, 26].

Homozygous loss-of-function mutations in *TREM2* (or in its adaptor protein DAP12) are responsible for Nasu-Hakola disease, a rare neurodegenerative disease characterised by an early-onset frontal syndrome and concurrent bone cysts [27, 28]. Additionally, low-frequency coding variants in *TREM2* have been linked to a two- to fivefold increased risk of AD [29–31]. These variants have also been associated with increased risk of other neurodegenerative diseases [32–34]. In patients harbouring homozygous *TREM2* mutations, both cerebrospinal fluid (CSF) and blood levels of sTREM2 are notably low or undetectable [16], suggesting a loss-of-function mechanism. Conversely, CSF sTREM2 levels in individuals carrying one copy of a *TREM2* risk variant exhibit variability; they may be elevated, reduced, or similar to those observed in AD patients lacking *TREM2* risk variants [35–37]. This variability suggests the involvement of a more intricate mechanism in *TREM2* variants associated with AD risk.

The soluble form of TREM2, sTREM2, can be quantified in CSF and blood [16, 38]. CSF sTREM2 may reflect the TREM2 signalling competence on the surface of microglia, thereby potentially serving as a marker for the activation of full-length TREM2-mediated signalling. This concept finds support in the rapid constitutive cleavage of full-length TREM2 on the cell surface, with a half-life of less than one hour [39]. Furthermore, the levels of sTREM2 in the brain of a mouse model of Aβ pathology correlate with the 18-kDa translocator protein (TSPO) small animal positron emission tomography (μPET) signal and the knock-in mouse model carrying the *TREM2* p.T66M mutation displays reduced microglial reactivity.

Most studies have consistently reported elevated levels of CSF sTREM2 in AD [36, 37, 40–46]. However, these results were not found in all studies, and there is a notable degree of variability in CSF sTREM2 levels [16, 47]. Further studies have revealed dynamic fluctuations in CSF sTREM2 levels along the AD *continuum*, peaking during later asymptomatic and early symptomatic stages of both sporadic AD and autosomal-dominant AD [37, 44–46]. Variability in CSF sTREM2 levels is more apparent during the dementia stage, likely reflecting the heterogeneous nature of this response. Additionally, an initial decline in CSF sTREM2 levels has been observed in early disease stages, particularly in Aβ-positive individuals that are Tau-negative (A+T−) [37]; however, this decrease disappears upon adjusting for CSF Aβ40 levels [48]. Furthermore, increases in CSF sTREM2 levels have also been documented in multiple sclerosis (MS) and other neuroinflammatory diseases [38, 49], HIV-1 infection [50], and dementia patients with delirium [51], which may suggest that TREM2-mediated microglial reactivity is not exclusive of AD and also occurs in other neurodegenerative diseases. However, increases in CSF sTREM2 levels have not been observed in frontotemporal dementia (FTD) [16, 52], Huntington's disease [53], or repetitive head impacts [54]. While alterations in CSF sTREM2 levels have been observed in AD and other neurological diseases, there remains a significant overlap in values between patients and controls, limiting its utility as a diagnostic marker. Investigations into blood sTREM2 have been relatively limited. No significant changes have been detected in plasma or serum sTREM2 levels in MS, neuroinflammatory diseases, AD, FTD, or Parkinson's disease (PD) [36, 37, 40, 55]. Intriguingly, elevated serum sTREM2 levels have been associated with a higher risk of developing dementia in the Japanese population [56]. CSF sTREM2 levels increase with age, and some reports suggest that they may be higher in males than in females [36, 45, 46, 57].

In individuals with autosomal-dominant AD, longitudinal studies have revealed a significant early increase in CSF sTREM2 levels [58]. Those carrying a pathogenic mutation exhibit significantly higher CSF sTREM2 levels compared to non-carriers, a difference observed up to 21 years before the anticipated onset of symptoms. Among carriers of pathogenic mutations, lower CSF Aβ42 levels were associated with longitudinal increases in CSF sTREM2. Additionally, longitudinal increases in CSF sTREM2 were linked to an improvement in Aβ burden over time, as evidenced by a slower decline in CSF Aβ42 levels and a slower increase in Aβ PET uptake. Independent investigations in individuals with sporadic AD have underscored that elevated baseline CSF sTREM2 is associated with reduced rates of increase in amyloid-PET signal [59]. These findings collectively suggest that the TREM2-mediated microglial response is intricately linked with early Aβ pathophysiological changes in AD and plays a role in modulating the balance between soluble and insoluble Aβ. Early alterations in soluble Aβ may trigger a TREM2-mediated microglial response, which in turn mitigates the deposition of amyloid plaques. These findings, although observational in nature and conducted in humans, should be interpreted cautiously. However, they align with findings from murine models of Aβ pathology, demonstrating that the TREM2-mediated microglial response restrains amyloid seeding [18, 60].

Research has also explored the connection between CSF sTREM2 and tau pathology. Several cross-sectional studies have consistently demonstrated a correlation between sTREM2 and levels of phosphorylated tau (p-tau) and total tau (t-tau) in the CSF. It is important to note that this correlation does not imply causality; rather, it probably suggests that, following early Aβ changes, there is a simultaneous response involving both TREM2 and tau pathology, thus explaining their correlation. Longitudinal studies have provided more insightful findings. In individuals with autosomal-dominant AD, high longitudinal increases in CSF sTREM2 were associated with a slowdown in the rise of CSF p-tau181 levels [58]. Similarly, in individuals at the preclinical stage of sporadic AD, higher baseline CSF sTREM2 was linked to a deceleration in tau deposition, as measured by tau-PET [61].

A pivotal question in the field is whether the TREM2-mediated microglial response exerts a beneficial or detrimental influence. Evidence supporting a protective role of TREM2 includes the observation that AD-associated risk variants in the *TREM2* gene induce partial loss of protein function. Moreover, complete loss of TREM2 function locks microglia in a homeostatic state, hindering their response to pathological stimuli such as amyloid deposition or neuronal death [62]. Initial studies in animal models investigating the consequences of TREM2 loss of function yielded inconsistent results. However, there is growing consensus that the absence of TREM2 function impedes microglial migration and the encapsulation of amyloid plaques, resulting in an increase in diffuse amyloid plaques and neuritic pathology [18, 20, 63–65]. Nonetheless, this effect may vary depending on the age of the mice [64, 66, 67]. In animal models of tau pathology and neuronal death, the impact of

TREM2 loss is less definitive, with some studies suggesting a protective role [20, 68, 69], while others lean towards a detrimental one [70]. Notably, TREM2-dependent microglial reactivity has been shown to impede tau accumulation and brain atrophy only if amyloid pathology is present [71, 72].

CSF sTREM2 serves as an illustrative example of a biomarker that currently lacks clear clinical utility but has significantly contributed to our understanding of TREM2 role in AD in humans. In this regard, CSF sTREM2 has complemented findings from animal models, providing support for its potential protective role. Initial neuroimaging cross-sectional and longitudinal investigations in humans linked higher CSF sTREM2 with increased grey matter volume and reduced diffusivity in early AD [73, 74]. Notably, a longitudinal study within the Alzheimer's Disease Neuroimaging Initiative (ADNI) cohort demonstrated that individuals with biomarker evidence of Aβ and tau pathology exhibit attenuated decline in memory and global cognition at follow-up when baseline CSF sTREM2 levels are higher [75]. This seminal study has been corroborated by subsequent longitudinal investigations, which have demonstrated that higher CSF sTREM2 is linked to a slower rate of hippocampal atrophy and lower rates of cognitive decline in the early clinical stages of the disease, and a reduced likelihood of progression from mild cognitive impairment (MCI) to dementia [76–78]. Additionally, common variants of the *MS4A* gene associated with increased CSF sTREM2 concentrations are also linked with reduced AD risk and delayed age at disease onset [55]. Collectively, these findings suggest a beneficial effect of TREM2 function. Aligned with these discoveries, therapeutic strategies targeting TREM2 are under consideration, and several TREM2 agonistic antibodies are being investigated in clinical trials [79, 80].

2.2 *Progranulin*

Progranulin (PGRN), a glycoprotein composed of seven and a half tandem repeats, known as granulin domains, undergoes secretion into the extracellular space through the secretory pathway. The full-length protein may undergo proteolytic processing, resulting in the release of individual granulin peptides [81–84]. Within the CNS, PGRN is predominantly expressed by microglia [85, 86]. Studies in PGRN-deficient mice have revealed an enhanced inflammatory response in these animals [87–89], with their microglia exhibiting suppression of genes associated with the homeostatic state and activation of the microglial neurodegenerative phenotype (MGnD) signature [62].

Its close association with FTD is well-established. Heterozygous mutations in the progranulin gene (*GRN*) result in FTLD-TDP due to haploinsufficiency. Homozygous mutations in the *GRN* gene lead to neuronal ceroid lipofuscinosis [90]. Reduced levels of PGRN in blood or CSF are observed in individuals carrying *GRN* mutations, serving as a diagnostic biomarker [91–93]. A recent systematic review involving over 7000 individuals established cut-off values in both CSF and plasma capable of discriminating between carriers and non-carriers of pathogenic *GRN* mutations [94].

Interestingly, certain *GRN* variants have been linked to an increased risk of AD [95, 96]. An AD-associated *GRN* variant (rs5848) also results in a reduction in PGRN levels in CSF and plasma [96–98], albeit not as pronounced as observed in FTD patients carrying a *GRN* mutation. In contrast to FTD, investigations have generally reported no significant differences in CSF PGRN in AD [97–100]. Nevertheless, a cross-sectional analysis within the DIAN (Dominantly Inherited Alzheimer's Network) and ADNI cohorts revealed an increase in CSF PGRN levels throughout the progression of both autosomal-dominant AD and sporadic AD [101]. Conversely, initial investigations into blood PGRN and AD have yielded less promising results, with no discernible changes in AD and poor correlation with brain PGRN levels [97, 98, 102]. However, a recent systematic review reported an increase in plasma PGRN levels in various clinical presentations of AD, as well as in FTD cases without *GRN* mutations [94].

Interestingly, sex differences have been observed in PGRN levels. While plasma PGRN levels tend to be higher in women, levels in CSF tend to be higher in men [94, 98, 101].

2.3 Galectin-3

Galectin-3 (Gal-3) is a β-galactosidase-binding lectin and consists of a single C-terminal carbohydrate recognition domain and a non-lectin collagen-like N-terminal region. Gal-3 is expressed in various glial cells within the CNS and is notably upregulated in microglia in response to CNS inflammatory conditions, such as experimental autoimmune encephalomyelitis, ischaemic injury, or traumatic brain injury (TBI) [103, 104]. In addition to its intracellular presence, Gal-3 is released into the extracellular space, where it can be detected in biological fluids.

In AD, Gal-3 is implicated in microglial response to Aβ, being upregulated in the brain cortex and hippocampus of both autosomal-dominant and sporadic AD patients. Notably, microglia expressing Gal-3 are found clustered around Aβ plaques, particularly those of larger size and irregular shape [105–107]. Deletion of Gal-3 in transgenic AD mouse models has been shown to attenuate the microglial response to Aβ, leading to reduced plaque burden and improved cognitive function [105, 108]. Consequently, Gal-3 is considered a mediator of microglial response with detrimental effects on Aβ clearance, making it a potential therapeutic target in AD.

In CSF, Gal-3 levels are elevated in AD patients and correlate with markers of tau pathology, neuronal injury, and synaptic dysfunction [106]. Increased levels of Gal-3 are observed in patients with sporadic FTD and symptomatic carriers of *MAPT* mutations, but not in carriers of *GRN* mutations or *C9orf72* expansions [109]. Gal-3 is increased in the serum or plasma of patients with AD [110, 111], PD [112, 113], Huntington's disease [114], and Amyotrophic Lateral Sclerosis (ALS)

[115], FTD [109], and TBI [116]. Furthermore, serum Gal-3 has also been proposed as a prognostic biomarker in ischaemic stroke [117].

2.4 PET Imaging of Microglial Reactivity

PET techniques enable the assessment of in vivo microglial and astrocyte reactivity using specific radioligands. Unlike fluid biomarkers, PET studies can provide topographical information as well. Microglial reactivity has been extensively studied with radioligands that bind TSPO, expressed in the outer mitochondrial membrane of reactive microglia under pathological conditions [118]. First-generation radioligands like [^{11}C]PK11195 have been widely used as TSPO ligands. Subsequently, second- and third-generation TSPO PET ligands with higher signal-to-noise ratios have been developed, such as [^{11}C]PBR28, [^{18}F]DPA-714, [^{18}F]GE-180, and [^{11}C]ER176. However, the binding of these next-generation TSPO tracers is influenced by a single nucleotide polymorphism (rs6971) of the *TSPO* gene, while [^{11}C]PK11195 binding is less affected by this polymorphism [119].

In AD, increased TSPO PET uptake has been documented in medial temporal regions, including the entorhinal cortex, temporoparietal cortex, including the precuneus, and the cingulate cortex [120–125]. Increased TSPO PET uptake is associated with reduced glucose brain metabolism [125, 126], and a more pronounced increase in TSPO binding correlates with cognitive worsening [122, 123].

TSPO PET uptake is also increased in progressive supranuclear palsy (PSP), particularly in areas such as the basal ganglia, striatum, midbrain, frontal cortex, and cerebellum [127–129], and this elevation correlates with the severity of clinical symptoms [130]. Elevated TSPO uptake has also been observed in corticobasal degeneration (CBD), particularly in the basal ganglia and forebrain regions [128]. Furthermore, TSPO PET has also been used to investigate microglial reactivity in FTD, revealing increased uptake in frontotemporal regions in both sporadic cases and patients with mutations in *MAPT* or *GRN* genes and *C9orf72* expansions [131–135]. Increased uptake in frontal regions has been associated with a more rapid cognitive decline [136].

3 Astrocyte Reactivity Biomarkers

3.1 GFAP

Reactive astrocytes overexpress GFAP and are typically found in post-mortem AD brain regions with Aβ and tau pathology [137–139]. Consequently, GFAP can be regarded as a marker of astrocyte reactivity, which can be measured in CSF as well as blood. While several studies have reported an elevation in CSF GFAP in AD

[140–145], the findings have not been consistently significant across all studies [146, 147]. However, a comprehensive meta-analysis revealed an overall increase in CSF GFAP in symptomatic AD patients [148]. CSF GFAP is also increased in FTD, in dementia with Lewy bodies (DLB), PD with dementia (PDD), and prion diseases [140, 142–144].

In recent years, several studies have indicated an increase in GFAP in plasma or serum [144, 149–153]. Notably, research has also demonstrated elevated blood GFAP in the early stages of AD, namely in individuals who are cognitively unimpaired but Aβ-positive [154–158]. This suggests that blood GFAP might serve as a promising candidate biomarker for this initial stage of the disease. Studies comparing GFAP measurements in CSF and plasma yielded surprising results [155, 158]. The effect size of the increases in plasma GFAP across the AD *continuum* is larger than that observed in CSF. Additionally, the accuracy of plasma GFAP in discriminating between Aβ-positive and Aβ-negative individuals is higher than that of CSF. These findings suggest that plasma GFAP provides additional information beyond serving as a mere surrogate for CSF measurements, as is common with most plasma biomarkers. Several reasons may account for this unexpected finding. Firstly, there could be a technical explanation; GFAP tends to be more stable in plasma than in CSF, particularly after freeze-thaw cycles [159, 160]. Furthermore, there may be biological explanations; astrocytes are integral to the blood-brain barrier (BBB), with their end-feet covering brain capillaries, potentially leading to the direct release of GFAP into the bloodstream following an increase in BBB permeability observed in AD [161]. This latter hypothesis is supported by the high levels of plasma GFAP found in intracranial haemorrhage, TBI, and other acute neurological conditions, where there is a structural disintegration of astrocytes and BBB [162–166].

There is still some debate regarding whether plasma GFAP is more indicative of astrocyte reactivity to Aβ pathology, tau pathology, or both. On the one hand, plasma GFAP increases in individuals with low Aβ burden, where there may be changes in soluble p-tau, but independently of Tau PET [155, 158]. Plasma GFAP is also associated with Tau PET burden, but only in Aβ-positive cognitively impaired individuals, and these associations are lost after adjusting for the effect of Aβ PET burden [158]. This data favours the idea that plasma GFAP probably reflects astrocyte reactivity to Aβ pathology, at least in early stages. However, a study addressing the neuropathological correlates of blood biomarkers found that plasma GFAP is mainly related to tau tangles [167]. Interestingly, some studies have shown that plasma GFAP mediates, at least partially, the association between Aβ deposition and tau deposition [158]. In line with this, a multicentre study demonstrated that in cognitively unimpaired individuals, Aβ pathology was associated with subsequent tau phosphorylation only in those individuals with high levels of plasma GFAP [168]. The role of astrocyte reactivity may even be earlier, as some evidence shows that plasma GFAP mediates the relationship between CSF Aβ42/40 and Aβ PET [169].

Furthermore, plasma or serum GFAP might serve as a prognostic marker. Plasma GFAP is associated with both longitudinal Aβ accumulation and cognitive decline over time [158] and predicts progression to dementia [170]. In cognitively normal individuals, serum GFAP is associated with the risk of incident dementia and cognitive decline at follow-up [171]. In a previous population study with a biracial design, plasma GFAP predicted the development of AD dementia and cognitive decline [151]. Notably, plasma GFAP decreases after treatment with anti-amyloid drugs [172] and may therefore be used as a pharmacodynamic biomarker in clinical trials, not only those targeting astrocytes but also Aβ pathology. Nevertheless, more evidence is needed to consider plasma GFAP as a surrogate biomarker in clinical trials.

Although plasma GFAP has mainly been studied in AD, several studies have found changes in other neurodegenerative diseases. GFAP in blood is increased in FTD [144, 173], particularly those carrying a *GRN* mutation [174], as well as in DLB/PDD [144] and CADASIL (cerebral autosomal dominant arteriopathy with subcortical infarcts and leukoencephalopathy) [175].

3.2 YKL-40

YKL-40, also known as chitinase 3-like protein 1 (CHI3L1), is a secreted 40-kDa glycoprotein of the human chitinase family that lacks hydrolase activity. It is upregulated in several inflammatory diseases, and it is implicated in inflammation-associated tissue remodelling [176, 177]. While it is expressed by various peripheral cells such as chondrocytes, macrophages, neutrophils, synoviocytes, and vascular smooth muscle cells, its predominant expression in the CNS is found in reactive astrocytes under various neuropathological conditions [176–178].

Since the initial observation of increased YKL-40 in CSF of AD patients [179], many subsequent studies have replicated this finding [180–184], as summarised in a recent review [185]. Furthermore, CSF YKL-40 increases in the preclinical stage of AD [181, 186–188] and continues to increase as the disease progresses [189–191], which has also been shown in longitudinal analyses [192, 193]. CSF YKL-40 is also increased in asymptomatic autosomal-dominant AD mutation carriers [194]. In AD, CSF YKL-40 is highly correlated with markers of tau pathology and neurodegeneration [180, 181, 184, 190]. Although YKL-40 levels in blood are moderately correlated with those in CSF and tend to be higher in AD compared to controls, their utility to detect AD pathology is limited, and they do not predict cognitive changes [179, 195].

CSF YKL-40 is associated with cognition and clinical outcomes. In the seminal paper by Craig-Schapiro et al., it was demonstrated that the CSF YKL-40/Aβ42 ratio predicted cognitive decline and was associated with progression from the cognitively unimpaired to the cognitively impaired stages [179]. Subsequent research showed that CSF YKL-40 correlates with greater longitudinal cognitive decline in non-demented patients [190, 196] and predicts progression from MCI to AD dementia [191]. However, the latter study also showed that in individuals with AD dementia, lower CSF YKL-40 levels were associated with a more rapid cognitive decline [191].

CSF YKL-40 is also associated with structural neuroimaging. In individuals in early stages of AD, higher CSF YKL40 is associated with cortical thinning and reduced grey matter volume in regions more vulnerable to AD [190, 197]. However, some studies have not observed these correlations [196], suggesting potential complexity in the structural relationships of CSF YKL-40. Among patients with MCI and mild dementia, a non-linear (quadratic) association has been reported between CSF YKL40 levels and grey matter volume in temporoparietal regions, insula, and cerebellum, indicating that intermediate CSF YKL40 levels are associated with greater volumes [189]. Additionally, these associations may be influenced by factors such as age [73], sex [198], and apolipoprotein E4 (*APOE* ε4)-carrier status [187].

Increases in CSF YKL-40 are not exclusive to AD but are also observed in a variety of chronic and acute neurological diseases, including MS, ALS, CBD, PSP, FTD, Creutzfeldt-Jakob disease (CJD), stroke, and TBI [177, 179, 180, 199–205]. Conversely, CSF YKL-40 levels are not increased (or are even decreased) in PD, DLB, and other synucleinopathies [182, 204, 206–208], unless concomitant AD pathology is present [209].

3.3 PET Imaging of Astrocyte Reactivity

Reactive astrogliosis has been investigated with radiotracers targeting monoamine oxidase B (MAO-B), an enzyme predominantly found on the mitochondrial membrane of a subtype of reactive astrocytes [210], but also present to a lesser extent in monoaminergic neurons. [^{11}C]deuterium-L-deprenyl (DED) has been the most widely used radiotracer and has facilitated the visualisation of reactive astrogliosis in the early stages of both sporadic and autosomal-dominant AD [3, 211–213]. Interestingly, plasma GFAP and [^{11}C]DED tracer follow divergent trajectories. While the former shows a continuous increase throughout the disease course, the latter demonstrates an early elevation in preclinical stages (notably, in autosomal-dominant AD, occurring 17 years before the estimated age of symptom onset) followed by a subsequent decrease over time [3, 212, 214]. As highlighted throughout this chapter, each biomarker, fluid- or neuroimaging-based, provides insight into distinct facets of the glial response.

4 Other Glial and Innate Immunity-Related Biomarkers

Various other glial and innate immunity biomarkers have been investigated. However, the results from these studies are inconsistent, particularly regarding cytokines. Moreover, not all biomarkers can be reliably detected in CSF or blood plasma and serum [215–218]. A few additional biomarkers are worth mentioning briefly.

S100B, a calcium-binding protein mainly expressed in astrocytes within the CNS, is elevated in CSF across several neurodegenerative diseases, including AD [219–221], FTD [221], PD [222], and CJD [143]. CSF levels of human interferon γ-inducible protein-10 (IP-10) have been reported to increase in individuals with MCI and mild AD dementia, although this finding is not consistent among studies [208, 223]. Monocyte chemoattractant protein-1 (MCP-1) is slightly increased in CSF of AD patients, while no differences are found in blood [147].

Recent studies have underscored the significance of SMOC1 in neurodegenerative diseases. SMOC1, SPARC-related modular calcium binding protein 1, is a protein primarily involved in cell-matrix interactions [224] and is enriched in oligodendrocyte precursor cells in the brain. SMOC1 colocalizes with Aβ plaques [225] and is increased in the CSF of AD patients [225–230], as well as in plasma [226]. CSF SMOC1 increases in preclinical AD and continues to rise throughout the *continuum* of sporadic AD [231]. In autosomal dominant AD, CSF SMOC1 increases in mutation carriers 29 years before symptom onset and continues to rise across the disease *continuum*, even preceding significant changes in CSF Aβ42 and p-tau [232]. Regarding other neurological diseases, a slight increase of plasma SMOC1 in PD has been described, but no changes were found in the CSF of PD, ALS, or FTD [226].

Studies that combine measurements of different immune-related biomarkers and correlate their levels with disease characteristics are highly needed. Brosseron et al. have conducted comprehensive investigations with panels of several innate immunity-related biomarkers, such as sAXL, sTyro3, MCP-1, IP-10, MIF, interleukin (IL)-6, IL-18, CRP, ApoE, and complement factors C1q, C3, C3b, C4, B, and H, besides sTREM2 and YKL-40 [216, 233]. In the DELCODE cohort, which focuses on subjective cognitive decline (SCD), they observed elevations of CSF sTREM2, YKL-40, sAXL, sTyro3, MIF, C1q, C4, and H during predementia stages in individuals with pathological levels of tau and neurodegeneration biomarkers. Notably, they found that soluble TAM receptors sAXL and sTyro3, along with ApoE, in CSF were associated with larger brain volumes and stable cognitive outcomes during longitudinal follow-up, indicating a potentially protective role [234]. In the same cohort, various inflammatory signatures based on combinations of CSF biomarkers were defined, and each of them was associated with different effects, either beneficial or detrimental, on brain structure and cognition [235]. In the ADNI cohort, higher levels of proteins related to the soluble tumour necrosis factor receptor 1 (sTNFR1) were associated with reduced risk of conversion to dementia [77]. An innovative approach to defining biomarker profiles reflecting microglial states has been conducting proteomics studies in animal models (e.g. *APP* transgenic or α-synuclein transgenic mice and *TREM2*- or *GRN*-knockout mice) and confirming the findings in human CSF [236, 237].

5 Future Perspectives

Both microglia and astrocytes dynamically sense and respond to physiological and pathological conditions, with different subpopulations exhibiting specific responses to various scenarios. Each of these responses may either limit the progression of the disease or, conversely, facilitate its advancement. Furthermore, these responses may vary depending on the type of stimuli, the microenvironment, and the region of the brain, as well as accumulated immunological memory, disease stage, sex, and age. Over the past few years, several studies have characterised numerous transcriptomic and functional signatures of microglia and astrocytes. It is important to note – once more – that each of the glial biomarkers described above only represents a limited aspect of glial reactivity. New biomarkers may be necessary to capture the diverse microglial and astrocytic responses, and signatures composed of a combination of glial biomarkers rather than individual biomarkers may more accurately define each predominant glial response.

Moreover, more studies are needed to understand the biological significance of these biomarkers. Biomarker-neuropathological correlation studies are a promising starting point [238]. A better understanding of the biology of glial reactivity and how it is reflected by biomarkers can facilitate the development of new drugs and aid in the identification of novel drug targets in conjunction with experimental procedures. As demonstrated in TREM2 research, the approach should be bidirectional, with animal model experiments validated by human biomarker data and the latter guiding animal model experiments. A pending issue is the development of blood-based biomarkers that reflect glial reactivity in the brain; while most of the biomarkers described above can be measured in blood, their origin is primarily peripheral, with the notable exception of plasma GFAP. On the other hand, investigating the role of peripheral innate immunity in the brain is certainly an extremely promising line of research to pursue [239].

Finally, research should not be limited to the mere description of whether a glial biomarker increases or decreases in a disease or whether it has diagnostic utility (which is rarely the case), but rather to understand the variables that trigger changes in these biomarkers, the interaction with pathological features, and their potential as prognostic or predictive biomarkers. Amid the development of new disease-modifying drugs in AD, glial and innate immunity biomarkers could serve to identify those patients more likely to respond to treatment or as target engagement biomarkers for treatments targeting glial cells.

Funding MS-C receives funding from the European Research Council (ERC) under the European Union's Horizon 2020 research and innovation programme (Grant agreement No. 948677); ERA PerMed-ERA NET and the Generalitat de Catalunya (Departament de Salut) through the project SLD077/21/000001; Project PI19/00155 and "PI22/00456, funded by Instituto de Salud Carlos III (ISCIII) and co-funded by the European Union; and from a fellowship from "la Caixa" Foundation (ID 100010434) and from the European Union's Horizon 2020 research and innovation programme under the Marie Skłodowska-Curie grant agreement No 847648 (LCF/BQ/PR21/11840004).

References

1. Heneka MT, Carson MJ, El Khoury J, Landreth GE, Brosseron F, Feinstein DL, et al. Neuroinflammation in Alzheimer's disease. Lancet Neurol. 2015;14:388–405. https://doi.org/10.1016/S1474-4422(15)70016-5.
2. Patani R, Hardingham GE, Liddelow SA. Functional roles of reactive astrocytes in neuroinflammation and neurodegeneration. Nat Rev Neurol. 2023;19:395–409. https://doi.org/10.1038/S41582-023-00822-1.
3. Schöll M, Carter SF, Westman E, Rodriguez-Vieitez E, Almkvist O, Thordardottir S, et al. Early astrocytosis in autosomal dominant Alzheimer's disease measured in vivo by multi-tracer positron emission tomography. Sci Rep. 2015;5:16404. https://doi.org/10.1038/srep16404.
4. Cummings J, Zhou Y, Lee G, Zhong K, Fonseca J, Cheng F. Alzheimer's disease drug development pipeline: 2024. Alzheimers Dement. 2024;10 https://doi.org/10.1002/trc2.12465.
5. Romero-Molina C, Garretti F, Andrews SJ, Marcora E, Goate AM. Microglial efferocytosis: diving into the Alzheimer's disease gene pool. Neuron. 2022;110:3513–33. https://doi.org/10.1016/j.neuron.2022.10.015.
6. Bieger A, Rocha A, Bellaver B, Machado L, Da Ros L, Borelli WV, et al. Neuroinflammation biomarkers in the AT(N) framework across the Alzheimer's disease continuum. J Prev Alzheimers Dis. 2023;10:401–17. https://doi.org/10.14283/jpad.2023.54.
7. Malpetti M, Franzmeier N, Brendel M. PET imaging to measure Neuroinflammation in vivo. Methods Mol Biol. 2024:177–93. https://doi.org/10.1007/978-1-0716-3774-6_12.
8. Escartin C, Galea E, Lakatos A, O'Callaghan JP, Petzold GC, Serrano-Pozo A, et al. Reactive astrocyte nomenclature, definitions, and future directions. Nat Neurosci. 2021;24:312–25. https://doi.org/10.1038/s41593-020-00783-4.
9. Paolicelli RC, Sierra A, Stevens B, Tremblay M-E, Aguzzi A, Ajami B, et al. Microglia states and nomenclature: a field at its crossroads. Neuron. 2022;110:3458–83. https://doi.org/10.1016/j.neuron.2022.10.020.
10. Hickman SE, El Khoury J. TREM2 and the neuroimmunology of Alzheimer's disease. Biochem Pharmacol. 2014;88:495–8. https://doi.org/10.1016/j.bcp.2013.11.021.
11. Schmid CD, Sautkulis LN, Danielson PE, Cooper J, Hasel KW, Hilbush BS, et al. Heterogeneous expression of the triggering receptor expressed on myeloid cells-2 on adult murine microglia. J Neurochem. 2002;83:1309–20. https://doi.org/10.1046/j.1471-4159.2002.01243.x.
12. Atagi Y, Liu CC, Painter MM, Chen XF, Verbeeck C, Zheng H, et al. Apolipoprotein E is a ligand for triggering receptor expressed on myeloid cells 2 (TREM2). J Biol Chem. 2015;290:26043–50. https://doi.org/10.1074/jbc.M115.679043.
13. Bailey CC, DeVaux LB, Farzan M. The triggering receptor expressed on myeloid cells 2 binds Apolipoprotein E. J Biol Chem. 2015;290:26033–42. https://doi.org/10.1074/jbc.M115.677286.
14. Keren-Shaul H, Spinrad A, Weiner A, Matcovitch-Natan O, Dvir-Szternfeld R, Ulland TK, et al. A unique microglia type associated with restricting development of Alzheimer's disease. Cell. 2017;169:1276–1290.e17. https://doi.org/10.1016/j.cell.2017.05.018.
15. Kleinberger G, Brendel M, Mracsko E, Wefers B, Groeneweg L, Xiang X, et al. The FTD-like syndrome causing TREM2 T66M mutation impairs microglia function, brain perfusion, and glucose metabolism. EMBO J. 2017;36:1837–53. https://doi.org/10.15252/embj.201796516.
16. Kleinberger G, Yamanishi Y, Suárez-Calvet M, Czirr E, Lohmann E, Cuyvers E, et al. TREM2 mutations implicated in neurodegeneration impair cell surface transport and phagocytosis. Sci Transl Med. 2014;6:243ra86. https://doi.org/10.1126/scitranslmed.3009093.
17. Mazaheri F, Snaidero N, Kleinberger G, Madore C, Daria A, Werner G, et al. TREM2 deficiency impairs chemotaxis and microglial responses to neuronal injury. EMBO Rep. 2017;18:1186–98. https://doi.org/10.15252/embr.201743922.

18. Parhizkar S, Arzberger T, Brendel M, Kleinberger G, Deussing M, Focke C, et al. Loss of TREM2 function increases amyloid seeding but reduces plaque-associated ApoE. Nat Neurosci. 2019;22:191–204. https://doi.org/10.1038/s41593-018-0296-9.
19. Takahashi K, Rochford CDP, Neumann H. Clearance of apoptotic neurons without inflammation by microglial triggering receptor expressed on myeloid cells-2. J Exp Med. 2005;201:647–57. https://doi.org/10.1084/jem.20041611.
20. Wang Y, Cella M, Mallinson K, Ulrich JD, Young KL, Robinette ML, et al. TREM2 lipid sensing sustains the microglial response in an Alzheimer's disease model. Cell. 2015b;160:1061–71. https://doi.org/10.1016/j.cell.2015.01.049.
21. Zhong L, Chen X-F, Wang T, Wang Z, Liao C, Wang Z, et al. Soluble TREM2 induces inflammatory responses and enhances microglial survival. J Exp Med. 2017;214:jem.20160844. https://doi.org/10.1084/jem.20160844.
22. Schlepckow K, Kleinberger G, Fukumori A, Feederle R, Lichtenthaler SF, Steiner H, et al. An Alzheimer-associated TREM2 variant occurs at the ADAM cleavage site and affects shedding and phagocytic function. EMBO Mol Med. 2017;9:1356–65. https://doi.org/10.15252/emmm.201707672.
23. Del-Aguila JL, Benitez BA, Li Z, Dube U, Mihindukulasuriya KA, Budde JP, et al. TREM2 brain transcript-specific studies in AD and TREM2 mutation carriers. Mol Neurodegener. 2019;14:18. https://doi.org/10.1186/s13024-019-0319-3.
24. Wunderlich P, Glebov K, Kemmerling N, Tien NT, Neumann H, Walter J. Sequential proteolytic processing of the triggering receptor expressed on myeloid cells-2 (TREM2) protein by ectodomain shedding and γ-secretase-dependent intramembranous cleavage. J Biol Chem. 2013;288:33027–36. https://doi.org/10.1074/jbc.M113.517540.
25. Filipello F, Goldsbury C, You SF, Locca A, Karch CM, Piccio L. Soluble TREM2: innocent bystander or active player in neurological diseases? Neurobiol Dis. 2022;165:105630. https://doi.org/10.1016/j.nbd.2022.105630.
26. Zhong L, Xu Y, Zhuo R, Wang T, Wang K, Huang R, et al. Soluble TREM2 ameliorates pathological phenotypes by modulating microglial functions in an Alzheimer's disease model. Nat Commun. 2019;10:1–16. https://doi.org/10.1038/s41467-019-09118-9.
27. Guerreiro R, Bilgic B, Guven G, Brás J, Rohrer J, Lohmann E, et al. A novel compound heterozygous mutation in TREM2 found in a Turkish frontotemporal dementia-like family. Neurobiol Aging. 2013a;34:2890.e1–5. https://doi.org/10.1016/j.neurobiolaging.2013.06.005.
28. Paloneva J, Manninen T, Christman G, Hovanes K, Mandelin J, Adolfsson R, et al. Mutations in two genes encoding different subunits of a receptor signaling complex result in an identical disease phenotype. Am J Hum Genet. 2002;71:656–62. https://doi.org/10.1086/342259.
29. Guerreiro R, Wojtas A, Bras J, Carrasquillo M, Rogaeva E, Majounie E, et al. TREM2 variants in Alzheimer's disease. N Engl J Med. 2013b;368:117–27. https://doi.org/10.1056/NEJMoa1211851.
30. Jonsson T, Stefansson H, Steinberg S, Jonsdottir I, Jonsson PV, Snaedal J, et al. Variant of TREM2 associated with the risk of Alzheimer's disease. N Engl J Med. 2013;368:107–16. https://doi.org/10.1056/NEJMoa1211103.
31. Lill CM, Rengmark A, Pihlstrøm L, Fogh I, Shatunov A, Sleiman PM, et al. The role of TREM2 R47H as a risk factor for Alzheimer's disease, frontotemporal lobar degeneration, amyotrophic lateral sclerosis, and Parkinson's disease. Alzheimers Dement. 2015;11:1407–16. https://doi.org/10.1016/J.JALZ.2014.12.009.
32. Borroni B, Ferrari F, Galimberti D, Nacmias B, Barone C, Bagnoli S, et al. Heterozygous TREM2 mutations in frontotemporal dementia. Neurobiol Aging. 2014;35:934.e7–10. https://doi.org/10.1016/j.neurobiolaging.2013.09.017.
33. Cady J, Koval ED, Benitez BA, Zaidman C, Jockel-Balsarotti J, Allred P, et al. TREM2 variant p.R47H as a risk factor for sporadic amyotrophic lateral sclerosis. JAMA Neurol. 2014;71:449–53. https://doi.org/10.1001/jamaneurol.2013.6237.
34. Rayaprolu S, Mullen B, Baker M, Lynch T, Finger E, Seeley WW, et al. TREM2 in neurodegeneration: evidence for association of the p.R47H variant with frontotemporal dementia and Parkinson's disease. Mol Neurodegener. 2013;8:19. https://doi.org/10.1186/1750-1326-8-19.

35. Ashton NJ, Suárez-Calvet M, Heslegrave A, Hye A, Razquin C, Pastor P, et al. Plasma levels of soluble TREM2 and neurofilament light chain in TREM2 rare variant carriers. Alzheimers Res Ther. 2019;11:94. https://doi.org/10.1186/s13195-019-0545-5.
36. Piccio L, Deming Y, Del-Águila JL, Ghezzi L, Holtzman DM, Fagan AM, et al. Cerebrospinal fluid soluble TREM2 is higher in Alzheimer disease and associated with mutation status. Acta Neuropathol. 2016;131:925–33. https://doi.org/10.1007/s00401-016-1533-5.
37. Suárez-Calvet M, Morenas-Rodríguez E, Kleinberger G, Schlepckow K, Araque Caballero MÁ, Franzmeier N, et al. Early increase of CSF sTREM2 in Alzheimer's disease is associated with tau related-neurodegeneration but not with amyloid-β pathology. Mol Neurodegener. 2019;14:1. https://doi.org/10.1186/s13024-018-0301-5.
38. Piccio L, Buonsanti C, Cella M, Tassi I, Schmidt RE, Fenoglio C, et al. Identification of soluble TREM-2 in the cerebrospinal fluid and its association with multiple sclerosis and CNS inflammation. Brain. 2008;131:3081–91. https://doi.org/10.1093/brain/awn217.
39. Thornton P, Sevalle J, Deery MJ, Fraser G, Zhou Y, Ståhl S, et al. TREM2 shedding by cleavage at the H157-S158 bond is accelerated for the Alzheimer's disease-associated H157Y variant. EMBO Mol Med. 2017;9:1366–78. https://doi.org/10.15252/emmm.201707673.
40. Bekris LM, Khrestian M, Dyne E, Shao Y, Pillai JA, Rao SM, et al. Soluble TREM2 and biomarkers of central and peripheral inflammation in neurodegenerative disease. J Neuroimmunol. 2018;319:19–27. https://doi.org/10.1016/j.jneuroim.2018.03.003.
41. Heslegrave A, Heywood W, Paterson R, Magdalinou N, Svensson J, Johansson P, et al. Increased cerebrospinal fluid soluble TREM2 concentration in Alzheimer's disease. Mol Neurodegener. 2016;11:3. https://doi.org/10.1186/s13024-016-0071-x.
42. Liu D, Cao B, Zhao Y, Huang H, McIntyre RS, Rosenblat JD, et al. Soluble TREM2 changes during the clinical course of Alzheimer's disease: a meta-analysis. Neurosci Lett. 2018;686:10–6. https://doi.org/10.1016/j.neulet.2018.08.038.
43. Nordengen K, Kirsebom B-E, Henjum K, Selnes P, Gísladóttir B, Wettergreen M, et al. Glial activation and inflammation along the Alzheimer's disease continuum. J Neuroinflammation. 2019;16:46. https://doi.org/10.1186/s12974-019-1399-2.
44. Rauchmann B-S, Schneider-Axmann T, Alexopoulos P, Perneczky R. CSF soluble TREM2 as a measure of immune response along the Alzheimer's disease continuum. Neurobiol Aging. 2019;74:182–90. https://doi.org/10.1016/j.neurobiolaging.2018.10.022.
45. Suárez-Calvet M, Araque Caballero MÁ, Kleinberger G, Bateman RJ, Fagan AM, Morris JC, et al. Early changes in CSF sTREM2 in dominantly inherited Alzheimer's disease occur after amyloid deposition and neuronal injury. Sci Transl Med. 2016a;8:369ra178. https://doi.org/10.1126/scitranslmed.aag1767.
46. Suárez-Calvet M, Kleinberger G, Araque Caballero MÁ, Brendel M, Rominger A, Alcolea D, et al. sTREM2 cerebrospinal fluid levels are a potential biomarker for microglia activity in early-stage Alzheimer's disease and associate with neuronal injury markers. EMBO Mol Med. 2016b;8:466–76. https://doi.org/10.15252/emmm.201506123.
47. Henjum K, Almdahl IS, Årskog V, Minthon L, Hansson O, Fladby T, et al. Cerebrospinal fluid soluble TREM2 in aging and Alzheimer's disease. Alzheimers Res Ther. 2016;8:17. https://doi.org/10.1186/s13195-016-0182-1.
48. Karlsson L, Vogel J, Arvidsson I, Åström K, Janelidze S, Blennow K, et al. Cerebrospinal fluid reference proteins increase accuracy and interpretability of biomarkers for brain diseases. bioRxiv. 2023:2023.06.08.544222. https://doi.org/10.1101/2023.06.08.544222.
49. Öhrfelt A, Axelsson M, Malmeström C, Novakova L, Heslegrave A, Blennow K, et al. Soluble TREM-2 in cerebrospinal fluid from patients with multiple sclerosis treated with natalizumab or mitoxantrone. Mult Scler. 2016;22:1587–95. https://doi.org/10.1177/1352458515624558.
50. Gisslén M, Heslegrave A, Veleva E, Yilmaz A, Andersson L-M, Hagberg L, et al. CSF concentrations of soluble TREM2 as a marker of microglial activation in HIV-1 infection. Neurol Neuroimmunol Neuroinflammation. 2019;6:e512. https://doi.org/10.1212/NXI.0000000000000512.
51. Henjum K, Quist-Paulsen E, Zetterberg H, Blennow K, Nilsson LNG, Watne LO. CSF sTREM2 in delirium—relation to Alzheimer's disease CSF biomarkers Aβ42, t-tau and p-tau. J Neuroinflammation. 2018;15:304. https://doi.org/10.1186/s12974-018-1331-1.

52. Woollacott IOC, Nicholas JM, Heslegrave A, Heller C, Foiani MS, Dick KM, et al. Cerebrospinal fluid soluble TREM2 levels in frontotemporal dementia differ by genetic and pathological subgroup. Alzheimers Res Ther. 2018;10:79. https://doi.org/10.1186/s13195-018-0405-8.
53. Byrne LM, Rodrigues FB, Johnson EB, De Vita E, Blennow K, Scahill R, et al. Cerebrospinal fluid neurogranin and TREM2 in Huntington's disease. Sci Rep. 2018;8:4260. https://doi.org/10.1038/s41598-018-21788-x.
54. Alosco ML, Tripodis Y, Fritts NG, Heslegrave A, Baugh CM, Conneely S, et al. Cerebrospinal fluid tau, Aβ, and sTREM2 in former National Football League Players: Modeling the relationship between repetitive head impacts, microglial activation, and neurodegeneration. Alzheimers Dement. 2018;14:1159–70. https://doi.org/10.1016/J.JALZ.2018.05.004.
55. Deming Y, Filipello F, Cignarella F, Cantoni C, Hsu S, Mikesell R, et al. The MS4A gene cluster is a key modulator of soluble TREM2 and Alzheimer's disease risk. Sci Transl Med. 2019;11:eaau2291. https://doi.org/10.1126/scitranslmed.aau2291.
56. Ohara T, Hata J, Tanaka M, Honda T, Yamakage H, Yoshida D, et al. Serum soluble triggering receptor expressed on myeloid cells 2 as a biomarker for incident dementia: the Hisayama study. Ann Neurol. 2019;85:47–58. https://doi.org/10.1002/ana.25385.
57. Wilson EN, Swarovski MS, Linortner P, Shahid M, Zuckerman AJ, Wang Q, Channappa D, Minhas PS, Mhatre SD, Plowey ED, Quinn JF, Zabetian CP, Tian L, Longo FM, Cholerton B, Montine TJ, Poston KL, Andreasson KI. Soluble TREM2 is elevated in Parkinson's disease subgroups with increased CSF tau. Brain. 2020;143(3):932–943. https://doi.org/10.1093/brain/awaa021. PMID: 32065223; PMCID: PMC7089668.
58. Morenas-Rodríguez E, Li Y, Nuscher B, Franzmeier N, Xiong C, Suárez-Calvet M, et al. Soluble TREM2 in CSF and its association with other biomarkers and cognition in autosomal-dominant Alzheimer's disease: a longitudinal observational study. Lancet Neurol. 2022;21:329–41. https://doi.org/10.1016/S1474-4422(22)00027-8.
59. Ewers M, Biechele G, Suárez-Calvet M, Sacher C, Blume T, Morenas-Rodriguez E, et al. Higher CSF sTREM2 and microglia activation are associated with slower rates of beta-amyloid accumulation. EMBO Mol Med. 2020;12:e12308. https://doi.org/10.15252/EMMM.202012308.
60. Zhao N, Qiao W, Li F, Ren Y, Zheng J, Martens YA, et al. Elevating microglia TREM2 reduces amyloid seeding and suppresses disease-associated microglia. J Exp Med. 2022b;219 https://doi.org/10.1084/JEM.20212479/213467.
61. Pereira JB, Janelidze S, Strandberg O, Whelan CD, Zetterberg H, Blennow K, et al. Microglial activation protects against accumulation of tau aggregates in nondemented individuals with underlying Alzheimer's disease pathology. Nat Aging. 2022;2:1138–44. https://doi.org/10.1038/s43587-022-00310-z.
62. Götzl JK, Brendel M, Werner G, Parhizkar S, Sebastian Monasor L, Kleinberger G, et al. Opposite microglial activation stages upon loss of PGRN or TREM2 result in reduced cerebral glucose metabolism. EMBO Mol Med. 2019;11:e9711. https://doi.org/10.15252/emmm.201809711.
63. Ulrich JD, Finn M, Wang Y, Shen A, Mahan TE, Jiang H, et al. Altered microglial response to Aβ plaques in APPPS1-21 mice heterozygous for TREM2. Mol Neurodegener. 2014;9:20. https://doi.org/10.1186/1750-1326-9-20.
64. Wang Y, Ulland TK, Ulrich JD, Song W, Tzaferis JA, Hole JT, et al. TREM2-mediated early microglial response limits diffusion and toxicity of amyloid plaques. J Exp Med. 2016;213:667–75. https://doi.org/10.1084/jem.20151948.
65. Yuan P, Condello C, Keene CDD, Wang Y, Bird TDD, Paul SMM, et al. TREM2 haplodeficiency in mice and humans impairs the microglia barrier function leading to decreased amyloid compaction and severe axonal dystrophy. Neuron. 2016;90:724–39. https://doi.org/10.1016/j.neuron.2016.05.003.
66. Jay TR, Hirsch AM, Broihier ML, Miller CM, Neilson LE, Ransohoff RM, et al. Disease progression-dependent effects of TREM2 deficiency in a mouse model of Alzheimer's disease. J Neurosci. 2017;37:637–47. https://doi.org/10.1523/JNEUROSCI.2110-16.2017.

67. Jay TR, Miller CM, Cheng PJ, Graham LC, Bemiller S, Broihier ML, et al. TREM2 deficiency eliminates TREM2 + inflammatory macrophages and ameliorates pathology in Alzheimer's disease mouse models. J Exp Med. 2015;212:287–95. https://doi.org/10.1084/jem.20142322.
68. Bemiller SM, McCray TJ, Allan K, Formica SV, Xu G, Wilson G, et al. TREM2 deficiency exacerbates tau pathology through dysregulated kinase signaling in a mouse model of tauopathy. Mol Neurodegener. 2017;12:74. https://doi.org/10.1186/s13024-017-0216-6.
69. Leyns CEG, Gratuze M, Narasimhan S, Jain N, Koscal LJ, Jiang H, et al. TREM2 function impedes tau seeding in neuritic plaques. Nat Neurosci. 2019;22:1217–22. https://doi.org/10.1038/s41593-019-0433-0.
70. Leyns CEG, Ulrich JD, Finn MB, Stewart FR, Koscal LJ, Remolina Serrano J, et al. TREM2 deficiency attenuates neuroinflammation and protects against neurodegeneration in a mouse model of tauopathy. Proc Natl Acad Sci. 2017;114:201710311. https://doi.org/10.1073/pnas.1710311114.
71. Gratuze M, Chen Y, Parhizkar S, Jain N, Strickland MR, Serrano JR, et al. Activated microglia mitigate Aβ-associated tau seeding and spreading. J Exp Med. 2021;218 https://doi.org/10.1084/JEM.20210542.
72. Lee SH, Meilandt WJ, Xie L, Gandham VD, Ngu H, Barck KH, et al. Trem2 restrains the enhancement of tau accumulation and neurodegeneration by β-amyloid pathology. Neuron. 2021;109:1283–1301.e6. https://doi.org/10.1016/J.NEURON.2021.02.010.
73. Falcon C, Monté-Rubio GC, Grau-Rivera O, Suárez-Calvet M, Sánchez-Valle R, Rami L, et al. CSF glial biomarkers YKL40 and sTREM2 are associated with longitudinal volume and diffusivity changes in cognitively unimpaired individuals. NeuroImage Clin. 2019;23:101801. https://doi.org/10.1016/j.nicl.2019.101801.
74. Gispert JD, Suárez-Calvet M, Monté GC, Tucholka A, Falcon C, Rojas S, et al. Cerebrospinal fluid sTREM2 levels are associated with gray matter volume increases and reduced diffusivity in early Alzheimer's disease. Alzheimers Dement. 2016b;12:1259–72. https://doi.org/10.1016/j.jalz.2016.06.005.
75. Ewers M, Franzmeier N, Suárez-Calvet M, Morenas-Rodriguez E, Caballero MAA, Kleinberger G, et al. Increased soluble TREM2 in cerebrospinal fluid is associated with reduced cognitive and clinical decline in Alzheimer's disease. Sci Transl Med. 2019;11:eaav6221. https://doi.org/10.1126/scitranslmed.aav6221.
76. Franzmeier N, Suárez-Calvet M, Frontzkowski L, Moore A, Hohman TJ, Morenas-Rodriguez E, et al. Higher CSF sTREM2 attenuates ApoE4-related risk for cognitive decline and neurodegeneration. Mol Neurodegener. 2020;15:57. https://doi.org/10.1186/s13024-020-00407-2.
77. Hu WT, Ozturk T, Kollhoff A, Wharton W, Christina Howell J, Weiner M, et al. Higher CSF sTNFR1-related proteins associate with better prognosis in very early Alzheimer's disease. Nat Commun. 2021;12:4001. https://doi.org/10.1038/s41467-021-24220-7.
78. Zhao A, Jiao Y, Ye G, Kang W, Tan L, Li Y, et al. Soluble TREM2 levels associate with conversion from mild cognitive impairment to Alzheimer's disease. J Clin Invest. 2022a;132 https://doi.org/10.1172/JCI158708.
79. Lewcock JW, Schlepckow K, Di Paolo G, Tahirovic S, Monroe KM, Haass C. Emerging microglia biology defines novel therapeutic approaches for Alzheimer's disease. Neuron. 2020;108:801–21. https://doi.org/10.1016/J.NEURON.2020.09.029.
80. Schlepckow K, Morenas-Rodríguez E, Hong S, Haass C. Stimulation of TREM2 with agonistic antibodies—an emerging therapeutic option for Alzheimer's disease. Lancet Neurol. 2023;22:1048–60. https://doi.org/10.1016/S1474-4422(23)00247-8.
81. Bateman A, Belcourt D, Bennett H, Lazure C, Solomon S. Granulins, a novel class of peptide from leukocytes. Biochem Biophys Res Commun. 1990;173:1161–8. https://doi.org/10.1016/S0006-291X(05)80908-8.
82. Bhandari V, Palfree RG, Bateman A. Isolation and sequence of the granulin precursor cDNA from human bone marrow reveals tandem cysteine-rich granulin domains. Proc Natl Acad Sci. 1992;89:1715–9. https://doi.org/10.1073/pnas.89.5.1715.
83. Holler CJ, Taylor G, Deng Q, Kukar T. Intracellular proteolysis of progranulin generates stable, lysosomal granulins that are haploinsufficient in patients with frontotemporal

dementia caused by GRN mutations. eNeuro. 2017;4:ENEURO.0100-17.2017. https://doi.org/10.1523/ENEURO.0100-17.2017.
84. Kleinberger G, Capell A, Haass C, Van Broeckhoven C. Mechanisms of granulin deficiency: lessons from cellular and animal models. Mol Neurobiol. 2013;47:337–60. https://doi.org/10.1007/s12035-012-8380-8.
85. Lui H, Zhang J, Makinson SR, Cahill MK, Kelley KW, Huang HY, et al. Progranulin deficiency promotes circuit-specific synaptic pruning by microglia via complement activation. Cell. 2016;165:921–35. https://doi.org/10.1016/j.cell.2016.04.001.
86. Zhang Y, Chen K, Sloan SA, Bennett ML, Scholze AR, O'Keeffe S, et al. An RNA-sequencing transcriptome and splicing database of glia, neurons, and vascular cells of the cerebral cortex. J Neurosci. 2014;34:11929–47. https://doi.org/10.1523/JNEUROSCI.1860-14.2014.
87. Martens LH, Zhang J, Barmada SJ, Zhou P, Kamiya S, Sun B, et al. Progranulin deficiency promotes neuroinflammation and neuron loss following toxin-induced injury. J Clin Invest. 2012;122:3955–9. https://doi.org/10.1172/JCI63113.
88. Tanaka Y, Matsuwaki T, Yamanouchi K, Nishihara M. Increased lysosomal biogenesis in activated microglia and exacerbated neuronal damage after traumatic brain injury in progranulin-deficient mice. Neuroscience. 2013;250:8–19. https://doi.org/10.1016/j.neuroscience.2013.06.049.
89. Yin F, Banerjee R, Thomas B, Zhou P, Qian L, Jia T, et al. Exaggerated inflammation, impaired host defense, and neuropathology in progranulin-deficient mice. J Exp Med. 2010;207:117–28. https://doi.org/10.1084/jem.20091568.
90. Smith KR, Damiano J, Franceschetti S, Carpenter S, Canafoglia L, Morbin M, et al. Strikingly different clinicopathological phenotypes determined by progranulin-mutation dosage. Am J Hum Genet. 2012;90:1102–7. https://doi.org/10.1016/j.ajhg.2012.04.021.
91. Finch N, Baker M, Crook R, Swanson K, Kuntz K, Surtees R, et al. Plasma progranulin levels predict progranulin mutation status in frontotemporal dementia patients and asymptomatic family members. Brain. 2009;132:583–91. https://doi.org/10.1093/brain/awn352.
92. Ghidoni R, Benussi L, Glionna M, Franzoni M, Binetti G. Low plasma progranulin levels predict progranulin mutations in frontotemporal lobar degeneration. Neurology. 2008;71:1235–9. https://doi.org/10.1212/01.wnl.0000325058.10218.fc.
93. Sleegers K, Brouwers N, Van Damme P, Engelborghs S, Gijselinck I, van der Zee J, et al. Serum biomarker for progranulin-associated frontotemporal lobar degeneration. Ann Neurol. 2009;65:603–9. https://doi.org/10.1002/ana.21621.
94. Swift IJ, Rademakers R, Finch NC, Baker M, Ghidoni R, Benussi L, et al. A systematic review of progranulin concentrations in biofluids in over 7,000 people-assessing the pathogenicity of GRN mutations and other influencing factors. Alzheimers Res Ther. 2024;16 https://doi.org/10.1186/S13195-024-01420-Z.
95. Brouwers N, Sleegers K, Engelborghs S, Maurer-Stroh S, Gijselinck I, van der Zee J, et al. Genetic variability in progranulin contributes to risk for clinically diagnosed Alzheimer disease. Neurology. 2008;71:656–64. https://doi.org/10.1212/01.wnl.0000319688.89790.7a.
96. Rademakers R, Eriksen JL, Baker M, Robinson T, Ahmed Z, Lincoln SJ, et al. Common variation in the miR-659 binding-site of GRN is a major risk factor for TDP43-positive frontotemporal dementia. Hum Mol Genet. 2008;17:3631–42. https://doi.org/10.1093/hmg/ddn257.
97. Morenas-Rodríguez E, Cervera-Carles L, Vilaplana E, Alcolea D, Carmona-Iragui M, Dols-Icardo O, et al. Progranulin protein levels in cerebrospinal fluid in primary neurodegenerative dementias. J Alzheimers Dis. 2015;50:539–46. https://doi.org/10.3233/JAD-150746.
98. Nicholson AM, Finch NA, Thomas CS, Wojtas A, Rutherford NJ, Mielke MM, et al. Progranulin protein levels are differently regulated in plasma and CSF. Neurology. 2014;82:1871–8. https://doi.org/10.1212/WNL.0000000000000445.
99. Körtvélyessy P, Gukasjan A, Sweeney-Reed CM, Heinze HJ, Thurner L, Bittner DM. Progranulin and amyloid-β levels: relationship to neuropsychology in frontotemporal and Alzheimer's disease. J Alzheimers Dis. 2015;46:375–80. https://doi.org/10.3233/JAD-150069.

100. Wilke C, Gillardon F, Deuschle C, Hobert MA, Jansen IE, Metzger FG, et al. Cerebrospinal fluid progranulin, but not serum progranulin, is reduced in GRN-negative frontotemporal dementia. Neurodegener Dis. 2017;17:83–8. https://doi.org/10.1159/000448896.
101. Suárez-Calvet M, Capell A, Araque Caballero MÁ, Morenas-Rodríguez E, Fellerer K, Franzmeier N, et al. CSF progranulin increases in the course of Alzheimer's disease and is associated with sTREM2, neurodegeneration and cognitive decline. EMBO Mol Med. 2018;10:e9712. https://doi.org/10.15252/emmm.201809712.
102. Wilke C, Gillardon F, Deuschle C, Dubois E, Hobert MA, Müller vom Hagen J, et al. Serum levels of progranulin do not reflect cerebrospinal fluid levels in neurodegenerative disease. Curr Alzheimer Res. 2016;13:654–62. https://doi.org/10.2174/1567205013666160314151247.
103. García-Revilla J, Boza-Serrano A, Espinosa-Oliva AM, Soto MS, Deierborg T, Ruiz R, et al. Galectin-3, a rising star in modulating microglia activation under conditions of neurodegeneration. Cell Death Dis. 2022;13:628. https://doi.org/10.1038/s41419-022-05058-3.
104. Shin T. The pleiotropic effects of galectin-3 in neuroinflammation: a review. Acta Histochem. 2013;115:407–11. https://doi.org/10.1016/J.ACTHIS.2012.11.010.
105. Boza-Serrano A, Ruiz R, Sanchez-Varo R, García-Revilla J, Yang Y, Jimenez-Ferrer I, et al. Galectin-3, a novel endogenous TREM2 ligand, detrimentally regulates inflammatory response in Alzheimer's disease. Acta Neuropathol. 2019;138:251–73. https://doi.org/10.1007/S00401-019-02013-Z/FIGURES/7.
106. Boza-Serrano A, Vrillon A, Minta K, Paulus A, Camprubí-Ferrer L, Garcia M, et al. Galectin-3 is elevated in CSF and is associated with Aβ deposits and tau aggregates in brain tissue in Alzheimer's disease. Acta Neuropathol. 2022;144:843–59. https://doi.org/10.1007/s00401-022-02469-6.
107. Krasemann S, Madore C, Cialic R, Baufeld C, Calcagno N, El Fatimy R, et al. The TREM2-APOE pathway drives the transcriptional phenotype of dysfunctional microglia in neurodegenerative diseases. Immunity. 2017;47:566–581.e9. https://doi.org/10.1016/j.immuni.2017.08.008.
108. Tao C-C, Cheng K-M, Ma Y-L, Hsu W-L, Chen Y-C, Fuh J-L, et al. Galectin-3 promotes Aβ oligomerization and Aβ toxicity in a mouse model of Alzheimer's disease. Cell Death Differ. 2020;27:192–209. https://doi.org/10.1038/s41418-019-0348-z.
109. Borrego-Écija S, Pérez-Millan A, Antonell A, Fort-Aznar L, Kaya-Tilki E, León-Halcón A, et al. Galectin-3 is upregulated in frontotemporal dementia patients with subtype specificity. Alzheimers Dement. 2024;20:1515–26. https://doi.org/10.1002/ALZ.13536.
110. Wang X, Zhang S, Lin F, Chu W, Yue S. Elevated galectin-3 levels in the serum of patients with Alzheimer's disease. Am J Alzheimer's Dis Other Dementiasr. 2015a;30:729–32. https://doi.org/10.1177/1533317513495107.
111. Yazar T, Olgun Yazar H, Cihan M. Evaluation of serum galectin-3 levels at Alzheimer patients by stages: a preliminary report. Acta Neurol Belg. 2021;121:949–54. https://doi.org/10.1007/s13760-020-01477-1.
112. Cengiz T, Türkboyları S, Gençler OS, Anlar Ö. The roles of galectin-3 and galectin-4 in the idiopatic Parkinson disease and its progression. Clin Neurol Neurosurg. 2019;184:105373. https://doi.org/10.1016/J.CLINEURO.2019.105373.
113. Yazar HO, Yazar T, Cihan M. A preliminary data: evaluation of serum galectin-3 levels in patients with idiopathic Parkinson's disease. J Clin Neurosci. 2019;70:164–8. https://doi.org/10.1016/j.jocn.2019.08.032.
114. Siew JJ, Chen H-M, Chen H-Y, Chen H-L, Chen C-M, Soong B-W, et al. Galectin-3 is required for the microglia-mediated brain inflammation in a model of Huntington's disease. Nat Commun. 2019;10:3473. https://doi.org/10.1038/s41467-019-11441-0.
115. Yan J, Xu Y, Zhang L, Zhao H, Jin L, Liu W-G, et al. Increased expressions of plasma galectin-3 in patients with amyotrophic lateral sclerosis. Chin Med J. 2016;129:2797–803. https://doi.org/10.4103/0366-6999.194656.
116. Shen YF, Yu WH, Dong XQ, Du Q, Yang DB, Wu GQ, et al. The change of plasma galectin-3 concentrations after traumatic brain injury. Clin Chim Acta. 2016;456:75–80. https://doi.org/10.1016/J.CCA.2016.02.029.

117. Wang A, Zhong C, Zhu Z, Xu T, Peng Y, Xu T, et al. Serum galectin-3 and poor outcomes among patients with acute ischemic stroke. Stroke. 2018;49:211–4. https://doi.org/10.1161/STROKEAHA.117.019084.
118. Banati RB. Visualising microglial activation in vivo. Glia. 2002;40:206–17. https://doi.org/10.1002/GLIA.10144.
119. Owen DR, Yeo AJ, Gunn RN, Song K, Wadsworth G, Lewis A, et al. An 18-kDa translocator protein (TSPO) polymorphism explains differences in binding affinity of the PET Radioligand PBR28. J Cereb Blood Flow Metab. 2012;32:1–5. https://doi.org/10.1038/jcbfm.2011.147.
120. Edison P, Archer HA, Gerhard A, Hinz R, Pavese N, Turkheimer FE, et al. Microglia, amyloid, and cognition in Alzheimer's disease: an [11C](R)PK11195-PET and [11C]PIB-PET study. Neurobiol Dis. 2008;32:412–9. https://doi.org/10.1016/j.nbd.2008.08.001.
121. Hamelin L, Lagarde J, Dorothé G, Leroy C, Labit M, Comley RA, et al. Early and protective microglial activation in Alzheimer's disease: a prospective study using 18 F-DPA-714 PET imaging. Brain. 2016;1252–1264 https://doi.org/10.1093/brain/aww017.
122. Hamelin L, Lagarde J, Dorothée G, Potier MC, Corlier F, Kuhnast B, et al. Distinct dynamic profiles of microglial activation are associated with progression of Alzheimer's disease. Brain. 2018;141:1855–70. https://doi.org/10.1093/brain/awy079.
123. Kreisl WC, Lyoo CH, Liow J-S, Wei M, Snow J, Page E, et al. 11C-PBR28 binding to translocator protein increases with progression of Alzheimer's disease. Neurobiol Aging. 2016;44:53–61. https://doi.org/10.1016/j.neurobiolaging.2016.04.011.
124. Kreisl WC, Lyoo CH, McGwier M, Snow J, Jenko KJ, Kimura N, et al. In vivo radioligand binding to translocator protein correlates with severity of Alzheimer's disease. Brain. 2013;136:2228–38. https://doi.org/10.1093/brain/awt145.
125. Tondo G, Iaccarino L, Caminiti SP, Presotto L, Santangelo R, Iannaccone S, et al. The combined effects of microglia activation and brain glucose hypometabolism in early-onset Alzheimer's disease. Alzheimers Res Ther. 2020;12:1–10. https://doi.org/10.1186/s13195-020-00619-0.
126. Fan Z, Okello AA, Brooks DJ, Edison P. Longitudinal influence of microglial activation and amyloid on neuronal function in Alzheimer's disease. Brain. 2015;138:3685–98. https://doi.org/10.1093/brain/awv288.
127. Gerhard A, Trender-Gerhard I, Turkheimer F, Quinn NP, Bhatia KP, Brooks DJ. In vivo imaging of microglial activation with [11C](R)-PK11195 PET in progressive supranuclear palsy. Mov Disord. 2006;21:89–93. https://doi.org/10.1002/MDS.20668.
128. Palleis C, Sauerbeck J, Beyer L, Harris S, Schmitt J, Morenas-Rodriguez E, et al. In vivo assessment of neuroinflammation in 4-repeat tauopathies. Mov Disord. 2021;36:883–94. https://doi.org/10.1002/MDS.28395.
129. Passamonti L, Rodríguez PV, Hong YT, Allinson KSJ, Bevan-Jones WR, Williamson D, et al. [11C]PK11195 binding in Alzheimer disease and progressive supranuclear palsy. Neurology. 2018;90:e1989–96. https://doi.org/10.1212/WNL.0000000000005610.
130. Malpetti M, Passamonti L, Rittman T, Jones PS, Vázquez Rodríguez P, Bevan-Jones WR, et al. Neuroinflammation and tau colocalize in vivo in progressive supranuclear palsy. Ann Neurol. 2020;88:1194–204. https://doi.org/10.1002/ANA.25911.
131. Bevan-Jones WR, Cope TE, Jones PS, Kaalund SS, Passamonti L, Allinson K, et al. Neuroinflammation and protein aggregation co-localize across the frontotemporal dementia spectrum. Brain. 2020;143:1010–26. https://doi.org/10.1093/BRAIN/AWAA033.
132. Cagnin A, Rossor M, Sampson EL, MacKinnon T, Banati RB. In vivo detection of microglial activation in frontotemporal dementia. Ann Neurol. 2004;56:894–7. https://doi.org/10.1002/ANA.20332.
133. Kim MJ, McGwier M, Jenko KJ, Snow J, Morse C, Zoghbi SS, et al. Neuroinflammation in frontotemporal lobar degeneration revealed by 11 C-PBR28 PET. Ann Clin Transl Neurol. 2019;6:1327–31. https://doi.org/10.1002/ACN3.50802.
134. Malpetti M, Rittman T, Jones PS, Cope TE, Passamonti L, Bevan-Jones WR, et al. In vivo PET imaging of neuroinflammation in familial frontotemporal dementia. J Neurol Neurosurg Psychiatry. 2021;92:319–22. https://doi.org/10.1136/JNNP-2020-323698

135. Pascual B, Funk Q, Zanotti-Fregonara P, Cykowski MD, Veronese M, Rockers E, et al. Neuroinflammation is highest in areas of disease progression in semantic dementia. Brain. 2021;144:1565–75. https://doi.org/10.1093/BRAIN/AWAB057.
136. Malpetti M, Cope TE, Street D, Jones PS, Hezemans FH, Mak E, et al. Microglial activation in the frontal cortex predicts cognitive decline in frontotemporal dementia. Brain. 2023;146:3221–31. https://doi.org/10.1093/brain/awad078.
137. Carter SF, Herholz K, Rosa-Neto P, Pellerin L, Nordberg A, Zimmer ER. Astrocyte biomarkers in Alzheimer's disease. Trends Mol Med. 2019;25:77–95. https://doi.org/10.1016/j.molmed.2018.11.006.
138. Marutle A, Gillberg P-G, Bergfors A, Yu W, Ni R, Nennesmo I, et al. 3H-deprenyl and 3H-PIB autoradiography show different laminar distributions of astroglia and fibrillar β-amyloid in Alzheimer brain. J Neuroinflammation. 2013;10:90. https://doi.org/10.1186/1742-2094-10-90.
139. Serrano-Pozo A, Mielke ML, Gómez-Isla T, Betensky RA, Growdon JH, Frosch MP, et al. Reactive glia not only associates with plaques but also parallels tangles in Alzheimer's disease. Am J Pathol. 2011;179:1373–84. https://doi.org/10.1016/j.ajpath.2011.05.047.
140. Abu-Rumeileh S, Steinacker P, Polischi B, Mammana A, Bartoletti-Stella A, Oeckl P, et al. CSF biomarkers of neuroinflammation in distinct forms and subtypes of neurodegenerative dementia. Alzheimers Res Ther. 2019;12:2. https://doi.org/10.1186/s13195-019-0562-4.
141. Fukuyama R, Izumoto T, Fushiki S. The cerebrospinal fluid level of glial fibrillary acidic protein is increased in cerebrospinal fluid from Alzheimer's disease patients and correlates with severity of dementia. Eur Neurol. 2001;46:35–8. https://doi.org/10.1159/000050753.
142. Ishiki A, Kamada M, Kawamura Y, Terao C, Shimoda F, Tomita N, et al. Glial fibrillar acidic protein in the cerebrospinal fluid of Alzheimer's disease, dementia with Lewy bodies, and frontotemporal lobar degeneration. J Neurochem. 2016;136:258–61. https://doi.org/10.1111/JNC.13399.
143. Jesse S, Steinacker P, Cepek L, Arnim CV, Tumani H, Lehnert S, et al. Glial fibrillary acidic protein and protein S-100B: different concentration pattern of glial proteins in cerebrospinal fluid of patients with Alzheimer's disease and Creutzfeldt-Jakob disease. J Alzheimers Dis. 2009;17:541–51. https://doi.org/10.3233/JAD-2009-1075.
144. Oeckl P, Halbgebauer S, Anderl-Straub S, Steinacker P, Huss AM, Neugebauer H, et al. Glial fibrillary acidic protein in serum is increased in Alzheimer's disease and correlates with cognitive impairment. J Alzheimers Dis. 2019;67:481–8. https://doi.org/10.3233/JAD-180325.
145. Wallin A, Blennow K, Rosengren LE. Glial fibrillary acidic protein in the cerebrospinal fluid of patients with dementia. Dementia. 1996;7:267–72. https://doi.org/10.1159/000106891.
146. Andreasen N, Gottfries J, Vanmechelen E, Vanderstichele H, Davidson P, Blennow K, et al. Evaluation of CSF biomarkers for axonal and neuronal degeneration, gliosis, and beta-amyloid metabolism in Alzheimer's disease. J Neurol Neurosurg Psychiatry. 2001;71:557–8.
147. Olsson B, Lautner R, Andreasson U, Öhrfelt A, Portelius E, Bjerke M, et al. CSF and blood biomarkers for the diagnosis of Alzheimer's disease: a systematic review and meta-analysis. Lancet Neurol. 2016;15:673–84. https://doi.org/10.1016/S1474-4422(16)00070-3.
148. Bellaver B, Ferrari-Souza JP, Uglione da Ros L, Carter SF, Rodriguez-Vieitez E, Nordberg A, et al. Astrocyte biomarkers in Alzheimer disease. Neurology. 2021;96:e2944–55. https://doi.org/10.1212/WNL.0000000000012109.
149. Asken BM, Elahi FM, La Joie R, Strom A, Staffaroni AM, Lindbergh CA, et al. Plasma glial fibrillary acidic protein levels differ along the spectra of amyloid burden and clinical disease stage. J Alzheimers Dis. 2020;78:265–76. https://doi.org/10.3233/JAD-200755.
150. Elahi FM, Casaletto KB, La Joie R, Walters SM, Harvey D, Wolf A, et al. Plasma biomarkers of astrocytic and neuronal dysfunction in early- and late-onset Alzheimer's disease. Alzheimers Dement. 2020;16:681–95. https://doi.org/10.1016/j.jalz.2019.09.004.
151. Rajan KB, Aggarwal NT, McAninch EA, Weuve J, Barnes LL, Wilson RS, et al. Remote blood biomarkers of longitudinal cognitive outcomes in a population study. Ann Neurol. 2020;88:1065–76. https://doi.org/10.1002/ana.25874.

152. Simrén J, Leuzy A, Karikari TK, Hye A, Benedet AL, Lantero-Rodriguez J, et al. The diagnostic and prognostic capabilities of plasma biomarkers in Alzheimer's disease. Alzheimers Dement. 2021;17:1145–56. https://doi.org/10.1002/alz.12283.
153. Verberk IMW, Thijssen E, Koelewijn J, Mauroo K, Vanbrabant J, de Wilde A, et al. Combination of plasma amyloid beta(1-42/1-40) and glial fibrillary acidic protein strongly associates with cerebral amyloid pathology. Alzheimers Res Ther. 2020;12:118. https://doi.org/10.1186/s13195-020-00682-7.
154. Ashton NJ, Janelidze S, Mattsson-Carlgren N, Binette AP, Strandberg O, Brum WS, et al. Differential roles of Aβ42/40, p-tau231 and p-tau217 for Alzheimer's trial selection and disease monitoring. Nat Med. 2022;28:2555–62. https://doi.org/10.1038/s41591-022-02074-w.
155. Benedet AL, Milà-Alomà M, Vrillon A, Ashton NJ, Pascoal TA, Lussier F, et al. Differences between plasma and cerebrospinal fluid glial fibrillary acidic protein levels across the Alzheimer disease continuum. JAMA Neurol. 2021;78:1471. https://doi.org/10.1001/jamaneurol.2021.3671.
156. Chatterjee P, Pedrini S, Stoops E, Goozee K, Villemagne VL, Asih PR. Plasma glial fi brillary acidic protein is elevated in cognitively normal older adults at risk of Alzheimer ' s disease. Transl Psychiatry. 2021;11:27. https://doi.org/10.1038/s41398-020-01137-1.
157. Milà-Alomà M, Ashton NJ, Shekari M, Salvadó G, Ortiz-Romero P, Montoliu-Gaya L, et al. Plasma p-tau231 and p-tau217 as state markers of amyloid-β pathology in preclinical Alzheimer's disease. Nat Med. 2022;28:1797–801. https://doi.org/10.1038/s41591-022-01925-w.
158. Pereira JB, Janelidze S, Smith R, Mattsson-Carlgren N, Palmqvist S, Teunissen CE, et al. Plasma GFAP is an early marker of amyloid-β but not tau pathology in Alzheimer's disease. Brain. 2021;144:3505–16. https://doi.org/10.1093/brain/awab223.
159. Ashton NJ, Suárez-Calvet M, Karikari TK, Lantero-Rodriguez J, Snellman A, Sauer M, et al. Effects of pre-analytical procedures on blood biomarkers for Alzheimer's pathophysiology, glial activation, and neurodegeneration. Alzheimers Dement. 2021;13:e12168. https://doi.org/10.1002/DAD2.12168.
160. Simrén J, Weninger H, Brum WS, Khalil S, Benedet AL, Blennow K, et al. Differences between blood and cerebrospinal fluid glial fibrillary acidic protein levels: the effect of sample stability. Alzheimers Dement. 2022;18:1988–92. https://doi.org/10.1002/ALZ.12806.
161. Sweeney MD, Sagare AP, Zlokovic BV. Blood-brain barrier breakdown in Alzheimer disease and other neurodegenerative disorders. Nat Rev Neurol. 2018;14:133–50. https://doi.org/10.1038/nrneurol.2017.188.
162. Mayer CA, Brunkhorst R, Niessner M, Pfeilschifter W, Steinmetz H, Foerch C. Blood levels of glial fibrillary acidic protein (GFAP) in patients with neurological diseases. PLoS One. 2013;8:e62101. https://doi.org/10.1371/journal.pone.0062101.
163. Nylén K, Csajbok LZ, Öst M, Rashid A, Blennow K, Nellgård B, et al. Serum glial fibrillary acidic protein is related to focal brain injury and outcome after aneurysmal subarachnoid hemorrhage. Stroke. 2007;38:1489–94. https://doi.org/10.1161/STROKEAHA.106.478362.
164. Nylén K, Öst M, Csajbok LZ, Nilsson I, Blennow K, Nellgård B, et al. Increased serum-GFAP in patients with severe traumatic brain injury is related to outcome. J Neurol Sci. 2006;240:85–91. https://doi.org/10.1016/j.jns.2005.09.007.
165. Papa L, Brophy GM, Welch RD, Lewis LM, Braga CF, Tan CN, et al. Time course and diagnostic accuracy of glial and neuronal blood biomarkers GFAP and UCH-L1 in a large cohort of trauma patients with and without mild traumatic brain injury. JAMA Neurol. 2016;73:551–60. https://doi.org/10.1001/jamaneurol.2016.0039.
166. Papa L, Lewis LM, Falk JL, Zhang Z, Silvestri S, Giordano P, et al. Elevated levels of serum glial fibrillary acidic protein breakdown products in mild and moderate traumatic brain injury are associated with intracranial lesions and neurosurgical intervention. Ann Emerg Med. 2012;59:471–83. https://doi.org/10.1016/j.annemergmed.2011.08.021.

167. Salvadó G, Ossenkoppele R, Ashton NJ, Beach TG, Serrano GE, Reiman EM, et al. Specific associations between plasma biomarkers and postmortem amyloid plaque and tau tangle loads. EMBO Mol Med. 2023;15 https://doi.org/10.15252/emmm.202217123.
168. Bellaver B, Puig-Pijoan A, Ferrari-Souza JP, Leffa DT, Lussier FZ, Ferreira PCL, et al. Blood-brain barrier integrity impacts the use of plasma amyloid-β as a proxy of brain amyloid-β pathology. Alzheimers Dement. 2023;19:3815–25. https://doi.org/10.1002/ALZ.13014.
169. Pelkmans W, Shekari M, Brugulat-Serrat A, Sánchez-Benavides G, Minguillón C, Fauria K, et al. Astrocyte biomarkers GFAP and YKL-40 mediate early Alzheimer's disease progression. Alzheimers Dement. 2024;20:483–93. https://doi.org/10.1002/alz.13450.
170. Cicognola C, Janelidze S, Hertze J, Zetterberg H, Blennow K, Mattsson-Carlgren N, et al. Plasma glial fibrillary acidic protein detects Alzheimer pathology and predicts future conversion to Alzheimer dementia in patients with mild cognitive impairment. Alzheimers Res Ther. 2021;13 https://doi.org/10.1186/s13195-021-00804-9.
171. Verberk IMW, Laarhuis MB, van den Bosch KA, Ebenau JL, van Leeuwenstijn M, Prins ND, et al. Serum markers glial fibrillary acidic protein and neurofilament light for prognosis and monitoring in cognitively normal older people: a prospective memory clinic-based cohort study. Lancet Heal Longev. 2021;2:e87–95. https://doi.org/10.1016/S2666-7568(20)30061-1.
172. Pontecorvo MJ, Lu M, Burnham SC, Schade AE, Dage JL, Shcherbinin S, et al. Association of donanemab treatment with exploratory plasma biomarkers in early symptomatic Alzheimer disease: a secondary analysis of the TRAILBLAZER-ALZ randomized clinical trial. JAMA Neurol. 2022;79:1250–9. https://doi.org/10.1001/JAMANEUROL.2022.3392.
173. Benussi A, Ashton NJ, Karikari TK, Gazzina S, Premi E, Benussi L, et al. Serum glial fibrillary acidic protein (GFAP) is a marker of disease severity in frontotemporal lobar degeneration. J Alzheimers Dis. 2020;77:1129–41. https://doi.org/10.3233/JAD-200608.
174. Heller C, Foiani MS, Moore K, Convery R, Bocchetta M, Neason M, et al. Plasma glial fibrillary acidic protein is raised in progranulin-associated frontotemporal dementia. J Neurol Neurosurg Psychiatry. 2020;91 https://doi.org/10.1136/jnnp-2019-321954.
175. Chen C-H, Cheng Y-W, Chen Y-F, Tang S-C, Jeng J-S. Plasma neurofilament light chain and glial fibrillary acidic protein predict stroke in CADASIL. J Neuroinflammation. 2020;17:124. https://doi.org/10.1186/s12974-020-01813-5.
176. Bonneh-Barkay D, Bissel SJ, Kofler J, Starkey A, Wang G, Wiley CA. Astrocyte and macrophage regulation of YKL-40 expression and cellular response in neuroinflammation. Brain Pathol. 2012;22:530–46. https://doi.org/10.1111/J.1750-3639.2011.00550.X.
177. Bonneh-Barkay D, Wang G, Starkey A, Hamilton RL, Wiley CA. In vivo CHI3L1 (YKL-40) expression in astrocytes in acute and chronic neurological diseases. J Neuroinflammation. 2010a;7:34. https://doi.org/10.1186/1742-2094-7-34.
178. Querol-Vilaseca M, Colom-Cadena M, Pegueroles J, San Martín-Paniello C, Clarimon J, Belbin O, et al. YKL-40 (Chitinase 3-like I) is expressed in a subset of astrocytes in Alzheimer's disease and other tauopathies. J Neuroinflammation. 2017;14:118. https://doi.org/10.1186/s12974-017-0893-7.
179. Craig-Schapiro R, Perrin RJ, Roe CM, Xiong C, Carter D, Cairns NJ, et al. YKL-40: a novel prognostic fluid biomarker for preclinical Alzheimer's disease. Biol Psychiatry. 2010;68:903–12. https://doi.org/10.1016/j.biopsych.2010.08.025.
180. Alcolea D, Carmona-Iragui M, Suárez-Calvet M, Sánchez-Saudinós MB, Sala I, Antón-Aguirre S, et al. Relationship between β-secretase, inflammation and core cerebrospinal fluid biomarkers for Alzheimer's disease. J Alzheimers Dis. 2014;42:157–67. https://doi.org/10.3233/JAD-140240.
181. Antonell A, Mansilla A, Rami L, Lladó A, Iranzo A, Olives J, et al. Cerebrospinal fluid level of YKL-40 protein in preclinical and prodromal Alzheimer's disease. J Alzheimers Dis. 2014;42:901–8. https://doi.org/10.3233/JAD-140624.
182. Janelidze S, Hertze J, Zetterberg H, Landqvist Waldö M, Santillo A, Blennow K, et al. Cerebrospinal fluid neurogranin and YKL-40 as biomarkers of Alzheimer's disease. Ann Clin Transl Neurol. 2016;3:12–20. https://doi.org/10.1002/acn3.266.

183. Mattsson N, Tabatabaei S, Johansson P, Hansson O, Andreasson U, Månsson J-E, et al. Cerebrospinal fluid microglial markers in Alzheimer's disease: elevated chitotriosidase activity but lack of diagnostic utility. NeuroMolecular Med. 2011;13:151–9. https://doi.org/10.1007/s12017-011-8147-9.
184. Olsson B, Hertze J, Lautner R, Zetterberg H, Nägga K, Höglund K, et al. Microglial markers are elevated in the prodromal phase of Alzheimer's disease and vascular dementia. J Alzheimers Dis. 2013b;33:45–53. https://doi.org/10.3233/JAD-2012-120787.
185. Connolly K, Lehoux M, O'Rourke R, Assetta B, Erdemir GA, Elias JA, et al. Potential role of chitinase-3-like protein 1 (CHI3L1/YKL-40) in neurodegeneration and Alzheimer's disease. Alzheimers Dement. 2023;19:9–24. https://doi.org/10.1002/ALZ.12612.
186. Alcolea D, Martínez-Lage P, Sánchez-Juan P, Olazarán J, Antúnez C, Izagirre A, et al. Amyloid precursor protein metabolism and inflammation markers in preclinical Alzheimer disease. Neurology. 2015a;85:626–33.
187. Gispert JD, Monté GC, Suárez-Calvet M, Falcon C, Tucholka A, Rojas S, et al. The APOE ε4 genotype modulates CSF YKL-40 levels and their structural brain correlates in the continuum of Alzheimer's disease but not those of sTREM2. Alzheimers Dement. 2017;6:50–9. https://doi.org/10.1016/j.dadm.2016.12.002.
188. Milà-Alomà M, Salvadó G, Gispert JD, Vilor-Tejedor N, Grau-Rivera O, Sala-Vila A, et al. Amyloid beta, tau, synaptic, neurodegeneration, and glial biomarkers in the preclinical stage of the Alzheimer's continuum. Alzheimers Dement. 2020;alz.12131 https://doi.org/10.1002/alz.12131.
189. Gispert JD, Monté GC, Falcon C, Tucholka A, Rojas S, Sánchez-Valle R, et al. CSF YKL-40 and pTau181 are related to different cerebral morphometric patterns in early AD. Neurobiol Aging. 2016a;38:47–55. https://doi.org/10.1016/j.neurobiolaging.2015.10.022.
190. Janelidze S, Mattsson N, Stomrud E, Lindberg O, Palmqvist S, Zetterberg H, et al. CSF biomarkers of neuroinflammation and cerebrovascular dysfunction in early Alzheimer disease. Neurology. 2018;91:e867–77. https://doi.org/10.1212/WNL.0000000000006082.
191. Kester MI, Teunissen CE, Sutphen C, Herries EM, Ladenson JH, Xiong C, et al. Cerebrospinal fluid VILIP-1 and YKL-40, candidate biomarkers to diagnose, predict and monitor Alzheimer's disease in a memory clinic cohort. Alzheimers Res Ther. 2015;7:59. https://doi.org/10.1186/s13195-015-0142-1.
192. Lleó A, Alcolea D, Martínez-Lage P, Scheltens P, Parnetti L, Poirier J, et al. Longitudinal cerebrospinal fluid biomarker trajectories along the Alzheimer's disease continuum in the BIOMARKAPD study. Alzheimers Dement. 2019;0 https://doi.org/10.1016/j.jalz.2019.01.015.
193. Sutphen CL, McCue L, Herries EM, Xiong C, Ladenson JH, Holtzman DM, et al. Longitudinal decreases in multiple cerebrospinal fluid biomarkers of neuronal injury in symptomatic late onset Alzheimer's disease. Alzheimers Dement. 2018;14:869–79. https://doi.org/10.1016/j.jalz.2018.01.012.
194. Schindler SE, Li Y, Todd KW, Herries EM, Henson RL, Gray JD, et al. Emerging cerebrospinal fluid biomarkers in autosomal dominant Alzheimer's disease. Alzheimers Dement. 2019;15:655–65. https://doi.org/10.1016/j.jalz.2018.12.019.
195. Choi J, Lee H-W, Suk K. Plasma level of chitinase 3-like 1 protein increases in patients with early Alzheimer's disease. J Neurol. 2011;258:2181–5. https://doi.org/10.1007/s00415-011-6087-9.
196. Sala-Llonch R, Idland A-V, Borza T, Watne LO, Wyller TB, Brækhus A, et al. Inflammation, amyloid, and atrophy in the aging brain: relationships with longitudinal changes in cognition. J Alzheimers Dis. 2017;58:829–40. https://doi.org/10.3233/JAD-161146.
197. Alcolea D, Vilaplana E, Pegueroles J, Montal V, Sánchez-Juan P, González-Suárez A, et al. Relationship between cortical thickness and cerebrospinal fluid YKL-40 in predementia stages of Alzheimer's disease. Neurobiol Aging. 2015b;36:2018–23. https://doi.org/10.1016/j.neurobiolaging.2015.03.001.

198. Falcon C, Grau-Rivera O, Suárez-Calvet M, Bosch B, Sánchez-Valle R, Arenaza-Urquijo EM, et al. Sex differences of longitudinal brain changes in cognitively unimpaired adults. J Alzheimers Dis. 2020;76:1413–22. https://doi.org/10.3233/JAD-200293.
199. Alcolea D, Irwin DJ, Illán-Gala I, Muñoz L, Clarimón J, McMillan CT, et al. Elevated YKL-40 and low sAPPβ:YKL-40 ratio in antemortem cerebrospinal fluid of patients with pathologically confirmed FTLD. J Neurol Neurosurg Psychiatry. 2019;90:180–6. https://doi.org/10.1136/jnnp-2018-318993.
200. Bonneh-Barkay D, Zagadailov P, Zou H, Niyonkuru C, Figley M, Starkey A, et al. YKL-40 expression in traumatic brain injury: an initial analysis. J Neurotrauma. 2010b;27:1215–23. https://doi.org/10.1089/NEU.2010.1310.
201. Canto E, Tintore M, Villar LM, Costa C, Nurtdinov R, Alvarez-Cermeno JC, et al. Chitinase 3-like 1: prognostic biomarker in clinically isolated syndromes. Brain. 2015;138:918–31. https://doi.org/10.1093/BRAIN/AWV017.
202. Comabella M, Fernández M, Martin R, Rivera-Vallvé S, Borrás E, Chiva C, et al. Cerebrospinal fluid chitinase 3-like 1 levels are associated with conversion to multiple sclerosis. Brain. 2010;133:1082–93. https://doi.org/10.1093/brain/awq035.
203. Illán-Gala I, Alcolea D, Montal V, Dols-Icardo O, Muñoz L, de Luna N, et al. CSF sAPPβ, YKL-40, and NfL along the ALS-FTD spectrum. Neurology. 2018;91:e1619–28. https://doi.org/10.1212/WNL.0000000000006383.
204. Llorens F, Thüne K, Tahir W, Kanata E, Diaz-Lucena D, Xanthopoulos K, et al. YKL-40 in the brain and cerebrospinal fluid of neurodegenerative dementias. Mol Neurodegener. 2017;12:83. https://doi.org/10.1186/s13024-017-0226-4.
205. Teunissen CE, Elias N, Koel-Simmelink MJA, Durieux-Lu S, Malekzadeh A, Pham TV, et al. Novel diagnostic cerebrospinal fluid biomarkers for pathologic subtypes of frontotemporal dementia identified by proteomics. Alzheimer's Dement. 2016;2:86–94. https://doi.org/10.1016/j.dadm.2015.12.004.
206. Hall S, Janelidze S, Surova Y, Widner H, Zetterberg H, Hansson O. Cerebrospinal fluid concentrations of inflammatory markers in Parkinson's disease and atypical parkinsonian disorders. Sci Rep. 2018;8:13276. https://doi.org/10.1038/s41598-018-31517-z.
207. Olsson B, Constantinescu R, Holmberg B, Andreasen N, Blennow K, Zetterberg H. The glial marker YKL-40 is decreased in synucleinopathies. Mov Disord. 2013a;28:1882–5. https://doi.org/10.1002/mds.25589.
208. Wennström M, Surova Y, Hall S, Nilsson C, Minthon L, Hansson O, et al. The inflammatory marker YKL-40 is elevated in cerebrospinal fluid from patients with Alzheimer's but not Parkinson's disease or dementia with Lewy bodies. PLoS One. 2015;10:e0135458. https://doi.org/10.1371/journal.pone.0135458.
209. Morenas-Rodríguez E, Alcolea D, Suárez-Calvet M, Muñoz-Llahuna L, Vilaplana E, Sala I, et al. Different pattern of CSF glial markers between dementia with Lewy bodies and Alzheimer's disease. Sci Rep. 2019;9:7803. https://doi.org/10.1038/s41598-019-44173-8.
210. Ekblom J, Jossan SS, Oreland L, Walum E, Aquilonius SM. Reactive gliosis and monoamine oxidase B. J Neural Transm Suppl. 1994;41:253–8. https://doi.org/10.1007/978-3-7091-9324-2_33.
211. Carter SF, Schöll M, Almkvist O, Wall A, Engler H, Långström B, et al. Evidence for astrocytosis in prodromal alzheimer disease provided by 11C-deuterium-L-deprenyl: a multitracer PET paradigm combining 11C-Pittsburgh compound B and 18F-FDG. J Nucl Med. 2012;53:37–46. https://doi.org/10.2967/jnumed.110.087031.
212. Rodriguez-Vieitez E, Saint-Aubert L, Carter SF, Almkvist O, Farid K, Schöll M, et al. Diverging longitudinal changes in astrocytosis and amyloid PET in autosomal dominant Alzheimer's disease. Brain. 2016;139:922–36. https://doi.org/10.1093/brain/awv404.
213. Santillo AF, Gambini JP, Lannfelt L, Långström B, Ulla-Marja L, Kilander L, et al. In vivo imaging of astrocytosis in Alzheimer's disease: an 11C-L-deuteriodeprenyl and PIB PET study. Eur J Nucl Med Mol Imaging. 2011;38:2202–8. https://doi.org/10.1007/S00259-011-1895-9.
214. Chiotis K, Johansson C, Rodriguez-Vieitez E, Ashton NJ, Blennow K, Zetterberg H, et al. Tracking reactive astrogliosis in autosomal dominant and sporadic Alzheimer's disease with

multi-modal PET and plasma GFAP. Mol Neurodegener. 2023;18:60. https://doi.org/10.1186/s13024-023-00647-y.
215. Brosseron F, Krauthausen M, Kummer M, Heneka MT. Body fluid cytokine levels in mild cognitive impairment and Alzheimer's disease: a comparative overview. Mol Neurobiol. 2014;50:534–44. https://doi.org/10.1007/s12035-014-8657-1.
216. Brosseron F, Traschütz A, Widmann CN, Kummer MP, Tacik P, Santarelli F, et al. Characterization and clinical use of inflammatory cerebrospinal fluid protein markers in Alzheimer's disease. Alzheimers Res Ther. 2018;10:25. https://doi.org/10.1186/s13195-018-0353-3.
217. Casaletto KB, Elahi FM, Fitch R, Walters S, Fox E, Staffaroni AM, et al. A comparison of biofluid cytokine markers across platform technologies: correspondence or divergence? Cytokine. 2018;111:481–9. https://doi.org/10.1016/J.CYTO.2018.05.032.
218. Shen X-N, Niu L-D, Wang Y-J, Cao X-P, Liu Q, Tan L, et al. Inflammatory markers in Alzheimer's disease and mild cognitive impairment: a meta-analysis and systematic review of 170 studies. J Neurol Neurosurg Psychiatry. 2019;90:590–8. https://doi.org/10.1136/jnnp-2018-319148.
219. Christl J, Verhülsdonk S, Pessanha F, Menge T, Seitz RJ, Kujovic M, et al. Association of cerebrospinal fluid S100B protein with Core biomarkers and cognitive deficits in prodromal and mild Alzheimer's disease. J Alzheimers Dis. 2019;72:1119–27. https://doi.org/10.3233/JAD-190550.
220. Peskind ER, Griffin WST, Akama KT, Raskind MA, Van Eldik LJ. Cerebrospinal fluid S100B is elevated in the earlier stages of Alzheimer's disease. Neurochem Int. 2001;39:409–13. https://doi.org/10.1016/S0197-0186(01)00048-1.
221. Petzold A, Jenkins R, Watt HC, Green AJE, Thompson EJ, Keir G, et al. Cerebrospinal fluid S100B correlates with brain atrophy in Alzheimer's disease. Neurosci Lett. 2003;336:167–70. https://doi.org/10.1016/S0304-3940(02)01257-0.
222. Papuć E, Rejdak K. Increased cerebrospinal fluid S100B and NSE reflect neuronal and glial damage in Parkinson's disease. Front Aging Neurosci. 2020;12 https://doi.org/10.3389/FNAGI.2020.00156.
223. Galimberti D, Schoonenboom N, Scheltens P, Fenoglio C, Bouwman F, Venturelli E, et al. Intrathecal chemokine synthesis in mild cognitive impairment and Alzheimer disease. Arch Neurol. 2006;63:538. https://doi.org/10.1001/archneur.63.4.538.
224. Vannahme C, Smyth N, Miosge N, Gösling S, Frie C, Paulsson M, et al. Characterization of SMOC-1, a novel modular calcium-binding protein in basement membranes. J Biol Chem. 2002;277:37977–86. https://doi.org/10.1074/JBC.M203830200.
225. Bai B, Wang X, Li Y, Chen PC, Yu K, Dey KK, et al. Deep multilayer brain proteomics identifies molecular networks in Alzheimer's disease progression. Neuron. 2020;105:975–991.e7. https://doi.org/10.1016/j.neuron.2019.12.015.
226. Dammer EB, Ping L, Duong DM, Modeste ES, Seyfried NT, Lah JJ, et al. Multi-platform proteomic analysis of Alzheimer's disease cerebrospinal fluid and plasma reveals network biomarkers associated with proteostasis and the matrisome. Alzheimers Res Ther. 2022;14:174. https://doi.org/10.1186/s13195-022-01113-5.
227. Dayon L, Núñez Galindo A, Wojcik J, Cominetti O, Corthésy J, Oikonomidi A, et al. Alzheimer disease pathology and the cerebrospinal fluid proteome. Alzheimers Res Ther. 2018;10:66. https://doi.org/10.1186/s13195-018-0397-4.
228. Higginbotham L, Ping L, Dammer EB, Duong DM, Zhou M, Gearing M, et al. Integrated proteomics reveals brain-based cerebrospinal fluid biomarkers in asymptomatic and symptomatic Alzheimer's disease. Sci Adv. 2020;6 https://doi.org/10.1126/sciadv.aaz9360.
229. Wang H, Dey KK, Chen P-C, Li Y, Niu M, Cho J-H, et al. Integrated analysis of ultra-deep proteomes in cortex, cerebrospinal fluid and serum reveals a mitochondrial signature in Alzheimer's disease. Mol Neurodegener. 2020;15:43. https://doi.org/10.1186/s13024-020-00384-6.

230. Zhou M, Haque RU, Dammer EB, Duong DM, Ping L, Johnson ECB, et al. Targeted mass spectrometry to quantify brain-derived cerebrospinal fluid biomarkers in Alzheimer's disease. Clin Proteomics. 2020;17 https://doi.org/10.1186/S12014-020-09285-8.
231. Watson CM, Dammer EB, Ping L, Duong DM, Modeste E, Carter EK, et al. Quantitative mass spectrometry analysis of cerebrospinal fluid protein biomarkers in Alzheimer's disease. Sci Data. 2023;10:261. https://doi.org/10.1038/s41597-023-02158-3.
232. Johnson ECB, Bian S, Haque RU, Carter EK, Watson CM, Gordon BA, et al. Cerebrospinal fluid proteomics define the natural history of autosomal dominant Alzheimer's disease. Nat Med. 2023;29:1979–88. https://doi.org/10.1038/s41591-023-02476-4.
233. Brosseron F, Kolbe C, Santarelli F, Carvalho S, Antonell A, Castro-Gomez S, et al. Multicenter Alzheimer's and Parkinson's disease immune biomarker verification study. Alzheimers Dement. 2020;16:292–304. https://doi.org/10.1016/j.jalz.2019.07.018.
234. Brosseron F, Maass A, Kleineidam L, Ravichandran KA, González PG, McManus RM, et al. Soluble TAM receptors sAXL and sTyro3 predict structural and functional protection in Alzheimer's disease. Neuron. 2022;110:1009–1022.e4. https://doi.org/10.1016/J.NEURON.2021.12.016.
235. Hayek D, Ziegler G, Kleineidam L, Brosseron F, Nemali A, Vockert N, et al. Different inflammatory signatures based on CSF biomarkers relate to preserved or diminished brain structure and cognition. Mol Psychiatry. 2024; https://doi.org/10.1038/S41380-023-02387-3.
236. Eninger T, Müller SA, Bacioglu M, Schweighauser M, Lambert M, Maia LF, et al. Signatures of glial activity can be detected in the CSF proteome. Proc Natl Acad Sci. 2022;119:e2119804119. https://doi.org/10.1073/pnas.2119804119.
237. Pesämaa I, Müller SA, Robinson S, Darcher A, Paquet D, Zetterberg H, et al. A microglial activity state biomarker panel differentiates FTD-granulin and Alzheimer's disease patients from controls. Mol Neurodegener. 2023;18:70. https://doi.org/10.1186/s13024-023-00657-w.
238. Sánchez-Juan P, Valeriano-Lorenzo E, Ruiz-González A, Pastor AB, Rodrigo Lara H, López-González F, et al. Serum GFAP levels correlate with astrocyte reactivity, post-mortem brain atrophy and neurofibrillary tangles. Brain. 2024;139:16–7. https://doi.org/10.1093/brain/awae035.
239. Bettcher BM, Tansey MG, Dorothée G, Heneka MT. Peripheral and central immune system crosstalk in Alzheimer disease – a research prospectus. Nat Rev Neurol. 2021;17:689–701. https://doi.org/10.1038/S41582-021-00549-X.

Fluid and Imaging Markers of Vascular Disease

Gary A. Rosenberg

Abstract Cognitive impairment is a major concern in the aging population. The advent of a number of disease-modifying treatments for dementia of the Alzheimer's type that need to be used at an early stage has emphasized the importance of making a correct diagnosis. The overlap among clinical phenotypes associated with cognitive impairment has led to the use of biomarkers to improve disease classification. Biomarkers can come from neuroimaging, cerebrospinal fluid, blood, and neuropsychological testing. Initially, the main biomarkers were amyloid and tau proteins in the CSF and on PET studies. Recently, the use of CSF and PET has been replaced with plasma-based biomarkers as measured by ultra-low quantities of these proteins in the blood. These measurements can be done with commercially available instruments such as the Quanterix HD-X Simoa and the MesoScale Discovery platform. Simoa can be used to measure neurofilament light (NfL) to detect axonal injury and glial fibrillary acidic protein (GFAP) for astrocyte activation as a measure of inflammation. By bypassing an invasive lumbar puncture or an expensive PET study, these blood measurements open the possibility of large-scale screening for early dementia populations that would otherwise be inaccessible. Adding neuroimaging of the white matter with diffusion tensor imaging indicates vascular damage. Combining blood-based biomarkers with imaging using a clustering machine learning approach expands the classification process to include vascular cognitive impairment and multi-etiology dementia. The use of biomarkers for early detection will improve the selection of patients for disease-modifying therapies and make them accessible to underrepresented populations.

Keywords CSF · PET · Trichotomy classification · Neurofilament light · Glial fibrillary acidic protein

G. A. Rosenberg (✉)
Department of Neurology, Center for Memory and Aging, University of New Mexico, Albuquerque, NM, USA
e-mail: grosenberg@salud.unm.edu

R. Tarawneh (ed.), *Biomarkers of Neurodegenerative Disorders*,
https://doi.org/10.1007/978-3-031-80849-4_7

1 Introduction

Multiple pathologies can impair cognition, leading to dementia. Clinically, the various syndromes resulting from these pathologies have overlapping clinical presentations that complicate diagnosis. This is particularly true for patients with Alzheimer disease (AD) and related dementias (ADRD). To improve the diagnostic accuracy of the clinical syndromes, biomarkers have been used to indicate the underlying pathophysiology. Biomarkers reflect the altered physiological processes in the pathogenesis of a disorder, making them helpful for diagnosis, therapeutic management, and monitoring of disease progression. The discovery of disease-modifying treatments (DMT) for AD and the need to use these treatments in the early stages of the disease have provided a strong impetus to identify potential disease biomarkers.

AD was the first type of dementia that adopted biomarkers for diagnosis; this was driven by the discovery of amyloid in plaques of AD and the ability to measure pathological forms of amyloid in the cerebrospinal fluid (CSF) [1, 2]. However, the recognition that other pathological changes also cause cognitive impairment has expanded the search for novel biomarkers that can be used for diagnosis. The original emphasis on clinical diagnosis based on clinical phenomenology has been replaced with a symptom-agnostic, biomarker-driven classification approach called the amyloid beta (Aβ)/tau/neurodegeneration (AT(N)) system [3]. This AD-centric formula left out other major causes of cognitive loss, and a new formulation, ATX(N), that incorporates multiple causes has been proposed, where X is added to allow for new biomarkers to be incorporated as they are discovered [4]. Biological diagnoses using biomarkers are being used in research and in the clinic.

A major breakthrough occurred with the discovery that these biomarkers could be detected in blood with the use of ultrasensitive assay instruments. The ability to perform ultrasensitive assays of proteins in the blood for use as biomarkers made it possible to bypass the invasive lumbar puncture to measure AD proteins. In addition, it obviates the need to do the expensive and scarce positron emission tomography (PET) to identify amyloid and tau in the brain. This ability to use blood as the biomarker for ADRD will facilitate large-scale screening of populations to select those that will need further evaluation in a more economical way.

While AD is considered the most common form of dementia, there is growing awareness that most patients have multiple pathological changes in the brain. It is estimated that 70% of patients with amyloid in the brain at autopsy also have vascular pathology, making the combination of AD and vascular disease (mixed dementia) the most common form [5]. When both are present, the cognitive decline is accelerated [6]. Vascular disease does not currently have biomarkers in blood or CSF but can be identified by the identification of white matter ischemic injury with magnetic resonance imaging (MRI). When vascular disease is present, there is damage to the white matter that can be qualitatively assessed in the MRI image and quantified by changes in the diffusion tensor image (DTI). Combining the fluid measurements in blood and CSF of neurodegeneration with white matter injury on MRI provides a powerful tool to improve patient classification through the mathematical method of cluster analysis. In this way, those with pure AD and pure vascular contributions to cognitive impairment and dementia (VCID) can be separated from mixed dementia and normal cognition. Such an approach will be important now that several treatments are available for ADRD and there is a need to correctly diagnose patients early before the abnormal accumulation of pathological proteins has resulted in cell death.

2 Biomarkers Present in the CSF in Dementia

CSF is the most useful in the detection of AD proteins. The continuity of the CSF with brain interstitial fluid that is contiguous with brain tissue makes it more likely to represent molecules associated with brain activity than those measured in plasma. There are multiple forms of amyloid that can be measured in the CSF. Levels of $A\beta_{1-42}$ are reduced in patients with AD, most likely due to the sink action of the amyloid plaques. Confirmation of the CSF levels has been obtained by studies of PET and from autopsy. Measurements of amyloid in body fluids are more variable for a variety of reasons, and forms of tau protein in CSF or plasma have been shown to be a more reliable indicator of AD where they indicate amyloid pathology. CSF levels of the phosphorylated tau protein are a more reliable biomarker of brain AD pathology than amyloid. The tau protein can be phosphorylated at multiple sites. Currently, the most used is tau phosphorylated at threonine 181 (pTau181), which also reflects levels of amyloid in the brain. Ptau181 was one of the first to be extensively studied and found to be a good indicator of both amyloid and tau pathology in the brain. While early studies were done with pTau181, more recently, the pTau217 assays are being shown to be more reliable measures of brain AD pathology. Most research studies are switching to pTau217 instead of pTau181 [7, 8]. Recently, a new tau assay was reported that measures the microtubule binding region of tau, showing brain-derived forms of tau [9]. Forming a ratio of pTau with $A\beta_{1-42}$ increases the accuracy of the test. While total tau is also elevated in AD, pTau is generally preferred as the biomarker since total tau is elevated in other diseases, such as stroke and head trauma [10].

Although $A\beta_{1-42}$ is commonly measured in CSF and PET as a biomarker for amyloid in the brain, it is not optimal for diagnosis because of the variability between laboratories. Some of the variability observed in earlier studies has been resolved with the use of ultrasensitive assay systems that are more reliable. Because of these major advances in the science of biomarkers, there is an increase in their use for "biological diagnoses" as opposed to the clinical diagnoses that have been the standard for many years. As the accuracy of the biomarkers increases, they are becoming important in early diagnosis of AD and selection of patients for the new anti-amyloid treatments.

A formula has been proposed that includes several biomarkers to form a biological diagnosis. This is referred to as the A/T/N framework, with A being amyloid, T being pTau, and N being neurodegeneration [N is shown by neurofilament light

(NfL) and total tau]. This formula is only useful for confirmation of a diagnosis of AD. It fails to include the other causes of cognitive loss, such as vascular disease. A major limitation of the A/T/N formula is the need for performing an invasive lumbar puncture to obtain CSF or a costly PET scan to detect brain amyloid and tau. This has been a strong impetus to find substitute biomarkers in the more readily accessible blood. As the accuracy of the blood-based biomarkers improves, they could replace CSF and PET studies.

3 Biomarkers for Amyloid in the Blood in Dementia

The movement away from the measurement of biomarkers in CSF and PET has been greatly accelerated by ultrasensitive assay instruments capable of measuring ultralow levels of molecules in the blood. Several of these machines are commercially available and are rapidly changing the biomarker field. Large databanks of CSF with concomitant PET and autopsy studies have facilitated the adoption of the blood studies. Early studies were done with mass spectrometry, which is only available in large research laboratories, limiting the use of these instruments [11]. Single-molecule array (SIMOA) technology is being broadly adopted since it allows studies in blood. SIMOA is an ultrasensitive enzyme-linked immunosorbent assay (ELISA) that uses antibody-coated beads, a fluorescently conjugated detection antibody, and a digital readout. After the beads, sample, and detection antibody have been mixed, the solution is applied to a cartridge with over 235,000 microwells. Each microwell can hold only one bead. The digital readout is highly reproducible from site to site. Other highly sensitive assay systems include the commercially available Meso Scale Discovery (MSD) and O-Link instruments. Forming the ratio of $A\beta_{1-42}/A\beta_{1-40}$ is more accurate than used individually.

4 Importance of Mixed Dementia

An important report appeared in the 1960s: A religious order of uncloistered sisters that were teachers and working in communities consented to be followed over multiple years with neuropsychological testing and agreed to autopsy. This important longitudinal data showed that the combination of AD and vascular disease accelerated cognitive decline compared to those with AD pathology alone [6]. This was confirmed in a reexamination of neuropathological tissue from the National Alzheimer's Coordinating Center (NACC) brain bank, which on re-review showed up to 70 percent had both AD and vascular disease [5].

It is important to have diversified groups of subjects to increase the validity of the biomarker studies. Although most studies have been done in Caucasian populations, studies are now being done in more diverse populations. In a study of elderly American Indians, lower levels of Aβ and higher levels of pTau181 suggested more

AD pathology than in majority populations. A study of Caribbean Hispanic individuals from the Dominican Republic and northern New York City was done with a panel of plasma biomarkers, including pTau181 and the ratio of pTau181 to $A\beta_{1-42}$, which identified biological evidence of AD. A proportion of asymptomatic individuals without dementia had biomarker evidence of AD, suggesting they were presymptomatic, while a proportion of affected individuals with dementia lacked biomarker evidence of AD and may have had other dementia disorders [12]. Another study was done in a population-based cohort of older adults from medically underserved small towns of relatively low socio-economic status in western Pennsylvania. They used a novel clustering approach to classify participants into groups of distinct plasma $A\beta_{1-42}/A\beta_{1-40}$ profiles. They hypothesized that associations between plasma biomarkers and memory could identify individuals at risk of ADRD in the community. Clustering identified two distinct plasma Aβ42/40 modes that were further divided into normal, uncertain, and abnormal. They showed that in different groups, plasma pTau181, NfL, and glial fibrillary acidic protein (GFAP) were inversely correlated with $A\beta_{1-42}/A\beta_{1-40}$, Clinical Dementia Rating®, and memory composite scores, with the strongest associations in the abnormal group [13]. The next step was to develop panels of plasma biomarkers that can accurately classify patients [14]. As the number of such studies grows, confidence in the use of plasma-based biomarkers will increase, and their use will be solidified.

5 Neuroimaging Biomarkers

Ischemia, which is the hallmark of white matter injury, can be detected by magnetic resonance spectroscopy (MRS) and imaging (MRI). Commonly used methods to bring out the changes in white matter are the qualitative fluid-attenuated inversion recovery (FLAIR) sequences and quantitative DTI [15]. Early studies with biochemical methods, such as proton MRS (^{1}H-MRS), can accurately detect ischemia using the abundant signal from N-acetylaspartate (NAA), whose role in the brain is uncertain but which decreases with ischemia [16–18].

FLAIR pulse sequences null the water signal in the sulci and ventricles, revealing lesions in the white matter because the water in the tissue remains in the signal. The size of the white matter hyperintense regions can be measured, and a volume of white matter hyperintensities (WMH) can be used as a qualitative measure of white matter injury. DTI is also sensitive to changes and can be quantified, making it more useful for research studies. The two most commonly used DTI sequences are mean free water and peak width of skeletonized mean diffusivity (PSMD) [19–22]. MRS is less frequently used because it is not automated on most MRI scanners and requires an MRI engineer for implementation and interpretation.

The NAA, mean free water, and PSMD signals correlate with neuropsychological testing, indicating their usefulness in research [17, 21, 22]. By combining the neuroimaging to show white matter injury and fluid biomarkers from either CSF or blood, a more complete picture of the underlying pathophysiology can be obtained than from either alone. Clustering methods have been developed that show the results of a combination of biomarkers. A recent clustering study involved the vascular injury factor formed from a composite of MRI findings on the x-axis and the AD injury factor obtained from amyloid and tau in the CSF on the y-axis [23]. This created four quadrants with pure AD, pure subcortical ischemic vascular disease (SIVD), mixed AD and SIVD, and normal. This "dichotomy" approach was subsequently expanded to include results of neuropsychological testing, resulting in a 3-axis "trichotomy," which produced eight categories [24].

Using the clustering approach, it is possible to determine the major factors present at the onset of dementia and the way it progresses. For example, AD may begin as a single entity when the patient is 40 or 50 years old. As the subject ages and vascular disease begins to appear, usually by age 60 and beyond, it would be possible to map this change in two or three dimensions. In this scenario, if amyloid is found in the fluids or on PET scans, anti-amyloid treatment could be started. With time, as vascular disease appears, an aggressive approach to control vascular risk factors could be started.

6 Biomarkers of Inflammation and Blood-Brain Barrier Permeability

A healthy blood-brain barrier (BBB) is critical for normal brain cell function. The anatomical basis for the BBB is a series of interfaces between the systemic circulation and the brain cells. Anatomically, the most important component of the BBB is the tight junction proteins between endothelial cells. Other components of the BBB include the basal lamina around the blood vessels, the pericytes embedded in the extracellar matrix, and the astrocytic foot processes that wrap around the blood vessels. This complex cellular structure is referred to as the neurovascular unit (NVU) [25]. During an insult to the brain caused by infection, trauma, immunological reactions, and other major insults, a molecular cascade is initiated that leads to activation of the intrinsic inflammatory cells, the microglia, and the influx of extrinsic inflammatory cells, the macrophages. These inflammatory cells release cytokines and proteases that attack the NVU, breaking down tight junctions and extracellular matrix, causing brain edema.

While many methods can be used in animals, mainly two methods are used in humans: (1) measurements of albumin in blood and CSF, and (2) dynamic contrast-enhanced MRI. The albumin index (Q_{alb}) depends on the separation of the albumin in the blood from that in the CSF that is maintained by the large size of the albumin molecule which is excluded to a large part from the CSF space. When CSF proteins were used as a diagnostic test for multiple sclerosis (MS), it was important to separate the proteins produced in the brain that appeared in the CSF from those entering the brain from the blood, which had a high concentration of albumin. The albumin index indicated if the BBB was disrupted and the proteins in the CSF came from the blood. This was important for the measurement of immunoglobulins, such as immunoglobulin G (IgG). Albumin index is routinely measured in MS patients undergoing lumbar puncture. The earliest use of Q_{alb} in patients with dementia was to compare the BBB in patients with AD and vascular dementia. These studies showed that the disruption of the BBB was seen in vascular dementia but not in AD [26]. When compared with controls, the AD group showed a higher mean albumin ratio, indicating a decline in BBB function. However, no significant reduction in BBB function was found in AD patients without vascular factors. These findings support the hypothesis that BBB disturbance is related to vascular factors coexisting with AD. When infarcts were present, they appeared to be endpoint manifestations of the vascular pathology rather than the cause of the disease, as had been suggested at that time when thromboembolism with multiple cerebral infarcts was considered the only pathogenic substrate of vascular dementia, which was called multi-infarct dementia (MID). They suggested that subcortical white matter changes are another important cause of vascular dementia [27]. These early studies were unable to support the commonly held opinion at the time that multiple strokes were the cause of MID. Furthermore, they linked the disruption of the BBB as shown by elevated Q_{alb} with white matter damage.

With the introduction of MRI into dementia research, a contrast-based method of measuring the BBB in humans was described first in animals [28]. Dynamic contrast enhanced MRI (DCEMRI) required a single injection of gadolinium with repeated T1 measurements over 20 minutes. The Patlak equations were used to calculate transfer constants (K_{trans}). DCEMRI has been adapted to human studies [29]. The variability in results from different groups has limited studies in humans, and an attempt to standardize protocols has been made [30].

Albumin index was compared to DCEMRI in a group of patients from one center using the biomarkers from the Biomarkers in Vascular Cognitive Impairment and Dementia (MarkVCID) consortium [31, 32]. They found that Q_{alb} and K_{trans} did not correlate with each other, suggesting that they measured different aspects of the BBB function. However, Q_{alb} correlated with a group of MarkVCID biomarkers that were proinflammatory. These included the angiogenic factors, vascular endothelial growth factor-C (VEGF-C), placental growth factor (PlGF), matrix metalloproteinases (MMP-2, MMP-3, and MMP-10), and the cytokines (interleukin-[IL]-6, IL-8, and tumor necrosis factor [TNF]-α). On the other hand, K_{trans} was associated with the diffusion measures, mean free water, and PSMD, which indicate white matter injury. The MMPs are important in disrupting the BBB by attacking the extracellular matrix and the tight junction proteins [33]. Mild cognitive impairment subjects with increased CSF levels of MMP-10 showed a higher probability of progressing to dementia of the Alzheimer's type and a faster cognitive decline [34].

7 Other Biomarkers to Delineate Pathophysiology

The National Institutes of Health (NIH) have funded a consortium to identify biomarkers in vascular cognitive dementia called the MarkVCID. A series of fluid and imaging biomarkers were selected to be studied by the group. Initially, CSF was selected as the optimal fluid, but blood was also collected, while multiple forms of imaging were used. The biomarkers initially selected by the group were commercially available in a series of kits that included angiogenic factors, cytokines, and proteases. The assays were made by the MSD platform. Initially, the assays were performed at multiple sites in the consortium, with the results stored in a central database. Imaging ranged from analysis of FLAIR structural imaging to detection of injury with DTI. The long-term goal was to use the biomarkers as surrogates for clinical trials [32].

8 New Biomarkers in CSF and Blood

Two new biomarkers have engendered great interest; NfL and GFAP, and these are elevated in several neurological diseases. NfL is a component of the axon that is released when the axon is injured. This can occur in MS, amyotrophic lateral sclerosis, frontotemporal dementia, vascular dementia, and brain trauma. The nonspecific nature of these biomarkers makes them unsuitable for diagnosis. However, they indicate disease progression and improvement with treatments. In a multicenter Alzheimer's Disease Neuroimaging Initiative (ADNI) study, longitudinal plasma NfL levels increased in association with several baseline and longitudinal hallmarks of AD. The increase in plasma NfL levels was associated with changes in other established measures of neurodegeneration in AD [35]. Reactive astrocytes express GFAP, which can be detected in CSF and blood. Since astrocytes are only found in the brain, elevated levels of GFAP are a more specific indicator of release by the brain. This makes GFAP a useful biomarker for brain inflammation. It has been shown that GFAP levels in plasma and CSF are higher in AD and correlate with cognitive impairment [36]. Use of NfL in biomarker-based diagnosis is not perfect because the biomarker measurements in older controls are similar to those in younger clinically diagnosed AD cases (which likely represents increased levels with aging in older controls) [37]. The continuing effort to develop novel blood-based biomarkers is important in the goal of substituting them for CSF and PET.

Inflammasomes, which are molecules that promote inflammation by activating proinflammatory cytokines, have been identified in the CSF of dementia patients [38]. Studies of CSF and blood in dementia patients show upregulation of two potential biomarkers of inflammation at the cellular level, MMP-10 and apoptosis-associated speck-like protein containing a CARD (ASC). Stimulation of microglia cultures with purified insoluble ASC speck aggregates and MMP-10 resulted in the release of MMP-3 and MMP-10. Furthermore, MMP-10-stimulated microglia released additional MMP-10 along with the inflammatory cytokines, TNF-α, IL-6, and CXC motif chemokine ligand 1 (CXCL1). How they interact with the matrix metalloproteinases to damage the endothelial tight junctions needs to be studied.

9 Multimodal Biomarkers Are Important to Show Etiologies Other Than AD

Although when cognitive impairment is suspected, AD is the first and often only diagnosis that is considered; it is rarely found in isolation and usually occurs in combination with other types of dementia, particularly vascular disease. Using cluster analysis with both neurodegenerative and vascular biomarkers, it is possible to identify multiple forms of dementia in a single patient. When the vascular disease as shown by DTI is plotted on the x-axis and neurodegeneration factors on the y-axis, mixed dementia can be determined. Clustering classification methods, such as the "dichotomy and trichotomy approaches" will be important in the future in early detection of the different forms of dementia in the elderly [24, 39]. The factors selected can be modified as new information is obtained on the optimal biomarkers to use and the cut-points can be recalculated. The usefulness of this approach will need to be validated in longitudinal studies of larger cohorts. The main problem with biomarkers is the identification of cut points to separate normal from abnormal values. In addition, new biomarkers are being discovered, and the optimal set to use for cluster analysis is evolving.

10 Future Biomarkers for Use in Diagnosis, Classification, and Patient Selection for Treatment

The biomarkers described in the literature represent only a small number of potential indicators of disease and pathophysiology. The use of ultra-sensitive assay instruments, such as SIMOA and O-Link, has only been available for several years and has rapidly replaced other technologies. These new ultrasensitive analysis systems can create very large datasets that require the use of computers capable of analysis of large datasets. For example, O-link instruments can measure over 5000 molecules in small samples of fluid. Genetic studies have begun to identify potential biomarkers, and the ability to combine modalities will offer to further improve diagnostic accuracy. These developments in biomarkers will lead to the ability to form a biological diagnosis that can be used in precision medicine with computational methods and statistical programs.

References

1. Glenner GG, Wong CW. Alzheimer's disease: initial report of the purification and characterization of a novel cerebrovascular amyloid protein. Biochem Biophys Res Commun. 1984;120(3):885–90.
2. Seubert P, et al. Isolation and quantification of soluble Alzheimer's beta-peptide from biological fluids. Nature. 1992;359(6393):325–7.
3. Jack CR Jr, et al. NIA-AA research framework: toward a biological definition of Alzheimer's disease. Alzheimers Dement. 2018;14(4):535–62.
4. Hampel H, et al. Developing the ATX(N) classification for use across the Alzheimer disease continuum. Nat Rev Neurol. 2021;17(9):580–9.
5. Toledo JB, et al. Contribution of cerebrovascular disease in autopsy confirmed neurodegenerative disease cases in the National Alzheimer's coordinating Centre. Brain. 2013;136(Pt 9):2697–706.
6. Snowdon DA, et al. Brain infarction and the clinical expression of Alzheimer disease. The nun study [see comments]. JAMA. 1997;277(10):813–7.
7. Gonzalez-Ortiz F, et al. Plasma phospho-tau in Alzheimer's disease: towards diagnostic and therapeutic trial applications. Mol Neurodegener. 2023;18(1):18.
8. Ossenkoppele R, et al. Tau biomarkers in Alzheimer's disease: towards implementation in clinical practice and trials. Lancet Neurol. 2022;21(8):726–34.
9. Horie K, et al. Change in cerebrospinal fluid tau microtubule binding region detects symptom onset, cognitive decline, tangles, and atrophy in dominantly inherited Alzheimer's disease. Ann Neurol. 2023;93(6):1158–72.
10. Hampel H, et al. Blood-based biomarkers for Alzheimer's disease: current state and future use in a transformed global healthcare landscape. Neuron. 2023;111(18):2781–99.
11. Paterson RW, et al. SILK studies - capturing the turnover of proteins linked to neurodegenerative diseases. Nat Rev Neurol. 2019;15(7):419–27.
12. Honig LS, et al. Evaluation of plasma biomarkers for a/T/N classification of Alzheimer disease among adults of Caribbean Hispanic ethnicity. JAMA Netw Open. 2023;6(4):e238214.
13. Ferreira PCL, et al. Plasma biomarkers identify older adults at risk of Alzheimer's disease and related dementias in a real-world population-based cohort. Alzheimers Dement. 2023;19(10):4507–19.
14. Palmqvist S, et al. An accurate fully automated panel of plasma biomarkers for Alzheimer's disease. Alzheimers Dement. 2023;19(4):1204–15.
15. Raja R, et al. MRI measurements of blood-brain barrier function in dementia: a review of recent studies. Neuropharmacology. 2017;
16. Brooks WM, et al. 1H-MRS differentiates white matter hyperintensities in subcortical arteriosclerotic encephalopathy from those in normal elderly. Stroke. 1997;28(10):1940–3.
17. Gasparovic C, et al. 1H-MR spectroscopy metabolite levels correlate with executive function in vascular cognitive impairment. J Neurol Neurosurg Psychiatry. 2013;84(7):715–21.
18. Sappey-Marinier D, et al. Proton magnetic resonance spectroscopy of human brain: applications to normal white matter, chronic infarction, and MRI white matter signal hyperintensities. Magn Reson Med. 1992;26(2):313–27.
19. Duering MBG, Brodtmann A, Chen C, Cordonnier C, de Leeuw FE, Debette S, Frayne R, Jouvent E, Rost NS, Ter Telgte A, Al-Shahi Salman R, Backes WH, Bae HJ, Brown R, Chabriat H, De Luca A, de Carli C, Dewenter A, Doubal FN, Ewers M, Field TS, Ganesh A, Greenberg S, Helmer KG, Hilal S, ACC J, Jokinen H, Kuijf H, BYK L, Lebenberg J, BJ MI, Maillard P, VCT M, Pantoni L, Rudilosso S, Satizabal CL, Schirmer MD, Schmidt R, Smith C, Staals J, Thrippleton MJ, van Veluw SJ, Vemuri P, Wang Y, Werring D, Zedde M, Akinyemi RO, Del Brutto OH, Markus HS, Zhu YC, Smith EE, Dichgans M, Wardlaw JM. Neuroimaging standards for research into small vessel disease-advances since 2013. Lancet Neurol. 2023:602–18.
20. Maillard P, et al. FLAIR and diffusion MRI signals are independent predictors of white matter hyperintensities. AJNR Am J Neuroradiol. 2013;34(1):54–61.

21. Maillard P, et al. MRI free water as a biomarker for cognitive performance: validation in the MarkVCID consortium. Alzheimers Dement (Amst). 2022a;14(1):e12362.
22. Maillard P, et al. Instrumental validation of free water, peak-width of skeletonized mean diffusivity, and white matter hyperintensities: MarkVCID neuroimaging kits. Alzheimers Dement (Amst). 2022b;14(1):e12261.
23. Caprihan A, et al. A double-dichotomy clustering of dual pathology dementia patients. Cereb Circ Cogn Behav. 2021;2
24. Caprihan A, et al. A trichotomy method for defining homogeneous subgroups in a dementia population. Ann Clin Transl Neurol. 2023;10(10):1802–15.
25. Iadecola C. The neurovascular unit coming of age: a journey through neurovascular coupling in health and disease. Neuron. 2017;96(1):17–42.
26. Blennow K, et al. Blood-brain barrier disturbance in patients with Alzheimer's disease is related to vascular factors. Acta Neurol Scand. 1990;81(4):323–6.
27. Wallin A, Blennow K. Pathogenetic basis of vascular dementia. Alzheimer Dis Assoc Disord. 1991;5(2):91–102.
28. Ewing JR, et al. Patlak plots of Gd-DTPA MRI data yield blood-brain transfer constants concordant with those of 14C-sucrose in areas of blood-brain opening. Magn Reson Med. 2003;50(2):283–92.
29. Taheri S, et al. Blood-brain barrier permeability abnormalities in vascular cognitive impairment. Stroke. 2011;42(8):2158–63.
30. Heye AK, et al. Assessment of blood-brain barrier disruption using dynamic contrast-enhanced MRI. A systematic review. Neuroimage Clin. 2014;6:262–74.
31. Hillmer L, et al. Blood-brain barrier disruption measured by albumin index correlates with inflammatory fluid biomarkers. J Cereb Blood Flow Metab. 2023;43(5):712–21.
32. Wilcock D, et al. MarkVCID cerebral small vessel consortium: I. Enrollment, clinical, fluid protocols. Alzheimers Dement. 2021;17(4):704–15.
33. Yang Y, et al. Matrix metalloproteinase-mediated disruption of tight junction proteins in cerebral vessels is reversed by synthetic matrix metalloproteinase inhibitor in focal ischemia in rat. J Cereb Blood Flow Metab. 2007;27(4):697–709.
34. Martino Adami PV, et al. Matrix metalloproteinase 10 is linked to the risk of progression to dementia of the Alzheimer's type. Brain. 2022;145(7):2507–17.
35. Mattsson N, et al. Association between longitudinal plasma neurofilament light and neurodegeneration in patients with Alzheimer disease. JAMA Neurol. 2019;76(7):791–9.
36. Arslan B, et al. Blood-based biomarkers in Alzheimer's disease - moving towards a new era of diagnostics. Clin Chem Lab Med. 2024;
37. Stevenson-Hoare J, et al. Plasma biomarkers and genetics in the diagnosis and prediction of Alzheimer's disease. Brain. 2023;146(2):690–9.
38. Sanchez KE, et al. Apoptosis-associated speck-like protein containing a CARD-mediated release of matrix metalloproteinase 10 stimulates a change in microglia phenotype. Front Mol Neurosci. 2022;15:976108.
39. Rosenberg GA, et al. A multimodal approach to stratification of patients with dementia: selection of mixed dementia patients prior to autopsy. Brain Sci. 2019;9(8).

Part II
Fluid and Imaging Markers of Neurodegenerative Disorders

Fluid Markers of Alzheimer Disease

Courtney L. Sutphen, Katherine Volluz, and Rawan Tarawneh

Abstract Fluid biomarkers provide a quantifiable means to model and monitor disease progression in Alzheimer disease (AD). As usage of markers in the clinical realm expands, there is an increased need for novel, sensitive, and specific measures to accurately differentiate and distinguish stages of AD and other neurodegenerative diseases. Improvement in biomarker quality by selective assay platforms can allow for early detection or development of new markers, particularly when ultrasensitive or automated platforms are considered. Additionally, cerebrospinal fluid (CSF) biomarkers provide a backbone for the development and discovery of less invasive blood-based tests, while biomarker usage guidelines help bridge the gap between research and clinical usage. In order to produce robust biomarker data, it is important to consider and minimize the effects of pre-analytical factors (such as sample processing and handling) as well as analytical effects, including batch-to-batch variation within manufacturers, differences between assay platforms, and deviation from manufacturer protocols. Gold standard diagnostic assays, as defined by the US FDA, may also aid in clinical diagnosis.

Biomarkers discussed in this chapter encompass traditional CSF markers of amyloid and tau pathology (Aβ40, Aβ42, total tau, and various species of hyperphosphorylated tau) as well as emerging CSF and blood-based markers of neurodegeneration (i.e., NfL, UCHL1, FABP and VILIP-1), neuroinflammation (i.e., GFAP and YKL-40), and synaptic integrity (i.e., Ng, SNAP-25 and synaptotagmin-1). Still, improvements in assay performance, standardization of sample

C. L. Sutphen
Wake Forest Alzheimer Disease Research Center, Wake Forest School of Medicine, Winston-Salem, NC, USA
e-mail: csutphen@wakehealth.edu

K. Volluz
Knight Alzheimer Disease Research Center, Washington University School of Medicine, St. Louis, MO, USA

R. Tarawneh (✉)
Department of Neurology, University of New Mexico, Albuquerque, NM, USA
e-mail: raltarawneh@salud.unm.edu

R. Tarawneh (ed.), *Biomarkers of Neurodegenerative Disorders*,
https://doi.org/10.1007/978-3-031-80849-4_8

handling and data analysis, and expanded biomarker assessments in diverse cohorts are necessary to advance our knowledge of fluid biomarkers in AD for broader applicability as the field moves forward.

Keywords CSF · Plasma · Amyloid · Tau · Biomarker

1 Introduction

Amyloid beta and tau proteins are cemented as crucial fluid biomarkers in the field of Alzheimer disease (AD) research and now play a pivotal role in both diagnosis and monitoring of disease progression. The journey to establishing their importance began with confirmation of significant relationships between postmortem amyloid plaques and neurofibrillary tangles, composed of aggregated amyloid beta and hyperphosphorylated tau, respectively, and levels of amyloid beta 42 peptide (Aβ42) and tau in cerebrospinal fluid (CSF) [1, 2] in individuals with AD. Diagnosis in living individuals is the primary aim for AD biomarkers, and immense effort has shown clear associations between in vivo amyloid and tau neuroimaging agents, CSF biomarkers, and postmortem neuropathology [3–5]. A decades-long evolution of continuous improvement and refinement in CSF immunoassays for amyloid, tau, and emerging biomarkers of AD-adjacent processes has resulted in a small number of FDA-approved diagnostic tests in individuals with early cognitive impairment, and recent, rapid advancement in immunoassay sensitivity has allowed their use in plasma and serum. This chapter aims to provide a history of the evolution of fluid biomarkers of AD, their status in research and clinical applications, and current well-established CSF biomarker data that spans the time-course of AD. Improvement in biomarker quality and expanded availability offers hope for timely diagnosis, more effective treatments, and ultimately, a better quality of life for those affected by AD.

2 Assay Platforms

Careful consideration of assay platforms remains a topic of importance when discussing fluid biomarkers of AD. Ability to detect proteins in CSF, plasma, and serum, cost of implementation, and assay performance within and between laboratories are concerns every group investigating AD biomarkers must account for. We will briefly discuss a number of assays and their current usage in the AD fluid biomarker landscape, with a particular focus on CSF.

2.1 Early AD Assays

Enzyme-linked immunosorbent assays (ELISAs) were the original platform for AD CSF biomarker determination. The INNOTEST assays (Fujirebio, Europe), single-analyte ELISAs for Aβ42, Aβ40, total tau (tTau), and tau phosphorylated at threonine 181 (pTau181) were among the most widely used to determine biomarker levels in CSF. A number of important discoveries were made using these early assays, including the effect of pre-analytical handling events such as freeze/thaw cycles [6] and assay drift over long periods of analysis [7, 8]. However, early studies of plasma biomarkers in matched samples did not correlate with CSF, amyloid positron emission tomography (PET) neuroimaging, or cognitive decline in multiple cohorts using either the INNOTEST assays or other early ELISAs [3, 9–12]. A few studies found significant differences between AD and cognitively normal participants in plasma; however, changes were not in a consistent direction: Mayeux et al. found elevated levels of Aβ40 or Aβ42 in AD [13], while Pesaresi et al. found a reduction in AD [14].

Multiple Analyte Profiling (xMAP) (Luminex, USA) is a bead-based immunoassay using principles similar to traditional ELISAs. These assays allow simultaneous measurement of Aβ42, Aβ40, tTau, and pTau181 in various combinations, thus reducing the effort, time, and sample volume required to generate data [15]. Similar to traditional ELISAs, levels of amyloid and tau in CSF differed in AD groups compared with non-AD, and both assay types performed similarly when diagnostic accuracy was compared with pathologically-confirmed dementia [16], amyloid PET [17], and AD vs. controls [18]. Also, similar to traditional ELISAs, these findings did not translate well to plasma [19]. Importantly, steps towards automation on the xMAP platform showed a significant improvement in assay precision, which in turn revealed modest associations between CSF and plasma biomarkers but not group differences in plasma Aβ between AD vs. healthy controls [20]. One early meta-analysis of plasma amyloid levels, as measured by ELISA and xMAP, hinted at possible plasma diagnostic or prognostic utility: longitudinal analysis of Aβ in cognitively normal individuals who later converted to AD showed elevated levels of Aβ42 and Aβ40, while cross-sectional analysis of AD vs. normal participants showed lower, though not significantly, levels of Aβ42 [21].

2.2 Ultrasensitive Assays

A number of newer assay platforms are now available to measure AD biomarkers. Built upon lessons learned from earlier iterations, the majority of current assays use varying combinations of multi-analyte measurement, ultra-sensitive technology, and/or automation. Automation, in particular, has been a major investment resulting in improved precision beyond even highly experienced assay technicians [22]. The combination of automation and ultrasensitivity has resulted in greatly expedited development of plasma biomarkers compared to that in CSF.

The single molecule array (SIMOA) platform (Quanterix, USA) is a bead-based immunoassay utilizing ELISA principles, but the final reaction takes place in a disc containing an array of reaction chambers sized to fit one bead each, allowing detection as low as a single molecule [23]. Assays may be single- or multi-analyte, can be performed in CSF or plasma, and are either fully automated or with a transfer step to a 96-well plate that preserves the sample.

Electrochemiluminescent and chemiluminescent platforms comprise the bulk of current ultra-sensitive AD assays. MesoScale Discovery (MesoScale Diagnostics, USA) is an electrochemiluminescent platform capable of measuring AD biomarkers in both CSF and plasma. With easily accessible assay development features and a large commercially available assay inventory, the MesoScale Discovery platform is most commonly used for novel biomarker development [24] or assessment of inflammatory or other AD-associated biomarkers [25]. The Cobas® system (Roche Diagnostics, Switzerland) is a fully automated electrochemiluminescent platform with accompanying single-analyte Elecsys assays for amyloid and tau. The Cobas ratio of CSF pTau181/Aβ42 was the second in-vitro diagnostic biomarker approved for use by the United States Food and Drug Administration (FDA) in individuals with cognitive impairment [26]. Lastly, the Lumipulse® G1200 system (Fujirebio, Japan) is a fully automated, chemiluminescent platform and can perform multiple single-analyte assays from a sample, now in both CSF and plasma [27, 28]. The Lumipulse® G ratio of CSF Aβ42/Aβ40 was the first in vitro diagnostic biomarker to be FDA-approved for use in the US [29].

2.3 Mass Spectrometry

Mass spectrometry (MS) is a complex platform with high sensitivity and multiplex capabilities. Unlike the above assays, samples must go through extensive preparation steps, which may include a denaturing step to prevent aggregation of Aβ isoforms and ensure monomeric isoforms are available for analysis, lyophilization, immunoprecipitation, or solid-phase extraction [30]. Preparation can then be followed by either matrix-assisted laser desorption ionization-time of flight (MALDI-ToF-MS) or liquid chromatography tandem MS (LC-MS/MS). MALDI-ToF-MS is a fast procedure that measures the mass-to-charge ratio of peptides in a sample, from which the presence of the target molecule can be confirmed, although this method can only indicate the relative quantity of peptides unlike quantitative methods [31]; however, improvements in quantitation are possible by spiking in stable-isotope-labelled peptides of interest [32]. LC-MS/MS is more expensive and effort-intensive, but can be used for peptide quantification with selected reaction monitoring if the mass-to-charge ratio of the peptide of interest is known. Peptides in a sample are separated by size on a liquid chromatography column, followed by ionization targeted to the mass-to-charge ratio of the peptide of interest. The ionized product is then fragmented and moved to a second ionization step targeting the mass-to-charge ratio only of fragments of the peptide of interest. These fragments travel to a detector, which counts the fragments, and internal standard curves and controls enable accurate quantification of the amount of peptide in a sample. This tandem MS procedure allows sensitive measurement because even peptides that elute from liquid chromatography in close proximity and have similar ionization properties will separate during the second ionization step based on their specific gas-phase ion chemistry. Additionally, with modern equipment, LC-MS/MS is a platform capable of multiplex analysis [33]. An additional format of MS—isotope dilution-MS—was used as a Reference Measurement Procedure to validate the first Certified Reference Material (CRM) for CSF Aβ42, a significant step towards aligning the measured concentrations of all the above immunoassay methods, an ongoing process being undertaken by many major manufacturers of Aβ42 assays [34].

3 Gold Standard CSF and Blood-Based AD Biomarkers

3.1 What Makes an AD Biomarker a Gold Standard?

Gold Standard biomarkers in AD reflect the underlying disease pathology. Without biomarkers, the only way to diagnose AD is by autopsy confirmation of amyloid plaques and tau tangles—however, autopsy is still the most accurate form of diagnosis because reliable biomarkers for common neurodegenerative copathologies, such as α-synuclein or non-AD tauopathies, are not yet available. It was recognized in the earliest immunoassays for amyloid and tau that levels of these proteins differed significantly in research participants diagnosed with dementia from cognitively normal participants [35, 36], and estimates of diagnostic validity in a meta-analysis of AD and control participants were reported at 78% specificity and 85% sensitivity across all included tests and study populations [37]. Further studies confirmed that levels of CSF Aβ42, tTau, and pTau181 accurately reflected amyloid and tau neuropathology as assessed at autopsy [1, 38]. The advent of amyloid PET neuroimaging spurred comparisons between CSF and amyloid brain pathology in living participant cohorts, where it was found that low CSF amyloid (i.e. low CSF Aβ42 or CSF Aβ42/Aβ40) corresponded almost completely with high amyloid PET ligand binding in the brain [3, 39]. These validations of links between CSF biomarkers, postmortem pathology, and in vivo PET imaging have firmly placed biomarkers of amyloid and tau as antemortem Gold Standard identifiers of the pathology driving incident cognitive impairment in individuals with AD. Adding to the value of these biomarkers is evidence that they can not only predict clinical progression [40], but also reliably detect the presence of pathology in cognitively normal individuals [41] (i.e., those with pre-symptomatic AD) and can increasingly differentiate AD from other types of dementia [42].

3.2 Evolution of Biomarker Usage Guidelines

The earliest established framework for the identification of AD was published in 1984 from the US National Institute of Neurological and Communicative Disorders and Stroke and the Alzheimer's Disease and Related Disorders Association (NINCDS-ADRDA) and was based on the cognitive symptomatology of AD rather than a pathologically anchored assessment of disease [43]. The International Working Group (IWG) for New Research Criteria for the Diagnosis of AD established recommendations in 2007 requiring the presence of subtle changes in episodic memory, called predementia or prodromal disease, alongside one or more abnormal biomarkers of amyloid, tau, or AD-related structural brain changes [44]. In 2011, a novel framework for AD identification was proposed by a joint group of researchers from the National Institute on Aging and the Alzheimer's Association (NIA-AA) [45]. This proposal retained and refined much of the core clinical and biomarker criteria identified between 1984 and 2010 [44, 46]. However, an integral part of these new criteria was the proposal—for research use only—of distinct disease stages, including preclinical AD, where cognitive signs or symptoms are absent but biomarker identification of pathological change is evident [47, 48]. Updates to the IWG and NIA-AA frameworks continued as it became broadly recognized that biomarkers of amyloid, tau, and other processes occurred in a temporal pattern even in cognitively normal individuals [49–51].

A notable update to the NIA-AA framework, again for research use only, was recommended in 2018, which focused solely on biomarkers of amyloid (A), tau (T), and neurodegeneration (N), where cognitive decline was proposed to assist in assessing the clinical severity of AD but not to define the presence of pathology [52]. Referred to as the AT(N) framework, this system is designed to describe AD as a purely biological construct rather than a syndromic one, with 6 stages encompassing normal AD biomarkers with normal cognition through abnormal biomarkers with severe dementia. This report emphasized that the framework was flexible with respect to the addition of new biomarkers; viewed AD as pathological change with additional cognitive decline across a temporal continuum; and finally, not yet ready for clinical use. However, because some individuals with frank amyloid and tau neuropathology remain or pass away cognitively intact despite decades of biomarker positivity [53, 54], the appropriate use of biomarkers as diagnostic implements, particularly in preclinical AD, in clinical settings versus research settings is widely debated [55–57]. Recent 2021 publications from both the NIA-AA group and the IWG agree on a key point: particularly in preclinical individuals, biomarkers should not be used to clinically define AD as a basis for establishing a treatment plan. The NIA-AA group applied the 2018 6-stage recommendations to the Mayo Clinic Study of Aging, a population-based study of aging in a geographically defined area of the US, and concluded that the staging of individuals varied based on which biomarkers and cognitive parameters were used and that the criteria may need further refinement before they are ready for use in clinical trials [58]. The IWG focused mainly on limitations in the usage of biomarkers for AD diagnosis, ultimately recommending that a clinical diagnosis be restricted to individuals with both abnormal biomarkers *and* evidence of cognitive decline, while cognitively normal individuals with abnormal biomarkers only be considered *at risk* for progression to AD rather than receiving a diagnosis of AD [59].

A revision of the 2018 recommendations from the Alzheimer's Association is now available, which aims to bridge the transition between biomarkers as research-use-only tools and biomarkers as clinically available diagnostic tests in a world where disease-modifying treatments are available [60]. Three major shifts in the field drove these revised guidelines: the advent of regulatory approval for treatments targeting AD pathology, advances in plasma biomarker performance, and recognition that biomarker modalities may not be interchangeable for diagnosis and disease staging. These new guidelines propose a pairing of biological and clinical definitions of AD as disease staging tools, which may help account for potential comorbid pathologies and/or cognitive reserve or resilience. Instead of the AT(N) framework, biomarker groupings are defined as "Core 1" indicative of amyloid pathology (including A, Aβ proteinopathy such as Aβ42, and T1, phosphorylated and secreted AD tau such as pTau181, pTau217, and pTau231), "Core 2" indicative of AD-associated tau pathology (T2, AD tau proteinopathy; in the CSF, largely novel Tau measures not specifically associated with amyloid pathology such as the microtubule binding region tau beginning at amino acid 243 [MTBR-tau243] and pTau205), and additional non-Core categories for "non-specific processes" such as neurodegeneration and neuroinflammation and "non-AD copathology" such as α-synuclein and vascular brain injury. Unlike previous guidelines, the presence of Core 1 biomarker positivity is enough to diagnose an individual with AD; in CSF, reduced Aβ42/40, elevated pTau181/Aβ42, or elevated tTau/Aβ42 are suggested. This recommendation in particular was hotly debated during public open comment periods for a number of reasons including: incomplete validation of biomarkers in diverse groups or individuals with potentially confounding comorbidities, exacerbation of medical inequality based on socioeconomic status, and emotional distress in individuals who are biomarker positive with no cognitive changes given the current requirement that cognitive decline must be present to receive approved disease-modifying therapies [61]. Additionally, the guidelines state that disease staging using fluid biomarkers is not equivalent to PET biomarkers, where the spatial uptake of Tau PET can be used to identify disease stages. The proposal is to use Core 1 CSF, plasma, or Amyloid PET positivity as Stage A, medial temporal Tau PET as Stage B, moderate neocortical Tau PET as Stage C, and high neocortical Tau PET as Stage D. Clinical staging is defined as Stage 0, no evidence of cognitive or biomarker change; Stage 1, biomarker evidence with no cognitive change; Stage 2, mild detectable cognitive change; Stage 3, cognitive impairment and early functional change; Stages 4–6 are dementia with mild, moderate, and severe functional impairment. See Table 1 for the proposed integrated staging scheme.

The final Alzheimer's Association Workgroup report illustrates how far the field of biomarker research in AD has come from the initial 2007 IWG recommendations regarding biomarker usage. Until a disease-modifying treatment is widely available, it seems likely that the clinical usage of AD biomarkers, particularly as a broadly used tool for AD detection, will remain a passionately contested topic. However, it is clear that the accuracy of many biomarkers is at or near the level of validation

Table 1 Proposed diagnostic groups using combined biological and clinical staging schemes. Note that CSF biomarkers are only included here as identifiers of amyloid pathology; all further biological staging is defined by spatial Tau PET uptake

Diagnostic group	Combined biomarker and clinical staging
Stage 0—healthy aging	No biomarker positivity, no cognitive change
Typical AD trajectory	Positive amyloid biomarker (A), Stage 1 clinical Medial temporal Tau PET (B), Stage 2 clinical Moderate neocortical Tau PET (C), Stage 3 clinical High neocortical Tau PET (D), Stage 4–6 clinical
AD with possible comorbid pathology (worse cognitive performance than biomarkers indicate)	Amyloid positive biomarker only (A), Stage 2–6 clinical Medial temporal Tau PET (B), Stage 3–6 clinical Moderate neocortical Tau PET (C), Stage 4–6 clinical
AD with possible cognitive reserve (better cognitive performance than biomarkers indicate)	Clinical Stage 1, biomarker groups B–D Clinical Stage 2, biomarker groups C and D Clinical Stage 3, biomarker group D

where transition into a real-world setting is increasingly possible. This is largely due to technical advances made in measuring biomarkers, which drives the need for updated guidelines and recommendations.

3.3 Development and Refinement of Biomarker Assays

Alongside evolving biomarker usage guidelines, the analytical capability to measure AD biomarkers has greatly expanded, in part due to (1) the increased use of biomarkers in research and clinical trial settings; (2) updated and refined assays; and (3) concerted worldwide efforts to harmonize fluid biomarkers of disease across cohorts and assay platforms. Biomarkers of AD, particularly Aβ42, are prone to a surplus of pre-analytical and analytical influences that, to this day, make the measurement of this peptide difficult to compare across assay platforms, studies, and laboratories [62]. Pre-analytical factors include time of day at sample collection, adsorption due to contact with tubing or non-polypropylene storage tubes, sample processing, handling, and storage, and more—though some factors like gravity drip vs. aspiration for CSF collection do not seem to have an effect on biomarker levels [63–67]. Analytical factors include assay platform, deviation from manufacturer protocols, and batch-to-batch variation within manufacturers [7, 8, 62, 68, 69].

In 2004 and 2008, laboratories worldwide participated in a multicenter comparison of assays for Aβ42, tTau, and pTau181 [70]. A set of three CSF pools was processed in a single laboratory to limit pre-analytical effects and sent to participating groups, and the coefficient of variation (%CV) was calculated from analyte values reported by each group. A total of eight laboratories participated in both years using the INNOTEST ELISA platform for each analyte, yet even with these provisions, between-laboratory sample CVs from 2004 and 2008 were 30% and 21% for Aβ42, 21% and 15% for tTau, and 13% and 9% for pTau181. Within laboratory CVs were similar at 25%, 18%, and 7% for Aβ42, tTau, and pTau181, respectively. Sources of variability were proposed to be manufacturer lot-to-lot and/or differences in assay protocol, level of pipetting automation, or other laboratory-specific factors [70]. Reports between 2010 and 2015 emphasized the need for standardization of sample processing and assay procedures in both CSF and blood, highlighting areas where improved documentation and standardization could aid in reduction of between-laboratory variability [71, 72]. Numerous individual laboratories and small groups assessed pre-analytical and analytical variation, reporting: statistical modeling to test the limits of biomarker variability on clinical diagnostic accuracy [73], novel assay platform (Elecsys and Lumipulse) variability [74, 75], in-depth assessment of between-laboratory pre-analytical procedures [76], and strengthened recommendations for unified sample handling protocols [77, 78].

Among these efforts, one of the largest and longest-standing has been the Alzheimer's Association Global Biomarker Standardization Consortium (GBSC) to aid in the improvement of measures within and between laboratories [79]. In the first year of the program, forty laboratories participated using INNOTEST ELISA, xMAP, or MesoScale assays for amyloid and tau, and pools of CSF were prepared in a single laboratory to limit pre-analytical confounds. Variation differed by assay platform, but overall, the between-laboratory CV for all tested analytes ranged from 13% to 36%, while well-established routine CSF measures, such as albumin, typically have between-laboratory CVs of 10% and 15% or less. The authors proposed the wide range of variances to result from a combination of laboratory-specific sample handling or analytic procedures, assay lot variability, assay manufacturer variability, and the lack of AD-biomarker-specific certified reference materials, meaning an "anchor" value to adjust the raw analyte values across platforms did not exist. A 2013 GBSC update reported between-laboratory CVs of 20–30% across 84 participating laboratories using INNOTEST, xMAP, and MesoScale platforms, with the majority of variability still explainable by between-laboratory effects such as analytic procedures or sample handling, but with a significant contribution from assay lot-to-lot variability [80]. The GBSC is continually collecting data, with 44 rounds of worldwide assay comparisons across now more than 100 laboratories [81].

4 Gold Standard Diagnostic Assays

4.1 Regulatory Approval

The United States FDA and National Institute of Health (i.e., the FDA-NIH Biomarker Working Group) maintains a resource for **B**iomarkers, **E**ndpoint**S**, and other **T**ools (BEST) that guides biomarker usage over a range of applications, from diagnosis to prediction of future change or disease risk, through drug monitoring or response [82]. In turn, assay manufacturers may undertake rigorous validation processes to apply for approval through one of many pathways to allow biomarker availability for clinical use: In vitro diagnostic device (IVD), companion diagnostic, or laboratory-developed test. Briefly, if an assay is being evaluated for use as a diagnostic biomarker it begins with a designation of Research Use Only—available to researchers or clinicians to provide additional context for biomarker development but not allowed for clinical use. IVD designation includes tools that are used for diagnosis or to guide patient treatment; however, these are not tied to a specific therapeutic product (e.g., a drug or medical device), while a companion diagnostic is an IVD deemed essential for the safe and effective use of a therapy. Biomarker assays may qualify for 510(k) approval, a premarket submission to demonstrate substantial equivalence to a similar, already legally marketed, test. Laboratory-developed tests are limited to measurement in a single laboratory with Clinical Laboratory Improvement Amendments (CLIA) certification using assays from a single source; typically, these tests are not as highly scrutinized as those obtaining IVD status [83].

A testament to the thorough investigation and ongoing standardization efforts noted above, three CSF biomarker tests now have sufficient validation to achieve IVD status in the USA. The first, approved in May 2022 for the Lumipulse® G assay platform, uses the Aβ42/Aβ40 ratio and is not intended for use as a stand-alone diagnostic but for confirmation of AD pathology in individuals 55 years of age or older with cognitive impairment [29]. The second and third approvals were the Roche Elecsys® pTau181/Aβ42 and tTau/Aβ42 ratios, intended to – again – be used in individuals 55 years of age or older with cognitive impairment [26]. The Athena Diagnostics ADmark Alzheimer's Evaluation includes apolipoprotein E (APOE) genotyping combined with CSF INNOTEST ELISA assays for Aβ42, tTau, and pTau181 and is a Laboratory Developed Test available to clinicians in the US [84]. In the European Union, a number of additional assays have been approved by the European Medicines Agency. All current approved assays require a combination of Aβ42, Aβ40, tTau, and pTau181, and validation of these assays has occurred over more than two decades. As each assay has a specific cutoff delineating a biomarker as positive or negative, so too does each laboratory, company, or collaborative group have specific and evolving means of determining what a positive biomarker means. Therefore, the ultimate standard of truth for AD diagnosis remains the postmortem neuropathological assessment.

4.2 CSF Amyloid and Tau Biomarkers

As the primary component of amyloid plaques, Aβ42 is one of the most widely studied CSF biomarkers in AD. This peptide is produced via sequential cleavage of the Amyloid Precursor Protein (APP), a transmembrane protein of – as yet – unknown physiological function [85]. Cleavage by α- and γ-secretase enzymes results in non-amyloidogenic processing of the less aggregation-prone Aβ40 peptide, while cleavage with the β- and γ-secretases results in amyloidogenic processing and the creation of Aβ42 [86]. During the course of AD, the balance of Aβ42 concentration in the brain shifts, though whether through increased amyloidogenic processing, decreased Aβ42 clearance, and/or other mechanisms is still a key point of study [85]. The end result is increasing deposition of Aβ42 into amyloid plaques in the brain and a concomitant, marked reduction in the CSF. By contrast, the Aβ40 peptide remains largely stable across the disease continuum and seems to be indicative of the overall production rate of amyloid peptides within a given individual [87, 88]. As a result, many AD biomarker investigators utilize the ratio of Aβ42/Aβ40 in place of Aβ42 concentration alone, which helps to balance differing rates of inter-individual amyloid production with the AD-driven reduction in Aβ42 [89]. This characteristic has become particularly pertinent as it also helps mitigate some effects of pre-analytical variability, such as differences in storage tube manufacturers, freeze thaws, or other sample handling variables [90].

The primary component of neurofibrillary tangles is microtubule-associated protein tau, a component of microtubule stability in neuronal axons. Tau plays a major role in a number of neurodegenerative diseases—a large variety of mutations, splicing alternatives, post-translational modifications, and even the location and specific shapes of tau aggregates can determine which tau-associated neurodegenerative disease an individual has [91]. Alternative splicing results in the roughly equal production of 6 different Tau isoforms, with differential inclusion of exons 2 and 3 (neither, only exon 2, or both) and, based on exon 10 splicing, with either three microtubule binding domains (3R) or four (4R). Post-translational modifications can include ubiquitination, acetylation, and phosphorylation at myriad sites—some of which are disease specific [92]. In AD, neurofibrillary tangles contain both 3R and 4R Tau, and the most studied aspect of the protein is phosphorylation. In particular, Tau phosphorylated at threonine 181, 231, or 217 are closely linked to AD pathogenic processes and are largely specific for AD compared with other neurodegenerative diseases [93]. In general, Tau protein measurements are less susceptible to pre-analytical variability compared to Aβ [90].

As central biomarkers for AD, measures of CSF amyloid and tau have been well validated across a spectrum of disease stages. Autopsy confirmation of AD as the standard of truth is the best test of CSF biomarker validity, though published reports are limited largely because research participants who both pass away and donate their brain for neuropathological assessment are limited within any given cohort. Early manuscripts often had either very small cohorts and, specifically with respect to CSF biomarkers, a large amount of heterogeneity in assays and pre-analytical handling as global efforts for assay standardization had not yet begun [37]. Even so, these studies reported a strong correlation between the levels of amyloid plaques in the neocortex and hippocampus at autopsy [1] and reduced levels of CSF Aβ42 alongside significant elevations in tTau and pTau181, which, when modeled together, produced an Area Under the Receiver Operating Characteristic Curve (AUC) of 0.95 when differentiating autopsy-confirmed AD versus non-AD and healthy controls [94]. Other autopsy cohorts revealed Aβ42 alone to be a highly sensitive biomarker with an AUC of 0.91 in AD vs. healthy controls, and when modeled via logistic regression with tTau and the *APOE* genotype, the AUC increased to 0.94 [95]. These CSF biomarkers also have added value in distinguishing AD from non-AD dementia pathology, even in cases where the clinical diagnosis was incorrect or where an additional neurodegenerative co-pathology was present [96]. A recent study of multiple biomarker modalities in an autopsy-confirmed cohort showed that CSF Aβ42 alone had a reasonable discriminatory value, AUC = 0.84, and when modeled with the *APOE* genotype and demographics including age at death and sex, the discriminatory value between AD and non-AD (either healthy controls or non-AD dementia) increased to AUC = 0.92, whereas the ratio of pTau181/Aβ42 had an AUC of 0.95, increasing to only AUC = 0.96 when modeled with *APOE* genotype and demographics [5]; however, the authors stress that replication in large, diverse, and independent cohorts of autopsy-confirmed diagnosis is still necessary.

Across the vast majority of CSF biomarker analyses, some combination of amyloid and tau biomarkers consistently provides the best discrimination between AD and either non-AD or healthy control groups. The ratio of CSF pTau/Aβ42 or tTau/Aβ42 (or vice versa) is a well-studied combination biomarker and in many cases is better at identifying AD than either biomarker alone, including in autopsy comparisons [5, 97], cognitive diagnosis [98, 99], and prediction of future cognitive decline [100, 101]. However, as the field moves forward with disease-modifying therapies, the identification of the earliest stages of disease will become even more important—a potential pitfall of Tau(s)/Aβ42 ratios is that they may miss the earliest stages of disease before the development of significant tau pathology [97]. Another commonly assessed ratio is Aβ42/Aβ40. Individuals produce varying amounts of amyloid peptides, thus using a raw biomarker value like Aβ42 concentration alone may result in misinterpretation of disease status in a person who simply produces very little amyloid [89], and while levels of CSF Aβ42 decline in AD, CSF Aβ40 remains relatively stable, which allows usage of the Aβ42/Aβ40 ratio as a control for inter-individual differences in total amyloid concentration and even pre-analytical sample handling differences [90]. A number of studies have shown that the Aβ42/Aβ40 ratio begins to change as early as 20 years before the onset of cognitive changes and is significantly lower in AD compared to healthy controls [48, 87, 102]. While Aβ42/Aβ40 may not necessarily provide superior diagnostic information to Aβ42 alone in discriminating AD from healthy groups, in individuals with intermediate or discrepant pTau181 values, the addition of the Aβ42/Aβ40 ratio improved diagnostic accuracy for AD vs. healthy controls [103, 104]. A recent working group convened to assess the value of CSF Aβ42/Aβ40 and concluded that it should always be analyzed regardless of whether an individual is considered Aβ42 positive or negative because it improves the percentage of appropriately diagnosed patients whether considering AD vs. healthy controls or AD vs. non-AD dementias [90] and may change even earlier than Amyloid PET in the disease process [105]. With respect to disease staging, once individuals are identified as amyloid positive, the ratio of Aβ42/Aβ40 does not provide additional value in discriminating preclinical, early, and late AD [106]. This is the premise of classification systems such as AT(N), where change in amyloid is then followed by changes in biomarkers of tau accumulation and neurodegeneration [102, 107].

Tau accumulation in AD is currently best identified via phosphorylated species of Tau, in particular pTau181, which is preferentially elevated in AD and not in non-AD dementias [108]. Though changes in CSF tau measures occur after changes in amyloid, elevations still begin during the preclinical stage of disease and can predict future cognitive decline [109–111]. Similarly, CSF pTau181 and tTau can be used together with CSF amyloid to accurately predict future cognitive decline in mildly impaired individuals and even differentiate cohorts with stable mild cognitive impairment (MCI) from those who transition from MCI to AD dementia [112, 113]. Unlike tTau, pTau181 is considered disease-specific for AD as elevations of tTau occur after acute stroke and in some non-AD dementias where pTau181 remains stable [114, 115].

Based on more than two decades of strong evidence, CSF Aβ42, Aβ40, pTau181, and tTau are rightly considered the Gold Standard fluid biomarkers for AD. They are not, however, the only measures of amyloid and tau to be investigated. Tau phosphorylated at residues other than 181, particularly pTau217, has recently shown promise as an AD diagnostic biomarker, perhaps with even higher accuracy than pTau181 in differentiating AD from non-AD dementias [116, 117]. Staging of disease from a single timepoint will likely become a necessity to enroll the best candidates for clinical trials as more disease-modifying therapeutics are studied, or allowing patients to receive a drug that will have the greatest impact on their disease trajectory—and it is probable that emerging measures of AD-related tau modifications may provide this capability. A recent study modeled 5 stages of CSF biomarker classification defined by changes in Aβ42/Aβ40, pTau217, pTau205, MTBR-tau243 and tTau. In two independent cohorts, the authors found that CSF Aβ42/Aβ40 was the earliest biomarker to change, followed by pTau217 and pTau205, which were associated with abnormal amyloid PET, MTBR-tau243 with tau PET and cognitive impairment, and tTau with magnetic resonance imaging (MRI) measures of neurodegeneration [118]. Further studies are needed, to validate this model, but alongside PET, MRI, and cognitive measures, it is a step forward in AD staging based on CSF. A number of other studies indicate that pTau217, pTau205, and/or pTau231 reflect AD-related change better than pTau181 and, perhaps most importantly, perform equally well in plasma as they do in CSF, which increases the potential for wide-spread usage [116, 119, 120]. Each pTau measurement may also be related to different aspects of AD—for instance, CSF pTau205 may be more closely related to tau neurofibrillary tangles, while pTau217 and pTau231 may be more closely related to amyloid pathology, further increasing their potential as disease staging biomarkers [121]. Forms of MTBR-tau may also aid in differential diagnosis between primary tauopathies, including AD from, for example, frontotemporal dementia subtypes (FTD), based on which amino acid residue MTBR-tau begins [122, 123].

To a lesser extent, other isoforms of Aβ or Aβ-associated proteins have been analyzed as well. Early studies indicated that the Aβ38 isoform was elevated in AD CSF [124], though a recent meta-analysis of eight studies including Aβ38 concentrations did not show differences between AD and control groups [113]. Recent studies are similarly inconclusive, one showing that high Aβ38 levels were associated with slower decline in cognitive measures and a lower risk of conversion to AD dementia [125] and another showing that high Aβ38 levels were associated with elevated pTau and tTau, though cognition was not assessed [126]. There may be benefit in profiling a panel of Aβ isoforms for differential diagnosis between AD and cerebral amyloid angiopathy (CAA), as one study suggests peptides shorter than Aβ42 were decreased in CAA compared with AD-like individuals [127], though another suggested that adding Aβ38 and Aβ43 to the Gold Standard AD biomarkers did not improve differential diagnosis between AD and CAA [128]. Unlike the expansive exploration of different forms of tau in AD, investigation of Aβ38 and other isoforms is progressing at a much slower pace and – to date – has not shown incredible promise as biomarkers of AD, though perhaps Aβ isoforms are merely understudied.

Other less-studied measures are Aβ-adjacent, such as concentrations of β-site amyloid precursor protein cleaving enzyme 1 (BACE1), or sAPP-α and -β. Both sAPPα and sAPPβ are soluble cleavage products of APP, the precursor protein of Aβ. Meta-analysis of nine and ten cohorts, respectively, showed no difference in either cleavage product between AD and control groups [113]. The enzyme BACE1 is the primary β-secretase responsible for Aβ42 production, and BACE1 activity or concentration can be measured in the CSF. It is also considered an attractive drug target for AD but, thus far, has not shown efficacy in clinical trials [129]. In a small study of AD and non-AD controls, BACE1 activity was higher in the AD group, but the authors concluded that the complexity of BACE1 species in CSF, including truncated forms, should be accounted for when assessing activity [130] and a study of BACE1 in amyloid-positive individuals subdivided by CSF pTau181 quartiles showed significantly lower BACE1 at the lowest quartile of pTau181 and significantly higher levels in each subsequent quartile compared with controls [131]. Another group assessed BACE1 concentration across AT(N) profiles and found that levels remained stable across all stages [132]. At present, the utility of BACE1 as a biomarker for AD may be limited, but further study is warranted.

With clear relationships to AD pathology at autopsy, diagnosis, differential diagnosis, and disease staging, it is no surprise that measures of Aβ42, Aβ40, tTau, and pTau181 are considered the Gold Standard in fluid biomarkers for AD. Perhaps pTau217 or other Tau phospho-sites may be considered burgeoning Gold Standard markers given the immense additions they have made in diagnosis, prognosis, and staging. There is still more work to be done, though—assessment of these biomarkers in diverse cohorts, real-world cohorts, and in populations world-wide where they are yet to be extensively assessed (such as across Africa) is necessary to ensure the best quality of diagnosis and prognosis is equitably available to all.

5 Novel and Emerging CSF Biomarkers

5.1 Importance of Novel Biomarkers

There is much more to AD than just amyloid and tau aggregation—which means there are many possibilities for biomarkers of disease-adjacent processes. Among the most widely studied are biomarkers of neurodegeneration, neuroinflammation, and synaptic function—all processes that play major roles in the course of AD. With lessons learned from the development of amyloid and tau fluid biomarkers, the addition of new biomarkers in AD has become a more streamlined, targeted process, meaning some of these novel or emerging biomarkers already have significant published evidence of their utility in AD.

5.2 Neurodegeneration Biomarkers

Neurofilaments are a class of proteins found primarily in myelinated axons, where they play a crucial role in axon stability and are released in high volumes during axonal injury or degradation. As early as 1996, the light subunit of neurofilament (neurofilament light, or NfL) was identified as a biomarker of neuronal degeneration: in the CSF of individuals with AD, amyotrophic lateral sclerosis (ALS), and "miscellaneous" neurodegenerative diseases, NfL was significantly increased compared with controls [133]. In fact, NfL is likely better classified as a highly validated fluid biomarker in AD rather than "novel" or "emerging." Hundreds of published reports have since followed the 1996 report, cementing NfL as a widely-used biomarker across myriad neurodegenerative diseases, particularly elevated in ALS, FTD, and atypical parkinsonian disorders [134] and as such is currently at a high level of validation and considered close to clinical implementation as a diagnostic, differential, or prognostic biomarker. For instance, in healthy individuals, it is recognized that levels of CSF NfL increase significantly with age [135] and efforts are already underway to identify pre-analytical or analytical variables that may interfere with routine clinical use [136]. In AD, a meta-analysis confirmed elevated NfL in MCI, and even preclinical AD, compared with healthy control groups without biomarker evidence of AD pathology [137]. In disease staging studies, CSF NfL performs highly when included with Gold Standard AD biomarkers and aids in differentiating AD from controls across all disease stages [109, 138]. More granular studies have been done as well, which is where non-Gold Standard biomarkers may truly have a chance to shine, and CSF NfL has been shown to: predict cortical amyloid load and brain atrophy [139], predict decline in attention but not episodic memory in preclinical disease [140], associate negatively with cortical thickness during preclinical disease in some of the earliest brain areas affected by tau pathology [141] and change as early as 9 years before the onset of cognitive symptoms [142]. Studies have also shown promise in CSF NfL as a differential diagnosis biomarker in AD vs. FTD [143, 144], and as an exclusionary diagnostic when combined with "AD Gold Standard" markers to rule out Creutzfeldt-Jakob disease (CJD) in suspected cases [145]. Excitingly, plasma NfL correlates closely with CSF NfL and may be an easily accessible, minimally invasive biomarker with wide use in screening for AD, including preclinically [146–148].

Visinin-like protein-1 (VILIP-1) is a neuronal calcium sensor protein that modulates intracellular calcium-mediated signaling and can be located in either the pre- or post-synaptic compartments, but is mostly localized in the neuronal perikarya and proximal dendrites. Due to its abundant expression in the neuronal perikarya, VILIP-1 is generally considered a marker of neuronal injury [149], and has demonstrated utility as a diagnostic and prognostic marker in large longitudinal cohorts of AD and controls who were followed over time. CSF VILIP-1 levels are elevated in even the earliest preclinical stages of AD, reliably predict brain amyloid positivity on PET, and predict future cognitive decline, potentially to a better extent than CSF tTau, pTau181, and Aβ42 [149]. Further, the VILIP-1/Aβ42 ratio was a stronger predictor of future cognitive decline in cognitively normal individuals than the tau/Aβ42 or p-tau181/Aβ42 ratios. Importantly, combining CSF VILIP-1 with CSF tTau, pTau181, and Aβ42 improved the collective ability of these markers to predict future cognitive decline in cognitively normal individuals [149]. Preclinically, elevations in VILIP-1 are also associated with future risk of AD in cognitively normal *APOE4* carriers [87, 150]. In symptomatic disease, VILIP-1 is also elevated compared to healthy controls [151–153] and scales with dementia severity, with higher baseline levels predicting higher rates of future global and domain-specific cognitive decline and whole brain or regional atrophy in AD-vulnerable regions such as the hippocampus and entorhinal cortex, to a potentially better extent than CSF tTau, pTau181, and Aβ42 [154, 155]. Data regarding the utility of VILIP-1 in the differential diagnosis of AD is currently limited but suggests a relative specificity to AD-related neurodegeneration. In AD compared with Dementia with Lewy bodies (DLB) [156, 157], CSF concentrations are higher in AD and similarly may be higher in AD compared with FTD and Parkinson's disease (PD) [149]. In CJD, CSF levels of VILIP-1 are significantly higher, likely reflecting the significant and rapid neuronal loss associated with this rapidly progressive dementia [158]. Translation from the CSF to blood has only been published in two studies: in one study, serum VILIP-1, while significantly elevated in AD and CJD vs. controls, had an overall performance that was relatively inferior to CSF VILIP-1 values [149, 158]. An earlier study had also shown that plasma VILIP-1 levels, measured using the single-molecule counting technology Erenna (Singulex, USA), were increased in early symptomatic AD compared to controls, although the group differences were smaller than those observed in the CSF [149].

Ubiquitin C-terminal hydrolase L1 (UCHL1) is a novel biomarker that can be considered indicative of neurodegeneration and, similar to NfL, is located primarily in neurons and axons in the central nervous system. This protein is critical for the degradation of unnecessary, misfolded, or damaged proteins but also associated with tau and is important for microtubule stabilization, and in AD brain tissue, UCHL1 is significantly reduced compared to healthy tissue [159]. Particularly in CSF, UCHL1 is not a highly studied biomarker, but it is significantly elevated in AD

CSF and does differentiate between AD and control groups [160]. Another study found elevated levels of UCHL1 in AD CSF compared to controls, even during preclinical disease as defined by positive AT(N) biomarkers in the absence of cognitive change [161], though when assessed in conjunction with NfL, it did not provide additional diagnostic value [109]. Fatty Acid Binding Proteins (FABP) are also promising biomarkers of AD and may be considered neurodegenerative biomarkers due to significant correlations with tau [162], though they may be more accurately classified as metabolic biomarkers, rather than neurodegeneration markers, as their main function in the brain is regulation of the lipid composition of cellular membranes. Similar to UCHL1, FABPs are elevated in AD as early as preclinical disease [161, 163]. A combination of elevated heart-FABP and pTau showed promise as a differential diagnosis biomarker between AD and DLB, and elevated heart-FABP alone was discriminatory between AD, PD, and other neurodegenerative diseases [164]. Additionally, in individuals with low CSF Aβ42, elevated heart-FABP was associated with MRI measures of brain atrophy in AD-specific regions, regardless of tau status [165].

5.3 Neuroinflammatory Biomarkers

A number of promising biomarkers indicate processes of neuroinflammation over the course of disease. The most highly validated – to date – is glial fibrillary acidic protein (GFAP), which is likely an indicator of reactive astrocytes. Surprisingly, despite being relatively little studied until around 2010, GFAP was identified as a CSF biomarker of inflammation in AD and other neurodegenerative processes in 1986 [166]. Similar to NfL, GFAP concentrations in CSF are reported to increase with aging regardless of AD status [167] and initial investigations into the pre-analytical effects on CSF and plasma GFAP concentrations have begun [168]. In cognitively intact amyloid positive-individuals, GFAP was elevated in comparison to amyloid-negative individuals [169], correlated with the Preclinical Alzheimer's Cognitive Composite, a composite score of subtle preclinical cognitive changes [138], was positively associated with amyloid PET load but mediated by CSF pTau levels [170], and may change up to 12–14 years before development of cognitive symptoms [102]. Symptomatic AD studies are similar: GFAP is elevated in symptomatic AD compared to healthy controls [171–173] and correlates with dementia severity [174]. With respect to differential dementia diagnosis, one study reported that GFAP is significantly more elevated in FTD than both AD and DLB, but with no difference between AD and DLB [171] and another reported significant elevations in AD compared to normal controls and CJD [175]. Perhaps most interesting, though, is that blood-based GFAP may be a better measure of AD-related processes than CSF GFAP. Multiple studies have reported greater changes in magnitude compared to CSF [169, 176], elevations earlier in the disease process compared to CSF [109], and even apparent resilience to an important pre-analytical variable: number of freeze-thaw cycles [168]. Many AD biomarkers perform better in CSF than plasma, though a growing number like pTau217 and NfL are equivalently valuable in both fluid types, which makes GFAP an oddity in need of more investigation [177].

Another AD-associated neuroinflammatory protein is chitinase 3-like protein 1 (YKL-40), which appears to play a role in tissue remodeling during inflammatory processes and is upregulated in myriad inflammatory conditions such as Crohn's disease, some cancers, and neurodegenerative disease [178]. In CSF, elevations in YKL-40 appear to occur with increasing age regardless of disease status [87] and the effects of pre-analytical sample handling are minimal [179]. In preclinical disease, elevations in CSF YKL-40 were associated with risk of developing cognitive impairment [180, 181] and negatively correlated with MRI cortical thickness in normal and MCI participants with positive CSF Aβ42 [182], though these results are not always replicated—one study showed significant elevations in prodromal, but not preclinical, AD compared with normal controls [183]. In MCI and AD dementia, compared with cognitively normal groups, YKL-40 is consistently shown to be elevated [151, 184, 185]. In a number of other types of dementia, YKL-40 is also elevated, which may limit its potential as a differential biomarker. One study found no difference in YKL-40 between vascular dementia (VaD) and AD [186] while another found increased YKL-40 in AD and CJD but not VaD or DLB [187] and other groups reported increases in AD, but not DLB or PD [188], and in both AD and FTD [182], compared with cognitively normal controls. Notably, YKL-40 is at an earlier stage of development compared to other biomarkers discussed here and is assessed in a number of relatively small cohorts, which may help explain the comparative lack of concordance discussed above.

5.4 Synaptic Integrity Biomarkers

The importance of synaptic integrity cannot be overstated as a necessity for a well-functioning nervous system. In AD, even more than amyloid and tau, loss of synapses correlates closely with the level of cognitive impairment [189–191]. The usage of synaptic proteins as fluid biomarkers of AD may provide additional granular measures of cognitive decline across the disease continuum, making them prime candidates for disease staging, monitoring of treatment response in clinical trials, or prognostic indicators of the rate of future cognitive decline [192].

Neurogranin (Ng) is the most highly investigated CSF synaptic biomarker and the only post-synaptic biomarker to date. It is a calmodulin-binding protein enriched in dendritic spines that plays an important role in long-term potentiation. In cognitively normal individuals, Ng is elevated in amyloid positive versus amyloid negative groups [170, 193, 194]. Early disease stages are also characterized by elevated Ng in biomarker-positive MCI compared with normal control groups [194–196] and these early elevations correlate with future cognitive decline [195, 197]. Studies

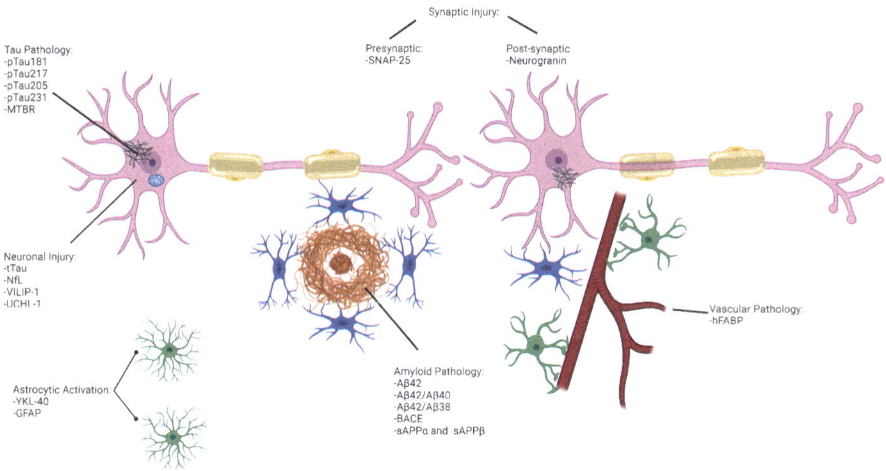

Fig. 1 Most highly validated biomarkers in the CSF indicating different pathologies and/or disease processes involved in AD. Pathologies include amyloid plaques and tau tangles; processes include neuronal injury, inflammation, synaptic injury, and vascular injury. Abbreviations: Aβ amyloid-beta peptides, BACE1 β-secretase, GFAP glial fibrillary acidic protein, hFABP heart fatty acid binding protein, MTBR tau microtubule binding region, NfL neurofilament light chain, Ng neurogranin, pTau hyperphosphorylated tau, sAPP soluble amyloid precursor protein fragment, SNAP-25 synaptosomal-associated protein-25, tTau total tau, UCHL-1 ubiquitin C-terminal hydrolase L1, VILIP-1 visinin-like protein-1, YKL-40 chitinase 3-like protein 1. (Adapted with permission from Tarawneh R. Biomarkers: Our Path Towards a Cure for Alzheimer Disease. Biomarker Insights. 2020;15 [215]. Copyright © 2020 by SAGE Publications Ltd. Reprinted by Permission of Sage Publications; permission obtained through the Copyright Clearance Center, Inc). Created with Biorender.com

have also shown that CSF Ng is closely associated with whole brain and regional atrophy in AD-vulnerable regions [194]. Interestingly, Ng appears to be an AD-specific CSF biomarker. In a study of FTD, DLB, ALS, and five other neurodegenerative diseases, elevations in Ng were specific to AD [197] and another study found elevations to be specific to AD in comparison to some FTD subtypes, Lewy body dementias (LBD), PD, and three other non-AD cognitive impairment groups [198].

Pre-synaptic protein biomarkers of AD with reasonable experimental evidence include synaptosomal-associated protein 25 (SNAP-25) and synaptotagmin-1. Part of the pre-synaptic SNARE protein complex, SNAP-25 is an essential component of the cellular machinery that mediates the release of synaptic vesicles containing neurotransmitters. In AD, CSF SNAP-25 is elevated in preclinical AD [102, 109], and clinical stages of AD [199, 200] compared with cognitively normal controls. Differential diagnosis is also aided by CSF SNAP-25 evaluation: in AD vs. FTD [201], atypical AD vs. FTD [202], and AD vs. ALS, PD, and FTD but not AD vs. CJD [203]. Interestingly, in cognitively normal *APOE4* carriers who are considered at risk for development of AD, SNAP-25 is elevated, suggesting potential presynaptic disruption may be a very early disease process [204]. Synaptotagmin-1 (Syt-1) is a pre-synaptic calcium-sensor protein necessary for the release of neurotransmitters. In AD, evidence indicates CSF Syt-1 is elevated in MCI and AD compared to healthy controls [205, 206] and it may have utility as a differential biomarker for FTD and AD [202]. The translation of these pre-synaptic markers from CSF to plasma is still in relative infancy, with no published data for Syt-1 and a single for SNAP-25, where the SNAP-25 assay lacked analytical sensitivity [207].

Tracking these AD-adjacent processes of neurodegeneration, neuroinflammation, and synaptic health through all stages of AD has the potential to improve diagnostic and prognostic potential above and beyond that offered by amyloid and tau biomarkers. Well-validated non-Gold Standard biomarkers also provide excellent targets for monitoring target engagement or slowing of decline in AD clinical trials. Eventually, it may be possible to assess a panel of biomarkers that not only would allow diagnosis and prognosis but also differentiate AD from other dementias in clinical settings and potentially guide treatment options based on an individual's disease stage. Figure 1 demonstrates the main emerging fluid markers in AD and the different pathologies they measure.

6 Conclusions and Future Directions

Within the last 15 years, assessment of AD-related biomarkers has grown dramatically. The identification of pre-analytical variables that affect measured levels of analytes was an incredibly important step forward, and, as protocols and assay adjustments using certified reference materials become more widely adopted, it is likely that the array of already robust assays that are used in research settings will transition into routine clinical use, especially as disease-modifying therapies become more widely available. Assays that are less robust will benefit from similar improvements and may begin, when analyzed in conjunction with Gold Standard markers, to offer highly granular information about disease status within a given individual instead of at group levels.

Even within the Gold Standard biomarkers, though, there is much work to be done. The influence of factors like sex and gender, race, and socio- and economic disadvantage may have significant effects on biomarker measures of AD processes and, in turn, may result in reduced accuracy of diagnosis, prognosis, or disease staging in diverse groups. For instance, studies in Europe and Asia have reported sex-specific differences in disease prevalence, while studies from North and South America typically do not [208]. Whether these world-wide differences can be attributed to differences in biology, sex-specific comorbidities, or sociocultural constructs that lead to gender inequality is yet to be elucidated [209]. A recent systematic review of sex differences in relation to progression of MCI to dementia noted that *APOE4* status and depression were associated with an increased risk of progression in women, but changes in MRI and CSF markers were associated with an increased risk in men—though the authors noted that due to heterogeneity and underpowered statistical analyses, much more work is needed to parse out differences in AD conversion risk and sex [210]. Studies on neighborhood disadvantage have consistently shown connections between the level of disadvantage and AD. In one autopsy cohort, neighborhood disadvantage was significantly associated with an increased odds of having AD neuropathology [211]. Similarly, in older veterans, greater disadvantage was associated with elevated AD risk over 11 years even after adjusting for factors such as sex, race, and comorbid medical conditions [212]. Studies of diverse racial and ethnic groups highlight the need for characterization in larger cohorts. One study of plasma Aβ42/Aβ40 in Black and White research participants reported that Black individuals had on average higher levels of Aβ42/Aβ40, indicating less AD neuropathology, but longitudinal change between the two groups was similar, suggesting that accumulation of amyloid pathology was similar [213]. In another study of Black and White individuals with MCI, pTau181 and NfL in the CSF were consistently lower in Black participants, and differences could potentially exacerbate the under-representation of Black individuals in AD clinical trials [214]. Studies able to robustly assess a combination of sex, disadvantage, and race combined in relation to CSF biomarkers are difficult – but exceedingly necessary – to ensure equitable access to clinical trials and approved disease-modifying therapeutics for all.

Fortunately, the increase in collaborative world-wide research efforts and highly characterized Gold Standard and emerging CSF biomarkers of AD are combining to allow such research to flourish. The landscape of fluid biomarkers in AD is certainly rapidly changing. The next decade of improvements in assays, sample handling, data analysis, and expansion of biomarker assessments in ever more diverse groups is set to be just as fast-paced and informative as the last.

Funding This work was supported by the National Institute of Health- National Institute of Aging Grant RF1AG083744 (RT).

References

1. Strozyk D, Blennow K, White LR, Launer LJ. CSF Abeta 42 levels correlate with amyloid-neuropathology in a population-based autopsy study. Neurology. 2003;60(4):652–6. https://doi.org/10.1212/01.wnl.0000046581.81650.d0.
2. Tapiola T, Alafuzoff I, Herukka SK, et al. Cerebrospinal fluid {beta}-amyloid 42 and tau proteins as biomarkers of Alzheimer-type pathologic changes in the brain. Arch Neurol. 2009;66(3):382–9. https://doi.org/10.1001/archneurol.2008.596.
3. Fagan AM, Mintun MA, Mach RH, et al. Inverse relation between in vivo amyloid imaging load and cerebrospinal fluid Abeta42 in humans. Ann Neurol. 2006;59(3):512–9. https://doi.org/10.1002/ana.20730.
4. Forsberg A, Engler H, Almkvist O, et al. PET imaging of amyloid deposition in patients with mild cognitive impairment. Neurobiol Aging. 2008;29(10):1456–65. https://doi.org/10.1016/j.neurobiolaging.2007.03.029.
5. Wang ZB, Tan L, Gao PY, et al. Associations of the A/T/N profiles in PET, CSF, and plasma biomarkers with Alzheimer's disease neuropathology at autopsy. Alzheimers Dement

J Alzheimers Assoc. Published online July 28. 2023;19:4421. https://doi.org/10.1002/alz.13413.

6. Zimmermann R, Lelental N, Ganslandt O, Maler JM, Kornhuber J, Lewczuk P. Preanalytical sample handling and sample stability testing for the neurochemical dementia diagnostics. J Alzheimers Dis. 2011;25(4):739–45. https://doi.org/10.3233/JAD-2011-110212.
7. Se S, Cl S, CT, et al. Upward drift in cerebrospinal fluid amyloid β 42 assay values for more than 10 years. Alzheimers Dement J Alzheimers Assoc. 2018;14(1):62. https://doi.org/10.1016/j.jalz.2017.06.2264.
8. Simonsen AH, Musaeus CS, Christensen GL, Hasselbalch SG, Waldemar G. Upwards drift of cerebrospinal fluid amyloid-β 42 over twelve years in a consecutive clinical cohort. J Alzheimers Dis JAD. 2021;81(4):1369–73. https://doi.org/10.3233/JAD-201506.
9. Mehta PD, Pirttilä T, Mehta SP, Sersen EA, Aisen PS, Wisniewski HM. Plasma and cerebrospinal fluid levels of amyloid β proteins 1-40 and 1-42 in Alzheimer disease. Arch Neurol. 2000;57(1):100–5. https://doi.org/10.1001/archneur.57.1.100.
10. Vanderstichele H, Kerschaver EV, Hesse C, et al. Standardization of measurement of β-amyloid(1-42) in cerebrospinal fluid and plasma. Amyloid. 2000;7(4):245–58. https://doi.org/10.3109/13506120009146438.
11. Mehta PD, Pirttila T, Patrick BA, Barshatzky M, Mehta SP. Amyloid β protein 1–40 and 1–42 levels in matched cerebrospinal fluid and plasma from patients with Alzheimer disease. Neurosci Lett. 2001;304(1):102–6. https://doi.org/10.1016/S0304-3940(01)01754-2.
12. Freeman SH, Raju S, Hyman BT, Frosch MP, Irizarry MC. Plasma Aβ levels do not reflect brain Aβ levels. J Neuropathol Exp Neurol. 2007;66(4):264–71. https://doi.org/10.1097/NEN.0b013e31803d3ae4.
13. Mayeux R, Honig LS, Tang MX, et al. Plasma Aβ40 and Aβ42 and Alzheimer's disease: relation to age, mortality, and risk. Neurology. 2003;61(9):1185–90. https://doi.org/10.1212/01.WNL.0000091890.32140.8F.
14. Pesaresi M, Lovati C, Bertora P, et al. Plasma levels of beta-amyloid (1–42) in Alzheimer's disease and mild cognitive impairment. Neurobiol Aging. 2006;27(6):904–5. https://doi.org/10.1016/j.neurobiolaging.2006.03.004.
15. Kang JH, Vanderstichele H, Trojanowski JQ, Shaw LM. Simultaneous analysis of cerebrospinal fluid biomarkers using microsphere-based xMAP multiplex technology for early detection of Alzheimer's disease. Methods. 2012;56(4):484–93. https://doi.org/10.1016/j.ymeth.2012.03.023.
16. Le Bastard N, Coart E, Vanderstichele H, Vanmechelen E, Martin JJ, Engelborghs S. Comparison of two analytical platforms for the clinical qualification of Alzheimer's disease biomarkers in pathologically-confirmed dementia. J Alzheimers Dis JAD. 2013;33(1):117–31. https://doi.org/10.3233/JAD-2012-121246.
17. Fagan AM, Shaw LM, Xiong C, et al. Comparison of analytical platforms for cerebrospinal fluid measures of β-amyloid 1-42, total tau, and P-tau181 for identifying Alzheimer disease amyloid plaque pathology. Arch Neurol. 2011;68(9):1137–44. https://doi.org/10.1001/archneurol.2011.105.
18. Wang LS, Leung YY, Chang SK, et al. Comparison of xMAP and ELISA assays for detecting cerebrospinal fluid biomarkers of Alzheimer's disease. J Alzheimers Dis. 2012;31(2):439–45. https://doi.org/10.3233/JAD-2012-120082.
19. Hansson O, Zetterberg H, Vanmechelen E, et al. Evaluation of plasma Aβ40 and Aβ42 as predictors of conversion to Alzheimer's disease in patients with mild cognitive impairment. Neurobiol Aging. 2010;31(3):357–67. https://doi.org/10.1016/j.neurobiolaging.2008.03.027.
20. Figurski MJ, Waligórska T, Toledo J, et al. Improved protocol for measurement of plasma β-amyloid in longitudinal evaluation of Alzheimer's Disease Neuroimaging Initiative study patients. Alzheimers Dement J Alzheimers Assoc. 2012;8(4):250–60. https://doi.org/10.1016/j.jalz.2012.01.001.
21. Song F, Poljak A, Valenzuela M, Mayeux R, Smythe GA, Sachdev PS. Meta-analysis of plasma amyloid-β levels in Alzheimer's disease. J Alzheimers Dis. 2011;26(2):365–75. https://doi.org/10.3233/JAD-2011-101977.

22. Kollhoff AL, Howell JC, Hu WT. Automation vs. experience: measuring Alzheimer's beta-amyloid 1–42 peptide in the CSF. Front Aging Neurosci. 2018;10:253. https://doi.org/10.3389/fnagi.2018.00253.
23. Rissin DM, Kan CW, Campbell TG, et al. Single-molecule enzyme-linked immunosorbent assay detects serum proteins at subfemtomolar concentrations. Nat Biotechnol. 2010;28(6):595–9. https://doi.org/10.1038/nbt.1641.
24. Palmqvist S, Tideman P, Cullen N, et al. Prediction of future Alzheimer's disease dementia using plasma phospho-tau combined with other accessible measures. Nat Med. 2021;27(6):1034–42. https://doi.org/10.1038/s41591-021-01348-z.
25. Sajjad MU, Blennow K, Knapskog AB, et al. Cerebrospinal fluid levels of interleukin-8 in delirium, dementia, and cognitively healthy patients. J Alzheimers Dis JAD. 2020;73(4):1363–72. https://doi.org/10.3233/JAD-190941.
26. Roche Alzheimer's disease Cerebrospinal Fluid (CSF) assays receive FDA clearance, supporting more accurate and timely diagnosis. Accessed June 26, 2023. https://diagnostics.roche.com/us/en/news-listing/2022/roche-alzheimers-disease-cerebrospinal-fluid-assays-receive-fda-clearance.html
27. Zecca C, Brescia V, Piccininni M, et al. Comparative evaluation of two immunoassays for cerebrospinal fluid β-amyloid1–42 measurement. Clin Chim Acta. 2019;493:107–11. https://doi.org/10.1016/j.cca.2019.02.033.
28. Wilson EN, Young CB, Ramos Benitez J, et al. Performance of a fully-automated Lumipulse plasma phospho-tau181 assay for Alzheimer's disease. Alzheimers Res Ther. 2022;14(1):172. https://doi.org/10.1186/s13195-022-01116-2.
29. Lumipulse® G β-Amyloid Ratio (1–42/1–40) in vitro diagnostic test receives FDA marketing authorization for the assessment of Alzheimer's disease in the United States | Fujirebio. Accessed June 26, 2023. https://www.fujirebio.com/en/news-events/lumipulser-g-bamyloid-ratio-142140-in-vitro-diagnostic-test-receives-fda-marketing
30. Korecka M, Shaw LM. Mass spectrometry-based methods for robust measurement of Alzheimer's disease biomarkers in biological fluids. J Neurochem. 2021;159(2):211–33. https://doi.org/10.1111/jnc.15465.
31. Bros P, Delatour V, Vialaret J, et al. Quantitative detection of amyloid-β peptides by mass spectrometry: state of the art and clinical applications. Clin Chem Lab Med CCLM. 2015;53(10):1483–93. https://doi.org/10.1515/cclm-2014-1048.
32. Nakamura A, Kaneko N, Villemagne VL, et al. High performance plasma amyloid-β biomarkers for Alzheimer's disease. Nature. 2018;554(7691):249–54. https://doi.org/10.1038/nature25456.
33. The use of selected reaction monitoring in quantitative proteomics | Bioanalysis. Accessed 30 Sept 2023. https://www.future-science.com/doi/10.4155/bio.12.126
34. Boulo S, Kuhlmann J, Andreasson U, et al. First amyloid β1-42 certified reference material for re-calibrating commercial immunoassays. Alzheimers Dement. 2020;16(11):1493–503. https://doi.org/10.1002/alz.12145.
35. Motter R, Vigo-Pelfrey C, Kholodenko D, et al. Reduction of beta-amyloid peptide42 in the cerebrospinal fluid of patients with Alzheimer's disease. Ann Neurol. 1995;38(4):643–8. https://doi.org/10.1002/ana.410380413.
36. Blennow K, Wallin A, Agren H, Spenger C, Siegfried J, Vanmechelen E. Tau protein in cerebrospinal fluid: a biochemical marker for axonal degeneration in Alzheimer disease? Mol Chem Neuropathol. 1995;26(3):231–45. https://doi.org/10.1007/BF02815140.
37. Cure S, Abrams K, Belger M, dell'agnello G, Happich M. Systematic literature review and meta-analysis of diagnostic test accuracy in Alzheimer's disease and other dementia using autopsy as standard of truth. J Alzheimers Dis. 2014;42(1):169–82. https://doi.org/10.3233/JAD-131559.
38. Blennow K, Hampel H, Weiner M, Zetterberg H. Cerebrospinal fluid and plasma biomarkers in Alzheimer disease. Nat Rev Neurol. 2010;6(3):131–44. https://doi.org/10.1038/nrneurol.2010.4.

39. Blennow K, Mattsson N, Schöll M, Hansson O, Zetterberg H. Amyloid biomarkers in Alzheimer's disease. Trends Pharmacol Sci. 2015;36(5):297–309. https://doi.org/10.1016/j.tips.2015.03.002.
40. Hansson O, Seibyl J, Stomrud E, et al. CSF biomarkers of Alzheimer's disease concord with amyloid-β PET and predict clinical progression: A study of fully automated immunoassays in BioFINDER and ADNI cohorts. Alzheimers Dement. 2018;14(11):1470–81. https://doi.org/10.1016/j.jalz.2018.01.010.
41. Long JM, Coble DW, Xiong C, et al. Preclinical Alzheimer's disease biomarkers accurately predict cognitive and neuropathological outcomes. Brain J Neurol. 2022;145(12):4506–18. https://doi.org/10.1093/brain/awac250.
42. Bjerke M, Engelborghs S. Cerebrospinal fluid biomarkers for early and differential Alzheimer's disease diagnosis. J Alzheimers Dis JAD. 2018;62(3):1199–209. https://doi.org/10.3233/JAD-170680.
43. McKhann G, Drachman D, Folstein M, Katzman R, Price D, Stadlan EM. Clinical diagnosis of Alzheimer's disease: report of the NINCDS-ADRDA work group under the auspices of Department of Health and Human Services Task Force on Alzheimer's disease. Neurology. 1984;34(7):939–44. https://doi.org/10.1212/wnl.34.7.939.
44. Dubois B, Feldman HH, Jacova C, et al. Research criteria for the diagnosis of Alzheimer's disease: revising the NINCDS-ADRDA criteria. Lancet Neurol. 2007;6(8):734–46. https://doi.org/10.1016/S1474-4422(07)70178-3.
45. McKhann GM, Knopman DS, Chertkow H, et al. The diagnosis of dementia due to Alzheimer's disease: recommendations from the National Institute on Aging-Alzheimer's association workgroups on diagnostic guidelines for Alzheimer's disease. Alzheimers Dement. 2011;7(3):263–9. https://doi.org/10.1016/j.jalz.2011.03.005.
46. Dubois B, Feldman HH, Jacova C, et al. Revising the definition of Alzheimer's disease: a new lexicon. Lancet Neurol. 2010;9(11):1118–27. https://doi.org/10.1016/S1474-4422(10)70223-4.
47. Albert MS, DeKosky ST, Dickson D, et al. The diagnosis of mild cognitive impairment due to Alzheimer's disease: recommendations from the National Institute on Aging-Alzheimer's association workgroups on diagnostic guidelines for Alzheimer's disease. Alzheimers Dement. 2011;7(3):270–9. https://doi.org/10.1016/j.jalz.2011.03.008.
48. Sperling RA, Aisen PS, Beckett LA, et al. Toward defining the preclinical stages of Alzheimer's disease: recommendations from the National Institute on Aging-Alzheimer's association workgroups on diagnostic guidelines for Alzheimer's disease. Alzheimers Dement J Alzheimers Assoc. 2011;7(3):280–92. https://doi.org/10.1016/j.jalz.2011.03.003.
49. Cummings JL, Dubois B, Molinuevo JL, Scheltens P. International work group criteria for the diagnosis of Alzheimer disease. Med Clin North Am. 2013;97(3):363–8. https://doi.org/10.1016/j.mcna.2013.01.001.
50. Dubois B, Feldman HH, Jacova C, et al. Advancing research diagnostic criteria for Alzheimer's disease: the IWG-2 criteria. Lancet Neurol. 2014;13(6):614–29. https://doi.org/10.1016/S1474-4422(14)70090-0.
51. Dubois B, Hampel H, Feldman HH, et al. Preclinical Alzheimer's disease: definition, natural history, and diagnostic criteria. Alzheimers Dement. 2016;12(3):292–323. https://doi.org/10.1016/j.jalz.2016.02.002.
52. Jack CR, Bennett DA, Blennow K, et al. NIA-AA research framework: toward a biological definition of Alzheimer's disease. Alzheimers Dement J Alzheimers Assoc. 2018;14(4):535–62. https://doi.org/10.1016/j.jalz.2018.02.018.
53. Iacono D, Resnick SM, O'Brien R, et al. Mild cognitive impairment and asymptomatic Alzheimer disease subjects: equivalent β-amyloid and tau loads with divergent cognitive outcomes. J Neuropathol Exp Neurol. 2014;73(4):295–304. https://doi.org/10.1097/NEN.0000000000000052.
54. Jansen WJ, Ossenkoppele R, Knol DL, et al. Prevalence of cerebral amyloid pathology in persons without dementia: a meta-analysis. JAMA. 2015;313(19):1924–38. https://doi.org/10.1001/jama.2015.4668.

55. Langa KM, Burke JF. Preclinical Alzheimer disease—early diagnosis or overdiagnosis? JAMA Intern Med. 2019;179(9):1161–2. https://doi.org/10.1001/jamainternmed.2019.2629.
56. Frisoni GB, Ritchie C, Carrera E, et al. Re-aligning scientific and lay narratives of Alzheimer's disease. Lancet Neurol. 2019;18(10):918–9. https://doi.org/10.1016/S1474-4422(19)30323-0.
57. Rabinovici GD, Carrillo MC. Biomarker-informed treatment decisions in cognitively impaired patients do not apply to preclinical Alzheimer disease. JAMA Intern Med. 2019;179(12):1736–7. https://doi.org/10.1001/jamainternmed.2019.5114.
58. Petersen RC, Wiste HJ, Weigand SD, et al. NIA-AA Alzheimer's disease framework: clinical characterization of stages. Ann Neurol. 2021;89(6):1145–56. https://doi.org/10.1002/ana.26071.
59. Dubois B, Villain N, Frisoni GB, et al. Clinical diagnosis of Alzheimer's disease: recommendations of the international working group. Lancet Neurol. 2021;20(6):484–96. https://doi.org/10.1016/S1474-4422(21)00066-1.
60. Jack CR Jr, Andrews JS, Beach TG, et al. Revised criteria for diagnosis and staging of Alzheimer's disease: Alzheimer's association workgroup. Alzheimers Dement. 2024;20:5143. https://doi.org/10.1002/alz.13859.
61. Alzheimer's Association Workgroup. Revised Criteria for the Diagnosis and Staging of Alzheimer's Disease: Alzheimer's Association Workgroup. https://aaic.alz.org/diagnostic-criteria.asp#drafts
62. Mattsson N, Blennow K, Zetterberg H. Inter-laboratory variation in cerebrospinal fluid biomarkers for Alzheimer's disease: united we stand, divided we fall. Clin Chem Lab Med. 2010;48(5):603–7. https://doi.org/10.1515/CCLM.2010.131.
63. Bjerke M, Portelius E, Minthon L, et al. Confounding factors influencing amyloid beta concentration in cerebrospinal fluid. Int J Alzheimers Dis. 2010;2010:986310. https://doi.org/10.4061/2010/986310.
64. Ja D, T K, Y H, et al. Diurnal patterns of soluble amyloid precursor protein metabolites in the human central nervous system. PLoS One. 2014;9(3) https://doi.org/10.1371/journal.pone.0089998.
65. Leitão MJ, Baldeiras I, Herukka SK, et al. Chasing the effects of pre-analytical confounders—a multicenter study on CSF-AD biomarkers. Front Neurol. 2015;6:153. https://doi.org/10.3389/fneur.2015.00153.
66. A F, E P, H Z, K B, I Q, A PL. Pre-analytical and analytical factors influencing Alzheimer's disease cerebrospinal fluid biomarker variability. Clin Chim Acta Int J Clin Chem. 2015;449:449. https://doi.org/10.1016/j.cca.2015.05.024.
67. Rembach A, Evered LA, Li QX, et al. Alzheimer's disease cerebrospinal fluid biomarkers are not influenced by gravity drip or aspiration extraction methodology. Alzheimers Res Ther. 2015;7(1):71. https://doi.org/10.1186/s13195-015-0157-7.
68. Willemse EAJ, van Uffelen KWJ, van der Flier WM, Teunissen CE. Effect of long-term storage in biobanks on cerebrospinal fluid biomarker Aβ1-42, T-tau, and P-tau values. Alzheimers Dement Amst Neth. 2017;8:45–50. https://doi.org/10.1016/j.dadm.2017.03.005.
69. Tijms BM, Willemse EAJ, Zwan MD, et al. Unbiased approach to counteract upward drift in cerebrospinal fluid amyloid-β 1-42 analysis results. Clin Chem. 2018;64(3):576–85. https://doi.org/10.1373/clinchem.2017.281055.
70. Verwey NA, van der Flier WM, Blennow K, et al. A worldwide multicentre comparison of assays for cerebrospinal fluid biomarkers in Alzheimer's disease. Ann Clin Biochem. 2009;46(Pt 3):235–40. https://doi.org/10.1258/acb.2009.008232.
71. Teunissen CE, Verwey NA, Kester MI, van Uffelen K, Blankenstein MA. Standardization of assay procedures for analysis of the CSF biomarkers amyloid β (1-42), tau, and phosphorylated tau in Alzheimer's disease: report of an international workshop. Int J Alzheimers Dis. 2010;2010:635053. https://doi.org/10.4061/2010/635053.

72. Ad W, Ka P, Ar R, Cl M, Vl V, Kj B. Variability in blood-based amyloid-beta assays: the need for consensus on pre-analytical processing. J Alzheimers Dis JAD. 2012;30(2):323. https://doi.org/10.3233/JAD-2012-120058.
73. Niemantsverdriet E, Goossens J, Struyfs H, et al. Diagnostic impact of cerebrospinal fluid biomarker (pre-)analytical variability in Alzheimer's disease. J Alzheimers Dis JAD. 2016;51(1):97–106. https://doi.org/10.3233/JAD-150953.
74. Rozga M, Bittner T, Höglund K, Blennow K. Accuracy of cerebrospinal fluid Aβ1-42 measurements: evaluation of pre-analytical factors using a novel Elecsys immunosassay. Clin Chem Lab Med. 2017;55(10):1545–54. https://doi.org/10.1515/cclm-2016-1061.
75. Leitão MJ, Silva-Spínola A, Santana I, et al. Clinical validation of the Lumipulse G cerebrospinal fluid assays for routine diagnosis of Alzheimer's disease. Alzheimers Res Ther. 2019;11(1):91. https://doi.org/10.1186/s13195-019-0550-8.
76. Delaby C, Teunissen CE, Blennow K, et al. Clinical reporting following the quantification of cerebrospinal fluid biomarkers in Alzheimer's disease: an international overview. Alzheimers Dement J Alzheimers Assoc. 2022;18(10):1868–79. https://doi.org/10.1002/alz.12545.
77. Janelidze S, Stomrud E, Brix B, Hansson O. Towards a unified protocol for handling of CSF before β-amyloid measurements. Alzheimers Res Ther. 2019;11(1):63. https://doi.org/10.1186/s13195-019-0517-9.
78. Hansson O, Batrla R, Brix B, et al. The Alzheimer's association international guidelines for handling of cerebrospinal fluid for routine clinical measurements of amyloid β and tau. Alzheimers Dement J Alzheimers Assoc. 2021;17(9):1575–82. https://doi.org/10.1002/alz.12316.
79. Mattsson N, Andreasson U, Persson S, et al. The Alzheimer's association external quality control program for cerebrospinal fluid biomarkers. Alzheimers Dement J Alzheimers Assoc. 2011;7(4):386–395.e6. https://doi.org/10.1016/j.jalz.2011.05.2243.
80. Mattsson N, Andreasson U, Persson S, et al. CSF biomarker variability in the Alzheimer's association quality control program. Alzheimers Dement J Alzheimers Assoc. 2013;9(3):251–61. https://doi.org/10.1016/j.jalz.2013.01.010.
81. The Alzheimer's Association QC program for CSF and blood biomarkers. University of Gothenburg Institute of Neuroscience and Physiology. https://www.gu.se/en/neuroscience-physiology/the-alzheimers-association-qc-program-for-csf-and-blood-biomarkers
82. FDA-NIH Biomarker Working Group. BEST (biomarkers, endpoints, and other tools) resource. Food and Drug Administration (US); 2016. Accessed 1 Oct 2023. http://www.ncbi.nlm.nih.gov/books/NBK326791/
83. Cummings J, Kinney J. Biomarkers for Alzheimer's disease: context of use, qualification, and roadmap for clinical implementation. Medicina (Mex). 2022;58(7):952. https://doi.org/10.3390/medicina58070952.
84. Iaccarino L, Burnham SC, Dell'Agnello G, Dowsett SA, Epelbaum S. Diagnostic biomarkers of amyloid and tau pathology in Alzheimer's disease: an overview of tests for clinical practice in the United States and Europe. J Prev Alzheimers Dis. 2023;10(3):426–42. https://doi.org/10.14283/jpad.2023.43.
85. O'Brien RJ, Wong PC. Amyloid precursor protein processing and Alzheimer's disease. Annu Rev Neurosci. 2011;34:185–204. https://doi.org/10.1146/annurev-neuro-061010-113613.
86. Papaliagkas V, Kalinderi K, Vareltzis P, Moraitou D, Papamitsou T, Chatzidimitriou M. CSF biomarkers in the early diagnosis of mild cognitive impairment and Alzheimer's disease. Int J Mol Sci. 2023;24(10):8976. https://doi.org/10.3390/ijms24108976.
87. Sutphen CL, Jasielec MS, Shah AR, et al. Longitudinal cerebrospinal fluid biomarker changes in preclinical Alzheimer disease during middle age. JAMA Neurol. 2015;72(9):1029–42. https://doi.org/10.1001/jamaneurol.2015.1285.
88. Lleó A, Alcolea D, Martínez-Lage P, et al. Longitudinal cerebrospinal fluid biomarker trajectories along the Alzheimer's disease continuum in the BIOMARKAPD study. Alzheimers Dement. 2019;15(6):742–53. https://doi.org/10.1016/j.jalz.2019.01.015.

89. Wiltfang J, Esselmann H, Bibl M, et al. Amyloid beta peptide ratio 42/40 but not A beta 42 correlates with phospho-Tau in patients with low- and high-CSF A beta 40 load. J Neurochem. 2007;101(4):1053–9. https://doi.org/10.1111/j.1471-4159.2006.04404.x.
90. Hansson O, Lehmann S, Otto M, Zetterberg H, Lewczuk P. Advantages and disadvantages of the use of the CSF amyloid β (Aβ) 42/40 ratio in the diagnosis of Alzheimer's disease. Alzheimers Res Ther. 2019;11(1):34. https://doi.org/10.1186/s13195-019-0485-0.
91. Arendt T, Stieler JT, Holzer M. Tau and tauopathies. Brain Res Bull. 2016;126:238–92. https://doi.org/10.1016/j.brainresbull.2016.08.018.
92. Creekmore BC, Watanabe R, Lee EB. Neurodegenerative disease tauopathies. Annu Rev Pathol. Published online October 13, 2023. 2023;13:345. https://doi.org/10.1146/annurev-pathmechdis-051222-120750.
93. Milà-Alomà M, Ashton NJ, Shekari M, et al. Plasma p-tau231 and p-tau217 as state markers of amyloid-β pathology in preclinical Alzheimer's disease. Nat Med. 2022;11:1–5. https://doi.org/10.1038/s41591-022-01925-w. Published online August 11 2022
94. Engelborghs S, De Vreese K, Van de Casteele T, et al. Diagnostic performance of a CSF-biomarker panel in autopsy-confirmed dementia. Neurobiol Aging. 2008;29(8):1143–59. https://doi.org/10.1016/j.neurobiolaging.2007.02.016.
95. Shaw LM, Vanderstichele H, Knapik-Czajka M, et al. Cerebrospinal fluid biomarker signature in Alzheimer's disease neuroimaging initiative subjects. Ann Neurol. 2009;65(4):403–13. https://doi.org/10.1002/ana.21610.
96. Niemantsverdriet E, Feyen BFE, Le Bastard N, et al. Added diagnostic value of cerebrospinal fluid biomarkers for differential dementia diagnosis in an autopsy-confirmed cohort. J Alzheimers Dis. 63(1):373–81. https://doi.org/10.3233/JAD-170927.
97. Reimand J, Boon BDC, Collij LE, et al. Amyloid-β PET and CSF in an autopsy-confirmed cohort. Ann Clin Transl Neurol. 2020;7(11):2150–60. https://doi.org/10.1002/acn3.51195.
98. Palmqvist S, Zetterberg H, Mattsson N, et al. Detailed comparison of amyloid PET and CSF biomarkers for identifying early Alzheimer disease. Neurology. 2015;85(14):1240–9. https://doi.org/10.1212/WNL.0000000000001991.
99. Wang L, Fagan AM, Shah AR, et al. Cerebrospinal fluid proteins predict longitudinal hippocampal degeneration in early-stage dementia of the Alzheimer type. Alzheimer Dis Assoc Disord. 2012;26(4):314–21. https://doi.org/10.1097/WAD.0b013e31823c0cf4.
100. Fagan AM, Roe CM, Xiong C, Mintun MA, Morris JC, Holtzman DM. Cerebrospinal fluid tau/beta-amyloid(42) ratio as a prediction of cognitive decline in nondemented older adults. Arch Neurol. 2007;64(3):343–9. https://doi.org/10.1001/archneur.64.3.noc60123.
101. Landau SM, Harvey D, Madison CM, et al. Comparing predictors of conversion and decline in mild cognitive impairment. Neurology. 2010;75(3):230–8. https://doi.org/10.1212/WNL.0b013e3181e8e8b8.
102. Li Y, Yen D, Hendrix RD, et al. Timing of biomarker changes in sporadic Alzheimer's disease in estimated years from symptom onset. Ann Neurol. 2024;24:951. https://doi.org/10.1002/ana.26891. Published online February 24 2024
103. Nutu M, Zetterberg H, Londos E, et al. Evaluation of the cerebrospinal fluid amyloid-β1-42/amyloid-β1-40 ratio measured by alpha-LISA to distinguish Alzheimer's disease from other dementia disorders. Dement Geriatr Cogn Disord. 2013;36(1–2):99–110. https://doi.org/10.1159/000353442.
104. Sauvée M, DidierLaurent G, Latarche C, Escanyé MC, Olivier JL, Malaplate-Armand C. Additional use of Aβ42/Aβ40 ratio with cerebrospinal fluid biomarkers P-tau and Aβ42 increases the level of evidence of Alzheimer's disease pathophysiological process in routine practice. J Alzheimers Dis JAD. 2014;41(2):377–86. https://doi.org/10.3233/JAD-131838.
105. Palmqvist S, Mattsson N, Hansson O. For the Alzheimer's Disease Neuroimaging Initiative. Cerebrospinal fluid analysis detects cerebral amyloid-β accumulation earlier than positron emission tomography. Brain. 2016;139(4):1226–36. https://doi.org/10.1093/brain/aww015.
106. Wojdała AL, Bellomo G, Toja A, Gaetani L, Parnetti L, Chiasserini D. CSF and plasma Aβ42/40 across Alzheimer's disease continuum: comparison of two ultrasensitive Simoa®

assays targeting distinct amyloid regions. Clin Chem Lab Med CCLM. 2024;62(2):332–40. https://doi.org/10.1515/cclm-2023-0659.
107. Jack CR, Bennett DA, Blennow K, et al. A/T/N: an unbiased descriptive classification scheme for Alzheimer disease biomarkers. Neurology. 2016;87(5):539–47. https://doi.org/10.1212/WNL.0000000000002923.
108. Hampel H, Buerger K, Zinkowski R, et al. Measurement of phosphorylated tau epitopes in the differential diagnosis of Alzheimer disease: a comparative cerebrospinal fluid study. Arch Gen Psychiatry. 2004;61(1):95–102. https://doi.org/10.1001/archpsyc.61.1.95.
109. Wojdała AL, Bellomo G, Gaetani L, et al. Trajectories of CSF and plasma biomarkers across Alzheimer's disease continuum: disease staging by NF-L, p-tau181, and GFAP. Neurobiol Dis. 2023;189:106356. https://doi.org/10.1016/j.nbd.2023.106356.
110. Tan MS, Ji X, Li JQ, et al. Longitudinal trajectories of Alzheimer's ATN biomarkers in elderly persons without dementia. Alzheimers Res Ther. 2020;12(1):55. https://doi.org/10.1186/s13195-020-00621-6.
111. Ebenau JL, Timmers T, Wesselman LMP, et al. ATN classification and clinical progression in subjective cognitive decline: the SCIENCe project. Neurology. 2020;95(1):e46–58. https://doi.org/10.1212/WNL.0000000000009724.
112. Blennow K, Shaw LM, Stomrud E, et al. Predicting clinical decline and conversion to Alzheimer's disease or dementia using novel Elecsys Aβ(1-42), pTau and tTau CSF immunoassays. Sci Rep. 2019;9(1):19024. https://doi.org/10.1038/s41598-019-54204-z.
113. Olsson B, Lautner R, Andreasson U, et al. CSF and blood biomarkers for the diagnosis of Alzheimer's disease: a systematic review and meta-analysis. Lancet Neurol. 2016;15(7):673–84. https://doi.org/10.1016/S1474-4422(16)00070-3.
114. Hesse C, Rosengren L, Andreasen N, et al. Transient increase in total tau but not phospho-tau in human cerebrospinal fluid after acute stroke. Neurosci Lett. 2001;297(3):187–90. https://doi.org/10.1016/s0304-3940(00)01697-9.
115. Skillbäck T, Rosén C, Asztely F, Mattsson N, Blennow K, Zetterberg H. Diagnostic performance of cerebrospinal fluid total tau and phosphorylated tau in Creutzfeldt-Jakob disease: results from the Swedish mortality registry. JAMA Neurol. 2014;71(4):476–83. https://doi.org/10.1001/jamaneurol.2013.6455.
116. Janelidze S, Stomrud E, Smith R, et al. Cerebrospinal fluid p-tau217 performs better than p-tau181 as a biomarker of Alzheimer's disease. Nat Commun. 2020;11(1):1683. https://doi.org/10.1038/s41467-020-15436-0.
117. Hanes J, Kovac A, Kvartsberg H, et al. Evaluation of a novel immunoassay to detect p-tau Thr217 in the CSF to distinguish Alzheimer disease from other dementias. Neurology. 2020;95(22):e3026–35. https://doi.org/10.1212/WNL.0000000000010814.
118. Salvadó G, Horie K, Barthélemy NR, et al. Novel CSF tau biomarkers can be used for disease staging of sporadic Alzheimer's disease. medRxiv. Published online July 16, 2023:2023.07.14.23292650. https://doi.org/10.1101/2023.07.14.23292650.
119. Ashton NJ, Puig-Pijoan A, Milà-Alomà M, et al. Plasma and CSF biomarkers in a memory clinic: Head-to-head comparison of phosphorylated tau immunoassays. Alzheimers Dement J Alzheimers Assoc. 2023;19(5):1913–24. https://doi.org/10.1002/alz.12841.
120. Lantero-Rodriguez J, Montoliu-Gaya L, Benedet AL, et al. CSF p-tau205: a biomarker of tau pathology in Alzheimer's disease. Acta Neuropathol (Berl). 2024;147(1):12. https://doi.org/10.1007/s00401-023-02659-w.
121. Wennström M, Schultz N, Gallardo PM, et al. The relationship between p-tau217, p-tau231, and p-tau205 in the human brain is affected by the cellular environment and Alzheimer's disease pathology. Cells. 2024;13(4):331. https://doi.org/10.3390/cells13040331.
122. Horie K, Barthélemy NR, Spina S, et al. CSF tau microtubule-binding region identifies pathological changes in primary tauopathies. Nat Med. 2022;24:1–8. https://doi.org/10.1038/s41591-022-02075-9. Published online November 24 2022
123. Horie K, Barthélemy NR, Sato C, Bateman RJ. CSF tau microtubule binding region identifies tau tangle and clinical stages of Alzheimer's disease. Brain. 2021;144(2):515–27. https://doi.org/10.1093/brain/awaa373.

124. Wiltfang J, Esselmann H, Bibl M, et al. Highly conserved and disease-specific patterns of carboxyterminally truncated Aβ peptides 1–37/38/39 in addition to 1–40/42 in Alzheimer's disease and in patients with chronic neuroinflammation. J Neurochem. 2002;81(3):481–96. https://doi.org/10.1046/j.1471-4159.2002.00818.x.
125. Cullen N, Janelidze S, Palmqvist S, Stomrud E, Mattsson-Carlgren N, Hansson O. Association of CSF Aβ38 levels with risk of Alzheimer disease–related decline. Neurology. 2022;98(9):e958–67. https://doi.org/10.1212/WNL.0000000000013228.
126. Povala G, Bellaver B, De Bastiani MA, et al. Soluble amyloid-beta isoforms predict downstream Alzheimer's disease pathology. Cell Biosci. 2021;11(1):204. https://doi.org/10.1186/s13578-021-00712-3.
127. van den Berg E, Kersten I, Brinkmalm G, et al. Profiling amyloid-β peptides as biomarkers for cerebral amyloid angiopathy. J Neurochem. 2024;16:1254. https://doi.org/10.1111/jnc.16074. Published online February 16 2024
128. Dargvainiene J, Jensen-Kondering U, Bender B, et al. Aβ38 and Aβ43 do not differentiate between Alzheimer's disease and cerebral amyloid angiopathy. Ann Clin Transl Neurol. 2024;7:806. https://doi.org/10.1002/acn3.51987. Published online January 7 2024
129. Bazzari FH, Bazzari AH. BACE1 inhibitors for Alzheimer's disease: the past, present and any future? Molecules. 2022;27(24):8823. https://doi.org/10.3390/molecules27248823.
130. Lopez-Font I, Boix CP, Zetterberg H, Blennow K, Sáez-Valero J. Characterization of cerebrospinal fluid BACE1 species. Mol Neurobiol. 2019;56(12):8603–16. https://doi.org/10.1007/s12035-019-01677-8.
131. Wesenhagen KEJ, Tijms BM, Boonkamp L, et al. P-tau subgroups in AD relate to distinct amyloid production and synaptic integrity profiles. Alzheimers Res Ther. 2022;14:95. https://doi.org/10.1186/s13195-022-01038-z.
132. Kirsebom BE, Richter G, Nordengen K, et al. Stable cerebrospinal fluid neurogranin and β-site amyloid precursor protein cleaving enzyme 1 levels differentiate predementia Alzheimer's disease patients. Brain Commun. 2022;4(5):fcac244. https://doi.org/10.1093/braincomms/fcac244.
133. Rosengren LE, Karlsson JE, Karlsson JO, Persson LI, Wikkelsø C. Patients with amyotrophic lateral sclerosis and other neurodegenerative diseases have increased levels of neurofilament protein in CSF. J Neurochem. 1996;67(5):2013–8. https://doi.org/10.1046/j.1471-4159.1996.67052013.x.
134. Hansson O. Biomarkers for neurodegenerative diseases. Nat Med. 2021;27(6):954–63. https://doi.org/10.1038/s41591-021-01382-x.
135. Simrén J, Andreasson U, Gobom J, et al. Establishment of reference values for plasma neurofilament light based on healthy individuals aged 5–90 years. Brain Commun. 2022;4(4):fcac174. https://doi.org/10.1093/braincomms/fcac174.
136. Arslan B, Zetterberg H. Neurofilament light chain as neuronal injury marker—what is needed to facilitate implementation in clinical laboratory practice? Clin Chem Lab Med CCLM. 2023;61(7):1140–9. https://doi.org/10.1515/cclm-2023-0036.
137. Gaur A, Rivet L, Mah E, et al. Novel fluid biomarkers for mild cognitive impairment: a systematic review and meta-analysis. Ageing Res Rev. 2023;91:102046. https://doi.org/10.1016/j.arr.2023.102046.
138. Johnson SC, Suárez-Calvet M, Suridjan I, et al. Identifying clinically useful biomarkers in neurodegenerative disease through a collaborative approach: the NeuroToolKit. Alzheimers Res Ther. 2023;15:25. https://doi.org/10.1186/s13195-023-01168-y.
139. Dhiman K, Gupta VB, Villemagne VL, et al. Cerebrospinal fluid neurofilament light concentration predicts brain atrophy and cognition in Alzheimer's disease. Alzheimers Dement Amst Neth. 2020;12(1):e12005. https://doi.org/10.1002/dad2.12005.
140. Aschenbrenner AJ, Gordon BA, Fagan AM, et al. Neurofilament light predicts decline in attention but not episodic memory in preclinical Alzheimer's disease. J Alzheimers Dis JAD. 2020;74(4):1119–29. https://doi.org/10.3233/JAD-200018.

141. Salvadó G, Shekari M, Falcon C, et al. Brain alterations in the early Alzheimer's continuum with amyloid-β, tau, glial and neurodegeneration CSF markers. Brain Commun. 2022;4(3):fcac134. https://doi.org/10.1093/braincomms/fcac134.
142. Jia J, Ning Y, Chen M, et al. Biomarker changes during 20 years preceding Alzheimer's disease. N Engl J Med. 2024;390(8):712–22. https://doi.org/10.1056/NEJMoa2310168.
143. Chaudhry A, Houlden H, Rizig M. Novel fluid biomarkers to differentiate frontotemporal dementia and dementia with Lewy bodies from Alzheimer's disease: A systematic review. J Neurol Sci. 2020;415:116886. https://doi.org/10.1016/j.jns.2020.116886.
144. Giuffrè GM, Quaranta D, Costantini EM, et al. Cerebrospinal fluid neurofilament light chain and total-tau as biomarkers of neurodegeneration in Alzheimer's disease and frontotemporal dementia. Neurobiol Dis. 2023;186:106267. https://doi.org/10.1016/j.nbd.2023.106267.
145. Wang Z, Lewis V, Stehmann C, et al. Alzheimer's disease biomarker utilization at first referral enhances differential diagnostic precision with simultaneous exclusion of Creutzfeldt-Jakob disease. Alzheimers Dement Amst Neth. 2024;16(1):e12548. https://doi.org/10.1002/dad2.12548.
146. Hu Q, Shi M, Li Y, Zhao X. Elevated plasma neurofilament light was associated with multimodal neuroimaging features in Alzheimer's disease signature regions and predicted future tau deposition. Res Sq. 2024;19:rs.3.rs-3946421. https://doi.org/10.21203/rs.3.rs-3946421/v1. Published online February 19, 2024
147. Garcia MJ, Leadley R, Ross J, et al. Prognostic and predictive factors in early Alzheimer's disease: A systematic review. J Alzheimers Dis Rep. 2024;8(1):203–40. https://doi.org/10.3233/ADR-230045.
148. Vrillon A, Ashton NJ, Karikari TK, et al. Comparison of CSF and plasma NfL and pNfH for Alzheimer's disease diagnosis: a memory clinic study. J Neurol. 2024;271(3):1297–310. https://doi.org/10.1007/s00415-023-12066-6.
149. Tarawneh R, D'Angelo G, Macy E, et al. Visinin-like protein-1: diagnostic and prognostic biomarker in Alzheimer disease. Ann Neurol. 2011;70(2):274–85. https://doi.org/10.1002/ana.22448.
150. Wang L, Zhang M, Wang Q, Jiang X, Li K, Liu J. APOE ε4 allele is associated with elevated levels of CSF VILIP-1 in preclinical Alzheimer's disease. Neuropsychiatr Dis Treat. 2020;16:923–31. https://doi.org/10.2147/NDT.S235395.
151. Kester MI, Teunissen CE, Sutphen C, et al. Cerebrospinal fluid VILIP-1 and YKL-40, candidate biomarkers to diagnose, predict and monitor Alzheimer's disease in a memory clinic cohort. Alzheimers Res Ther. 2015;7(1):59. https://doi.org/10.1186/s13195-015-0142-1.
152. Lee JM, Blennow K, Andreasen N, et al. The brain injury biomarker, VLP-1, is increased in the CSF of Alzheimer's disease patients. Clin Chem. 2008;54(10):1617–23. https://doi.org/10.1373/clinchem.2008.104497.
153. Mroczko B, Groblewska M, Zboch M, et al. Evaluation of visinin-like protein 1 concentrations in the cerebrospinal fluid of patients with mild cognitive impairment as a dynamic biomarker of Alzheimer's disease. J Alzheimers Dis JAD. 2015;43(3):1031–7. https://doi.org/10.3233/JAD-141050.
154. Tarawneh R, Lee JM, Ladenson JH, Morris JC, Holtzman DM. CSF VILIP-1 predicts rates of cognitive decline in early Alzheimer disease. Neurology. 2012;78(10):709–19. https://doi.org/10.1212/WNL.0b013e318248e568.
155. Tarawneh R, Head D, Allison S, et al. Cerebrospinal fluid markers of neurodegeneration and rates of brain atrophy in early Alzheimer disease. JAMA Neurol. 2015;72(6):656–65. https://doi.org/10.1001/jamaneurol.2015.0202.
156. Babić Leko M, Borovečki F, Dejanović N, Hof PR, Šimić G. Predictive value of cerebrospinal fluid visinin-like Protein-1 levels for Alzheimer's disease early detection and differential diagnosis in patients with mild cognitive impairment. J Alzheimers Dis JAD. 2016;50(3):765–78. https://doi.org/10.3233/JAD-150705.
157. Luo X, Hou L, Shi H, et al. CSF levels of the neuronal injury biomarker visinin-like protein-1 in Alzheimer's disease and dementia with Lewy bodies. J Neurochem. 2013;127(5):681–90. https://doi.org/10.1111/jnc.12331.

158. Halbgebauer S, Steinacker P, Riedel D, et al. Visinin-like protein 1 levels in blood and CSF as emerging markers for Alzheimer's and other neurodegenerative diseases. Alzheimers Res Ther. 2022;14(1):175. https://doi.org/10.1186/s13195-022-01122-4.
159. Butterfield DA. Ubiquitin carboxyl-terminal hydrolase L-1 in brain: focus on its oxidative/nitrosative modification and role in brains of subjects with Alzheimer disease and mild cognitive impairment. Free Radic Biol Med. 2021;177:278–86. https://doi.org/10.1016/j.freeradbiomed.2021.10.036.
160. Öhrfelt A, Johansson P, Wallin A, et al. Increased cerebrospinal fluid levels of ubiquitin carboxyl-terminal hydrolase L1 in patients with Alzheimer's disease. Dement Geriatr Cogn Disord Extra. 2016;6(2):283–94. https://doi.org/10.1159/000447239.
161. Paciotti S, Wojdała AL, Bellomo G, et al. Potential diagnostic value of CSF metabolism-related proteins across the Alzheimer's disease continuum. Alzheimers Res Ther. 2023;15(1):124. https://doi.org/10.1186/s13195-023-01269-8.
162. Chiasserini D, Parnetti L, Andreasson U, et al. CSF levels of heart fatty acid binding protein are altered during early phases of Alzheimer's disease. J Alzheimers Dis JAD. 2010;22(4):1281–8. https://doi.org/10.3233/JAD-2010-101293.
163. Trombetta BA, Wu CY, Kuo E, et al. Cerebrospinal fluid biomarker profiling of diverse pathophysiological domains in Alzheimer's disease. Alzheimers Dement N Y N. 2024;10(1):e12440. https://doi.org/10.1002/trc2.12440.
164. Chiasserini D, Biscetti L, Eusebi P, et al. Differential role of CSF fatty acid binding protein 3, α-synuclein, and Alzheimer's disease core biomarkers in Lewy body disorders and Alzheimer's dementia. Alzheimers Res Ther. 2017;9:52. https://doi.org/10.1186/s13195-017-0276-4.
165. Desikan RS, Thompson WK, Holland D, et al. Heart fatty acid binding protein and Aβ-associated Alzheimer's neurodegeneration. Mol Neurodegener. 2013;8:39. https://doi.org/10.1186/1750-1326-8-39.
166. Noppe M, Crols R, Andries D, Lowenthal A. Determination in human cerebrospinal fluid of glial fibrillary acidic protein, S-100 and myelin basic protein as indices of non-specific or specific central nervous tissue pathology. Clin Chim Acta Int J Clin Chem. 1986;155(2):143–50. https://doi.org/10.1016/0009-8981(86)90275-5.
167. Milà-Alomà M, Salvadó G, Gispert JD, et al. Amyloid beta, tau, synaptic, neurodegeneration, and glial biomarkers in the preclinical stage of the Alzheimer's continuum. Alzheimers Dement. 2020;16(10):1358–71. https://doi.org/10.1002/alz.12131.
168. Simrén J, Weninger H, Brum WS, et al. Differences between blood and cerebrospinal fluid glial fibrillary acidic protein levels: the effect of sample stability. Alzheimers Dement J Alzheimers Assoc. 2022;18(10):1988–92. https://doi.org/10.1002/alz.12806.
169. Benedet AL, Milà-Alomà M, Vrillon A, et al. Differences between plasma and cerebrospinal fluid glial fibrillary acidic protein levels across the Alzheimer disease continuum. JAMA Neurol. 2021;78(12):1471–83. https://doi.org/10.1001/jamaneurol.2021.3671.
170. Salvadó G, Milà-Alomà M, Shekari M, et al. Cerebral amyloid-β load is associated with neurodegeneration and gliosis: mediation by p-tau and interactions with risk factors early in the Alzheimer's continuum. Alzheimers Dement. 2021;17(5):788–800. https://doi.org/10.1002/alz.12245.
171. Ishiki A, Kamada M, Kawamura Y, et al. Glial fibrillar acidic protein in the cerebrospinal fluid of Alzheimer's disease, dementia with Lewy bodies, and frontotemporal lobar degeneration. J Neurochem. 2016;136(2):258–61. https://doi.org/10.1111/jnc.13399.
172. McGrowder DA, Miller F, Vaz K, et al. Cerebrospinal fluid biomarkers of Alzheimer's disease: current evidence and future perspectives. Brain Sci. 2021;11(2):215. https://doi.org/10.3390/brainsci11020215.
173. Teitsdottir UD, Jonsdottir MK, Lund SH, Darreh-Shori T, Snaedal J, Petersen PH. Association of glial and neuronal degeneration markers with Alzheimer's disease cerebrospinal fluid profile and cognitive functions. Alzheimers Res Ther. 2020;12(1):92. https://doi.org/10.1186/s13195-020-00657-8.

174. Fukuyama R, Izumoto T, Fushiki S. The cerebrospinal fluid level of glial fibrillary acidic protein is increased in cerebrospinal fluid from Alzheimer's disease patients and correlates with severity of dementia. Eur Neurol. 2001;46(1):35–8. https://doi.org/10.1159/000050753.
175. Jesse S, Steinacker P, Cepek L, et al. Glial fibrillary acidic protein and protein S-100B: different concentration pattern of glial proteins in cerebrospinal fluid of patients with Alzheimer's disease and Creutzfeldt-Jakob disease. J Alzheimers Dis. 2009;17(3):541–51. https://doi.org/10.3233/JAD-2009-1075.
176. Honey MIJ, Wesenhagen KEJ, Willemse EAJ, et al. Comparison between plasma, serum and cerebrospinal fluid glial fibrillary acidic protein in Alzheimer's disease and dementia with Lewy bodies and the effect of age and sex on diagnostic performance. Alzheimers Dement. 2022;18(S5):e067313. https://doi.org/10.1002/alz.067313.
177. Leuzy A, Cullen NC, Mattsson-Carlgren N, Hansson O. Current advances in plasma and cerebrospinal fluid biomarkers in Alzheimer's disease. Curr Opin Neurol. 2021;34(2):266. https://doi.org/10.1097/WCO.0000000000000904.
178. Bonneh-Barkay D, Wang G, Starkey A, Hamilton RL, Wiley CA. In vivo CHI3L1 (YKL-40) expression in astrocytes in acute and chronic neurological diseases. J Neuroinflammation. 2010;7:34. https://doi.org/10.1186/1742-2094-7-34.
179. Willemse EAJ, Vermeiren Y, Garcia-Ayllon MS, et al. Pre-analytical stability of novel cerebrospinal fluid biomarkers. Clin Chim Acta Int J Clin Chem. 2019;497:204–11. https://doi.org/10.1016/j.cca.2019.07.024.
180. Craig-Schapiro R, Perrin RJ, Roe CM, et al. YKL-40: A novel prognostic fluid biomarker for preclinical Alzheimer's disease. Biol Psychiatry. 2010;68(10):903–12. https://doi.org/10.1016/j.biopsych.2010.08.025.
181. Alcolea D, Martínez-Lage P, Sánchez-Juan P, et al. Amyloid precursor protein metabolism and inflammation markers in preclinical Alzheimer disease. Neurology. 2015;85(7):626–33. https://doi.org/10.1212/WNL.0000000000001859.
182. Alcolea D, Vilaplana E, Pegueroles J, et al. Relationship between cortical thickness and cerebrospinal fluid YKL-40 in predementia stages of Alzheimer's disease. Neurobiol Aging. 2015;36(6):2018–23. https://doi.org/10.1016/j.neurobiolaging.2015.03.001.
183. Antonell A, Mansilla A, Rami L, et al. Cerebrospinal fluid level of YKL-40 protein in preclinical and prodromal Alzheimer's disease. J Alzheimers Dis JAD. 2014;42(3):901–8. https://doi.org/10.3233/JAD-140624.
184. Wang L, Gao T, Cai T, et al. Cerebrospinal fluid levels of YKL-40 in prodromal Alzheimer's disease. Neurosci Lett. 2020;715:134658. https://doi.org/10.1016/j.neulet.2019.134658.
185. Hellwig K, Kvartsberg H, Portelius E, et al. Neurogranin and YKL-40: independent markers of synaptic degeneration and neuroinflammation in Alzheimer's disease. Alzheimers Res Ther. 2015;7:74. https://doi.org/10.1186/s13195-015-0161-y.
186. Llorens F, Schmitz M, Knipper T, et al. Cerebrospinal fluid biomarkers of Alzheimer's disease show different but partially overlapping profile compared to vascular dementia. Front Aging Neurosci. 2017;9:289. https://doi.org/10.3389/fnagi.2017.00289.
187. Llorens F, Thüne K, Tahir W, et al. YKL-40 in the brain and cerebrospinal fluid of neurodegenerative dementias. Mol Neurodegener. 2017;12(1):83. https://doi.org/10.1186/s13024-017-0226-4.
188. Wennström M, Surova Y, Hall S, et al. The inflammatory marker YKL-40 is elevated in cerebrospinal fluid from patients with Alzheimer's but not Parkinson's disease or dementia with Lewy bodies. PLoS One. 2015;10(8):e0135458. https://doi.org/10.1371/journal.pone.0135458.
189. Terry RD, Masliah E, Salmon DP, et al. Physical basis of cognitive alterations in Alzheimer's disease: synapse loss is the major correlate of cognitive impairment. Ann Neurol. 1991;30(4):572–80. https://doi.org/10.1002/ana.410300410.
190. Scheff SW, Price DA, Schmitt FA, Mufson EJ. Hippocampal synaptic loss in early Alzheimer's disease and mild cognitive impairment. Neurobiol Aging. 2006;27(10):1372–84. https://doi.org/10.1016/j.neurobiolaging.2005.09.012.

191. DeKosky ST, Scheff SW. Synapse loss in frontal cortex biopsies in Alzheimer's disease: correlation with cognitive severity. Ann Neurol. 1990;27(5):457–64. https://doi.org/10.1002/ana.410270502.
192. Colom-Cadena M, Spires-Jones T, Zetterberg H, et al. The clinical promise of biomarkers of synapse damage or loss in Alzheimer's disease. Alzheimers Res Ther. 2020;12(1):21. https://doi.org/10.1186/s13195-020-00588-4.
193. Bos I, Vos S, Verhey F, et al. Cerebrospinal fluid biomarkers of neurodegeneration, synaptic integrity, and astroglial activation across the clinical Alzheimer's disease spectrum. Alzheimers Dement. 2019;15(5):644–54. https://doi.org/10.1016/j.jalz.2019.01.004.
194. Tarawneh R, D'Angelo G, Crimmins D, et al. Diagnostic and prognostic utility of the synaptic marker neurogranin in Alzheimer disease. JAMA Neurol. 2016;73(5):561–71. https://doi.org/10.1001/jamaneurol.2016.0086.
195. Kvartsberg H, Duits FH, Ingelsson M, et al. Cerebrospinal fluid levels of the synaptic protein neurogranin correlates with cognitive decline in prodromal Alzheimer's disease. Alzheimers Dement J Alzheimers Assoc. 2015;11(10):1180–90. https://doi.org/10.1016/j.jalz.2014.10.009.
196. Nilsson J, Pichet Binette A, Palmqvist S, et al. Cerebrospinal fluid biomarker panel for synaptic dysfunction in a broad spectrum of neurodegenerative diseases. Brain J Neurol. Published online February 7, 2024:awae032. https://doi.org/10.1093/brain/awae032.
197. Portelius E, Zetterberg H, Skillbäck T, et al. Cerebrospinal fluid neurogranin: relation to cognition and neurodegeneration in Alzheimer's disease. Brain. 2015;138(11):3373–85. https://doi.org/10.1093/brain/awv267.
198. Wellington H, Paterson RW, Portelius E, et al. Increased CSF neurogranin concentration is specific to Alzheimer disease. Neurology. 2016;86(9):829–35. https://doi.org/10.1212/WNL.0000000000002423.
199. Brinkmalm A, Brinkmalm G, Honer WG, et al. SNAP-25 is a promising novel cerebrospinal fluid biomarker for synapse degeneration in Alzheimer's disease. Mol Neurodegener. 2014;9(1):53. https://doi.org/10.1186/1750-1326-9-53.
200. Kivisäkk P, Carlyle BC, Sweeney T, et al. Increased levels of the synaptic proteins PSD-95, SNAP-25, and neurogranin in the cerebrospinal fluid of patients with Alzheimer's disease. Alzheimers Res Ther. 2022;14(1):58. https://doi.org/10.1186/s13195-022-01002-x.
201. Das S, van Engelen MPE, Goossens J, et al. The use of synaptic biomarkers in cerebrospinal fluid to differentiate behavioral variant of frontotemporal dementia from primary psychiatric disorders and Alzheimer's disease. Alzheimers Res Ther. 2024;16(1):34. https://doi.org/10.1186/s13195-024-01409-8.
202. Clarke MTM, Brinkmalm A, Foiani MS, et al. CSF synaptic protein concentrations are raised in those with atypical Alzheimer's disease but not frontotemporal dementia. Alzheimers Res Ther. 2019;11(1):105. https://doi.org/10.1186/s13195-019-0564-2.
203. Halbgebauer S, Steinacker P, Hengge S, et al. CSF levels of SNAP-25 are increased early in Creutzfeldt-Jakob and Alzheimer's disease. J Neurol Neurosurg Psychiatry. Published online August 22, 2022:jnnp-2021-328646. https://doi.org/10.1136/jnnp-2021-328646.
204. Butt OH, Long JM, Henson RL, et al. Cognitively normal APOE ε4 carriers have specific elevation of CSF SNAP-25. Neurobiol Aging. 2021;102:64–72. https://doi.org/10.1016/j.neurobiolaging.2021.02.008.
205. Tible M, Sandelius Å, Höglund K, et al. Dissection of synaptic pathways through the CSF biomarkers for predicting Alzheimer disease. Neurology. 2020;95(8):e953–61. https://doi.org/10.1212/WNL.0000000000010131.
206. Öhrfelt A, Brinkmalm A, Dumurgier J, et al. The pre-synaptic vesicle protein synaptotagmin is a novel biomarker for Alzheimer's disease. Alzheimers Res Ther. 2016;8:41. https://doi.org/10.1186/s13195-016-0208-8.
207. Nilsson J, Ashton NJ, Benedet AL, et al. Quantification of SNAP-25 with mass spectrometry and Simoa: a method comparison in Alzheimer's disease. Alzheimers Res Ther. 2022;14(1):78. https://doi.org/10.1186/s13195-022-01021-8.

208. Vila-Castelar C, Udeh-Momoh C, Aggarwal NT, Mielke MM. Sex and gender considerations in dementia: a call for global research. Nat Aging. 2023;3(5):463–5. https://doi.org/10.1038/s43587-023-00374-5.
209. Shang X, Roccati E, Zhu Z, et al. Leading mediators of sex differences in the incidence of dementia in community-dwelling adults in the UK biobank: a retrospective cohort study. Alzheimers Res Ther. 2023;15(1):7. https://doi.org/10.1186/s13195-022-01140-2.
210. Martin J, Reid N, Ward DD, King S, Hubbard RE, Gordon EH. Investigating sex differences in risk and protective factors in the progression of mild cognitive impairment to dementia: a systematic review. J Alzheimers Dis JAD. 2024;97(1):101–19. https://doi.org/10.3233/JAD-230700.
211. Powell WR, Buckingham WR, Larson JL, et al. Association of neighborhood-level disadvantage with Alzheimer disease neuropathology. JAMA Netw Open. 2020;3(6):e207559. https://doi.org/10.1001/jamanetworkopen.2020.7559.
212. Dintica CS, Bahorik A, Xia F, Kind A, Yaffe K. Dementia risk and disadvantaged neighborhoods. JAMA Neurol. 2023;80(9):903–9. https://doi.org/10.1001/jamaneurol.2023.2120.
213. Xiong C, Schindler S, Luo J, et al. Baseline levels and longitudinal rates of change in plasma Aβ42/40 among self-identified Black/African American and White individuals. Res Sq. Published online January 8, 2024:rs.3.rs-3783571. https://doi.org/10.21203/rs.3.rs-3783571/v1.
214. Hajjar I, Yang Z, Okafor M, et al. Association of plasma and cerebrospinal fluid Alzheimer disease biomarkers with race and the role of genetic ancestry, vascular comorbidities, and neighborhood factors. JAMA Netw Open. 2022;5(10):e2235068. https://doi.org/10.1001/jamanetworkopen.2022.35068.
215. Tarawneh R. Biomarkers: our path towards a cure for Alzheimer disease. Biomark Insights. 2020;15:1177271920976367. https://doi.org/10.1177/1177271920976367.

Fluid and Imaging Markers of Frontotemporal Lobar Degeneration

Fieke A. M. Prinse, Julie F. H. De Houwer, Elise G. P. Dopper, and Harro Seelaar

Authors Elise G.P. Dopper and Harro Seelaar are the shared last authors and they have contributed equally to this chapter.

Abstract Frontotemporal lobar degeneration (FTLD) consists of a heterogeneous spectrum of pathologies and genetic underpinnings, associated with a spectrum of clinical phenotypes. As a result, reaching a definitive diagnosis, differentiating FTLD from other neurodegenerative disorders with overlapping features, tracking disease progression, and evaluating outcomes in therapeutic trials remain challenging. In this chapter, we review the current knowledge of fluid and imaging biomarkers for FTLD. We discuss markers that have been implemented in clinical practice, including anatomical T1-weighted magnetic resonance imaging (MRI) to evaluate grey matter atrophy, 18F-fluorodeoxyglucose positron emission tomography (PET) imaging which measures regional hypometabolism, and cerebrospinal fluid (CSF) or blood levels of neurofilament light chain as a nonspecific marker for neurodegeneration. We also discuss how combinations of biomarkers may be used to differentiate FTLD from Alzheimer's disease (AD) or distinguish among different FTLD

Authors Fieke A. M. Prinse and Julie F. H. De Houwer are the shared first authors and they have contributed equally to this chapter.

F. A. M. Prinse
Department of Neurology, Alzheimer Center Erasmus MC, Erasmus University Medical Center, Rotterdam, The Netherlands

Department of Radiology, C. J. Gorter Center for MRI, Leiden University Medical Center, Leiden, The Netherlands

J. F. H. De Houwer · E. G. P. Dopper · H. Seelaar (✉)
Department of Neurology, Alzheimer Center Erasmus MC, Erasmus University Medical Center, Rotterdam, The Netherlands
e-mail: h.seelaar@erasmusmc.nl

R. Tarawneh (ed.), *Biomarkers of Neurodegenerative Disorders*,
https://doi.org/10.1007/978-3-031-80849-4_9

pathological subtypes. In addition, we discuss several biomarkers that may offer potential as diagnostic or predictive markers in FTLD but need more thorough investigation, such as molecular markers that can distinguish between different types of FTLD pathology or those that reflect common mechanisms in neurodegenerative disorders such as neuroinflammation and synaptic injury. Finally, we review promising developments in the field of FTLD biomarkers and propose the need for a multimodality approach in differential diagnosis and prognostication.

Keywords Frontotemporal lobar degeneration · Frontotemporal dementia · Fluid · Neuroimaging · Biomarkers

1 Background

Frontotemporal lobar degeneration (FTLD) refers to a heterogeneous spectrum of neuropathological changes characterized by relatively selective degeneration of the frontal and/or temporal lobes [1]. FTLD is the second most common cause of dementia under the age of 65, with an incidence estimate of ~10.8 cases per 100,000 and an almost equal distribution between women and men. The mortality rate is 1.56/100,000 person per year [2].

Clinical presentations of FTLD include the behavioral variant of frontotemporal dementia (bvFTD) [4], primary progressive aphasia (PPA), and the atypical parkinsonian syndromes of progressive supranuclear palsy (PSP) and corticobasal syndrome (CBS). Individuals with FTLD may also develop motor neuron symptoms, sometimes fulfilling criteria for amyotrophic lateral sclerosis (ALS) (Fig. 1). For an extensive and recent review of FTLD, see Grossman et al. [3].

1.1 Clinical Spectrum

BvFTD is characterized by progressive behavioral problems, including disinhibition, apathy, lack of empathy and sympathy, compulsions, and executive dysfunction with relative sparing of memory and visuospatial functions [4]. Individuals with PPA present predominantly with language disturbances. Two subvariants of PPA are commonly associated with FTLD pathology: nonfluent variant PPA (nfvPPA) and semantic variant PPA (svPPA). Individuals with nfvPPA present with slow effortful speech, speech apraxia, and/or agrammatism, while individuals with svPPA present with anomia and impaired object knowledge [5]. PSP typically manifests with postural instability, early recurrent falls, oculomotor dysfunction, parkinsonism, and executive dysfunction [6]. CBS presents with asymmetric motor symptoms, including parkinsonism, myoclonus, and/or dystonia combined with higher cortical dysfunctions such as apraxia, alien limb phenomenon, aphasia, and cortical sensory deficits [7].

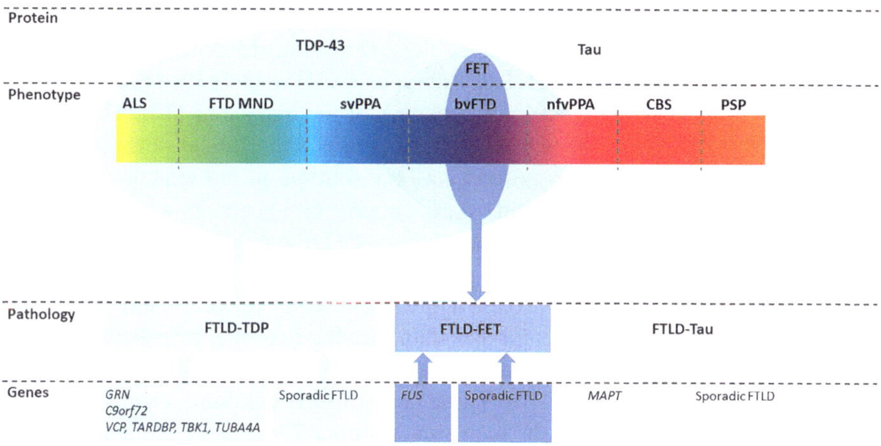

Fig. 1 Clinical, pathological, and genetic spectrum of FTLD. Genetic forms have predictable pathology: *GRN* mutations and *C9orf72* hexanucleotide repeat expansions (HRE) show TDP-43 pathology, whereas *MAPT* mutations show tau pathology. The clinical FTLD syndromes have more variable underlying pathological and genetic causes. However, ALS and svPPA are generally associated with TDP-43 pathology; while PSP, CBS, and nfvPPA are usually associated with tau-pathology. Abbreviations: ALS amyotrophic lateral sclerosis, bvFTD behavioral variant of FTD, C9orf72 chromosome 9 open reading frame 72, CBS corticobasal syndrome, FET FUS (fused in sarcoma) EWS (Ewing sarcoma) and TAF15 (TATA-binding associated factor 15), FTD frontotemporal dementia, FUS fused in sarcoma, GRN progranulin, MAPT microtubule-associated protein tau, MND motor neuron disease, nfvPPA nonfluent variant of primary progressive aphasia, PSP progressive supranuclear palsy, svPPA semantic variant of primary progressive aphasia, TARDBP TAR DNA-binding protein, tau tubulin associated unit, TBK1 TANK-binding kinase 1, TDP-43 transactive response DNA-binding protein, TUBA4A tubulin alpha 4a, VCP valosin-containing protein. (Adapted with permission from Meeter et al. 2017 [211]. Copyright © 2017, Springer Nature Limited; permission obtained through the Copyright Clearance Center, Inc).

1.2 Pathology

The pathological substrate of FTLD is the aggregation of insoluble proteins forming pathological inclusions within neurons. The pathological classification of FTLD is based on the immunohistochemical identification of those proteins (Fig. 1). The vast majority of FTLD can be divided into two major groups: FTLD-TAR DNA-binding protein (FTLD-TDP; approximately 50%) and FTLD-tubulin-associated unit (FTLD-tau; 45%). Less than 5% of FTLD cases are FET (FUS (Fused in sarcoma), EWS (Ewing sarcoma) and TAF15 (TATA-binding associated factor 15)) proteinopathies and a minority cannot be explained by these proteinopathies and are called FTLD with ubiquitin-positive inclusions (FTLD-UPS) [8].

FTLD-TDP is characterized by accumulation of TDP-43, a DNA- and RNA-binding protein which is involved in various aspects of RNA metabolism. TDP-43 regulates DNA and RNA splicing, stability, turnover, and trafficking [9]. Pathological forms of TDP-43 become hyperphosphorylated, ubiquitinated, and cleaved at the N-terminus [10–13] and accumulate in the cytoplasm, dendrites, axons, and the nucleus. Four established FTLD-TDP subtypes (A–D) can be distinguished based on the morphology and the distribution pattern of the inclusions [14]. These different subtypes are associated with distinct mutations and clinical phenotypes [15]. Recently, TDP type E was proposed as a new subtype in individuals with FTLD who had rapid clinical deterioration over the course of 3 years. However, it is debatable whether TDP type E is a separate subtype of TDP pathology or a continuum of TDP type B pathology [16].

In FTLD-tau, intraneuronal protein aggregates consist predominantly of tau. Tau is a microtubule-associated protein that stabilizes the neuronal cytoskeleton, which is encoded by *MAPT* gene, and is present in six isoforms in the brain due to alternative splicing in exons 2, 3, and 10. These six isoforms exhibit distinctions in their N-terminus, with variances in the inclusion of either 29 or 58 amino acids, as well as the composition of exon 10, which comprises either three (3R) or four (4R) consecutive tandem repeat sequences of 31 or 32 amino acids [17]. These 3R and 4R variants are almost equally abundant in the healthy human brain. Tau is involved in microtubule assembly, stabilization of the cytoskeleton, and axonal transport [18, 19]. Acetylation or hyperphosphorylation can cause aggregation of tau and neurotoxicity as it impairs microtubule stability and axonal transport [20–24]. Accumulation of 3R-tau most often leads to clinical syndromes of bvFTD or PPA [25, 26], whereas CBS and PSP are the most common clinical phenotypes associated with 4R-tauopathies [27, 28]. 4R-tauopathies can be further subdivided based on pathological hallmarks, such as the presence of tufted astrocytes in PSP pathology or astrocytic plaques in corticobasal degeneration (CBD) [17].

A subset of FTLD exhibits abnormal accumulation of FUS, EWS, or TAF-15 proteins, a protein family collectively called FET. This subtype is classified as FTLD-FET, although some articles refer to it as FTLD-FUS, the first identified protein of this cluster [29–32]. Most cases of FTLD-FET are sporadic [3]. Nearly all individuals with FTLD-FET with an underlying FUS pathology present with early onset sporadic bvFTD (<40 years), and demonstrate severe and rapidly progressive behavioral decline associated with a high prevalence of psychotic symptoms [33, 34].

1.3 Genetics

Although most cases of FTLD are sporadic, 20–30% of the cases are familial with an autosomal dominant pattern of inheritance (Fig. 1) [8]. We refer in this chapter to genetic FTLD when a pathogenic variant has been found in the patient. The three most common genes associated with FTLD include: *microtubule-associated protein tau* (*MAPT*) and *progranulin* (*GRN*) mutations, and the *chromosome 9 open reading frame* 72 hexanucleotide repeat expansion (*C9orf72* HRE). Other less commonly associated genes include *TAR DNA-binding protein* (*TARDBP*), *tubulin alpha 4a* (*TUBA4A*), *charged multivesicular body protein 2B* (*CHMP2B*), *valosin-containing*

Table 1 Overview of the most promising fluid biomarkers in FTLD

	Unspecified FTD	FTLD-GRN		FTLD-C9orf72		FTLD-MAPT		Potential value as a biomarker	Ref.
		P	S	P	S	P	S		
NfL	↑	↑	↑	↑	↑	↑	↑	Diagnosis of FTLD Proximity to symptom onset in genetic FTLD Prognosis Evaluation of therapeutic response	65-91
Tau	-	-	↑	-	-	-	↑	Differential diagnosis with AD	
DPR	-			↑	↑			Differentiate mutation carriers and non-carriers Target engagement	114-118
PGRN	-	↓	↓	-	-			Differentiate mutation carriers and non-carriers Target engagement	122-132
GFAP	↑	-	↑	-	-	-	-	Differential diagnosis with other neurodegenerative diseases	155, 157-158
NPTX2	↓	↓	↓	↓	↓	↓	↓	Prognosis Proximity to symptom onset	193-194

– indicates unchanged levels; ↑ indicates increased levels; ↓ indicates decreased levels; all compared with controls. The size of the arrow indicates the magnitude of change. Yellow, in blood and cerebrospinal fluid; blue, in cerebrospinal fluid.

Abbreviations: *NfL* neurofilament light chain, *DPR* dipeptide repeat proteins, *PGRN* progranulin, *GFAP* glial fibrillary acidic protein, *NPTX2* neuronal pentraxin-2, *P* presymptomatic mutation carriers, *S* symptomatic mutation carriers, *FTLD* frontotemporal lobar degeneration, *AD* Alzheimer's disease, *FTD* frontotemporal dementia

protein (*VCP*), *TANK-binding kinase 1* (*TBK1*), and *Fused in sarcoma* (*FUS*). Mutations in these latter genes contribute to the clinical and pathological heterogeneity observed in FTLD cases (Table 1).

The most common genetic cause of FTLD is a hexanucleotide repeat expansion (HRE) of GGGGCC in the first intron of *C9orf72*, explaining approximately 25% of familial FTLD [35, 36] and 6% of sporadic FTLD [37, 38]. Healthy individuals have a repeat length of 24 or less, with the majority only having two to eight repeats [35, 36]. The presence of more than 30 hexanucleotide repeats is considered a pathological expansion. Nevertheless, in most carriers, a substantially larger expansion is observed, spanning from hundreds to thousands of repeats. Three different pathophysiological mechanisms by which *C9orf72* HRE contributes to neurodegeneration have been proposed: (i) loss-of-function of *C9orf72*; (ii) formation of toxic RNA foci from repeat expansion transcripts; and (iii) formation and accumulation of dipeptide repeat proteins (DPR) via repeat-associated non-ATG-initiated translation (RAN-translation) [39]. At the pathologic level, *C9orf72* HREs are often associated with FTLD-TDP type B but sometimes TDP type A or a mixture of types A and B [40, 41]. Individuals may, however, present with various clinical phenotypes, most commonly bvFTD, ALS, or a psychiatric disorder [42, 43]. The mean age of onset is 58.2 ± 9.8 years with a mean disease duration of 6.4 years and a mean age at death of 65 years [44]. However, the age of symptom onset varies considerably across individuals [45]. This is reflected by a penetrance of 80–90% by the age of 80 years [44].

GRN mutations are responsible for 15–20% of familial FTLD cases [46]. *GRN* encodes progranulin, a growth factor involved in neuronal survival, inflammation, axonal growth, and lysosomal function [47, 48]. The mechanism by which *GRN* gene variations cause FTLD remains to be elucidated; however, it is thought that reduced progranulin levels lead to TDP-43 pathology, usually FTLD-TDP type

A. *GRN*-associated FTLD typically presents as bvFTD but may also manifest as nfvPPA or CBS [8, 49, 50]. *GRN* families display remarkable inter- and intra-familial variability in the age of symptom onset and the associated clinical phenotype [44, 51]. The mean age of onset is 61.3 ± 8.8 years, with an average disease duration of 7.1 years and age at death of 69 years [44].

MAPT mutations are implicated in 20% of familial FTLD [52]. *MAPT* encodes the tau protein and mutations are thus associated with FTLD-tau pathology. A mutation in the *MAPT* gene alters the physiological balance of the two tau isoforms (i.e., 3R and 4R), leading to hyperphosphorylation and impaired binding to the axonal cytoskeleton [53, 54]. Both 3R- and 4R-tau pathologies and their related clinical phenotypes can be caused by a *MAPT* mutation, depending on the location of the mutation [55]. The mean age of onset in individuals with a *MAPT* mutation is 49.5 ± 10 years, with a mean age at death of 59 years and a disease duration of 9.3 years [44].

We here summarize the current literature regarding the utility of fluid and imaging markers in assisting with the diagnosis of various FTLD subtypes, differentiating FTLD from AD and other neurodegenerative disorders, predicting age of symptom onset in asymptomatic FTLD gene mutation carriers, evaluating disease severity, and their potential use as stratification or outcome variables in clinical trials of disease-modifying therapies in FTLD.

2 Fluid Biomarkers

2.1 Introduction

Advances in fluid biomarkers are rapidly emerging as they are an objective indicator of normal and pathological biological processes in the brain. This makes biomarkers promising tools not only to better understand the underlying pathophysiology of a neurological disease but also to improve patient care. Up to this point, fluid biomarkers are not yet used in routine clinical practice in FTLD. Generally, fluid biomarkers are measured in cerebrospinal fluid (CSF) and blood (serum and plasma). More noninvasive ways, such as urine or saliva, are beyond the scope of this chapter. We will discuss the prospects of candidate biomarkers according to the following classification [56]: (i) diagnostic markers, aiding in the differential diagnosis between FTLD and other neurodegenerative and nonneurodegenerative disorders; (ii) prognostic markers, to support prediction of disease course and survival; (iii) staging markers, of specific importance in genetic FTLD to forecast proximity to symptom onset; and (iv) markers of therapeutic response, evaluating target engagement and efficacy. This section focuses on markers of neurodegeneration, markers for the underlying proteinopathy as well as markers for other major underlying pathological pathways (e.g., neuro-inflammation, synaptic dysfunction, and gene-specific markers).

2.2 Neurofilament

Over the past years, neurofilaments sparked interest as a promising biomarker for FTLD. Neurofilaments are class IV intermediate filaments that are abundantly present in the central and peripheral nervous system and are an integral part of the cytoskeleton. They consist of four subunits, classified according to their molecular weight: light chain (NfL), medium chain (NfM), heavy chain (NfH), and α-internexin. Neurofilaments are thought to play a role in nerve conduction velocity by modulating cell shape and axonal diameter [57–59]. Upon axonal injury, neurofilaments are released into the extracellular space and eventually into the CSF and the bloodstream. Thus, neurofilaments serve as an indirect marker of axonal injury, irrespective of the underlying cause. Therefore, elevated levels can be detected in multiple neurological conditions, such as multiple sclerosis, neurodegenerative disorders, stroke, and traumatic brain injury [58, 60, 61]. NfL is the most studied subunit, due to its abundance and solubility [57]. The development of a single molecule array (Simoa) has greatly improved the accuracy of NfL blood measurements [57, 58]. Ever since, multiple studies have demonstrated a close correlation between NfL levels in serum or plasma and levels in CSF [58, 62, 63].

Strongly elevated CSF or blood levels of NfL in individuals with FTLD have consistently been reported, with a diagnostic accuracy of over 90% to distinguish patients from healthy controls [62, 64–66]. Blood NfL concentrations are similar across the clinical phenotypes of FTLD, although concentrations are the highest in patients with concomitant ALS and lower levels are seen in patients with slow progression or *C9orf72* HRE carriers with a psychiatric phenotype [67, 68]. NfL can aid in the differentiation between FTLD and nonneurodegenerative disease, such as primary psychiatric disorders (PPD), a clinically relevant issue given the substantial symptomatic overlap. Generally, PPD patients have CSF and blood NfL levels that are comparable to healthy controls [69–73] while being elevated among various subtypes of FTLD. Therefore, high NFL levels, in the absence of Alzheimer's disease (AD) biomarkers, reflect a non-AD neurodegenerative disorder (e.g., FTLD) rather than PPD. NfL levels in CSF and blood are also higher in FTLD compared to other forms of dementia, such as AD or dementia with Lewy bodies (DLB) [73–77]. However, due to a considerable overlap in concentrations, the diagnostic accuracy to discriminate FTLD from other neurodegenerative disorders is low [73–77]. This limits the use of NfL as a disease-specific marker. Importantly, NfL in CSF and blood increases with age in healthy individuals [73, 78], with a steeper increase after the age of 60. It is hypothesized that this increase is due to age-related degenerative processes without overt clinical symptoms. Furthermore, multiple studies have shown a negative correlation between renal function and blood-based NfL, especially in older individuals [79–81]. To aid clinicians, a Shiny app has been developed with age- and disease-specific reference values [82].

Multiple studies have shown a strong correlation between CSF and blood concentrations of NfL and various clinical, neuropsychological, and imaging variables

[66, 67, 74, 75, 83–85]. Individuals with higher NfL levels have a shorter disease duration, more functional and cognitive impairment, more brain atrophy, and a shorter overall survival. This underlines the possibility of NfL as a prognostic marker.

An accurate estimation of symptom onset in presymptomatic mutation carriers would be extremely valuable for participant enrollment and screening in clinical trials. Reports of higher cross-sectional, baseline, blood NfL levels in the presymptomatic phase of disease in converters versus nonconverters highlight the potential of NfL as a marker for symptom onset [66, 85–88]. Elevated blood NfL levels are seen 1–2 years before symptom onset (i.e., conversion) in mutation carriers. Furthermore, this phase is marked by the highest rate of change in NfL levels [85, 87]. Over the past few years, efforts have been made to further elucidate the exact trajectory of NfL changes around conversion. For example, one study found higher serum NfL more than 3 years before conversion to the prodromal phase, with a steeper increase shortly after conversion [88]. Importantly the longitudinal trajectory depends on the causative gene and phenotype [67, 88, 89]. *GRN* carriers have higher baseline plasma levels and higher annualized rate changes (ARC) compared to *C9orf72* HRE carriers [67]. Recently, higher plasma NfL levels were observed in *GRN* carriers with, compared to those without, atypical white matter hyperintensities [90]. Moreover, among *C9orf72* HRE carriers, there is a large variability in baseline values and ARC depending on the phenotype, with higher levels in patients with ALS and lower values in patients with a psychiatric presentation [67, 68]. Current evidence demonstrates the potential of blood NfL to reflect disease progression, yet gene-specific cutoff thresholds will be imperative. To further demarcate the conversion period, a recent study investigated the added benefit of measuring NfH levels in serum [87]. This study suggests that the onset of conversion to the symptomatic stage is marked by increased NfL levels, but normal NfH levels, whereas its completion is marked by increased NfH levels. Further research will have to determine the clinical relevance of these findings.

Furthermore, NfL has potential as a marker of treatment effect. In multiple sclerosis, a decrease in NfL has been observed with disease-modifying drugs [91]. However, in genetic FTLD, it is still unknown if a decrease in prodromal levels or prevention of a further increase is possible and clinically meaningful [92].

In summary, NfL is a promising biomarker for several purposes. First, as a diagnostic marker to differentiate FTLD patients from healthy controls or patients with nonneurodegenerative diseases such as PPD. Second, as a marker of symptom onset in genetic FTLD, as evidenced by the fact that individuals who are close to conversion have higher NfL levels. Third, the baseline measurement can predict disease intensity and therefore, prognosis. Last, as a potential marker for therapeutic response in investigational treatment trials, as a decrease in NfL reflects a decrease in the rate of neurodegeneration.

2.3 *Gene-Specific Biomarkers*

2.3.1 Tau

Tau sparks interest as a possible biomarker, as underlying tau pathology represents 45% of all FTLD cases. However, in contrast to AD, CSF levels of phosphorylated tau_{181} ($p\text{-}tau_{181}$) are generally normal and levels of total tau (t-tau) are normal to elevated [93]. Nonetheless, in rare R406W *MAPT* mutations, elevated p-tau levels are reported [94]. This specific mutation has a mixed 3R/4R tau accumulation, similar to AD, which is reflected by the elevated p-tau levels at residues 181 or 217. Furthermore, elevated levels of t-tau can be variably seen in *GRN* mutation carriers [95, 96]. To date, the most promising use of tau in clinical practice is the differential diagnosis with AD. The core biomarkers for AD, for example, $A\beta_{42}$, $p\text{-}tau_{181}$, and t-tau, can exclude or confirm underlying AD pathology, yet they cannot confirm or rule out FTLD pathology. Importantly, co-pathology of AD and FTLD also occurs, which is particularly prevalent in older patients. Therefore, the underlying pathology cannot be determined solely by biomarker levels, and those levels should be interpreted in the context of the clinical and radiological presentation.

The $p\text{-}tau_{181}$/t-tau ratio has been suggested as a biomarker to differentiate FTLD-TDP from FTLD-tau. This ratio is reported to be lower in FTLD-TDP than FTLD-tau, possibly driven by the presence of concomitant ALS in some patients with FTLD-TDP [97–100].

In recent years, significant advances have been made in measuring p-tau isoforms in plasma. Concentrations of $p\text{-}tau_{181}$, $p\text{-}tau_{231}$, and $p\text{-}tau_{217}$ are higher in patients with AD relative to controls; however, an increase is not observed in FTLD [101–105]. These markers are specific to AD and can discriminate between AD and FTLD with nearly 100% accuracy [103–105]. Currently, studies are ongoing to determine the potential added benefit of measuring other blood-based biomarkers in clinical practice beyond those that reflect the core AD pathologies of amyloid and tau (i.e., Aβ42 and $p\text{-}tau_{181}$).

Thus far, pathological tau-isoforms specific for 3R- and 4R-tauopathies are yet to be identified. Assays based on real-time quaking-induced conversion (RT-QuIC), with promising results in Pick's disease and the 4R-tauopathies, are currently under investigation [106–109].

2.3.2 Dipeptide Repeat Proteins in C9orf72-related Disease

As described earlier, accumulation of dipeptide repeat proteins (DPR) is one of the three postulated pathophysiological mechanisms by which *C9orf72* HRE contributes to neurodegeneration. The HRE is bidirectionally transcribed into RNA, which forms sense and antisense RNA foci. Five distinct DPRs are formed, poly(GA), poly(GP), poly(GR), poly(PR), and poly(PA), via RAN-translation. DPRs are abundantly found in the brains of patients with FTLD and ALS associated with

C9orf72 HRE. They are mainly seen in cytoplasmic inclusions. However, their presence does not correlate with the degree of neurodegeneration, making the contribution of DPR-associated toxicity to disease progression and the relative pathogenic contribution of each individual DPR a topic of debate. The arginine containing DPRs, poly(GR) and poly(PR), are thought to be the most toxic. The abundantly present poly(GA) might be toxic to a lesser extent [110].

Poly(GP) is the most studied DPR due to its abundance and solubility. Poly(GP) has been consistently shown to be elevated in the CSF of individuals with a *C9orf72* HRE, but not detectable in healthy controls or individuals with sporadic FTLD [111–114]. CSF levels of poly(GP) are already increased in the presymptomatic phase of disease [111–114]. A trend towards higher levels is seen in symptomatic carriers compared to presymptomatic *C9orf72* HRE carriers. However, only one study found significantly higher levels after adjusting for age and gender [113]. Based on these results, poly(GP) could serve as a diagnostic marker to differentiate *C9orf72* HRE carriers from noncarriers with high sensitivity and specificity. Nonetheless, given the lack of correlation with clinical markers and markers of neurodegeneration, the potential prognostic value of poly(GP) is limited [111–114]. To date, only one study has investigated other DPRs, namely, poly(GR) and poly(GA), which reported similar results as for poly(GP) [115]. Further, DPRs have the potential to be used as markers of target engagement as one study showed decreased levels of poly(GP) in a cell and mouse model after treatment with an antisense oligonucleotide (ASO) targeting the *C9orf72* HRE [111]. Additionally, one patient treated with an ASO, showed a significant decrease in poly(GP), poly(GR), and poly(GA) 6 months after treatment [115, 116].

2.3.3 Progranulin

Progranulin (PGRN) is a glycoprotein that is encoded by the *GRN* gene and is ubiquitously expressed by multiple cell types in the brain and the periphery. In the brain, PGRN is mainly found in neurons and microglia, where it is involved in multiple cellular processes, including axonal growth, lysosomal function, and neuroinflammation [117]. Heterozygous mutations in *GRN* are a common cause of genetic FTLD [118]. Most mutations are frameshift, nonsense, or splice site mutations and introduce a premature stop codon, resulting in reduction in PGRN levels through haploinsufficiency.

Immunoassays can measure PGRN in the blood and CSF. Lower PGRN levels are consistently reported in *GRN* carriers with a diagnostic accuracy of nearly 100% in differentiating mutation carriers from sporadic FTLD patients [119–126]. Reduced PGRN levels are already observed in the presymptomatic stage in mutation carriers, as early as the late teens [122, 124–126], which suggests that levels may be low at birth. Moreover, stable PGRN levels have been reported over a 4-year longitudinal follow-up [126]. PGRN has potential as a diagnostic marker, to accurately detect a pathogenic *GRN* variant prior to genetic testing or to determine the pathogenicity of a variant of unknown significance. Nonetheless, given the lack of correlation with the extent of neurodegeneration or clinical measures, PGRN has limited use as a prognostic or disease stage marker [125, 126]. PGRN is currently used as a marker of target engagement in clinical trials. A phase 2 trial of a monoclonal antibody against sortilin showed plasma and CSF levels of PGRN restored back to the normal range [127]. Importantly, a weak correlation between blood and CSF levels is seen, with only 6–29% of variability in CSF levels explained by plasma levels [64, 124, 128, 129]. This warrants caution when using plasma PGRN as a pharmacodynamic marker, since an increase in plasma levels does not necessarily reflect an increase in CSF levels. Furthermore, considerable variability exists in peripheral PGRN levels within individuals. Various factors, such as metabolic syndrome, auto-immune disease, pregnancy, and cancer, influence CSF levels [130, 131]. In addition, multiple genetic factors influencing PGRN levels have been identified, namely *GRN* rs5848, *SORT1* rs646776, and possibly *TMEM106b* rs1990622 polymorphisms [35, 124, 132–134].

2.3.4 TDP-43

There have been numerous studies investigating the potential of TDP-43 as a biomarker in FTLD-TDP, albeit with inconsistent results. For example, most studies observed elevated CSF levels in sporadic ALS [135–138], yet decreased CSF levels are reported in a neuropathologically confirmed study of FTLD-TDP [139]. Similarly, variable results are seen in the blood [135, 140–143]. One study found lower serum total TDP-43 levels in *C9orf72* HRE carriers and FTLD-ALS patients, but not in *GRN* mutation carriers. It has been suggested that part of the variation can be explained by differences in the predominance of various subtypes of TDP-43 pathology (e.g., type B versus type A) [142]. Thus far, detection of phosphorylated TDP-43 in CSF has not been successful [139, 141]. In plasma, however, higher levels of phosphorylated TDP-43 are found in *GRN* and *C9orf72* HRE carriers compared to controls and individuals with sporadic FTLD [141].

So far, it has been difficult to differentiate physiological from pathological forms of TDP-43. Recently, a study found higher levels of TDP-43 oligomers in plasma in a small cohort of patients diagnosed with svPPA which is associated with FTLD-TDP type C, compared with other neurodegenerative diseases and healthy controls [143]. However, the role of TDP-43 oligomers in physiology versus pathology is still under debate [144].

A promising technological development is RT-QuIC. A protocol was developed using a recombinant C-terminal fragment [145] and validated in a small, genetic cohort. A positive result was obtained in TDP-43 positive samples with a sensitivity of 94% and specificity of 85%. Nonetheless, the investigators reported a number of false positive results, possibly due to the presence of low levels of TDP-43 aggregates in healthy controls or the heterogeneous presence of contaminants in the CSF. Optimization of the protocol could lead to a breakthrough in the detection of pathological forms of TDP-43 in CSF, with the possibility to serve as a biomarker to distinguish FTLD-TDP from FTLD-tau.

2.4 Markers of Neuroinflammation

Neuroinflammation is being recognized as a pivotal component in the pathophysiology of FTLD (reviewed in [146]). It is a complex and multistage process that involves innate immune resident central nervous system glial cells and pro- and anti-inflammatory factors. To date, it is debated whether neuro-inflammation is a primary or secondary event, with beneficial or detrimental effects. In this section, we will discuss possible markers related to these processes.

2.4.1 Glial Markers

Glial cells, such as microglia and astrocytes, are fundamental to the inflammatory response in the brain. Glial activation leads to the production and release of inflammatory cytokines and chemokines, which modulate a pro- or anti-inflammatory environment. In the next segment, we will discuss the following markers of glial activation: glial fibrillary acidic protein (GFAP), YKL-40 (otherwise known as chitinase-3-like protein 1 or CHI3L1), chitotriosidase (CHIT1) and soluble triggering receptor expressed on myeloid cells 2 (sTREM2).

GFAP is an intermediate filament and part of the cytoskeleton in astrocytes [147]. It is a marker of astrogliosis as it is produced and released upon neuronal injury. In AD, GFAP is a promising diagnostic and prognostic biomarker [148]. In FTLD, however, less consistent results have been reported. In mixed FTLD cohorts, elevated CSF levels are measured compared to controls [149–151]. In blood, some studies report higher levels [152–154], yet others found no differences [151]. Nonetheless, levels are not increased to the extent seen in AD. Interestingly, several studies suggest higher serum or plasma levels in symptomatic *GRN* carriers compared to symptomatic *MAPT* and *C9orf72* HRE carriers [152, 154, 155]. Serum or plasma GFAP has potential as a marker for the differential diagnosis with other neurodegenerative disorders, with a diagnostic accuracy of around 80% to differentiate AD and FTLD [154]. One study suggested that FTLD-tau and FTLD-TDP can be discriminated by the ratio of plasma GFAP and NfL [156]. Furthermore, GFAP is also a possible prognostic biomarker as some studies showed a correlation with functional impairment, disease severity, and brain atrophy [152, 157].

YKL-40, a chitinase, is a pro-inflammatory molecule and is mainly produced by activated astrocytes [158, 159]. Elevated levels of YKL-40 have been observed in multiple acute and chronic neurological disorders [151, 160–163]. In FTLD, however, conflicting results are reported. For example, higher levels are seen in patients with a clinical diagnosis of bvFTD or PPA [163]. Importantly, within this group, patients with a CSF profile compatible with underlying AD pathology had the highest levels. Nonetheless, in a genetic FTLD cohort, there was no difference in YKL-40 compared to controls [164]. To date, YKL-40 appears to have limited value as a biomarker for FTLD.

CHIT1 is part of the chitinases and is secreted by activated microglia. Elevated levels are reported in ALS [151]. In FTLD, however, inconsistent results are noted. CHIT1 levels in CSF were found to be increased in a cohort with sporadic FTLD, compared to controls [150, 163], yet another study did not find a difference between genetic FTLD and controls [151]. In a large, well-defined genetic cohort, higher levels are reported in symptomatic *GRN* carriers, although not in symptomatic *C9orf72* HRE or *MAPT* carriers nor in presymptomatic carriers [164]. Interestingly, there is a negative correlation with the Mini-Mental State Examination (MMSE) scores in presymptomatic *GRN* carriers, suggesting that CHIT1 increases in proximity to symptom onset [164]. Of note, a polymorphic 24-bp duplication in exon 10 of the *CHIT1* gene may reduce activity of CHIT1 [151, 165]. This polymorphism is highly prevalent in European populations (35–50%) [166], which potentially limits the use of CHIT1 as a potential biomarker.

TREM2 is a transmembrane protein with a secreted extracellular ectodomain (sTREM2), exclusively expressed in myeloid cells such as microglia. Genetic variants of *TREM2* are associated with an increased risk of neurodegenerative diseases [146]. Nonetheless, several studies suggest that sTREM2 levels in CSF do not differ between FTLD patients and healthy controls [164, 167, 168]. Interestingly, a large variability in individual concentrations is noted, especially among *GRN* carriers [164, 168]. It is suggested that, comparable to AD [169], high sTREM2 levels in the presymptomatic stage correlate with a delayed disease onset [164, 168]. Therefore, sTREM2 has potential as a prognostic biomarker in genetic FTLD.

2.4.2 Cytokines and Chemokines

Upon exposure to pathogens or inflammatory stimuli, microglia become activated and produce different molecules such as cytokines and chemokines [146, 159]. There have been numerous studies to investigate various cytokines and chemokines, including interleukin (IL)-6 and tumor necrosis factor (TNF)-α, in CSF and blood with conflicting results [170–178]. A recent report investigated the prognostic value of multiple inflammatory markers in a cohort with genetic FTLD. They report an association between higher levels of TNF-α with disease severity and cognitive impairment in *MAPT* and *C9orf72* HRE carriers [179], suggesting the potential of TNF-α as a prognostic marker. The current evidence underlines the role of peripheral inflammation in the pathogenesis of FTLD. Nonetheless, convincing evidence for the application of cytokines and chemokines as biomarkers is lacking.

2.4.3 Complement

The complement system consists of multiple plasma proteins that are activated by pathogens, leading to an inflammatory response to help fight infections. C1q is an essential component of the classical complement pathway and leads to activation of C3b, which eventually leads to a pro-inflammatory state. It is thought that the accumulation of C3b on synapses contributes to the synaptic loss observed in several neurodegenerative disorders [180].

Elevated CSF and plasma levels of complement are seen in various neurodegenerative diseases [181–183]. A small number of studies have investigated complement levels in FTLD, showing elevated CSF levels of C1q and C3b [180, 184, 185]. In a genetic FTLD cohort, complement levels were found to correlate with regional grey matter volume and plasma NfL levels in presymptomatic mutation carriers [185], pointing to complement as a possible marker for proximity to symptom onset. Further research is necessary to elucidate the potential of complement as a biomarker. Nonetheless, given the substantial overlap in concentrations between patients and controls [185] and increase in other neurodegenerative diseases, its use as a diagnostic marker is limited.

2.5 Synaptic Markers

Synaptic dysfunction is a key pathological feature among multiple neurodegenerative diseases which has led to the investigation of many synaptic proteins as potential markers in FTLD.

Neuronal pentraxins consist of three proteins, namely, neuronal pentraxin 1 (NPTX1), neuronal pentraxin 2 (NPTX2), and neuronal pentraxin receptor (NPTXR). They are part of a family of extracellular scaffolding proteins that modulate synaptic function and plasticity [186]. Thus far, neuronal pentraxins can only be measured in CSF. Proteomics identified the neuronal pentraxins as candidate biomarkers [187, 188], showing reduced CSF levels of NPTXR in patients in genetic and sporadic FTLD. Decreased NPTX2 levels have been consistently shown in symptomatic mutation carriers (*GRN, C9orf72, MAPT)* and individuals with sporadic FTLD [189–191]. Similar results are seen for NPTX1 and NPTXR in genetic FTLD [189, 190]. Furthermore, NPTX2 concentrations are associated with several clinical measures of disease severity, such as the Clinical Dementia Rating (CDR)® plus NACC-FTLD [189, 190]. In presymptomatic *GRN* mutation and *C9orf72* HRE carriers, but not in *MAPT* carriers, levels of neuronal pentraxins correlate with brain volumes, suggesting that their levels may change in close proximity to symptom onset [190]. One study constructed a disease progression timeline of genetic FTLD, using discriminative event-based modeling, and found that CSF NPTX2 was the first biomarker to become abnormal, followed by NfL [192]. These findings support the potential value of CSF NPTX2 and other neuronal pentraxins as predictive markers for symptom onset in genetic FTLD.

Limited studies have investigated other synaptic proteins, such as synaptosomal-associated protein-25 (SNAP-25), synaptotagmin-1, vesicle-associated membrane protein-2 (VAMP2), and neurogranin, although with conflicting results [191, 193, 194]. More research is needed to determine their role in the pathophysiology of FTLD and their potential as FTLD biomarkers.

3 Imaging Biomarkers

3.1 Introduction

Neuroimaging techniques, particularly magnetic resonance imaging (MRI) and positron emission tomography (PET), play a crucial role in the diagnosis of FTLD in clinical practice and are part of the clinical criteria for bvFTD, PPA, CBS, and PSP [5, 6]. MRI scans reveal characteristic atrophy in the frontal and temporal lobes, while PET scans using [^{18}F]-Fluorodeoxyglucose (also referred to as fluorodeoxyglucose-F-18) can assess the glucose metabolism in the brain, a marker for regional cortical hypometabolism in the brain. These specific patterns of atrophy and hypometabolism help differentiate FTLD from other causes of cognitive disturbances (e.g., AD, Parkinson's disease dementia, vascular dementia) or psychiatric disorders. In light of recent developments in investigational treatments that target specific proteins implicated in FTLD pathogenesis, it has become increasingly important to differentiate between the various pathological subtypes of FTLD. Moreover, the FTLD brain begins to change many years before the first clinical symptoms appear, which could be used to predict symptom onset. These brain changes appear at the molecular (tau and TDP-43 deposition), cellular (hypometabolic changes due to loss of neurons and synapses or reduced neuronal activity), and tissue level (changes in grey and white matter connectivity or atrophy) [195]. It will be important to develop neuroimaging biomarkers that can help accurately diagnose and predict FTLD subtypes and monitor disease progression. In this section, we will first describe patterns of brain atrophy in grey and white matter associated with various FTLD subtypes using structural imaging. We will then review functional connectivity in the FTLD brain, and propose several emerging imaging biomarkers for FTLD.

3.2 Structural Imaging

3.2.1 Grey Matter

The gold standard to evaluate changes in grey matter structure is visual assessment of a volumetric T1-weighted MRI by a neuroradiologist. Different scales for visual assessment, such as GCA for global cortical atrophy [196, 197] and MTA for mediotemporal atrophy, [198] can aid in this assessment. To date, various automated quantitative methods or post-processing image analysis tools can be used to measure brain volume and the rate of brain atrophy in specific brain regions in research and clinical settings [199–203]. These automated methods generate regional brain volumes for the study participant and compare these to norms generated from a healthy population of the same age group. The use of quantitative software improves diagnostic accuracy and improves inter-rater reliability for visual assessments

[204]. A combination of three volumetric MRI biomarkers is able to discriminate FTLD and non-FTLD, for example, AD, vascular dementia, mild cognitive impairment, and dementia with Lewy bodies, and to separate FTLD subtypes: (1) the anterior-posterior index, (2) the asymmetry index, and (3) the volume of the left temporal pole. The results showed that these biomarkers can be used to improve differential diagnosis of FTLD, such as nvPPA and svPPA [205].

FTLD Clinical Syndromes

FTLD is characterized by degeneration of the frontal and/or temporal lobes, which can be visualized on T1-weighted MRI. Depending on the location of the atrophy, different symptoms and clinical subtypes can arise. These specific patterns of atrophy can support clinical diagnoses within the FTLD spectrum. BvFTD is typically associated with bilateral atrophy in the prefrontal and anterior temporal lobes and, in most cases, involvement of the anterior cingulate, anterior insula, orbital and medial frontal lobes [4]. Therefore, bvFTD is characterized by atrophy of brain regions involved in the basal and limbic networks. Conversely, individuals with nfvPPA have prominent atrophy in the left posterior-inferior frontal regions, for example, Broca's area and the superior premotor cortex, which then spreads to the prefrontal cortex and basal ganglia over time. In contrast, individuals with svPPA present with atrophy in the anteromedial temporal lobe(s) and later in the disease, also in the insula, and orbitofrontal cortex [5]. Individuals with PSP have mild atrophy in the frontal lobes, and characteristic flattening or concavity of the superior aspect of the midbrain, which is referred to as the "hummingbird sign" (as it resembles a side view of a hummingbird's elongated peak on midline sagittal T1 sequences) [6, 206–209]. The posterior frontal and anterior parietal (the peri-Rolandic) lobes can show substantial asymmetric atrophy in CBS with additional involvement of the basal ganglia [7]. The rate with which the brain atrophy progresses in individuals with CBD has a steeper decline compared to those with PSP [214].

Pathological and Genetic FTLD Subtypes

The different genetic and pathological subtypes of FTLD have distinct patterns of grey matter atrophy at the group level, which spread uniquely throughout the brain and progress differentially over time. However, rates of regional brain atrophy do not typically aid in identifying the underlying pathology at an individual level [210, 211].

In FTLD-TDP type A, brain atrophy has been found to be widespread and asymmetric, with either left or right predominance, involving the frontal, temporal, and parietal lobes [212, 213]. The most significantly affected regions include the dorsal frontal, anterior, medial, and posterior temporal and the inferior parietal lobes, orbitofrontal cortex, insula, frontal operculum, caudate, and thalamus (Fig. 2) [25,

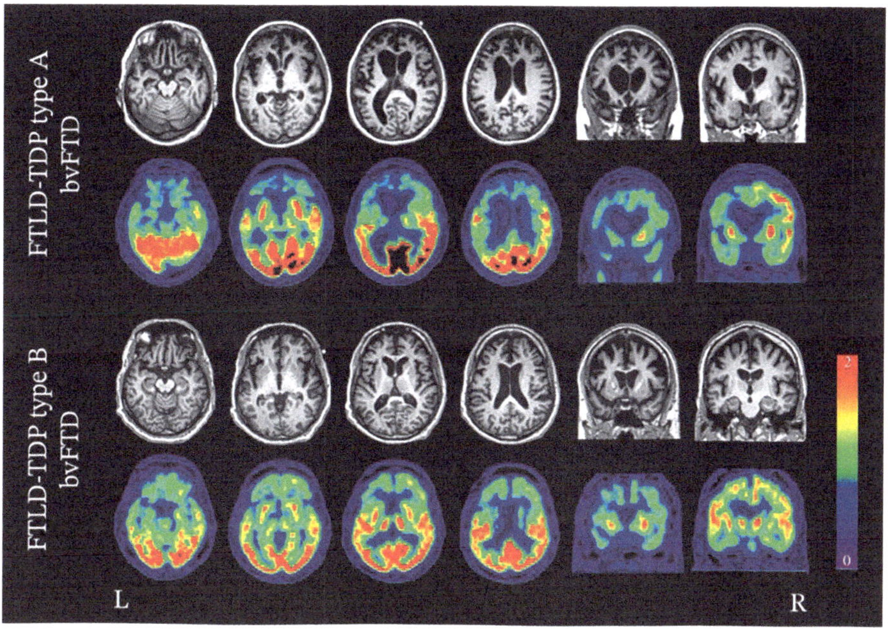

Fig. 2 Atrophy patterns of bvFTD associated with FTLD-TDP type A (upper panel) and type B (lower panel) visible on structural MRI (top row) and FDG-PET (bottom row). TDP type A is associated with degeneration and hypometabolism of the bilateral frontal lobes extending to the parietal lobes with the left being more impacted than the right. TDP type B is associated with relatively symmetrical degeneration and hypometabolism in the bilateral frontal lobes; however, the parietal lobes are less affected compared to TDP type A. (Taken with permission from Peet et al. 2021 [224]; Open Access through a Creative Commons license. https://creativecommons.org/licenses/by/4.0/). Abbreviations: *FTLD* frontotemporal lobar degeneration, *TDP* TAR DNA-binding protein, *bvFTD* behavioral variant of frontotemporal dementia

212–214]. In particular, involvement of the anterior insula, frontal operculum, and parietal lobe can help differentiate FTLD-TDP type A from the other FTLD-TDP subtypes [25, 213, 214]. Individuals with TDP type A due to an underlying *GRN* mutation have severe atrophy of the lateral temporal and medial parietal lobes, which can help differentiate *GRN* mutations from other genetic causes of FTLD-TDP [215–218]. In FTLD-TDP type B, atrophy is relatively symmetric and involves the medial and lateral frontal lobes. In addition, limbic and subcortical structures are also significantly involved, mostly affecting the anterior cingulate gyrus, anterior insula, and hippocampi (Fig. 2) [25, 213, 214]. The largest neuroimaging studies involving individuals with FTLD-TDP type B included those with *C9orf72* HRE. These cases usually present with mild cortical atrophy, which is widespread and symmetric [213]. Additionally, atrophy in these individuals is also present in the thalamus, inferior and superior parietal lobe, and the cerebellum [218–220]. FTLD-TDP type C pathology typically presents with asymmetric degeneration of the medial aspect of the anterior temporal lobes, specifically the temporal lobe pole,

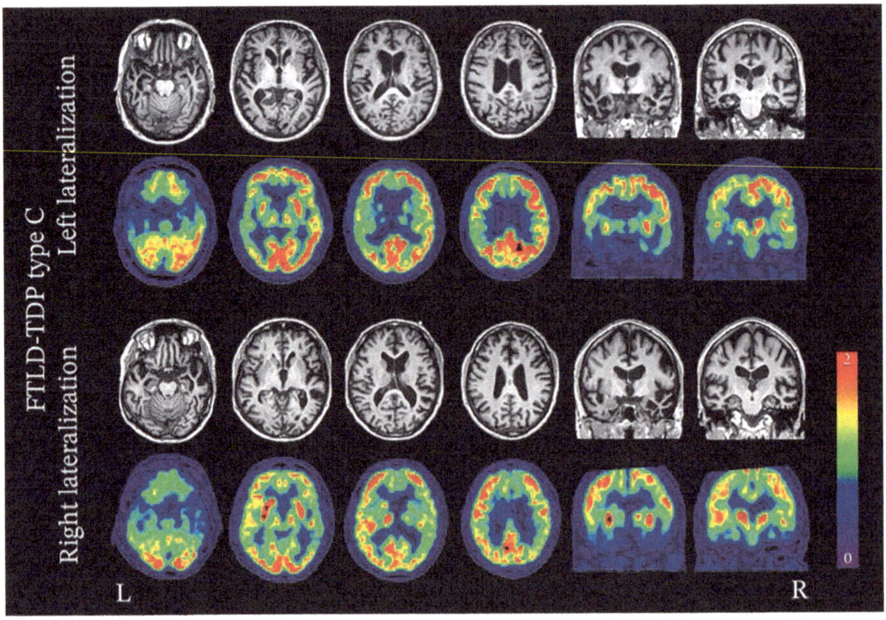

Fig. 3 Atrophy (top row) and hypometabolism (bottom row) patterns associated with FTLD-TDP type C. TDP type C exhibits anterior temporal lobe degeneration either isolated in the left hemisphere (upper panel), associated with svPPA or the right hemisphere (lower panel), associated with bvFTD. (Taken with permission from Peet et al. 2021 [224]; Open Access through a Creative Commons license. https://creativecommons.org/licenses/by/4.0/). Abbreviations: *FTLD* frontotemporal lobar degeneration, *TDP* TAR DNA-binding protein

fusiform gyrus, lingual gyrus, and amygdala-hippocampal area (Fig. 3) [221, 222], that is, the language-dominant temporal lobe in svPPA [21, 199]. Atrophy may also be present in the orbitofrontal cortex, but does not typically involve the parietal lobe (Fig. 3) [223]. Atrophy progresses in a stereotypical pattern, despite initial lateralization of the disease to the right or left hemisphere, as it typically begins in the anterior temporal lobes, then progresses to both the ipsilateral posterior temporal areas and the contralateral temporal pole, and subsequently involves the orbitofrontal cortex. The anterior temporal lobes demonstrate the greatest degree of atrophy [222]. Accordingly, FTLD-TDP type C is associated with the clinical phenotypes of svPPA or temporal variant of bvFTD. Specific atrophy patterns for TDP type D and E remain to be elucidated.

Structural MRI in tauopathies often reveals severe symmetric atrophy affecting the bilateral temporal lobes with early involvement of the anterior cingulate gyrus, orbitofrontal, and fronto-insular regions. In addition, early degeneration of the hippocampal formation is common. In 3R-tauopathies, neuroimaging shows striking atrophy, often described as knife-edge (GCA 3), affecting the frontal and temporal lobes, specifically the dorsolateral prefrontal cortex, ventromedial prefrontal cortex, temporal poles, and the anteromedial and –lateral regions of the temporal lobes (Fig. 4) [224]. Other cortical and subcortical structures including the anterior

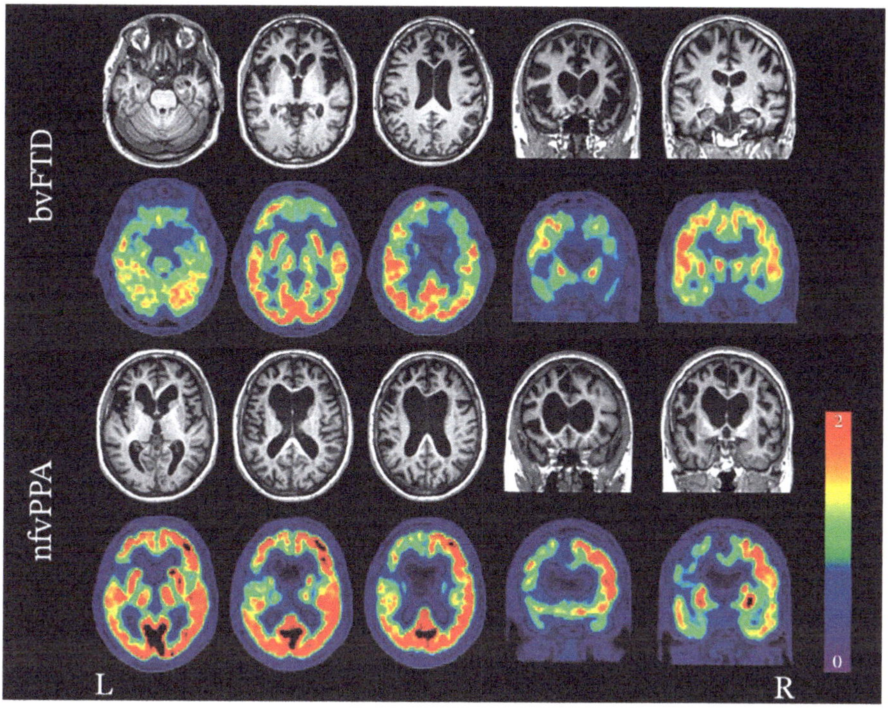

Fig. 4 Neuroimaging patterns associated with 3R FTLD-tau. BvFTD (upper panel) demonstrates bilateral frontal lobe atrophy (upper row) and hypometabolism (lower row). nfvPPA shows lateralized left frontal lobe atrophy (upper row) and hypometabolism (lower row). (Taken with permission from Peet et al. 2021 [224]; Open Access through a Creative Commons license. https://creativecommons.org/licenses/by/4.0/). Abbreviations: *FTLD* frontotemporal lobar degeneration, *tau* tubule associated unit, *bvFTD* behavioral variant of frontotemporal dementia, *nfvPPA* non-fluent variant of primary progressive aphasia

cingulate cortex, anterior insula, posterior insula, and caudate are affected as well, while the substantia nigra and the parietal cortex are fairly spared from neurodegeneration in comparison to 4R-tauopaties [225–227]. In most cases, atrophy in 3R-tauopathies is bilateral, but some patients show a predominance of atrophy on the left side of the brain [225]. Aside from severe temporal atrophy, 4R-tauopathies have a more heterogeneous pattern of atrophy, comparable to the described patterns for the clinical syndromes of PSP and CBS above (Fig. 5) [214]. Additionally, the epicenters of neuropathology can be the brainstem with associated pronounced atrophy of the midbrain and the superior cerebellar peduncles [195]. Patients with an underlying *MAPT* mutation usually show symmetric atrophy in the anteromedial temporal lobes [214, 228, 229], a feature that can already be found in an early disease stage [230] but can vary depending on the specific type of mutation [231]. Some *MAPT* mutations present with additional atrophy in the basal ganglia and brainstem, which may explain the additional parkinsonian features observed in these patients [232–234].

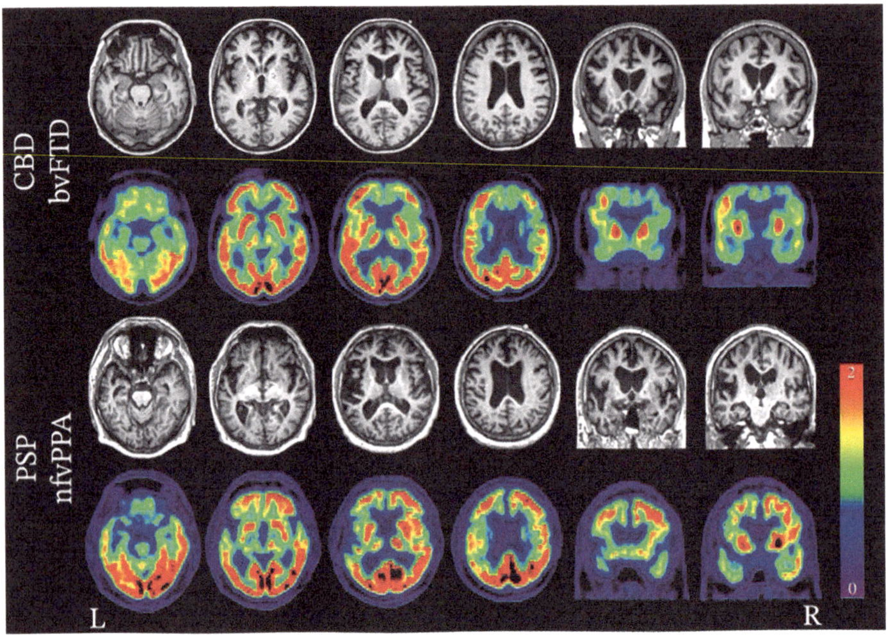

Fig. 5 Neuroimaging patterns associated with 4R FTLD-tau. CBD pathology in a patient with a clinical presentation of bvFTD (upper panel) exhibits dorsolateral and medial frontal atrophy (upper row) and hypometabolism (lower row). A pathologically-confirmed PSP case with a clinical presentation of nfvPPA shows heterogeneous atrophy (upper row) and hypometabolism (lower row) in the brainstem, midbrain, and superior cerebellar peduncles. (Taken with permission from Peet et al. 2021 [224]; Open Access through a Creative Commons license. https://creativecommons.org/licenses/by/4.0/). Abbreviations: *FTLD* frontotemporal lobar degeneration; tau, tubule associated unit, *CBD* cortical basal degeneration, *bvFTD* behavioral variant of frontotemporal dementia, *PSP* progressive supranuclear palsy, *nfvPPA* non-fluent variant of primary progressive aphasia

FTLD-FET is the least prevalent pathological subtype but has a characteristic pattern of grey matter atrophy in the brain. There appears to be some frontotemporal atrophy, but the most notable is severe caudate atrophy, which is disproportional compared to other regions in the brain. This caudate atrophy allows the distinction of the FTLD-FET subtype from the other FTLD pathological subtypes on brain MRI [33, 235, 236].

Presymptomatic Genetic FTLD

Presymptomatic carriers of genetic mutations are highly valuable in the evaluation of the FTLD disease process, as there are already structural changes visible before symptom onset, progressing afterwards. Following those at-risk individuals over time can provide useful information regarding disease pathogenesis and factors that can predict symptom onset, which could be a valuable time-point to start in therapeutic trials. Several consortia, for example, The Genetic Frontotemporal Dementia Initiative (GENFI), a European and Canadian consortium, and ARTFL-LEFFTDS Longitudinal Frontotemporal Lobar Degeneration (ALLFTD), their counterpart in the United States, follow mutation carriers who are at-risk for future symptomatic conversion to better understand the molecular and pathological changes that precede symptom onset.

Asymptomatic *C9orf72* HRE carriers already present with smaller total brain volumes at a mean age of 26 years, supporting the hypothesis that *C9orf72* might be a neurodevelopmental disorder [237]. Additional research in young presymptomatic at-risk individuals is necessary to further test this hypothesis. In addition, *C9orf72* HRE carriers show a diffuse pattern of atrophy including frontal, temporal, parietal, insular, and posterior cortical regions, as well as lower subcortical volumes of the thalamus, hippocampus, and the cerebellum [203, 218, 238–242]. This early involvement of subcortical regions is specific for *C9orf72* HRE carriers, distinguishing them from other FTLD mutation carriers [243].

In *GRN* mutation carriers, there is already asymmetric atrophy present at 5 years prior to expected symptom onset [244]. This atrophy has been found to be widespread with reductions not only in frontotemporal, but also in the parietal, posterior cingulate, precuneus, and occipital grey matter [245]. In addition, asymptomatic *GRN* mutation carriers have more rapid changes in grey matter around symptom onset compared to asymptomatic *MAPT* mutation carriers [246].

In presymptomatic carriers with a *MAPT* mutation, the cortical volume appears to decline in the temporal lobe early in the asymptomatic stage, up to 15 years before symptom onset, with an acceleration of atrophy rates along with the development of symptoms followed by atrophy in the frontal and parietal lobes [244, 247]. Additional atrophy was found in the anteromedial temporal lobe, orbitofrontal cortex, and hippocampus.

To conclude, the clinical, pathological, and genetic subtypes have distinct patterns of grey matter atrophy, and different rates of progression, which can be used for an accurate diagnosis or tracking disease progression. In addition, the changes that precede clinical diagnosis could help predict symptom onset. Thus, structural grey matter MRI represents an early biomarker allowing for patient stratification for clinical trials and monitoring of therapeutic effects.

3.2.2 White Matter

Diffusion tensor imaging (DTI) is a valuable noninvasive MRI technique for the assessment of the integrity and structure of brain white matter. This technique measures the microstructural integrity of white matter tracts by determining the motion of water molecules in different directions, described as diffusion, and can reliably detect microstructural abnormalities in white matter tracts. Fiber tracts restrict water diffusion along a particular axis. Fractional anisotropy (FA) is a measure of directionality of this restricted water flow and is associated with axonal integrity. If the FA value is high, then there is restricted diffusion along tracts, which can be interpreted as intact axons, while a low value reflects free diffusion in all directions [248]. Mean diffusivity (MD) is a metric that reflects the average diffusion measurements in all directions, characterizing the structural properties of the tissue. High MD values reflect that water molecules can move more freely in the examined tissue. If there is degeneration of white matter fibers, water molecules can diffuse more freely and in more directions compared to nonaffected tissue. Affected white matter is thus reflected by a low FA and high MD [249].

In FTLD, degeneration in the anterior corpus callosum, anterior cingulate cortex, and uncinate fasciculus tracts, areas that engage in the executive, language, and motor domains, has been described [250]. White matter degradation coincides with grey matter atrophy in the frontal, temporal, and insular regions and probably precedes grey matter atrophy [251]. Until now, research into white matter changes has mostly focused on genetic FTLD and very little on the pathological subtypes or sporadic FTLD.

Although FTLD-TDP is predominantly associated with grey matter atrophy compared to white matter pathology, white matter is still affected in this subtype. Especially, individuals with a *GRN* mutation can show severe white matter loss. Two studies demonstrated decreased FA in the left uncinate fasciculus, the left inferior occipitofrontal fasciculus, and the genu of the corpus callosum. In addition, reduced FA and increased MD in the internal capsule was found. *GRN* mutation carriers also show left-right asymmetry in most white matter tracts [252, 253] and white matter hyperintensities [254]. Individuals with a *C9orf72* HRE have dorsal white matter tract pathology located in the cingulum, corpus callosum, and the superior cerebellar peduncles [255, 256]. Other studies reported reduced white matter integrity in the uncinate fasciculus and the inferior longitudinal fasciculus (tracts that connect the frontal lobes to other brain regions), the corticospinal tracts, the corona radiata, and the internal and external capsules (tracts related to motor function) [240, 253, 257, 258]. In asymptomatic *C9orf72* HRE carriers, lower FA values are observed in the frontotemporal tracts and higher MD values in white matter tracts throughout the brain [242].

In FTLD-tau, there is severe loss of white matter integrity compared to FTLD-TDP [250, 259]. In symptomatic individuals, this loss is visible in the anterior corpus callosum, the white matter connections of the limbic system, the subcortical white matter in the temporal and frontal lobes, and the connections of the frontal and occipital lobes [260]. A post-mortem study showed that tau pathology is further associated with axonal loss and tau inclusions in glial cells [250]. Individuals with *MAPT* mutations have degeneration of the uncinate fasciculus and right parahippocampal cingulum [255, 256]. Furthermore, other studies found reduced FA and increased MD in the bilateral uncinate fasciculus and cingulum [253, 261]. *MAPT* mutation carriers show an increased MD in the entorhinal white matter even prior to symptom onset [260].

Thus, white matter changes are more pronounced in FTLD-tau than in FTLD-TDP, but the various subtypes are associated with distinct patterns of microstructural white matter damage [251]. McMillan et al. showed that the two pathological groups (FTLD-tau and FTLD-TDP) can be differentiated using DTI with a sensitivity of 96% and a specificity of 100%, which was higher than differentiation based on grey matter thickness (64% and 60%, respectively) [250]. Moreover, white matter abnormalities precede grey matter atrophy. These results can be used to predict conversion or symptom onset in genetic FTLD.

3.3 Functional Imaging

3.3.1 FDG-PET

PET with ^{18}F-Fluorodeoxyglucose (FDG-PET) is a marker for cerebral glucose metabolism, providing a surrogate of neuronal and synaptic activity. For the clinical syndromes associated with FTLD, hypometabolism is found in the same areas as the typical grey matter atrophy [262–269]; for example, in bvFTD, the hypometabolism shows in the orbitofrontal and anterior temporal lobes; in nfvPPA, in the left peri-insular regions; in svPPA, in the left anterior temporal lobes; in PSP, in the medial frontal and anterior cingulate cortices; and in CBS, in an asymmetric fronto-parietal pattern. The hypometabolism seen with PET precedes structural atrophy in the grey matter, making it a useful biomarker in differentiating FTLD from other dementia syndromes [266]. Moreover, these patterns of hypometabolism are distinct in the pathological subtypes on a group level [224], and the rate of hypometabolism depends on the type of disease [266]. These findings could help differentiate the various subtypes for clinical diagnosis or to stratify groups for trials. FTLD has distinct patterns of hypometabolism compared to AD (parietotemporal lobes) or DLB (parieto-occipital lobes, but relatively preserved posterior cingulate metabolism) [270], making it a useful technique to exclude pathology based on these two diseases [249]. FDG-PET is an FDA-approved tool to differentiate FTLD from AD [271]. Another useful application of FDG-PET in the diagnosis of bvFTD is the differentiation from psychiatric disorders [197, 272, 273].

FDG-PET is the preferred technique to measure hypometabolism in the clinic due to various reasons, despite the fact that there are noninvasive MRI sequences. First, the analysis is easier compared to that of MRI techniques. Second, where the MRI sequences only have found differences at the group level, FDG-PET has been able to discover distinct patterns of regional hypometabolism that enable an accurate clinical diagnosis of individual patients, both by visual inspection and, especially, by quantitative assessment [211].

Some studies have shown that FDG-PET can already point out changes in the presymptomatic phase of FTLD, which could serve as a proxy for FTLD in future trials. For example, presymptomatic *GRN* mutation carriers present with hypometabolism in the frontal and temporal lobes [274, 275]. Other studies tracked the disease progression in the form of longitudinal studies, providing additional information about the patterns and speed of pathological spread [211]. FDG-PET has the potential to be used to identify presymptomatic gene carriers which can be included in drug studies. FDG-PET can also be used in combination with MRI to monitor drug effects.

3.3.2 Arterial Spin Labeling

Arterial spin labeling (ASL) is an MRI technique that measures brain perfusion by labeling arterial water as an endogenous tracer to quantify cerebral blood flow [249]. Brain perfusion measurements correlate with the metabolism evaluated with FDG-PET and thus, can be used as an indirect measurement for neuronal activity. Studies have shown similar results as FDG-PET with hypoperfusion in bilateral frontal lobes, the anterior cingulate cortex, insula, and thalamus compared with controls [276–281].

ASL is a promising biomarker for the assessment of hypoperfusion in FTLD for multiple reasons. ASL can be easily combined with other MRI techniques in a single session, is noninvasive, involves no radiation exposure unlike FDG-PET, is widely available, and is less expensive than FDG-PET. As hypoperfusion generally occurs before grey matter atrophy, ASL can be informative in the presymptomatic stages [280–282]. However, it should be noted, that stand-alone ASL is less sensitive as a diagnostic imaging tool compared to FDG-PET [283]. In addition, compared to PET data, the analysis of ASL data is more complex, and more user-friendly analyses are needed to facilitate analyses of ASL imaging data and make it more readily accessible in the clinic. Due to these limitations, ASL remains a research imaging tool in FTLD and is not typically incorporated into clinical MRI assessments.

3.3.3 Resting State fMRI

Resting state functional MRI (rs-fMRI) measures intrinsic functional connectivity between brain regions to detect synchronous patterns of low-frequency fluctuations in blood oxygen level dependent (BOLD) signals during rest [249]. These alterations are based on the concept that neuronal activity leads to deoxygenated hemoglobin in draining veins, which is more paramagnetic (i.e., aligned with the magnetic field, thus increasing the local magnetic field) than oxygenated hemoglobin. Due to the effect of neurovascular coupling, neuronal activation is assumed to lead to an increased flow of oxygenated blood in the corresponding brain regions, causing a change in the local magnetic field. Rs-fMRI can therefore be used to detect functional connectivity between different brain areas, because communicating brain areas demonstrate slow synchronized oscillations of BOLD signals that are correlated over time [284].

In FTLD, rs-fMRI shows altered connectivity in networks that closely follow the atrophy patterns. The most common finding is reduced functional connectivity within the salience network [285–287], which encompasses various brain structures, such as the anterior cingulate cortex and the ventral anterior insular cortices. The network is activated in response to communication, social behavior, and self-awareness, functions that are diminished in some variants of FTLD [288]. Conversely, other studies have shown that there is an increase in the connectivity in the default network [289]. The default network is a group of brain regions that are active during rest but deactivate during cognitively demanding tasks [290]. In FTLD, connectivity in the default network is increased in the striatum, motor cortex, cerebellum, and basal ganglia. Specific network alterations are also found in the different genetic variants of FTLD [291, 292]. Small studies demonstrated alterations of the salience network in *C9orf72* HRE carriers, the frontoparietal network in *GRN* mutation carriers, and the default mode network in *MAPT* mutation carriers.

Rs-fMRI is a promising biomarker for multiple reasons. First, it is noninvasive and is easily applicable. Second, rs-fMRI studies have shown some preliminary results that differentiation between various subtypes or diseases is possible. For example, bvFTD showed elevated apathy scores which correlated with a more diffuse pattern of hyperconnectivity in the prefrontal cortex compared to svPPA [286]. Additionally, one study showed different functional brain connectivity between bvFTD and AD [293]. Lastly, rs-fMRI might be able to measure altered connectivity in presymptomatic mutation carriers [261, 291, 294, 295]. However, these studies were performed in small sample sizes, and validation of these results in a larger cohort is recommended [296]. Another disadvantage is the various methods used to analyze fMRI data. All the abovementioned studies have used different analyses which makes the interpretation or comparison of the results more difficult. To make this technique more accessible in the clinic and to correlate study findings, consensus for the analysis is needed. Additionally, the findings of these studies are based on group level differences, which cannot yet be translated to the individual level.

3.4 Potential Neuroimaging Biomarkers

3.4.1 MR Spectroscopy

MR spectroscopy (MRS) is a noninvasive MRI technique that provides an estimation of brain biochemistry. The outcome measure is not an anatomical image, but the concentration of certain metabolites in a predefined region of interest. These metabolites are associated with either neurons and/or glia such as astrocytes. N-acetylaspartate (NAA) is a measure for neuronal integrity, as it is only present in neurons [297–300]. The metabolite myo-inositol resides in astrocytes, and since astrocytes are involved with the neuroinflammatory response in the brain, it is thought to be a marker for inflammation in the brain [301]. Other metabolites are creatine, representing energy metabolism in both neurons and astrocytes, and used as a reference, and choline, which indicates membrane turnover. Small studies in symptomatic FTLD subjects have shown an increase in myo-inositol to creatine ratio and a decrease in NAA to creatine ratio, which suggests an increase in the neuroinflammatory response and a decline in neuronal integrity, or atrophy of the brain [302–305]. However, the underlying pathology or genetic mutation in these studies is not mentioned, thus, the results are difficult to translate to the clinic. Only two studies have investigated metabolite changes in a specific genetic subgroup. In presymptomatic *MAPT* mutation carriers, an elevation in the myo-inositol to creatine level was found, suggesting that neuroinflammation precedes neuronal degeneration [306]. A small study with symptomatic *MAPT* mutation carriers shows metabolic alterations in the anterior temporal lobe compared to non-carriers [307]. Unfortunately, these studies were only exploratory, as the study population was small. Moreover, MR spectroscopy has not yet been performed in *GRN* mutation or *C9orf72* HRE carriers or in the various proteinopathies. For future investigations, these studies should be repeated with a larger sample size to replicate these findings and include the different pathological and genetic subtypes in order to examine the differences in metabolite concentrations and compositions.

3.4.2 Molecular Imaging

Tau-PET

PET tracers that could image tau protein deposits in vivo and therefore differentiate patients with FTLD-tau from other FTLD subtypes or other diseases such as AD would be helpful in the clinic. However, most currently available tau-tracers are mainly developed for AD. One current available tracer, ^{18}F-flortaucipir, can distinguish AD from FTLD with a specificity of 90.6% and a sensitivity of 89.9% due to different uptake patterns [283, 308]. In FTLD, ^{18}F-flortaucipir uptake was found in temporal, insular, and frontal regions in a selection of symptomatic *MAPT* mutation carriers, and to a lesser extent, in presymptomatic carriers [309–316]. However, participants of this study mainly had mixed 3R/4R tauopathies, indicating the flaw of this specific tracer as it mostly binds to AD-related tauopathies. AD tauopathies have a mix of 3R and 4R tau, while most FTLD-tau have either 3R or 4R to which this tracer does not bind well [224]. Moreover, multiple other studies have found signal in *GRN*-related or *C9orf72*-related FTLD and in svPPA, which is also related to TDP-43 pathology, questioning the specificity of the binding of these first-generation tracers to tau [317–320]. Additionally, these results come from either singular cases or small-size cross-sectional studies. Studies with larger study populations or longitudinal data are necessary to evaluate the efficacy of ^{18}F-flortaucipir in FTLD-tau.

Another major difficulty in this context is found in the different tau-isoforms detected in the different FTLD subtypes and the yet unknown binding behavior of the available tau-tracers to these isoforms in vivo [316, 321]. To selectively image the different tau isoforms and their intracellular aggregation, this requires the application of different tau ligands [322]. Many studies are ongoing to find other such suitable tracers for imaging tau deposition in FTLD, for example, ^{11}C-PBB3 and ^{18}F-PM-PBB3, but they come with conflicting results. ^{11}C-PBB3 was able to bind to 3R/4R and 4R pathology, but not to 3R. ^{18}F-PM-PBB3 also showed increased binding affinities to 4R, but not to 3R. In both studies, there was widespread binding in participants with rapid progression, but low affinity for those with slow progression [323, 324]. If tau-PET tracers are developed, they could become effective for diagnosis of underlying tau pathology in FTLD and differentiation of the various pathological subgroups.

TDP-PET

Development of TDP-43 tracers for clinical use would be of major importance to diagnose and predict the underlying pathology during life, in particular in sporadic FTLD. TDP-PET would also be insightful in the presymptomatic phase of genetic FTLD, including mutation carriers, when TDP-43 accumulation starts. Currently, tracers are being developed and the first human studies are in progress [325].

4 Concluding Remarks and Future Directions

In recent years, great strides have been made in identifying novel fluid and neuroimaging biomarkers, yet the implementation in clinical practice remains limited. At present, CSF analysis is utilized to rule out AD; however, it cannot affirmatively diagnose FTLD. Additionally, in tertiary care centers, NfL is incorporated in the clinical work-up to support the diagnosis of neurodegenerative diseases. Furthermore, anatomical T1-weighted MRI, to assess grey matter atrophy, and FDG-PET, for hypometabolism, aid clinicians in diagnosing clinical syndromes of FTLD. However, as the diagnosis can only be definite postmortem or by genetic testing, the need for additional biomarkers is pressing. Such biomarkers are needed for supporting the clinical diagnosis, predicting disease onset, disease monitoring, and evaluation of therapeutic trials. Additionally, considering upcoming clinical trials, the development of a marker that can predict the underlying neuropathological substrate in sporadic FTLD is imperative.

This chapter described promising fluid biomarkers, for example, fluid markers reflecting the various pathophysiological processes that go beyond the different proteinopathies, such as inflammatory markers. Interestingly, despite the convincing evidence supporting the role of neuroinflammation in FTLD [146], conflicting results are reported. It is suggested that this variability might be related to disease state and severity. If so, those markers could be of important prognostic value. A propitious strategy for the identification of novel fluid biomarkers is the use of large omics studies. Thus far, most studies have focused on proteomics, mainly using unbiased mass spectrometry [187, 326]. Recently, highly sensitive, high-throughput protein arrays have been developed, for example, aptamer-based technologies, immuno-based proximity extension assays, and antibody suspension bead arrays. These techniques allow for the measurement of large sets of protein libraries in CSF and blood samples. In AD, the use of immuno-based proximity extension assays has produced promising results [327, 328]. Similar studies in FTLD are yet to be conducted.

In addition, multiple, promising imaging techniques have been investigated and developed in the past decade. Two promising techniques are DTI, assessing white matter tract degeneration [250], and ASL, measuring hypoperfusion [276]. These techniques are safe, noninvasive, repeatable, and easily combined in a single MRI session. However, DTI and ASL have only been assessed at the group level; thus, most of the outcomes are not yet applicable on a single-subject level. Therefore, validation in well-characterized genetic and sporadic cohorts is crucial before introducing them in clinical practice or as a measure of therapeutic response in investigational trials.

Another interesting approach is the combination of various biomarkers within a multimodal approach. To date, several studies have investigated the possible advantage of combining multiple imaging techniques [329–334] or fluid biomarkers [76, 335–337]. For example, a study combining T1-weighted MRI and FDG-PET imaging was able to distinguish between bvFTD, svPPA, nfvPPA, and other neurodegenerative disorders [334]. Multiple other studies combining structural MRI, DTI, and functional connectivity scans could predict the conversion of presymptomatic mutation carriers to symptomatic patients [331, 333]. In the field of fluid biomarkers, one study reported an improved diagnostic accuracy in differentiating FTLD from other neurodegenerative diseases with a model based on sex, age, CSF NPTX2, serum NfL, and serum GFAP [336]. The improved accuracy is feasible given that such models are a better reflection of the complex pathophysiology underlying neurodegenerative diseases. Moreover, the future focus should not only be on combining fluid or imaging biomarkers but should also extend to other modalities, such as neuropsychological assessments and new data science methods, including machine and deep learning.

References

1. Cairns NJ, Bigio EH, Mackenzie IR, Neumann M, Lee VM, Hatanpaa KJ, et al. Neuropathologic diagnostic and nosologic criteria for frontotemporal lobar degeneration: consensus of the Consortium for Frontotemporal Lobar Degeneration. Acta Neuropathol. 2007;114(1):5–22.
2. Coyle-Gilchrist IT, Dick KM, Patterson K, Vázquez Rodríquez P, Wehmann E, Wilcox A, et al. Prevalence, characteristics, and survival of frontotemporal lobar degeneration syndromes. Neurology. 2016;86(18):1736–43.
3. Grossman M, Seeley WW, Boxer AL, Hillis AE, Knopman DS, Ljubenov PA, et al. Frontotemporal lobar degeneration. Nat Rev Dis Primers. 2023;9(1):40.
4. Rascovsky K, Hodges JR, Knopman D, Mendez MF, Kramer JH, Neuhaus J, et al. Sensitivity of revised diagnostic criteria for the behavioural variant of frontotemporal dementia. Brain. 2011;134(Pt 9):2456–77.

5. Gorno-Tempini ML, Hillis AE, Weintraub S, Kertesz A, Mendez M, Cappa SF, et al. Classification of primary progressive aphasia and its variants. Neurology. 2011;76(11):1006–14.
6. Höglinger GU, Respondek G, Stamelou M, Kurz C, Josephs KA, Lang AE, et al. Clinical diagnosis of progressive supranuclear palsy: The movement disorder society criteria. Mov Disord. 2017;32(6):853–64.
7. Armstrong MJ, Litvan I, Lang AE, Bak TH, Bhatia KP, Borroni B, et al. Criteria for the diagnosis of corticobasal degeneration. Neurology. 2013;80(5):496–503.
8. Sellami L, Saracino D, Le Ber I. Genetic forms of frontotemporal lobar degeneration: Current diagnostic approach and new directions in therapeutic strategies. Rev Neurol (Paris). 2020;176(7–8):571–81.
9. Buratti E, Baralle FE. The multiple roles of TDP-43 in pre-mRNA processing and gene expression regulation. RNA Biol. 2010;7(4):420–9.
10. Arai T, Hasegawa M, Akiyama H, Ikeda K, Nonaka T, Mori H, et al. TDP-43 is a component of ubiquitin-positive tau-negative inclusions in frontotemporal lobar degeneration and amyotrophic lateral sclerosis. Biochem Biophys Res Commun. 2006;351(3):602–11.
11. Neumann M, Sampathu DM, Kwong LK, Truax AC, Micsenyi MC, Chou TT, et al. Ubiquitinated TDP-43 in frontotemporal lobar degeneration and amyotrophic lateral sclerosis. Science. 2006;314(5796):130–3.
12. Neumann M, Kwong LK, Lee EB, Kremmer E, Flatley A, Xu Y, et al. Phosphorylation of S409/410 of TDP-43 is a consistent feature in all sporadic and familial forms of TDP-43 proteinopathies. Acta Neuropathol. 2009;117(2):137–49.
13. Hasegawa M, Arai T, Nonaka T, Kametani F, Yoshida M, Hashizume Y, et al. Phosphorylated TDP-43 in frontotemporal lobar degeneration and amyotrophic lateral sclerosis. Ann Neurol. 2008;64(1):60–70.
14. Mackenzie IR, Neumann M, Baborie A, Sampathu DM, Du Plessis D, Jaros E, et al. A harmonized classification system for FTLD-TDP pathology. Acta Neuropathol. 2011;122(1):111–3.
15. Neumann M, Mackenzie IRA. Review: Neuropathology of non-tau frontotemporal lobar degeneration. Neuropathol Appl Neurobiol. 2019;45(1):19–40.
16. Neumann M, Lee EB, Mackenzie IR. Frontotemporal lobar degeneration TDP-43-immunoreactive pathological subtypes: clinical and mechanistic significance. Adv Exp Med Biol. 2021;1281:201–17.
17. Kovacs GG. Invited review: neuropathology of tauopathies: principles and practice. Neuropathol Appl Neurobiol. 2015;41(1):3–23.
18. Weingarten MD, Lockwood AH, Hwo SY, Kirschner MW. A protein factor essential for microtubule assembly. Proc Natl Acad Sci USA. 1975;72(5):1858–62.
19. Lee G, Leugers CJ. Tau and tauopathies. Prog Mol Biol Transl Sci. 2012;107:263–93.
20. Iqbal K, Liu F, Gong CX, Alonso Adel C, Grundke-Iqbal I. Mechanisms of tau-induced neurodegeneration. Acta Neuropathol. 2009;118(1):53–69.
21. Kovacs GG, Botond G, Budka H. Protein coding of neurodegenerative dementias: the neuropathological basis of biomarker diagnostics. Acta Neuropathol. 2010;119(4):389–408.
22. Irwin DJ, Cohen TJ, Grossman M, Arnold SE, Xie SX, Lee VM, et al. Acetylated tau, a novel pathological signature in Alzheimer's disease and other tauopathies. Brain. 2012;135(Pt 3):807–18.
23. Grinberg LT, Wang X, Wang C, Sohn PD, Theofilas P, Sidhu M, et al. Argyrophilic grain disease differs from other tauopathies by lacking tau acetylation. Acta Neuropathol. 2013;125(4):581–93.
24. Noble W, Hanger DP, Miller CC, Lovestone S. The importance of tau phosphorylation for neurodegenerative diseases. Front Neurol. 2013;4:83.
25. Perry DC, Brown JA, Possin KL, Datta S, Trujillo A, Radke A, et al. Clinicopathological correlations in behavioural variant frontotemporal dementia. Brain. 2017;140(12):3329–45.

26. Spinelli EG, Mandelli ML, Miller ZA, Santos-Santos MA, Wilson SM, Agosta F, et al. Typical and atypical pathology in primary progressive aphasia variants. Ann Neurol. 2017;81(3):430–43.
27. Arai T, Ikeda K, Akiyama H, Tsuchiya K, Yagishita S, Takamatsu J. Intracellular processing of aggregated tau differs between corticobasal degeneration and progressive supranuclear palsy. Neuroreport. 2001;12(5):935–8.
28. Munoz DG, Dickson DW, Bergeron C, Mackenzie IR, Delacourte A, Zhukareva V. The neuropathology and biochemistry of frontotemporal dementia. Ann Neurol. 2003;54(Suppl 5):S24–8.
29. Neumann M, Bentmann E, Dormann D, Jawaid A, DeJesus-Hernandez M, Ansorge O, et al. FET proteins TAF15 and EWS are selective markers that distinguish FTLD with FUS pathology from amyotrophic lateral sclerosis with FUS mutations. Brain. 2011;134(Pt 9):2595–609.
30. Neumann M, Rademakers R, Roeber S, Baker M, Kretzschmar HA, Mackenzie IR. A new subtype of frontotemporal lobar degeneration with FUS pathology. Brain. 2009;132(Pt 11):2922–31.
31. Tan AY, Manley JL. The TET family of proteins: functions and roles in disease. J Mol Cell Biol. 2009;1(2):82–92.
32. Mackenzie IR, Neumann M. Molecular neuropathology of frontotemporal dementia: insights into disease mechanisms from postmortem studies. J Neurochem. 2016;138(Suppl 1):54–70.
33. Seelaar H, Klijnsma KY, de Koning I, van der Lugt A, Chiu WZ, Azmani A, et al. Frequency of ubiquitin and FUS-positive, TDP-43-negative frontotemporal lobar degeneration. J Neurol. 2010;257(5):747–53.
34. Urwin H, Josephs KA, Rohrer JD, Mackenzie IR, Neumann M, Authier A, et al. FUS pathology defines the majority of tau- and TDP-43-negative frontotemporal lobar degeneration. Acta Neuropathol. 2010;120(1):33–41.
35. DeJesus-Hernandez M, Mackenzie IR, Boeve BF, Boxer AL, Baker M, Rutherford NJ, et al. Expanded GGGGCC hexanucleotide repeat in noncoding region of C9ORF72 causes chromosome 9p-linked FTD and ALS. Neuron. 2011;72(2):245–56.
36. Renton AE, Majounie E, Waite A, Simón-Sánchez J, Rollinson S, Gibbs JR, et al. A hexanucleotide repeat expansion in C9ORF72 is the cause of chromosome 9p21-linked ALS-FTD. Neuron. 2011;72(2):257–68.
37. van Blitterswijk M, Mullen B, Nicholson AM, Bieniek KF, Heckman MG, Baker MC, et al. TMEM106B protects C9ORF72 expansion carriers against frontotemporal dementia. Acta Neuropathol. 2014;127(3):397–406.
38. Boeve BF, Boylan KB, Graff-Radford NR, DeJesus-Hernandez M, Knopman DS, Pedraza O, et al. Characterization of frontotemporal dementia and/or amyotrophic lateral sclerosis associated with the GGGGCC repeat expansion in C9ORF72. Brain. 2012;135(Pt 3):765–83.
39. Cruts M, Gijselinck I, Van Langenhove T, van der Zee J, Van Broeckhoven C. Current insights into the C9orf72 repeat expansion diseases of the FTLD/ALS spectrum. Trends Neurosci. 2013;36(8):450–9.
40. Troakes C, Maekawa S, Wijesekera L, Rogelj B, Siklós L, Bell C, et al. An MND/ALS phenotype associated with C9orf72 repeat expansion: abundant p62-positive, TDP-43-negative inclusions in cerebral cortex, hippocampus and cerebellum but without associated cognitive decline. Neuropathology. 2012;32(5):505–14.
41. Mackenzie IR, Frick P, Neumann M. The neuropathology associated with repeat expansions in the C9ORF72 gene. Acta Neuropathol. 2014;127(3):347–57.
42. Le Ber I, Camuzat A, Guillot-Noel L, Hannequin D, Lacomblez L, Golfier V, et al. C9ORF72 repeat expansions in the frontotemporal dementias spectrum of diseases: a flow-chart for genetic testing. J Alzheimers Dis. 2013;34(2):485–99.
43. Snowden JS, Adams J, Harris J, Thompson JC, Rollinson S, Richardson A, et al. Distinct clinical and pathological phenotypes in frontotemporal dementia associated with MAPT, PGRN and C9orf72 mutations. Amyotroph Lateral Scler Frontotemporal Degener. 2015;16(7–8):497–505.

44. Moore KM, Nicholas J, Grossman M, McMillan CT, Irwin DJ, Massimo L, et al. Age at symptom onset and death and disease duration in genetic frontotemporal dementia: an international retrospective cohort study. Lancet Neurol. 2020;19(2):145–56.
45. Murphy NA, Arthur KC, Tienari PJ, Houlden H, Chiò A, Traynor BJ. Age-related penetrance of the C9orf72 repeat expansion. Sci Rep. 2017;7(1):2116.
46. Rademakers R, Neumann M, Mackenzie IR. Advances in understanding the molecular basis of frontotemporal dementia. Nat Rev Neurol. 2012;8(8):423–34.
47. Ahmed Z, Mackenzie IR, Hutton ML, Dickson DW. Progranulin in frontotemporal lobar degeneration and neuroinflammation. J Neuroinflammation. 2007;4:7.
48. Paushter DH, Du H, Feng T, Hu F. The lysosomal function of progranulin, a guardian against neurodegeneration. Acta Neuropathol. 2018;136(1):1–17.
49. Grossman M. Biomarkers in the primary progressive aphasias. Aphasiology. 2014;28(8–9):922–40.
50. Kim G, Ahmadian SS, Peterson M, Parton Z, Memon R, Weintraub S, et al. Asymmetric pathology in primary progressive aphasia with progranulin mutations and TDP inclusions. Neurology. 2016;86(7):627–36.
51. Le Ber I, Camuzat A, Hannequin D, Pasquier F, Guedj E, Rovelet-Lecrux A, et al. Phenotype variability in progranulin mutation carriers: a clinical, neuropsychological, imaging and genetic study. Brain. 2008;131(Pt 3):732–46.
52. Pickering-Brown S, Baker M, Bird T, Trojanowski J, Lee V, Morris H, et al. Evidence of a founder effect in families with frontotemporal dementia that harbor the tau +16 splice mutation. Am J Med Genet B Neuropsychiatr Genet. 2004;125B(1):79–82.
53. Pickering-Brown SM. The complex aetiology of frontotemporal lobar degeneration. Exp Neurol. 2007;206(1):1–10.
54. Ghetti B, Oblak AL, Boeve BF, Johnson KA, Dickerson BC, Goedert M. Invited review: Frontotemporal dementia caused by microtubule-associated protein tau gene (MAPT) mutations: a chameleon for neuropathology and neuroimaging. Neuropathol Appl Neurobiol. 2015;41(1):24–46.
55. Rohrer JD, Warren JD. Phenotypic signatures of genetic frontotemporal dementia. Curr Opin Neurol. 2011;24(6):542–9.
56. Swift IJ, Sogorb-Esteve A, Heller C, Synofzik M, Otto M, Graff C, et al. Fluid biomarkers in frontotemporal dementia: past, present and future. J Neurol Neurosurg Psychiatry. 2021;92(2):204–15. https://doi.org/10.1136/jnnp-2020-323520.
57. Petzold A. Neurofilament phosphoforms: surrogate markers for axonal injury, degeneration and loss. J Neurol Sci. 2005;233(1–2):183–98.
58. Khalil M, Teunissen CE, Otto M, Piehl F, Sormani MP, Gattringer T, et al. Neurofilaments as biomarkers in neurological disorders. Nat Rev Neurol. 2018;14(10):577–89.
59. Verde F, Otto M, Silani V. Neurofilament light chain as biomarker for amyotrophic lateral sclerosis and frontotemporal dementia. Front Neurosci. 2021;15:679199.
60. Bridel C, van Wieringen WN, Zetterberg H, Tijms BM, Teunissen CE, the NFLG, et al. Diagnostic value of cerebrospinal fluid neurofilament light protein in neurology: a systematic review and meta-analysis. JAMA Neurol. 2019;76(9):1035–48.
61. Gaetani L, Blennow K, Calabresi P, Di Filippo M, Parnetti L, Zetterberg H. Neurofilament light chain as a biomarker in neurological disorders. J Neurol Neurosurg Psychiatry. 2019;90(8):870–81.
62. Meeter LH, Dopper EG, Jiskoot LC, Sanchez-Valle R, Graff C, Benussi L, et al. Neurofilament light chain: a biomarker for genetic frontotemporal dementia. Ann Clin Transl Neurol. 2016;3(8):623–36.
63. Wilke C, Preische O, Deuschle C, Roeben B, Apel A, Barro C, et al. Neurofilament light chain in FTD is elevated not only in cerebrospinal fluid, but also in serum. J Neurol Neurosurg Psychiatry. 2016;87(11):1270–2.

64. Wilke C, Gillardon F, Deuschle C, Dubois E, Hobert MA, Müller vom Hagen J, et al. Serum levels of progranulin do not reflect cerebrospinal fluid levels in neurodegenerative disease. Curr Alzheimer Res. 2016;13(6):654–62.
65. Abu-Rumeileh S, Mometto N, Bartoletti-Stella A, Polischi B, Oppi F, Poda R, et al. Cerebrospinal fluid biomarkers in patients with frontotemporal dementia spectrum: a single-center study. J Alzheimers Dis. 2018;66(2):551–63.
66. Rojas JC, Wang P, Staffaroni AM, Heller C, Cobigo Y, Wolf A, et al. Plasma neurofilament light for prediction of disease progression in familial frontotemporal lobar degeneration. Neurology. 2021;96(18):e2296–e312.
67. Saracino D, Dorgham K, Camuzat A, Rinaldi D, Rametti-Lacroux A, Houot M, et al. Plasma NfL levels and longitudinal change rates in C9orf72 and GRN-associated diseases: from tailored references to clinical applications. J Neurol Neurosurg Psychiatry. 2021;92(12):1278–88.
68. Ooi S, Patel SK, Eratne D, Kyndt C, Reidy N, Lewis C, et al. Plasma neurofilament light chain and clinical diagnosis in frontotemporal dementia syndromes. J Alzheimers Dis. 2022;89(4):1221–31.
69. Vijverberg EG, Dols A, Krudop WA, Del Campo MM, Kerssens CJ, Gossink F, et al. Cerebrospinal fluid biomarker examination as a tool to discriminate behavioral variant frontotemporal dementia from primary psychiatric disorders. Alzheimers Dement (Amst). 2017;7:99–106.
70. Al Shweiki MR, Steinacker P, Oeckl P, Hengerer B, Danek A, Fassbender K, et al. Neurofilament light chain as a blood biomarker to differentiate psychiatric disorders from behavioural variant frontotemporal dementia. J Psychiatr Res. 2019;113:137–40.
71. Katisko K, Cajanus A, Jääskeläinen O, Kontkanen A, Hartikainen P, Korhonen VE, et al. Serum neurofilament light chain is a discriminative biomarker between frontotemporal lobar degeneration and primary psychiatric disorders. J Neurol. 2020;267(1):162–7.
72. Eratne D, Loi SM, Walia N, Farrand S, Li QX, Varghese S, et al. A pilot study of the utility of cerebrospinal fluid neurofilament light chain in differentiating neurodegenerative from psychiatric disorders: A 'C-reactive protein' for psychiatrists and neurologists? Aust N Z J Psychiatry. 2020;54(1):57–67.
73. Ashton NJ, Janelidze S, Al Khleifat A, Leuzy A, van der Ende EL, Karikari TK, et al. A multicentre validation study of the diagnostic value of plasma neurofilament light. Nat Commun. 2021;12(1):3400.
74. Skillbäck T, Farahmand B, Bartlett JW, Rosén C, Mattsson N, Nägga K, et al. CSF neurofilament light differs in neurodegenerative diseases and predicts severity and survival. Neurology. 2014;83(21):1945–53.
75. Steinacker P, Anderl-Straub S, Diehl-Schmid J, Semler E, Uttner I, von Arnim CAF, et al. Serum neurofilament light chain in behavioral variant frontotemporal dementia. Neurology. 2018;91(15):e1390–e401.
76. Baiardi S, Quadalti C, Mammana A, Dellavalle S, Zenesini C, Sambati L, et al. Diagnostic value of plasma p-tau181, NfL, and GFAP in a clinical setting cohort of prevalent neurodegenerative dementias. Alzheimers Res Ther. 2022;14(1):153.
77. Götze K, Vrillon A, Bouaziz-Amar E, Mouton-Liger F, Hugon J, Martinet M, et al. Plasma neurofilament light chain in memory clinic practice: Evidence from a real-life study. Neurobiol Dis. 2023;176:105937.
78. Khalil M, Pirpamer L, Hofer E, Voortman MM, Barro C, Leppert D, et al. Serum neurofilament light levels in normal aging and their association with morphologic brain changes. Nat Commun. 2020;11(1):812.
79. Tang R, Panizzon MS, Elman JA, Gillespie NA, Hauger RL, Rissman RA, et al. Association of neurofilament light chain with renal function: mechanisms and clinical implications. Alzheimers Res Ther. 2022;14(1):189.

80. Wu J, Xiao Z, Wang M, Wu W, Ma X, Liang X, et al. The impact of kidney function on plasma neurofilament light and phospho-tau 181 in a community-based cohort: the Shanghai Aging Study. Alzheimers Res Ther. 2024;16(1):32.
81. Zhang B, Zhang C, Wang Y, Chen L, Qiao Y, Wang Y, et al. Effect of renal function on the diagnostic performance of plasma biomarkers for Alzheimer's disease. Front Aging Neurosci. 2023;15:1150510.
82. Vermunt L, Otte M, Verberk IMW, Killestein J, Lemstra AW, van der Flier WM, et al. Age- and disease-specific reference values for neurofilament light presented in an online interactive support interface. Ann Clin Transl Neurol. 2022;9(11):1832–7.
83. Rohrer JD, Woollacott IO, Dick KM, Brotherhood E, Gordon E, Fellows A, et al. Serum neurofilament light chain protein is a measure of disease intensity in frontotemporal dementia. Neurology. 2016;87(13):1329–36.
84. Cajanus A, Katisko K, Kontkanen A, Jääskeläinen O, Hartikainen P, Haapasalo A, et al. Serum neurofilament light chain in FTLD: association with C9orf72, clinical phenotype, and prognosis. Ann Clin Transl Neurol. 2020;7(6):903–10.
85. Gendron TF, Heckman MG, White LJ, Veire AM, Pedraza O, Burch AR, et al. Comprehensive cross-sectional and longitudinal analyses of plasma neurofilament light across FTD spectrum disorders. Cell Rep Med. 2022;3(4):100607.
86. van der Ende EL, Meeter LH, Poos JM, Panman JL, Jiskoot LC, Dopper EGP, et al. Serum neurofilament light chain in genetic frontotemporal dementia: a longitudinal, multicentre cohort study. Lancet Neurol. 2019;18(12):1103–11.
87. Wilke C, Reich S, van Swieten JC, Borroni B, Sanchez-Valle R, Moreno F, et al. Stratifying the Presymptomatic Phase of Genetic Frontotemporal Dementia by Serum NfL and pNfH: A Longitudinal Multicentre Study. Ann Neurol. 2022;91(1):33–47.
88. Giannini LAA, Seelaar H, van der Ende EL, Poos JM, Jiskoot LC, Dopper EGP, et al. Clinical value of longitudinal serum neurofilament light chain in prodromal genetic frontotemporal dementia. Neurology. 2023;101(10):e1069–e82.
89. Staffaroni AM, Quintana M, Wendelberger B, Heuer HW, Russell LL, Cobigo Y, et al. Temporal order of clinical and biomarker changes in familial frontotemporal dementia. Nat Med. 2022;28(10):2194–206.
90. Vítor J, Saracino D, Ströer S, Camuzat A, Dorgham K, Clot F, et al. Atypical white matter hyperintensities markedly impact plasma neurofilament light chain variability in GRN patients. J Alzheimers Dis. 2023;94:1351.
91. Varhaug KN, Torkildsen Ø, Myhr KM, Vedeler CA. Neurofilament light chain as a biomarker in multiple sclerosis. Front Neurol. 2019;10:338.
92. Zetterberg H, Teunissen C, van Swieten J, Kuhle J, Boxer A, Rohrer JD, et al. The role of neurofilament light in genetic frontotemporal lobar degeneration. Brain Commun. 2023;5(1):fcac310
93. Paterson RW, Slattery CF, Poole T, Nicholas JM, Magdalinou NK, Toombs J, et al. Cerebrospinal fluid in the differential diagnosis of Alzheimer's disease: clinical utility of an extended panel of biomarkers in a specialist cognitive clinic. Alzheimers Res Ther. 2018;10(1):32.
94. Sato C, Mallipeddi N, Ghoshal N, Wright BA, Day GS, Davis AA, et al. MAPT R406W increases tau T217 phosphorylation in absence of amyloid pathology. Ann Clin Transl Neurol. 2021;8(9):1817–30.
95. Sha SJ, Miller ZA, Min SW, Zhou Y, Brown J, Mitic LL, et al. An 8-week, open-label, dose-finding study of nimodipine for the treatment of progranulin insufficiency from GRN gene mutations. Alzheimers Dement (N Y). 2017;3(4):507–12.
96. Goossens J, Bjerke M, Van Mossevelde S, Van den Bossche T, Goeman J, De Vil B, et al. Diagnostic value of cerebrospinal fluid tau, neurofilament, and progranulin in definite frontotemporal lobar degeneration. Alzheimers Res Ther. 2018;10(1):31.
97. Hu WT, Watts K, Grossman M, Glass J, Lah JJ, Hales C, et al. Reduced CSF p-Tau181 to Tau ratio is a biomarker for FTLD-TDP. Neurology. 2013;81(22):1945–52.

98. Borroni B, Benussi A, Archetti S, Galimberti D, Parnetti L, Nacmias B, et al. Csf p-tau181/tau ratio as biomarker for TDP pathology in frontotemporal dementia. Amyotroph Lateral Scler Frontotemporal Degener. 2015;16(1–2):86–91.
99. Pijnenburg YA, Verwey NA, van der Flier WM, Scheltens P, Teunissen CE. Discriminative and prognostic potential of cerebrospinal fluid phosphoTau/tau ratio and neurofilaments for frontotemporal dementia subtypes. Alzheimers Dement (Amst). 2015;1(4):505–12.
100. Meeter LHH, Vijverberg EG, Del Campo M, Rozemuller AJM, Donker Kaat L, de Jong FJ, et al. Clinical value of neurofilament and phospho-tau/tau ratio in the frontotemporal dementia spectrum. Neurology. 2018;90(14):e1231–e9.
101. Benussi A, Karikari TK, Ashton N, Gazzina S, Premi E, Benussi L, et al. Diagnostic and prognostic value of serum NfL and p-Tau(181) in frontotemporal lobar degeneration. J Neurol Neurosurg Psychiatry. 2020;91(9):960–7.
102. Karikari TK, Pascoal TA, Ashton NJ, Janelidze S, Benedet AL, Rodriguez JL, et al. Blood phosphorylated tau 181 as a biomarker for Alzheimer's disease: a diagnostic performance and prediction modelling study using data from four prospective cohorts. Lancet Neurol. 2020;19(5):422–33.
103. Thijssen EH, La Joie R, Wolf A, Strom A, Wang P, Iaccarino L, et al. Diagnostic value of plasma phosphorylated tau181 in Alzheimer's disease and frontotemporal lobar degeneration. Nat Med. 2020;26(3):387–97.
104. Thijssen EH, La Joie R, Strom A, Fonseca C, Iaccarino L, Wolf A, et al. Plasma phosphorylated tau 217 and phosphorylated tau 181 as biomarkers in Alzheimer's disease and frontotemporal lobar degeneration: a retrospective diagnostic performance study. Lancet Neurol. 2021;20(9):739–52.
105. Ashton NJ, Pascoal TA, Karikari TK, Benedet AL, Lantero-Rodriguez J, Brinkmalm G, et al. Plasma p-tau231: a new biomarker for incipient Alzheimer's disease pathology. Acta Neuropathol. 2021;141(5):709–24.
106. Kraus A, Saijo E, Metrick MA 2nd, Newell K, Sigurdson CJ, Zanusso G, et al. Seeding selectivity and ultrasensitive detection of tau aggregate conformers of Alzheimer disease. Acta Neuropathol. 2019;137(4):585–98.
107. Metrick MA 2nd, Ferreira NDC, Saijo E, Kraus A, Newell K, Zanusso G, et al. A single ultrasensitive assay for detection and discrimination of tau aggregates of Alzheimer and Pick diseases. Acta Neuropathol Commun. 2020;8(1):22.
108. Saijo E, Ghetti B, Zanusso G, Oblak A, Furman JL, Diamond MI, et al. Ultrasensitive and selective detection of 3-repeat tau seeding activity in Pick disease brain and cerebrospinal fluid. Acta Neuropathol. 2017;133(5):751–65.
109. Saijo E, Metrick MA 2nd, Koga S, Parchi P, Litvan I, Spina S, et al. 4-Repeat tau seeds and templating subtypes as brain and CSF biomarkers of frontotemporal lobar degeneration. Acta Neuropathol. 2020;139(1):63–77.
110. Schmitz A, Pinheiro Marques J, Oertig I, Maharjan N, Saxena S. Emerging perspectives on dipeptide repeat proteins in C9ORF72 ALS/FTD. Front Cell Neurosci. 2021;15:637548.
111. Gendron TF, Chew J, Stankowski JN, Hayes LR, Zhang YJ, Prudencio M, et al. Poly(GP) proteins are a useful pharmacodynamic marker for C9ORF72-associated amyotrophic lateral sclerosis. Sci Transl Med. 2017;9(383)
112. Lehmer C, Oeckl P, Weishaupt JH, Volk AE, Diehl-Schmid J, Schroeter ML, et al. Poly-GP in cerebrospinal fluid links C9orf72-associated dipeptide repeat expression to the asymptomatic phase of ALS/FTD. EMBO Mol Med. 2017;9(7):859–68.
113. Meeter LHH, Gendron TF, Sias AC, Jiskoot LC, Russo SP, Donker Kaat L, et al. Poly(GP), neurofilament and grey matter deficits in C9orf72 expansion carriers. Ann Clin Transl Neurol. 2018;5(5):583–97.
114. Wilson KM, Katona E, Glaria I, Carcolé M, Swift IJ, Sogorb-Esteve A, et al. Development of a sensitive trial-ready poly(GP) CSF biomarker assay for C9orf72-associated frontotemporal dementia and amyotrophic lateral sclerosis. J Neurol Neurosurg Psychiatry. 2022;93(7):761–71.

115. Krishnan G, Raitcheva D, Bartlett D, Prudencio M, McKenna-Yasek DM, Douthwright C, et al. Poly(GR) and poly(GA) in cerebrospinal fluid as potential biomarkers for C9ORF72-ALS/FTD. Nat Commun. 2022;13(1):2799.
116. Tran H, Moazami MP, Yang H, McKenna-Yasek D, Douthwright CL, Pinto C, et al. Suppression of mutant C9orf72 expression by a potent mixed backbone antisense oligonucleotide. Nat Med. 2022;28(1):117–24.
117. Cenik B, Sephton CF, Kutluk Cenik B, Herz J, Yu G. Progranulin: a proteolytically processed protein at the crossroads of inflammation and neurodegeneration. J Biol Chem. 2012;287(39):32298–306.
118. Baker M, Mackenzie IR, Pickering-Brown SM, Gass J, Rademakers R, Lindholm C, et al. Mutations in progranulin cause tau-negative frontotemporal dementia linked to chromosome 17. Nature. 2006;442(7105):916–9.
119. Ghidoni R, Benussi L, Glionna M, Franzoni M, Binetti G. Low plasma progranulin levels predict progranulin mutations in frontotemporal lobar degeneration. Neurology. 2008;71(16):1235–9.
120. Finch N, Baker M, Crook R, Swanson K, Kuntz K, Surtees R, et al. Plasma progranulin levels predict progranulin mutation status in frontotemporal dementia patients and asymptomatic family members. Brain. 2009;132(Pt 3):583–91.
121. Schofield EC, Halliday GM, Kwok J, Loy C, Double KL, Hodges JR. Low serum progranulin predicts the presence of mutations: a prospective study. J Alzheimers Dis. 2010;22(3):981–4.
122. Ghidoni R, Stoppani E, Rossi G, Piccoli E, Albertini V, Paterlini A, et al. Optimal plasma progranulin cutoff value for predicting null progranulin mutations in neurodegenerative diseases: a multicenter Italian study. Neurodegener Dis. 2012;9(3):121–7.
123. Almeida MR, Baldeiras I, Ribeiro MH, Santiago B, Machado C, Massano J, et al. Progranulin peripheral levels as a screening tool for the identification of subjects with progranulin mutations in a Portuguese cohort. Neurodegener Dis. 2014;13(4):214–23.
124. Meeter LH, Patzke H, Loewen G, Dopper EG, Pijnenburg YA, van Minkelen R, et al. Progranulin levels in plasma and cerebrospinal fluid in granulin mutation carriers. Dement Geriatr Cogn Dis Extra. 2016;6(2):330–40.
125. Galimberti D, Fumagalli GG, Fenoglio C, Cioffi SMG, Arighi A, Serpente M, et al. Progranulin plasma levels predict the presence of GRN mutations in asymptomatic subjects and do not correlate with brain atrophy: results from the GENFI study. Neurobiol Aging. 2018;62(245):e9–e12.
126. Sellami L, Rucheton B, Ben Younes I, Camuzat A, Saracino D, Rinaldi D, et al. Plasma progranulin levels for frontotemporal dementia in clinical practice: a 10-year French experience. Neurobiol Aging. 2020;91(167):e1–9.
127. Kurnellas M, Mitra A, Schwabe T, Paul R, Arrant AE, Roberson ED, et al. Latozinemab, a novel progranulin-elevating therapy for frontotemporal dementia. J Transl Med. 2023;21(1):387.
128. Nicholson AM, Finch NA, Thomas CS, Wojtas A, Rutherford NJ, Mielke MM, et al. Progranulin protein levels are differently regulated in plasma and CSF. Neurology. 2014;82(21):1871–8.
129. Feneberg E, Steinacker P, Volk AE, Weishaupt JH, Wollmer MA, Boxer A, et al. Progranulin as a candidate biomarker for therapeutic trial in patients with ALS and FTLD. J Neural Transm (Vienna). 2016;123(3):289–96.
130. Han JJ, Yu M, Houston N, Steinberg SM, Kohn EC. Progranulin is a potential prognostic biomarker in advanced epithelial ovarian cancers. Gynecol Oncol. 2011;120(1):5–10.
131. Todoric J, Handisurya A, Perkmann T, Knapp B, Wagner O, Tura A, et al. Circulating progranulin levels in women with gestational diabetes mellitus and healthy controls during and after pregnancy. Eur J Endocrinol. 2012;167(4):561–7.
132. Cruchaga C, Graff C, Chiang HH, Wang J, Hinrichs AL, Spiegel N, et al. Association of TMEM106B gene polymorphism with age at onset in granulin mutation carriers and plasma granulin protein levels. Arch Neurol. 2011;68(5):581–6.

133. Finch N, Carrasquillo MM, Baker M, Rutherford NJ, Coppola G, Dejesus-Hernandez M, et al. TMEM106B regulates progranulin levels and the penetrance of FTLD in GRN mutation carriers. Neurology. 2011;76(5):467–74.
134. Hsiung GY, Fok A, Feldman HH, Rademakers R, Mackenzie IR. rs5848 polymorphism and serum progranulin level. J Neurol Sci. 2011;300(1–2):28–32.
135. Kasai T, Kojima Y, Ohmichi T, Tatebe H, Tsuji Y, Noto YI, et al. Combined use of CSF NfL and CSF TDP-43 improves diagnostic performance in ALS. Ann Clin Transl Neurol. 2019;6(12):2489–502.
136. Kasai T, Tokuda T, Ishigami N, Sasayama H, Foulds P, Mitchell DJ, et al. Increased TDP-43 protein in cerebrospinal fluid of patients with amyotrophic lateral sclerosis. Acta Neuropathol. 2009;117(1):55–62.
137. Noto Y, Shibuya K, Sato Y, Kanai K, Misawa S, Sawai S, et al. Elevated CSF TDP-43 levels in amyotrophic lateral sclerosis: specificity, sensitivity, and a possible prognostic value. Amyotroph Lateral Scler. 2011;12(2):140–3.
138. Junttila A, Kuvaja M, Hartikainen P, Siloaho M, Helisalmi S, Moilanen V, et al. Cerebrospinal Fluid TDP-43 in Frontotemporal Lobar Degeneration and Amyotrophic Lateral Sclerosis Patients with and without the C9ORF72 Hexanucleotide Expansion. Dement Geriatr Cogn Dis Extra. 2016;6(1):142–9.
139. Kuiperij HB, Versleijen AA, Beenes M, Verwey NA, Benussi L, Paterlini A, et al. Tau Rather than TDP-43 Proteins are Potential Cerebrospinal Fluid Biomarkers for Frontotemporal Lobar Degeneration Subtypes: A Pilot Study. J Alzheimers Dis. 2017;55(2):585–95.
140. Foulds PG, Davidson Y, Mishra M, Hobson DJ, Humphreys KM, Taylor M, et al. Plasma phosphorylated-TDP-43 protein levels correlate with brain pathology in frontotemporal lobar degeneration. Acta Neuropathol. 2009;118(5):647–58.
141. Suárez-Calvet M, Dols-Icardo O, Lladó A, Sánchez-Valle R, Hernández I, Amer G, et al. Plasma phosphorylated TDP-43 levels are elevated in patients with frontotemporal dementia carrying a C9orf72 repeat expansion or a GRN mutation. J Neurol Neurosurg Psychiatry. 2014;85(6):684–91.
142. Katisko K, Huber N, Kokkola T, Hartikainen P, Krüger J, Heikkinen AL, et al. Serum total TDP-43 levels are decreased in frontotemporal dementia patients with C9orf72 repeat expansion or concomitant motoneuron disease phenotype. Alzheimers Res Ther. 2022;14(1):151.
143. Jamerlan AM, Shim KH, Youn YC, Teunissen C, An SSA, Scheltens P, et al. Increased oligomeric TDP-43 in the plasma of Korean frontotemporal dementia patients with semantic dementia. Alzheimers Dement. 2023;19:4020.
144. Lye YS, Chen YR. TAR DNA-binding protein 43 oligomers in physiology and pathology. IUBMB Life. 2022;74(8):794–811.
145. Scialò C, Tran TH, Salzano G, Novi G, Caponnetto C, Chiò A, et al. TDP-43 real-time quaking induced conversion reaction optimization and detection of seeding activity in CSF of amyotrophic lateral sclerosis and frontotemporal dementia patients. Brain Commun. 2020;2(2):fcaa142.
146. Bright F, Werry EL, Dobson-Stone C, Piguet O, Ittner LM, Halliday GM, et al. Neuroinflammation in frontotemporal dementia. Nat Rev Neurol. 2019;15(9):540–55. https://doi.org/10.1038/s41582-019-0231-z.
147. Colangelo AM, Alberghina L, Papa M. Astrogliosis as a therapeutic target for neurodegenerative diseases. Neurosci Lett. 2014;565:59–64.
148. Kim KY, Shin KY, Chang KA. GFAP as a potential biomarker for Alzheimer's disease: a systematic review and meta-analysis. Cells. 2023;12(9)
149. Ishiki A, Kamada M, Kawamura Y, Terao C, Shimoda F, Tomita N, et al. Glial fibrillar acidic protein in the cerebrospinal fluid of Alzheimer's disease, dementia with Lewy bodies, and frontotemporal lobar degeneration. J Neurochem. 2016;136(2):258–61. https://doi.org/10.1111/jnc.13399.

150. Abu-Rumeileh S, Steinacker P, Polischi B, Mammana A, Bartoletti-Stella A, Oeckl P, et al. CSF biomarkers of neuroinflammation in distinct forms and subtypes of neurodegenerative dementia. Alzheimers Res Ther. 2019;12(1):2. https://doi.org/10.1186/s13195-019-0562-4.
151. Oeckl P, Weydt P, Steinacker P, Anderl-Straub S, Nordin F, Volk AE, et al. Different neuroinflammatory profile in amyotrophic lateral sclerosis and frontotemporal dementia is linked to the clinical phase. J Neurol Neurosurg Psychiatry. 2019;90(1):4–10.
152. Benussi A, Ashton NJ, Karikari TK, Gazzina S, Premi E, Benussi L, et al. Serum Glial Fibrillary Acidic Protein (GFAP) is a marker of disease severity in frontotemporal lobar degeneration. J Alzheimers Dis. 2020;77(3):1129–41.
153. Marelli C, Hourregue C, Gutierrez LA, Paquet C, Menjot de Champfleur N, De Verbizier D, et al. Cerebrospinal fluid and plasma biomarkers do not differ in the presenile and late-onset behavioral variants of frontotemporal dementia. J Alzheimers Dis. 2020;74(3):903–11.
154. Oeckl P, Anderl-Straub S, Von Arnim CAF, Baldeiras I, Diehl-Schmid J, Grimmer T, et al. Serum GFAP differentiates Alzheimer's disease from frontotemporal dementia and predicts MCI-to-dementia conversion. J Neurol Neurosurg Psychiatry. 2022;93:659.
155. Heller C, Foiani MS, Moore K, Convery R, Bocchetta M, Neason M, et al. Plasma glial fibrillary acidic protein is raised in progranulin-associated frontotemporal dementia. J Neurol Neurosurg Psychiatry. 2020;91(3):263–70.
156. Cousins KAQ, Shaw LM, Chen-Plotkin A, Wolk DA, Van Deerlin VM, Lee EB, et al. Distinguishing frontotemporal lobar degeneration Tau from TDP-43 using plasma biomarkers. JAMA Neurol. 2022;79(11):1155–64.
157. Katisko K, Cajanus A, Huber N, Jääskeläinen O, Kokkola T, Kärkkäinen V, et al. GFAP as a biomarker in frontotemporal dementia and primary psychiatric disorders: diagnostic and prognostic performance. J Neurol Neurosurg Psychiatry. 2021;92(12):1305–12.
158. Querol-Vilaseca M, Colom-Cadena M, Pegueroles J, San Martín-Paniello C, Clarimon J, Belbin O, et al. YKL-40 (Chitinase 3-like I) is expressed in a subset of astrocytes in Alzheimer's disease and other tauopathies. J Neuroinflammation. 2017;14(1):118.
159. Russo C, Valle MS, Casabona A, Malaguarnera L. Chitinase signature in the plasticity of neurodegenerative diseases. Int J Mol Sci. 2023;24(7)
160. Alcolea D, Irwin DJ, Illán-Gala I, Muñoz L, Clarimón J, McMillan CT, et al. Elevated YKL-40 and low sAPPβ:YKL-40 ratio in antemortem cerebrospinal fluid of patients with pathologically confirmed FTLD. J Neurol Neurosurg Psychiatry. 2019;90(2):180–6.
161. Illán-Gala I, Alcolea D, Montal V, Dols-Icardo O, Muñoz L, de Luna N, et al. CSF sAPPβ, YKL-40, and NfL along the ALS-FTD spectrum. Neurology. 2018;91(17):e1619–e28.
162. Antonell A, Tort-Merino A, Rios J, Balasa M, Borrego-Ecija S, Auge JM, et al. Synaptic, axonal damage and inflammatory cerebrospinal fluid biomarkers in neurodegenerative dementias. Alzheimers Dement'. 2020;16(2):262–72.
163. Woollacott IOC, Nicholas JM, Heller C, Foiani MS, Moore KM, Russell LL, et al. Cerebrospinal Fluid YKL-40 and chitotriosidase levels in frontotemporal dementia vary by clinical, genetic and pathological subtype. Dement Geriatr Cogn Disord. 2020;49(1):56–76. https://doi.org/10.1159/000506282.
164. Woollacott IOC, Swift IJ, Sogorb-Esteve A, Heller C, Knowles K, Bouzigues A, et al. CSF glial markers are elevated in a subset of patients with genetic frontotemporal dementia. Ann Clin Transl Neurol. 2022;9(11):1764–77. https://doi.org/10.1002/acn3.51672.
165. Boot RG, Renkema GH, Verhoek M, Strijland A, Bliek J, de Meulemeester TM, et al. The human chitotriosidase gene. Nature of inherited enzyme deficiency. J Biol Chem. 1998;273(40):25680–5.
166. Malaguarnera L, Simporè J, Prodi DA, Angius A, Sassu A, Persico I, et al. A 24-bp duplication in exon 10 of human chitotriosidase gene from the sub-Saharan to the Mediterranean area: role of parasitic diseases and environmental conditions. Genes Immun. 2003;4(8):570–4.
167. Woollacott IOC, Nicholas JM, Heslegrave A, Heller C, Foiani MS, Dick KM, et al. Cerebrospinal fluid soluble TREM2 levels in frontotemporal dementia differ by genetic

and pathological subgroup. Alzheimers Res Ther. 2018;10(1):79. https://doi.org/10.1186/s13195-018-0405-8.
168. van der Ende EL, Morenas-Rodriguez E, McMillan C, Grossman M, Irwin D, Sanchez-Valle R, et al. CSF sTREM2 is elevated in a subset in GRN-related frontotemporal dementia. Neurobiol Aging. 2021;103(158):e1–5.
169. Ewers M, Franzmeier N, Suárez-Calvet M, Morenas-Rodriguez E, Caballero MAA, Kleinberger G, et al. Increased soluble TREM2 in cerebrospinal fluid is associated with reduced cognitive and clinical decline in Alzheimer's disease. Sci Transl Med. 2019;11(507)
170. Sjögren M, Folkesson S, Blennow K, Tarkowski E. Increased intrathecal inflammatory activity in frontotemporal dementia: pathophysiological implications. J Neurol Neurosurg Psychiatry. 2004;75(8):1107–11.
171. Galimberti D, Bonsi R, Fenoglio C, Serpente M, Cioffi SM, Fumagalli G, et al. Inflammatory molecules in Frontotemporal Dementia: cerebrospinal fluid signature of progranulin mutation carriers. Brain Behav Immun. 2015;49:182–7.
172. Galimberti D, Schoonenboom N, Scheltens P, Fenoglio C, Venturelli E, Pijnenburg YA, et al. Intrathecal chemokine levels in Alzheimer disease and frontotemporal lobar degeneration. Neurology. 2006;66(1):146–7.
173. Galimberti D, Venturelli E, Fenoglio C, Guidi I, Villa C, Bergamaschini L, et al. Intrathecal levels of IL-6, IL-11 and LIF in Alzheimer's disease and frontotemporal lobar degeneration. J Neurol. 2008;255(4):539–44.
174. Miller ZA, Rankin KP, Graff-Radford NR, Takada LT, Sturm VE, Cleveland CM, et al. TDP-43 frontotemporal lobar degeneration and autoimmune disease. J Neurol Neurosurg Psychiatry. 2013;84(9):956–62.
175. Gibbons L, Rollinson S, Thompson JC, Robinson A, Davidson YS, Richardson A, et al. Plasma levels of progranulin and interleukin-6 in frontotemporal lobar degeneration. Neurobiol Aging. 2015;36(3):1603e1–4.
176. Fraga VG, Magalhães CA, Loures CMG, de Souza LC, Guimarães HC, Zauli DAG, et al. Inflammatory and pro-resolving mediators in frontotemporal dementia and Alzheimer's disease. Neuroscience. 2019;421:123–35.
177. Vieira ÉLM, Caramelli P, Rocha NP, Freitas Cardoso MDG, de Miranda AS, Teixeira AL, et al. Tumor necrosis factor superfamily molecules are increased in behavioral variant frontotemporal dementia and correlate with cortical atrophy: An exploratory investigation. J Neuroimmunol. 2021;354:577531.
178. Chu M, Wen L, Jiang D, Liu L, Nan H, Yue A, et al. Peripheral inflammation in behavioural variant frontotemporal dementia: associations with central degeneration and clinical measures. J Neuroinflammation. 2023;20(1):65.
179. Asken BM, Ljubenkov PA, Staffaroni AM, Casaletto KB, Vandevrede L, Cobigo Y, et al. Plasma inflammation for predicting phenotypic conversion and clinical progression of autosomal dominant frontotemporal lobar degeneration. J Neurol Neurosurg Psychiatry. 2023;94:541.
180. Zhou J, Wade SD, Graykowski D, Xiao MF, Zhao B, Giannini LAA, et al. The neuronal pentraxin Nptx2 regulates complement activity and restrains microglia-mediated synapse loss in neurodegeneration. Sci Transl Med. 2023;15(689):eadf0141.
181. Daborg J, Andreasson U, Pekna M, Lautner R, Hanse E, Minthon L, et al. Cerebrospinal fluid levels of complement proteins C3, C4 and CR1 in Alzheimer's disease. J Neural Transm (Vienna). 2012;119(7):789–97.
182. Bonham LW, Desikan RS, Yokoyama JS, Alzheimer's Disease Neuroimaging I. The relationship between complement factor C3, APOE ε4, amyloid and tau in Alzheimer's disease. Acta Neuropathol Commun. 2016;4(1):65.
183. Morgan AR, Touchard S, Leckey C, O'Hagan C, Nevado-Holgado AJ, Consortium N, et al. Inflammatory biomarkers in Alzheimer's disease plasma. Alzheimers Dement. 2019;15(6):776–87.

184. Lui H, Zhang J, Makinson SR, Cahill MK, Kelley KW, Huang HY, et al. Progranulin deficiency promotes circuit-specific synaptic pruning by microglia via complement activation. Cell. 2016;165(4):921–35.
185. van der Ende EL, Heller C, Sogorb-Esteve A, Swift IJ, McFall D, Peakman G, et al. Elevated CSF and plasma complement proteins in genetic frontotemporal dementia: results from the GENFI study. J Neuroinflammation. 2022;19(1):217.
186. de San G, José N, Massa F, Halbgebauer S, Oeckl P, Steinacker P, Otto M. Neuronal pentraxins as biomarkers of synaptic activity: from physiological functions to pathological changes in neurodegeneration. J Neural Transm (Vienna). 2022;129(2):207–30.
187. van der Ende EL, Meeter LH, Stingl C, van Rooij JGJ, Stoop MP, Nijholt DAT, et al. Novel CSF biomarkers in genetic frontotemporal dementia identified by proteomics. Ann Clin Transl Neurol. 2019;6(4):698–707. https://doi.org/10.1002/acn3.745.
188. Remnestål J, Öijerstedt L, Ullgren A, Olofsson J, Bergström S, Kultima K, et al. Altered levels of CSF proteins in patients with FTD, presymptomatic mutation carriers and non-carriers. Transl Neurodegener. 2020;9(1):27.
189. van der Ende EL, Xiao M, Xu D, Poos JM, Panman JL, Jiskoot LC, et al. Neuronal pentraxin 2: a synapse-derived CSF biomarker in genetic frontotemporal dementia. J Neurol Neurosurg Psychiatry. 2020;91(6):612–21.
190. Sogorb-Esteve A, Nilsson J, Swift IJ, Heller C, Bocchetta M, Russell LL, et al. Differential impairment of cerebrospinal fluid synaptic biomarkers in the genetic forms of frontotemporal dementia. Alzheimers Res Ther. 2022;14(1):118.
191. Das S, Goossens J, Jacobs D, Dewit N, Pijnenburg YAL, In 't Veld S, et al. Synaptic biomarkers in the cerebrospinal fluid associate differentially with classical neuronal biomarkers in patients with Alzheimer's disease and frontotemporal dementia. Alzheimers Res Ther. 2023;15(1):62.
192. van der Ende EL, Bron EE, Poos JM, Jiskoot LC, Panman JL, Papma JM, et al. A data-driven disease progression model of fluid biomarkers in genetic frontotemporal dementia. Brain. 2022;145(5):1805–17.
193. Goetzl EJ, Kapogiannis D, Schwartz JB, Lobach IV, Goetzl L, Abner EL, et al. Decreased synaptic proteins in neuronal exosomes of frontotemporal dementia and Alzheimer's disease. FASEB J. 2016;30(12):4141–8.
194. Clarke MTM, Brinkmalm A, Foiani MS, Woollacott IOC, Heller C, Heslegrave A, et al. CSF synaptic protein concentrations are raised in those with atypical Alzheimer's disease but not frontotemporal dementia. Alzheimers Res Ther. 2019;11(1):105.
195. Niessen WJ. MR brain image analysis in dementia: From quantitative imaging biomarkers to ageing brain models and imaging genetics. Med Image Anal. 2016;33:107–13.
196. Pasquier F, Leys D, Weerts JG, Mounier-Vehier F, Barkhof F, Scheltens P. Inter- and intraobserver reproducibility of cerebral atrophy assessment on MRI scans with hemispheric infarcts. Eur Neurol. 1996;36(5):268–72.
197. Sheikh-Bahaei N, Sajjadi SA, Pierce AL. Current role for biomarkers in clinical diagnosis of alzheimer disease and frontotemporal dementia. Curr Treat Options Neurol. 2017;19(12):46.
198. Scheltens P, Leys D, Barkhof F, Huglo D, Weinstein HC, Vermersch P, et al. Atrophy of medial temporal lobes on MRI in "probable" Alzheimer's disease and normal ageing: diagnostic value and neuropsychological correlates. J Neurol Neurosurg Psychiatry. 1992;55(10):967–72.
199. Fischl B, Salat DH, Busa E, Albert M, Dieterich M, Haselgrove C, et al. Whole brain segmentation: automated labeling of neuroanatomical structures in the human brain. Neuron. 2002;33(3):341–55.
200. Vrooman HA, Cocosco CA, van der Lijn F, Stokking R, Ikram MA, Vernooij MW, et al. Multi-spectral brain tissue segmentation using automatically trained k-Nearest-Neighbor classification. NeuroImage. 2007;37(1):71–81.

201. Fortunati V, Verhaart RF, Niessen WJ, Veenland JF, Paulides MM, van Walsum T. Automatic tissue segmentation of head and neck MR images for hyperthermia treatment planning. Phys Med Biol. 2015;60(16):6547–62.
202. Matsuda H. MRI morphometry in Alzheimer's disease. Ageing Res Rev. 2016;30:17–24.
203. Poos JM, Grandpierre LDM, van der Ende EL, Panman JL, Papma JM, Seelaar H, et al. Longitudinal brain atrophy rates in presymptomatic carriers of genetic frontotemporal dementia. Neurology. 2022;99(24):e2661–71.
204. Calloni SF, Vezzulli PQ, Castellano A, Leone R, Basaia S, von Loon A, et al. Combining semi-quantitative rating and automated brain volumetry in MRI evaluation of patients with probable behavioural variant of fronto-temporal dementia: an added value for clinical practise? Neuroradiology. 2023;65(6):1025–35.
205. Bruun M, Koikkalainen J, Rhodius-Meester HFM, Baroni M, Gjerum L, van Gils M, et al. Detecting frontotemporal dementia syndromes using MRI biomarkers. Neuroimage Clin. 2019;22:101711.
206. Righini A, Antonini A, De Notaris R, Bianchini E, Meucci N, Sacilotto G, et al. MR imaging of the superior profile of the midbrain: differential diagnosis between progressive supranuclear palsy and Parkinson disease. AJNR Am J Neuroradiol. 2004;25(6):927–32.
207. Gröschel K, Kastrup A, Litvan I, Schulz JB. Penguins and hummingbirds: midbrain atrophy in progressive supranuclear palsy. Neurology. 2006;66(6):949–50.
208. Massey LA, Jäger HR, Paviour DC, O'Sullivan SS, Ling H, Williams DR, et al. The midbrain to pons ratio: a simple and specific MRI sign of progressive supranuclear palsy. Neurology. 2013;80(20):1856–61.
209. Massey LA, Micallef C, Paviour DC, O'Sullivan SS, Ling H, Williams DR, et al. Conventional magnetic resonance imaging in confirmed progressive supranuclear palsy and multiple system atrophy. Mov Disord. 2012;27(14):1754–62.
210. Whitwell JL, Boeve BF, Weigand SD, Senjem ML, Gunter JL, Baker MC, et al. Brain atrophy over time in genetic and sporadic frontotemporal dementia: a study of 198 serial magnetic resonance images. Eur J Neurol. 2015;22(5):745–52.
211. Meeter LH, Donker Kaat L, Rohrer JD, van Swieten JC. Imaging and fluid biomarkers in frontotemporal dementia. Nat Rev Neurol. 2017;13(7):406–19.
212. Rohrer JD, Geser F, Zhou J, Gennatas ED, Sidhu M, Trojanowski JQ, et al. TDP-43 subtypes are associated with distinct atrophy patterns in frontotemporal dementia. Neurology. 2010;75(24):2204–11.
213. Whitwell JL, Jack CR Jr, Parisi JE, Senjem ML, Knopman DS, Boeve BF, et al. Does TDP-43 type confer a distinct pattern of atrophy in frontotemporal lobar degeneration? Neurology. 2010;75(24):2212–20.
214. Rohrer JD, Lashley T, Schott JM, Warren JE, Mead S, Isaacs AM, et al. Clinical and neuroanatomical signatures of tissue pathology in frontotemporal lobar degeneration. Brain. 2011;134(Pt 9):2565–81.
215. Spina S, Murrell JR, Huey ED, Wassermann EM, Pietrini P, Grafman J, et al. Corticobasal syndrome associated with the A9D Progranulin mutation. J Neuropathol Exp Neurol. 2007;66(10):892–900.
216. Beck J, Rohrer JD, Campbell T, Isaacs A, Morrison KE, Goodall EF, et al. A distinct clinical, neuropsychological and radiological phenotype is associated with progranulin gene mutations in a large UK series. Brain. 2008;131(Pt 3):706–20.
217. Rohrer JD, Warren JD, Omar R, Mead S, Beck J, Revesz T, et al. Parietal lobe deficits in frontotemporal lobar degeneration caused by a mutation in the progranulin gene. Arch Neurol. 2008;65(4):506–13.
218. Cash DM, Bocchetta M, Thomas DL, Dick KM, van Swieten JC, Borroni B, et al. Patterns of gray matter atrophy in genetic frontotemporal dementia: results from the GENFI study. Neurobiol Aging. 2018;62:191–6.

219. Sha SJ, Takada LT, Rankin KP, Yokoyama JS, Rutherford NJ, Fong JC, et al. Frontotemporal dementia due to C9ORF72 mutations: clinical and imaging features. Neurology. 2012;79(10):1002–11.
220. Häkkinen S, Chu SA, Lee SE. Neuroimaging in genetic frontotemporal dementia and amyotrophic lateral sclerosis. Neurobiol Dis. 2020;145:105063.
221. Bocchetta M, Iglesias Espinosa MDM, Lashley T, Warren JD, Rohrer JD. In vivo staging of frontotemporal lobar degeneration TDP-43 type C pathology. Alzheimers Res Ther. 2020;12(1):34.
222. Borghesani V, Battistella G, Mandelli ML, Welch A, Weis E, Younes K, et al. Regional and hemispheric susceptibility of the temporal lobe to FTLD-TDP type C pathology. Neuroimage Clin. 2020;28:102369.
223. Whitwell JL, Josephs KA. Neuroimaging in frontotemporal lobar degeneration—predicting molecular pathology. Nat Rev Neurol. 2012;8(3):131–42.
224. Peet BT, Spina S, Mundada N, La Joie R. Neuroimaging in frontotemporal dementia: heterogeneity and relationships with underlying neuropathology. Neurotherapeutics. 2021;18(2):728–52.
225. Whitwell JL, Jack CR Jr, Parisi JE, Knopman DS, Boeve BF, Petersen RC, et al. Imaging signatures of molecular pathology in behavioral variant frontotemporal dementia. J Mol Neurosci. 2011;45(3):372–8.
226. Whitwell JL, Josephs KA, Rossor MN, Stevens JM, Revesz T, Holton JL, et al. Magnetic resonance imaging signatures of tissue pathology in frontotemporal dementia. Arch Neurol. 2005;62(9):1402–8.
227. Rankin KP, Mayo MC, Seeley WW, Lee S, Rabinovici G, Gorno-Tempini ML, et al. Behavioral variant frontotemporal dementia with corticobasal degeneration pathology: phenotypic comparison to bvFTD with Pick's disease. J Mol Neurosci. 2011;45(3):594–608.
228. van Swieten JC, Stevens M, Rosso SM, Rizzu P, Joosse M, de Koning I, et al. Phenotypic variation in hereditary frontotemporal dementia with tau mutations. Ann Neurol. 1999;46(4):617–26.
229. Ghetti B, Spina S, Murrell JR, Huey ED, Pietrini P, Sweeney B, et al. In vivo and postmortem clinicoanatomical correlations in frontotemporal dementia and parkinsonism linked to chromosome 17. Neurodegener Dis. 2008;5(3–4):215–7.
230. Boeve BF, Tremont-Lukats IW, Waclawik AJ, Murrell JR, Hermann B, Jack CR Jr, et al. Longitudinal characterization of two siblings with frontotemporal dementia and parkinsonism linked to chromosome 17 associated with the S305N tau mutation. Brain. 2005;128(Pt 4):752–72.
231. Whitwell JL, Jack CR Jr, Boeve BF, Senjem ML, Baker M, Ivnik RJ, et al. Atrophy patterns in IVS10+16, IVS10+3, N279K, S305N, P301L, and V337M MAPT mutations. Neurology. 2009;73(13):1058–65.
232. Arima K, Kowalska A, Hasegawa M, Mukoyama M, Watanabe R, Kawai M, et al. Two brothers with frontotemporal dementia and parkinsonism with an N279K mutation of the tau gene. Neurology. 2000;54(9):1787–95.
233. Soliveri P, Rossi G, Monza D, Tagliavini F, Piacentini S, Albanese A, et al. A case of dementia parkinsonism resembling progressive supranuclear palsy due to mutation in the tau protein gene. Arch Neurol. 2003;60(10):1454–6.
234. Slowinski JL, Schweitzer KJ, Imamura A, Uitti RJ, Strongosky AJ, Dickson DW, et al. Brainstem atrophy on routine MR study in pallidopontonigral degeneration. J Neurol. 2009;256(5):827–9.
235. Josephs KA, Whitwell JL, Parisi JE, Petersen RC, Boeve BF, Jack CR Jr, et al. Caudate atrophy on MRI is a characteristic feature of FTLD-FUS. Eur J Neurol. 2010;17(7):969–75.
236. Rohrer JD, Lashley T, Holton J, Revesz T, Urwin H, Isaacs AM, et al. The clinical and neuroanatomical phenotype of FUS associated frontotemporal lobar degeneration. J Neurol Neurosurg Psychiatry. 2011;82(12):1405–7.

237. Finger E, Malik R, Bocchetta M, Coleman K, Graff C, Borroni B, et al. Neurodevelopmental effects of genetic frontotemporal dementia in young adult mutation carriers. Brain. 2023;146(5):2120–31.
238. Walhout R, Schmidt R, Westeneng HJ, Verstraete E, Seelen M, van Rheenen W, et al. Brain morphologic changes in asymptomatic C9orf72 repeat expansion carriers. Neurology. 2015;85(20):1780–8.
239. Lee EB, Porta S, Michael Baer G, Xu Y, Suh E, Kwong LK, et al. Expansion of the classification of FTLD-TDP: distinct pathology associated with rapidly progressive frontotemporal degeneration. Acta Neuropathol. 2017;134(1):65–78.
240. Bertrand A, Wen J, Rinaldi D, Houot M, Sayah S, Camuzat A, et al. Early cognitive, structural, and microstructural changes in presymptomatic C9orf72 carriers younger than 40 years. JAMA Neurol. 2018;75(2):236–45.
241. Popuri K, Dowds E, Beg MF, Balachandar R, Bhalla M, Jacova C, et al. Gray matter changes in asymptomatic C9orf72 and GRN mutation carriers. Neuroimage Clin. 2018;18:591–8.
242. Panman JL, Jiskoot LC, Bouts M, Meeter LHH, van der Ende EL, Poos JM, et al. Gray and white matter changes in presymptomatic genetic frontotemporal dementia: a longitudinal MRI study. Neurobiol Aging. 2019;76:115–24.
243. Chen Q, Kantarci K. Imaging biomarkers for neurodegeneration in presymptomatic familial frontotemporal lobar degeneration. Front Neurol. 2020;11:80.
244. Rohrer JD, Nicholas JM, Cash DM, van Swieten J, Dopper E, Jiskoot L, et al. Presymptomatic cognitive and neuroanatomical changes in genetic frontotemporal dementia in the Genetic Frontotemporal dementia Initiative (GENFI) study: a cross-sectional analysis. Lancet Neurol. 2015;14(3):253–62.
245. Olm CA, McMillan CT, Irwin DJ, Van Deerlin VM, Cook PA, Gee JC, et al. Longitudinal structural gray matter and white matter MRI changes in presymptomatic progranulin mutation carriers. Neuroimage Clin. 2018;19:497–506.
246. Jiskoot LC, Panman JL, Meeter LH, Dopper EGP, Donker Kaat L, Franzen S, et al. Longitudinal multimodal MRI as prognostic and diagnostic biomarker in presymptomatic familial frontotemporal dementia. Brain. 2019;142(1):193–208.
247. Pengo M, Premi E, Borroni B. Dissecting the many faces of frontotemporal dementia: an imaging perspective. Int J Mol Sci. 2022;23(21)
248. Diehl-Schmid J, Onur OA, Kuhn J, Gruppe T, Drzezga A. Imaging frontotemporal lobar degeneration. Curr Neurol Neurosci Rep. 2014;14(10):489.
249. Dev SI, Dickerson BC, Touroutoglou A. Neuroimaging in frontotemporal lobar degeneration: research and clinical utility. Adv Exp Med Biol. 2021;1281:93–112.
250. McMillan CT, Irwin DJ, Avants BB, Powers J, Cook PA, Toledo JB, et al. White matter imaging helps dissociate tau from TDP-43 in frontotemporal lobar degeneration. J Neurol Neurosurg Psychiatry. 2013;84(9):949–55.
251. Zhang Y, Tartaglia MC, Schuff N, Chiang GC, Ching C, Rosen HJ, et al. MRI signatures of brain macrostructural atrophy and microstructural degradation in frontotemporal lobar degeneration subtypes. J Alzheimers Dis. 2013;33(2):431–44.
252. Borroni B, Alberici A, Premi E, Archetti S, Garibotto V, Agosti C, et al. Brain magnetic resonance imaging structural changes in a pedigree of asymptomatic progranulin mutation carriers. Rejuvenation Res. 2008;11(3):585–95.
253. Jiskoot LC, Bocchetta M, Nicholas JM, Cash DM, Thomas D, Modat M, et al. Presymptomatic white matter integrity loss in familial frontotemporal dementia in the GENFI cohort: A cross-sectional diffusion tensor imaging study. Ann Clin Transl Neurol. 2018;5(9):1025–36.
254. Sudre CH, Bocchetta M, Heller C, Convery R, Neason M, Moore KM, et al. White matter hyperintensities in progranulin-associated frontotemporal dementia: a longitudinal GENFI study. Neuroimage Clin. 2019;24:102077.
255. Mahoney CJ, Ridgway GR, Malone IB, Downey LE, Beck J, Kinnunen KM, et al. Profiles of white matter tract pathology in frontotemporal dementia. Hum Brain Mapp. 2014;35(8):4163–79.

256. Mahoney CJ, Simpson IJ, Nicholas JM, Fletcher PD, Downey LE, Golden HL, et al. Longitudinal diffusion tensor imaging in frontotemporal dementia. Ann Neurol. 2015;77(1):33–46.
257. Lee SE, Sias AC, Mandelli ML, Brown JA, Brown AB, Khazenzon AM, et al. Network degeneration and dysfunction in presymptomatic C9ORF72 expansion carriers. Neuroimage Clin. 2017;14:286–97.
258. Papma JM, Jiskoot LC, Panman JL, Dopper EG, den Heijer T, Donker Kaat L, et al. Cognition and gray and white matter characteristics of presymptomatic C9orf72 repeat expansion. Neurology. 2017;89(12):1256–64.
259. Agosta F, Galantucci S, Magnani G, Marcone A, Martinelli D, Antonietta Volontè M, et al. MRI signatures of the frontotemporal lobar degeneration continuum. Hum Brain Mapp. 2015;36(7):2602–14.
260. Chen Q, Boeve BF, Schwarz CG, Reid R, Tosakulwong N, Lesnick TG, et al. Tracking white matter degeneration in asymptomatic and symptomatic MAPT mutation carriers. Neurobiol Aging. 2019;83:54–62.
261. Dopper EG, Rombouts SA, Jiskoot LC, den Heijer T, de Graaf JR, de Koning I, et al. Structural and functional brain connectivity in presymptomatic familial frontotemporal dementia. Neurology. 2014;83(2):e19–26.
262. Grimmer T, Diehl J, Drzezga A, Förstl H, Kurz A. Region-specific decline of cerebral glucose metabolism in patients with frontotemporal dementia: a prospective 18F-FDG-PET study. Dement Geriatr Cogn Disord. 2004;18(1):32–6.
263. Ishii K. PET approaches for diagnosis of dementia. AJNR Am J Neuroradiol. 2014;35(11):2030–8.
264. Iaccarino L, Crespi C, Della Rosa PA, Catricalà E, Guidi L, Marcone A, et al. The semantic variant of primary progressive aphasia: clinical and neuroimaging evidence in single subjects. PLoS One. 2015;10(3):e0120197.
265. Matias-Guiu JA, Cabrera-Martín MN, Moreno-Ramos T, García-Ramos R, Porta-Etessam J, Carreras JL, et al. Clinical course of primary progressive aphasia: clinical and FDG-PET patterns. J Neurol. 2015;262(3):570–7.
266. Buhour MS, Doidy F, Laisney M, Pitel AL, de La Sayette V, Viader F, et al. Pathophysiology of the behavioral variant of frontotemporal lobar degeneration: a study combining MRI and FDG-PET. Brain Imaging Behav. 2017;11(1):240–52.
267. Cerami C, Dodich A, Greco L, Iannaccone S, Magnani G, Marcone A, et al. The role of single-subject brain metabolic patterns in the early differential diagnosis of primary progressive aphasias and in prediction of progression to dementia. J Alzheimers Dis. 2017;55(1):183–97.
268. Walker Z, Gandolfo F, Orini S, Garibotto V, Agosta F, Arbizu J, et al. Clinical utility of FDG PET in Parkinson's disease and atypical parkinsonism associated with dementia. Eur J Nucl Med Mol Imaging. 2018;45(9):1534–45.
269. Pardini M, Huey ED, Spina S, Kreisl WC, Morbelli S, Wassermann EM, et al. FDG-PET patterns associated with underlying pathology in corticobasal syndrome. Neurology. 2019;92(10):e1121–e35.
270. Lim SM, Katsifis A, Villemagne VL, Best R, Jones G, Saling M, et al. The 18F-FDG PET cingulate Island sign and comparison to 123I-beta-CIT SPECT for diagnosis of dementia with Lewy bodies. J Nucl Med. 2009;50(10):1638–45.
271. Medicare National Coverage Determinations. Centers for Medicare and Medicaid Services Web site. Manual: Chapter 1, Part 4 (Sections 200-310.1)—Coverage Determinations. http://www.cms.gov/Regulations-and-Guidance/Guidance/Manuals/Downloads/ncd103c1_Part4.pdf
272. Kerklaan BJ, van Berckel BN, Herholz K, Dols A, van der Flier WM, Scheltens P, et al. The added value of 18-fluorodeoxyglucose-positron emission tomography in the diagnosis of the behavioral variant of frontotemporal dementia. Am J Alzheimers Dis Other Dement. 2014;29(7):607–13.

273. Krudop WA, Dols A, Kerssens CJ, Eikelenboom P, Prins ND, Möller C, et al. The pitfall of behavioral variant frontotemporal dementia mimics despite multidisciplinary application of the FTDC criteria. J Alzheimers Dis. 2017;60(3):959–75.
274. Jacova C, Hsiung GY, Tawankanjanachot I, Dinelle K, McCormick S, Gonzalez M, et al. Anterior brain glucose hypometabolism predates dementia in progranulin mutation carriers. Neurology. 2013;81(15):1322–31.
275. Caroppo P, Habert MO, Durrleman S, Funkiewiez A, Perlbarg V, Hahn V, et al. Lateral temporal lobe: an early imaging marker of the presymptomatic GRN disease? J Alzheimers Dis. 2015;47(3):751–9.
276. Dopper EG, Chalos V, Ghariq E, den Heijer T, Hafkemeijer A, Jiskoot LC, et al. Cerebral blood flow in presymptomatic MAPT and GRN mutation carriers: a longitudinal arterial spin labeling study. Neuroimage Clin. 2016;12:460–5.
277. Du AT, Jahng GH, Hayasaka S, Kramer JH, Rosen HJ, Gorno-Tempini ML, et al. Hypoperfusion in frontotemporal dementia and Alzheimer disease by arterial spin labeling MRI. Neurology. 2006;67(7):1215–20.
278. Fällmar D, Haller S, Lilja J, Danfors T, Kilander L, Tolboom N, et al. Arterial spin labeling-based Z-maps have high specificity and positive predictive value for neurodegenerative dementia compared to FDG-PET. Eur Radiol. 2017;27(10):4237–46.
279. Steketee RM, Bron EE, Meijboom R, Houston GC, Klein S, Mutsaerts HJ, et al. Early-stage differentiation between presenile Alzheimer's disease and frontotemporal dementia using arterial spin labeling MRI. Eur Radiol. 2016;26(1):244–53.
280. Verfaillie SC, Adriaanse SM, Binnewijzend MA, Benedictus MR, Ossenkoppele R, Wattjes MP, et al. Cerebral perfusion and glucose metabolism in Alzheimer's disease and frontotemporal dementia: two sides of the same coin? Eur Radiol. 2015;25(10):3050–9.
281. Tosun D, Schuff N, Rabinovici GD, Ayakta N, Miller BL, Jagust W, et al. Diagnostic utility of ASL-MRI and FDG-PET in the behavioral variant of FTD and AD. Ann Clin Transl Neurol. 2016;3(10):740–51.
282. Mutsaerts H, Mirza SS, Petr J, Thomas DL, Cash DM, Bocchetta M, et al. Cerebral perfusion changes in presymptomatic genetic frontotemporal dementia: a GENFI study. Brain. 2019;142(4):1108–20.
283. Ward J, Ly M, Raji CA. Brain PET imaging: frontotemporal dementia. PET Clin. 2023;18(1):123–33.
284. Logothetis NK, Pfeuffer J. On the nature of the BOLD fMRI contrast mechanism. Magn Reson Imaging. 2004;22(10):1517–31.
285. Zhou J, Greicius MD, Gennatas ED, Growdon ME, Jang JY, Rabinovici GD, et al. Divergent network connectivity changes in behavioural variant frontotemporal dementia and Alzheimer's disease. Brain. 2010;133(Pt 5):1352–67.
286. Farb NA, Grady CL, Strother S, Tang-Wai DF, Masellis M, Black S, et al. Abnormal network connectivity in frontotemporal dementia: evidence for prefrontal isolation. Cortex. 2013;49(7):1856–73.
287. Filippi M, Agosta F, Scola E, Canu E, Magnani G, Marcone A, et al. Functional network connectivity in the behavioral variant of frontotemporal dementia. Cortex. 2013;49(9):2389–401.
288. Seeley WW, Menon V, Schatzberg AF, Keller J, Glover GH, Kenna H, et al. Dissociable intrinsic connectivity networks for salience processing and executive control. J Neurosci. 2007;27(9):2349–56.
289. Hohenfeld C, Werner CJ, Reetz K. Resting-state connectivity in neurodegenerative disorders: is there potential for an imaging biomarker? Neuroimage Clin. 2018;18:849–70.
290. Bonnelle V, Ham TE, Leech R, Kinnunen KM, Mehta MA, Greenwood RJ, et al. Salience network integrity predicts default mode network function after traumatic brain injury. Proc Natl Acad Sci USA. 2012;109(12):4690–5.
291. Borroni B, Alberici A, Cercignani M, Premi E, Serra L, Cerini C, et al. Granulin mutation drives brain damage and reorganization from preclinical to symptomatic FTLD. Neurobiol Aging. 2012;33(10):2506–20.

292. Lee SE, Khazenzon AM, Trujillo AJ, Guo CC, Yokoyama JS, Sha SJ, et al. Altered network connectivity in frontotemporal dementia with C9orf72 hexanucleotide repeat expansion. Brain. 2014;137(Pt 11):3047–60.
293. Möller C, Hafkemeijer A, Pijnenburg YA, Rombouts SA, van der Grond J, Dopper E, et al. Joint assessment of white matter integrity, cortical and subcortical atrophy to distinguish AD from behavioral variant FTD: a two-center study. Neuroimage Clin. 2015;9:418–29.
294. Premi E, Cauda F, Costa T, Diano M, Gazzina S, Gualeni V, et al. Looking for neuroimaging markers in frontotemporal lobar degeneration clinical trials: a multi-voxel pattern analysis study in granulin disease. J Alzheimers Dis. 2016;51(1):249–62.
295. Premi E, Cauda F, Gasparotti R, Diano M, Archetti S, Padovani A, et al. Multimodal FMRI resting-state functional connectivity in granulin mutations: the case of fronto-parietal dementia. PLoS One. 2014;9(9):e106500.
296. Greaves CV, Rohrer JD. An update on genetic frontotemporal dementia. J Neurol. 2019;266(8):2075–86.
297. Bates TE, Strangward M, Keelan J, Davey GP, Munro PM, Clark JB. Inhibition of N-acetylaspartate production: implications for 1H MRS studies in vivo. Neuroreport. 1996;7(8):1397–400.
298. Petroff OA, Errante LD, Kim JH, Spencer DD. N-acetyl-aspartate, total creatine, and myo-inositol in the epileptogenic human hippocampus. Neurology. 2003;60(10):1646–51.
299. Valenzuela MJ, Sachdev P. Magnetic resonance spectroscopy in AD. Neurology. 2001;56(5):592–8.
300. Murray ME, Przybelski SA, Lesnick TG, Liesinger AM, Spychalla A, Zhang B, et al. Early Alzheimer's disease neuropathology detected by proton MR spectroscopy. J Neurosci. 2014;34(49):16247–55.
301. Brand A, Richter-Landsberg C, Leibfritz D. Multinuclear NMR studies on the energy metabolism of glial and neuronal cells. Dev Neurosci. 1993;15(3–5):289–98.
302. Ernst T, Chang L, Melchor R, Mehringer CM. Frontotemporal dementia and early Alzheimer disease: differentiation with frontal lobe H-1 MR spectroscopy. Radiology. 1997;203(3):829–36.
303. Kizu O, Yamada K, Ito H, Nishimura T. Posterior cingulate metabolic changes in frontotemporal lobar degeneration detected by magnetic resonance spectroscopy. Neuroradiology. 2004;46(4):277–81.
304. Coulthard E, Firbank M, English P, Welch J, Birchall D, O'Brien J, et al. Proton magnetic resonance spectroscopy in frontotemporal dementia. J Neurol. 2006;253(7):861–8.
305. Chawla S, Wang S, Moore P, Woo JH, Elman L, McCluskey LF, et al. Quantitative proton magnetic resonance spectroscopy detects abnormalities in dorsolateral prefrontal cortex and motor cortex of patients with frontotemporal lobar degeneration. J Neurol. 2010;257(1):114–21.
306. Kantarci K, Boeve BF, Wszolek ZK, Rademakers R, Whitwell JL, Baker MC, et al. MRS in presymptomatic MAPT mutation carriers: a potential biomarker for tau-mediated pathology. Neurology. 2010;75(9):771–8.
307. Chen Q, Boeve BF, Tosakulwong N, Lesnick T, Brushaber D, Dheel C, et al. Brain MR spectroscopy changes precede frontotemporal lobar degeneration phenoconversion in mapt mutation carriers. J Neuroimaging. 2019;29(5):624–9.
308. Ossenkoppele R, Rabinovici GD, Smith R, Cho H, Schöll M, Strandberg O, et al. Discriminative accuracy of [18F] flortaucipir positron emission tomography for alzheimer disease vs other neurodegenerative disorders. JAMA. 2018;320(11):1151–62.
309. Bevan Jones WR, Cope TE, Passamonti L, Fryer TD, Hong YT, Aigbirhio F, et al. [(18) F]AV-1451 PET in behavioral variant frontotemporal dementia due to MAPT mutation. Ann Clin Transl Neurol. 2016;3(12):940–7.
310. Smith R, Puschmann A, Schöll M, Ohlsson T, van Swieten J, Honer M, et al. 18F-AV-1451 tau PET imaging correlates strongly with tau neuropathology in MAPT mutation carriers. Brain. 2016;139(Pt 9):2372–9.

311. Spina S, Schonhaut DR, Boeve BF, Seeley WW, Ossenkoppele R, O'Neil JP, et al. Frontotemporal dementia with the V337M MAPT mutation: Tau-PET and pathology correlations. Neurology. 2017;88(8):758–66.
312. Tsai RM, Bejanin A, Lesman-Segev O, LaJoie R, Visani A, Bourakova V, et al. (18) F-flortaucipir (AV-1451) tau PET in frontotemporal dementia syndromes. Alzheimers Res Ther. 2019;11(1):13.
313. Convery RS, Jiao J, Clarke MTM, Moore KM, Koriath CAM, Woollacott IOC, et al. Longitudinal ((18)F)AV-1451 PET imaging in a patient with frontotemporal dementia due to a Q351R MAPT mutation. J Neurol Neurosurg Psychiatry. 2020;91(1):106–8.
314. Soleimani-Meigooni DN, Iaccarino L, La Joie R, Baker S, Bourakova V, Boxer AL, et al. 18F-flortaucipir PET to autopsy comparisons in Alzheimer's disease and other neurodegenerative diseases. Brain. 2020;143(11):3477–94.
315. Malpetti M, Kaalund SS, Tsvetanov KA, Rittman T, Briggs M, Allinson KSJ, et al. In vivo (18)F-flortaucipir PET does not accurately support the staging of progressive supranuclear palsy. J Nucl Med. 2022;63(7):1052–7.
316. Wolters EE, Papma JM, Verfaillie SCJ, Visser D, Weltings E, Groot C, et al. [(18)F] flortaucipir PET across various MAPT mutations in presymptomatic and symptomatic carriers. Neurology. 2021;97(10):e1017–e30.
317. Bevan-Jones WR, Cope TE, Jones PS, Passamonti L, Hong YT, Fryer TD, et al. [(18) F]AV-1451 binding in vivo mirrors the expected distribution of TDP-43 pathology in the semantic variant of primary progressive aphasia. J Neurol Neurosurg Psychiatry. 2018;89(10):1032–7.
318. Josephs KA, Martin PR, Botha H, Schwarz CG, Duffy JR, Clark HM, et al. [(18) F]AV-1451 tau-PET and primary progressive aphasia. Ann Neurol. 2018;83(3):599–611.
319. Makaretz SJ, Quimby M, Collins J, Makris N, McGinnis S, Schultz A, et al. Flortaucipir tau PET imaging in semantic variant primary progressive aphasia. J Neurol Neurosurg Psychiatry. 2018;89(10):1024–31.
320. Cho H, Kim HJ, Choi JY, Ryu YH, Lee MS, Na DL, et al. (18)F-flortaucipir uptake patterns in clinical subtypes of primary progressive aphasia. Neurobiol Aging. 2019;75:187–97.
321. Murray ME, Kouri N, Lin WL, Jack CR Jr, Dickson DW, Vemuri P. Clinicopathologic assessment and imaging of tauopathies in neurodegenerative dementias. Alzheimers Res Ther. 2014;6(1):1.
322. Villemagne VL, Fodero-Tavoletti MT, Masters CL, Rowe CC. Tau imaging: early progress and future directions. Lancet Neurol. 2015;14(1):114–24.
323. Ikeda A, Shimada H, Nishioka K, Takanashi M, Hayashida A, Li Y, et al. Clinical heterogeneity of frontotemporal dementia and Parkinsonism linked to chromosome 17 caused by MAPT N279K mutation in relation to tau positron emission tomography features. Mov Disord. 2019;34(4):568–74.
324. Su Y, Fu J, Yu J, Zhao Q, Guan Y, Zuo C, et al. Tau PET imaging with [18F]PM-PBB3 in frontotemporal dementia with MAPT mutation. J Alzheimers Dis. 2020;76(1):149–57.
325. Rao PPN, Shakeri A, Zhao Y, Calon F. Strategies in the design and development of (TAR) DNA-binding protein 43 (TDP-43) binding ligands. Eur J Med Chem. 2021;225:113753.
326. Teunissen CE, Elias N, Koel-Simmelink MJ, Durieux-Lu S, Malekzadeh A, Pham TV, et al. Novel diagnostic cerebrospinal fluid biomarkers for pathologic subtypes of frontotemporal dementia identified by proteomics. Alzheimers Dement (Amst). 2016;2:86–94.
327. Del Campo M, Peeters CFW, Johnson ECB, Vermunt L, Hok AHYS, van Nee M, et al. CSF proteome profiling across the Alzheimer's disease spectrum reflects the multifactorial nature of the disease and identifies specific biomarker panels. Nat Aging. 2022;2(11):1040–53.
328. van der Ende EL, In 't Veld S, Hanskamp I, van der Lee S, Dijkstra JIR, Hok AHYS, et al. CSF proteomics in autosomal dominant Alzheimer's disease highlights parallels with sporadic disease. Brain. 2023;19
329. Bron EE, Smits M, Papma JM, Steketee RME, Meijboom R, de Groot M, et al. Multiparametric computer-aided differential diagnosis of Alzheimer's disease and frontotemporal dementia using structural and advanced MRI. Eur Radiol. 2017;27(8):3372–82.

330. Vijverberg EG, Wattjes MP, Dols A, Krudop WA, Möller C, Peters A, et al. Diagnostic accuracy of MRI and additional [18F] FDG-PET for behavioral variant frontotemporal dementia in patients with late onset behavioral changes. J Alzheimers Dis. 2016;53(4):1287–97.
331. Feis RA, Bouts M, de Vos F, Schouten TM, Panman JL, Jiskoot LC, et al. A multimodal MRI-based classification signature emerges just prior to symptom onset in frontotemporal dementia mutation carriers. J Neurol Neurosurg Psychiatry. 2019;90(11):1207–14.
332. Feis RA, Bouts M, Panman JL, Jiskoot LC, Dopper EGP, Schouten TM, et al. Single-subject classification of presymptomatic frontotemporal dementia mutation carriers using multimodal MRI. Neuroimage Clin. 2019;22:101718.
333. Feis RA, van der Grond J, Bouts M, Panman JL, Poos JM, Schouten TM, et al. Classification using fractional anisotropy predicts conversion in genetic frontotemporal dementia, a proof of concept. Brain Commun. 2020;2(2):fcaa079.
334. Tahmasian M, Shao J, Meng C, Grimmer T, Diehl-Schmid J, Yousefi BH, et al. Based on the network degeneration hypothesis: separating individual patients with different neurodegenerative syndromes in a preliminary hybrid PET/MR study. J Nucl Med. 2016;57(3):410–5.
335. Zhu N, Santos-Santos M, Illán-Gala I, Montal V, Estellés T, Barroeta I, et al. Plasma glial fibrillary acidic protein and neurofilament light chain for the diagnostic and prognostic evaluation of frontotemporal dementia. Transl Neurodegener. 2021;10(1):50.
336. Bolsewig K, Hok AHYS, Sepe FN, Boonkamp L, Jacobs D, Bellomo G, et al. A combination of neurofilament light, glial fibrillary acidic protein, and neuronal pentraxin-2 discriminates between frontotemporal dementia and other dementias. J Alzheimers Dis. 2022;90(1):363–80.
337. Chouliaras L, Thomas A, Malpetti M, Donaghy P, Kane J, Mak E, et al. Differential levels of plasma biomarkers of neurodegeneration in Lewy body dementia, Alzheimer's disease, frontotemporal dementia and progressive supranuclear palsy. J Neurol Neurosurg Psychiatry. 2022;93(6):651–8.

Fluid and Imaging Markers of Lewy Body Dementias

Kayla Julio, Harm Jan van der Horn, and Sephira G. Ryman

Abstract Lewy body dementias encompass Parkinson's disease dementia (PDD) and dementia with Lewy bodies (DLB), both characterized by the presence of Lewy bodies and Lewy neurites resulting from α-synuclein aggregation. Clinical variations in these diseases stem from differing pathological locations and severity, along with the co-occurrence of Alzheimer's disease pathology. Consequently, there is a pressing need for biomarkers capable of early detection, distinguishing individuals with Lewy body dementias from other neurodegenerative disorders, and capturing co-existing pathologies. The chapter begins by exploring clinical diagnostic categories within the Lewy body disease spectrum and evaluating existing biomarkers integrated into diagnostic criteria. The discussion then delves into ongoing research on these biomarkers and emerging alternatives. Future directions include identifying early disease markers and predicting progression in line with recent prodromal PD and DLB research criteria. Furthermore, research suggests that factors such as inflammation, gut-brain-axis dysfunction, and cerebrovascular issues play a role in Lewy body diseases, and quantifying these markers may be an important future direction.

Keywords Lewy body · Neurodegeneration · Dopamine · Fluid biomarkers · Imaging biomarkers

1 Introduction

Lewy body dementias include Parkinson's disease dementia (PDD) and Dementia with Lewy bodies (DLB) [1]. Both disorders are neuropathologically defined by the presence of Lewy bodies and Lewy neurites, which are the end result of α-synuclein aggregation [1–3]. Substantial heterogeneity in the clinical phenotypes exists due to

K. Julio · H. J. van der Horn · S. G. Ryman (✉)
The Mind Research Network, Albuquerque, NM, USA
e-mail: sryman@mrn.org

R. Tarawneh (ed.), *Biomarkers of Neurodegenerative Disorders*,
https://doi.org/10.1007/978-3-031-80849-4_10

(1) the location and severity of the pathology [4–6] as well as (2) the presence of co-pathology, such as Alzheimer's disease (AD) [7, 8]. As such, there is an enormous clinical need to develop biomarkers that can not only detect early disease processes and differentiate these disorders from other neurodegenerative conditions, but also capture co-occurring pathologies in Lewy body dementias. As α-synuclein biomarkers are covered in detail in another chapter, we briefly discuss this work and expand on non-synuclein-based biomarkers. We begin our discussion by reviewing clinical diagnostic categories that fall within the spectrum of Lewy body diseases (LBD), examining which biomarkers are currently incorporated into present diagnostic criteria. We then discuss the current research on each of these as well as emerging biomarkers.

2 Spectrum of Lewy Body Diseases

The spectrum of LBD includes the established clinical phenotypes, from Parkinson's disease (PD) with varying degrees of cognitive impairment to Dementia with Lewy bodies (DLB) at the most severe end [9]. Additionally, research criteria have recently been developed to identify both prodromal PD [10] and DLB [11] (Fig. 1). The incidence of PD is approximately 15 per 100,000 annually, with prevalence ranging between 100 and 200 per 100,000 people [12]. PD is diagnosed based on the presence of motor symptoms (bradykinesia, rigidity, and tremor), which is considered an essential criterion [13]. Supportive criteria for the diagnosis only include one biomarker, cardiac sympathetic denervation on ^{123}I-metaiodobenzylguanidine (MIBG) imaging, in addition to response to dopaminergic therapy, the presence of levodopa-induced dyskinesia, and olfactory loss. Clinically established PD is diagnosed when individuals exhibit bradykinesia in combination with the presence of at least either rest tremor and/or rigidity, and there is an absence of absolute exclusion criteria, the presence of at least two supportive criteria, and no red flags. Absolute exclusion criteria suggest other etiologies, such as cerebellar abnormalities, downward vertical supranuclear gaze palsy suggestive of progressive supranuclear palsy (PSP), symptoms of frontotemporal dementia, and drug-induced parkinsonism, among others. If present, these features rule out a diagnosis of PD. A diagnosis of clinically probable PD is made when there is an absence of absolute exclusion criteria, and the presence of red flags is counterbalanced by supportive criteria, which is the primary way biomarkers are leveraged for diagnosis beyond MIBG imaging (i.e., magnetic resonance imaging [MRI] or dopaminergic neuroimaging).

There is increased appreciation that cognitive dysfunction is common and debilitating, as a large number of individuals with PD progress to dementia [14]. Approximately 20% of individuals with PD meet criteria for mild cognitive impairment (PD-MCI) at diagnosis, and 80% of those who survive two decades develop PDD [15, 16]. Cognitive decline in PD is variable in terms of the cognitive domains impacted as well as the temporal course of impairments [17, 18]. MCI is the stage between normal aging and dementia that was initially used to characterize the

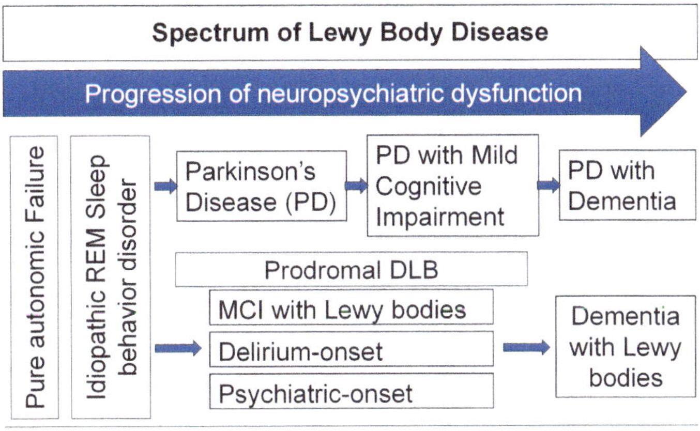

Fig. 1 Spectrum of Lewy body diseases. Parkinson's disease dementia (PDD) and Dementia with Lewy bodies (DLB) all represent an end stage of the disease progression. The figure illustrates potential paths by which patients may exhibit prodromal phenotypes and how they may progress towards dementia. The paths are not exhaustive in that patients do not need to progress in the illustrated order. Further, prodromal Parkinson's disease (PD) and DLB diagnoses are only used for research purposes at this time

amnestic deficits that precede AD dementia and predict an emerging dementia syndrome [19]. Initially, these same criteria were adapted for use in PD [20, 21] though different cognitive tests and cut-offs were used across studies [15, 22]. The Movement Disorders Society developed criteria for PD-MCI [23] to ensure consistent diagnostic practices in the field. Diagnostic criteria for PD-MCI include a diagnosis of PD and evidence of a gradual decline in cognition in the context of established PD, reported by the patient or informant or observed by the clinician. Additionally, cognitive deficits on either formal neuropsychological testing (Level II comprehensive assessment) or a scale of global cognitive abilities (Level I abbreviated assessment, e.g., Montreal Cognitive Assessment) are required, and these should not yet interfere significantly with functional independence [23]. When cognitive impairment is severe enough to significantly interfere with activities of daily living in the context of established PD, PDD is diagnosed.

Currently, biomarkers are not used in the diagnosis of PD-MCI or PDD, though future validation of biomarkers is highly desired as individuals with PD-MCI exhibit significant heterogeneity in terms of cognitive impairment, pathological processes, and progression to PDD [24–26]. Unlike AD, in which most individuals typically exhibit a clear amnestic memory impairment early in the course of the disorder that predicts decline, cognitive impairment in PD exhibits a more variable course. For instance, individuals with PD commonly exhibit executive impairment early in the disease course but may or may not experience further decline over time [24, 27, 28]. Rather, the emergence of visuospatial and memory impairments are more consistent predictors of an emerging dementia syndrome [24, 28–30].

While individuals with PD may experience the core clinical features of DLB (detailed below), if cognitive symptoms develop first, DLB should be diagnosed based on the one-year rule. DLB is likely underdiagnosed, and the prevalence estimates indicate point and period prevalence between 0.02 and 63.5 per 1000 persons [31–33]. Core clinical features of DLB include (1) fluctuating cognition, (2) recurrent visual hallucinations, (3) REM sleep behavior disorder (RBD), and (4) one or more of the motor features of parkinsonism [34]. Probable DLB is diagnosed if two or more core clinical features are present, *or* only one core clinical feature is present with one or more indicative biomarkers, *and* the patient experiences cognitive impairment sufficient to impair activities of daily living. Possible DLB is diagnosed if only one core clinical feature is present, *or* one or more indicative biomarkers are present, but there are no core clinical features, *and* the patient experiences cognitive impairment sufficient to impair activities of daily living. Indicative biomarkers include reduced dopamine transporter uptake in basal ganglia demonstrated by single-photon emission computed tomography (SPECT) or positron emission tomography (PET), reduced MIBG uptake, and polysomnographic (PSG) confirmation of REM sleep without atonia (i.e., RBD). Supportive biomarkers are also provided to help the diagnostic evaluation, though they are not explicitly listed in the criteria. These include relative preservation of the medial temporal lobe structures on computed tomography (CT) or MRI, [^{18}F]-Fluorodeoxyglucose PET (FDG-PET) occipital hypometabolism or ^{99m}Tc-hexamethylpropyleneamine oxime (HMPAO) SPECT hypoperfusion, relative preservation of posterior or midcingulate metabolism, and prominent posterior slow-wave activity on electroencephalogram (EEG) with periodic fluctuations in the pre-alpha/theta range.

As significant neurodegeneration occurs prior to the clinical diagnoses of either PD or DLB [35], there are increased efforts to identify prodromal and at-risk cohorts that provide earlier opportunities for intervention. Research criteria to identify prodromal PD [10, 36] and DLB have been published [11]. One of the most specific and earliest prodromal indicators of Lewy body diseases (LBD) is isolated/idiopathic iRBD (referred to as idiopathic/isolated when it occurs in the absence of PD or DLB). Approximately 33% of individuals with iRBD convert to an α-synucleinopathy (inclusive of LBD as well as multiple system atrophy [MSA]) within five years, 82.4% at 10.5 years, and 96.6% at 14 years [37–39]. Of the converters, it is estimated that 25% will convert to DLB, 43% will convert to PD, and 7% will convert to MSA [40]. Pure autonomic failure (PAF) is a clinical diagnosis that also exhibits a high rate of phenoconversion to an α-synucleinopathy [41]. Specifically, PAF is diagnosed based on severe autonomic dysfunction in the absence of central nervous system symptoms. As high as 34% of individuals with PAF convert to an α-synucleinopathy within four years, with 52% converting to DLB, 24% converting to PD, and 24% converting to MSA [41]. Of the patients that convert, 96% exhibit iRBD [41].

The prodromal PD research criteria use an evidence-based conceptual framework that integrates risk and prodromal markers to identify probable prodromal PD. The majority of risk and prodromal markers have predictive values between one and two (e.g., pesticide exposure, nonuse of caffeine, no smoking history), whereas the following markers exhibit significantly greater predictive values: substantia nigra hyperechogenicity (3.4), PSG-proven RBD (130), dopaminergic PET/SPECT abnormalities (43.3), subthreshold parkinsonism (9.6), olfactory loss (18.5), and orthostatic hypotension (18.5). In contrast, the prodromal DLB criteria adopt a traditional list of symptoms and specifications required for the diagnosis. They specify three different subtypes: MCI with Lewy bodies (MCI-LB), psychiatric-onset DLB, and delirium-onset DLB [11].

Importantly, despite a broad range of clinical presentations, both individuals with LBD and MSA exhibit misfolding and aggregation of α-synuclein. However, LBD exhibit α-synuclein aggregates (which ultimately form Lewy bodies) in the neurons, whereas individuals with MSA exhibit α-synuclein aggregates in the glial cells [42, 43]. There is significant variation in the location of the initiation and progression of disease pathology, which leads to the different clinical phenotypes [4–6]. While the diagnosis of PD versus DLB has been useful in clinical practice, there is considerable overlap in the disorders [9]. For example, while PD is diagnosed based on the presence of motor symptoms, these individuals can – and often do – develop cognitive fluctuations, RBD, and visual hallucinations. As such, the use of biomarkers can help identify individuals at risk, characterize the location and extent of neuropathological processes, help inform diagnosis, monitor disease processes, and predict clinical symptoms or progression.

3 Pathogenesis

At autopsy, Lewy bodies and Lewy neurites are the defining neuropathological characteristics of PD and DLB [1–3]. Alpha (α)-synuclein was determined to be the principal component of Lewy bodies following the discovery that point mutations in the gene encoding α-synuclein (*SNCA*) are pathogenic for familial forms of PD [42, 44]. An important area of research is currently underway to understand what factors contribute to the misfolding and aggregation of α-synuclein. Post-translational modifications (such as phosphorylation, ubiquitination, nitration, and truncation) have been implicated in abnormal protein folding [45], resulting in multiple α-synuclein conformers, including oligomers, protofibrils, fibrils, and mature Lewy bodies [42, 46, 47]. It is unknown whether the different protein structures are indicative of distinct pathologies or a continuum. Of note, there is some controversy regarding whether the α-synuclein aggregates are in themselves driving pathogenesis or rather are upstream mechanisms or compensatory responses [48].

Numerous potential biological mechanisms may contribute to α-synuclein aggregation, including gut dysbiosis [49, 50], head trauma [51], environmental exposures [52, 53], infections [54–56], and cerebrovascular dysfunction [57]. Each of these alone is likely not the sole source of α-synuclein aggregation sufficient to lead to PD or DLB [58], which has led some to posit that diseases such as PD require triggers, facilitators, and aggravators that influence the disease [59]. In other words, these factors may initiate pathological processes, but it is only in the context of facilitators such as peripheral inflammation, gut barrier dysfunction, and aggravators such as neuroinflammation and oxidative stress that the disease persists and progresses.

As individuals with PD are diagnosed based on their motor symptoms, they exhibit abnormal deposition of α-synuclein in the substantia nigra and brainstem that leads to degeneration of the monoamine neurotransmitter systems (dopaminergic, noradrenergic, and cholinergic) [60, 61]. Frontoparietal network dysfunction due to degeneration of the substantia nigra plays a significant role in executive impairments in PD in addition to dysfunction of the mesocortical and noradrenergic networks due to degeneration of the ventral tegmental area and locus coeruleus, respectively [62]. The severity of degeneration in these regions corresponds to the severity of executive dysfunction. Memory and visuospatial impairments, in contrast, are likely related to broader degeneration, possibly reflecting progression of α-synuclein to the limbic and cortical regions or development of co-occurring neuropathological processes, including amyloid-beta and tau abnormalities [24, 63] and cerebrovascular dysfunction [64, 65]. These co-occurring neuropathological processes may significantly contribute to the development and progression of cognitive impairment and indicate a more malignant course of decline [66–68]. Individuals with DLB exhibit midbrain pathology as well as limbic and cortical α-synuclein pathology and often present with cognitive impairment in attention, executive, visuospatial, and memory functions [69].

AD neuropathological change commonly occurs in Lewy body dementias and has been recently thought to be an important factor driving cognitive impairment [67]. Estimates as high as 35% of individuals with PDD and 70% of individuals with DLB exhibit co-occurring AD amyloid-beta plaques and tau-positive neurofibrillary tangles based on postmortem assessments [8, 70, 71]. Co-occurring AD pathology influences the onset of dementia in PD [70, 71] and leads to a more severe decline in domain-specific dysfunction [72], highlighting the importance of detecting co-pathology, particularly when evaluating the efficacy of disease-modifying treatments.

While cerebrovascular pathology is less common in LBD based on post-mortem neuropathological assessments, there is increasing evidence that cerebrovascular dysfunction may contribute to disease initiation or progression. Individuals with LBD exhibit unique cardiovascular risk factors relative to other dementia syndromes given autonomic cardiovascular dysfunction is common and optimal cerebral perfusion is crucial to optimal brain function. Clinically, individuals may experience abnormal heart rate variability, supine hypertension, postprandial hypotension, non-dipping blood pressure in the nocturnal periods, and orthostatic hypotension severe enough to result in end organ hypoperfusion and hypoxia [73–75]. While estimates vary, a large proportion of individuals with PD experience autonomic-related orthostatic hypotension (30.1% [76]–49.7% with hypotension [77]), and individuals with orthostatic hypotension may exhibit greater cerebrovascular disease and cognitive impairment [77–79]. Currently, there is limited understanding of the impact of these factors on cognitive outcomes and potential underlying mechanisms. Due to this, we include a brief section below that discusses our recently developed framework that can guide future work in this area.

4 Conceptualization of Biomarkers in Lewy Body Dementias

Above, we noted the relatively established biomarkers that can be used for diagnosis. However, there are numerous categories of biomarkers (e.g., susceptibility/risk biomarker, diagnostic biomarker, monitoring biomarker, and prognostic biomarker [80]) that can serve to answer different clinical and research questions. For example, as individuals with PD are typically seen by Movement Disorders neurologists, a large area of research is developing diagnostic biomarkers for PD, which are most commonly intended to help differentiate PD from atypical parkinsonian disorders, such as MSA, PSP, and corticobasal degeneration (CBD). A separate line of work has focused on identifying biomarkers that are indicative of cognitive impairment in PD. DLB, in contrast, is often seen by behavioral neurologists and a key differential diagnosis is other dementia syndromes, such as AD. Additionally, as noted, PD is diagnosed based on the presence of motor symptoms and known corresponding neuroanatomical locus of pathology (e.g., the substantia nigra). However, both PDD and DLB occur when substantial progression of neuropathology has occurred and has impacted broader regions of the brain. Given these contexts and the heterogeneity in neuropathological factors, there is substantial complexity in the discussion of biomarkers of Lewy body dementias. To organize our discussion, we start with biomarkers that capture disease processes that occur early in the course, such as RBD, sympathetic denervation, and midbrain/brainstem degeneration. We then move on to discuss how other techniques capture broader degeneration, including MRI, PET, and SPECT imaging techniques. Finally, we discuss biomarkers that detect co-occurring neuropathological processes, primarily AD, as this has received the most attention, though we also discuss the future need to investigate the role of cerebrovascular dysfunction in LBD.

5 REM Sleep Behavior Disorder

The detection of REM sleep without atonia and/or the diagnosis of iRBD can be used to identify prodromal individuals and help differentiate individuals with LBD relative to other neurodegenerative etiologies and identify individuals with PD at risk of cognitive impairment. RBD is a parasomnia in which individuals act out their dreams while sleeping. The physical movements occur during REM phases of sleep and the absence of atonia (sleep muscle paralysis). Patients may not be aware they are acting out their dreams, but it can be significantly disruptive to bed partners, particularly if the patient's dreams are action-oriented and include running or fighting [81]. As noted, iRBD is a prodromal indicator of α-synucleinopathies, including PD, DLB, and MSA [37–40].

RBD is diagnosed based on the presence of the following criteria described by the International Classification of Sleep Disorders (ICSD-3): (1) repeated episodes of sleep-related vocalization and/or complex motor behaviors; (2) these behaviors are documented by PSG to occur during REM sleep or, based on clinical history of dream enactment, are presumed to occur during REM sleep; (3) polysomnographic recording demonstrates REM sleep without atonia; and (4) the disturbance is not better explained by other sleep disorders, mental disorders, medication, or substance abuse. Based on these criteria, the diagnosis of RBD can only be made with PSG. However, it is often a burden for individuals to undergo sleep studies and screening questionnaires, such as the Mayo Sleep questionnaire [82] or use of a screening question ("Have you ever been told, or suspected yourself, that you seem to 'act out your dreams' while asleep (for example, punching, flailing your arms in the air, making running movements, etc.)?") are often performed [83]. The highest risk of phenotypic conversion to LBD in iRBD is present for PSG-confirmed cases, leading to its increased use as a biomarker.

The recent DLB criteria have increased the diagnostic weighting given to RBD, based on the added sensitivity to differentiate autopsy-confirmed cases when RBD was included as a criterion [34]. Specifically, when clinically probable DLB was determined based on two or more of the three core features without RBD as a criterion (i.e., two of the three cardinal features of parkinsonism, cognitive fluctuations, and visual hallucinations), sensitivity was 85% and specificity was 73%. When RBD was added as a core clinical criterion and DLB was defined as two or more of the four core features, sensitivity improved to 88% [84]. In the same study, RBD occurred in 76% of autopsy-confirmed DLB cases, relative to 4% in the non-DLB comparison cohort. This is consistent with prior clinicopathologic correlations, which indicated that of the individuals with RBD evaluated, 160/170 (94%) exhibited an α-synucleinopathy at autopsy [85]. While a clinical diagnosis of RBD is a core clinical feature, polysomnographic confirmation of REM sleep without atonia is an indicative biomarker.

In contrast to DLB, RBD is not included as a diagnostic criterion in PD, though it occurs in approximately 42.3% of individuals with PD [86] and is an important risk factor and predictor of cognitive impairment in PD [87]. Specifically, individuals with PD with RBD have a significantly higher frequency (42%) of dementia than individuals with PD without RBD (7.3%) [88]. Conversely, that same study indicated that the frequency of RBD in the PDD group (77%) was significantly higher than in the PD normal cognition group (27%). MCI diagnosis frequency is three times higher in individuals with PD with RBD (66%) relative to individuals with PD without RBD (23%) [89]. The same study also indicated that subjective cognitive decline was significantly higher in individuals with PD with RBD (89%) relative to individuals with PD without RBD (58%).

Taken together, the use of PSG to identify individuals with RBD serves as an important biomarker to identify individuals at risk of an α-synucleinopathy (susceptibility/risk biomarker) and to differentiate individuals with α-synucleinopathies from other neurodegenerative conditions (diagnostic biomarker). Lastly, in the context of established PD, it can also serve to identify individuals at risk for cognitive impairment (prognostic biomarker).

6 Imaging Markers in LBD

6.1 ^{123}I-metaiodobenzylguanidine (MIBG) Scans

MIBG myocardial scintigraphy is used to quantify myocardial sympathetic nerve damage [90]. MIBG and noradrenaline (norepinephrine) have the same mechanisms for uptake, storage, and release by postganglionic, presynaptic cardiac sympathetic nerves that innervate the heart. As such, it can be measured to assess the integrity of these neurons. It is routinely used in different heart conditions, such as congestive heart failure, ischemic heart disease, cardiomyopathy, as well as diabetic neuropathy. In contrast to RBD, which is presumed to be due to central nervous system pathology, sympathetic nerve damage occurs in the periphery. This, in combination with the increased appreciation of prodromal gastrointestinal and autonomic symptoms in PD and DLB, has raised support for the notion of a "body-first" pathogenesis of PD [6]. The combination of iRBD and decreased MIBG uptake are considered key biomarkers indicative of an emerging body-first PD phenotype [5]. While the body-first conceptualization of PD pathogenesis is largely a theory at this stage, as noted above, the use of MIBG myocardial scintigraphy is given increased diagnostic weighting in both PD and DLB relative to other biomarkers [13, 34]. Similar to RBD, it occurs in the prodromal phase and thus, can serve as a susceptibility/risk biomarker and diagnostic biomarker.

The MIBG procedure includes the intravenous injection of MIBG followed by SPECT imaging during an early (15–30 min) and delayed (3–4 h) imaging phase. The uptake can be quantified by calculating the heart to mediastinum (HM) ratio. The early imaging phase captures the density of the presynaptic cardiac sympathetic nerve endings. In contrast, the presynaptic functional tone of the cardiac sympathetic nerve is captured by the delayed phase. It has long been known that prodromal cohorts, such as iRBD [91] and PAF [92] individuals, exhibit reduced cardiac MIBG uptake relative to healthy controls. In a detailed review by Sakakibara et al. (2014) [93], the authors demonstrated that many of the prodromal non-motor symptoms associated with PD (autonomic dysfunction based on the presence of constipation and postural hypotension, depression/anxiety, visual hallucinations/psychosis in older adults, and RBD), with the exception of impaired olfaction, were accompanied by low MIBG uptake. Comprehensive imaging to quantify gut innervation (^{11}C-donepezil PET), cardiac sympathetic innervation (MIBG), noradrenergic nuclei in the brainstem and their projections (^{11}C-methylreboxetine [MeNER] PET) and, degeneration of the locus coeruleus (neuromelanin-sensitive MRI), and nigrostriatal dopamine storage capacity (^{18}F-dihydroxyphenylalanine [DOPA]) demonstrated that individuals with iRBD exhibited fully developed pathology in the peripheral autonomic nervous system and the locus coeruleus (e.g., to the same degree as observed in PD) but lacked the loss of the nigrostriatal dopamine storage capacity observed in PD [94]. This is consistent with a separate study that highlights how MIBG imaging can identify premotor individuals with PD/DLB during a negative DAT scan [95].

While iRBD is relatively specific to α-synucleinopathies (inclusive of MSA), an added benefit of MIBG is that it has demonstrated the ability to differentiate PD from MSA (noted below). The combination of iRBD and reduced MIBG uptake, therefore, has the potential to increase diagnostic specificity. Individuals with iRBD exhibited reduced MIBG relative to both MSA and PSP cohorts [96]. Studies have indicated that reduced MIBG uptake is an important predictor of phenoconversion to LBD in iRBD [97]. A separate study has also demonstrated a significant reduction in MIBG in individuals with PAF relative to CBD [98]. A meta-analysis of thirteen MIBG studies including 625 individuals with PD and 220 individuals with atypical parkinsonism syndromes (MSA, PSP, and CBD) has indicated delayed HM ratios on MIBG and demonstrated a 94.1% sensitivity and 80.2% specificity in discriminating PD from atypical parkinsonism when only including individuals with PD who were early in the disease course [99]. The ability of the early and late HM ratios was slightly less precise when all individuals with PD were included, exhibiting sensitivities and specificities of 89.7% and 82.6%, respectively. A meta-analysis specific to differentiating PD from MSA indicated that MIBG shows a high sensitivity to detect PD (89%) and an acceptable specificity (77%) to discriminate PD from MSA, with an area under the receiver operating characteristic curve (ROC) of 0.93 [100]. This confirms that MIBG is an accurate test to help detect PD and differentiate it from atypical parkinsonism.

Low HM ratios are consistently observed in both individuals with PD and DLB, with an early study highlighting that 90.0% of individuals with PD and 94.1% of individuals with DLB exhibit reduced HM ratios, with a sensitivity of 91.3% and a specificity of 89.1% to differentiate LBD from other movement or cognitive disorders [101]. MIBG HM ratios are associated with cognitive functioning in PD [102], though this has received relatively limited investigation. MIBG has been used to differentiate DLB from AD (sensitivity = 69%; specificity = 87%; with increased discriminatory accuracy in mild cases—sensitivity = 77% and specificity =94%) [103]. A 2016 study comparing MIBG and SPECT found a 93% sensitivity and 100% specificity of MIBG in differentiating DLB from non-LBD. Both MIBG and [^{123}I]-2β-carbomethoxy-3β-(4-iodophenyl)-N-(3-fluoropropyl) nortropane SPECT (^{123}I-FP-CIT DAT-SPECT) demonstrated similar sensitivities for identifying DLB (93% vs. 90%); however, MIBG was far more specific (100% vs. 76%), highlighting that MIBG may be a more reliable and accurate method of excluding non-LB dementias, especially when paired with other diagnostic criteria [104].

6.2 MRI Biomarkers of Substantia Nigra and Locus Coeruleus Degeneration

As PD motor symptoms are associated with degeneration of the substantia nigra and corresponding dysfunction of the fronto-subcortical systems [105], substantial efforts have focused on developing biomarkers to detect midbrain degeneration, though these markers are not currently incorporated in diagnostic criteria. We have recently reviewed these in PD, highlighting three classes of MRI sequences that have been used to detect early changes in the substantia nigra, which include iron sensitive imaging (referred to as nigrosome imaging below, typically used to detect nigrosome degeneration, also known as the "swallow tail" sign), neuromelanin sensitive imaging, and diffusion MRI (most commonly evaluating measures of fractional anisotropy and free water diffusion) [106].

As noted elsewhere, individuals with iRBD may not exhibit degeneration of the substantia nigra, which is consistent with recent studies. For example, only 27.5% of individuals with RBD exhibited nigrosome degeneration, compared with 7.7% of control subjects and 96% of individuals with PD [107]. However, neuromelanin-sensitive imaging demonstrated that individuals with iRBD exhibited a reduction in the substantia nigra volume, signal intensity, and fractional anisotropy relative to controls. Neuromelanin-sensitive imaging volume and signal intensities demonstrated a diagnostic accuracy of 0.86 and 0.79, respectively, to differentiate iRBD from controls [108]. Additionally, free water in the substantia nigra was increased in iRBD, though not to the same extent observed in PD [109].

Quantification of the signal in the substantia nigra using the above techniques is increasingly recognized as a reliable and valid biomarker of degeneration in this region, with high sensitivity to differentiate individuals with PD from healthy controls [106]. Similar to the issues faced with dopaminergic imaging, discussed below, atypical parkinsonism syndromes often exhibit dopaminergic deficits, and it is currently unclear whether these new biomarkers can effectively differentiate individuals with PD from atypical parkinsonism disorders, and further research is warranted [106].

In terms of cognitive functioning in PD, neuromelanin-sensitive imaging may have an added benefit to the above approaches as it facilitates the quantification of the locus coeruleus signal. Neuromelanin-sensitive imaging of the substantia nigra is correlated with disease duration, whereas the locus coeruleus is associated with cognitive dysfunction [110]. This is consistent with an earlier study that indicated reduced neuromelanin-sensitive locus coeruleus signal in PD-MCI relative to healthy controls (with PD-normal cognition exhibiting an intermediate between the healthy controls and PD-MCI [111]).

As individuals with AD do not exhibit significant degeneration of the substantia nigra, these measures are useful to differentiate DLB from AD. For example, 63% of individuals with DLB exhibited the "swallow tail" sign, showing a sensitivity and specificity of 63% and 76%, respectively, to differentiate DLB from AD [112]. A separate study, however, indicated that the evaluation of nigrosome-1 degeneration in DLB exhibited a similar but increased diagnostic accuracy (93% sensitivity, 87% specificity) to differentiate DLB from AD/amnestic MCI relative to ^{123}I-FP-CIT DAT-SPECT (93% sensitivity, 84% specificity) [113].

Taken together, these MRI biomarkers may be informative, though research is needed to evaluate how they are altered across individuals with PD with varying degrees of cognitive impairment and patients with DLB. While not yet included in any diagnostic criteria, the benefit of MRI is that it does not require the use of radionuclide tracers and is already routinely used and relatively more cost-effective.

Figure 2 demonstrates examples of brain MRI measures used in PD.

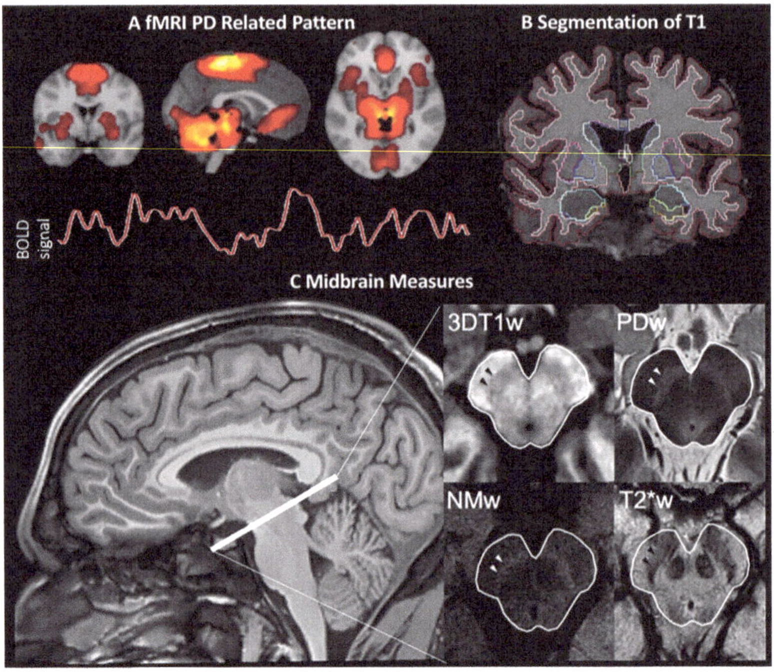

Fig. 2 Common MRI methods in Parkinson's disease. Panel A shows the Parkinson's Disease Related Pattern (PDRP) identified through resting-state functional MRI (rs-fMRI). It features increased activity in the basal ganglia, thalamus, cerebellum/pons, anterior cingulate cortex (ACC), and supplementary motor area (SMA). Panel B illustrates an example of cortical and subcortical segmentation from a T1-weighted MRI image. Panel C displays normal anatomy of the Substantia Nigra (SN) at 3 T. The left image is a parasagittal T1-weighted view indicating the level of the axial slice, while the right image is an axial slice through the midbrain and SN (marked with arrowheads). The SN appears hypointense in three-dimensional T1-weighted MP2RAGE (3DT1w) and two-dimensional T2*-weighted images (T2*w), and hyperintense in neuromelanin T1-weighted spin-echo images (NMw) and proton density-weighted images (PDw). The contours of the midbrain are outlined in white. (Figure 2A-B is adapted with permission from Ryman et al. [106] (Copyright of © 2019 Elsevier Ltd). Figure 2C is adapted with permission from Lehéricy et al. [243] (Copyright of John Wiley and Sons, © 2014 International Parkinson and Movement Disorder Society). Permissions were obtained through the Copyright Clearance Center, Inc

6.3 *Dopamine Transporter Uptake*

Reduced dopamine transporter uptake in the basal ganglia demonstrated by SPECT or PET is considered an indicative biomarker in DLB [34] and is used to counteract red flags in the diagnosis of PD [13]. One of the most widely studied approaches in PD is the DAT-SPECT. Two radiopharmaceuticals are available for DAT-SPECT, including ^{123}I-FP-CIT (mentioned above) and [^{123}I]2β-carboxymethoxy-3β-(4-iodophenyl) tropane (subsequently referred to as ^{123}Iβ-CIT) [114], which is used to detect dopaminergic deficits and clinically approved to differentiate PD from essential tremor [115]. DAT-SPECT uses ligands for the presynaptic monoamine transporter, which quantifies the dysfunction or loss of striatal dopaminergic terminals [114]. PET (F-DOPA PET) has also been used and measures the uptake of dopamine precursors (levodopa decarboxylase activity and trapping of F-dopamine in synaptic vesicles) [116]. A large proportion of individuals with iRBD (40–58.6%) exhibit reduced striatal dopamine uptake on ^{123}I-FP-CIT DAT-SPECT [117, 118]. Imaging deficits of DAT-SPECT were associated with an increased risk of an incident α-synucleinopathy [118]. DAT-SPECT exhibited relatively low diagnostic value to differentiate patients with iRBD from controls (sensitivity = 29.3% and specificity = 100%), though there was a high prognostic value in identifying individuals who convert to an α-synucleinopathy [119]. A recent study identified a prodromal DLB cohort, defined as patients with non-motor symptoms associated with LBD such as RBD, olfactory dysfunction, autonomic dysfunction, and depression in conjunction with diffuse occipital hypometabolism (based on FDG-PET). The prodromal cohort exhibited reduced DAT-SPECT uptake relative to patients with AD [120]. A second study evaluated whether visually rated DAT-SPECT could differentiate possible and probable MCI-LB relative to MCI-AD. This study indicated relatively poor sensitivity of 54.2% [95% confidence interval (CI) 39.2–68.6] and good specificity of 89.0% (95% CI 70.8–97.6) [121]. As noted above, DAT-SPECT can be helpful to rule out a non-dopaminergic etiology in the context of parkinsonism; however, despite substantial research, there is limited use beyond this in PD, given the limited sensitivity and specificity of DAT-SPECT and F-DOPA PET to differentiate between PD and MSA, PSP, and CBD, as these diseases also involve some presynaptic dopaminergic loss [122–124]. However, DAT-SPECT can be useful to differentiate DLB from other neurodegenerative conditions, primarily AD (sensitivity and specificity ranging from 78–87% and 90–94%, respectively) [125–127], and semi-quantitative assessments have shown promise to differentiate DLB from frontotemporal degeneration [128]. A meta-analysis with a total of 419 subjects estimated the pooled sensitivity and specificity of DAT-SPECT to differentiate DLB versus "no DLB" were 86.5% [95% CI: 72–94.1%] and 93.6% (95% CI: 88.5–96.6%), respectively [129].

6.4 *Metabolic and Perfusion PET Imaging*

Unlike the prior approaches, which focus on the nigrostriatal system, metabolic imaging captures the broader degeneration that occurs in PDD and DLB. As noted, low occipital uptake on FDG-PET/HMPAO-SPECT is used as a supportive biomarker for the diagnosis of DLB. FDG-PET quantifies patterns of cerebral glucose metabolism and has indicated that patients with DLB exhibit occipital hypometabolism. FDG-PET imaging can be evaluated in a variety of ways, including comparison of regional metabolism to a normative database, comparison of regional group differences, and advanced multivariate approaches that can quantify broader network dysfunction [130]. The latter approach can be used to obtain disease-related patterns of dysfunction. Examination of FDG-PET in individuals with iRBD indicated a subset of participants exhibited reduced occipital metabolic activity, ranging from 44% to 90% of the samples examined [131, 132]. An iRBD-related metabolic network pattern indicated that these individuals exhibited relatively increased activities in the pons, thalamus, medial frontal and sensorimotor areas, hippocampus, supramarginal and inferior temporal gyri, and posterior cerebellum, with decreased activity in the occipital and superior temporal regions [133]. The topography of the iRBD pattern partially overlapped with the PD-related pattern, and the iRBD pattern was significantly expressed in individuals with PD relative to healthy controls, suggesting that the iRBD pattern is an early manifestation of the PD-related pattern [134]. Additionally, individuals with iRBD that exhibited a significant baseline PD-related pattern or exhibited a higher rate of change in the PD-related pattern exhibited an increased risk for conversion to an α-synucleinopathy [135].

FDG-PET has demonstrated significant promise as a biomarker for the differential diagnosis of parkinsonism [124], with pooled sensitivities and specificities to differentiate PD from atypical parkinsonism of 91.4% (71.6–97.8%) and 90.6% (70.1–97.5%), respectively, when evaluating observer-dependent visual reads supported by voxel-based statistical analyses [136]. The PD-related pattern is characterized by relatively increased activity in the pallidothalamic and pontine regions and reduced activity in the premotor cortex, supplementary motor area, and parietal association regions [137]. Additional patterns have been identified for atypical parkinsonism syndromes, which have also been useful for the differential diagnosis of PD [138, 139].

Cortical hypometabolism in the posterior temporoparietal and occipital areas is common in nondemented individuals with PD, though it has been found to significantly progress as an individual converts to PDD [136, 140–142]. For example, individuals with PD who exhibited progression to PDD exhibited significant metabolic reductions in the occipital and posterior cingulate regions before progressing to PDD. At the time they had progressed to PDD, they exhibited a significant reduction in the thalamus, posterior cingulate, occipital, parietal, and frontal cortices (ranging from 7% to 11.4%) [143]. Notably, hypometabolism has been demonstrated to precede the regionally corresponding atrophy changes on MRI [144], suggesting that this may contribute to atrophy. Additionally, regional glucose metabolism in posterior temporal and parietal regions correlated with memory and visuospatial functioning, highlighting that the impact on these regions accounts for the cognitive impairments that are predictive of PDD [141, 145].

A PD-related cognitive pattern (PDCP) has also been developed and demonstrates metabolic reductions in frontal and parietal association areas [146]. Of note, the PDCP largely overlaps with networks commonly referred to as the frontoparietal network in addition to broader regions. The expression of this network is a replicable imaging biomarker of PD-cognitive dysfunction [147]. PDCP expression increases over time, corresponding to gradual cognitive decline, slightly less than the expression of the PD-related pattern (PDRP), which corresponds to motor progression [148]. PDD and DLB exhibit widespread lateral frontal, temporal, and parietal hypometabolism in conjunction with occipital hypometabolism. The occipital hypometabolism is best able to differentiate PDD and DLB from AD [149, 150]. A metabolic brain pattern has been identified for DLB, which shares topographic similarities with the AD-related pattern and the PDRP but not with the PDCP [151]. The diagnostic and prognostic accuracy of the FDG-PET hypometabolism pattern represents a reliable biomarker of DLB, and it has been suggested that it should be included in the diagnostic criteria (beyond a supportive biomarker) [152].

6.5 Functional MRI Network Approaches

Functional MRI (fMRI) approaches are not currently included in any diagnostic criteria. Research approaches leveraging these networks have largely focused on understanding the mechanisms of dysfunction in LBD and are only reviewed briefly here. Numerous studies have highlighted reductions in resting state connectivity [153–155] and reductions in frontostriatal engagement during executive functioning tasks [156, 157] in PD. Additionally, fMRI has been used to identify [158] and validate [159] a PD-related pattern.

Additionally, the default mode network (DMN), which consists of the medial prefrontal cortex, posterior cingulate cortex, hippocampi, and inferior parietal lobes [160], is initially normal early in individuals with PD without evidence of cognitive impairment and exhibits a decrement with the onset of cognitive decline [137, 161]. Default-mode network activity is notably reduced in AD [162] and associated with cerebrospinal fluid (CSF) Aβ42/pTau-181 [163]. Meta-analysis of resting state fMRI studies in PD indicates that cognitive impairment in PD is most consistently associated with a reduction in DMN connectivity [164]. Examination of PD cohorts with limited cognitive impairment suggests CSF α-synuclein is associated with sensorimotor network connectivity and longitudinal change in network connectivity between the dorsal attention and frontoparietal control networks [165, 166]. In a cohort that differentiated cognitive status, however, cognitively impaired groups (PD-MCI and PD dementia) exhibited a reduction in the CSF Aβ42/Aβ40 ratio in addition to reduced DMN connectivity [167], suggesting these biomarkers may be an indicator of DMN decline.

6.6 Volumetric MRI Biomarkers

As LBD exhibits relatively subtle atrophy in systems impacted early in the disease course (e.g., fronto-subcortical systems), they may appear normal on routine clinical MRI scans. In the diagnosis of PD based on current clinical criteria, MRIs are only used to identify and rule out alternate etiologies, such as cerebrovascular lesions (i.e., lacunes, white matter hyperintensities, and enlarged perivascular spaces) and normal pressure hydrocephalus [13]. In DLB, the relative preservation of the medial temporal lobe structures on CT/MRI is considered a supportive biomarker, as degeneration of the medial temporal lobe structures is indicative of AD. Patients with DLB do have significantly greater medial temporal lobe atrophy relative to healthy controls, but significantly less than those with AD [168].

There have been substantial research efforts to identify whether quantitative volumetric approaches can be helpful biomarkers. Due to the relatively subtle volumetric changes and heterogeneity in neuroanatomical systems impacted, group differences are observed (e.g., iRBD or PD cohorts exhibit reduced volumes), but the ability of atrophy/volumetric quantification to differentiate patient groups from controls is infrequently reported. Atrophy is more apparent at later disease stages and may be more helpful to differentiate patient groups. Methods that quantify neuronal, vascular, and metabolic dysfunction are more helpful at earlier disease stages. Patients with iRBD have structural changes consistent with atrophy of the fronto-subcortical systems [169]. For example, analysis of cortical thickness revealed that individuals with iRBD exhibited thinning in the bilateral medial superior frontal cortex, orbitofrontal cortex, anterior cingulate cortex, and the right dorsolateral primary motor cortex. Additionally, voxel-based morphometry (VBM) results indicated a reduction in gray matter volume in individuals with iRBD, particularly in the frontal lobes, anterior cingulate gyri, and the caudate nucleus [170]. iRBD brain atrophy patterns are suggestive of a prion-like and regional vulnerability hypothesis of α-synuclein [171]. Individuals with MCI-LB exhibit atrophy in the bilateral insula, precuneus, medial frontal structures, right claustrum, left anterior cingulate, left frontal, right superior and inferior frontal regions, and right cingulate cortex [172]. Individuals with PD exhibit atrophy of the fronto-subcortical systems. Progression of cortical atrophy in the caudate, putamen, temporal/hippocampal, and frontal and parietal regions is observed as the disease progresses. Regions specific to the progression of cognitive symptoms include the hippocampus, frontotemporal areas, caudate, thalamus, and nucleus accumbens [173], which are generally consistent with a recent meta-analysis [174]. The pattern of atrophy in DLB differs from AD, with relatively little cortical gray matter loss relative to the widespread pattern of gray matter loss observed in AD [175]. Unfortunately, these studies do not report the sensitivity or specificity of these measures to differentiate patient groups, raising challenges to evaluate the potential of these measures to serve as biomarkers. As discussed above, MRI biomarkers of the midbrain and brainstem are likely more promising. Additionally, whole-brain susceptibility weighted imaging has demonstrated relatively more robust imaging characteristics than MRI volumetric approaches [176].

6.7 Cerebrovascular-Related Biomarkers

Current diagnostic criteria do not incorporate information obtained from imaging that quantifies cerebrovascular dysfunction in DLB, though there are numerous studies indicating that cerebrovascular-related dysfunction (e.g., measured by cerebral blood flow or cerebrovascular reactivity) is observed in LBD [177, 178]. As

LBD exhibits relatively unique cardiovascular risk factors and dysfunction, there is currently no clear framework to conceptualize whether these are co-occurring disease processes or central to the disease pathogenesis. We have recently reviewed the literature [57] and discuss how cerebrovascular function is vital not just for ensuring optimal blood flow to the brain but also facilitating the clearance of interstitial waste, such as α-synuclein. Given LBD patients experience unique cardiovascular risk factors, including fluctuations/low blood pressure and autonomic dysfunction that may occur in the prodromal phase of the disorder, developing cerebrovascular biomarkers may be an important area of future research. Evaluation of individuals with DLB relative to individuals with AD has indicated high sensitivities (87–100%) and specificities (90–95%) of cerebral blood flow metrics to differentiate these patient groups [179]. Direct comparison with FDG-PET, however, indicates that FDG-PET (area under the curve = 0.91 [95% CI: 0.65–0.95]) still outperforms perfusion measures (area under the curve = 0.80 [95% CI: 0.79–1.00]) [180]. The pattern of hypoperfusion on MRI-arterial spin labeling (ASL) is similar to the hypometabolism of FDG-PET. Future methodological development may help improve the performance of MRI-ASL imaging. Given that MRI-ASL does not require radiotracers and is noninvasive, it has the potential to serve as an important biomarker in the future.

Cerebrovascular reactivity paradigms in conjunction with MRI (which evaluate changes in blood-oxygen-level dependent (BOLD) or cerebral blood flow in response to a hypercapnia challenge [181]) have been recently evaluated in PD. Initial results have been inconsistent and based on small sample sizes [182–184]. Additionally, prior approaches failed to model the latency of the cerebrovascular response. Our recent analyses demonstrated reduced amplitude and increased latencies in the cerebrovascular response in PD relative to controls [178]. Our initial approach quantified summary metrics (e.g., average gray and white matter amplitude and latency), though we have recently identified a PD cerebrovascular reactivity pattern using the latency measure that exhibited high sensitivity (85%) and specificity (75%) to differentiate individuals with PD from controls [185].

7 Fluid Biomarkers

While fluid biomarkers are not yet included in current diagnostic criteria, the potential impact of their use cannot be understated given their ease of use in clinical practice. Imaging approaches provide anatomical specificity, whereas fluid biomarkers are more easily obtained and cost-effective. Identification of α-synuclein and its aggregates has been evaluated most extensively and is reviewed in detail in another chapter in this book. Here, we focus our efforts on a brief discussion of α-synuclein biomarkers followed by dopamine metabolites. As AD pathology commonly occurs in both PD and DLB (more so in the latter), we discuss biomarkers of co-occurring AD. As there are likely several upstream mechanisms that may occur prior to α-synuclein aggregation, we organize our subsequent discussion into contributing disease processes.

7.1 Alpha-Synuclein Fluid Biomarkers

Initial research focused on quantifying total, phosphorylated, and oligomeric α-synuclein in CSF and serum. However, despite significant efforts, these approaches never reached acceptable diagnostic performances [186]. Seeded aggregation assays (SAAs) and immunohistochemical detection of pathologic α-synuclein have demonstrated promise as they amplify small amounts of aggregated protein. Specific approaches include protein misfolding cyclic amplification (PMCA) and real-time quaking-induced conversion (RT-QuIC) and have demonstrated high diagnostic utility (e.g., PMCA demonstrated an area under the curve of 0.93 and RT-QuIC demonstrated an area under the curve of 0.89 to differentiate PD from healthy controls) [187]. The specificity and sensitivity of PMCA assays are high (often greater than 90%) in comparing CSF samples of individuals with PD to healthy controls [187], individuals with AD [188], and other neurological conditions [189]. The seeding activity is even greater in individuals with DLB, resulting in high specificity and sensitivity (greater than 90%) to differentiate DLB from non-synucleinopathies (that included individuals with AD) [188, 190] and healthy controls [191]. Together, this highlights the significant promise of the seeded aggregation assays. In addition to seeded aggregation, a protease that degrades α-synuclein, CSF neurosin, is reduced in DLB and PD relative to AD and healthy controls [192], which may also help identify individuals with LBD.

As noted, the substantia nigra pars compacta (SNc) is particularly susceptible to α-synuclein aggregation in LBD. Additionally, excess α-synuclein can interact with dopamine synthesis and its metabolites [193, 194], potentially facilitating α-synuclein aggregation. Alternatively, altered dopamine and corresponding metabolites may be due to neurodegeneration in the SNc. Either mechanism would result in altered dopamine and its metabolites. There have been recent reports highlighting that DOPA decarboxylase (DDC), which is essential for the formation of dopamine, can be detected in CSF, plasma, and urine, and is upregulated in PD relative to healthy controls (sensitivity = 0.83; specificity = 0.83) [195] and AD [196]. Similar results were observed in de novo cohorts in these studies. Other studies have demonstrated that the dopamine metabolite, CSF homovanillic acid, is reduced in LBD and AD relative to healthy controls [197, 198]. Homovanillic acid is also correlated with the DAT-SPECT signal in individuals with PD [199].

7.2 Co-occurring AD Pathological Processes

As noted, there is a high rate of co-occurring AD pathology in both PDD and DLB, highlighting the utility of AD fluid biomarkers in PDD/DLB. Evidence thus far highlights that AD Aβ and tau co-pathologies strongly influence the clinical presentation and progression of PD, supported by converging evidence from studies of neuropathology and ante-mortem biomarkers [8, 67, 72]. As amyloid and tau CSF biomarkers were developed and validated within AD populations, many of the initial evaluations of these biomarkers in LBD were to understand how these biomarkers influence clinical presentations in Lewy body dementias. For example, a large cohort of de novo individuals with PD exhibited lower CSF β-amyloid$_{1-42}$ (Aβ_{42}), phosphorylated tau (p-tau), and total tau (t-tau) than healthy controls at baseline assessments, with 31.5% of individuals exhibiting pathologically low levels of CSF Aβ_{42} at baseline relative to 27.7% of control individuals. Individuals with PD exhibited a greater decline in CSF Aβ_{42} and p-tau relative to healthy controls over the course of three years [200]. Individuals with PDD and DLB that exhibited CSF Aβ_{42}, p-tau, and t-tau concentrations suggestive of co-occurring AD (based on an autopsy-validated CSF cutoff: t-tau/Aβ_{42} ratio > 0.3, $n = 43$) performed worse on the Boston Naming Test than individuals with DLB/PDD without evidence of co-occurring AD [201].

Efforts to evaluate the cutoffs for CSF Aβ_{42}, p-tau, and t-tau concentrations in PD using PET amyloid imaging as the gold standard have indicated that the optimal cutoffs likely differ significantly from cutoffs derived from AD populations. Specifically, the cut-off of Aβ_{42} to predict amyloid-PET positivity was higher than the AD cutoff. The cut-offs of t-tau/Aβ_{42} and p-tau/Aβ_{42} ratios to predict cognitive impairment were lower than the AD cutoffs. This suggests that the presence of other pathologies, such as α-synuclein, leads to disease-specific CSF biomarker characteristics. This highlights the complexity of developing biomarkers in a heterogeneous disorder. However, it is notable that a recent meta-analysis of CSF biomarkers in PD indicated that the combination of total α-synuclein, Aβ_{42}, and neurofilament light chain (NfL) are the most promising diagnostic biomarkers to differentiate PD from atypical parkinsonism [202].

7.3 *Neuronal and Synaptic Injury Markers*

NfL is considered a component of the neuronal cytoskeleton and is released when neuronal damage and axonal injury occur. Individuals with PD do not exhibit significant differences in NfL early in the disease course, whereas atypical parkinsonian disorders have demonstrated increased NfL, highlighting the potential of NfL to differentiate PD from these disorders using both CSF NfL [203, 204] and blood-based NfL [205, 206]. NfL, however, is associated with both motor and cognitive progression in PD, highlighting its potential use as a progression biomarker [207–209]. NfL is elevated in many neurodegenerative conditions, limiting its differential diagnostic ability once cognitive decline has already occurred. Rather, it can be better conceptualized as a marker of progressive neurodegeneration, as the highest levels are observed in rapidly progressing neurodegenerative conditions (e.g., CBD and PSP as well as Amyotrophic Lateral Sclerosis and Frontotemporal Dementia) [210, 211]. Visinin-like protein-1 (VILIP-1) is a marker of calcium-mediated neuronal injury. It is significantly increased in AD relatively to healthy controls and DLB, supporting a potential role of differentiating AD relative to DLB, as it appears it is altered in AD only [212, 213].

Proteomic approaches have found reduced synaptic markers when evaluating postmortem brain samples, including Ras-related protein Rab-3A, synaptosomal-associated protein-25 (SNAP-25), and neurogranin in DLB, PDD, and AD relative to healthy controls [214]. While useful in AD, the authors suggest that the use of these measures should not be extended to Lewy body diseases. Neurogranin is a dendritic protein that can be measured in the CSF and is considered a marker for synaptic injury. In contrast to AD, neurogranin is decreased in PD, PDD, MSA, and PSP relative to both healthy controls and AD [215], further indicating that neurogranin may not be useful to diagnose LBD but can be useful to differentiate LBD from AD.

7.4 Biomarkers Capturing Contributing Disease Processes

As noted, there are many mechanisms that increase risk for PD or DLB and may contribute to α-synuclein aggregation or interact with it. For example, infections [54–56], gut dysbiosis [49, 50], head trauma [51], environmental exposures [52, 53], and cerebrovascular dysfunction [57] may either trigger or worsen α-synuclein pathology. It has been suggested that these, in combination with facilitators and aggravators (e.g., inflammation, gut barrier dysfunction, mitochondrial and lysosomal dysfunction, and oxidative stress), contribute to disease pathophysiology and progression [59]. Based on this, it is not surprising that biomarkers measuring aspects of these processes have demonstrated promise, as discussed below.

7.4.1 Inflammatory Markers

Inflammation may contribute to the initiation or exacerbation of α-synuclein aggregation. Microglia are key innate immune cells that mediate phagocytosis and inflammatory responses. They can be activated by misfolded α-synuclein or other early disease processes, which then may contribute to immune responses, ultimately resulting in damage to neurons [216]. Much of the current work has focused on inflammatory markers that can be detected in the blood. A meta-analysis has identified that individuals with PD exhibit significantly increased blood interleukin (IL)-6, tumor necrosis factor, IL-1β, IL-2, IL-10, C-reactive protein, and RANTES (regulated on activation, normal T-expressed, and presumably secreted) compared with healthy controls [217]. However, a systematic review of the literature spanning PD, PDD, and DLB indicated an increase in blood-based markers of interleukin-1β (IL-1β), tumor necrosis factor (TNF)-α, IL-6, and IL-10 [218]. The same systematic review indicated that CSF IL-1β, IL-6, transforming growth factor-β1, and C-reactive protein levels were increased while interferon-γ was decreased in individuals with

LBD relative to healthy controls, though the authors acknowledge that there are complex changes in the individual cytokines. A recent study evaluated potential differences in DLB-MCI, AD-MCI, DLB, and AD and found elevated plasma levels of IL-10, IL-1β, IL-4, and IL-2 in DLB-MCI and AD-MCI, with no significant differences in inflammatory markers between DLB and AD relative to healthy controls, which suggests that inflammation is highest earlier in the disease course [219]; although other studies have indicated that individuals with DLB and AD exhibit elevated serum levels of IL-6 and TNF-α relative to healthy controls in small sample sizes [220]. Evaluation of CSF glial proteins YKL-40, soluble triggering receptor expressed on myeloid cells 2 (sTREM2), and progranulin in prodromal DLB, DLB, and AD found that individuals with DLB did not exhibit increased sTREM2 and progranulin levels, whereas YKL-40 was only increased in individuals with DLB who exhibited an AD biomarker profile [221].

7.4.2 Gut Dysbiosis

There is increasing evidence indicating individuals with LBD exhibit altered gut microbiota in conjunction with altered gut permeability and systemic inflammation, which has sparked great interest in understanding whether these mechanisms may occur early in the disease process and contribute to disease pathogenesis [50]. While the etiological role of gut dysbiosis in disease pathogenesis has yet to be examined, α-synuclein aggregates are detected in the gut and have been proposed as useful biomarkers [222]. Further, PD patients exhibit an elevation in gram-negative bacteria and a corresponding increase in endotoxin lipopolysaccharide (LPS) content [223], which can be detected in the periphery. Similarly, a bloom in rare bacteria, such as sulfate-reducing bacteria, may also be a potential biomarker for LBD, given observed elevations in the *Desulfovibrionaceae* family in individuals with PD [224, 225], which also correlates with the severity of clinical symptoms [225]. Depending on the specificity of the microbiome composition to LBD, further work incorporating microbiome/gut permeability biomarkers may be an important area of future research.

7.4.3 Autophagy, Mitochondria, and Oxidative Stress

Dysfunction of the autophagy-lysosomal pathway is implicated in intracellular aggregation of α-synuclein [226], in part due to the increased risk for PD and DLB in carriers of the glucocerebrosidase gene (*GBA*) mutation [227]. The failure of this pathway contributes to diminished ability to degrade misfolded α-synuclein. Individuals with PD and DLB [228, 229] exhibit reduced beta-glucocerebrosidase (GCase) activity relative to healthy controls. However, diagnostic accuracy of GCase was suboptimal, with a sensitivity of 67% and specificity of 77% when differentiating patients with PD from healthy controls [229]. A combination of GCase and the oligomeric/total α-synuclein ratio improved the discriminatory ability to a sensitivity of 82% and a specificity of 71% to differentiate PD from healthy controls [230].

There are numerous potential sources of oxidative stress in LBD, including mitochondrial dysfunction, inflammation, and environmental exposure [231]. Oxidative stress has been implicated in dopamine cell degeneration, highlighting that early detection may capture early disease processes [232]. A recent meta-analysis [233] indicated that individuals with PD exhibited increases in blood 8-OhdG, MDA, nitrite, and ferritin relative to healthy controls. Additionally, patients with PD and DLB exhibit altered mitochondrial functioning [234], with notable Complex I deficiencies of the respiratory chain, among others. There has been a strong link between oxidative stress and disrupted mitochondrial respiration and a potential unique role of the antioxidants, such as neuropeptide cocaine- and amphetamine-regulated transcript (CART) [235]. CART is localized to a greater degree in the mitochondria and has been significantly reduced in patients with DLB relative to AD and healthy controls [236]. This work has largely focused on understanding disease mechanisms, and future work is necessary to evaluate the potential for the use of these measures as biomarkers.

7.5 Differentiating DLB and AD

Herein, we have discussed the broad landscape of biomarkers relevant to LBD, emphasizing non-α-synuclein biomarkers and discussing both established and emerging biomarkers based on the biological system involved. However, recent efforts have more practically summarized which CSF biomarkers differ between DLB and AD, recreated in Table 1. Those noted in the table but not discussed above include elevations of calcium (Ca) and magnesium (Mg) in DLB compared to controls as well as AD and vascular dementia (VaD) [237], and elevated amino acids, asparagine and glycine, in DLB relative to controls [238]. There is rapid development and identification of candidate biomarkers using proteomics, metabolomics, and transcriptomics in LBD. For example, the following six CSF proteins were lower in participants with DLB relative to controls: neurosecretory protein (VGF), secretogranin-2 (SCG2), neuronal pentraxin-2 (NPTX2), neuronal pentraxin receptor (NPTXR), proenkephalin-B (PDYN), and proSAAS (PCSK1N) [239]. An additional study found CSF levels of VGF and PDYN were lower in DLB compared to AD. VGF, PDYN, and SCG2, when combined, best differentiate DLB from non-DLB with a specificity of 83% and sensitivity of 69% [240]. Much of this work is largely exploratory, with limited validation at this stage. As such, we have included some notable findings but deferred an exhaustive review until further validation.

Table 1 Presents the potential biomarkers in LBD CSF; ↑ denotes an increase relative to controls; ↓ denotes a decrease relative to controls; – denotes no change

Fluid biomarkers	DLB	AD	References
YKL-40	↑ in DLB + AD profile, not in prodromal stages	↑ in prodromal stages	[221]
Neurogranin	↓ in DLB and PDD	↑	[214, 215]
VILIP-1	–	↑ in AD compared to DLB	[212, 213]
Neurosin	↓	–	[192]
IL-6	↑	–	[217–220]
CART	↓	–	[236]
HVA	↓	↓	[197–199]
RT-QuIC	↑	–	[187, 190, 191]
Magnesium, calcium, copper	↑ compared to AD	–	[237]
Asparagine, glycine	↑	–	[238]
PDYN	↓	–	[239]
NPTX2, VGF, SCG2	↓	↓	[239, 240]

Source: Adapted from Bousiges et al. 2022, *International Journal of Molecular Sciences* [241]. Abbreviations: *YKL-40* chitinase 3-like protein 1 or human cartilage glycoprotein 39, *VILIP-1* Visinin-like protein-1, *IL-6* Interleukin-6, *CART* Cocaine-and-amphetamine-regulated transcript, *HVA* homovanillic acid, *RT-QuIC* Real-Time Quaking-Induced Conversion, *PDYN* Proenkephalin-B, *NPTX2* Neuronal pentraxin-2, *VGF* nerve growth factor inducible (a neurosecretory protein), *SCG2* Secretogranin-2

8 Summary and Future Directions

Over time, diagnostic criteria for LBD are increasingly incorporating biomarkers, each of which is reviewed herein. There are numerous future directions to pursue. As noted, a significant benefit of biomarkers is identifying individuals before significant degeneration has occurred. Given the recent development of research criteria for prodromal PD and DLB, there is a need for additional research to understand which biomarkers are present early in the disease course and predict progression. In parallel, several lines of research have highlighted other potential disease mechanisms, such as inflammation, gut-brain-axis dysfunction, and cerebrovascular dysfunction, that are important to LBD pathogenesis. Quantification of these different disease mechanism markers will likely lead to new biomarker developments [217, 242].

Funding This work was supported by NIH grant 3R01NS133569-01S1 (SR).

References

1. Okazaki H, Lipkin LE, Aronson SM. Diffuse intracytoplasmic ganglionic inclusions (Lewy type) associated with progressive dementia and quadriparesis in flexion. J Neuropathol Exp Neurol. 1961;20:237–44.
2. Kosaka K, Oyanagi S, Matsushita M, Hori A. Presenile dementia with Alzheimer-, pick-and Lewy-body changes. Acta Neuropathol. 1976;36:221–33.
3. Ince PG, Perry EK, Morris CM. Dementia with Lewy bodies. A distinct non-Alzheimer dementia syndrome? Brain Pathol. 1998;8:299–324.
4. Just MK, Gram H, Theologidis V, Jensen PH, Nilsson KPR, Lindgren M, et al. Alpha-Synuclein strain variability in body-first and brain-first Synucleinopathies. Front Aging Neurosci. 2022;14
5. Horsager J, Andersen KB, Knudsen K, Skjærbæk C, Fedorova TD, Okkels N, et al. Brain-first versus body-first Parkinson's disease: a multimodal imaging case-control study. Brain. 2020;143:3077–88.
6. Borghammer P. The α-synuclein origin and connectome model (SOC model) of Parkinson's disease: explaining motor asymmetry, non-motor phenotypes, and cognitive decline. J Parkinsons Dis. 2021;11:455–74.
7. Irwin DJ, Hurtig HI. The contribution of tau, amyloid-beta and alpha-synuclein pathology to dementia in Lewy body disorders. J Alzheimer Dis Parkinson. 2018:8.
8. Irwin DJ, Lee VM-Y, Trojanowski JQ. Parkinson's disease dementia: convergence of α-synuclein, tau and amyloid-β pathologies. Nat Rev Neurosci. 2013;14:626.
9. Jellinger KA, Korczyn AD. Are dementia with Lewy bodies and Parkinson's disease dementia the same disease? BMC Med. 2018;16:34.
10. Heinzel S, Berg D, Gasser T, Chen H, Yao C, Postuma RB, et al. Update of the MDS research criteria for prodromal Parkinson's disease. Mov Disord. 2019;34:1464–70.
11. McKeith IG, Ferman TJ, Thomas AJ, Blanc F, Boeve BF, Fujishiro H, et al. Research criteria for the diagnosis of prodromal dementia with Lewy bodies. Neurology. 2020;94:743–55.
12. Tysnes O-B, Storstein A. Epidemiology of Parkinson's disease. J Neural Transm. 2017;124:901–5.
13. Postuma RB, Berg D, Stern M, Poewe W, Olanow CW, Oertel W, et al. MDS clinical diagnostic criteria for Parkinson's disease. Mov Disord. 2015;30:1591–601.
14. Pfeiffer RF. Non-motor symptoms in Parkinson's disease. Parkinsonism Relat Disord. 2016;22:S119–22.
15. Aarsland D, Bronnick K, Williams-Gray C, Weintraub D, Marder K, Kulisevsky J, et al. Mild cognitive impairment in Parkinson disease: a multicenter pooled analysis. Neurology. 2010;75:1062–9.
16. Aarsland D, Kurz MW. The epidemiology of dementia associated with Parkinson's disease. Brain Pathol. 2010;20:633–9.
17. Biundo R, Weis L, Antonini A. Cognitive decline in Parkinson's disease: the complex picture. npj Parkinson's Dis. 2016;2:1–7.
18. Winer JR, Maass A, Pressman P, Stiver J, Schonhaut DR, Baker SL, et al. Associations between tau, β-amyloid, and cognition in parkinson disease. JAMA Neurol. 2018;75:227–35.
19. Petersen RC, Doody R, Kurz A, Mohs RC, Morris JC, Rabins PV, et al. Current concepts in mild cognitive impairment. Arch Neurol. 2001;58:1985–92.
20. Winblad B, Palmer K, Kivipelto M, Jelic V, Fratiglioni L, Wahlund L, et al. Mild cognitive impairment–beyond controversies, towards a consensus: report of the international working group on mild cognitive impairment. J Intern Med. 2004;256:240–6.
21. Petersen RC. Mild cognitive impairment as a diagnostic entity. J Intern Med. 2004;256:183–94.
22. Litvan I, Aarsland D, Adler CH, Goldman JG, Kulisevsky J, Mollenhauer B, et al. MDS task force on mild cognitive impairment in Parkinson's disease: critical review of PD-MCI. Mov Disord. 2011;26:1814–24.

23. Litvan I, Goldman JG, Tröster AI, Schmand BA, Weintraub D, Petersen RC, et al. Diagnostic criteria for mild cognitive impairment in Parkinson's disease: Movement Disorder Society task force guidelines. Mov Disord. 2012;27:349–56.
24. Kehagia AA, Barker RA, Robbins TW. Neuropsychological and clinical heterogeneity of cognitive impairment and dementia in patients with Parkinson's disease. The Lancet Neurology. 2010;9:1200–13.
25. Kang JH, Mollenhauer B, Coffey CS, Toledo JB, Weintraub D, Galasko DR, et al. CSF biomarkers associated with disease heterogeneity in early Parkinson's disease: the Parkinson's progression markers initiative study. Acta Neuropathol. 2016;131:935–49.
26. Adler CH, Caviness JN, Sabbagh MN, Shill HA, Connor DJ, Sue L, et al. Heterogeneous neuropathological findings in Parkinson's disease with mild cognitive impairment. Acta Neuropathol. 2010;120:827–8.
27. McKinlay A, Grace RC, Dalrymple-Alford JC, Roger D. Characteristics of executive function impairment in Parkinson's disease patients without dementia. J Int Neuropsychol Soc. 2010;16:268–77.
28. Williams-Gray CH, Evans JR, Goris A, Foltynie T, Ban M, Robbins TW, et al. The distinct cognitive syndromes of Parkinson's disease: 5 year follow-up of the CamPaIGN cohort. Brain. 2009;132:2958–69.
29. Levin BE, Llabre MM, Reisman S, Weiner WJ, Sanchez-Ramos J, Singer C, et al. Visuospatial impairment in Parkinson's disease. Neurology. 1991;41:365.
30. Levy G, Jacobs DM, Tang M, Côté LJ, Louis ED, Alfaro B, et al. Memory and executive function impairment predict dementia in Parkinson's disease. Mov Disorder. 2002;17:1221–6.
31. Hogan DB, Fiest KM, Roberts JI, Maxwell CJ, Dykeman J, Pringsheim T, et al. The prevalence and incidence of dementia with Lewy bodies: a systematic review. Can J Neurol Sci. 2016;43:S83–95.
32. Mok W, Chow TW, Zheng L, Mack WJ, Miller C. Clinicopathological concordance of dementia diagnoses by community versus tertiary care clinicians. Am J Alzheimers Dis Other Dement. 2004;19:161–5.
33. Jones SAV, O'brien JT. The prevalence and incidence of dementia with Lewy bodies: a systematic review of population and clinical studies. Psychol Med. 2014;44:673–83.
34. McKeith IG, Boeve BF, Dickson DW, Halliday G, Taylor J-P, Weintraub D, et al. Diagnosis and management of dementia with Lewy bodies: fourth consensus report of the DLB consortium. Neurology. 2017;89:88–100.
35. Brooks DJ. The early diagnosis of Parkinson's disease. Ann Neurol. 1998;44:S10–8.
36. Berg D, Postuma RB, Adler CH, Bloem BR, Chan P, Dubois B, et al. MDS research criteria for prodromal Parkinson's disease. Mov Disord. 2015;30:1600–11.
37. Iranzo A, Tolosa E, Gelpi E, Molinuevo JL, Valldeoriola F, Serradell M, et al. Neurodegenerative disease status and post-mortem pathology in idiopathic rapid-eye-movement sleep behaviour disorder: an observational cohort study. Lancet Neurol. 2013;12:443–53.
38. Postuma RB, Iranzo A, Hu M, Högl B, Boeve BF, Manni R, et al. Risk and predictors of dementia and parkinsonism in idiopathic REM sleep behaviour disorder: a multicentre study. Brain. 2019;142:744–59.
39. Postuma RB, Berg D. Prodromal Parkinson's disease: the decade past, the decade to come. Mov Disord. 2019;34:665–75.
40. Galbiati A, Verga L, Giora E, Zucconi M, Ferini-Strambi L. The risk of neurodegeneration in REM sleep behavior disorder: a systematic review and meta-analysis of longitudinal studies. Sleep Med Rev. 2019;43:37–46.
41. Kaufmann H, Norcliffe-Kaufmann L, Palma J, Biaggioni I, Low PA, Singer W, et al. Natural history of pure autonomic failure: AU nited S tates prospective cohort. Ann Neurol. 2017;81:287–97.
42. Spillantini MG, Schmidt ML, Lee VM-Y, Trojanowski JQ, Jakes R, Goedert M. α-Synuclein in Lewy bodies. Nature. 1997;388:839–40.

43. Wakabayashi K, Yoshimoto M, Tsuji S, Takahashi H. α-Synuclein immunoreactivity in glial cytoplasmic inclusions in multiple system atrophy. Neurosci Lett. 1998;249:180–2.
44. Polymeropoulos MH, Lavedan C, Leroy E, Ide SE, Dehejia A, Dutra A, et al. Mutation in the α-synuclein gene identified in families with Parkinson's disease. Science. 1997;276:2045–7.
45. Fares MB, Jagannath S, Lashuel HA. Reverse engineering Lewy bodies: how far have we come and how far can we go? Nat Rev Neurosci. 2021;22:111–31.
46. Iwatsubo T, Yamaguchi H, Fujimuro M, Yokosawa H, Ihara Y, Trojanowski JQ, et al. Purification and characterization of Lewy bodies from the brains of patients with diffuse Lewy body disease. Am J Pathol. 1996;148:1517.
47. Roberts RF, Wade-Martins R, Alegre-Abarrategui J. Direct visualization of alpha-synuclein oligomers reveals previously undetected pathology in Parkinson's disease brain. Brain. 2015;138:1642–57.
48. Espay AJ, Vizcarra JA, Marsili L, Lang AE, Simon DK, Merola A, et al. Revisiting protein aggregation as pathogenic in sporadic Parkinson and Alzheimer diseases. Neurology. 2019;92:329–37.
49. Nishiwaki H, Ito M, Ishida T, Hamaguchi T, Maeda T, Kashihara K, et al. Meta-analysis of gut dysbiosis in Parkinson's disease. Mov Disord. 2020;35:1626–35.
50. Ryman S, Vakhtin AA, Richardson SP, Lin HC. Microbiome–gut–brain dysfunction in prodromal and symptomatic Lewy body diseases. J Neurol. 2023;270:746–58.
51. Camacho-Soto A, Warden MN, Searles Nielsen S, Salter A, Brody DL, Prather H, et al. Traumatic brain injury in the prodromal period of Parkinson's disease: a large epidemiological study using Medicare data. Ann Neurol. 2017;82:744–54.
52. Gunnarsson L-G, Bodin L. Parkinson's disease and occupational exposures: a systematic literature review and meta-analyses. Scand J Work Environ Health. 2017;43:197–209.
53. Dusek P, Roos PM, Litwin T, Schneider SA, Flaten TP, Aaseth J. The neurotoxicity of iron, copper and manganese in Parkinson's and Wilson's diseases. J Trace Elem Med Biol. 2015;31:193–203.
54. Wijarnpreecha K, Chesdachai S, Jaruvongvanich V, Ungprasert P. Hepatitis C virus infection and risk of Parkinson's disease: a systematic review and meta-analysis. Eur J Gastroenterol Hepatol. 2018;30:9–13.
55. Stolzenberg E, Berry D, Yang DE, Lee EY, Kroemer A, Kaufman S, et al. A role for neuronal alpha-synuclein in gastrointestinal immunity. J Innate Immun. 2017;9:456–63.
56. Shannon KM. Infections and changes in commensal bacteria and the pathogenesis of Parkinson's disease. J Parkinsons Dis. 2022;12:1–7.
57. Ryman SG, Vakhtin AA, Mayer AR, van der Horn HJ, Shaff NA, Nitschke S, et al. Abnormal cerebrovascular activity, perfusion and glymphatic clearance in Lewy body diseases, vol. 39. Movement Disorders; 2024. p. 1258.
58. Parkkinen L, Pirttilä T, Alafuzoff I. Applicability of current staging/categorization of α-synuclein pathology and their clinical relevance. Acta Neuropathol. 2008;115:399–407.
59. Johnson ME, Stecher B, Labrie V, Brundin L, Brundin P. Triggers, facilitators, and aggravators: redefining Parkinson's disease pathogenesis. Trends Neurosci. 2019;42:4–13.
60. Owen AM, James M, Leigh PN, Summers BA, Marsden CD, Quinn NP, et al. Fronto-striatal cognitive deficits at different stages of Parkinson's disease. Brain. 1992;115:1727–51.
61. Christopher L, Marras C, Duff-Canning S, Koshimori Y, Chen R, Boileau I, et al. Combined insular and striatal dopamine dysfunction are associated with executive deficits in Parkinson's disease with mild cognitive impairment. Brain. 2014;137:565–75.
62. Del Tredici K, Braak H. Dysfunction of the locus coeruleus–norepinephrine system and related circuitry in Parkinson's disease-related dementia. J Neurol Neurosurg Psychiatry. 2013;84:774–83.
63. Kehagia AA, Barker RA, Robbins TW. Cognitive impairment in Parkinson's disease: the dual syndrome hypothesis. Neurodegener Dis. 2013;11:79–92.

64. Chahine LM, Dos Santos C, Fullard M, Scordia C, Weintraub D, Erus G, et al. Modifiable vascular risk factors, white matter disease and cognition in early Parkinson's disease. Eur J Neurol. 2019;26:246–e18.
65. Shibata K, Sugiura M, Nishimura Y, Sakura H. The effect of small vessel disease on motor and cognitive function in Parkinson's disease. Clin Neurol Neurosurg. 2019;182:58–62.
66. Buchman AS, Yu L, Wilson RS, Leurgans SE, Nag S, Shulman JM, et al. Progressive parkinsonism in older adults is related to the burden of mixed brain pathologies. Neurology. 2019;92:e1821–30.
67. Coughlin DG, Hurtig HI, Irwin DJ. Pathological influences on clinical heterogeneity in Lewy body diseases. Mov Disord. 2020;35:5–19.
68. Celebi O, Temuçin ÇM, Elibol B, Saka E. Short latency afferent inhibition in Parkinson's disease patients with dementia. Mov Disord. 2012;27:1052–5.
69. Tsuboi Y, Uchikado H, Dickson DW. Neuropathology of Parkinson's disease dementia and dementia with Lewy bodies with reference to striatal pathology. Parkinson Relat Disord. 2007;13:S221–4.
70. Irwin DJ, Grossman M, Weintraub D, Hurtig HI, Duda JE, Xie SX, et al. Neuropathological and genetic correlates of survival and dementia onset in synucleinopathies: a retrospective analysis. Lancet Neurol. 2017;16:55–65.
71. Jellinger KA, Seppi K, Wenning GK, Poewe W. Impact of coexistent Alzheimer pathology on the natural history of Parkinson's disease. J Neural Transm. 2002;109:329–39.
72. Ryman SG, Yutsis M, Tian L, Henderson VW, Montine TJ, Salmon DP, et al. Cognition at each stage of Lewy body disease with co-occurring Alzheimer's disease pathology. J Alzheimers Dis. 2021;80:1–14.
73. Nanhoe-Mahabier W, De Laat KF, Visser JE, Zijlmans J, de Leeuw F-E, Bloem BR. Parkinson disease and comorbid cerebrovascular disease. Nat Rev Neurol. 2009;5:533.
74. Kaufmann H, Norcliffe-Kaufmann L, Palma J-A. Baroreflex dysfunction. N Engl J Med. 2020;382:163–78.
75. Chen Z, Li G, Liu J. Autonomic dysfunction in Parkinson's disease: implications for pathophysiology, diagnosis, and treatment. Neurobiol Dis. 2020;134:104700.
76. Velseboer DC, de Haan RJ, Wieling W, Goldstein DS, de Bie RMA. Prevalence of orthostatic hypotension in Parkinson's disease: a systematic review and meta-analysis. Parkinsonism Relat Disord. 2011;17:724–9.
77. Allcock LM, Kenny RA, Mosimann UP, Tordoff S, Wesnes KA, Hildreth AJ, et al. Orthostatic hypotension in Parkinson's disease: association with cognitive decline? Int. J Geriatr Psychiatr. 2006;21:778–83.
78. Peralta C, Stampfer-Kountchev M, Karner E, Köllensperger M, Geser F, Wolf E, et al. Orthostatic hypotension and attention in Parkinson's disease with and without dementia. J Neural Transm. 2007;114:585–8.
79. Hohler AD, Zuzuárregui J-RP, Katz DI, DePiero TJ, Hehl CL, Leonard A, et al. Differences in motor and cognitive function in patients with Parkinson's disease with and without orthostatic hypotension. Int J Neurosci. 2012;122:233–6.
80. FDA-NIH Biomarker Working Group. BEST (Biomarkers, EndpointS, and other Tools) resource. 2016;
81. Howell M, Avidan AY, Foldvary-Schaefer N, Malkani RG, During EH, Roland JP, et al. Management of REM sleep behavior disorder: an American Academy of sleep medicine clinical practice guideline. J Clin Sleep Med. 2023;19:759–68.
82. Boeve BF, Molano JR, Ferman TJ, Smith GE, Lin SC, Bieniek K, et al. Validation of the Mayo sleep questionnaire to screen for REM sleep behavior disorder in an aging and dementia cohort. Sleep Med. 2011;12:445.
83. Postuma RB, Arnulf I, Hogl B, Iranzo A, Miyamoto T, Dauvilliers Y, et al. A single-question screen for rapid eye movement sleep behavior disorder: a multicenter validation study. Mov Disord. 2012;27:913–6.

84. Ferman TJ, Boeve BF, Smith GE, Lin S-C, Silber MH, Pedraza O, et al. Inclusion of RBD improves the diagnostic classification of dementia with Lewy bodies. Neurology. 2011;77:875–82.
85. Boeve BF, Silber MH, Ferman TJ, Lin SC, Benarroch EE, Schmeichel AM, et al. Clinicopathologic correlations in 172 cases of rapid eye movement sleep behavior disorder with or without a coexisting neurologic disorder. Sleep Med. 2013;14:754–62.
86. Zhang X, Sun X, Wang J, Tang L, Xie A. Prevalence of rapid eye movement sleep behavior disorder (RBD) in Parkinson's disease: a meta and meta-regression analysis. Neurol Sci. 2017;38:163–70.
87. Lin Y-Q, Chen S-D. RBD: a red flag for cognitive impairment in Parkinson's disease? Sleep Med. 2018;44:38–44.
88. Marion M-H, Qurashi M, Marshall G, Foster O. Is REM sleep behaviour disorder (RBD) a risk factor of dementia in idiopathic Parkinson's disease? J Neurol. 2008;255:192–6.
89. Jozwiak N, Postuma RB, Montplaisir J, Latreille V, Panisset M, Chouinard S, et al. REM sleep behavior disorder and cognitive impairment in Parkinson's disease. Sleep. 2017;40:zsx101.
90. Orimo S, Ozawa E, Nakade S, Sugimoto T, Mizusawa H. 123I-metaiodobenzylguanidine myocardial scintigraphy in Parkinson's disease. J Neurol Neurosurg Psychiatry. 1999;67:189–94.
91. Miyamoto T, Miyamoto M, Inoue Y, Usui Y, Suzuki K, Hirata K. Reduced cardiac 123I-MIBG scintigraphy in idiopathic REM sleep behavior disorder. Neurology. 2006;67:2236–8.
92. Kashihara K, Imamura T, Shinya T. Cardiac 123I-MIBG uptake is reduced more markedly in patients with REM sleep behavior disorder than in those with early stage Parkinson's disease. Parkinsonism Relat Disord. 2010;16:252–5.
93. Sakakibara R, Tateno F, Kishi M, Tsuyusaki Y, Terada H, Inaoka T. MIBG myocardial scintigraphy in pre-motor Parkinson's disease: a review. Parkinsonism Relat Disord. 2014;20:267–73.
94. Knudsen K, Fedorova TD, Hansen AK, Sommerauer M, Otto M, Svendsen KB, et al. In-vivo staging of pathology in REM sleep behaviour disorder: a multimodality imaging case-control study. The Lancet Neurology. 2018;17:618–28.
95. Sakakibara R, Tateno F, Aiba Y, Ogata T, Kishi M, Terada H, et al. MIBG myocardial scintigraphy identifies premotor PD/DLB during a negative DAT scan period: second report. Mov Disord Clin Pract. 2019;6:46–50.
96. Miyamoto T, Miyamoto M, Suzuki K, Nishibayashi M, Iwanami M, Hirata K. 123I-MIBG cardiac scintigraphy provides clues to the underlying neurodegenerative disorder in idiopathic REM sleep behavior disorder. Sleep. 2008;31:717–23.
97. Park DG, Kim JY, Kim MS, Kim MH, An Y-S, Chang J, et al. Neurofilament light chain and cardiac MIBG uptake as predictors for phenoconversion in isolated REM sleep behavior disorder. J Neurol. 2023;1–10:4393.
98. Kashihara K, Ohno M, Kawada S, Okumura Y. Reduced cardiac uptake and enhanced washout of 123I-MIBG in pure autonomic failure occurs conjointly with Parkinson's disease and dementia with Lewy bodies. J Nucl Med. 2006;47:1099–101.
99. Orimo S, Suzuki M, Inaba A, Mizusawa H. 123I-MIBG myocardial scintigraphy for differentiating Parkinson's disease from other neurodegenerative parkinsonism: a systematic review and meta-analysis. Parkinson Relat Disord. 2012;18:494–500.
100. Treglia G, Cason E, Stefanelli A, Cocciolillo F, Di Giuda D, Fagioli G, et al. MIBG scintigraphy in differential diagnosis of parkinsonism: a meta-analysis. Clin Auton Res. 2012;22:43–55.
101. Tateno F, Sakakibara R, Kishi M, Ogawa E, Terada H, Ogata T, et al. Sensitivity and specificity of metaiodobenzylguanidine (MIBG) myocardial accumulation in the diagnosis of Lewy body diseases in a movement disorder clinic. Parkinson Relat Disord. 2011;17:395–7.
102. Kim J-S, Shim Y-S, Song I-U, Yoo J-Y, Kim H-T, Kim Y-I, et al. Cardiac sympathetic denervation and its association with cognitive deficits in Parkinson's disease. Parkinson Relat Disord. 2009;15:706–8.

103. Yoshita M, Arai H, Arai H, Arai T, Asada T, Fujishiro H, et al. Diagnostic accuracy of 123I-meta-iodobenzylguanidine myocardial scintigraphy in dementia with Lewy bodies: a multicenter study. PLoS One. 2015;10:e0120540.
104. Tiraboschi P, Corso A, Guerra UP, Nobili F, Piccardo A, Calcagni ML, et al. 123I-2β-carbomethoxy-3β-(4-iodophenyl)-N-(3-fluoropropyl) nortropane single photon emission computed tomography and 123I-metaiodobenzylguanidine myocardial scintigraphy in differentiating dementia with lewy bodies from other dementias: a comparative study. Ann Neurol. 2016;80:368–78.
105. Obeso JA, Rodriguez-Oroz MC, Stamelou M, Bhatia KP, Burn DJ. The expanding universe of disorders of the basal ganglia. Lancet. 2014;384:523–31.
106. Ryman SG, Poston KL. MRI biomarkers of motor and non-motor symptoms in Parkinson's disease. Parkinson Relat Disord. 2019;73:85–93.
107. Barber TR, Griffanti L, Bradley KM, McGowan DR, Lo C, Mackay CE, et al. Nigrosome 1 imaging in REM sleep behavior disorder and its association with dopaminergic decline. Ann Clin Transl Neurol. 2020;7:26–35.
108. Pyatigorskaya N, Gaurav R, Arnaldi D, Leu-Semenescu S, Yahia-Cherif L, Valabregue R, et al. Magnetic resonance imaging biomarkers to assess substantia nigra damage in idiopathic rapid eye movement sleep behavior disorder. Sleep. 2017;40:zsx149.
109. Zhou L, Li G, Zhang Y, Zhang M, Chen Z, Zhang L, et al. Increased free water in the substantia nigra in idiopathic REM sleep behaviour disorder. Brain. 2021;144:1488–97.
110. Prasuhn J, Prasuhn M, Fellbrich A, Strautz R, Lemmer F, Dreischmeier S, et al. Association of locus coeruleus and substantia nigra pathology with cognitive and motor functions in patients with Parkinson disease. Neurology. 2021;97:e1007–16.
111. Li Y, Wang C, Wang J, Zhou Y, Ye F, Zhang Y, et al. Mild cognitive impairment in de novo Parkinson's disease: a neuromelanin MRI study in locus coeruleus. Mov Disord. 2019;34:884–92.
112. Shams S, Fällmar D, Schwarz S, Wahlund L-O, Van Westen D, Hansson O, et al. MRI of the swallow tail sign: a useful marker in the diagnosis of Lewy body dementia? Am J Neuroradiol. 2017;38:1737–41.
113. Kamagata K, Nakatsuka T, Sakakibara R, Tsuyusaki Y, Takamura T, Sato K, et al. Diagnostic imaging of dementia with Lewy bodies by susceptibility-weighted imaging of nigrosomes versus striatal dopamine transporter single-photon emission computed tomography: a retrospective observational study. Neuroradiology. 2017;59:89–98.
114. Scherfler C, Schwarz J, Antonini A, Grosset D, Valldeoriola F, Marek K, et al. Role of DAT-SPECT in the diagnostic work up of parkinsonism. Mov Disord. 2007;22:1229–38.
115. Benamer HTS, Patterson J, Grosset DG, Booij J, De Bruin K, Van Royen E, et al. Accurate differentiation of parkinsonism and essential tremor using visual assessment of [123I]-FP-CIT SPECT imaging: the [123I]-FP-CIT study group. Mov Disord. 2000;15:503–10.
116. Leenders KL, Palmer AJ, Quinn N, Clark JC, Firnau G, Garnett ES, et al. Brain dopamine metabolism in patients with Parkinson's disease measured with positron emission tomography. J Neurol Neurosurg Psychiatry. 1986;49:853–60.
117. Iranzo A, Lomeña F, Stockner H, Valldeoriola F, Vilaseca I, Salamero M, et al. Decreased striatal dopamine transporter uptake and substantia nigra hyperechogenicity as risk markers of synucleinopathy in patients with idiopathic rapid-eye-movement sleep behaviour disorder: a prospective study. Lancet Neurol. 2010;9:1070–7.
118. Iranzo A, Santamaría J, Valldeoriola F, Serradell M, Salamero M, Gaig C, et al. Dopamine transporter imaging deficit predicts early transition to synucleinopathy in idiopathic rapid eye movement sleep behavior disorder. Ann Neurol. 2017;82:419–28.
119. Arnaldi D, Chincarini A, Hu MT, Sonka K, Boeve B, Miyamoto T, et al. Dopaminergic imaging and clinical predictors for phenoconversion of REM sleep behaviour disorder. Brain. 2021;144:278–87.

120. Kasanuki K, Iseki E, Ota K, Kondo D, Ichimiya Y, Sato K, et al. 123 I-FP-CIT SPECT findings and its clinical relevance in prodromal dementia with Lewy bodies. Eur J Nucl Med Mol Imaging. 2017;44:358–65.
121. Thomas AJ, Donaghy P, Roberts G, Colloby SJ, Barnett NA, Petrides G, et al. Diagnostic accuracy of dopaminergic imaging in prodromal dementia with Lewy bodies. Psychol Med. 2019;49:396–402.
122. Tolosa E, Garrido A, Scholz SW, Poewe W. Challenges in the diagnosis of Parkinson's disease. Lancet Neurol. 2021;20:385–97.
123. Eshuis SA, Jager PL, Maguire RP, Jonkman S, Dierckx RA, Leenders KL. Direct comparison of FP-CIT SPECT and F-DOPA PET in patients with Parkinson's disease and healthy controls. Eur J Nucl Med Mol Imaging. 2009;36:454–62.
124. Thobois S, Prange S, Scheiber C, Broussolle E. What a neurologist should know about PET and SPECT functional imaging for parkinsonism: a practical perspective. Parkinson Relat Disord. 2019;59:93–100.
125. Costa DC, Walker Z, Walker RWH, Fontes FRG. Dementia with Lewy bodies versus Alzheimer's disease: role of dopamine transporter imaging. Mov Disord. 2003;18:S34–8.
126. McKeith I, O'Brien J, Walker Z, Tatsch K, Booij J, Darcourt J, et al. Sensitivity and specificity of dopamine transporter imaging with 123I-FP-CIT SPECT in dementia with Lewy bodies: a phase III, multicentre study. Lancet Neurol. 2007;6:305–13.
127. Thomas AJ, Attems J, Colloby SJ, O'Brien JT, McKeith I, Walker R, et al. Autopsy validation of 123I-FP-CIT dopaminergic neuroimaging for the diagnosis of DLB. Neurology. 2017;88:276–83.
128. Jreige M, Kurian GK, Perriraz J, Potheegadoo J, Bernasconi F, Stampacchia S, et al. The diagnostic performance of functional dopaminergic scintigraphic imaging in the diagnosis of dementia with Lewy bodies: an updated systematic review. Eur J Nucl Med Mol Imaging. 2023;50:1988–2035.
129. Papathanasiou ND, Boutsiadis A, Dickson J, Bomanji JB. Diagnostic accuracy of 123I-FP-CIT (DaTSCAN) in dementia with Lewy bodies: a meta-analysis of published studies. Parkinsonism Relat Disord. 2012;18:225–9.
130. Woo C-W, Chang LJ, Lindquist MA, Wager TD. Building better biomarkers: brain models in translational neuroimaging. Nat Neurosci. 2017;20:365–77.
131. Kashihara K, Ban M, Baba T, Takeda A. Occipital hypometabolism on [18F]-FDG PET scans in patients with idiopathic REM sleep behavior disorder is less prominent than those of patients with Parkinson's disease. Parkinsonism Relat Disord. 2018;46:e72.
132. Fujishiro H, Iseki E, Murayama N, Yamamoto R, Higashi S, Kasanuki K, et al. Diffuse occipital hypometabolism on [18F]-FDG PET scans in patients with idiopathic REM sleep behavior disorder: prodromal dementia with Lewy bodies? Psychogeriatrics. 2010;10:144–52.
133. Wu P, Yu H, Peng S, Dauvilliers Y, Wang J, Ge J, et al. Consistent abnormalities in metabolic network activity in idiopathic rapid eye movement sleep behaviour disorder. Brain. 2014;137:3122–8.
134. Meles SK, Renken RJ, Janzen A, Vadasz D, Pagani M, Arnaldi D, et al. The metabolic pattern of idiopathic REM sleep behavior disorder reflects early-stage Parkinson disease. J Nucl Med. 2018;59:1437–44.
135. Kogan RV, Janzen A, Meles SK, Sittig E, Renken RJ, Gurvits V, et al. Four-year follow-up of [18F] Fluorodeoxyglucose positron emission tomography–based Parkinson's disease–related pattern expression in 20 patients with isolated rapid eye movement sleep behavior disorder shows prodromal progression. Mov Disord. 2021;36:230–5.
136. Meyer PT, Frings L, Rücker G, Hellwig S. 18F-FDG PET in parkinsonism: differential diagnosis and evaluation of cognitive impairment. J Nucl Med. 2017;58:1888–98.
137. Schindlbeck KA, Eidelberg D. Network imaging biomarkers: insights and clinical applications in Parkinson's disease. Lancet Neurol. 2018;17:629–40.

138. Hellwig S, Amtage F, Kreft A, Buchert R, Winz OH, Vach W, et al. [18F] FDG-PET is superior to [123I] IBZM-SPECT for the differential diagnosis of parkinsonism. Neurology. 2012;79:1314–22.
139. Bohnen NI, Minoshima S. FDG-PET and molecular brain imaging in the movement disorders clinic. Neurology. 2012;79:1306–7.
140. Baba T, Hosokai Y, Nishio Y, Kikuchi A, Hirayama K, Suzuki K, et al. Longitudinal study of cognitive and cerebral metabolic changes in Parkinson's disease. J Neurol Sci. 2017;372:288–93.
141. Firbank MJ, Yarnall AJ, Lawson RA, Duncan GW, Khoo TK, Petrides GS, et al. Cerebral glucose metabolism and cognition in newly diagnosed Parkinson's disease: ICICLE-PD study. J Neurol Neurosurg Psychiatry. 2017;88:310–6.
142. Tard C, Demailly F, Delval A, Semah F, Defebvre L, Dujardin K, et al. Hypometabolism in posterior and temporal areas of the brain is associated with cognitive decline in Parkinson's disease. J Parkinsons Dis. 2015;5:569–74.
143. Bohnen NI, Koeppe RA, Minoshima S, Giordani B, Albin RL, Frey KA, et al. Cerebral glucose metabolic features of Parkinson disease and incident dementia: longitudinal study. J Nucl Med. 2011;52:848–55.
144. Gonzalez-Redondo R, García-García D, Clavero P, Gasca-Salas C, García-Eulate R, Zubieta JL, et al. Grey matter hypometabolism and atrophy in Parkinson's disease with cognitive impairment: a two-step process. Brain. 2014;137:2356–67.
145. Garcia-Garcia D, Clavero P, Gasca Salas C, Lamet I, Arbizu J, Gonzalez-Redondo R, et al. Posterior parietooccipital hypometabolism may differentiate mild cognitive impairment from dementia in Parkinson's disease. Eur J Nucl Med Mol Imaging. 2012;39:1767–77.
146. Huang C, Mattis P, Tang C, Perrine K, Carbon M, Eidelberg D. Metabolic brain networks associated with cognitive function in Parkinson's disease. NeuroImage. 2007;34:714–23.
147. Meles SK, Tang CC, Teune LK, Dierckx RA, Dhawan V, Mattis PJ, et al. Abnormal metabolic pattern associated with cognitive impairment in Parkinson's disease: a validation study. J Cereb Blood Flow Metab. 2015;35:1478–84.
148. Mattis PJ, Niethammer M, Sako W, Tang CC, Nazem A, Gordon ML, et al. Distinct brain networks underlie cognitive dysfunction in Parkinson and Alzheimer diseases. Neurology. 2016;87:1925–33.
149. Minoshima S, Foster NL, Sima AAF, Frey KA, Albin RL, Kuhl DE. Alzheimer's disease versus dementia with Lewy bodies: cerebral metabolic distinction with autopsy confirmation. Ann Neurol. 2001;50:358–65.
150. Vander Borght T, Minoshima S, Giordani B, Foster NL, Frey KA, Berent S, et al. Cerebral metabolic difference s in Parkinson's and Alzheimer's disease s matched for dementia severity. J Nucl Med. 1997;38:797–802.
151. Perovnik M, Tomše P, Jamšek J, Tang C, Eidelberg D, Trošt M. Metabolic brain pattern in dementia with Lewy bodies: relationship to Alzheimer's disease topography. NeuroImage: Clinical. 2022;35:103080.
152. Caminiti SP, Sala A, Iaccarino L, Beretta L, Pilotto A, Gianolli L, et al. Brain glucose metabolism in Lewy body dementia: implications for diagnostic criteria. Alzheimers Res Ther. 2019;11:1–14.
153. Rektorová I, Krajcovicova L, Marecek R, Mikl M. Default mode network and extrastriate visual resting state network in patients with Parkinson's disease dementia. Neurodegener Dis. 2012;10:232–7.
154. Dubbelink KTEO, Schoonheim MM, Deijen JB, Twisk JWR, Barkhof F, Berendse HW. Functional connectivity and cognitive decline over 3 years in Parkinson disease. Neurology. 2014;83:2046–53.
155. Baggio H, Segura B, Sala-Llonch R, Marti M, Valldeoriola F, Compta Y, et al. Cognitive impairment and resting-state network connectivity in P arkinson's disease. Hum Brain Mapp. 2015;36:199–212.

156. Lewis SJG, Dove A, Robbins TW, Barker RA, Owen AM. Cognitive impairments in early Parkinson's disease are accompanied by reductions in activity in frontostriatal neural circuitry. J Neurosci. 2003;23:6351–6.
157. Monchi O, Petrides M, Doyon J, Postuma RB, Worsley K, Dagher A. Neural bases of set-shifting deficits in Parkinson's disease. J Neurosci. 2004;24:702–10.
158. Vo A, Sako W, Fujita K, Peng S, Mattis PJ, Skidmore FM, et al. Parkinson's disease-related network topographies characterized with resting state functional MRI. Hum Brain Mapp. 2017;38:617–30.
159. Rommal A, Vo A, Schindlbeck KA, Greuel A, Ruppert MC, Eggers C, et al. Parkinson's disease-related pattern (PDRP) identified using resting-state functional MRI: validation study. Neuroimage: Reports. 2021;1:100026.
160. Buckner RL, Andrews-Hanna JR, Schacter DL. The brain's default network: anatomy, function, and relevance to disease, vol. 1124; 2008. p. 1.
161. Spetsieris PG, Ko JH, Tang CC, Nazem A, Sako W, Peng S, et al. Metabolic resting-state brain networks in health and disease. Proc Natl Acad Sci. 2015;112:2563–8.
162. Greicius MD, Srivastava G, Reiss AL, Menon V. Default-mode network activity distinguishes Alzheimer's disease from healthy aging: evidence from functional MRI. Proc Natl Acad Sci. 2004;101:4637–42.
163. Li X, Li T-Q, Andreasen N, Wiberg MK, Westman E, Wahlund L-O. Ratio of Aβ42/P-tau 181p in CSF is associated with aberrant default mode network in AD. Sci Rep. 2013;3:1–5.
164. Wolters AF, van de Weijer SCF, Leentjens AFG, Duits AA, Jacobs HIL, Kuijf ML. Resting-state fMRI in Parkinson's disease patients with cognitive impairment: a meta-analysis. Parkinsonism Relat Disord. 2018;62:16.
165. Campbell MC, Jackson JJ, Koller JM, Snyder AZ, Kotzbauer PT, Perlmutter JS. Proteinopathy and longitudinal changes in functional connectivity networks in Parkinson disease. Neurology. 2020;94:e718–28.
166. Campbell MC, Koller JM, Snyder AZ, Buddhala C, Kotzbauer PT, Perlmutter JS. CSF proteins and resting-state functional connectivity in Parkinson disease. Neurology. 2015;84:2413–21.
167. Zarifkar P, Kim J, La C, Zhang K, YorkWilliams S, Levine TF, et al. Cognitive impairment in Parkinson's disease is associated with Default Mode Network subsystem connectivity and cerebrospinal fluid Aβ. Parkinsonism Relat Disord. 2021;83:71–8.
168. Barber R, Gholkar A, Scheltens P, Ballard C, McKeith IG, O'Brien JT. Medial temporal lobe atrophy on MRI in dementia with Lewy bodies. Neurology. 1999;52:1153.
169. Campabadal A, Segura B, Junque C, Iranzo A. Structural and functional magnetic resonance imaging in isolated REM sleep behavior disorder: a systematic review of studies using neuroimaging software. Sleep Med Rev. 2021;59:101495.
170. Rahayel S, Postuma RB, Montplaisir J, Bedetti C, Brambati S, Carrier J, et al. Abnormal gray matter shape, thickness, and volume in the motor cortico-subcortical loop in idiopathic rapid eye movement sleep behavior disorder: association with clinical and motor features. Cereb Cortex. 2018;28:658–71.
171. Rahayel S, Tremblay C, Vo A, Zheng YQ, Lehéricy S, Arnulf I, et al. Brain atrophy in prodromal synucleinopathy is shaped by structural connectivity and gene expression. Brain. 2022;145:3162–78.
172. Durcan R, Donaghy P, Osborne C, Taylor J, Thomas AJ. Imaging in prodromal dementia with Lewy bodies: where do we stand? Int J Geriatr Psychiatry. 2019;34:635–46.
173. Sarasso E, Agosta F, Piramide N, Filippi M. Progression of grey and white matter brain damage in Parkinson's disease: a critical review of structural MRI literature. J Neurol. 2021;268:3144–79.
174. He H, Liang L, Tang T, Luo J, Wang Y, Cui H. Progressive brain changes in Parkinson's disease: a meta-analysis of structural magnetic resonance imaging studies. Brain Res. 2020;1740:146847.

175. Whitwell JL, Weigand SD, Shiung MM, Boeve BF, Ferman TJ, Smith GE, et al. Focal atrophy in dementia with Lewy bodies on MRI: a distinct pattern from Alzheimer's disease. Brain. 2007;130:708–19.
176. Acosta-Cabronero J, Cardenas-Blanco A, Betts MJ, Butryn M, Valdes-Herrera JP, Galazky I, et al. The whole-brain pattern of magnetic susceptibility perturbations in Parkinson's disease. Brain. 2017;140:118–31.
177. Joshi D, Prasad S, Saini J, Ingalhalikar M. Role of arterial spin labeling (ASL) images in Parkinson's disease (PD): a systematic review. Acad Radiol. 2022;30:1695.
178. Ryman SG, Shaff N, Dodd A, Nitschke S, Wertz C, Julio K, et al. Reduced and delayed cerebrovascular reactivity in patients with Parkinson's disease. Mov Disord. 2023;38:1262.
179. Roquet D, Sourty M, Botzung A, Armspach J-P, Blanc F. Brain perfusion in dementia with Lewy bodies and Alzheimer's disease: an arterial spin labeling MRI study on prodromal and mild dementia stages. Alzheimers Res Ther. 2016;8:1–13.
180. Nedelska Z, Senjem ML, Przybelski SA, Lesnick TG, Lowe VJ, Boeve BF, et al. Regional cortical perfusion on arterial spin labeling MRI in dementia with Lewy bodies: associations with clinical severity, glucose metabolism and tau PET. NeuroImage: Clinical. 2018;19:939–47.
181. Liu P, Li Y, Pinho M, Park DC, Welch BG, Lu H. Cerebrovascular reactivity mapping without gas challenges. NeuroImage. 2017;146:320–6.
182. Krainik A, Maillet A, Fleury V, Sahin M, Troprès I, Lamalle L, et al. Levodopa does not change cerebral vasoreactivity in Parkinson's disease. Mov Disord. 2013;28:469–75.
183. Al-Bachari S, Parkes LM, Vidyasagar R, Hanby MF, Tharaken V, Leroi I, et al. Arterial spin labelling reveals prolonged arterial arrival time in idiopathic Parkinson's disease. NeuroImage: Clinical. 2014;6:1–8.
184. Pelizzari L, Laganà MM, Rossetto F, Bergsland N, Galli M, Baselli G, et al. Cerebral blood flow and cerebrovascular reactivity correlate with severity of motor symptoms in Parkinson's disease. Ther Adv Neurol Disord. 2019;12:1756286419838354.
185. van der Horn HJ, Vakhtin AA, Julio K, Nitschke S, Shaff NA, Dodd AB, et al. Parkinson's disease cerebrovascular reactivity pattern: a feasibility study. J Cereb Blood Flow Metab. 15:895.
186. Simonsen AH, Kuiperij B, El-Agnaf OMA, Engelborghs S, Herukka S-K, Parnetti L, et al. The utility of α-synuclein as biofluid marker in neurodegenerative diseases: a systematic review of the literature. Biomark Med. 2016;10:19–34.
187. Kang UJ, Boehme AK, Fairfoul G, Shahnawaz M, Ma TC, Hutten SJ, et al. Comparative study of cerebrospinal fluid α-synuclein seeding aggregation assays for diagnosis of Parkinson's disease. Mov Disord. 2019;34:536–44.
188. Fairfoul G, McGuire LI, Pal S, Ironside JW, Neumann J, Christie S, et al. Alpha-synuclein RT-Qu IC in the CSF of patients with alpha-synucleinopathies. Ann Clinic Transl Neurol. 2016;3:812–8.
189. Shahnawaz M, Tokuda T, Waragai M, Mendez N, Ishii R, Trenkwalder C, et al. Development of a biochemical diagnosis of Parkinson disease by detection of α-synuclein misfolded aggregates in cerebrospinal fluid. JAMA Neurol. 2017;74:163–72.
190. Groveman BR, Orrù CD, Hughson AG, Raymond LD, Zanusso G, Ghetti B, et al. Rapid and ultra-sensitive quantitation of disease-associated α-synuclein seeds in brain and cerebrospinal fluid by αSyn RT-QuIC. Acta Neuropathol Commun. 2018;6:1–10.
191. Bargar C, Wang W, Gunzler SA, LeFevre A, Wang Z, Lerner AJ, et al. Streamlined alpha-synuclein RT-QuIC assay for various biospecimens in Parkinson's disease and dementia with Lewy bodies. Acta Neuropathol Commun. 2021;9:1–13.
192. Wennström M, Surova Y, Hall S, Nilsson C, Minthon L, Boström F, et al. Low CSF levels of both α-synuclein and the α-synuclein cleaving enzyme neurosin in patients with synucleinopathy. PLoS One. 2013;8:e53250.
193. Galvin JE. Interaction of alpha-synuclein and dopamine metabolites in the pathogenesis of Parkinson's disease: a case for the selective vulnerability of the substantia nigra. Acta Neuropathol. 2006;112:115–26.

194. Wise RM, Wagener A, Fietzek UM, Klopstock T, Mosharov EV, Zucca FA, et al. Interactions of dopamine, iron, and alpha-synuclein linked to dopaminergic neuron vulnerability in Parkinson's disease and neurodegeneration with brain iron accumulation disorders. Neurobiol Dis. 2022;175:105920.
195. Pereira JB, Kumar A, Hall S, Palmqvist S, Stomrud E, Bali D, et al. DOPA decarboxylase is an emerging biomarker for parkinsonian disorders including preclinical Lewy body disease. Nat Aging. 2023;3:1201–9.
196. Rutledge J, Lehallier B, Zarifkar P, Losada PM, Shahid-Besanti M, Western D, et al. Comprehensive proteomics of CSF, plasma, and urine identify DDC and other biomarkers of early Parkinson's disease. Acta Neuropathol. 2024;147:1–18.
197. Lourenco MV, Ribeiro FC, Santos LE, Beckman D, Melo HM, Sudo FK, et al. Cerebrospinal fluid neurotransmitters, cytokines, and chemokines in Alzheimer's and Lewy body diseases. J Alzheimers Dis. 2021;82:1067–74.
198. Morimoto S, Takao M, Hatsuta H, Nishina Y, Komiya T, Sengoku R, et al. Homovanillic acid and 5-hydroxyindole acetic acid as biomarkers for dementia with Lewy bodies and coincident Alzheimer's disease: An autopsy-confirmed study. PLoS One. 2017;12:e0171524.
199. Goto R, Kurihara M, Kameyama M, Komatsu H, Higashino M, Hatano K, et al. Correlations between cerebrospinal fluid homovanillic acid and dopamine transporter SPECT in degenerative parkinsonian syndromes. J Neural Transm. 2023;130:513–20.
200. Irwin DJ, Fedler J, Coffey CS, Caspell-Garcia C, Kang JH, Simuni T, et al. Evolution of Alzheimer's disease cerebrospinal fluid biomarkers in early Parkinson's disease. Ann Neurol. 2020;88:574–87.
201. Howard E, Irwin DJ, Rascovsky K, Nevler N, Shellikeri S, Tropea TF, et al. Cognitive profile and markers of Alzheimer disease–type pathology in patients with Lewy body dementias. Neurology. 2021;96:e1855–64.
202. Xiang C, Cong S, Tan X, Ma S, Liu Y, Wang H, et al. A meta-analysis of the diagnostic utility of biomarkers in cerebrospinal fluid in Parkinson's disease. npj Parkinson's Dis. 2022;8:165.
203. Bäckström DC, Domellöf ME, Linder J, Olsson B, Öhrfelt A, Trupp M, et al. Cerebrospinal fluid patterns and the risk of future dementia in early, incident Parkinson disease. JAMA Neurol. 2015;72:1175–82.
204. Magdalinou N, Paterson R, Schott J, Fox N, Mummery C, Blennow K, et al. A panel of nine cerebrospinal fluid biomarkers may identify patients with atypical parkinsonian syndromes. J Neurol Neurosurg Psychiatry. 2015;86:1240–7.
205. Hansson O, Janelidze S, Hall S, Magdalinou N, Lees AJ, Andreasson U, et al. Blood-based NfL: a biomarker for differential diagnosis of parkinsonian disorder. Neurology. 2017;88:930–7.
206. Morris HR. Blood based biomarkers for movement disorders. Acta Neurol Scand. 2022;146:353–61.
207. Lin C-H, Li C-H, Yang K-C, Lin F-J, Wu C-C, Chieh J-J, et al. Blood NfL: a biomarker for disease severity and progression in Parkinson disease. Neurology. 2019;93:e1104–11.
208. Aamodt WW, Waligorska T, Shen J, Tropea TF, Siderowf A, Weintraub D, et al. Neurofilament light chain as a biomarker for cognitive decline in Parkinson disease. Mov Disord. 2021;36:2945–50.
209. Pilotto A, Imarisio A, Carrarini C, Russo M, Masciocchi S, Gipponi S, et al. Plasma neurofilament light chain predicts cognitive progression in prodromal and clinical dementia with Lewy bodies. J Alzheimers Dis. 2021;82:913–9.
210. Delaby C, Alcolea D, Carmona-Iragui M, Illán-Gala I, Morenas-Rodríguez E, Barroeta I, et al. Differential levels of neurofilament light protein in cerebrospinal fluid in patients with a wide range of neurodegenerative disorders. Sci Rep. 2020;10:9161.
211. Abu-Rumeileh S, Capellari S, Stanzani-Maserati M, Polischi B, Martinelli P, Caroppo P, et al. The CSF neurofilament light signature in rapidly progressive neurodegenerative dementias. Alzheimers Res Ther. 2018;10:1–11.

212. Luo X, Hou L, Shi H, Zhong X, Zhang Y, Zheng D, et al. CSF levels of the neuronal injury biomarker visinin-like protein-1 in Alzheimer's disease and dementia with Lewy bodies. J Neurochem. 2013;127:681–90.
213. Babić Leko M, Borovečki F, Dejanović N, Hof PR, Šimić G. Predictive value of cerebrospinal fluid visinin-like protein-1 levels for Alzheimer's disease early detection and differential diagnosis in patients with mild cognitive impairment. J Alzheimers Dis. 2016;50:765–78.
214. Bereczki E, Francis PT, Howlett D, Pereira JB, Höglund K, Bogstedt A, et al. Synaptic proteins predict cognitive decline in Alzheimer's disease and Lewy body dementia. Alzheimers Dement. 2016;12:1149–58.
215. Hall S, Janelidze S, Zetterberg H, Brix B, Mattsson N, Surova Y, et al. Cerebrospinal fluid levels of neurogranin in parkinsonian disorders. Mov Disord. 2020;35:513–8.
216. Wang T, Shi C, Luo H, Zheng H, Fan L, Tang M, et al. Neuroinflammation in Parkinson's disease: triggers, mechanisms, and immunotherapies. Neuroscientist. 2022;28:364–81.
217. Qu Y, Li J, Qin Q, Wang D, Zhao J, An K, et al. A systematic review and meta-analysis of inflammatory biomarkers in Parkinson's disease. npj Parkinson's Dis. 2023;9:18.
218. King E, Thomas A. Systemic inflammation in Lewy body diseases: a systematic review. Alzheimer Dis Assoc Disord. 2017;31:346–56.
219. King E, O'Brien JT, Donaghy P, Morris C, Barnett N, Olsen K, et al. Peripheral inflammation in prodromal Alzheimer's and Lewy body dementias. J Neurol Neurosurg Psychiatry. 2018;89:339–45.
220. Costantini E, Carrarini C, Calisi D, De Rosa M, Simone M, Di Crosta A, et al. Search in the periphery for potential inflammatory biomarkers of dementia with Lewy bodies and Alzheimer's disease. J Alzheimer Dis. 1:1–12.
221. Morenas-Rodríguez E, Alcolea D, Suárez-Calvet M, Muñoz-Llahuna L, Vilaplana E, Sala I, et al. Different pattern of CSF glial markers between dementia with Lewy bodies and Alzheimer's disease. Sci Rep. 2019;9:7803.
222. Ruffmann C, Bengoa-Vergniory N, Poggiolini I, Ritchie D, Hu M, Alegre-Abarrategui J, et al. Detection of alpha-synuclein conformational variants from gastro-intestinal biopsy tissue as a potential biomarker for Parkinson's disease. Neuropathol Appl Neurobiol. 2018;44:722–36.
223. Raetz CRH, Whitfield C. Lipopolysaccharide endotoxins. Annu Rev Biochem. 2002;71:635.
224. Lin A, Zheng W, He Y, Tang W, Wei X, He R, et al. Gut microbiota in patients with Parkinson's disease in southern China. Parkinsonism Relat Disord. 2018;53:82–8.
225. Murros KE, Huynh VA, Takala TM, Saris PEJ. Desulfovibrio bacteria are associated with Parkinson's disease. Front Cell Infect Microbiol. 2021;11:652617.
226. Moors T, Paciotti S, Chiasserini D, Calabresi P, Parnetti L, Beccari T, et al. Lysosomal dysfunction and α-Synuclein aggregation in Parkinson's disease: diagnostic links. Mov Disord. 2016;31:791–801.
227. Creese B, Bell E, Johar I, Francis P, Ballard C, Aarsland D. Glucocerebrosidase mutations and neuropsychiatric phenotypes in Parkinson's disease and Lewy body dementias: review and meta-analyses. Am J Med Genet B Neuropsychiatr Genet. 2018;177:232–41.
228. Parnetti L, Balducci C, Pierguidi L, De Carlo C, Peducci M, D'Amore C, et al. Cerebrospinal fluid β-glucocerebrosidase activity is reduced in dementia with Lewy bodies. Neurobiol Dis. 2009;34:484–6.
229. Parnetti L, Paciotti S, Eusebi P, Dardis A, Zampieri S, Chiasserini D, et al. Cerebrospinal fluid β-glucocerebrosidase activity is reduced in Parkinson's disease patients. Mov Disord. 2017;32:1423–31.
230. Parnetti L, Chiasserini D, Persichetti E, Eusebi P, Varghese S, Qureshi MM, et al. Cerebrospinal fluid lysosomal enzymes and alpha-synuclein in Parkinson's disease. Mov Disord. 2014;29:1019–27.
231. Blesa J, Trigo-Damas I, Quiroga-Varela A, Jackson-Lewis VR. Oxidative stress and Parkinson's disease. Front Neuroanat. 2015;9:91.
232. Jenner P. Oxidative stress in Parkinson's disease. Ann Neurol. 2003;53:S26–38.

233. Wei Z, Li X, Li X, Liu Q, Cheng Y. Oxidative stress in Parkinson's disease: a systematic review and meta-analysis. Front Mol Neurosci. 2018;11:236.
234. Schapira AHV. Mitochondria in the aetiology and pathogenesis of Parkinson's disease. Lancet Neurol. 2008;7:97–109.
235. Mao P. Oxidative stress and its clinical applications in dementia. J Neurodegener Dis. 2013;2013:319898.
236. Schultz K, Wiehager S, Nilsson K, Nielsen JE, Lindquist SG, Hjermind LE, et al. Reduced CSF CART in dementia with Lewy bodies. Neurosci Lett. 2009;453:104–6.
237. Boström F, Hansson O, Gerhardsson L, Lundh T, Minthon L, Stomrud E, et al. CSF Mg and Ca as diagnostic markers for dementia with Lewy bodies. Neurobiol Aging. 2009;30:1265–71.
238. Molina J, Gómez P, Vargas C, Ortiz S, Pérez-Rial S, Urigüen L, et al. Neurotransmitter amino acid in cerebrospinal fluid of patients with dementia with Lewy bodies. J Neural Transm. 2005;112:557–63.
239. Van Steenoven I, Koel-Simmelink MJ, Vergouw LJ, Tijms BM, Piersma SR, Pham TV, et al. Identification of novel cerebrospinal fluid biomarker candidates for dementia with Lewy bodies: a proteomic approach. Mol Neurodegener. 2020;15:1–15.
240. Boiten WA, van Steenoven I, Xiao M-F, Worley PF, Noli B, Cocco C, et al. Pathologically decreased CSF levels of synaptic marker NPTX2 in DLB are correlated with levels of alpha-synuclein and VGF. Cells. 2020;10:38.
241. Bousiges O, Blanc F. Biomarkers of dementia with Lewy bodies: differential diagnostic with Alzheimer's disease. IJMS. 2022;23:6371.
242. Zimmermann M, Brockmann K. Blood and cerebrospinal fluid biomarkers of inflammation in Parkinson's disease. J Parkinsons Dis. 2022;12:S183–200.
243. Lehéricy S, Bardinet E, Poupon C, Vidailhet M, François C. 7 tesla magnetic resonance imaging: A closer look at substantia nigra anatomy in Parkinson's disease. Mov Disord. 2014;29:1574-1581

Fluid and Imaging Markers of Huntington's Disease

Mena Farag, Rachael I. Scahill, Edward J. Wild, and Sarah J. Tabrizi

Abstract Huntington's disease (HD) is an autosomal dominant neurodegenerative condition characterised by motor, cognitive and neuropsychiatric symptoms, resulting from an expanded CAG trinucleotide repeat in the huntingtin (*HTT*) gene, leading to the accumulation of toxic mutant huntingtin (mHTT) protein. Despite the established genetic aetiology, the pathophysiology of HD is complex, and the lack of disease-modifying treatments accentuates the pressing need for effective therapeutic strategies. Recent advancements in HD research are increasingly directed towards gene-targeted therapies such as antisense oligonucleotides and small molecule splicing modifiers, aimed at reducing levels of toxic mHTT protein. Essential for the success of these emerging therapies is the need for biomarkers that reliably and robustly correlate with disease progression. Effects can be determined via biomarkers to assess target engagement and response to intervention at a cellular level (e.g. via biofluids) and at a macrostructural level (e.g. via imaging).

Herein, this chapter provides a comprehensive summary of the current landscape of fluid and imaging biomarkers in HD and offers insights into the future trajectory of biomarker research in this era of HTT-lowering therapeutics.

Keywords Huntington's disease · Biomarkers · Cerebrospinal fluid · Plasma · Neuroimaging

M. Farag · R. I. Scahill · E. J. Wild · S. J. Tabrizi (✉)
Huntington's Disease Centre, Department of Neurodegenerative Disease, UCL Queen Square Institute of Neurology, London, UK
e-mail: s.tabrizi@ucl.ac.uk

R. Tarawneh (ed.), *Biomarkers of Neurodegenerative Disorders*,
https://doi.org/10.1007/978-3-031-80849-4_11

1 Introduction

1.1 Huntington's Disease

Huntington's disease is an autosomal dominant neurodegenerative disease caused by an expanded cytosine adenine guanine (CAG) trinucleotide repeat in exon 1 of the huntingtin gene (*HTT*) located on the short arm of chromosome 4, which encodes the ubiquitously expressed HTT protein [1]. Pathogenesis in HD arises from the accumulation of toxic mutant huntingtin (mHTT) protein, which contains an abnormally long polyglutamine tract that corresponds to the CAG expansion [2]. The expression of mHTT results in the formation of soluble protein oligomers and insoluble aggregates that cause progressive dysfunction of neuronal physiology and death of neurons. Clinically, HD manifests as a triad of progressive motor dysfunction, cognitive decline and neuropsychiatric disturbance, with death typically within 20 years from clinical onset [2].

The well-established genetic basis of HD enables predictive and genetic testing for the condition. Although HD can affect all populations, the condition occurs at much higher frequencies among individuals of European ancestry [2]. In the UK, the prevalence of HD was initially estimated to be 5.4 per 100,000 in 1990 but more recent work in 2010 suggested it was as high as 12.3 per 100,000 [3]. In Canada, the estimated prevalence of HD in the general population was reported at 13.7 per 100,000 [4].

To date, there are currently no approved disease-modifying treatments to delay the onset or slow progression of HD [5]. Management of HD integrates a multidisciplinary approach that combines supportive and symptomatic interventions directed towards alleviating behavioural, mood and motor symptoms [6].

Since the discovery of the gene in 1993 [1], significant strides in research have led to progress in the pursuit of disease-modifying treatments for HD. These include the development of antisense oligonucleotides (ASOs) targeting *HTT* messenger ribonucleic acid (mRNA) via intrathecal injection (ClinicalTrials.gov Identifiers: NCT03761849; NCT05686551), HTT-lowering via oral splicing modulators (ClinicalTrials.gov Identifier: NCT05358717) and direct intraparenchymal striatal delivery of HTT-lowering gene therapy via viral vectors (ClinicalTrials.gov Identifier: NCT05243017). The ability to objectively measure the effect of disease-modifying treatments for HD is crucial to the success of clinical trials. Effects can be determined via biomarkers to assess target engagement and response to intervention at a cellular level (e.g. via biofluids) and at a macrostructural level (e.g. via imaging). Herein, this chapter will provide a current exploration of fluid and imaging biomarkers in HD and provide insights into the future trajectory of biomarker research in this era of HTT-lowering therapeutics.

1.2 Genetics

As a monogenic neurodegenerative condition with autosomal dominant inheritance, the offspring of an affected individual with HD carry a 50% probability of inheriting the disease-causing allele. An expansion of ≥40 CAG repeats at the *HTT* locus on the short arm (p) of chromosome 4 (4p16.3) represents fully penetrant mutations, resulting in affected HD gene expanded (HDGE) individuals who invariably develop HD within a typical lifespan [2]. Individuals whose CAG repeat length falls within the range of 36–39 exhibit reduced penetrance and are at increased risk of developing HD. CAG repeat lengths within the range of 27–35 represent intermediate alleles, and while this falls below the affected CAG range for the individual, this has implications for offspring due to the potential for CAG repeat expansion with germline instability [2].

Genetic anticipation is a phenomenon observed in HD, whereby expansion of the unstable CAG repeat results in increasing severity of disease and earlier manifestation of symptoms in successive generations. Anticipation in HD is predominantly associated with paternal transmission and the majority of juvenile-onset cases of HD are a product of male intergenerational transmissions [7, 8].

The length of the CAG repeat expansion is pivotal as a significant determinant of the disease course and phenotypic variation in HD, correlating positively with disease severity while exhibiting an inverse relationship with the age of disease onset [9, 10] and survival after diagnosis [11]. However, the length of the CAG repeat expansion does not exclusively dictate age at onset of HD. There are additional modifying factors that exert influence on the pathogenic process.

Over the course of an affected individual's lifespan, the pathogenic polyglutamine length exhibits an intrinsic inclination towards instability, often in regions of neuropathological involvement in HD, notably the striatum [12]. Somatic expansion significantly influences age at onset of HD with greater increases in CAG repeat length associated with earlier onset of disease [13]. Genetic modifiers implicated in DNA maintenance and repair mechanisms can alter the trajectory of HD and modify the age at onset of HD [14].

Genome-wide association studies (GWAS) in HD have identified at least six DNA maintenance genes located on various loci, such as chromosomes 2 (PMS1), 5 (MSH3/DHFR), 7 (PMS2), 15 (FAN1) and 19 (LIG1) [15]. These genes associated with DNA repair are additional drivers for the timing of onset of HD, influencing both the rate of CAG repeat expansion within target cells and the mechanism of toxicity by which somatically expanded repeats induce cellular damage [15]. DNA repair pathways have since emerged as potential novel targets within the therapeutic landscape of HD. For example, MSH3 lowering through MSH3-targeting ASOs has been demonstrated to effectively halt somatic expansion in medium spiny neurons derived from induced pluripotent stem cells of HD patients [16].

1.3 Pathophysiology

The pathophysiology of HD is multifaceted, involving a combination of genetic, molecular and cellular abnormalities that culminate in progressive neurodegeneration and manifestation of symptoms. Animal study models, especially mice, have played a pivotal role in uncovering disrupted pathways and providing crucial insights into the pathogenic mechanisms underlying HD [17].

Fragmentation of *HTT* is a critical early step in the pathogenic process of HD, which results in the formation of the pathogenic exon1 HTT protein due to incomplete splicing [2, 18]. When the expanded polyglutamine sequence falls within the pathogenic range, the mHTT protein has a propensity to aggregate, leading to the formation of toxic protein aggregates in selective neurons [19]. This induces neuronal dysfunction and death, which underlies the clinical manifestations in HD [2]. Immunohistochemistry performed on post-mortem brain tissue from HD patients revealed that mHTT forms aggregates within neuronal intranuclear inclusions and dystrophic neurites in the striatum and cortex [20].

While HD is primarily driven by a toxic gain-of-function due to polyglutamine (polyQ) expansion in the *HTT* gene, evidence suggests that the loss of normal HTT function also significantly contributes to disease pathophysiology [21]. Wild-type HTT protein interacts with various effector proteins and is involved in essential cellular processes, including nuclear import, transcriptional regulation, apoptotic signalling and synaptic function [21]. Understanding the extent to which cells can tolerate a reduction in normal *HTT* expression is important, especially with non-selective therapeutic approaches lowering both wild-type and mHTT proteins. This requires balancing the benefits of reducing mHTT with the risks associated with decreasing normal HTT protein.

Energy metabolism defects are implicated in the degeneration of selective neurons in the pathogenesis of HD via mechanisms of excitotoxic processes, oxidative damage and altered regulation of genes [22]. It is hypothesised that mHTT protein and fragmentation initiate a complex cascade of intricate molecular processes and altered regulation of genes, resulting in abnormal transcriptional regulation and mitochondrial dysfunction [23]. Nevertheless, the question of whether these metabolic aberrations initiate cell death pathways or occur as a result of another cellular defect remains elusive, and the mechanism by which mHTT triggers these sequelae still remains unclear.

In addition to neurodegeneration, the pathophysiological landscape of HD also encompasses immune activation, induced by mHTT, with upregulation of cytokines in the innate immune system both centrally and peripherally [24]. Immunohistochemistry on post-mortem brain tissue from HD patients and controls revealed the presence of activated microglia in the neostriatum, globus pallidus, adjoining white matter and cortex of HD brains, but not controls [25]. This observation implicates an association between the presence of activated microglia and the neurodegenerative processes affecting neurons in HD, thereby highlighting activated microglia as a pathological correlate in the pathogenesis of HD.

The development and identification of biomarkers in HD are intricately linked to the pathophysiology of the disease. For example, being able to measure and quantify mHTT protein aggregates in biofluids, such as cerebrospinal fluid (CSF), or through imaging methods with positron emission tomography (PET), can offer valuable insights into the natural history of the disease, degree of severity and response to therapeutics.

1.4 Clinical Features

Clinically, HD manifests as a triad of motor disorder, cognitive impairment and neuropsychiatric symptoms with age at onset varying, though usually occurring in the fourth decade of life [26]. In approximately 5%, symptoms of HD appear before the age of twenty and are classified as juvenile-onset HD (JoHD) [27].

Chorea is the most common motor symptom in adult patients, whereas parkinsonism is more common in juvenile patients [2, 26]. As the disease advances, chorea tends to subside, while dystonia and parkinsonism become prevalent [2, 26].

Motor signs of HD can be assessed using the Unified Huntington's Disease Rating Scale (UHDRS) Total Motor Score (TMS) [28]. This is a standardised clinical assessment tool that incorporates rating scales to evaluate motor features of HD such as ocular pursuit movements, saccade initiation and velocity, speech, tongue protrusion, coordination, bradykinesia, tone, the presence of chorea and/or dystonia, gait and postural reflexes [28]. Eliciting motor signs is crucial for obtaining vital information about clinical status and staging. Subtle motor abnormalities have been detected in premanifest HDGE groups in cross-sectional [29, 30] and longitudinal [31–33] observational studies. The HDGE group in the HD Young Adult Study (HD-YAS), at baseline, had a median UHDRS TMS score of 0 [34], within ranges reported for genotype-negative controls [35]. This indicates the absence of HD-related motor signs in this far-from-onset HDGE cohort who were approximately 24 years before predicted clinical motor diagnosis.

While the UHDRS-TMS holds potential as a clinical endpoint and outcome measure in both clinical trials and observational studies, the scale is susceptible to inter-rater and intra-rater variability in ratings. Accounting for this predisposition, the motor phenotype working group of the European Huntington's Disease Network (EHDN) developed online certification and training for the UHDRS-TMS [36].

Features of cognitive impairment in HD include difficulties in maintaining attention, organisation and planning skills, emotion recognition, disinhibition resulting in impulsivity, socially inappropriate behaviour and lack of insight into deficits [2, 26]. Early cognitive changes can occur before onset of motor signs in premanifest HDGE groups [30, 37, 38].

Neuropsychiatric features such as apathy, depression and irritability are common and can result in functional decline and significant disability, compared to motor symptoms [2, 26]. Severe psychiatric disturbances, including affective disorder or psychosis with delusions and hallucinations, are less prevalent in HD but nonetheless still occur [39]. The risk of suicide-related mortality in HD is markedly higher than that of the general population. Incidence rates of suicide derived from Enroll-HD, a global longitudinal, observational study of HD, revealed an incidence rate of 72 suicides per 100,000 person-years among HDGE individuals, compared to a lower incidence rate of 8 suicides per 100,000 person-years among non-HDGE individuals [40]. Suicide, a preventable cause of early mortality in HD, was the third most prevalent cause of death of individuals with manifest HD, constituting up to 6.6% of the overall mortality [41].

1.5 Disease Staging

Disease staging in HD can be assessed using the Shoulson-Fahn system, which uses the UHDRS Total Functional Capacity (TFC) score to rank independence from stage 1 (full to near-full functional capacity) to 5 (total dependence on care) [42].

The 'prodromal' stage of HD is characterised by the emergence of subtle motor signs with UHDRS TMS diagnostic confidence level (DCL) 2 or 3 [28]. DCL 2 suggests motor abnormalities that may be signs of HD with 50–89% confidence and DCL 3 suggests motor abnormalities that are likely signs of HD with 90–98% confidence. In contrast, DCL 4 is suggestive of unequivocal motor signs of HD with ≥99% confidence [28].

While the staging of HD has traditionally focused on overt clinical phenotypes, this approach fails to account for underlying disease pathobiology and biomarkers in HD that become apparent decades before functional decline. The HD Integrated Staging System (HD-ISS) has since been developed, which incorporates an evidence-based research framework and staging that spans the natural course and progression of HD [43].

The HD-ISS comprises four stages with defined criteria [43]. HD-ISS stage 0 integrates the genetic definition of HD, encompassing premanifest HDGE individuals with ≥40 CAG repeats, signifying full penetrance for developing the disease. At stage 0, no detectable pathological changes are observed. At stage 1, measurable biomarkers of pathogenesis are evident, with caudate and putamen volume landmarks serving as imaging markers of pathogenesis. At stage 2, subtle clinical signs or symptoms are evident and incorporate the UHDRS TMS for motor signs and/or deficits on the Symbol Digit Modalities Test (SDMT) for cognitive changes necessary to fulfil criteria for this stage. At stage 3, functional changes are evident, and this stage is stratified into mild, moderate and severe, as defined by the UHDRS TFC and Independence Scale [43].

The premanifest stage of HD, prior to the emergence of motor signs, represents an ideal and opportune window to initiate disease-modifying treatments to delay the onset of symptoms or slow the rate of disease progression [30]. Thus, identifying robust, sensitive and stable fluid and imaging biomarkers capable of detecting biological changes in premanifest HDGE cohorts is required to direct and inform intervention studies investigating disease-modifying treatments for HD.

2 Fluid Biomarkers

2.1 *Prerequisites*

When evaluating fluid biomarkers in the HD biomarker field, certain prerequisites must be considered. First, a robust biomarker should be specific, accurately reflecting the presence and progression of HD, distinguishing affected HDGE groups at least from control groups and ideally from other neurodegenerative diseases. Although the CAG repeat expansion in HD is considered a definitive trait biomarker, it alone does not convey information about disease state or severity [44]. Ideally, the biomarker should differentiate between different stages of HD and aid in prognosis.

Second, the biomarker should be sensitive and capable of accurately identifying changes in disease stage or response to disease-modifying treatments. Essential for the success of future prevention trials are biomarkers that reliably correlate with disease progression and can demonstrate target engagement and efficacy, i.e. response to disease-modifying therapeutics [5].

Third, the biomarker should be readily accessible and minimally invasive, cost-effective, reproducible and consistently measurable in biofluids such as blood, CSF and saliva. In this section, we discuss key fluid biomarkers in the field of HD.

2.2 *Peripheral Blood*

Peripheral blood sampling offers the advantages of easy accessibility for precise and reliable quantification of biomarkers [44]. Plasma can be processed in large batches; modern ultrasensitive assays have quantification thresholds low enough to detect substances derived from the central nervous system (CNS) in blood samples and multiplex assays enable quantification of multiple biomarkers of interest from one sample [45, 46].

2.2.1 Neurofilament Light

Numerous substances have been identified in blood that are altered in HD (Table 1). However, only neurofilament light (NfL) protein has consistently demonstrated clear separation from controls, association with clinical measures and prognostic utility. NfL, a protein in the axonal cytoskeleton, is released from damaged axons into both blood and CSF and serves as a proxy biomarker of neuronal injury. It cannot distinguish between central and peripheral neuroaxonal damage, nor between causes, so the context in which it is measured, and its clinical and other biomarker correlates are crucial to consider.

Table 1 Peripheral blood biomarkers in Huntington's disease

Biomarker	Marker	Method/assay	Study	Number	Direction	Correlation				
						DBS	TMS	TFC	Cognitive	Imaging
NfL (plasma)	Axonal damage	NF-Light Assay (UmanDiagnostics, Umeå, Sweden); (Simoa; Quanterix Corp, Boston, MA)	[47]	$n = 298$	M > P > C	Yes (CAP)	Yes	Yes	Yes (SDMT, SW)	Yes: smaller caudate, putamen, grey matter and white matter volumes
NfL (plasma)	Axonal damage	1-Plex Simoa Kit (NF-Light, Quanterix)	[48]	$n = 80$	M > P > C	–	Yes	Yes	Yes (SDMT, SC, SW, VFC) after adjusting for age and CAG_n	Yes: grey matter and caudate volumes (after adjusting for age and CAG_n) and whole-brain volume (after adjusting for age)
NfL (plasma)	Axonal damage	Simoa HD-1 Analyser (Quanterix); Neurology B-Plex A	[49]	$n = 74$	M > P > C	–	Yes (rate of change) after adjusting for age and CAG_n	No (rate of change) after adjusting for age and CAG_n	No	No: (rate of change) after adjusting for age and CAG_n
NfL (plasma)	Axonal damage	Simoa HD-1 Analyser (Quanterix); Neurology 4-Plex A	[34]	$n = 130$	P > C	–	–	–	–	–
NfL (plasma)	Axonal damage	(Simoa, Norcross, GA, USA); (Quanterix, Billerica, MA, USA)	[50]	$n = 83$	M > P > C	Yes	Yes	Yes	Yes (SDMT, SW)	–
NfL (plasma)	Axonal damage	Simoa Human NF-Light Advantage Kit (Simoa, Quanterix) SR-X Biomarker detection system	[51]	$n = 35$	N/A (M only)	Yes	Yes	Yes	Yes (UHDRS-Cogscore, SDMT, SC, SW, SI)	Yes: temporo-occipital increases in intracortical mean diffusivity (microstructural; T1-MRI and DTI)

NfL (plasma)	Axonal damage	(MSD; Rockville, MD) R-Plex Assay (Cat. No. F217X)	[52]	$n = 112$	M > P > C	Yes (PIN-YTO – P)	Yes (within PIN score, P)	–	Yes (SDMT within PIN score, P)	–
NfH (plasma)	Axonal damage	ELISA (SMI35; Sternberger Monoclonals Inc., Lutherville, USA)	[53]	$n = 117$	No significant difference between C, P or M	No (CAG_n)	–	–	–	–
GFAP (plasma)	Astrocyte reactivity	(Simoa, Norcross, GA, USA); (Quanterix, Billerica, MA, USA)	[50]	$n = 83$	M > P > C	Yes	Yes	Yes	No (SDMT, SW)	–
TDP-43 (plasma)	TDP-43 pathology	TDP-43 Advantage Kit (Catalog No. 103293); (Simoa; Quanterix, Lexington, MA, USA) SR-X	[54]	$n = 36$	N/A – no controls	No	No	No	No (SDMT, SW)	Yes: cortical thinning, microstructural degeneration involving frontal and anterior temporal regions
BDNF (plasma)	Neuropeptide	ELISA (Emax Immunoassay System, Promega, Madison, WI, USA)	[55]	$n = 38$	C > M	–	–	–	–	–
BDNF (plasma)	Neuropeptide	ELISA (Cat. No. SEA011Hu; Cloud-Clone Corp., Houston, Tx, USA)	[56]	$n = 82$	C > M	–	Yes (advanced)	Yes (early-stage)	No	–
BDNF (plasma)	Neuropeptide	Simoa Human BDNF Discovery Kit (Quanterix, USA); HD-1 Analyser (Quanterix, USA)	[57]	$n = 80$	No significant difference between C, P or M	–	No	No	Yes (VFC)	No

(continued)

Table 1 (continued)

Biomarker	Marker	Method/assay	Study	Number	Direction	Correlation				
						DBS	TMS	TFC	Cognitive	Imaging
BDNF (serum)	Neuropeptide	BDNF Emax Immunoassay ELISA Kit (Promega, Madison WI, USA)	[58]	$n = 84$	C > M	–	Yes	No	Yes (UHDRS II)	–
BDNF (serum)	Neuropeptide	BDNF Emax Immunoassay ELISA Kit (Promega, Madison WI, USA)	[59]	$n = 73$	High variability across all groups	–	–	No	–	–
GDNF (plasma)	Neuropeptide	ELISA (Emax Immunoassay System, Promega, Madison, WI, USA)	[55]	$n = 38$	C > M	–	–	–	–	–
Oxytocin (plasma)	Neuropeptide	ELISA (Enzo Life Sciences, Rome, Italy)	[60]	$n = 23$	No significant difference between C or M	–	–	–	–	–
Oxytocin (plasma)	Neuropeptide	ELISA (ADI-900-153A, Enzo Life Sciences, Farmingdale, NY)	[61]	$n = 55$	No significant difference between C, P or M	–	No	–	Yes (SDMT, VFC) for P only	–
Oxytocin (plasma)	Neuropeptide	Radioimmunoassay (RIA, Phoenix Pharmaceuticals, California, USA)	[62]	$n = 146$	No significant difference between C, P or M	No (CAP)	No	–	No (MMSE, MoCA)	–

NGF (plasma)	Neuropeptide	ELISA (Emax Immunoassay System, Promega, Madison, WI, USA)	[55]	$n = 38$	C > M	–	–	–	–	–
8-OHdG (leucocyte)	Oxidative stress	LCECA	[63]	$n = 77$	Prodromal > C	Yes (CAP)	–	–	–	–
8-OHdG (leucocyte)	Oxidative stress	HPLC-ECD [64]	[65]	$n = 52$	M > C	No	No	No	–	–
8-oxoGua (serum)	Oxidative stress	ELISA (Assay Designs, Ann Arbor, MI)	[66]	$n = 28$	M > C	–	–	No	–	–
MDA (plasma)	Oxidative stress	HPLC (Agarwal and Chase, 2002)	[65]	$n = 52$	M > C	No	Yes	Yes (IS)	–	–
MDA (serum)	Oxidative stress	Bioxytech LPO-586 Colorimetric Assay (Biostat, Stockport, Cheshire, UK)	[67]	$n = 26$	M > C	–	–	–	–	–
MDA (serum)	Oxidative stress	Bioxytech LPO-586 Colorimetric Assay (Biostat, Stockport, Cheshire, UK)	[68]	$n = 26$	M > C	–	–	–	–	–
4-HNE (serum)	Oxidative stress	Bioxytech LPO-586 Colorimetric Assay (Biostat, Stockport, Cheshire, UK)	[67]	$n = 26$	M > C	–	–	–	–	–
4-HNE (serum)	Oxidative stress	Bioxytech LPO-586 Colorimetric Assay (Biostat, Stockport, Cheshire, UK)	[68]	$n = 26$	M > C	–	–	–	–	–
LPO (plasma)	Oxidative stress	(Ultraspec 1000 E Pharmacy Biotech spectrophotometer)	[69]	$n = 66$	M > C; P > C	No	Yes	–	Yes (VFC, SW)	–

(continued)

Table 1 (continued)

Biomarker	Marker	Method/assay	Study	Number	Direction	Correlation				
						DBS	TMS	TFC	Cognitive	Imaging
LPO (plasma)	Oxidative stress	PeroxiDetect (Sigma)	[70]	$n = 91$	M > C; no significant difference between C or P	No	No	No	–	–
Cu/Zn-SOD (erythrocyte)	Oxidative stress	RANSOD Kit (Randox Laboratory, UK)	[65]	$n = 52$	C > M	No (CAG_n)	No	No	–	–
Cu/Zn-SOD (serum)	Oxidative stress	ELISA (Bender MedSystems GmbH, Vienna, Austria)	[66]	$n = 28$	M > C	–	–	No	–	–
GPx (erythrocyte)	Oxidative stress	[71]	[65]	$n = 52$	C > M	No (CAG_n)	No	No	–	–
GSH (plasma)	Oxidative stress	(Ultraspec 1000 E Pharmacy Biotech spectrophotometer)	[69]	$n = 66$	C > M; C > P	No (CAG_n)	No	–	No	–
Lactate (plasma)	Oxidative stress	L-lactic Acid Enzymatic Bioanalysis (Roche) Kit	[70]	$n = 91$	M > C; no significant difference between P or C	No (CAG_n)	No	No	–	–
Nitrotyrosine (plasma)	Oxidative stress	ELISA (Oxis International (Portland, OR, USA)	[55]	$n = 38$	M > C	–	–	–	–	–

TrRd-1 (plasma)	Oxidative stress	ELISA (RLF-EK0122R; BioVendor-Laboratorni medicina a.s., Brno, Czech Republic)	[72]	$n = 23$	C > M	–	–	–	–	–
Trx-1 (plasma)	Oxidative stress	ELISA (RLF-EK0125R; BioVendor-Laboratorni medicina a.s., Brno, Czech Republic)	[72]	$n = 23$	C > M	–	–	–	–	–
Uric acid (plasma)	Oxidative stress	Colorimetric Enzymatic Reaction Kit (Catalog No. 1-3802, Salimetrics, CA, USA)	[73, 74]	$n = 107$	C > P > M	No	Yes (male and female M HD)	Yes (male M HD)	No (MMSE)	–
Tryptophan (serum)	Kynurenine pathway	HPLC	[75]	$n = 23$	C > M	–	–	–	Yes (MMS)	–
Tryptophan (serum)	Kynurenine pathway	Absorbance Detection Spectrometry	[76]	$n = 113$	C > P > M; only significant between C and M	–	–	–	–	–
Kynurenine (plasma)	Kynurenine pathway	HPLC; Waters HPLC System (Waters, Milford, MA, USA)	[67]	$n = 26$	M > C	–	–	–	–	–
3-HAA (plasma)	Kynurenine pathway	HPLC; Waters HPLC System (Waters, Milford, MA, USA)	[67]	$n = 26$	C > M	–	–	–	–	–

(continued)

Table 1 (continued)

Biomarker	Marker	Method/assay	Study	Number	Direction	Correlation				
						DBS	TMS	TFC	Cognitive	Imaging
3-HK (plasma)	Kynurenine pathway	HPLC; Waters HPLC System (Waters, Milford, MA, USA)	[67]	$n = 26$	C > M	–	–	–	–	–
KA (plasma)	Kynurenine pathway	LC-MS/MS (Charles River Laboratories, The Netherlands)	[77]	$n = 78$	P > C > M trend but only P > M significant (after adjusting for age and CAG_n)	–	No	No	No (SDMT, SW, VFC)	No: whole-brain, grey matter, white matter and caudate volumes
XA (plasma)	Kynurenine pathway	HPLC; Waters HPLC System (Waters, Milford, MA, USA)	[67]	$n = 26$	M > C	–	–	–	–	–
IL-1ra (plasma)	Cytokine	Quantikine ELISA Human Immunoassay (R&D Systems)	[78]	Timepoint 1, $n = 124$ Timepoint 2, $n = 90$	N/A—no controls	–	–	–	Yes (ExCog score)	–
IL-4 (plasma)	Cytokine	Multiplex ELISA	[24]	$n = 194$	M > P > C	–	No	No	–	–
IL-6 (plasma)	Cytokine	IL-6 ELISA Kit (EK-033-31, Phoenix Pharmaceuticals, Belmont, CA)	[79]	$n = 96$	M > P > C	–	–	–	–	–
IL-6 (plasma)	Cytokine	Multiplex ELISA	[24]	$n = 194$	M > P > C	–	No	No	–	–

IL-6 (plasma)	Cytokine	ELISA (Biosource IL-6 Kit, Nivelles, Belgium)	[72]	$n = 23$	M > C	–	–	–	–	–
IL-6 (plasma)	Cytokine	ELISA (R&D)	[80]	$n = 36$	M > C; no significant difference between C or P	–	No	Yes	–	–
IL-6 (plasma)	Cytokine	Quantikine ELISA Human Immunoassay (R&D Systems)	[78]	Timepoint 1, $n = 124$ Timepoint 2, $n = 90$	N/A – no controls	–	–	–	Yes (ExCog score)	–
IL-8 (plasma)	Cytokine	Multiplex ELISA	[24]	$n = 194$	M > P > C	–	Yes	Yes	–	–
IL-10 (plasma)	Cytokine	Multiplex ELISA	[24]	$n = 194$	M > P > C	–	No	No	–	–
IL-18 (plasma)	Cytokine	ELISA (R&D)	[80]	$n = 36$	C > M; no significant difference between C or P	–	No	No	–	–
IL-23 (serum)	Cytokine	Biosource Human IL-23 Immunoassay Kit (Invitrogen, Paisley, UK)	[76]	$n = 113$	M > P > C; only significant between C and M	–	–	–	–	–
MMP-9 (plasma)	Cytokine	ELISA (R&D)	[80]	$n = 36$	M > C; no significant difference between C or P	–	No	No	–	–

(continued)

Table 1 (continued)

Biomarker	Marker	Method/assay	Study	Number	Direction	Correlation				
						DBS	TMS	TFC	Cognitive	Imaging
sIL-2R (serum)	Cytokine	ELISA (T-Cell Sciences, Cambridge, MA, USA)	[75]	$n = 23$	M > C	–	–	–	No (MMS)	–
sTNF-R (serum)	Cytokine	ELISA (Bender, Vienna, Austria)	[75]	$n = 23$	M > C	–	–	–	No (MMS)	–
TGF-β1 (plasma)	Cytokine	ELISA (R&D)	[80]	$n = 36$	M > C; no significant difference between C or P	–	No	No	–	–
TNF-α (plasma)	Cytokine	Multiplex ELISA	[24]	$n = 194$	M > P > C	–	Yes	Yes	–	–
C3 (serum)	Immune response	Nephelometry & Indirect Immunofluorescence	[75]	$n = 23$	M > C	–	–	–	No (MMS)	–
CfH (plasma)	Immune response	Luminex Assays (Millipore, Billerica, MA)	[81]	Cohort 1: $n = 79$ Cohort 2: $n = 42$	Cohort 1: C > M > P (trend not significant); Cohort 2: P > M > C	Yes (DBS, CAG_n) P only	No	No	–	–
IgA (serum)	Immune response	Nephelometry & Indirect Immunofluorescence	[75]	$n = 23$	M > C	–	–	–	No (MMS)	–
IgG (serum)	Immune response	Nephelometry & Indirect Immunofluorescence	[75]	$n = 23$	No significant difference between C or M	–	–	–	No (MMS)	–

IgM (serum)	Immune response	Nephelometry & Indirect Immunofluorescence	[75]	$n = 23$	No significant difference between C or M	–	–	–	Yes (MMS)	–
sHLA-G (serum)	Immune response	ExBIO/BioVendor ELISA assay kit (AXXORA; Nottingham, UK)	[76])	$n = 113$	M > C	–	–	–	–	–
α-2 M (plasma)	Inflammation	Luminex Assays (Millipore, Billerica, MA)	[81]	Cohort 1: $n = 79$ Cohort 2: $n = 42$	No significant difference between C, P or M	No	Yes (early M)	No	–	–
α1-AT (plasma)	Inflammation	Luminex Assays (Millipore, Billerica, MA)	[81]	Cohort 1: $n = 79$ Cohort 2: $n = 42$	No significant difference between C, P or M	Yes (DBS, CAG_n) P only	No	No	–	–
ApoE (plasma)	Inflammation	Luminex Assays (Millipore, Billerica, MA)	[81]	Cohort 1: $n = 79$ Cohort 2: $n = 42$	No significant difference between C, P or M	Yes (DBS) M and P	No	No	–	–
Clusterin (plasma)	Inflammation	ELISA (RD194034200, BioVendor, Modrice, Czech Republic)	[79]	$n = 73$	M > C; M > P	–	–	–	–	–
Clusterin (plasma)	Inflammation	ELISA (RD194034200, BioVendor, Modrice, Czech Republic)	[79]	$n = 29$	M > C	–	–	–	–	–

(continued)

Table 1 (continued)

Biomarker	Marker	Method/assay	Study	Number	Direction	Correlation				
						DBS	TMS	TFC	Cognitive	Imaging
CRP (plasma)	Inflammation	Assay Kit (OSR6147, CRP; Olympus, Orlando, FL, USA)	[72]	$n = 23$	M > C	–	–	–	–	–
CRP (plasma)	Inflammation	Luminex Assays (Millipore, Billerica, MA)	[81]	Cohort 1: $n = 79$ Cohort 2: $n = 42$	C > P > M (only significant between early M and C)	No	No	No	–	–
CRP (plasma)	Inflammation	ELISA (CRP, R&D Systems)	[73, 74]	$n = 125$	No significant difference between C, P or M	–	Yes	No	Yes (MoCA)	–
CRP (serum)	Inflammation	Behring Turbitimeter	[67]	$n = 26$	M > C	–	–	–	–	–
MPO/WBC ratio (plasma)	Inflammation	Oxis International (Portland, OR, USA)	[72]	$n = 23$	M > C	–	–	–	–	–
Neopterin (serum)	Inflammation	ELISA (BRAHMS-Diagnostica, Berlin, Germany)	[75]	$n = 23$	M > C	–	–	–	Yes (MMS)	–
Neopterin (serum)	Inflammation	Immunoassay Kit (Immunological Laboratories, Hamburg, Germany)	[67]	$n = 26$	M > C	–	–	–	–	–
Neopterin (serum)	Inflammation	Immunoassay Kit (Immunobiological Laboratories, Hamburg, Germany)	[68]	$n = 26$	M > C	–	–	–	–	–
Osteonectin (plasma)	Inflammation	SECTOR 6000 Instrument (MSD)	[81]	Cohort 1: $n = 79$ Cohort 2: $n = 42$	No significant difference between C, P or M	Yes (CAG_n) P only	No	No	–	–

Transthyretin (plasma)	Inflammation	Luminex Assays (Millipore, Billerica, MA)	[81]	Cohort 1: $n = 79$ Cohort 2: $n = 42$	No significant difference between C, P or M	Yes (CAG_n) P only	No	No	–	–
LDH (plasma)	Cell damage	ELISA (Linear Chemicals. S. L; Montgat, Barcelona, Spain)	[55]	$n = 38$	M > C	–	–	–	–	–
24OHC (plasma)	Metabolic activity	IDMS; GC-MS	[82]	$n = 191$	C > M	–	–	No	–	Yes: caudate volumes between P and M and between P and C
24OHC (plasma)	Metabolic activity	IDMS	[83]	$n = 303$	C > M; P > M	–	Yes	No	–	–
24OHC (plasma)	Metabolic activity	IDMS; GC-MS	[84]	$n = 150$	C > M	–	Yes	–	Yes (SDMT, SW, SC, SI)	Yes: striatum volumes (putamen, caudate)
27OHC (plasma)	Metabolic activity	IDMS	[83]	$n = 303$	C > P	–	Yes	No	–	–
Asparagine (plasma)	Metabolic activity	HPLC	[85]	$n = 59$	C > M	–	–	–	–	–
Glycine (plasma)	Metabolic activity	AbsoluteIDQ p180 Kit (BIOCRATES Life Sciences, AG, Austria)	[86]	$n = 82$	M > C	No (CAG_n)	No	No	–	–
Histidine (plasma)	Metabolic activity	HPLC	[85]	$n = 59$	C > M	–	–	–	–	–
Isoleucine (plasma)	Metabolic activity	^{1}H NMR Spectroscopy & IEX	[87]	$n = 53$	C > M; P > M	Yes (CAG_n)	Yes	–	–	–

(continued)

Table 1 (continued)

Biomarker	Marker	Method/assay	Study	Number	Direction	Correlation				
						DBS	TMS	TFC	Cognitive	Imaging
Isoleucine (plasma)	Metabolic activity	AbsoluteIDQ p180 Kit (BIOCRATES Life Sciences, AG, Austria)	[86]	$n = 82$	C > M	No (CAG_n)	No	No	–	–
Lanosterol (plasma)	Metabolic activity	IDSM	[83]	$n = 303$	C > M	–	Yes	No	–	–
Lathosterol (plasma)	Metabolic activity	IDSM	[83]	$n = 303$	C > P; C > M	–	No	No	–	–
Leucine (plasma)	Metabolic activity	^{1}H NMR Spectroscopy	[87]	$n = 53$	C > M; P > M	No (CAG_n)	Yes	–	–	–
Leucine (plasma)	Metabolic activity	HPLC	[85]	$n = 59$	C > M	–	–	–	–	–
Lysophosphatidylcholine acyle C20:3 (plasma)	Metabolic activity	AbsoluteIDQ p180 Kit (BIOCRATES Life Sciences, AG, Austria)	[86]	$n = 82$	C > M	No (CAG_n)	No	No	–	–
Phosphatidylcholine acyl-alkyl C34:0 (plasma)	Metabolic activity	AbsoluteIDQ p180 Kit (BIOCRATES Life Sciences, AG, Austria)	[86]	$n = 82$	C > M	No (CAG_n)	No	No	–	–
Phosphatidylcholine acyl-alkyl C36:0 (plasma)	Metabolic activity	AbsoluteIDQ p180 Kit (BIOCRATES Life Sciences, AG, Austria)	[86]	$n = 82$	C > M	No (CAG_n)	No	No	–	–
Serine (plasma)	Metabolic activity	HPLC	[85]	$n = 59$	C > M	–	–	–	–	–

Serotonin (plasma)	Metabolic activity	AbsoluteIDQ p180 Kit (BIOCRATES Life Sciences, AG, Austria)	[86]	$n = 82$	C > M	No (CAG_n)	No	No	–	–
Taurine (plasma)	Metabolic activity	AbsoluteIDQ p180 Kit (BIOCRATES Life Sciences, AG, Austria)	[86]	$n = 82$	C > M	No (CAG_n)	No	No	–	–
Threonine (plasma)	Metabolic activity	HPLC	[85]	$n = 59$	C > M	–	–	–	–	–
Valine (plasma)	Metabolic activity	^{1}H NMR Spectroscopy	[87]	$n = 53$	C > M; P > M	Yes (CAG_n)	Yes	–	–	–
Valine (plasma)	Metabolic activity	AbsoluteIDQ p180 Kit (BIOCRATES Life Sciences, AG, Austria)	[86]	$n = 82$	C > M	No (CAG_n)	No	No	–	–
ACE2 (plasma)	Endocrine	ELISA (MB723213; My BioSource, San Diego, CA, USA)	[88]	$n = 57$	No significant difference between C, P or M	No (CAG_n)	No	–	Yes (VFC)	–
Adiponectin (plasma)	Endocrine	Radioimmunoassay (Linco Research, St. Charles, MO, USA)	[89]	$n = 18$	No significant difference in mean 24-hour levels between M or C	No (CAG_n)	Yes (mean 24-hour level)	Yes (mean 24-hour level)	–	–
Ang II (plasma)	Endocrine	ELISA (MBS084052; My BioSource, San Diego, CA, USA)	[88]	$n = 57$	No significant difference between C, P or M	No (CAG_n)	No	–	Yes (MMSE)	–
Cortisol (plasma)	Endocrine	ELECYS2010 Kit (Roche Diagnostics, Meylan, France)	[90]	$n = 290$	M > C	–	No	No	No	–

(continued)

Table 1 (continued)

Biomarker	Marker	Method/assay	Study	Number	Direction	Correlation				
						DBS	TMS	TFC	Cognitive	Imaging
Cortisol (serum)	Endocrine	Commercial Kit (CRT Abbot, Time-delayed Fluorescence)	[91]	$n = 26$	M > C	–	No (AIMS)	No	No (MMS)	–
Cortisol (serum)	Endocrine	ELISA (Cat. No. RE52611; IBL Hamburg)	[92]	$n = 21$	M > C (for late-stage)	–	–	–	–	–
GH (plasma)	Endocrine	ACCESS2 Kit (Beckman-Coulter, Villepinte, France)	[90]	$n = 290$	M > C (across all disease stages)	–	Yes	Yes	No	–
GH (plasma)	Endocrine	ACCESS2 Kit (Beckman-Coulter, Villepinte, France)	[93]	$n = 109$	N/A – no controls	–	–	–	Yes (SW, SC, SI, SDMT) in men	–
GH (serum)	Endocrine	Time-resolved Fluoroimmunoassay (DELFIA hGH; PerkinElmer Life and Analytical Sciences, Wallac Oy, Turku, Finland)	[94]	$n = 18$	No significant difference in mean levels between C or M	No (CAG_n)	Yes (total 24-hour secretion)	Yes (total 24-hour secretion)	–	–
Ghrelin (plasma)	Endocrine	Radioimmunoassay (Linco Research, St. Charles, MO, USA)	[94]	$n = 18$	No significant difference in mean 24-hour levels between C or M	Yes (CAG_n) (post-/pre-prandial ratio)	Yes (post-/pre-prandial ratio)	Yes (post–/pre-prandial ratio)	–	–
Ghrelin (serum)	Endocrine	8089-K Kit (Linco Research, St Louis, USA)	[87]	$n = 53$	M > C (men)	–	No	–	–	–
IGF-1 (plasma)	Endocrine	IMMULITE2500 Kit (DPC, LA Garenne Coombes, France)	[90]	$n = 290$	M > C	–	No	Yes	Yes ($n = 98$; SI, SDMT, VFC)	–

IGF-1 (plasma)	Endocrine	IMMULITE2500 Kit (DPC, LA Garenne Coombes, France)	[93]	$n = 109$	N/A – no controls	–	–	–	Yes (SW, SC, SI, SDMT, VFC)	–
Leptin (plasma)	Endocrine	Radioimmunoassay (Linco Research, St. Charles, MO, USA)	[89]	$n = 18$	No significant difference in mean 24-hour levels between C or M	Yes (CAG_n)	Yes (pulsatile secretion rate)	Yes (pulsatile secretion rate)	–	–
Leptin (serum)	Endocrine	HL-81HK Kit (Linco Research, Paris, France)	[87]	$n = 53$	C > M (men)	–	No	–	–	–
Melatonin (serum)	Endocrine	ELISA (Cat. No. RE54021; IBL Hamburg)	[92]	$n = 21$	C > M for late-stage	–	–	–	–	–

This table provides a systematic summary of relevant peripheral blood biomarkers in HD, including the type of biomarker, quantification methods or assays employed, source studies, study sample sizes (n), direction of significant results among control (C), premanifest HD gene expanded (P) and manifest HD (M) groups, and significant correlations with disease burden score (DBS), total motor score (TMS), total functional capacity (TFC), cognitive measures and imaging markers

Abbreviations: *^{1}H-NMR spectroscopy* proton nuclear magnetic resonance spectroscopy, *3-HAA* 3-hydroxyanthranilic acid, *3-HK* 3-hydroxykynurenine, *4-HNE* 4-hydroxynonenal, *8-OHdG* 8-hydroxy-2′-deoxyguanosine, *8-oxoGua* 8-Oxo-7,8-dihydroguanine, *α1-AT* alpha-1-antitrypsin, *α-2 M* alpha-2-macroglobulin, *γ-GT* γ-glutamyltranspeptidase, *ACE2* angiotensin-converting enzyme 2, *AIMS* Abnormal Involuntary Movement Scale, *Ang II* angiotensin II, *ApoE* apolipoprotein E, *BDNF* brain-derived neurotrophic factor, *C3* complement component 3, *CAGn*, CAG repeat length, *CRP* C-reactive protein, *Cu/Zn-SOD* copper/zinc superoxide dismutase, *DTI* diffusion tensor imaging, *ELISA* enzyme-linked immunosorbent assay, *GC-MS* gas chromatography-mass spectrometry, *GH* growth hormone, *HDGE* HD gene expanded, *HPLC* high-performance liquid chromatography, *HPLC-ECD* high-performance liquid chromatography electrochemical detector, *IDMS* isotope dilution mass spectrometry, *IGF-1* insulin-like growth factor 1, *IgA* immunoglobulin A, *IgG* immunoglobulin G, *IgM* immunoglobulin M, *IL-1ra* interleukin-1 receptor antagonist, *IL-4* interleukin-4, *IL-6* interleukin-6, *IL-8* interleukin-8, *IL-10* interleukin-10, *IL-18* interleukin-18, *IL-23* interleukin-23, *IS* Independence Scale, *KA* kynurenic acid, *LCECA* liquid chromatography electrochemical array, *LDH* lactate dehydrogenase, *Lp-PLA2* lipoprotein-associated phospholipase A2, *MDA* malondialdehyde, *MMP-9* matrix metalloproteinase-9, *MMS* mini-mental status, *MMSE* Mini-Mental State Examination, *MoCA* Montreal Cognitive Assessment, *MSD* Meso Scale Discovery, *NfL* neurofilament light, NfH neurofilament heavy, *NGF* nerve growth factor, *PIN-YTO* prognostic index (PIN) years to onset, *SC* Stroop colour, *Simoa* single molecule array immunoassay, *SI* Stroop interference, *SMC* single molecule counting, *sTNF-R* soluble tumour necrosis factor receptor, *sIL-2R* soluble interleukin-2 receptor, *sHLA-G* soluble human leukocyte antigen-G, *SWR* Stroop word reading, *T1-MRI* T1-weighted magnetic resonance imaging, *TTR* transthyretin, *Trx-1* thioredoxin-1, *TrRd-1* thioredoxin reductase-1, *TNF-α* tumour necrosis factor-alpha, *UHDRS* Unified Huntington's Disease Rating Scale, *UHDRS* II UHDRS cognitive assessment, *VFC* verbal fluency categorical, *XA* xanthurenic acid.

Using the ultrasensitive Simoa method, NfL concentrations in plasma were quantified in the TRACK-HD cohort [30, 33], which included control ($n = 97$), premanifest HDGE ($n = 104$) and early-stage manifest HDGE ($n = 97$) groups [47]. Plasma NfL concentrations were significantly higher in HDGE groups, compared to controls, with differences increasing with incrementing disease stage and increasing CAG repeat length [47]. Cross-sectionally, plasma NfL concentrations also correlated with clinical and imaging metrics: higher NfL was associated with poorer scores on cognitive assessments, higher motor scores assessed by UHDRS TMS and reductions in global and regional brain volumes on T1 volumetric MRI scans [47]. Longitudinally, baseline plasma NfL was a predictor of subsequent motor onset, and progression in clinical and imaging measures. This important study underscores the longitudinal potential of NfL in plasma as a prognostic biomarker, serving as a predictor of disease onset and progression of HD.

Further study in the TRACK-HD cohort utilised voxel-based morphometry to map the regional distribution and volumetric changes in brain regions linked to plasma NfL levels [95]. Higher plasma NfL levels at baseline were significantly associated with decreased volume in both the caudate and putamen bilaterally [95]. Higher baseline plasma NfL levels were also predictive of subsequent longitudinal cortical grey and white matter atrophy [95]. These results reinforce the region-specific predictive influence of plasma NfL and support NfL as a disease state biomarker.

Subsequent research with the longitudinal HD-CSF cohort, which comprised healthy control ($n = 20$), premanifest HDGE ($n = 20$) and manifest HDGE ($n = 40$) groups, also demonstrated that plasma NfL has the strongest cross-sectional association with all UHDRS clinical measures, surviving adjustment for age and number of CAG repeats (CAG_n)—which are known predictors of disease progression—compared to CSF NfL and CSF mHTT [48, 96]. Furthermore, a strong correlation between plasma and CSF NfL levels supports inference of a CNS origin for NfL detected in plasma [47, 48, 96].

Longitudinal 24-month follow-up of the HD-CSF cohort revealed that only the rate of change in plasma NfL levels was significantly associated with progression in UHDRS TMS, after adjustment for age and CAG_n [49]. In addition, the rate of change in plasma NfL levels demonstrated prognostic value in differentiating between fast and slow progressors [49]. Compared to baseline CSF NfL and CSF mHTT biofluid markers and volumetric imaging biomarkers, baseline plasma NfL emerged as the strongest predictor of subsequent clinical progression in HD [49].

2.3 Cerebrospinal Fluid

Originating as an ultrafiltrate of blood plasma, CSF, synthesised within the CNS, is enriched in brain-derived constituents. Hence, quantification of disease-associated biomarkers in CSF presents a compelling avenue for investigating the chemical milieu and elucidating neuropathological mechanisms in HD. The pursuit of robust biomarkers in HD has fuelled an increase in research lumbar puncture (LP) procedures. Despite the more invasive nature of LPs to obtain CSF, compared to venepuncture for blood sampling, a recent multi-study cross-sectional analysis of two observational studies in HD underscores the safety and tolerability of LP in both premanifest and manifest HDGE groups [97].

2.3.1 Mutant Huntingtin

Given the role of mHTT in HD pathogenesis, quantifying the protein in its various forms could fulfil a dual role: as a disease state biomarker, and as a pharmacodynamic biomarker, the latter representing an indicator of target engagement in response to HTT-lowering therapeutic interventions. Quantification of mHTT in CSF has proved challenging due to the low concentration of mHTT and the intracellular aggregation-prone nature of the protein, which is gradually released into CSF [98].

CSF mHTT can be quantified using a single molecule counting (SMC) immunoassay [99]. This ultrasensitive bead-based immunoassay employs the 2B7 antibody for capture, which binds to the first 17 amino acids of the N-terminus of mHTT, and the MW1 antibody for detection, providing specificity for the polyQ expansion (Fig. 1).

A validation study of the 2B7-MW1 SMC mHTT detection assay demonstrated that the performance of this assay is accurate, reproducible and specific for the relative quantification of mHTT in CSF, adhering to established guidelines set forth by the Food and Drug Administration (FDA) and the European Medicines Agency (EMA) [100].

CSF mHTT as a pharmacodynamic biomarker of target engagement has since been incorporated as primary and secondary biomarker endpoints in clinical trials in HD.

A dose-dependent decrease in CSF mHTT levels was successfully observed in a landmark phase 1/2a clinical trial evaluating IONIS-HTTRx (ClinicalTrials.gov Identifier: NCT02519036), an ASO targeting *HTT* mRNA delivered by intrathecal administration [101].

In the discontinued phase 3 GENERATION HD1 clinical trial (ClinicalTrials.gov Identifier: NCT03761849), manifest HDGE individuals treated with Tominersen, an ASO, also exhibited dose-dependent decreases in CSF mHTT levels, with more pronounced reductions from baseline observed in those receiving the treatment every 8 weeks, compared to those treated every 16 weeks and the placebo group [102].

Current clinical trials in HD are integrating CSF mHTT protein as key biomarker endpoints, both primary and secondary, within their study design. For example, the phase 2 dose-finding GENERATION HD2 clinical trial (ClinicalTrials.gov Identifier: NCT05686551) is assessing the percentage reduction from baseline in the geometric mean of CSF mHTT levels at the 9-month mark. Similarly, a phase 2a study of PTC518 (ClinicalTrials.gov Identifier: NCT05358717), which is exploring the efficacy of an oral splicing modulator, is assessing the change in baseline CSF mHTT protein levels after 12 months.

As a disease state biomarker, CSF mHTT levels have demonstrated a correlation with clinical measures of disease severity [48, 49, 99, 103] and increases in conjunction with disease progression in HD [100] (Table 2).

In HD-YAS, CSF mHTT was detectable at low levels in nearly all premanifest HDGE individuals, with the exception of three with a low disease burden score (DBS) [34]. Higher CSF mHTT levels were significantly associated with higher age and CAG repeat length, including the interaction between these two covariates. Over half of the premanifest HDGE group had CSF mHTT concentrations between the detectable and quantifiable range, where the output from the 2B7-MW1 immunoassay is known to lack linearity [34].

Lowering CSF mHTT quantifiably, or from detectable to undetectable levels, could demonstrate target engagement in future preventive trials. However, the detection threshold of current assays may fall short in detecting mHTT in CSF in far-from-onset premanifest HDGE groups. Additionally, there is a large variability in mHTT levels among individuals [5]. These challenges currently restrict the utility of CSF mHTT as a biomarker in early intervention trials for premanifest HDGE groups at HD-ISS stages 0 and 1, particularly when trying to measure dose-response effects.

2.3.2 Total Huntingtin

Total huntingtin (tHTT) represents a potential biomarker of target engagement for HTT-lowering therapies. Advances in assays technology have facilitated detection of HTT by enzyme-linked immunosorbent assay (ELISAs) and electrochemiluminescence-based signal amplification via HTT Meso Scale Discovery (MSD) assays [125]. These assays have been used to quantify HTT protein levels across diverse systems, including in vitro cell cultures, in vivo animal models and human HD biosamples, with sensitivity of detecting HTT proteins at the low picomolar range [125]. Using ELISA-based MSD electrochemiluminescence immunoassay platforms, levels of N-, C-terminal and mid-region HTT protein were measured in peripheral blood mononuclear cells isolated from the whole blood of

Fig. 1 Schematic illustration of mHTT epitopes recognised by 2B7 and MW1 antibodies employed in the mHTT SMC immunoassay. (Adapted from Wild et al., [99]. Abbreviations: NH2 N-terminus, N17 17 amino acid, polyQ polyglutamine expansion, PRD proline-rich domain, COOH C-terminus

Table 2 Cerebrospinal fluid biomarkers in Huntington's disease

Biomarker	Marker	Method/assay	Study	Number	Direction	Correlation				
						DBS	TMS	TFC	Cognitive	Imaging
mHTT	Toxic protein	HDB4/MW1 HTT IP-FCM	[103]	$n = 37$	M > C P > C	Yes (DBS, CAG_n)	Yes (except late-stage)	–	Yes (SDMT, SI, VFC) – except late-stage	–
mHTT	Toxic protein	2B7-MW1 immunoassay (SMC Erenna platform, Merck)	[99]	$n = 52$	M > P > C	Yes	Yes	–	Yes (SDMT, SC, SW) after adjusting for DBS	–
mHTT	Toxic protein	2B7-MW1 SMC mHTT immunoassay	[100]	$n = 22$	M > C	–	–	–	–	–
mHTT	Toxic protein	2B7-MW1 immunoassay (SMC Erenna platform, Merck)	[48]	$n = 80$	M > P	–	Yes	Yes	Yes (SDMT, SC, SW, VFC) after adjusting for age	No
mHTT	Toxic protein	2B7-MW1 immunoassay (SMC Erenna platform, Merck)	[49]	$n = 71$	M > P	–	Yes (rate of change) after adjusting for CAG_n	No (rate of change) after adjusting for age and CAG_n	No (rate of change) after adjusting for age and CAG_n	No: (rate of change) after adjusting for age and CAG_n
mHTT	Toxic protein	2B7-MW1 immunoassay (SMCTM Erenna platform by Singulex, Evotek)	[34]	$n = 109$	P > C	–	–	–	–	–
mHTT	Toxic protein	mHTT assay (pAB 2B7/MW, CHDI)	[104]	$n = 45$	M > P > C	Yes (CAP)	Yes	Yes	Yes (SDMT, SW)	–
tHTT	Total huntingtin	2B7-DF7 immunoassay (SMCTM Erenna platform by Singulex, IRBM)	[34]	$n = 109$	–	–	–	–	–	–

(continued)

Table 2 (continued)

Biomarker	Marker	Method/assay	Study	Number	Direction	Correlation				
						DBS	TMS	TFC	Cognitive	Imaging
NfL	Axonal damage	ELISA (University of Gothenburg, Sweden)	[105]	$n = 70$	M > C	No (CAG_n)	No	Yes	No (SDMT, SC, SW, VFC)	–
NfL	Axonal damage	ELISA (UmanDiagnostics, Umeå, Sweden)	[106]	$n = 104$	M > P > C	Yes (CAP)	Yes	Yes	–	–
NfL	Axonal damage	ELISA (UmanDiagnostics, Umeå, Sweden)	[47]	$n = 37$	M > P > C	–	Yes	–	–	–
NfL	Axonal damage	ELISA (NF-light®, UmanDiagnostics, Umeå, Sweden)	[107]	$n = 23$	M > P; no controls	Yes	Yes after adjusting for age	No after adjusting for age	Yes (SC, SI, VFC) after adjusting for age	–
NfL	Axonal damage	ELISA (NF-light®, UmanDiagnostics AB, Umeå, Sweden)	[108]	$n = 52$	M > P > C	Yes	Yes after adjustment for DBS	Yes after adjustment for DBS	No	–
NfL	Axonal damage	ELISA (NF-Light; UmanDiagnostics, Umeå, Sweden)	[48]	$n = 80$	M > P > C	–	Yes	Yes	Yes (SDMT, SC, SW, VFC) after adjusting for age	Yes: whole-brain, grey matter, white matter and caudate volumes after adjusting for age and CAG_n
NfL	Axonal damage	ELISA (UmanDiagnostics AB, Umeå, Sweden)	[109]	$n = 60$	M > P > C	Yes (CAG_n not DBS)	No	No	–	–
NfL	Axonal damage	Simoa HD-1 Analyser (Quanterix); Neurology 4-Plex B	[49]	$n = 71$	M > P > C	–	Yes (rate of change) after adjusting for age and CAG_n	No (rate of change) after adjusting for age and CAG_n	No after adjusting for age and CAG_n	No: (rate of change) after adjusting for age and CAG_n
NfL	Axonal damage	Simoa HD-1 Analyser (Quanterix, UCL); Neurology 4-Plex A	[34]	$n = 109$	P > C	–	–	–	–	–
NfL	Axonal damage	Not specified (CHDI)	[104]	$n = 45$	M > P > C	Yes (CAP)	Yes	Yes	Yes (SDMT, SW)	–

P-tau181	Microtubular damage	ELISA (Innotest, phosphotau[181P] Fujirebio Europe, Gent, Belgium)	[108]	$n = 52$	C > P (significant after adjusting for age); M > P (trend not significant after adjusting for age)	Yes	Yes (after adjusting for DBS)	No	Yes (SI) after adjusting for DBS	–
T-tau	Microtubular damage	ELISA (Innotest hTau-Ag, Innogenetics, Gent, Belgium)	[110]	$n = 70$	M > C	–	No	No	No (SDMT)	–
T-tau	Microtubular damage	ELISA (Innotest, Fujirebio, Gent, Belgium)	[111]	$n = 76$	HDGE> C	Yes (DBS)	Yes (after adjusting for age and DBS)	Yes (after adjusting for age and DBS)	Yes(SDMT, SC, SW) after adjusting for age and DBS	–
T-tau	Microtubular damage	(Tau) Human ELISA (Life Technologies, Nærum, Denmark)	[106]	$n = 104$	No significant difference between M, P or C	Yes (CAP)	Yes	No	–	–
T-tau	Microtubular damage	ELISA (Innotest, hTau-Ag, Fujirebio Europe, Belgium)	[107]	$n = 23$	No significant difference between M or P; no controls	Yes	No (after adjusting for age or DBS)	No (after adjusting for age or DBS)	No (after adjusting for age or DBS)	–
T-tau	Microtubular damage	ELISA (Innotest, hTau-Ag, Fujirebio Europe, Gent, Belgium)	[108]	$n = 52$	C > P (significant after adjusting for age) M > P (trend not significant after adjusting for age)	Yes	Yes (after adjusting for DBS)	No	No	–
T-tau	Microtubular damage	Simoa HD-1 Analyser (Quanterix, UCL); Neurology 4-Plex A	[34]	$n = 109$	No significant difference between P or C	–	–	–	–	–
MBP	Neuronal damage	ELISA (Becman coulter, Sharon Hill, PA)	[106]	$n = 104$	M > C M > P	Yes (CAP)	No	Yes	–	–
UCH-L1	Neuronal damage	Simoa HD-1 Analyser (Quanterix, UCL); Neurology 4-Plex A	[34]	$n = 109$	No significant difference between P or C	–	–	–	–	–

(continued)

Table 2 (continued)

Biomarker	Marker	Method/assay	Study	Number	Direction	Correlation				
						DBS	TMS	TFC	Cognitive	Imaging
YKL-40	Microglial marker	ELISA (Quidel, Santa, Clara, California)	[112]	$n = 68$	M > P > C (trend not statistically significant)	Yes (CAG_n)	–	–	–	–
YKL-40	Microglial marker	ELISA (R&D Systems Inc. Minneapolis, MN)	[113]	$n = 37$	HDGE > C M (early, moderate) > P	No (DBS)	Yes (after adjusting for age or DBS)	Yes	–	–
YKL-40	Microglial marker	Quantikine ELISA (R&D Systems, Minneapolis, MN)	[106]	$n = 104$	M > C; M > P	Yes (CAP)	No	No	–	–
YKL-40	Microglial marker	ELISA (R&D Systems Minneapolis, MN, USA)	[108]	$n = 52$	M > P > C (trend not significant after adjusting for age)	Yes	Yes (after adjusting for DBS)	Yes (after adjusting for DBS)	Yes (SI) after adjusting for DBS	–
YKL-40	Microglial marker	U-PLEX Human YKL-40 Assay (MSD, UCL)	[34]	$n = 109$	P > C	–	–	–	–	–
YKL-40	Microglial marker	Nano-LC-PRM-MS	[114]	$n = 32$	No significant difference between M, P or C	No (CAP)	No	No	No (SDMT, SW, VFC)	–
Chito-triosidase	Microglial marker	In-house Enzyme Activity Assay	[113]	$n = 37$	HDGE > C; no significant association with disease stage	–	No	–	–	–

TREM2	Microglial marker	In-house MSD-based ELISA	[96]	$n = 80$	No significant difference between C, P or M	No	No	No	No (SDMT, SW, SC, VFC)	No: volumes of whole-brain, caudate, grey matter, white matter and lateral ventricles
PENK	Striatal marker	LC-MS	[115]	$n = 63$	C > P > M	Yes (CAP)	Yes (without adjusting for age)	Yes (without adjusting for age)	Yes (SC) without adjusting for age	–
PENK	Striatal marker	LC-MS/MS	[116]	$n = 139$	C > M	No (CAG_n)	Yes	No	Yes (UHDRS Cognitive Score)	–
PENK	Striatal marker	LC-MS/MS	[114]	$n = 32$	C > P > M	Yes (CAP)	No	No	No (SDMT, SW, VF)	–
PDYN	Striatal marker	LC-MS	[117]	$n = 71$	C > M	–	–	–	–	–
PDYN	Striatal marker	LC-MS/MS	[116]	$n = 139$	C > M	No (CAG_n)	No	No	Yes (UHDRS cognitive score)	–
PDYN	Striatal marker	LC-MS/MS	[114]	$n = 32$	C > P > M	No (CAP)	No	Yes	Yes (SDMT, VF but not SW)	–
Neurogranin	Synaptic dysfunction	In-house ELISA (Anti-N antibodies NG22 + NG2)	[96]	$n = 32$	No significant difference between HDGE (M, P) or C	No	No	No	–	–
Neurogranin	Synaptic dysfunction	ELISA (Neurogranin assay, Euroimmun, UCL)	[34]	$n = 109$	No significant difference between P or C	–	–	–	–	–
GFAP	Astrocyte reactivity	Simoa HD-1 Analyser (Quanterix, UCL); Neurology 4-Plex A	[34]	$n = 109$	No significant difference between P or C	–	–	–	–	–
Clusterin	Inflammation	ELISA (RD194034200, BioVendor, Modrice, Czech Republic)	[79]	$n = 29$	M > C	–	–	–	–	–

(continued)

Table 2 (continued)

Biomarker	Marker	Method/assay	Study	Number	Direction	Correlation				
						DBS	TMS	TFC	Cognitive	Imaging
CXCL13	Inflammation	Quantikine ELISA Kit (R&D Systems, Minneapolis, MN)	[106]	$n = 104$	Only detected in $n = 7$	No	–	–	–	–
MMP9	Inflammation	Quantikine ELISA Kit (R&D Systems, Minneapolis, MN)	[106]	$n = 104$	Only detected in $n = 3$	No	–	–	–	–
OPN	Inflammation	Quantikine ELISA Kit (R&D Systems, Minneapolis, MN)	[106]	$n = 104$	C > M; C > P	No	No	No	–	–
TCC	Inflammation	ELISA (Hycult Biotech, Uden, Netherlands)	[106]	$n = 104$	No significant difference between M, P or C	No	No	No	–	–
sCD27	Innate immunity	ELISA (Human CD27/ TNFRSF7 DY382, R&D Systems, Minneapolis, MN)	[108]	$n = 52$	M > P > C (significant after adjusting for sex)	No	No	No	No	–
IL-1β	Cytokine	Antibody-based Tetra-plex Array (MSD, Gaithersburg, MD, USA)	[113]	$n = 37$	Below the limit of detection	–	–	–	–	–
IL-6	Cytokine	Antibody-based Tetra-plex Array (MSD, Gaithersburg, MD, USA)	[113]	$n = 37$	HDGE > C	–	No	–	–	–

IL-6	Cytokine	Simoa HD-1 Analyser (Quanterix, UCL); Neurology 4-Plex A	[34]	$n = 109$	No significant difference between P or C	–	–	–	–	–
IL-8	Cytokine	Antibody-based Tetra-plex Array (MSD, Gaithersburg, MD, USA)	[113]	$n = 37$	No significant difference between HDGE or C	–	No	–	–	–
IL-8	Cytokine	Simoa HD-1 Analyser (Quanterix, UCL); Neurology 4-Plex A	[34]	$n = 109$	No significant difference between P or C	–	–	–	–	–
TNF-α	Cytokine	Antibody-based Tetra-plex Array (MSD, Gaithersburg, MD, USA)	[113]	$n = 37$	Below the limit of detection	–	–	–	–	–
NO_x	Oxidative stress	Quantikine ELISA (R&D systems, Minneapolis, MN)	[106]	$n = 104$	M > P P > C	No	No	No	–	–
3-HK	Kynurenine pathway	LC-MS/MS (Charles River Laboratories, The Netherlands)	[77]	$n = 80$	C > P > M trend but only M > P significant (after adjusting for age and CAG_n)	No (DBS, CAG_n)	No	No	No (SDMT, SW, VFC)	No: whole-brain, grey matter, white matter and caudate volumes
AA	Kynurenine pathway	LC-MS/MS (Charles River Laboratories, The Netherlands)	[77]	$n = 80$	No significant difference between M, P or C	–	No	No	No (SDMT, SW, VFC)	No: whole-brain, grey matter, white matter and caudate volumes
KA	Kynurenine pathway	LC-MS/MS (Charles River Laboratories, The Netherlands)	[77]	$n = 80$	No significant difference between M, P or C	No (DBS, CAG_n)	No	No	No (SDMT, SW, VFC)	No: whole-brain, grey matter, white matter and caudate volumes
KA	Kynurenine pathway	HPLC	[118]	$n = 73$	M > C	–	–	–	–	–

(continued)

Table 2 (continued)

Biomarker	Marker	Method/assay	Study	Number	Direction	Correlation				
						DBS	TMS	TFC	Cognitive	Imaging
Kynurenine	Kynurenine pathway	LC-MS/MS (Charles River Laboratories, The Netherlands)	[77]	$n = 80$	No significant difference between M, P or C	–	No	No	No (SDMT, SW, VFC)	No: whole-brain, grey matter, white matter and caudate volumes
Tryptophan	Kynurenine pathway	LC-MS/MS (Charles River laboratories, the Netherlands)	[77]	$n = 80$	No significant difference between M, P or C	–	No	No	No (SDMT, SW, VFC)	No: whole-brain, grey matter, white matter and caudate volumes
QA	Kynurenine pathway	LC-MS/MS (Charles River Laboratories, The Netherlands)	[77]	$n = 80$	No significant difference between M, P or C	No (DBS, CAG_n)	No	No	No (SDMT, SW, VFC)	No: whole-brain, grey matter, white matter and caudate volumes
QA	Kynurenine pathway	GC/ECNI-MS	[119]	$n = 18$	No significant difference between M or C	–	–	–	–	–
BDNF	Neuropeptide	Simoa Human BDNF Discovery Kit (Quanterix, USA); HD-1 Analyser (Quanterix, USA)	[57]	$n = 80$	No significant difference between C, P or M	–	No	No	No	Yes: white matter volume
Orexin-A	Neuropeptide	Radioimmunoassay Kit (Phoenix Pharmaceuticals, Belmont, CA)	[120]	$n = 22$	No significant difference between M or C	No (CAG_n)	–	–	–	–
Orexin-A	Neuropeptide	Radioimmunoassay Kit (Phoenix Pharmaceuticals, Belmont, CA)	[121]	$n = 67$	No significant difference between M or C	–	–	–	–	–

Orexin-A	Neuropeptide	Radioimmunoassay Kit (Phoenix Pharmaceuticals, Belmont, CA)	[122]	$n = 14$	No significant difference between M or C	No (CAG_n)	–	–	–	–
Oxytocin	Neuropeptide	Radioimmunoassay (RIA, Phoenix Pharmaceuticals, California, USA)	[62])	$n = 146$	M < C; no significant difference between P or C	No (CAP)	No	–	No (MMSE, MoCA)	–
5-HIAA	Dopaminergic transmission	HPLC	[123]	$n = 23$	No significant difference between M or C	–	–	–	–	–
DA	Dopaminergic transmission	HPLC	[123]	$n = 11$	M > C	–	–	–	–	–
DOPAC	Dopaminergic transmission	HPLC	[123]	$n = 22$	M > C	–	–	–	–	–
HVA	Dopaminergic transmission	HPLC	[123]	$n = 22$	No significant difference between M or C	–	–	–	–	–
MHPG	Dopaminergic transmission	HPLC	[123]	$n = 23$	No significant difference between M or C	–	–	–	–	–
NA	Dopaminergic transmission	HPLC	[123]	$n = 23$	No significant difference between M or C	–	–	–	–	–
GABA	Neurotransmitter	Ion Exchange-Fluorometric Method	[124]	$n = 61$	C > M; no significant difference between P or C	–	–	–	–	–
Cu	Metal homeostasis	GFAAS (Varian AA240, Varian, Inc., Palo Alto, CA)	[104]	$n = 45$	P > C; M (late) > C; no significant difference between M or P	No (CAP, CAG_n)	No	No	No	–
Fe	Metal homeostasis	GFAAS (Varian AA240, Varian, Inc., Palo Alto, CA)	[104]	$n = 45$	M > C; no significant difference between P or C	No (CAP, CAG_n)	No	No	No	–

(continued)

Table 2 (continued)

Biomarker	Marker	Method/assay	Study	Number	Direction	Correlation				
						DBS	TMS	TFC	Cognitive	Imaging
Mn	Metal homeostasis	GFAAS (Varian AA240, Varian, Inc., Palo Alto, CA)	[104]	$n = 45$	P > C; no significant difference between M or C	No (CAP, CAG_n)	No	No	No	–
Zn	Metal homeostasis	GFAAS (Varian AA240, Varian, Inc., Palo Alto, CA)	[104]	$n = 45$	C > P; C > M (late); no significant difference between M or P	No (CAP, CAG_n)	No	No	No	–

This table provides a systematic summary of relevant CSF biomarkers in HD, including the type of biomarker, quantification methods or assays employed, source studies, study sample sizes (n), direction of significant results among control (C), premanifest HD gene expanded (P) and manifest HD (M) groups, and significant correlations with disease burden score (DBS), total motor score (TMS), total functional capacity (TFC), cognitive measures and imaging markers

Abbreviations: *3-HK* 3-hydroxykynurenine, *5-HIAA* 5-hydroxyindoleacetic acid, *AA* arachidonic acid, *BDNF* brain-derived neurotrophic factor, CAG_n CAG repeat length, *CAP* CAG-Age Product, *CSF* cerebrospinal fluid, CXCL13 C-X-C motif chemokine ligand 13, *Cu* copper, *DA* dopamine, *DOPAC* dihydroxyphenylacetic acid, *ELISA* enzyme-linked immunosorbent assay, *Fe* iron, *GABA* gamma-aminobutyric acid, *GC/ECNI-MS* gas chromatography/electron capture negative ion mass spectrometry, *GFAP* glial fibrillary acidic protein, *GFAAAS* graphite furnace atomic absorption spectroscopy, *HDGE* HD gene expanded, *HPLC* high-performance liquid chromatography, *HVA* homovanillic acid, *IL-1β* interleukin-1 beta, *IL-6* interleukin-6, *IL-8* interleukin-8, *IP-FCM* immunoprecipitation and flow cytometry, *KA* kynurenic acid, *LC-MS/MS* liquid chromatography with mass spectrometry, *MBP* myelin basic protein, *MHPG* 3-methoxy-4-hydroxyphenylglycol, *mHTT* mutant Huntingtin, *MMP9* matrix metalloproteinase 9, *MMSE* Mini-Mental State Examination, *Mn* manganese, *MoCA* Montreal Cognitive Assessment, *MSD* Meso Scale Discovery, *NA* noradrenaline, *Nano-LC-PRM-MS* nano-liquid chromatography-parallel reaction monitoring-mass spectrometry, *NfL* neurofilament light, *NOx* nitric oxide metabolites, *OPN* osteopontin, *orexin-A* hypocretin-1, *P-tau181* phosphorylated tau at amino acid 181, *PDYN* prodynorphin, *PENK* proenkephalin, *QA* quinolinic acid, *SC* Stroop colour, *sCD27* soluble CD27 receptor, *SDMT* Symbol Digit Modalities Test, *SI* Stroop interference, *SMC* single molecule counting, *SW* Stroop word, *TCC* terminal complement complex, *tHTT* total Huntingtin, *TNF-α* tumour necrosis factor alpha, *TREM2* triggering receptor expressed on myeloid cells 2, *T-tau* total tau, *UCH-L1* ubiquitin carboxy-terminal hydrolase L1, *UHDRS* unified Huntington's Disease Rating Scale, *VFC* verbal fluency categorical, *YKL-40* chitinase-3-like protein 1, *Zn* zinc.

HD patients [126], which underscores the utility of these platforms in detecting specific segments of the HTT protein.

Further advancement in detection methods using the Single Molecule Counting (SMC) immunoassay and a combination of immunoprecipitation with flow cytometry has pushed the boundaries of sensitivity into the femtomolar range for quantification of HTT in human CSF [99, 103].

In HD-YAS, the 2B7-D7F7 immunoassay on the SMC Erenna® platform was used to measure CSF tHTT concentrations in the cohort studied [34], with the D7F7 antibody specific for a region surrounding Pro1220 of human HTT protein [127]. However, there were no significant differences in CSF tHTT levels between control and premanifest HDGE groups, the latter group who were approximately 24 years before predicted clinical motor diagnosis [34].

Despite the progress made, quantifying tHTT in CSF remains challenging due to the inherent limitations in assay accuracy and consistency.

2.3.3 NfL

In parallel with plasma, levels of NfL in CSF have consistently shown distinct differentiation in HDGE groups from control groups and a strong correlation with clinical indicators of disease severity (Table 2).

HD-YAS aimed to characterise the earliest measurable biomarkers of neurodegeneration in a premanifest HDGE cohort who were approximately 24 years away from predicted clinical motor diagnosis [34]. No significant differences were observed in cognitive or neuropsychiatric measures between HDGE (n = 64) and control (n = 67) groups, with both groups being closely matched for age, education and sex. In the HDGE group, both plasma and CSF NfL levels were significantly elevated, compared to controls. However, 53% of the HDGE group had CSF NfL levels that fell within the normal control range, whereas 87% had plasma NfL levels within the normal control range [34].

These findings suggest that, in far-from-onset HDGE groups, CSF NfL may serve as a more sensitive biomarker when compared to plasma NfL. Furthermore, the findings underscore CSF NfL as a disease state biomarker in HD, as NfL becomes abnormal more than two decades before predicted clinical motor diagnosis.

The role of CSF NfL as a biomarker in HD expanded following unexpected results from the phase 1/2a clinical trial evaluating IONIS-HTTRx (ClinicalTrials.gov Identifier: NCT02519036), an ASO targeting *HTT* mRNA delivered by intrathecal administration [101]. In this study, a subset of trial participants receiving higher doses of the ASO had elevated CSF NfL levels shortly after cessation of treatment, without corresponding changes observed in MRI brain safety assessments. These NfL elevations were generally transient, returning to baseline levels subsequently. This clinical trial importantly elucidated a clear dose-response and temporal relationship in fluctuations in CSF NfL levels during the dosing period, which normalised after the treatment was ceased.

A transient spike in CSF NfL was also observed in the phase 3 GENERATION HD1 clinical trial (ClinicalTrials.gov Identifier: NCT03761849) [102]. The increase in CSF NfL in a subset of manifest HDGE individuals receiving bi-monthly ASO was accompanied by increased leukocyte and total protein levels in their CSF, which suggests an inflammatory response. Of note, CSF NfL levels in the placebo group and the subset of manifest HDGE individuals dosed every 16 weeks remained similar throughout the study [102].

These findings underscore the evolving significant role of CSF NfL as a safety biomarker in HD therapeutics, with fluctuations in CSF NfL levels signalling potential toxic effects and inflammatory responses associated with ASO therapies. The ongoing phase 2 GENERATION HD2 dose-finding trial (ClinicalTrials.gov Identifier: NCT05686551) will be crucial to follow in assessing the early fluctuations of CSF NfL and their clinical significance.

2.3.4 Tau

Tau is a microtubule-associated protein that is widely expressed in the CNS released from nerve cells into the interstitial fluid of the brain, and then subsequently CSF [128]. Intracellular aggregates of abnormally phosphorylated tau are present in a group of neurodegenerative diseases known as tauopathies, notably prominent in Alzheimer's disease [129]. The expression of mHTT has been associated with alterations in tau function, including influencing cellular localisation and patterns of phosphorylation [130]. Pathological inclusions containing abnormally phosphorylated tau have been observed in the cortex and striatum of post-mortem brains from patients with HD [131]. These observations imply a potential secondary tauopathy effect in HD and have prompted investigations into whether tau could serve as a biomarker for disease progression.

Studies investigating the role of tau as a fluid biomarker in HD have demonstrated less consistency in results, compared to NfL. In one study, mean total tau protein levels in CSF were significantly higher in individuals with manifest HD (n = 35), compared to matched controls (n = 35), though there were no significant correlations between tau levels in CSF and CAG_n, cognitive, functional or motor measures [110].

In another study, mean total tau levels in CSF were also significantly higher in a HDGE group (n = 52), which included premanifest HDGE individuals (n = 13), compared to healthy controls (n = 24) [111, 113]. Levels of total tau in CSF were found to be correlated with cognitive and motor measures in this study, retaining significance following adjustments for age and DBS, which suggests that CSF total tau harbours independent predictive power for clinical phenotypes, extending beyond the known influences of age and CAG_n [111, 113].

However, in a larger cohort (n = 104), there emerged no significant difference in concentrations of total tau levels in CSF between control (n = 20), premanifest (n = 32) or manifest (n = 48) HDGE groups [106]. Among the three groups studied, tau exhibited notably distinct age-related effects, thus adjustments were made solely for sex and not age in the subsequent analyses [106].

Moreover, tau exhibited correlations with psychiatric symptoms, demonstrating distinct effects contingent upon disease burden (derived from CAG-Age-Product (CAP) score) within two discrete cohorts: HDGE individuals with and without psychiatric symptoms [106]. Tau increased in tandem with disease burden solely within the subset of HDGE individuals without psychiatric symptoms, though within HDGE individuals with psychiatric symptoms, tau consistently remained elevated, irrespective of disease burden [106].

There was no significant difference in CSF total tau levels between premanifest (n = 11) and manifest (n = 12) HDGE groups in another study [107]. Additionally, no significant correlation was observed between CSF total tau levels and clinical measures, after adjusting for age or disease burden. In comparison, CSF NfL levels in the same cohort were significantly higher in manifest HDGE individuals, compared to premanifest HDGE individuals, with differences persisting after adjusting for age. Furthermore, CSF NfL levels robustly correlated with cognitive and motor measures after adjusting for age. Notably, CSF NfL levels also strongly correlated with a 5-year probability of disease onset in premanifest HDGE individuals, which CSF total tau did not demonstrate [107].

Thus, head-to-head comparison between levels of NfL and total tau in CSF in this cohort suggests that NfL is superior to tau as a biomarker for disease stage in HD [107]. This assertion is further supported by findings from HD-YAS, the earliest premanifest HDGE group examined to date [34]. This study revealed no significant differences in CSF total tau levels between closely matched control and HDGE groups, with the latter group being approximately 24 years away from predicted clinical motor diagnosis [34]. However, within this far-from-onset HDGE group, CSF NfL levels were significantly elevated compared to the control group [34]. This further reinforces the case for NfL as a superior biofluid marker in tracking the earliest stages of disease progression in HD.

2.3.5 YKL-40

Chitinase-3-like protein 1 (CHI3L1), also known as YKL-40, is a glycoprotein released from different cell types, including macrophages [132]. Although the physiological role of YKL-40 remains to be fully elucidated, YKL-40 is expressed by astrocytes and microglial cells and is upregulated in neurodegenerative diseases associated with inflammation [112, 132]. YKL-40, as a marker for glia activation, has thus garnered attention as a biomarker that could reflect central neuroinflammatory perturbations implicated in HD pathogenesis.

In comparison to studies employing assays measuring YKL-40 using electrochemiluminescence [34] or via ELISA [106, 112], nano liquid chromatography-parallel reaction monitoring-mass spectrometry (nano-LC-PRM-MS) has also been used for YKL-40 quantification with high sensitivity and specificity [114]. Using nano-LC-PRM-MS, there was a 25% increase in CSF YKL-40 levels in HDGE individuals, compared to controls, although this change did not reach statistical significance [114]. No significant correlation was found between CSF YKL-40 levels and clinical measures of disease severity, including CAP score, UHDRS TFC, TMS and cognitive assessments. In this cohort, YKL-40 showed only moderate discriminatory power in distinguishing between premanifest and manifest HDGE groups, accurately classifying 93.8% of manifest HDGE individuals but only 62.5% of premanifest HDGE individuals [114].

Compared with controls, levels of YKL-40 in CSF were significantly elevated in the far-from-onset premanifest HDGE group in HD-YAS [34]. The premanifest HDGE group had significantly higher concentrations of CSF YKL-40, alongside CSF NfL and CSF mHTT, compared with controls, which suggests astrocytic activation due to mHTT-induced neuronal injury or cell-autonomous effects of mHTT astrocytes. Nonetheless, YKL-40 was less sensitive than CSF and plasma NfL [34].

Notably, YKL-40 levels in CSF were reduced in HDGE group enrolled in an open-label phase 2a clinical trial investigating the efficacy of anti-C1q therapy (ANX005; Annexon) [133]. ANX005 demonstrated target engagement and exhibited a decline in clinical progression within a subset of the HDGE group characterised by high baseline complement activity [133].

Recent considerations highlight the involvement of the complement pathway as a neural substrate for cognitive flexibility, linked to corticostriatal synapse loss [134, 135]. Cognitive flexibility has emerged as a potential cognitive biomarker reflecting potential disruptions in frontostriatal circuitry in HD [136]. Mild impairment in set-shifting was detected in a subset of the HDGE group in HD-YAS using the Cambridge Neuropsychological Test Automated Battery (CANTAB) Intra-Extra Dimensional Set Shift task [34, 136]. This impairment in attentional set-shifting correlated with altered functional connectivity in frontostriatal circuitry on resting-state functional MRI [136]. Longitudinal assessment of cognitive flexibility and CSF YKL-40 levels may provide greater insights into the role of YKL-40 as a fluid biomarker.

2.3.6 Striatal Markers

Peptides expressed in striatal medium spiny neuron populations have emerged as exploratory surrogate markers of striatal neurodegeneration in HD. Levels of proenkephalin (PENK) and prodynorphin (PDYN) in CSF were significantly reduced in manifest HDGE individuals, compared to other neurodegenerative conditions [116]. Levels of PENK in CSF were found to be significantly reduced in a manifest HDGE group, compared to premanifest HDGE and control groups [114, 115]. All three studies utilised liquid chromatography-mass spectrometry (LC-MS) within the methodology.

In one study, levels of PENK in CSF demonstrated the highest discriminatory potential for stratifying HDGE from control groups with improved discrimination when combined with NfL and albumin as an exploratory multi-marker CSF panel [114].

Striatal markers stand to gain from rigorous validation in large, well-powered studies, especially within premanifest HDGE cohorts.

2.3.7 Other Markers

Table 2 summarizes other CSF Markers of HD.

2.4 *Saliva*

Recently, saliva as a source for biomarker research in HD has garnered interest. As a non-invasive and easily accessible biofluid, saliva offers a rapid and simpler alternative to venepuncture and CSF collection for obtaining samples.

Elevations in inflammatory biomarkers such as interleukin-6 (IL-6), interleukin-1β (IL-1β) and C-reactive protein (CRP) have been observed in HDGE groups [73]. In addition, salivary concentrations of IL-6 and IL-1β showed a significant positive correlation with the UHDRS TMS and an inverse relationship with TFC. While these biomarkers are not specific for HD, altered levels detected in saliva may be biologically relevant and reflect inflammatory processes underlying neurodegeneration in HD.

Total HTT protein, encompassing both normal and mutant forms of the protein, has been detected and reproducibly measured in saliva in HDGE groups [137]. Akin to IL-6 and IL-1β, salivary tHTT also showed a significant positive correlation with UHDRS TMS and negative correlation with TFC, but not with DBS [137].

More recently, premanifest, and manifest HDGE groups had detectable salivary mHTT levels, compared with controls [138]. Salivary mHTT showed significant correlations with age, CAG repeat length and CAP score, which highlights the potential of salivary mHTT as a fluid biomarker in tracking disease progression in HD.

Saliva as a potential source of biomarkers for HD remains explorative and is in its infancy. Despite the novel approach, its viability for practical application and translation in clinical and research settings, especially in comparison to more established biomarkers like blood and CSF, remains uncertain and warrants further investigation.

3 Imaging Biomarkers

3.1 *Neuroimaging*

Neuroimaging methods have played an important role in advancing our understanding of the natural history of HD. Imaging biomarkers have emerged as robust contenders for endpoints in clinical trials, providing objective measures of disease progression and treatment efficacy. To this point, imaging biomarkers have already been employed in HD clinical trials as part of primary endpoints (ClinicalTrials.gov

Identifiers: NCT05686551, GENERATION HD2; NCT04120493, AMT-130) and secondary endpoints (ClinicalTrials.gov Identifier: NCT05358717, PIVOT-HD). Moreover, neuroimaging measures strongly correlate with fluid biomarkers, and clinical and functional measures, as detailed in Tables 1 and 2.

An ideal imaging biomarker would conveniently be available, accessible, cost-effective and capable of delivering reliable and reproducible results across different sites, irrespective of scanner-related factors like field strength and manufacturer [139].

Four key neuroimaging techniques employed in HD research include structural, diffusion, functional and molecular imaging. Each of these techniques provides independent yet complementary insights into the architecture and function of the brain, from macrostructural aspects captured by structural volumetric imaging to the microstructural nuances revealed by diffusion techniques. Assessing the effect of potential disease-modifying therapies in HD hinges on reliably detecting changes at both the cellular and macrostructural levels [5]. Therefore, imaging biomarkers derived from these techniques are increasingly sought after for evaluating target engagement and therapeutic responses. This section will discuss advancements and the current status of structural, diffusion, functional and molecular imaging in developing robust imaging biomarkers for HD.

3.2 Structural Imaging

The neuropathology of HD is characterised by loss of striatal projection neurons and progressive atrophy involving the entire cortex [140, 141]. *HTT* is expressed ubiquitously, and exon 1-encoded protein neuronal inclusions are present in affected regions of the brain prior to the onset of symptoms before any detectable signs of neurodegeneration are evident [142]. Neurodegeneration as sequelae of mHTT protein can be detected in premanifest HDGE groups prior to the emergence of motor signs of the disease. Multiple observational studies have consistently demonstrated the widespread effect on brain structure and volume, with progressive atrophy affecting the striatum (Fig. 2) and cortex, decades before predicted clinical motor diagnosis [30, 34, 37].

Importantly, estimates of atrophy correlate with cognitive, functional and motor decline [30]. Baseline striatal and grey matter volumes have emerged as predictors for HD diagnosis, while the reduction in striatal volume and the extent of cortical atrophy correlate with CAG repeat length [143, 144].

Brain atrophy occurs in normal ageing at a rate ranging between 0.1% and 0.3% per year [145–147]. Previous studies of premanifest HDGE individuals, with a median of 10.8 years for estimated years to onset, reported a whole-brain atrophy rate of 0.31% annually over a 2-year interval [32].

A premanifest HDGE group approximately 24 years before predicted clinical motor diagnosis was found to have a subtle reduction in putamen volumes, compared to a closely matched control group [34]. Interestingly, this finding did not

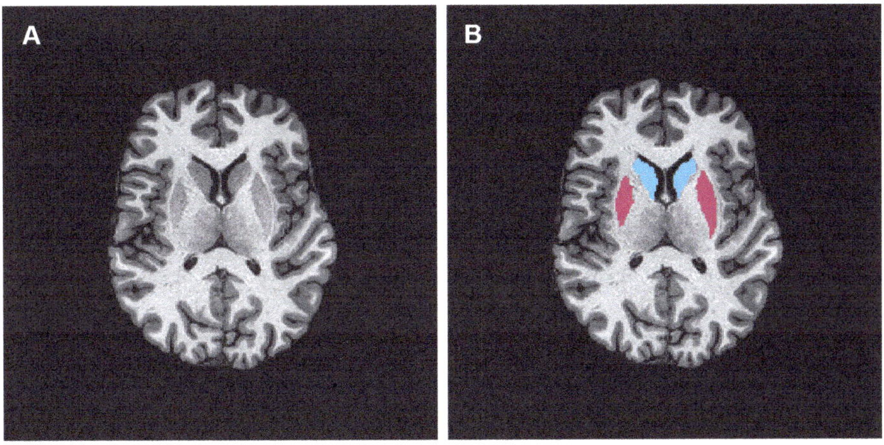

Fig. 2 Structural MRI for volumetric analysis. (**a**) Axial T1-weighted brain MRI, showing the basal ganglia as symmetrical deep grey matter structures. (**b**) The caudate (in blue) and the putamen (in pink) nuclei are highlighted as focal regions of interest susceptible to atrophy in HD, with striatal volume changes evident in premanifest HDGE groups decades before predicted clinical motor diagnosis.

correlate with predicted years to onset, which raises the question of a neurodevelopmental constitutive difference. Neurodevelopmental features associated with mHTT have been explored in an accelerated longitudinal study of HDGE children and adolescents [148]. Nevertheless, the reduction in putamen volumes observed in this far-from-onset HDGE group may reflect early neurodegenerative processes too nuanced to demonstrate robust correlation with disease burden at this time point. Longitudinal assessment of the HD-YAS cohort, integrating structural volumetric imaging and fluid biomarkers, holds promise in shedding light on the debate between early neurodegeneration and potential neurodevelopmental abnormalities.

A recent cross-study validation of structural imaging markers for HD highlighted the inclusion of caudate volume as a candidate imaging biomarker for future early-phase clinical trials [149]. This power analysis, which utilised datasets from the PREDICT-HD, TRACK-HD and IMAGE-HD cohorts, identified the caudate, pallidum and non-ventricular CSF as common regions showing significant effect sizes in comparisons between premanifest HDGE and control groups. Among these, caudate volume had the highest power [149].

To this point, clinical trials in HD are incorporating structural and volumetric MRI as imaging biomarkers within study designs. Tominersen (ClinicalTrials.gov Identifier: NCT05686551, GENERATION HD2), an ASO targeting *HTT* mRNA delivered via intrathecal injection, includes a change in baseline structural MRI as a primary outcome measure. PTC518 (ClinicalTrials.gov Identifier: NCT05358717, PIVOT-HD), an oral splicing modulator, has included change from baseline in caudate volume (via volumetric MRI) as a secondary outcome measure. This is expected, as MRI volumetric assessments of caudate volume and putamen volume have consistently shown the largest effect sizes for measurements at baseline and the rate of change longitudinally across all stages of HD [33].

3.3 Diffusion Imaging

Diffusion-weighted imaging (DWI) employs sequences that are sensitive to the motion of water molecules within the brain. In health, highly organised structures such as white matter fibres exhibit water diffusion in one direction, indicating anisotropy. However, when axonal damage occurs, leading to the breakdown of white matter fibres, there is an increase in diffusion in directions other than along the axons [139]. Thus, DWI has the potential to reveal pathological alterations in brain tissue architecture and provide insights into the selective targeting of neuronal connections in HD pathogenesis.

There are a number of techniques to process DWI data, the most widely used of which is diffusion tensor imaging (DTI). DTI markers, which quantify diffusion such as the apparent diffusion coefficient (ADC), fractional anisotropy (FA) and mean diffusivity (MD), provide surrogate measures of microstructural tissue change [150, 151]. Higher ADC values signify the tissue microstructure allows faster movement of water molecules. MD, which represents the average of ADC in three directions, typically increases in pathological tissue, particularly where there is a reduction in the density of cellular membranes [152]. In contrast, FA, which represents unidirectional diffusion, typically decreases with neural loss due to the diminished presence of cellular boundaries that impede diffusion, reflecting a decrease in microstructural organisation [150, 153].

White matter diffusivity changes related to HD pathology have been shown in the corpus callosum and around subcortical structures in the premanifest [154, 155] and manifest stages [156, 157] with later involvement of frontal, parietal and occipital white matter tracts as the disease progresses [158–160]. Another study failed to show significant differences in MD within the corpus callosum [161]. However, this study had a small sample size, and it is now well established that DTI is a sensitive tool for identifying the breakdown of white matter tracts across the disease span in HD [162].

Diffusion changes within grey matter are harder to interpret and imaging studies have provided conflicting results. Increased diffusivity in manifest HDGE individuals has been observed in the caudate and putamen [156, 161, 163] as well as the globus pallidus and thalamus [163]. However, Dumas and colleagues did not find significant differences between early manifest HDGE and control groups in diffusivity within the thalamus [156]. Interestingly, a number of studies have reported disease-related increases in FA within grey matter structures such as the striatum [160, 164], which paradoxically suggests *increased* tissue organisation as the disease progresses—a finding which may be difficult to understand biologically (see [162] for a discussion).

It is important to understand the relationship between DTI metrics and other measures of disease stage if this technique is to prove useful, for instance, as a biomarker for clinical trials. For example, Mascalchi and colleagues demonstrated increases in ADC values were associated with disease severity [157] and increased diffusion in the caudate and putamen correlates with CAG_n [163].

While inferences should exercise caution regarding causality, analyses correlating changes in DTI metrics with cognitive, neuropsychiatric and motor function provide insights into how distinct domains of HD may be associated with alterations in white matter pathways. Correlations have been observed between diffusivity measures in the corpus callosum and global cognitive performance [158] as well as more specifically performance on the Stroop task [160]. Additionally, correlations were found between MD values in the caudate and putamen and performance on the Mini-Mental State Examination (MMSE) [161]. MD changes in the corpus callosum have been associated with impaired motor performance [158] and increased diffusion in the striatum correlated inversely with UHDRS TFC [163]. Moreover, increased diffusivity in the corpus callosum, sensorimotor cortex and prefrontal cortex correlated with latency of antisaccade and percentage of errors in oculomotor function [156].

Regarding neuropsychiatric function, depression scores derived from the Hospital Anxiety and Depression Scale-Snaith Irritability Scale (HADS-SIS) were observed to correlate with reduced FA in the frontal cortex, anterior cingulate cortex, insula and cerebellum, without concurrent brain volume loss [165]. Diffusivity measures have also been shown to be associated with apathy [164]. This underscores the potential utility of diffusion measures as markers for the early detection of damage to white matter microstructure, which might not be evident on volumetric MRI scans.

Similar to cross-sectional studies of DTI in HD, disparities in findings persist in longitudinal studies. The first longitudinal study of DTI in HD utilised Tract-Based Spatial Statistics to compare diffusion metrics across whole-brain white matter and characterise changes in metrics over different disease stages [166]. In this study, a significant reduction of FA between baseline and follow-up DTI scans after a 1-year interval was observed in HDGE individuals ($n = 7$), compared to controls ($n = 7$) [166].

However, another longitudinal study reported no longitudinal changes in MD among HDGE individuals ($n = 18$), compared to controls ($n = 17$), in DWI acquired after a 1-year interval [161]. The discrepancies in longitudinal findings prompted exploration into whether progressive microstructural changes in HD may become evident in studies with larger sample sizes and with repeated measures over intervals longer than 1 year.

The IMAGE-HD study addressed this by examining a larger cohort of premanifest HDGE ($n = 40$), symptomatic HDGE ($n = 36$) and control ($n = 36$) groups and assessed longitudinal changes over 30 months using multimodal neuroimaging techniques [167]. In this study, only caudate FA demonstrated longitudinal differences between symptomatic HDGE and control groups. However, both premanifest and symptomatic HDGE groups demonstrated a significantly greater rate of change in caudate volume, compared to controls [167]. This suggests that diffusion measures may be less sensitive as imaging markers compared to volumetric measures in tracking disease progression in HD. Longitudinal neuroimaging analysis from the PADDINGTON study supports this inference, as effect sizes of microstructural (diffusion) metrics were smaller over shorter intervals, compared to macrostructural (volumetric) effect sizes [159, 168]. Over 6-, 9- and 15-months, effect sizes for longitudinal volumetric changes in caudate atrophy and ventricular expansion were larger in symptomatic HDGE, compared to control, groups [168].

Advancements in diffusion acquisition and modelling techniques hold promise for improving the sensitivity of diffusion measures in HD. Neurite orientation dispersion and density imaging (NODDI) is a diffusion MRI technique that can provide precise characterisation of axonal density and organisation by distinguishing intracellular, extracellular and CSF compartments within the microstructural environment [169].

In a premanifest HDGE group close to motor onset in TRACK-ON HD, NODDI revealed widespread reductions in axonal density and axonal reorganisation in the white matter surrounding the basal ganglia and appeared to be more sensitive than standard diffusion imaging [170]. No significant group differences were observed in standard diffusion and NODDI metrics between the far-from-onset premanifest HDGE group and the closely matched control group in HD-YAS [34]. This suggests that white matter integrity remains preserved in premanifest HDGE groups approximately 24 years before predicted clinical motor diagnosis. Compared to standard diffusion imaging techniques, NODDI appears more sensitive to HD-related white matter changes.

3.4 Functional Imaging

Functional MRI (fMRI) allows assessment of brain function during rest via resting-state fMRI (rs-fMRI), or with stimuli activating specific brain regions and networks via task-based fMRI. This imaging approach uses the blood oxygen level-dependent (BOLD) signal to measure brain activity—a biomarker generated by differences between oxyhaemoglobin and deoxyheamoglobin from changes in local blood flow associated with neuronal activity [171]. Compared to structural MRI, fMRI holds the potential advantage of detecting neurophysiological changes before structural alterations are evident [172].

Corticostriatal functional connectivity, assessed by synchrony of BOLD activity using rs-fMRI acquired on a 3-Tesla (3T) scanner, revealed reduced connectivity between the caudate nucleus and lateral premotor area in a prodromal HDGE group, compared to controls [173]. Subsequent fMRI studies, also acquired at 3T, have consistently demonstrated reduced functional connectivity in the striatal-motor pathway in HDGE groups [174–176]. Functional connectivity analyses during task-related fMRI studies have also uncovered alterations in frontostriatal circuits in premanifest HDGE groups [177, 178].

In another study using ultra-high field fMRI at 7-Tesla (7T), differential changes in functional connectivity were observed in the striatum-motor and striatum-prefrontal pathways among far-from-onset premanifest, close-to-onset premanifest and early manifest HDGE groups [179]. In all three HDGE groups, compared to controls, functional connectivity was altered between the premotor area and striatum as well as between the supplementary motor area and striatum [179].

In a premanifest HDGE group, there was evidence of reduced connectivity between the striatum and motor regions, alongside increased connectivity between the striatum and prefrontal regions [179]. This observation suggests a potential compensatory mechanism in the premanifest stage of HD in which additional cortical regions may be recruited to mitigate compromised connections and functions affected by HD pathology.

To this point, the investigation of compensatory processes in the premanifest stage of HD has been a particular area of interest in fMRI research endeavours. At this stage, premanifest HDGE individuals often maintain preserved levels of cognitive and motor function, despite subtle evidence of structural brain pathology. This preserved function is often attributed to the phenomenon of compensation; however, differentiating between functional changes due to HD-related pathology and changes due to compensatory behaviours is challenging [180–182].

Potential markers of compensation in premanifest HD have been investigated in the TRACK-ON HD cohort through a combination of rs-fMRI and task-based fMRI, while accounting for structural volumetric measures of the caudate, putamen, grey and white matter [182]. In this cohort, increased activation and connectivity in the right hemisphere were associated with better cognitive performance in premanifest HDGE individuals with increased atrophy and structural disease load [182]. However, despite this compensatory response to cognitive demands, no evidence of a similar compensatory mechanism was observed in motor function, as neither rs-fMRI nor task-based connectivity analyses demonstrated comparable patterns [182].

More recently, far-from-onset premanifest HDGE individuals in the HD-YAS cohort were found to have altered intrinsic functional connectivity between the caudate and dorsolateral prefrontal cortex during rs-fMRI [136]. Notably, these alterations were found to be associated with performance on attentional set-shifting in a task assessing cognitive flexibility, suggesting the possibility of a potential imaging biomarker for individual differences in cognitive flexibility in the premanifest stage of HD [136].

To date, fMRI biomarkers for application in clinical trials of HD have not yet been established. This is likely due to the variability in analytical methods across multiple study designs, limiting the standardisation of imaging acquisition parameters and the reproducibility and replicability of study findings. Lack of within-subject reproducibility and lack of longitudinal studies are limitations of using rs-fMRI [183]. With regard to task-based fMRI, variability in task performance poses a confounding factor that can influence activation patterns. In addition, ensuring consistency in task selection and instructions provided to study participants are additional factors warranting consideration. Addressing these limitations and establishing unified standardisation protocols in fMRI methodologies are essential steps required before fMRI can be incorporated as a reproducible—and therefore reliable—imaging biomarker in HD.

3.5 Molecular Imaging

3.5.1 PET

Positron emission tomography (PET) is a molecular imaging approach that employs radioactive tracers or radioligands, which can provide valuable insights into the molecular mechanisms underlying disease progression in HD [184]. By binding specifically to target molecules, for example, enzymes or receptors, PET enables visualisation and quantification of the distribution and concentration of radioligands in vivo.

PET studies in HD have employed a range of radioligands targeting phosphodiesterase 10A (PDE10A), an enzyme that is highly expressed in striatal medium spiny neurons [185]. At the molecular level, PDE10A plays a crucial role in modulating the downstream signalling pathways of cyclic adenosine monophosphate (cAMP) and cyclic guanosine monophosphate (cGMP), thereby influencing the regulation of striatal output and promoting neuronal survival [186]. Reductions in striatal PDE10A binding have been observed in premanifest HDGE groups, with further reductions as the disease progresses [186–189].

The radioligand [^{18}F]MNI-659 has been identified as a promising striatal imaging biomarker for in vivo assessment of PDE10A, with the potential for tracking disease progression in HD [188, 189]. In a cross-sectional analysis, reduced binding of [^{18}F]MNI-659 in the striatum was observed in premanifest and early manifest HDGE groups, exhibiting a robust correlation with disease severity assessed through UHDRS TMS, DBS and regional atrophy measures [188]. Furthermore, longitudinal follow-up PET scans approximately 1 year after the baseline scan revealed a decline in [^{18}F]MNI-659 uptake within the caudate nucleus and putamen among premanifest HDGE individuals [189].

Further investigation into the potential of [^{18}F]MNI-659 as an early disease state and pharmacodynamic biomarker has been undertaken in a larger cohort of premanifest and symptomatic HDGE groups [190]. Assessment of PDE10A availability in the striatum and globus pallidus, using [^{18}F]MNI-659 binding, revealed a significant reduction in HDGE groups across all disease stages, compared to controls, indicating diminished PDE10A availability in HD. Interestingly, a notable decline in PDE10A [^{18}F]MNI-659 binding was observed between the late premanifest and early symptomatic stages. These findings suggest that PDE10A holds potential as a molecular biomarker for capturing phenoconversion and the transition to symptomatic stages in HD. Moreover, the availability of PDE10A may represent an early neuroimaging indicator of change in HD progression [190].

Other PET studies in HD have explored alterations in dopaminergic function [191, 192], neuroinflammation [193–195] as well as metabolic and endocannabinoid neurotransmission [196–198], contributing to the molecular understanding of HD.

Using PET to directly measure mHTT in vivo is a more desirable approach for quantifying the therapeutic efficacy of HTT-lowering disease-modifying treatments, serving as a target engagement biomarker. Recent efforts have been directed towards developing radioligands with high affinity and selectivity for mHTT aggregates for clinical application. Leading novel candidates for PET imaging ligands with these properties, based on preclinical studies, include $[^{11}C]$CHDI-626 [199, 200] and ^{11}C ($[^{11}C]$CHDI-180R) [201]. While $[^{11}C]$CHDI-626 may preclude use in human in vivo PET imaging due to the risk of radiometabolite accumulation within the brain, ^{11}C ($[^{11}C]$CHDI-180R) has demonstrated encouraging kinetic properties [202]. These radioligands hold potential as PET imaging biomarkers to monitor the natural history of HD and as pharmacodynamic indicators to aid advances in therapeutic development.

While PET imaging offers the potential advantage of more sensitive biomarkers for target engagement, practical integration within the design and framework of studies and trials has logistical challenges. PET scanning, with limited accessibility and higher costs compared to MRI, along with the requirement for tracer isotope injection and exposure to ionising radiation, poses limitations on the feasibility of performing serial scans on the same subject.

3.5.2 MRS

Magnetic resonance spectroscopy (MRS) is a non-invasive imaging approach that has been studied as an imaging marker modality for tracking pathological changes at the biochemical level in HD. Alterations in key cerebral metabolites, such as N-acetylaspartate (NAA), indicative of neuronal integrity, creatine, reflecting brain energy metabolism and myo-inositol, a gliosis marker, have been demonstrated in regions of interest in MRS HD studies.

Studies have consistently shown reduced concentrations of NAA in the striatum of HDGE groups [203–210], highlighting its potential as an imaging biomarker of neuronal integrity.

Similarly, reductions in striatal creatine concentrations in HDGE groups have been observed [203–207], implicating alterations in energy metabolism in the brains of HDGE individuals.

Further MRS investigations into striatal metabolites in HDGE groups have shown elevated levels of lactate [204, 207], myo-inositol [203, 209, 210] and increased glutamate/glutamine activity in spectra localised to the striatum [207, 211].

Notably, reductions in NAA concentrations were found to correlate with the duration of symptoms [204] and UHDRS TMS [208]. Additionally, in both early manifest and premanifest HDGE groups, total NAA concentrations were correlated with DBS [210]. These findings suggest the potential utility of NAA as a metabolite in determining neurochemical responses to therapeutic interventions in HD.

MRS is evidently invaluable in identifying spectroscopic markers of the neurodegeneration in the striatum of HD, providing insights into the molecular mechanisms underlying HD-related neurodegeneration. Compared to structural and volumetric imaging, MRS provides a distinct advantage by allowing the quantification of dynamic biochemical processes implicated in the neuropathological cascades driving neurodegeneration. Moreover, compared to diffusion and functional imaging approaches, there appears to be a relatively established consensus with the methodologies used for MRS [212].

Nevertheless, MRS has received less investigation in HD, compared to other imaging approaches discussed. Future research on NAA, other cerebral metabolites of interest and the potential role of MRS as an imaging marker in HD requires further study with expansion across multiple sites and diverse, larger HDGE cohorts with longitudinal follow-up.

4 Conclusion

Advances in biofluid assay and imaging technologies have evidently improved the accuracy, reliability and sensitivity of biomarkers for HD. CSF mHTT, NfL and structural MRI have become pivotal in clinical trials, serving as critical measures for both primary and secondary outcomes. With the expansion of potentially transformative HD treatments on the horizon, the demand for reproducible, well-powered studies to validate efficacy biomarkers has never been more needed. These biomarkers must reliably correlate with disease progression and effectively indicate target engagement and response to disease-modifying treatments.

Striatal markers stand to gain from rigorous validation in large, well-powered studies, especially within premanifest HDGE cohorts. Additionally, integrating a combination of fluid biomarkers with imaging measures could lead to a more refined understanding of early neurodegeneration processes. Such an approach could help inform the timing for initiating disease-modifying treatments, ideally targeting HD-ISS stage 0 or 1, potentially altering the disease trajectory, and thereby preserving function.

References

1. The Huntington's Disease Collaborative Research Group. A novel gene containing a trinucleotide repeat that is expanded and unstable on Huntington's disease chromosomes. (1993). Available at: https://www.sciencedirect.com/science/article/pii/009286749390585E?via%3Dihub. Accessed 27 Sept 2022
2. Bates GP, et al. Huntington disease. In: Nature reviews disease primers. Nature Publishing Group; 2015. https://doi.org/10.1038/nrdp.2015.5.
3. Evans SJW, et al. Prevalence of adult Huntington's disease in the UK based on diagnoses recorded in general practice records. J Neurol Neurosurg Psychiatry. 2013;84(10):1156–60. https://doi.org/10.1136/jnnp-2012-304636.
4. Fisher ER, Hayden MR. Multisource ascertainment of Huntington disease in Canada: prevalence and population at risk. Mov Disord. 2014;29(1):105–14. https://doi.org/10.1002/mds.25717.

5. Tabrizi SJ, Estevez-Fraga C, et al. Potential disease-modifying therapies for Huntington's disease: lessons learned and future opportunities. Lancet Neurol. 2022b;21(7):645–58. https://doi.org/10.1016/S1474-4422(22)00121-1.
6. Farag M, et al. Advance care planning in Huntington's disease. J Huntington's Dis. 2023;12(1):77–82. https://doi.org/10.3233/JHD-220559.
7. Ridley RM, et al. Anticipation in Huntington's disease is inherited through the male line but may originate in the female. J Med Genet. 1988;25(9):589–95. https://doi.org/10.1136/jmg.25.9.589.
8. Wheeler VC, et al. Factors associated with HD CAG repeat instability in Huntington disease. J Med Genet. 2007;44(11):695–701. https://doi.org/10.1136/jmg.2007.050930.
9. Andrew SE, et al. The relationship between trinucleotide (CAG) repeat length and clinical features of Huntington's disease. Nat Genet. 1993;4:398. Available at: http://www.nature.com/naturegenetics
10. Snell RG, et al. Relationship between trinucleotide repeat expansion and phenotypic variation in Huntington's disease. Nat Genet. 1993;4:393. Available at: http://www.nature.com/naturegenetics
11. Langbehn DR, Registry Investigators of the European Huntington Disease Network. Longer CAG repeat length is associated with shorter survival after disease onset in Huntington disease. Am J Hum Genet. 2022;109(1):172–9. https://doi.org/10.1016/j.ajhg.2021.12.002.
12. Kennedy L, et al. Dramatic tissue-specific mutation length increases are an early molecular event in Huntington disease pathogenesis. Hum Mol Genet. 2003;12(24):3359–67. https://doi.org/10.1093/hmg/ddg352.
13. Swami M, et al. Somatic expansion of the Huntington's disease CAG repeat in the brain is associated with an earlier age of disease onset. Hum Mol Genet. 2009;18(16):3039–47. https://doi.org/10.1093/hmg/ddp242.
14. Lee JM, et al. Identification of genetic factors that modify clinical onset of Huntington's disease. Cell. 2015;162(3):516–26. https://doi.org/10.1016/j.cell.2015.07.003.
15. Lee JM, et al. CAG repeat not Polyglutamine length determines timing of Huntington's disease onset. Cell. 2019;178(4):887–900.e14. https://doi.org/10.1016/j.cell.2019.06.036.
16. Bunting E, et al. 'I01 Msh3-targeting antisense oligonucleotides halt CAG repeat expansions in Huntington's disease IPSC-derived neurons. In: I: Experimental therapeutics – preclinical. BMJ Publishing Group Ltd; 2022. p. A85.1-A85. https://doi.org/10.1136/jnnp-2022-ehdn.227.
17. Pouladi MA, Morton AJ, Hayden MR. Choosing an animal model for the study of Huntington's disease. Nat Rev Neurosci. 2013;14:708–21. https://doi.org/10.1038/nrn3570.
18. Neueder A, et al. The pathogenic exon 1 HTT protein is produced by incomplete splicing in Huntington's disease patients. Sci Rep. 2017;7(1):1307. https://doi.org/10.1038/s41598-017-01510-z.
19. Scherzinger E, et al. Huntingtin-encoded Polyglutamine expansions form amyloid-like protein aggregates in vitro and in vivo, cell, vol. 90; 1997. p. 549.
20. DiFiglia M, et al. Aggregation of huntingtin in neuronal Intranuclear inclusions and dystrophic neurites in brain. Science. 1997;277(5334):1990–3. https://doi.org/10.1126/science.277.5334.1990.
21. Schulte J, Littleton JT. The biological function of the huntingtin protein and its relevance to Huntington's disease pathology. Curr Trends Neurol. 2011;5:65–78. Available at: http://www.ncbi.nlm.nih.gov/pubmed/22180703
22. Browne SE, Flint Beal M. The energetics of Huntington's disease. Neurochem Res. 2004;29(3):531–46.
23. Kim J, et al. Mitochondrial loss, dysfunction and altered dynamics in Huntington's disease. Hum Mol Genet. 2010;19(20):3919–35. https://doi.org/10.1093/hmg/ddq306.

24. Björkqvist M, et al. A novel pathogenic pathway of immune activation detectable before clinical onset in Huntington's disease. J Exp Med. 2008;205(8):1869–77. https://doi.org/10.1084/jem.20080178.
25. Sapp E, et al. Early and progressive accumulation of reactive microglia in the Huntington disease brain. J Neuropathol Exp Neurol. 2001. Available at: https://academic.oup.com/jnen/article/60/2/161/2610064.
26. Edwards MJ, et al. 'Chorea', in OSH Parkinson's disease and other movement disorders. Oxford University Press; 2016. p. 159–74. https://doi.org/10.1093/med/9780198705062.003.0007.
27. Quarrell OWJ. Juvenile Huntington's disease. Oxford University Press; 2014. https://doi.org/10.1093/med/9780199929146.003.0004.
28. Huntington Study Group. Unified Huntington's disease rating scale: reliability and consistency. Mov Disord. 1996;11(2):136–42. https://doi.org/10.1002/mds.870110204.
29. Biglan KM, et al. Motor abnormalities in premanifest persons with Huntington's disease: the PREDICT-HD study. Mov Disord. 2009;24(12):1763–72. https://doi.org/10.1002/mds.22601.
30. Tabrizi SJ, et al. Articles biological and clinical manifestations of Huntington's disease in the longitudinal TRACK-HD study: cross-sectional analysis of baseline data. Lancet Neurol. 2009;8:791–801. https://doi.org/10.1016/S1474.
31. Dorsey ER, et al. Natural history of Huntington disease. JAMA Neurol. 2013;70(12):1520–30. https://doi.org/10.1001/jamaneurol.2013.4408.
32. Tabrizi SJ, et al. Potential endpoints for clinical trials in premanifest and early Huntington's disease in the TRACK-HD study: analysis of 24 month observational data. Lancet Neurol. 2012;11(1):42–53. https://doi.org/10.1016/S1474-4422(11)70263-0.
33. Tabrizi SJ, et al. Predictors of phenotypic progression and disease onset in premanifest and early-stage Huntington's disease in the TRACK-HD study: analysis of 36-month observational data. Lancet Neurol. 2013;12(7):637–49. https://doi.org/10.1016/S1474-4422(13)70088-7.
34. Scahill RI, et al. Biological and clinical characteristics of gene carriers far from predicted onset in the Huntington's disease Young Adult Study (HD-YAS): a cross-sectional analysis. Lancet Neurol. 2020;19(6):502–12. https://doi.org/10.1016/S1474-4422(20)30143-5.
35. Landwehrmeyer GB, et al. Data analytics from enroll-HD, a global clinical research platform for Huntington's disease. Mov Dis Clin Pract. 2017;4(2):212–24. https://doi.org/10.1002/mdc3.12388.
36. Reilmann R, et al. A teaching film, video library and online certification for the unified Huntington's disease rating scale Total motor score. Aktuelle Neurologie. 2009;36(S:02. https://doi.org/10.1055/s-0029-1238568.
37. Paulsen JS, et al. Detection of Huntington's disease decades before diagnosis: the Predict-HD study. J Neurol Neurosurg Psychiatry. 2008;79(8):874–80. https://doi.org/10.1136/jnnp.2007.128728.
38. Stout JC, et al. Neurocognitive signs in prodromal Huntington disease. Neuropsychology. 2011;25(1):1–14. https://doi.org/10.1037/a0020937.
39. Paulsen JS, et al. Neuropsychiatric aspects of Huntington's disease. J Neurol Neurosurg Psychiatry. 2001;71(3):310–4. https://doi.org/10.1136/jnnp.71.3.310.
40. van Duijn E, et al. Incidence of completed suicide and suicide attempts in a global prospective study of Huntington's disease. BJPsych Open. 2021;7(5):e158. https://doi.org/10.1192/bjo.2021.969.
41. Rodrigues FB, et al. Survival, mortality, causes and places of death in a European Huntington's disease prospective cohort. Movement Dis Clin Pract. 2017;4(5):737–42. https://doi.org/10.1002/mdc3.12502.
42. Shoulson I, Fahn S. Huntington disease: clinical care and evaluation. Neurology. 1979;29(1):1–1. https://doi.org/10.1212/WNL.29.1.1.
43. Tabrizi SJ, Schobel S, et al. A biological classification of Huntington's disease: the Integrated Staging System. Lancet Neurol. 2022a;21(7):632–44. https://doi.org/10.1016/S1474-4422(22)00120-X.

44. Weir DW, Sturrock A, Leavitt BR. Development of biomarkers for Huntington's disease. Lancet Neurol. 2011;10(6):573–90. https://doi.org/10.1016/S1474-4422(11)70070-9.
45. Rissin DM, et al. Single-molecule enzyme-linked immunosorbent assay detects serum proteins at subfemtomolar concentrations. Nat Biotechnol. 2010;28(6):595–9. https://doi.org/10.1038/nbt.1641.
46. Wilson DH, et al. The Simoa HD-1 analyzer: a novel fully automated digital immunoassay analyzer with single-molecule sensitivity and multiplexing. SLAS Technol. 2016;21(4):533–47. https://doi.org/10.1177/2211068215589580.
47. Byrne LM, et al. Neurofilament light protein in blood as a potential biomarker of neurodegeneration in Huntington's disease: a retrospective cohort analysis. Lancet Neurol. 2017;16(8):601–9. https://doi.org/10.1016/S1474-4422(17)30124-2.
48. Byrne LM, Rodrigues FB, Johnson EB, et al. Evaluation of mutant huntingtin and neurofilament proteins as potential markers in Huntington's disease. Sci Transl Med. 2018b;10(458):7108. https://doi.org/10.1126/scitranslmed.aat7108.
49. Rodrigues FB, et al. Mutant huntingtin and neurofilament light have distinct longitudinal dynamics in Huntington's disease. Sci Transl Med. 2020; Available at: https://www.science.org
50. You H, et al. Evaluation of blood glial fibrillary acidic protein as a potential marker in Huntington's disease. Front Neurol. 2021;12 https://doi.org/10.3389/fneur.2021.779890.
51. Sampedro F, et al. Cortical microstructural correlates of plasma neurofilament light chain in Huntington's disease. Parkinsonism Relat Dis. 2021;85:91–4. https://doi.org/10.1016/j.parkreldis.2021.03.008.
52. Parkin GM, et al. Associations between prognostic index scores and plasma neurofilament light in Huntington's disease. Parkinsonism Relat Dis. 2022;97:25–8. https://doi.org/10.1016/j.parkreldis.2022.02.023.
53. Wild EJ, et al. Plasma neurofilament heavy chain levels in Huntington's disease. Neurosci Lett. 2007;417(3):231–3. https://doi.org/10.1016/j.neulet.2007.02.053.
54. Sampedro F, et al. Plasma TDP-43 reflects cortical neurodegeneration and correlates with neuropsychiatric symptoms in Huntington's disease. Clin Neuroradiol. 2022;32(4):1077–85. https://doi.org/10.1007/s00062-022-01150-5.
55. Tasset I, et al. NGF and nitrosative stress in patients with Huntington's disease. J Neurol Sci. 2012;315(1–2):133–6. https://doi.org/10.1016/j.jns.2011.12.014.
56. Plinta K, et al. The utility of bdnf detection in assessing severity of huntington's disease. J Clin Med. 2021;10(21) https://doi.org/10.3390/jcm10215181.
57. Ou ZYA, et al. Brain-derived neurotrophic factor in cerebrospinal fluid and plasma is not a biomarker for Huntington's disease. Sci Rep. 2021;11(1):3481. https://doi.org/10.1038/s41598-021-83000-x.
58. Ciammola A, et al. Low brain-derived neurotrophic factor (BDNF) levels in serum of Huntington's disease patients. Am J Med Genet B Neuropsychiatr Genet. 2007;144(4):574–7. https://doi.org/10.1002/ajmg.b.30501.
59. Zuccato C, et al. Brain-derived neurotrophic factor in patients with Huntington's disease. PLoS One. 2011;6(8):e22966. https://doi.org/10.1371/journal.pone.0022966.
60. Unti E, et al. Social cognition and oxytocin in Huntington's disease: new insights. Brain Sci. 2018;8(9) https://doi.org/10.3390/brainsci8090161.
61. Fisher ER, et al. The relationship between plasma oxytocin and executive functioning in Huntington's disease: a pilot study. J Huntington's Dis. 2021;10(3):349–54. https://doi.org/10.3233/JHD-210467.
62. Hellem MNN, et al. Decreased CSF oxytocin relates to measures of social cognitive impairment in Huntington's disease patients. Parkinsonism Relat Disord. 2022;99:23–9. https://doi.org/10.1016/j.parkreldis.2022.05.003.
63. Long JD, et al. 8OHdG as a marker for Huntington disease progression. Neurobiol Dis. 2012;46(3):625–34. https://doi.org/10.1016/j.nbd.2012.02.012.

64. Cheng M-L, et al. Humic acid induces oxidative DNA damage, growth retardation, and apoptosis in human primary fibroblasts. Exp Biol Med (Maywood, N.J.). 2003;228(4):413–23. https://doi.org/10.1177/153537020322800412.
65. Chen CM, et al. Increased oxidative damage and mitochondrial abnormalities in the peripheral blood of Huntington's disease patients. Biochem Biophys Res Commun. 2007;359(2):335–40. https://doi.org/10.1016/j.bbrc.2007.05.093.
66. Ciancarelli I, et al. Peripheral biomarkers of oxidative stress and their limited potential in evaluation of clinical features of Huntington's patients. Biomarkers. 2014;19(6):452–6. https://doi.org/10.3109/1354750X.2014.935955.
67. Stoy N, et al. Tryptophan metabolism and oxidative stress in patients with Huntington's disease. J Neurochem. 2005;93(3):611–23. https://doi.org/10.1111/j.1471-4159.2005.03070.x.
68. Christofides J, et al. Blood 5-hydroxytryptamine, 5-hydroxyindoleacetic acid and melatonin levels in patients with either Huntington's disease or chronic brain injury. J Neurochem. 2006;97(4):1078–88. https://doi.org/10.1111/j.1471-4159.2006.03807.x.
69. Klepac N, et al. Oxidative stress parameters in plasma of Huntington's disease patients, asymptomatic Huntington's disease gene carriers and healthy subjects: a cross-sectional study. J Neurol. 2007;254(12):1676–83. https://doi.org/10.1007/s00415-007-0611-y.
70. Duran R, et al. Oxidative stress and plasma aminopeptidase activity in Huntington's disease. J Neural Transm. 2010;117(3):325–32. https://doi.org/10.1007/s00702-009-0364-0.
71. Paglia DE, Valentine WN. Studies on the quantitative and qualitative characterization of erythrocyte glutathione peroxidase. J Lab Clin Med. 1967;70(1):158–69.
72. Sánchez-López F, et al. Oxidative stress and inflammation biomarkers in the blood of patients with huntington's disease. Neurol Res. 2012;34(7):721–4. https://doi.org/10.1179/1743132812Y.0000000073.
73. Corey-Bloom J, Fischer RS, et al. Levels of interleukin-6 in saliva, but not plasma, correlate with clinical metrics in huntington's disease patients and healthy control subjects. Int J Mol Sci. 2020a;21(17):1–12. https://doi.org/10.3390/ijms21176363.
74. Corey-Bloom J, Haque A, et al. Uric acid as a potential peripheral biomarker for disease features in Huntington's patients. Front Neurosci. 2020b;14 https://doi.org/10.3389/fnins.2020.00073.
75. Leblhuber F, et al. Activated immune system in patients with Huntington's disease. Clin Chem Lab Med. 1998;36:747.
76. Forrest CM, et al. Blood levels of kynurenines, interleukin-23 and soluble human leucocyte antigen-G at different stages of Huntington's disease. J Neurochem. 2010;112(1):112–22. https://doi.org/10.1111/j.1471-4159.2009.06442.x.
77. Rodrigues FB, et al. Kynurenine pathway metabolites in cerebrospinal fluid and blood as potential biomarkers in Huntington's disease. J Neurochem. 2021;158(2):539–53. https://doi.org/10.1111/jnc.15360.
78. Bouwens JA, et al. Plasma cytokine levels in relation to neuropsychiatric symptoms and cognitive dysfunction in Huntington's disease. J Huntington's Dis. 2016;5(4):369–77. https://doi.org/10.3233/JHD-160213.
79. Dalrymple A, et al. Proteomic profiling of plasma in Huntington's disease reveals neuroinflammatory activation and biomarker candidates. J Proteome Res. 2007;6(7):2833–40. https://doi.org/10.1021/pr0700753.
80. Chang KH, et al. Plasma inflammatory biomarkers for Huntington's disease patients and mouse model. Brain Behav Immun. 2015;44:121–7. https://doi.org/10.1016/j.bbi.2014.09.011.
81. Silajdžić E, et al. A critical evaluation of inflammatory markers in huntingtons disease plasma. J Huntington's Dis. 2013;2(1):125–34. https://doi.org/10.3233/JHD-130049.
82. Leoni V, et al. Plasma 24S-hydroxycholesterol and caudate MRI in pre-manifest and early Huntington's disease. Brain. 2008;131(11):2851–9. https://doi.org/10.1093/brain/awn212.
83. Leoni V, et al. Whole body cholesterol metabolism is impaired in Huntington's disease. Neurosci Lett. 2011;494(3):245–9. https://doi.org/10.1016/j.neulet.2011.03.025.

84. Leoni V, et al. Plasma 24S-hydroxycholesterol correlation with markers of Huntington disease progression. Neurobiol Dis. 2013;55:37–43. https://doi.org/10.1016/j.nbd.2013.03.013.
85. Gruber BM, et al. Huntington' disease-imbalance of amino acid levels in plasma of patients and mutation carriers. Ann Agric Environ Med. Available at: www.aaem.pl. 2013;20:779.
86. Cheng ML, et al. Metabolic disturbances in plasma as biomarkers for Huntington's disease. J Nutr Biochem. 2016;31:38–44. https://doi.org/10.1016/j.jnutbio.2015.12.001.
87. Mochel F, et al. Early energy deficit in Huntingdon disease: identification of a plasma biomarker traceable during disease progression. PLoS One. 2007;2(7):e647. https://doi.org/10.1371/journal.pone.0000647.
88. Rocha NP, et al. Peripheral levels of renin-angiotensin system components are associated with cognitive performance in Huntington's disease. Front Neurosci. 2020;14 https://doi.org/10.3389/fnins.2020.594945.
89. Ahmad Aziz N, et al. Leptin secretion rate increases with higher CAG repeat number in Huntington's disease patients. Clin Endocrinol. 2010;73(2):206–11. https://doi.org/10.1111/j.1365-2265.2009.03661.x.
90. Saleh N, et al. Neuroendocrine disturbances in Huntington's disease. PLoS One. 2009;4(3):e4962. https://doi.org/10.1371/journal.pone.0004962.
91. Leblhuber F, et al. Serum dehydroepiandrosterone and cortisol measurements in Huntington's chorea. J Neurol Sci. 1995;132:76.
92. Adamczak-Ratajczak A, et al. Circadian rhythms of melatonin and cortisol in manifest Huntington's disease and in acute cortical ischemic stroke. J Physiol Pharmacol. 2017;68(4):539–46. Available at: http://www.ncbi.nlm.nih.gov/pubmed/29151070.
93. Saleh N, et al. High insulinlike growth factor I is associated with cognitive decline in Huntington disease On behalf of the Huntington French Speaking Group. Neurology. 2010; Available at: http://www.hdnetwork.org
94. Aziz NA, et al. Growth hormone and ghrelin secretion are associated with clinical severity in Huntington's disease. Eur J Neurol. 2010;17(2):280–8. https://doi.org/10.1111/j.1468-1331.2009.02798.x.
95. Johnson EB, et al. Neurofilament light protein in blood predicts regional atrophy in Huntington disease. Neurology. 2018;90(8):e717–23. https://doi.org/10.1212/WNL.0000000000005005.
96. Byrne LM, Rodrigues FB, Johnson EB, et al. Cerebrospinal fluid neurogranin and TREM2 in Huntington's disease. Sci Rep. 2018a;8(1):4260. https://doi.org/10.1038/s41598-018-21788-x.
97. Hassan YR, et al. Lumbar puncture safety and tolerability in premanifest and manifest Huntington's disease: a multi-analysis cross-sectional study. Sci Rep. 2022;12(1):18377. https://doi.org/10.1038/s41598-022-21934-6.
98. Byrne LM, Wild EJ. Cerebrospinal fluid biomarkers for Huntington's disease. J Huntington's Dis. 2016;5(1):1–13. https://doi.org/10.3233/JHD-160196.
99. Wild EJ, et al. Quantification of mutant huntingtin protein in cerebrospinal fluid from Huntington's disease patients. J Clin Invest. 2015;125(5):1979–86. https://doi.org/10.1172/JCI80743.
100. Fodale V, et al. Validation of ultrasensitive mutant huntingtin detection in human cerebrospinal fluid by single molecule counting immunoassay. J Huntington's Dis. 2017;6(4):349–61. https://doi.org/10.3233/JHD-170269.
101. Tabrizi SJ, et al. Targeting huntingtin expression in patients with Huntington's disease. N Engl J Med. 2019;380(24):2307–16. https://doi.org/10.1056/nejmoa1900907.
102. McColgan P, et al. Tominersen in adults with manifest Huntington's disease. N Engl J Med. 2023;389(23):2203–5. https://doi.org/10.1056/nejmc2300400.
103. Southwell AL, et al. Ultrasensitive measurement of huntingtin protein in cerebrospinal fluid demonstrates increase with Huntington disease stage and decrease following brain huntingtin suppression. Sci Rep. 2015;5 https://doi.org/10.1038/srep12166.
104. Pfalzer AC, et al. Alterations in metal homeostasis occur prior to canonical markers in Huntington disease. Sci Rep. 2022;12(1):10373. https://doi.org/10.1038/s41598-022-14169-y.

105. Constantinescu R, et al. Levels of the light subunit of neurofilament triplet protein in cerebrospinal fluid in Huntington's disease. Parkinsonism Relat Disord. 2009;15(3):245–8. https://doi.org/10.1016/j.parkreldis.2008.05.012.
106. Vinther-Jensen T, et al. Selected CSF biomarkers indicate no evidence of early neuroinflammation in Huntington disease. Neurol Neuroimmunol NeuroInflamm. 2016;3(6):e287. https://doi.org/10.1212/NXI.0000000000000287.
107. Niemelä V, et al. Tau or neurofilament light-which is the more suitable biomarker for Huntington's disease? PLoS One. 2017;12(2):e0172762. https://doi.org/10.1371/journal.pone.0172762.
108. Niemelä V, et al. Cerebrospinal fluid sCD27 levels indicate active T cell-mediated inflammation in premanifest Huntington's disease. PLoS One. 2018;13(2):e0193492. https://doi.org/10.1371/journal.pone.0193492.
109. Szejko N, et al. Quantification of the light subunit of Neurofilament protein in cerebrospinal fluid of Huntington's disease patients. PLoS Curr. 2018;10 https://doi.org/10.1371/currents.hd.280c8f9f7d9fa4f7f0c883d9f8e807da.
110. Constantinescu R, et al. Increased levels of total tau protein in the cerebrospinal fluid in Huntington's disease. Parkinsonism Relat Disord. 2011;17:714–5. https://doi.org/10.1016/j.parkreldis.2011.06.010.
111. Rodrigues FB, Byrne L, et al. Cerebrospinal fluid total tau concentration predicts clinical phenotype in Huntington's disease. J Neurochem. 2016b;139(1):22–5. https://doi.org/10.1111/jnc.13719.
112. Vinther-Jensen T, et al. YKL-40 in cerebrospinal fluid in Huntington's disease – a role in pathology or a nonspecific response to inflammation? Parkinsonism Relat Dis (Elsevier Ltd). 2014;20:1301–3. https://doi.org/10.1016/j.parkreldis.2014.08.011.
113. Rodrigues FB, Byrne LM, et al. Cerebrospinal fluid inflammatory biomarkers reflect clinical severity in huntington's disease. PLoS One. 2016a;11(9):e0163479. https://doi.org/10.1371/journal.pone.0163479.
114. Caron NS, et al. Cerebrospinal fluid biomarkers for assessing Huntington disease onset and severity. Brain Commun. 2022;4(6) https://doi.org/10.1093/braincomms/fcac309.
115. Niemela V, et al. Proenkephalin decreases in cerebrospinal fluid with symptom progression of Huntington's disease. Mov Disord. 2021;36(2):481–91. https://doi.org/10.1002/mds.28391.
116. Barschke P, et al. Cerebrospinal fluid levels of proenkephalin and prodynorphin are differentially altered in Huntington's and Parkinson's disease. J Neurol. 2022;269(9):5136–43. https://doi.org/10.1007/s00415-022-11187-8.
117. Al Shweiki MHDR, et al. Cerebrospinal fluid levels of Prodynorphin-derived peptides are decreased in Huntington's disease. Mov Disord. 2021;36(2):492–7. https://doi.org/10.1002/mds.28300.
118. Beal MF, et al. Kynurenine pathway measurements in Huntington's disease striatum: evidence for reduced formation of Kynurenic acid. J Neurochem. 1990;55(4):1327–39. https://doi.org/10.1111/j.1471-4159.1990.tb03143.x.
119. Heyes MP, et al. Regional brain and cerebrospinal fluid quinolinic acid concentrations in Huntington's disease. Neurosci Lett. 1991;122:265.
120. Meier A, et al. Normal hypocretin-1 (orexin-A) levels in the cerebrospinal fluid of patients with Huntington's disease. Brain Res. 2005;1063(2):201–3. https://doi.org/10.1016/j.brainres.2005.09.028.
121. Björkqvist M, et al. Cerebrospinal fluid levels of orexin-A are not a clinically useful biomarker for Huntington disease. Clin Genet. 2006;70(1):78–9. https://doi.org/10.1111/j.1399-0004.2006.00636.x.
122. Aziz A, et al. Hypocretin and melanin-concentrating hormone in patients with Huntington disease. Brain Pathol. 2008;18(4):474–83. https://doi.org/10.1111/j.1750-3639.2008.00135.x.
123. Garrett MC, Soares-da-Silva P. Increased cerebrospinal fluid dopamine and 3,4-Dihydroxyphenylacetic acid levels in Huntington's disease: evidence for an over-

active dopaminergic brain transmission. J Neurochem. 1992;58(1):101–6. https://doi.org/10.1111/j.1471-4159.1992.tb09283.x.
124. Manyam NV, et al. Levels of gamma-aminobutyric acid in cerebrospinal fluid in various neurologic disorders. Arch Neurol. 1980;37(6):352–5. https://doi.org/10.1001/archneur.1980.00500550054006.
125. Macdonald D, et al. Quantification assays for total and polyglutamine-expanded huntingtin proteins. PLoS One. 2014;9(5):e96854. https://doi.org/10.1371/journal.pone.0096854.
126. Hensman Moss DJ, et al. Quantification of huntingtin protein species in Huntington's disease patient leukocytes using optimised electrochemiluminescence immunoassays. PLoS One. 2017;12(12):e0189891. https://doi.org/10.1371/journal.pone.0189891.
127. Cariulo C, et al. Phosphorylation of huntingtin at residue T3 is decreased in Huntington's disease and modulates mutant huntingtin protein conformation. Proc Natl Acad Sci. 2017;114(50):E10809. https://doi.org/10.1073/pnas.1705372114.
128. Spillantini MG, Goedert M. Tau pathology and neurodegeneration. Lancet Neurol. 2013. www.thelancet.com/neurology. Available at: www.thelancet.com/neurology.
129. Grundke-Iqbal I, et al. Abnormal phosphorylation of the microtubule-associated protein X (tau) in Alzheimer cytoskeletal pathology (Alzheimer disease/neurofibrillary tangles/paired-helical filaments/microtubules), vol. 1. Med Sci; 1986. p. 202. Available at: https://www.pnas.org
130. Blum D, et al. Mutant huntingtin alters Tau phosphorylation and subcellular distribution. Hum Mol Genet. 2015;24(1):76–85. https://doi.org/10.1093/hmg/ddu421.
131. Vuono R, et al. The role of tau in the pathological process and clinical expression of Huntington's disease. Brain. 2015;138(7):1907–18. https://doi.org/10.1093/brain/awv107.
132. Llorens F, et al. YKL-40 in the brain and cerebrospinal fluid of neurodegenerative dementias. Mol Neurodegener. 2017;12(1):83. https://doi.org/10.1186/s13024-017-0226-4.
133. Kumar R, et al. A phase 2 open-label study to assess the safety, tolerability, pharmacokinetics, and pharmacodynamics of intravenous ANX005 in patients with, or at risk of, manifest Huntington's disease (HD) (S32.009). Neurology. 2023;100(17_supplement_2) https://doi.org/10.1212/WNL.0000000000203217.
134. Hamilton J, Farag M, Tabrizi SJ. Complementary insights into corticostriatal synapse loss and cognition in Huntington's disease'. Cell Rep Med (Cell Press). 2023;4 https://doi.org/10.1016/j.xcrm.2023.101314.
135. Wilton DK, et al. Microglia and complement mediate early corticostriatal synapse loss and cognitive dysfunction in Huntington's disease. Nat Med. 2023;29:2866. https://doi.org/10.1038/s41591-023-02566-3.
136. Langley C, et al. Fronto-striatal circuits for cognitive flexibility in far from onset Huntington's disease: evidence from the Young Adult Study. J Neurol Neurosurg Psychiatry. 2021;92(2):143–9. https://doi.org/10.1136/jnnp-2020-324104.
137. Corey-Bloom J, et al. Salivary levels of total huntingtin are elevated in Huntington's disease patients. Sci Rep. 2018;8(1):7371. https://doi.org/10.1038/s41598-018-25095-3.
138. Parkin GM, et al. Salivary Huntingtin protein is uniquely associated with clinical features of Huntington's disease. Sci Rep. 2023;13(1):1034. https://doi.org/10.1038/s41598-023-28019-y.
139. Zeun P, et al. Fluid and imaging biomarkers for Huntington's disease. Mol Cell Neurosci. 2019;97:67. https://doi.org/10.1016/j.mcn.2019.02.004.
140. Vonsattel JPG, DiFiglia M. Huntington disease. J Neuropathol Exp Neurol. 1998; 57(5):369–84. https://doi.org/10.1097/00005072-199805000-00001.
141. Vonsattel JPG, Keller C, Cortes Ramirez EP. Huntington's disease – neuropathology. Handb Clin Neurol. 2011;100:83–100. https://doi.org/10.1016/B978-0-444-52014-2.00004-5.
142. Sathasivam K, et al. Formation of polyglutamine inclusions in non-CNS tissue. Hum Mol Genet. 1999;8:813. Available at: https://academic.oup.com/hmg/article/8/5/813/662521
143. Bechtel N, et al. Tapping linked to function and structure in premanifest and symptomatic Huntington disease. Neurology. 2010;75:2150. Available at: www.neurology.org

144. Rosas HD, et al. Striatal volume loss in HD as measured by MRI and the influence of CAG repeat. Neurology. 2001;57(6):1025–8.
145. Fisher E, et al. Gray matter atrophy in multiple sclerosis: a longitudinal study. Ann Neurol. 2008;64(3):255–65. https://doi.org/10.1002/ana.21436.
146. Fotenos AF, et al. Normative estimates of cross-sectional and longitudinal brain volume decline in aging and AD. Neurology. 2005;64(6):1032–9. https://doi.org/10.1212/01.WNL.0000154530.72969.11.
147. Scahill RI, et al. A longitudinal study of brain volume changes in Normal aging using serial registered magnetic resonance imaging. Arch Neurol. 2003;60:989.
148. van der Plas E, et al. Abnormal brain development in child and adolescent carriers of mutant huntingtin. Neurology. 2019;93(10):e1021. https://doi.org/10.1212/WNL.0000000000008066.
149. Wijeratne PA, et al. Robust markers and sample sizes for multicenter trials of Huntington disease. Ann Neurol. 2020;87(5):751–62. https://doi.org/10.1002/ana.25709.
150. Beaulieu C. The basis of anisotropic water diffusion in the nervous system—a technical review. NMR Biomed. 2002;15(7–8):435–55. https://doi.org/10.1002/nbm.782.
151. Mamata H, Jolesz FA, Maier SE. Characterization of central nervous system structures by magnetic resonance diffusion anisotropy. Neurochem Int. 2004;45(4):553–60. https://doi.org/10.1016/j.neuint.2003.11.014.
152. Douaud G, et al. In vivo evidence for the selective subcortical degeneration in Huntington's disease. NeuroImage. 2009;46(4):958–66. https://doi.org/10.1016/j.neuroimage.2009.03.044.
153. Pierpaoli C, et al. Water diffusion changes in wallerian degeneration and their dependence on white matter architecture. NeuroImage. 2001;13(6):1174–85. https://doi.org/10.1006/nimg.2001.0765.
154. Poudel GR, et al. White matter connectivity reflects clinical and cognitive status in Huntington's disease. Neurobiol Dis. 2014;65:180–7. https://doi.org/10.1016/j.nbd.2014.01.013.
155. Rosas HD, et al. Altered white matter microstructure in the corpus callosum in Huntington's disease: implications for cortical "disconnection". NeuroImage. 2010;49(4):2995–3004. https://doi.org/10.1016/j.neuroimage.2009.10.015.
156. Dumas EM, et al. Early changes in white matter pathways of the sensorimotor cortex in premanifest Huntington's disease. Hum Brain Mapp. 2012;33(1):203–12. https://doi.org/10.1002/hbm.21205.
157. Mascalchi M, et al. Huntington disease: volumetric, diffusion-weighted, and magnetization transfer MR imaging of brain. Radiology. 2004;232(3):867–73. https://doi.org/10.1148/radiol.2322030820.
158. Bohanna I, et al. Diffusion Tensor Imaging in Huntington's disease reveals distinct patterns of white matter degeneration associated with motor and cognitive deficits. Brain Imaging Behav. 2011;5(3):171–80. https://doi.org/10.1007/s11682-011-9121-8.
159. Hobbs NZ, et al. Evaluation of multi-modal, multi-site neuroimaging measures in Huntington's disease: baseline results from the PADDINGTON study. NeuroImage Clin. 2013;2(1):204–11. https://doi.org/10.1016/j.nicl.2012.12.001.
160. Rosas HD, et al. Diffusion tensor imaging in presymptomatic and early Huntington's disease: selective white matter pathology and its relationship to clinical measures. Mov Disord. 2006;21(9):1317–25. https://doi.org/10.1002/mds.20979.
161. Sritharan A, et al. A longitudinal diffusion tensor imaging study in symptomatic Huntington's disease. J Neurol Neurosurg Psychiatry. 2010;81(3):257–62. https://doi.org/10.1136/jnnp.2007.142786.
162. Estevez-Fraga C, et al. Diffusion imaging in Huntington's disease: comprehensive review. J Neurol Neurosurg Psychiatry (BMJ Publishing Group). 2021;92:62–9. https://doi.org/10.1136/jnnp-2020-324377.
163. Seppi K, et al. Diffusion-weighted imaging in Huntington's disease. Mov Disord. 2006;21(7):1043–7. https://doi.org/10.1002/mds.20868.
164. Delmaire C, et al. The structural correlates of functional deficits in early huntington's disease. Hum Brain Mapp. 2013;34(9):2141–53. https://doi.org/10.1002/hbm.22055.

165. Sprengelmeyer R, et al. The neuroanatomy of subthreshold depressive symptoms in Huntington's disease: a combined diffusion tensor imaging (DTI) and voxel-based morphometry (VBM) study. Psychol Med. 2014;44(9):1867–78. https://doi.org/10.1017/S003329171300247X.
166. Weaver KE, et al. Longitudinal diffusion tensor imaging in Huntington's disease. Exp Neurol. 2009;216(2):525–9. https://doi.org/10.1016/j.expneurol.2008.12.026.
167. Dominguez JF, et al. Multimodal imaging biomarkers in premanifest and early Huntington's disease: 30-month IMAGE-HD data. Br J Psychiatry. 2016;208(6):571–8. https://doi.org/10.1192/bjp.bp.114.156588.
168. Hobbs NZ, et al. Short-interval observational data to inform clinical trial design in Huntington's disease. J Neurol Neurosurg Psychiatry. 2015;86(12):1291–8. https://doi.org/10.1136/jnnp-2014-309768.
169. Zhang H, et al. NODDI: practical in vivo neurite orientation dispersion and density imaging of the human brain. NeuroImage. 2012;61(4):1000–16. https://doi.org/10.1016/j.neuroimage.2012.03.072.
170. Zhang J, et al. In vivo characterization of white matter pathology in premanifest huntington's disease. Ann Neurol. 2018;84(4):497–504. https://doi.org/10.1002/ana.25309.
171. Johnson EB, Gregory S. Huntington's disease: brain imaging in Huntington's disease. In: Progress in molecular biology and translational science. Elsevier B.V.; 2019. p. 321–69. https://doi.org/10.1016/bs.pmbts.2019.04.004.
172. Bohanna I, et al. Magnetic resonance imaging as an approach towards identifying neuropathological biomarkers for Huntington's disease. Brain Res Rev. 2008;58:209–25. https://doi.org/10.1016/j.brainresrev.2008.04.001.
173. Unschuld PG, et al. Impaired cortico-striatal functional connectivity in prodromal Huntington's disease. Neurosci Lett. 2012;514(2):204–9. https://doi.org/10.1016/j.neulet.2012.02.095.
174. Odish OFF, et al. Longitudinal resting state fMRI analysis in healthy controls and premanifest Huntington's disease gene carriers: a three-year follow-up study. Hum Brain Mapp. 2015;36(1):110–9. https://doi.org/10.1002/hbm.22616.
175. Poudel GR, et al. Functional changes during working memory in Huntington's disease: 30-month longitudinal data from the IMAGE-HD study. Brain Struct Funct. 2015;220(1):501–12. https://doi.org/10.1007/s00429-013-0670-z.
176. Seibert TM, et al. Stability of resting fMRI interregional correlations analyzed in subject-native space: a one-year longitudinal study in healthy adults and premanifest Huntington's disease. NeuroImage. 2012;59(3):2452–63. https://doi.org/10.1016/j.neuroimage.2011.08.105.
177. Wolf RC, et al. Aberrant connectivity of lateral prefrontal networks in presymptomatic Huntington's disease. Exp Neurol. 2008a;213(1):137–44. https://doi.org/10.1016/j.expneurol.2008.05.017.
178. Wolf RC, et al. Altered frontostriatal coupling in pre-manifest Huntington's disease: effects of increasing cognitive load. Eur J Neurol. 2008b;15(11):1180–90. https://doi.org/10.1111/j.1468-1331.2008.02253.x.
179. Kronenbuerger M, et al. Differential changes in functional connectivity of striatum-prefrontal and striatum-motor circuits in Premanifest Huntington's disease. Neurodegener Dis. 2019;19(2):78–87. https://doi.org/10.1159/000501616.
180. Gregory S, et al. Operationalizing compensation over time in neurodegenerative disease. Brain. 2017;140(4):1158–65. https://doi.org/10.1093/brain/awx022.
181. Gregory S, et al. Testing a longitudinal compensation model in premanifest Huntington's disease. Brain. 2018;141(7):2156–66. https://doi.org/10.1093/brain/awy122.
182. Klöppel S, et al. Compensation in preclinical Huntington's disease: evidence from the track-on HD study. EBioMedicine. 2015;2(10):1420–9. https://doi.org/10.1016/j.ebiom.2015.08.002.
183. Hohenfeld C, Werner CJ, Reetz K. Resting-state connectivity in neurodegenerative disorders: is there potential for an imaging biomarker? NeuroImage: Clin (Elsevier Inc.). 2018;18:849–70. https://doi.org/10.1016/j.nicl.2018.03.013.

184. Wilson H, et al. Molecular imaging markers to track Huntington's disease pathology. Front Neurol Front Rese Found. 2017;8 https://doi.org/10.3389/fneur.2017.00011.
185. Coskran TM, et al. Immunohistochemical localization of phosphodiesterase 10A in multiple mammalian species. J Histochem Cytochem. 2006;54(11):1205–13. https://doi.org/10.1369/jhc.6A6930.2006.
186. Niccolini F, et al. Altered PDE10A expression detectable early before symptomatic onset in Huntington's disease. Brain. 2015;138(10):3016–29. https://doi.org/10.1093/brain/awv214.
187. Ahmad R, et al. PET imaging shows loss of striatal PDE10A in patients with Huntington disease. Neurology. 2014;82(3):279–81. https://doi.org/10.1212/WNL.0000000000000037.
188. Russell DS, et al. The phosphodiesterase 10 positron emission tomography tracer, [18F]MNI-659, as a novel biomarker for early Huntington disease. JAMA Neurol. 2014;71(12):1520–8. https://doi.org/10.1001/jamaneurol.2014.1954.
189. Russell DS, et al. Change in PDE10 across early Huntington disease assessed by [18 F]MNI-659 and PET imaging. Neurology. 2016;86(8):748–54.
190. Fazio P, et al. PET molecular imaging of phosphodiesterase 10A: an early biomarker of Huntington's disease progression. Mov Disord. 2020;35(4):606–15. https://doi.org/10.1002/mds.27963.
191. Pavese N, et al. Progressive striatal and cortical dopamine receptor dysfunction in Huntington's disease: a pet study. Brain. 2003;126(5):1127–35. https://doi.org/10.1093/brain/awg119.
192. Pavese N, et al. Cortical dopamine dysfunction in symptomatic and premanifest Huntington's disease gene carriers. Neurobiol Dis. 2010;37(2):356–61. https://doi.org/10.1016/j.nbd.2009.10.015.
193. Pavese N, et al. Microglial activation correlates with severity in Huntington disease a clinical and PET study. Neurology. 2006;66:1638.
194. Simmons DA, et al. TSPO-PET imaging using [18F]PBR06 is a potential translatable biomarker for treatment response in Huntington's disease: preclinical evidence with the p75NTR ligand LM11A-31. Hum Mol Genet. 2018;27(16):2893–912. https://doi.org/10.1093/hmg/ddy202.
195. Tai YF, et al. Microglial activation in presymptomatic Huntington's disease gene carriers. Brain. 2007;130(7):1759–66. https://doi.org/10.1093/brain/awm044.
196. Casteels C, et al. Metabolic and Type 1 cannabinoid receptor imaging of a transgenic rat model in the early phase of Huntington disease. Exp Neurol. 2011;229(2):440–9. https://doi.org/10.1016/j.expneurol.2011.03.014.
197. Van Laere K, et al. Widespread decrease of type 1 cannabinoid receptor availability in Huntington disease in vivo. J Nucl Med. 2010;51(9):1413–7. https://doi.org/10.2967/jnumed.110.077156.
198. Ooms M, et al. Early decrease of type 1 cannabinoid receptor binding and phosphodiesterase 10A activity invivo in R6/2 Huntington mice. Neurobiol Aging. 2014;35(12):2858–69. https://doi.org/10.1016/j.neurobiolaging.2014.06.010.
199. Bertoglio D, Verhaeghe J, et al. Longitudinal preclinical evaluation of the novel radioligand [11C]CHDI-626 for PET imaging of mutant huntingtin aggregates in Huntington's disease. Eur J Nucl Med Mol Imaging. 2022b;49(4):1166–75. https://doi.org/10.1007/s00259-021-05578-8.
200. Liu L, et al. CHDI-626, a PET tracer candidate for imaging mutant huntingtin aggregates with reduced binding to AD pathological proteins. J Med Chem. 2021;64(16):12003–21. https://doi.org/10.1021/acs.jmedchem.1c00667.
201. Bertoglio D, Bard J, et al. Development of a ligand for in vivo imaging of mutant huntingtin in Huntington's disease, vol. 14. Sci Trans Med; 2022a. p. eabm3682. Available at: https://www.science.org
202. Delva A, et al. Biodistribution and dosimetry in human healthy volunteers of the PET radioligands [11C]CHDI-00485180-R and [11C]CHDI-00485626, designed for quantification of cerebral aggregated mutant huntingtin. Eur J Nucl Med Mol Imaging. 2022;50(1):48–60. https://doi.org/10.1007/s00259-022-05945-z.

203. Hoang TQ, et al. Quantitative proton-decoupled 31P MRS and 1H MRS in the evaluation of Huntington's and Parkinson's diseases. Neurology. 1998;50(4):1033–40. https://doi.org/10.1212/wnl.50.4.1033.
204. Jenkins BG, et al. 1H NMR spectroscopy studies of Huntington's disease: correlations with CAG repeat numbers. Neurology. 1998;50(5):1357–65. https://doi.org/10.1212/wnl.50.5.1357.
205. Van Den Bogaard SJA, et al. Exploratory 7-Tesla magnetic resonance spectroscopy in Huntington's disease provides in vivo evidence for impaired energy metabolism. J Neurol. 2011;258(12):2230–9. https://doi.org/10.1007/s00415-011-6099-5.
206. Van Den Bogaard SJA, et al. Longitudinal metabolite changes in huntington's disease during disease onset. J Huntington's Dis. 2014;3(4):377–86. https://doi.org/10.3233/JHD-140117.
207. Reynolds NC, Prost RW, Mark LP. Heterogeneity in 1H-MRS profiles of presymptomatic and early manifest Huntington's disease. Brain Res. 2005;1031(1):82–9. https://doi.org/10.1016/j.brainres.2004.10.030.
208. Sánchez-Pernaute R, et al. Clinical correlation of striatal 1H MRS changes in Huntington's disease. Neurology. 1999;53(4):806–12. https://doi.org/10.1212/wnl.53.4.806.
209. Sturrock A, et al. Magnetic resonance spectroscopy biomarkers in premanifest and early Huntington disease. Neurology. 2010;75(19):1702–10. https://doi.org/10.1212/WNL.0b013e3181fc27e4.
210. Sturrock A, et al. A longitudinal study of magnetic resonance spectroscopy Huntington's disease biomarkers. Mov Disord. 2015;30(3):393–401. https://doi.org/10.1002/mds.26118.
211. Taylor-Robinson SD, et al. Proton magnetic resonance spectroscopy in Huntington's disease: evidence in favour of the glutamate excitotoxic theory. Movement Dis. 1996;11(2):167–73. https://doi.org/10.1002/mds.870110209.
212. Wilson M, et al. Methodological consensus on clinical proton MRS of the brain: review and recommendations. Magn Reson Med. 2019;82(2):527–50. https://doi.org/10.1002/mrm.27742.

Part III
Single and Integrated Omics Studies in AD Biomarker Discovery

Proteomics for Novel Biomarker and Drug Target Discovery in Alzheimer Disease

Huiming Zhu, Minqi Cai, Dezhu Chen, Jiaojiao Sha, Liangjia Du, Yiqiang Chen, Siyuan Liu, Qian Zheng, Xuefei Wei, and Bing Bai

Abstract With major advances in the technology of liquid chromatography coupled to tandem mass spectrometry (LC-MS/MS), the proteomics analyses in Alzheimer disease (AD) have achieved unprecedented depth and breadth in the protein landscape, providing insights into mechanisms, biomarkers, and therapeutic targets of this disease. Here, we introduce progress of this technology, review major reported AD proteomics analyses, summarize potential biomarkers and drug targets, and discuss challenges and possible solutions in this LC-MS/MS-based proteomics approach. With further technical improvements and more targeted experimental designs, this approach will be a powerful tool for the discovery of biomarkers and drug targets for this devastating disease.

H. Zhu
Chemistry and Biomedicine Innovation Center, Medical School of Nanjing University, Nanjing, China

M. Cai
Department of Laboratory Medicine, Nanjing Drum Tower Hospital Clinical College of Jiangsu University, Nanjing, China, Jiangsu

Department of Nuclear Medicine, Nanjing Drum Tower Hospital, Affiliated Hospital of Medical School, Nanjing University, Nanjing, Jiangsu, China

D. Chen · X. Wei
Department of Nuclear Medicine, Nanjing Drum Tower Hospital, Affiliated Hospital of Medical School, Nanjing University, Nanjing, Jiangsu, China

J. Sha
Department of Laboratory Medicine, Nanjing Drum Tower Hospital Clinical College of Nanjing University of Chinese Medicine, Nanjing, Jiangsu, China

L. Du
Department of Laboratory Medicine, Nanjing Drum Tower Hospital Clinical College of Nanjing Normal University, Nanjing, Jiangsu, China

Y. Chen · S. Liu
Center for Precision Medicine, Nanjing Drum Tower Hospital, Affiliated Hospital of Medical School, Nanjing University, Nanjing, Jiangsu, China

R. Tarawneh (ed.), *Biomarkers of Neurodegenerative Disorders*,
https://doi.org/10.1007/978-3-031-80849-4_12

Keywords Alzheimer disease · LC-MS/MS · Mass spectrometry · Proteomics · Biomarker · Drug target · Netrin-1

Abbreviations

AD	Alzheimer disease
MS	mass spectrometry
LC/LC-MS/MS	two-dimensional liquid chromatography coupled with tandem mass spectrometry

1 Background

While Alzheimer disease (AD) is the most common form of dementia in the aging population with a clear definition in diagnostic criteria, other dementias such as frontotemporal lobar degeneration, dementia with Lewy bodies, vascular dementia, and multifactorial dementias share many cognitive and pathological features with AD, and thus, can be difficult to distinguish from AD. These dementias are clinically referred to as AD and related dementias (ADRDs). However, in research, each of these dementias has a clear definition with unique pathology and distinct molecular signatures. So far, tremendous work has been focused on AD [1].

AD is an irreversible neurodegenerative disorder, the prevalence of which increases with age, and has increasingly become a global socioeconomic challenge with the expanding senior population in the current era [2]. Although lecanemab and other drugs have shown promising therapeutic effectiveness in slowing clinical disease progression [3–5], a cure for this devastating disease is not yet available.

AD dementia is associated with progressive memory and cognitive impairment, and affected individuals typically die within an average of 7–10 years from diagnosis. The pathological hallmarks of AD are extracellular amyloid plaques and intracellular neurofibrillary tangles (NFTs) in brain tissues [6]. The Aβ amyloid pathology initially occurs in basal parts of the neocortex and then gradually spreads to involve limbic and subcortical regions [7]. In contrast, the NFTs start early in the transentorhinal region of the medial temporal lobe, which can begin during middle age. Later, these tangles progress more severely into both entorhinal and transentorhinal regions until their full manifestation in limbic and neocortical areas [7]. This

Q. Zheng
Health Management Center, Nanjing Drum Tower Hospital, Affiliated Hospital of Medical School, Nanjing University, Nanjing, Jiangsu, China

B. Bai (✉)
Department of Laboratory Medicine, Nanjing Drum Tower Hospital, Affiliated Hospital of Medical School, Nanjing University, Nanjing, Jiangsu, China
e-mail: bing.bai@nju.edu.cn

suggests that Aβ plaques and NFTs do not follow the same temporal or anatomic pattern during aging and the development of AD. Besides, NFTs are almost universally present in the medial temporal lobe in subjects older than 70 years, while Aβ plaques are found only in a proportion of the older population [8]. In addition, the Aβ plaque burden or its peptide levels in the brain show no correlation with dementia severity, while the number of NFTs demonstrates a significant correlation with the extent of dying neurons and dementia [8]. Therefore, it is highly likely that Aβ and Tau proteinopathies initiate separately but might interact later during AD pathogenesis.

The core components of amyloid plaques and NFTs are Aβ peptides and hyperphosphorylated protein Tau, respectively [9]. The Aβ peptides are produced through the sequential cleavage of the amyloid precursor protein (APP) by the β-secretase (i.e., BACE1) and the γ-secretase enzymes [10]. The Aβ peptides are prone to aggregation, and their soluble forms, such as dimers, tetramers, and oligomers, demonstrate neuronal toxicity [11]. On the protein Tau, more than 40 amino acid sites are phosphorylated by about 20 kinases that include GSK3β, CDK5, MAPK, among others. Nearly 30 of these phosphorylated sites are involved in AD [12–14].

Genetic studies provide causative clues to the etiology of diseases. So far, there are three familial AD genes: *APP*, presenilin-1 (*PSEN1*), and presenilin-2 (*PSEN2*). The latter two encode presenilin proteins that are major components of the γ-secretase complex [15]. Therefore, all these three genes are directly involved in APP cleavage and Aβ generation. Moreover, mutations of *APP* that cause overproduction of Aβ promote the development of AD, while those that inhibit Aβ production protect people from AD [16]. Incorporating evidence from studies in pathology, biochemistry, cell biology, and animal work, the APP/Aβ amyloidogenic pathway is generally believed to be an initiator, and Tau a mediator, of AD pathogenesis [17].

In addition to Aβ and Tau, inflammation also plays a critical role in AD pathogenesis [18]. Among all AD genetic risk factors, almost half are related to inflammatory processes [15]. In AD brain cortices, complement proteins, such as C1q, C3b, C3c, C3d, and C4, localize within the amyloid plaques [19]. The fibrillar Aβ peptides and NFTs can directly activate the complement in vitro [20]. AD risk genes such as *TREM2*, *CR1*, *CD33*, and *CLU* are close regulators of the complement system. Besides the complement proteins, activated microglia almost universally colocalize with amyloid plaques in AD brains and correlate with both neuropathological and clinical stages of AD [21]. The progression of activated microglia closely parallels the evolution of neuritic plaques in AD brains [22]. Moreover, pro-inflammatory mediators, including interleukins, prostaglandins, reactive oxygen and nitrogen species, and other factors, are found in brain tissues, consistent with the observation of microglial activation [23, 24]. In fact, epidemiological and observational studies have reported that patients with inflammatory diseases who have received long-term treatments with nonsteroidal anti-inflammatory drugs (NSAIDs) show about a 50% reduction in the risk of developing AD [25]. It is generally considered that inflammation in AD might exert a dual function, neuroprotective for clearing Aβ at the early stage but detrimental later due to overactivation [21].

Besides Aβ, Tau, and inflammation, AD pathogenesis also involves alterations in many other biological processes and cellular signaling pathways, including cholinergic neuronal loss [26, 27], imbalanced calcium homeostasis [28, 29], mitochondrial oxidative stress [30, 31], membrane trafficking dysfunction [32, 33], meningeal lymphatic dysfunction [34, 35], microorganism infection [36–38], and endosomal abnormalities [39, 40]. Although these alterations might be evident in AD, the fact remains that amyloid pathology may be essential, but – alone – not sufficient, to cause AD [41, 42].

2 The Development of LC-MS/MS-Based Proteomics Technology

Proteomics provides a protein landscape at the systemic level of a biological sample. This can be achieved by high-throughput approaches by antibody-based methodologies such as protein arrays, tissue immunohistochemistry microarrays, oligonucleotide-linked proximity extension assays, aptamer-based assays, or the rising nanopore-based single-protein sequencing technology [43]. Besides specificity issues, the major caveat of the antibody- and aptamer-based approaches is that they can only identify proteins determined by the coated antibodies or aptamers. Novel proteins beyond recognition of these probes will be missed. Nanopore for protein sequencing is still under development, in which both the identification and quantification of proteins need major improvements.

The development of next-generation mass spectrometry-based quantitative proteomics has led to major advances in profiling the entire proteins of a biological sample at a system level. In the traditional LC-MS/MS analysis pipeline, proteins in samples are first digested by enzymes such as trypsin into peptides to be nearly uniform in biochemical properties, so that they can be well resuspended in the acidic solution and charged by the hydrogen ion (H^+), and then be eluted sequentially through the reverse phase liquid chromatography (RPLC) by the increasing gradient of an organic solvent. The eluted peptides are injected by electrospray into the mass spectrometer for scanning (MS1) to screen for peptide ions. An expected peptide ion will then be selected and fragmented for scanning again to generate a spectrum (MS2). Data information from MS1 and MS2 is derived to match their theoretical spectral data in the reference database. Thus, peptide and protein identification and quantification data are finally carried out.

Nowadays, significant improvements have been made in each procedure of the LC-MS/MS analysis pipeline. In this enhanced protocol (Fig. 1), with the multiplexed tandem isobaric mass tags (TMT), up to 18 biological protein samples can be analyzed in a single batch, greatly improving the sample analysis throughput and also increasing the quantification accuracy. The digested peptides are first separated by RPLC at basic pH into nearly 100 fractions. This greatly reduces the complexity of the sample, improves sensitivity due to concentrated peptides after fractionation,

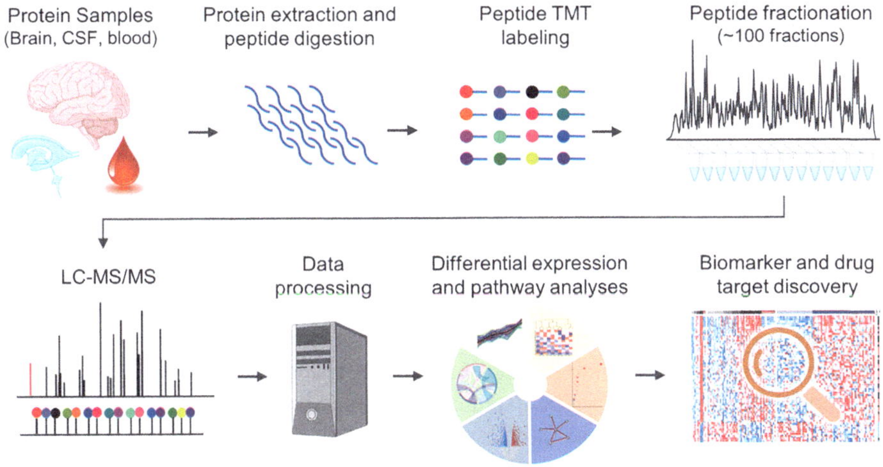

Fig. 1 General workflow of TMT-LC-LC-MS/MS-based quantitative proteomics analysis of AD human protein samples

and minimizes the interference from ion co-isolation. Besides, the use of dimethylsulfoxide (DMSO) in liquid chromatography increases the mass spectrometric detection sensitivity by several folds [44]. With significant improvements and optimizations, the TMT-LC-LC-MS/MS analysis pipeline emerges as a mainstream approach for profiling proteins as well as post-translational modifications in biological protein samples.

3 LC-MS/MS-Based Proteomics Analysis in AD

Prior to the full elucidation of the human genome, only incomplete protein databases were available, and therefore, mass spectrometry could only be used to analyze particular proteins specifically, identifying the Aβ peptide species in both soluble and insoluble forms [45, 46], and identifying the protein tau with its post-translational modifications (e.g. phosphorylation, acetylation, and deamidation) in AD brain tissues [47]. These have verified the core of the composites of the amyloid plaques and tau NFTs.

In 2004, Liao et al. collected the amyloid plaques from postmortem AD brain tissues by the laser capture microdissection (LCM) technique followed by protein extraction and MS analysis [48], which yielded identification of 488 proteins among which 26 showed enrichment in the plaques. We mentioned this as "amyloidome" [49]. Later, under a similar approach, more than 900 proteins in the brain amyloid plaques from rapidly progressive and sporadic AD were profiled, reporting unique protein changes in these different AD subtypes [50]. In 2019, Xiong et al. re-analyzed the AD amyloid plaques using the current TMT-LC/LC-MS/MS pipeline, increasing the proteome coverage to an unprecedented scale: 4000+ proteins, among which 40 proteins were differentially expressed, including apolipoprotein E (APOE) and complement proteins [51].

Insoluble proteome is another approach to looking into potentially aggregated proteins in AD brain tissues by biochemical differential extraction, in which proteins resistant to harsh detergents (e.g., sarkosyl) were collected for MS analysis. Gozal et al. analyzed the detergent-insoluble proteome of brain tissues, identifying 512 proteins, including 11 increased proteins in the AD samples [52]. We later expanded the coverage of the AD brain insoluble proteome, profiling 4216 proteins with 36 enriched proteins in AD [53]. This study has revealed proteopathy of U1 spliceosomal proteins and the concomitant RNA splicing deficiency in AD. Recently, recapitulating this aberrancy in mice has caused reduction of the GABA receptor subunit and post-synaptic hyperexcitability, implicating the importance of U1 in maintaining normal neuronal activities [54].

Synaptic loss is a feature of AD brain pathology. It occurs early and is highly correlated with cognitive decline [55], and thus might provide critical molecular clues to AD pathogenesis. Using ultracentrifugation and differential extraction, Zhou et al. isolated the post-synaptic density (PSD) samples from AD brains and identified 494 proteins [56]. More recently, about 5000 proteins were profiled in the synaptic subproteome by the accurate and multiplexed approach of TMT [57, 58]. It is notable that Carlyle et al. examined 100 brains in different disease stages from individuals with AD, revealing 123 unique proteins significantly associated with disease subtypes and enriched in serotonin release and oxidative phosphorylation pathways [59].

With the sophisticated LC-MS/MS analysis pipeline in recent years, deep proteome sequencing of the entire brain tissues has become possible. In 2015, an advanced LC/LC-MS/MS approach profiled over 10,000 proteins in the AD brain [60]. In 2020, we quantitatively identified 14,513 proteins in nearly 100 brain tissue samples from AD and four other different groups as controls that included nondemented controls with low or high amyloid plaques, mild cognitive impairment, and progressive supranuclear palsy [61]. Among these proteins, 173 were found to be differentially expressed in AD and enriched in 17 altered pathways: including Aβ cascade, Wnt, and TGF-β/BMP. The comprehensive whole proteome analyses of human AD brains demonstrate a wide and dynamic molecular fluctuation of proteins during AD progression [61].

So far, three research groups utilized the advanced TMT-LC/LC-MS/MS pipeline to generate seven deep proteomic datasets (each contains 8000+ proteins) from a total of 192 AD and control cortical specimens [61–64]. These datasets, as valuable resources, have enabled a meta-analysis to generate a list of 12,017 unique genes and yield 2698 significantly changed proteins that can be assigned to neurons, microglia, astrocytes, oligodendrocytes, and endothelium, implicating a wide contribution of many cell types to AD development [49].

Besides the whole proteome analysis, the LC-MS/MS pipeline was also utilized for the analysis of protein post-translational modifications (PTM) in AD. Tau proteopathy is associated with extensive PTMs, including phosphorylation, ubiquitination, acetylation, methylation, glycosylation, sumoylation, oxidation, and cleavage [65–67]. Some of these modifications have been analyzed by MS and cryo-electron microscopy [68]. More recently, Wesseling et al. have completed the most comprehensive tau PTM analysis in AD brains by multiple MS strategies, profiling a total of 95 modification events: 55 in phosphorylation, 17 in ubiquitination, 19 in acetylation, and four in methylation, reflecting the complexity in biological processing during AD progression [69].

In 2020, we identified 46,612 phosphopeptides with 34,173 phosphosites from 7083 proteins in AD and a set of controls, which was enabled by phosphopeptide enrichment and TMT-LC/LC-MS/MS analysis. Among these, 873 differentially expressed phosphopeptides in 398 proteins were identified, including 56 phosphosites on Tau [61]. This phosphoproteome revealed 28 differentially expressed kinase activities, covering all currently known 11 tau kinases [67]. Integration of kinase activities and levels collectively suggested the activation of MAP kinase signaling in AD [61]. Using a similar approach, Ping et al. have reported another deep dataset of AD phosphoproteomes (33,652 phosphosites on 8415 proteins) [70]. Besides phosphorylation, a global analysis of protein ubiquitination in AD was also performed, covering 4291 ubiquitinated sites on 1682 proteins [71].

Overall, current MS platforms and enrichment strategies have enabled proteome-wide analysis of PTMs in AD clinical samples. Similar deeper analyses will be continued in the near future. These comprehensive PTM datasets will be valuable resources for mining the fundamental mechanisms of AD pathogenesis.

4 Insights for Biomarker Discovery from AD Proteomics

Proteomics of cerebrospinal fluid (CSF) and serum/plasma provides great approaches for validation of proteins differentially expressed in the brain tissue as potential biomarkers. AD CSF proteomics has been tried in several studies but not in depth due to technical limitations. An early deep AD CSF proteomics analysis was performed with the TMT-LC/LC-MS/MS pipeline by Sathe et al., which profiled 2327 proteins including 139 differentially expressed proteins, such as MAPT, NPTX2, and VGF [72]. Higginbotham et al. also used a similar approach and profiled 2875 proteins from AD and control CSF samples among which 528 proteins were significantly changed, including MAPT, NEFL, VGF, and SMOC1 [63]. In addition, Bader et al. developed an advanced label-free strategy to analyze CSF samples from three independent cohorts of similar size (109 control and 88 AD cases in total), profiling 1445, 1478 and 1483 proteins, respectively, revealing the differentially expressed proteins of Tau, SOD1, and PARK7, which also bear genetic associations to aging-related neurodegeneration [73].

An ultradeep CSF proteome analysis has been reported by Wang et al., who, by extensive LC fractionation and long gradient LC-MS/MS analysis, profiled 5941 proteins, reaching an unprecedented scale in CSF proteome coverage [61, 74]. This study found 355 differentially expressed proteins, including SMOC1 and TGFB2, which showed consistent changes with previous reports.

The meta-analysis of these CSF datasets resulted in a list of 5939 proteins, including 311 upregulated and 165 downregulated proteins, in AD [49]. Overlapping these differentially expressed proteins with the AD brain proteome yielded a core table of 65 upregulated proteins (e.g., MAPT, SMOC1, HTRA1) and 44 downregulated proteins (e.g., NPTX2, VGF, PDHA1) in AD [49]. This provides a valuable reference for AD CSF biomarker discovery.

Directly sequencing all proteins in the serum/plasma is an ideal approach to finding biomarkers; however, this is very challenging. Concentrations of proteins in peripheral blood are in an extremely large dynamic range, ranging from mg/ml (such as albumin) to pg/ml (such as cytokines), spanning more than nine orders of magnitude. Those abundant proteins interfere with the LC-MS/MS system in analyzing those trace amounts of, but biologically important, proteins. Depletion of these abundant proteins with immunoaffinity or other approaches is often applied, but many proteins in low levels are depleted simultaneously and fail to be detected by the LC-MS/MS.

Using an extensive fractionation approach, Dey et al. separated the protein peptides into 180 fractions, each of which was analyzed by a long gradient (3 h) of LC-MS/MS analysis [75]. In this exhaustive study, 4826 proteins were profiled in serum samples from AD and controls, including 26 decreased proteins (e.g., PCK2, AK2, HSPA9) and four increased proteins (PUS10, BRF1, RC3H2, and CLIP1) [75].

A potential AD biomarker is ideally expected to meet these criteria: (1) large changes; (2) consistently changed in disease cases compared to controls; (3) consistently changed in brain tissues, CSF, and serum/plasma; (4) highly AD-specific; (5) brain (especially the affected region)-specific; (6) with evidence linked to AD pathogenesis, such as genetics and pathology; (7) secretable; (8) stable; and (9) detectable. Following these simple rules, we integrate the proteomics, genetics datasets, and the human protein atlas and generate a list of proteins (Table 1). It is admitted that none of these proteins can fulfill all these conditions. To establish a potential AD biomarker in blood is still challenging. However, considerations should be taken with the data. For example, if a protein expression level is about a ten-fold change in a subset of cells, it will only show about a two-fold change if this protein in these cells only occupies 10% of the entire tissue sample analyzed. Underestimations of protein level changes are common in samples analyzed as a whole.

Extracellular vesicles (EV) in CSF or blood are another interesting source for proteomics-based biomarker discovery. EVs are secreted from neurons, astrocytes, and microglia [76], carrying AD-related molecules including proteins of APP, Aβ peptides, and Tau [77, 78]. Analyses of EVs isolated from the plasma, CSF, and brain cells of AD patients and mouse models demonstrated higher levels of $A\beta_{42}$/$A\beta_{40}$ ratio, APP, Aβ monomeric and oligomeric forms, and Tau and Tau phosphorylation as reported in [76, 79, 80]. With increasing sophistication of the EV extraction methodology and development of other related analysis procedures, CSF or blood EV proteomics will be a new direction in AD biomarker discovery.

Table 1 Novel potential biomarkers indicated from major AD proteomic datasets

	Protein description	Log_2 (AD/Ctl)		
		Cortex	CSF	Serum
AD risk factor				
APOA2	Apolipoprotein A-II	0.11	0.04	0.57
Cortex-enriched				
PIANP	PILR alpha-associated neural protein	−0.08	−0.10	−2.97
PRKCG	Protein kinase C gamma type	−0.04	−0.09	−1.59
OLFM3	Noelin-3	−0.44	−0.10	−1.26
PTPRZ1	Receptor-type tyrosine-protein phosphatase zeta	0.05	0.17	0.84
PRRT2	Proline-rich transmembrane protein 2	−0.23	−0.32	
RPH3A	Rabphilin-3A	−0.51	−0.70	
Probably aggregated in AD brain tissues				
GLUD2	Glutamate dehydrogenase 2, mitochondrial	−0.02	−1.07	−1.37
EPB41L2	Band 4.1-like protein 2	0.23	−0.24	−0.90
DDX46	Probable ATP-dependent RNA helicase DDX46	0.06	−0.59	−0.84

5 Insights for Drug Targets from AD Proteomics

Discovery of potential drug targets from AD proteomics data is challenging and largely unexploited. Evidence for a protein to be considered as a potential drug target might include: (1) it is a familial gene or a genetic risk factor of this disease; (2) it demonstrates pathology in the affected tissue; (3) its pathological alteration occurs early and correlates with disease severity; (4) it is a hub protein in a cellular pathway or a biological process altered at the early stages of disease; (5) it links to critical proteins in the etiology or pathogenesis of this disease; or (6) activation or inhibition of this protein in cellular or mouse models shows a significant protective effect on this disease.

According to these conditions, one protein is particularly interesting, and it is Netrin-1. Netrin-1 was initially discovered as an axon guidance cue and was later found to be involved in inflammation, tumorigenesis, and other disorders. Recently, a monoclonal antibody (NP137) that targets Netrin-1 to disrupt its interaction with the UNC5B receptor has been developed and is currently being examined in a clinical phase 1 trial (NCT02977195) [81, 82].

In the ultradeep AD brain proteome analyses that we have reported, Netrin-1 shows extremely high correlation with Aβ, which is not only seen in human samples but also in AD mice that overexpress Aβ [14]. In addition, Netrin-1 colocalizes with Aβ plaques in both AD human and mouse brain cortices, and it can directly bind Aβ peptides in vitro [14]. Therefore, one might speculate that Aβ can possibly induce protein expression of Netrin-1, which in turn promotes Aβ aggregation through direct binding, forming a vicious cycle.

The importance of netrins in AD pathogenesis is highlighted by genetic evidence from their common receptor, UNC5C. A rare mutation T835M in the coding region of UNC5C was found in two AD families in an autosomal dominant pattern [83].

T835 is a conserved site in the hinge region of UNC5C, and this mutation increases cell death and potentiates Aβ neurotoxicity in vitro [83]. Besides, UNC5C cleavage by δ-secretase increases caspase-3 activation, promoting apoptosis and thus neurodegeneration in AD [84]. As UNC5H is reported to be a dependence receptor that will induce apoptosis when the netrins are absent [85], UNC5C in the presence of Netrin-1 in AD might be considered protective. It is also notable that Netrin-5 is a genomic risk locus identified from 90,338 (46,613 proxy) cases and 1,036,225 (318,246 proxy) controls, demonstrating a strong association with AD [95]. These lines of genetic evidence suggest causative roles of netrins in AD pathogenesis.

In fact, Netrin-1 can directly bind to APP to modulate Aβ production, aggregation, and toxicity. Netrin-1 can be coimmunoprecipitated with APP from cultured cells, be recruited to the plasma membrane of APP-expressing cells, and colocalize with APP in the growth cones of cortical neurons [86]. This binding is mediated through the Aβ region of APP, consistent with our report [14].

APP normally undergoes nontoxic cleavage by the α-secretase ADAM10. It is found that Aβ peptides are substrates of ADAM10 as inhibitors, shifting the α-cleavage of APP to β-cleavage, exerting a self-amplification effect [87]. However, this effect is inhibited by Netrin-1 in cultured cells and neurons, and Netrin-1 expression suppresses both Aβ40 and Aβ42 levels in mice, probably due to the binding of Netrin-1 in the Aβ region of APP, preventing its β-cleavage [87]. This provides a potential therapeutic approach to control Aβ overproduction in AD.

Indeed, Netrin-1 restores memory performance impaired by Aβ in mice, in which repeated intracerebroventricular injections of Netrin-1 rescued long-term potentiation reduction and all cognitive behavioral tasks [88]. This might be supported by the fact that selective homozygous deletion of Netrin-1 or, its receptor DCC, significantly affects memory consolidation [89]. This suggests that Netrin-1 is necessary in maintaining synaptic plasticity and normal memory.

Netrin-1 protects neurons from Aβ insults, possibly through inhibition of inflammation, oxidation, and apoptosis. Treating SH-SY5Y cells exposed to $A\beta_{42}$ peptides with Netrin-1 increased cell viability and partially suppressed the elevation of the inflammatory factors (TNFα and NF-κB) and the oxidation marker nuclear factor erythroid 2–like 2 (Nrf2) [90]. The Netrin-1 treatment also reduced caspase-3/7 activities in mice that received intrahippocampal injections of Aβ42 [91].

Besides functional links to APP/Aβ, Netrin-1 might also be related to a prominent AD pathology: amyloid plaques are often surrounded by activated microglia. It is reported that Netrin-1 can arrest macrophages by inactivating their egress from atherosclerotic plaques [92]. Netrin-1 exerts this effect by binding to its receptor UNC5b to inhibit activation of the actin-remodeling GTPase Rac1 and actin polymerization, making macrophages anergic [92]. The same effect can be seen in obesity, where Netrin-1 acts through UNC5b to retain macrophages in the adipose tissues [93]. Therefore, as microglia are similar to macrophages, they might also be retained around amyloid plaques where Netrin-1 is localized.

Although the effects of Netrin-1 on Aβ toxicity and microglia activation are complex, lines of evidence from genetics, pathology, biochemistry, cell biology, and animal models suggest that Netrin-1 is a promising target for early therapeutic intervention in AD.

6 Conclusions and Perspectives

It is estimated that ~16,000 genes are expressed in human brains [60], generating probably millions of protein species resulting from RNA alternative splicing, proteolytic cleavages, post-translational, and other biochemical modifications [94]. It is admitted that although current ultradeep LC-MS/MS-based proteomics studies have already profiled the brain and biofluids at an unprecedented scale, they only detect about 12,000 gene products of the brain tissues. Current large-scale AD proteomics data highlight mainly the mechanistic proteins and cellular pathway alterations. Deriving evidence of potential biomarkers from these data is still challenging, and finding potential drug targets is even harder.

To increase the chances of successful discovery of biomarkers and drug targets by proteomics, several strategies might be considered. First, the affected regions of the diseased tissue should be the key source for biomarkers to exploit where differentially expressed proteins (e.g., Aβ peptides, phosphorylated Tau) are enriched. This will not only significantly reduce the protein complexity of the sample but also enrich proteins of interest in a particular fraction. Moreover, sample sequential biochemical extraction should be considered. Proteins released to CSF and blood as biomarkers are generally secretable. These proteins can usually be enriched in the extraction buffer with mild detergents, making them easier to detect. Besides, for drug target discovery, interactome analyses of critical proteins can be performed. Protein exerts its functions always through interaction with its partners. Finding these partners in the affected tissues during early and later disease stages by coimmunoprecipitation and the subsequent proteomics analysis is an efficient way to reveal molecular mechanisms and therapeutic targets.

In summary, current deep proteomics studies have already profiled the brain tissues and biofluids at an unprecedented scale, revealing novel clues to mechanisms, biomarkers, and drug targets. It is notable that AD dementia reflects irreversible neurodegeneration and a nearly end-stage disease, in which many cellular pathways and biological processes are altered. With further technological development in the proteomics approach, it is not surprising to see even more molecular alterations and characteristic proteins discovered. It is also notable that many proteomics performed so far have focused on the late stage when AD is already fully manifested and extensive molecules are altered. To find biomarkers and drug targets, selection of samples from early – including presymptomatic – disease stages, performance of biochemical extraction, and interactome analysis of key proteins are some of the keys to successful discoveries.

Acknowledgements The authors thank all other members in the Bai lab and the department of clinical laboratory for helpful discussion.

Funding This work was supported by the National Natural Science Foundation of China (82172354, to BB), the Research Foundation of Jiangsu Provincial Commission of Health and Family Planning (M2021012 and LK2021028, to BB), the Nanjing Medical Science and Technology Development Foundation (ZKX22013, to BB), and the funding for clinical trials and new clinical technology from the Affiliated Drum Tower Hospital, Medical School of Nanjing University (2022-LCYJ-MS-02, XJSFZLX202331, to BB).

References

1. Majoka MA, Schimming C. Effect of social determinants of health on cognition and risk of Alzheimer disease and related dementias. Clin Ther. 2021;43:922–9. https://doi.org/10.1016/j.clinthera.2021.05.005.
2. Tahami Monfared AA, Byrnes MJ, White LA, Zhang Q. Alzheimer's disease: epidemiology and clinical progression. Neurol Ther. 2022;11:553–69. https://doi.org/10.1007/s40120-022-00338-8.
3. van Dyck CH, et al. Lecanemab in early Alzheimer's disease. N Engl J Med. 2023;388:9–21. https://doi.org/10.1056/NEJMoa2212948.
4. Burke JF, Kerber KA, Langa KM, Albin RL, Kotagal V. Lecanemab: looking before we leap. Neurology. 2023;101:661–5. https://doi.org/10.1212/WNL.0000000000207505.
5. Couzin-Frankel J. Side effects loom over Alzheimer's drugs. Science. 2023;381:466–7. https://doi.org/10.1126/science.adk0830.
6. Masters CL, et al. Alzheimer's disease. Nat Rev Dis Primers. 2015;1:15056. https://doi.org/10.1038/nrdp.2015.56.
7. Braak H, Braak E. Evolution of the neuropathology of Alzheimer's disease. Acta Neurol Scand Suppl. 1996;165:3–12.
8. Nelson PT, et al. Correlation of Alzheimer disease neuropathologic changes with cognitive status: a review of the literature. J Neuropathol Exp Neurol. 2012;71:362–81. https://doi.org/10.1097/NEN.0b013e31825018f7.
9. Duyckaerts C, Delatour B, Potier MC. Classification and basic pathology of Alzheimer disease. Acta Neuropathol. 2009;118:5–36. https://doi.org/10.1007/s00401-009-0532-1.
10. Greenfield JP, et al. Endoplasmic reticulum and trans-Golgi network generate distinct populations of Alzheimer beta-amyloid peptides. Proc Natl Acad Sci USA. 1999;96:742–7. https://doi.org/10.1073/pnas.96.2.742.
11. Haass C, Selkoe DJ. Soluble protein oligomers in neurodegeneration: lessons from the Alzheimer's amyloid beta-peptide. Nat Rev Mol Cell Biol. 2007;8:101–12. https://doi.org/10.1038/nrm2101.
12. Martin L, et al. Tau protein kinases: involvement in Alzheimer's disease. Ageing Res Rev. 2013;12:289–309. https://doi.org/10.1016/j.arr.2012.06.003.
13. Tan H, et al. Refined phosphopeptide enrichment by phosphate additive and the analysis of human brain phosphoproteome. Proteomics. 2015;15:500–7. https://doi.org/10.1002/pmic.201400171.
14. Bai B, et al. Deep multilayer brain proteomics identifies molecular networks in Alzheimer's disease progression. Neuron. 2020;105:975–991 e977. https://doi.org/10.1016/j.neuron.2019.12.015.
15. Yu M, Sporns O, Saykin AJ. The human connectome in Alzheimer disease – relationship to biomarkers and genetics. Nat Rev Neurol. 2021;17:545–63. https://doi.org/10.1038/s41582-021-00529-1.
16. Jonsson T, et al. A mutation in APP protects against Alzheimer's disease and age-related cognitive decline. Nature. 2012;488:96–9. https://doi.org/10.1038/nature11283.

17. Bloom GS. Amyloid-beta and tau: the trigger and bullet in Alzheimer disease pathogenesis. JAMA Neurol. 2014;71:505–8. https://doi.org/10.1001/jamaneurol.2013.5847.
18. Heppner FL, Ransohoff RM, Becher B. Immune attack: the role of inflammation in Alzheimer disease. Nat Rev Neurosci. 2015;16:358–72. https://doi.org/10.1038/nrn3880.
19. Eikelenboom P, Stam FC. Immunoglobulins and complement factors in senile plaques. An immunoperoxidase study. Acta Neuropathol. 1982;57:239–42. https://doi.org/10.1007/BF00685397.
20. Tenner AJ. Complement-mediated events in Alzheimer's disease: mechanisms and potential therapeutic targets. J Immunol. 2020;204:306–15. https://doi.org/10.4049/jimmunol.1901068.
21. Leng F, Edison P. Neuroinflammation and microglial activation in Alzheimer disease: where do we go from here? Nat Rev Neurol. 2021;17:157–72. https://doi.org/10.1038/s41582-020-00435-y.
22. Mrak RE. Microglia in Alzheimer brain: a neuropathological perspective. Int J Alzheimers Dis. 2012;2012:165021. https://doi.org/10.1155/2012/165021.
23. Mrak RE, Griffin WS. Glia and their cytokines in progression of neurodegeneration. Neurobiol Aging. 2005;26:349–54. https://doi.org/10.1016/j.neurobiolaging.2004.05.010.
24. Gyengesi E, Munch G. In search of an anti-inflammatory drug for Alzheimer disease. Nat Rev Neurol. 2020;16:131–2. https://doi.org/10.1038/s41582-019-0307-9.
25. Kinney JW, et al. Inflammation as a central mechanism in Alzheimer's disease. Alzheimers Dement (NY). 2018;4:575–90. https://doi.org/10.1016/j.trci.2018.06.014.
26. Davies P, Maloney AJ. Selective loss of central cholinergic neurons in Alzheimer's disease. Lancet. 1976;2:1403. https://doi.org/10.1016/s0140-6736(76)91936-x.
27. Francis PT, Palmer AM, Snape M, Wilcock GK. The cholinergic hypothesis of Alzheimer's disease: a review of progress. J Neurol Neurosurg Psychiatry. 1999;66:137–47. https://doi.org/10.1136/jnnp.66.2.137.
28. Mattson MP, et al. beta-Amyloid peptides destabilize calcium homeostasis and render human cortical neurons vulnerable to excitotoxicity. J Neurosci. 1992;12:376–89.
29. Mattson MP, Chan SL. Neuronal and glial calcium signaling in Alzheimer's disease. Cell Calcium. 2003;34:385–97. https://doi.org/10.1016/s0143-4160(03)00128-3.
30. Swerdlow RH, Khan SM. A "mitochondrial cascade hypothesis" for sporadic Alzheimer's disease. Med Hypotheses. 2004;63:8–20. https://doi.org/10.1016/j.mehy.2003.12.045.
31. Wang W, Zhao F, Ma X, Perry G, Zhu X. Mitochondria dysfunction in the pathogenesis of Alzheimer's disease: recent advances. Mol Neurodegener. 2020;15:30. https://doi.org/10.1186/s13024-020-00376-6.
32. Pimplikar SW, Nixon RA, Robakis NK, Shen J, Tsai LH. Amyloid-independent mechanisms in Alzheimer's disease pathogenesis. J Neurosci. 2010;30:14946–54. https://doi.org/10.1523/JNEUROSCI.4305-10.2010.
33. Jiang S, et al. Trafficking regulation of proteins in Alzheimer's disease. Mol Neurodegener. 2014;9:6. https://doi.org/10.1186/1750-1326-9-6.
34. Deane R, et al. LRP/amyloid beta-peptide interaction mediates differential brain efflux of Abeta isoforms. Neuron. 2004;43:333–44. https://doi.org/10.1016/j.neuron.2004.07.017.
35. Da Mesquita S, et al. Functional aspects of meningeal lymphatics in ageing and Alzheimer's disease. Nature. 2018;560:185–91. https://doi.org/10.1038/s41586-018-0368-8.
36. Kumar DK, et al. Amyloid-beta peptide protects against microbial infection in mouse and worm models of Alzheimer's disease. Sci Transl Med. 2016;8:340ra372. https://doi.org/10.1126/scitranslmed.aaf1059.
37. Readhead B, et al. Multiscale analysis of independent Alzheimer's cohorts finds disruption of molecular, genetic, and clinical networks by human herpesvirus. Neuron. 2018;99:64–82 e67. https://doi.org/10.1016/j.neuron.2018.05.023.
38. Eimer WA, et al. Alzheimer's disease-associated beta-amyloid is rapidly seeded by Herpesviridae to protect against brain infection. Neuron. 2018;100:1527–32. https://doi.org/10.1016/j.neuron.2018.11.043.

39. Nixon RA. Amyloid precursor protein and endosomal-lysosomal dysfunction in Alzheimer's disease: inseparable partners in a multifactorial disease. FASEB J. 2017;31:2729–43. https://doi.org/10.1096/fj.201700359.
40. Young JE, Holstege H, Andersen OM, Petsko GA, Small SA. On the causal role of retromer-dependent endosomal recycling in Alzheimer's disease. Nat Cell Biol. 2023;25:1394–7. https://doi.org/10.1038/s41556-023-01245-2.
41. Musiek ES, Holtzman DM. Three dimensions of the amyloid hypothesis: time, space and 'wingmen'. Nat Neurosci. 2015;18:800–6. https://doi.org/10.1038/nn.4018.
42. Hampel H, et al. The amyloid-beta pathway in Alzheimer's disease. Mol Psychiatry. 2021;26:5481–503. https://doi.org/10.1038/s41380-021-01249-0.
43. Cui M, Cheng C, Zhang L. High-throughput proteomics: a methodological mini-review. Lab Investig. 2022;102:1170–81. https://doi.org/10.1038/s41374-022-00830-7.
44. Hahne H, et al. DMSO enhances electrospray response, boosting sensitivity of proteomic experiments. Nat Methods. 2013;10:989–91. https://doi.org/10.1038/nmeth.2610.
45. Mori H, Takio K, Ogawara M, Selkoe DJ. Mass spectrometry of purified amyloid beta protein in Alzheimer's disease. J Biol Chem. 1992;267:17082.
46. Wang R, Sweeney D, Gandy SE, Sisodia SS. The profile of soluble amyloid beta protein in cultured cell media. Detection and quantification of amyloid beta protein and variants by immunoprecipitation-mass spectrometry. J Biol Chem. 1996;271:31894–902. https://doi.org/10.1074/jbc.271.50.31894.
47. Hasegawa M, et al. Protein sequence and mass spectrometric analyses of tau in the Alzheimer's disease brain. J Biol Chem. 1992;267:17047–54.
48. Liao et al. Proteomic characterization of postmortem amyloid plaques isolated by laser capture microdissection. J Biol Chem. 2004;279(35):37061–8.
49. Bai B, et al. Proteomic landscape of Alzheimer's disease: novel insights into pathogenesis and biomarker discovery. Mol Neurodegener. 2021;16:55. https://doi.org/10.1186/s13024-021-00474-z.
50. Drummond E, et al. Proteomic differences in amyloid plaques in rapidly progressive and sporadic Alzheimer's disease. Acta Neuropathol. 2017;133:933–54. https://doi.org/10.1007/s00401-017-1691-0.
51. Xiong F, Ge W, Ma C. Quantitative proteomics reveals distinct composition of amyloid plaques in Alzheimer's disease. Alzheimers Dement. 2019;15:429–40. https://doi.org/10.1016/j.jalz.2018.10.006.
52. Gozal YM, et al. Proteomics analysis reveals novel components in the detergent-insoluble subproteome in Alzheimer's disease. J Proteome Res. 2009;8:5069–79. https://doi.org/10.1021/pr900474t.
53. Bai B, et al. U1 small nuclear ribonucleoprotein complex and RNA splicing alterations in Alzheimer's disease. Proc Natl Acad Sci USA. 2013;110:16562–7. https://doi.org/10.1073/pnas.1310249110.
54. Chen PC, et al. Alzheimer's disease-associated U1 snRNP splicing dysfunction causes neuronal hyperexcitability and cognitive impairment. Nat Aging. 2022;2:923–40. https://doi.org/10.1038/s43587-022-00290-0.
55. Selkoe DJ. Alzheimer's disease is a synaptic failure. Science. 2002;298:789–91.
56. Zhou JY, et al. Proteomic analysis of postsynaptic density in Alzheimer disease. Clin Chim Acta. 2013;420:62–8.
57. Hesse R, et al. Comparative profiling of the synaptic proteome from Alzheimer's disease patients with focus on the APOE genotype. Acta Neuropathol Commun. 2019;7:214. https://doi.org/10.1186/s40478-019-0847-7.
58. Carlyle BC, et al. Multiplexed fractionated proteomics reveals synaptic factors associated with cognitive resilience in Alzheimer's disease. BioRxiv. 2020; https://doi.org/10.1101/2020.07.31.230680.

59. Carlyle BC, et al. Synaptic proteins associated with cognitive performance and neuropathology in older humans revealed by multiplexed fractionated proteomics. Neurobiol Aging. 2021;105:99–114. https://doi.org/10.1016/j.neurobiolaging.2021.04.012.
60. Wang H, et al. Systematic optimization of long gradient chromatography mass spectrometry for deep analysis of brain proteome. J Proteome Res. 2015;14:829–38. https://doi.org/10.1021/pr500882h.
61. Bai B, et al. Deep multilayer brain proteomics identifies molecular networks in Alzheimer's disease progression. Neuron. 2020;105:975–91.
62. Wang Z, et al. 27-plex tandem mass tag mass spectrometry for profiling brain proteome in Alzheimer's disease. Anal Chem. 2020; https://doi.org/10.1021/acs.analchem.0c00655.
63. Higginbotham L, et al. Integrated proteomics reveals brain-based cerebrospinal fluid biomarkers in asymptomatic and symptomatic Alzheimer's disease. Sci Adv. 2020;6:eaaz9360. https://doi.org/10.1126/sciadv.aaz9360.
64. Sathe G, et al. Quantitative proteomic analysis of the frontal cortex in Alzheimer's disease. J Neurochem. 2020;156:988. https://doi.org/10.1111/jnc.15116.
65. Gong CX, Liu F, Grundke-Iqbal I, Iqbal K. Post-translational modifications of tau protein in Alzheimer's disease. J Neural Transm. 2005;112:813–38.
66. Martin L, Latypova X, Terro F. Post-translational modifications of tau protein: implications for Alzheimer's disease. Neurochem Int. 2011;58:458–71. https://doi.org/10.1016/j.neuint.2010.12.023.
67. Wang Y, Mandelkow E. Tau in physiology and pathology. Nat Rev Neurosci. 2016;17:5–21. https://doi.org/10.1038/nrn.2015.1.
68. Arakhamia T, et al. Posttranslational modifications mediate the structural diversity of tauopathy strains. Cell. 2020;180:633–644. e612.
69. Wesseling H, et al. Tau PTM profiles identify patient heterogeneity and stages of Alzheimer's disease. Cell. 2020;183:1699–1713. e1613.
70. Ping L, et al. Global quantitative analysis of the human brain proteome and phosphoproteome in Alzheimer's disease. Sci Data. 2020;7:315. https://doi.org/10.1038/s41597-020-00650-8.
71. Abreha MH, et al. Quantitative analysis of the brain Ubiquitylome in Alzheimer's disease. Proteomics. 2018;18:e1800108. https://doi.org/10.1002/pmic.201800108.
72. Sathe G, et al. Quantitative proteomic profiling of cerebrospinal fluid to identify candidate biomarkers for Alzheimer's disease. Proteomics Clin Appl. 2019;13:e1800105. https://doi.org/10.1002/prca.201800105.
73. Bader JM, et al. Proteome profiling in cerebrospinal fluid reveals novel biomarkers of Alzheimer's disease. Mol Syst Biol. 2020;16:e9356. https://doi.org/10.15252/msb.20199356.
74. Wang H, et al. Integrated analysis of ultra-deep proteomes in cortex, cerebrospinal fluid and serum reveals a mitochondrial signature in Alzheimer's disease. Mol Neurodegener. 2020;15:43. https://doi.org/10.1186/s13024-020-00384-6.
75. Dey KK, et al. Deep undepleted human serum proteome profiling toward biomarker discovery for Alzheimer's disease. Clin Proteomics. 2019;16:16. https://doi.org/10.1186/s12014-019-9237-1.
76. Soares Martins T, et al. Diagnostic and therapeutic potential of exosomes in Alzheimer's disease. J Neurochem. 2021;156:162–81. https://doi.org/10.1111/jnc.15112.
77. Perez-Gonzalez R, Gauthier SA, Kumar A, Levy E. The exosome secretory pathway transports amyloid precursor protein carboxyl-terminal fragments from the cell into the brain extracellular space. J Biol Chem. 2012;287:43108–15. https://doi.org/10.1074/jbc.M112.404467.
78. Muraoka S, et al. Enrichment of neurodegenerative microglia signature in brain-derived extracellular vesicles isolated from Alzheimer's disease mouse models. J Proteome Res. 2021;20:1733–43. https://doi.org/10.1021/acs.jproteome.0c00934.
79. Rosas-Hernandez H, et al. Characterization of serum exosomes from a transgenic mouse model of Alzheimer's disease. Curr Alzheimer Res. 2019;16:388–95. https://doi.org/10.2174/1567205016666190321155422.

80. Saman S, et al. Exosome-associated tau is secreted in tauopathy models and is selectively phosphorylated in cerebrospinal fluid in early Alzheimer disease. J Biol Chem. 2012;287:3842–9. https://doi.org/10.1074/jbc.M111.277061.
81. Grandin M, et al. Structural decoding of the Netrin-1/UNC5 interaction and its Therapeutical implications in cancers. Cancer Cell. 2016;29:173–85. https://doi.org/10.1016/j.ccell.2016.01.001.
82. Cassier PA, et al. Netrin-1 blockade inhibits tumour growth and EMT features in endometrial cancer. Nature. 2023;620:409–16. https://doi.org/10.1038/s41586-023-06367-z.
83. Wetzel-Smith MK, et al. A rare mutation in UNC5C predisposes to late-onset Alzheimer's disease and increases neuronal cell death. Nat Med. 2014;20:1452–7. https://doi.org/10.1038/nm.3736.
84. Chen G, et al. Netrin-1 receptor UNC5C cleavage by active delta-secretase enhances neurodegeneration, promoting Alzheimer's disease pathologies. Sci Adv. 2021;7 https://doi.org/10.1126/sciadv.abe4499.
85. Llambi F, Causeret F, Bloch-Gallego E, Mehlen P. Netrin-1 acts as a survival factor via its receptors UNC5H and DCC. EMBO J. 2001;20:2715–22. https://doi.org/10.1093/emboj/20.11.2715.
86. Lourenco FC, et al. Netrin-1 interacts with amyloid precursor protein and regulates amyloid-beta production. Cell Death Differ. 2009;16:655–63. https://doi.org/10.1038/cdd.2008.191.
87. Spilman PR, et al. Netrin-1 interrupts amyloid-beta amplification, increases sAbetaPPalpha in vitro and in vivo, and improves cognition in a mouse model of Alzheimer's disease. J Alzheimers Dis. 2016;52:223–42. https://doi.org/10.3233/JAD-151046.
88. Shabani M, et al. Netrin-1 improves the amyloid-beta-mediated suppression of memory and synaptic plasticity. Brain Res Bull. 2017;131:107–16. https://doi.org/10.1016/j.brainresbull.2017.03.015.
89. Glasgow SD, Ruthazer ES, Kennedy TE. Guiding synaptic plasticity: novel roles for netrin-1 in synaptic plasticity and memory formation in the adult brain. J Physiol. 2021;599:493–505. https://doi.org/10.1113/JP278704.
90. Zamani E, et al. Netrin-1 protects the SH-SY5Y cells against amyloid beta neurotoxicity through NF-kappaB/Nrf2 dependent mechanism. Mol Biol Rep. 2020;47:9271–7. https://doi.org/10.1007/s11033-020-05996-1.
91. Zamani E, Parviz M, Roghani M, Mohseni-Moghaddam P. Key mechanisms underlying netrin-1 prevention of impaired spatial and object memory in Abeta(1-42) CA1-injected rats. Clin Exp Pharmacol Physiol. 2019;46:86–93. https://doi.org/10.1111/1440-1681.13020.
92. van Gils JM, et al. The neuroimmune guidance cue netrin-1 promotes atherosclerosis by inhibiting the emigration of macrophages from plaques. Nat Immunol. 2012;13:136–43. https://doi.org/10.1038/ni.2205.
93. Ramkhelawon B, et al. Netrin-1 promotes adipose tissue macrophage retention and insulin resistance in obesity. Nat Med. 2014;20:377–84. https://doi.org/10.1038/nm.3467.
94. Aebersold R, et al. How many human proteoforms are there? Nat Chem Biol. 2018;14:206–14. https://doi.org/10.1038/nchembio.2576.
95. Wightman DP, et al. Largest GWAS (N=1,126,563) of Alzheimer's Disease Implicates Microglia and Immune Cells. medRxiv 2020.11.20.20235275. https://doi.org/10.1101/2020.11.20.20235275.

Genomics and Transcriptomics of Alzheimer Disease

Natalia Vilor-Tejedor, Blanca Rodriguez-Fernandez, Patricia Genius, Emma S. Luckett, Federica Anastasi, Luigi Lorenzini, Hollydawn Murray, Caroline Bull, Matthew H. S. Clement, Andrew Morris, Arcadi Navarro, Junhao Wen, and Tavia E. Evans

Abstract Alzheimer disease (AD) and Alzheimer disease-related dementias (ADRD) are driven by complex interactions between genomic, transcriptomic, and neuroimaging features that contribute to neurodegeneration. This chapter discusses recent advancements, from identifying rare, high-impact mutations linked to familial AD to large-scale genome-wide association studies (GWAS) highlighting the polygenic nature of late-onset AD. Key genetic risk factors, notably the *APOE* ε4 allele, and dysregulated pathways, such as mitochondrial dysfunction, immune response, and synaptic plasticity, are emphasized. Integrative multi-omics approaches, combining genomics, transcriptomics, and brain imaging data, offer valuable insights into the heterogeneous molecular and neuroanatomical landscape of AD, supporting the development of precision medicine approaches. Future directions underscore the need for diverse genetic studies, research into gene-

N. Vilor-Tejedor (✉)
Alzheimer's Prevention Programme, Barcelonaβeta Brain Research Center (BBRC), Pasqual Maragall Foundation, Barcelona, Spain

Genetics, Radboud University, Nijmegen, The Netherlands

Computational Biology and Health Genomics, Centre for Genomic Regulation (CRG), The Barcelona Institute for Science and Technology, Barcelona, Spain

Neurosciences Programme, IMIM – Hospital del Mar Medical Research Institute, Barcelona, Spain
e-mail: nvilor@barcelonabeta.org

B. Rodriguez-Fernandez
Alzheimer's Prevention Programme, Barcelonaβeta Brain Research Center (BBRC), Pasqual Maragall Foundation, Barcelona, Spain

Neurosciences Programme, IMIM – Hospital del Mar Medical Research Institute, Barcelona, Spain

Bioinformatics and Genomics Programme, Centre for Genomic Regulation (CRG), The Barcelona Institute for Science and Technology, Barcelona, Spain

Biochemistry and Molecular Biology Department, Autonomous University of Barcelona (UAB), Barcelona, Spain

R. Tarawneh (ed.), *Biomarkers of Neurodegenerative Disorders*,
https://doi.org/10.1007/978-3-031-80849-4_13

environment interactions, and accessible, ethically governed data-sharing frameworks to enhance our understanding of AD pathogenesis and facilitate early intervention strategies.

Keywords Alzheimer disease · Genomics · Transcriptomics · Brain imaging genetics · Neurodegeneration

P. Genius
Alzheimer's Prevention Programme, Barcelonaβeta Brain Research Center (BBRC), Pasqual Maragall Foundation, Barcelona, Spain

Neurosciences Programme, IMIM – Hospital del Mar Medical Research Institute, Barcelona, Spain

Bioinformatics and Genomics Programme, Centre for Genomic Regulation (CRG), The Barcelona Institute for Science and Technology, Barcelona, Spain

E. S. Luckett · L. Lorenzini
Radiology and Nuclear Medicine, Amsterdam UMC, Amsterdam, The Netherlands

F. Anastasi
Alzheimer's Prevention Programme, Barcelonaβeta Brain Research Center (BBRC), Pasqual Maragall Foundation, Barcelona, Spain

Neurosciences Programme, IMIM – Hospital del Mar Medical Research Institute, Barcelona, Spain

Proteomics Unit, Centre for Genomic Regulation (CRG), The Barcelona Institute for Science and Technology, Barcelona, Spain

H. Murray · C. Bull · A. Morris
Health Data Research UK, London, UK

M. H. S. Clement
Alzheimer's Disease Data Initiative, Kirkland, WA, USA

A. Navarro
Alzheimer's Prevention Programme, Barcelonaβeta Brain Research Center (BBRC), Pasqual Maragall Foundation, Barcelona, Spain

Computational Biology and Health Genomics, Centre for Genomic Regulation (CRG), The Barcelona Institute for Science and Technology, Barcelona, Spain

Department of Medicine and Life Sciences, Universitat Pompeu Fabra, Barcelona, Spain

Institució Catalana de Recerca i Estudis Avançats (ICREA), Barcelona, Spain

Experimental and Health Sciences, Institute of Evolutionary Biology (CSIC-UPF), Barcelona, Spain

J. Wen
Department of Radiology, Columbia University, New York, NY, USA
e-mail: jw4745@cumc.columbia.edu

T. E. Evans
Department(s) of Human Genetics, Radboud University Medical Center, Nijmegen, Netherlands, Netherlands
e-mail: Tavia.Evans@radboudumc.nl

1 Introduction to the Genomic Era Focusing on Neurodegenerative Diseases

Neurodegenerative diseases, particularly Alzheimer disease (AD) and related dementias (ADRD), manifest as progressive and irreversible declines in cognitive and neurological function, leading to significant morbidity and mortality. AD is distinguished by its neuropathological hallmarks: the deposition of amyloid-β (Aβ) peptides into plaques and the aggregation of hyperphosphorylated tau protein into neurofibrillary tangles within the brain [1]. This protein deposition serves as the current endpoint for diagnosis, along with cognitive measures, which are also used to measure drug responses in AD clinical trials. Alongside these brain changes, the presence of Aβ and tau isoforms in biofluids such as cerebrospinal fluid or blood aids in determining clinical trial success and diagnosing the disease. Although these protein isoforms can be successfully detected in the brain and biofluids, and increase the risk for developing symptomatic AD, they are not always sufficient to develop clinical symptoms of the disease. Therefore, advances in omics-based research have been crucial to further understand what drives disease development and the onset of cognitive impairment.

The elucidation of the genetic landscape underlying ADRD has revealed a complex aetiology that includes both monogenic determinants and polygenic risk factors [2–4]. At one end of the genetic spectrum, rare, highly penetrant mutations in genes such as amyloid precursor protein (*APP*), presenilin 1 (*PSEN1*), and presenilin 2 (*PSEN2*) have been identified, which confer almost a certain risk for the development of familial AD with an early onset [5]. Conversely, the more common, late-onset form of AD is predominantly influenced by a combination of genetic variants, each conferring a modest additive risk for the disease. The identification of these variants, primarily through Genome-wide Association Studies (GWAS) [6–9], has highlighted a polygenic model of AD, wherein the cumulative effect of multiple alleles contributes to the overall genetic susceptibility to the disease. Among these, the apolipoprotein E (*APOE*) gene, and specifically its epsilon (ε4) allele, stands out as the major significant genetic risk factor for AD [10].

The genetic insights gained from the study of ADRD have profound implications for the field of neurodegenerative research. First, advancements in genetics offer the promise of personalized medicine, tailoring interventions to the genetic risk profile of each individual [11]. This approach forms the basis for preventive strategies, such as personalized lifestyle modifications and risk reduction interventions for individuals at increased genetic risk [12]. Second, genetic exploration of ADRD highlights the broader pathophysiological mechanisms underlying neurodegeneration, suggesting common pathways being shared between different neurodegenerative diseases [13, 14]. This has broadened the scope of research, encouraging a more integrated approach to the study of neurodegeneration that may uncover synergistic pathways for therapeutic intervention. Overall, the emergence of genetic insights heralds new steps for diagnostic precision. The potential to detect AD risk before the appearance of symptoms is moving from being an aspirational goal to a practical possibility [15–17]. Imagine a scenario in which a straightforward genetic test can determine a person's likelihood of developing dementia, enabling early interventions to prevent its onset. This concept is gradually becoming more concrete as research efforts identify and amalgamate genetic and biomolecular indicators, providing a predictive model for vulnerability to neurodegenerative diseases. This progress marks a significant turn towards pre-emptive management of such conditions, highlighted by the accuracy of genetic testing.

As we advance our exploration of the genetic underpinnings of ADRD, we explore multiple dimensions of genomic research. This chapter integrates diverse scientific insights, from basic genetic markers to advanced genomic interactions, and emphasizes their relevance in clinical applications and research directions. Our exploration begins with an introduction to the genomic era, highlighting its impact on our understanding of neurodegenerative diseases. We then delve into the genetics of AD, exploring individual genetic markers, the broader implications of GWAS and post-discovery GWAS analyses, and the potential of genetic and pathway scores beyond *APOE*. Additionally, we examine the scope of transcriptomics in AD, along with its integration with other omics layers, such as proteomics. This chapter also discusses genetic studies of AD endophenotypes that connect genetic findings with AD biomarkers and neuroimaging phenotypes to enhance our understanding of AD. Finally, we consider future directions in AD research, focusing on exploring gene-environment and gene–gene interactions, increasing diversity in genetic studies, and addressing ethical considerations surrounding data access and sharing.

2 Genetics of AD

Small changes in genes, referred to as genetic variants, can increase the likelihood of developing illnesses, including AD. While some genetic variants are common, and most do not cause disease, there are some that do. For instance, a person might inherit a gene variant that nearly guarantees the onset of a specific disease (i.e., disease-causing mutations). However, other variants may only alter the likelihood of developing a disease, either increasing (i.e., genetic risk factors) or decreasing the risk. Finding and understanding these variants is very important in disease treatment or prevention efforts.

The potential of genetic biomarkers in AD exemplifies the core of precision medicine: delivering the right treatment to the right patient at the right time, based on a comprehensive understanding of an individual's genetic, environmental, and lifestyle factors. As we continue to unravel the genetic complexities of AD, the integration of genetic biomarkers into clinical practice is set to transform the diagnosis, treatment, and prevention of this challenging condition, ushering in an era of more predictive, preventive, and personalized medicine.

2.1 Key Genetic Markers in Alzheimer Disease

Genetics contribute to both familial (early-onset) and sporadic (late-onset) forms of AD [3, 4]. In familial AD, mutations in specific genes directly cause the disease, typically manifesting at a younger age. Conversely, sporadic AD, which accounts for most cases, arises from a combination of genetic, environmental, and lifestyle factors, with age being the most significant risk factor.

Early-Onset AD (EOAD) represents a less common form of AD, typically affecting individuals under the age of 65. The genetic underpinnings of EOAD highlight the critical role of specific mutations [18, 19] in three genes—*APP*, *PSEN1*, and *PSEN2*—that lead to the development of the disease at a younger age (i.e., causative genes). The *APP* gene encodes the amyloid precursor protein, a key player in the production of Aβ peptides. Certain mutations in the *APP* gene alter the proteolytic processing of the amyloid precursor protein, favouring the formation of Aβ42, which is more pathogenic compared to other Aβ peptides. Aβ42 is more prone to aggregation, which accelerates the formation of amyloid plaques—a hallmark of AD pathology. Interestingly, certain mutations in the *APP* gene, located near the secretase cleavage sites, directly influence enzymatic activity, thereby increasing the production of Aβ42. The *PSEN1* and *PSEN2* genes encode the presenilin proteins, which are integral components of the gamma-secretase complex. This complex is responsible for the final cleavage of the amyloid precursor protein, producing Aβ peptides. Mutations in *PSEN1* are the most common cause of familial EOAD, with over 200 mutations associated with EOAD. Similarly, mutations in *PSEN2*, while less common, follow the same pattern of increasing the ratio of Aβ42 to Aβ40, promoting plaque formation. These genetic mutations demonstrate autosomal dominant inheritance patterns, meaning the presence of a single copy of the mutated allele from an affected parent is almost always associated with developing AD. Such findings underscore the importance of genetic counselling and testing for families with a history of EOAD, offering critical insights into the hereditary nature of the disease. Notably, while EOAD has a considerable genetic basis with a heritability of over 90%, variation in known EOAD genes accounts for only 5–10% of cases. Most cases are either sporadic or follow a non-Mendelian pattern of inheritance but are expected to be enriched for causative genetic factors [20].

Late-Onset AD (LOAD), which occurs after the age of 65, constitutes the majority of AD cases and showcases a complex genetic and environmental landscape. It involves multiple genes, each contributing a small increase in disease predisposition. The most significant genetic risk factor for sporadic AD is the *APOE* gene [10, 21, 22], particularly its ε4 allele. *APOE* plays a crucial role in lipid metabolism and is expressed in the brain and other tissues implicating it in neuronal repair processes [23]. It has three common alleles: ε2, ε3, and ε4. The ε4 allele is associated with an increased risk and earlier onset of AD, while the ε2 allele seems to have a protective effect. The exact mechanisms by which *APOE* ε4 influences AD risk are not fully understood but they are believed to involve processes such as Aβ aggregation and clearance, as well as neuroinflammation [24]. *APOE* ε4 is less efficient in amyloid clearance compared to the other alleles, leading to an increased accumulation of Aβ plaques. Additionally, *APOE* ε4 has been linked to neuroinflammation and tau pathology, further contributing to AD progression [21]. However, despite the strong association between *APOE* ε4 and AD risk, it is important to note that inheriting one or even two ε4 alleles does not necessarily imply the development of the disease. Conversely, individuals without an ε4 allele can still develop AD, highlighting the complex interplay of genetic and non-genetic factors in disease pathogenesis [25].

2.2 Key Markers from Genome-Wide Association Studies

Genome-wide association studies (GWAS) are a research method designed to identify genetic variations linked to specific diseases or traits across the genome [26, 27]. This method involves scanning the genomes of large cohorts, often by comparing individuals with a particular disease or trait to those without, to find genetic markers—typically single nucleotide polymorphisms (SNPs) —that occur more frequently in those with the condition. GWAS are particularly valuable in studying complex diseases such as AD, where multiple genes may contribute to a person's risk [28]. The approach can help researchers understand the biological pathways involved in these diseases and may lead to new preventive measures, diagnostic tools, and treatments.

GWAS focusing on ADRD have identified an increasing number of genetic loci associated with these conditions as sample sizes in these studies have expanded. The two most recent GWAS on AD have made significant advancements in understanding the genetic landscape of this condition [7, 8]. Wightman et al. significantly expanded their previous GWAS, including 1,126,563 participants (43,725 cases, 46,613 proxy-cases, 1,036,225 controls/proxy-controls), from new biobank and population-based dementia datasets. Meanwhile, Wightman et al. and Bellenguez et al. [7, 8] enhanced their GWAS by incorporating additional clinical case/control cohorts primarily from the European Alzheimer's & Dementia Biobank, resulting in a total of 788,989 participants (64,498 cases, 46,828 proxy-cases, and 677,643 controls/proxy-controls). These two studies collectively discovered 90 unique variants situated within 75 different loci (38 and 71, respectively) linked to AD. Of these, 34 loci were concurrently identified in both studies, and 42 of the loci are newly identified.

Since the first GWAS of AD was published in 2013 [29], subsequent AD GWAS have identified 101 independent genetic variants across 81 different loci, with heritability estimated at 79% [28]. These studies highlight genes involved in immune response, lipid metabolism, and cell signalling [6–9, 30]. Each of these genes contributes a small increase to the risk of developing AD, suggesting that AD is a polygenic disease influenced by the cumulative effects of many genetic variants (see Sect. 2.5). Additionally, although the X chromosome constitutes 5% of the genome and contains a substantial portion of brain-expressed genes, its contribution to AD remains relatively unexplored. Recent advancements, such as large-scale X chromosome-wide association studies, offer promising avenues for unravelling the genetic underpinnings of AD on this chromosome [31].

2.3 Whole-Genome Sequencing and Whole-Exome Genotype Analyses

Most GWAS performed for LOAD utilize genome-wide microarrays. These microarrays primarily capture previously identified genetic variants and are generally regarded as cost-effective compared to next-generation sequencing techniques, such as Whole Exome Sequencing (WES) and Whole Genome Sequencing (WGS). Although targeted microarrays have been instrumental in increasing GWAS sample sizes, they are limited in their ability to detect unknown genetic variants.

Targeted microarrays are designed to detect only known genetic variants, which are typically common and are estimated to account for merely 33% of the genetic variance in LOAD [32]. In contrast, genotyping-by-sequencing methods offer significant advantages for identifying less common, but potentially more impactful, genetic variants that standard GWAS, which use SNP arrays, might miss.

These methodologies, initially applied in family studies, have facilitated the identification of rare mutations in genes previously overlooked by GWAS due to their relatively low frequency. The role of the *TREM2* gene serves as an illustrative example of how WES and WGS have unveiled new pathological pathways in the disease. Initially known for causing Nasu-Hakola disease—a rare condition characterized by early-onset dementia and bone abnormalities in Finnish populations [33]—WES later identified new mutations in *TREM2* among highly consanguineous families with early-onset frontotemporal dementia but without bone involvement [34]. Subsequently, other mutations in *TREM2* were linked to an increased risk of LOAD [35].

While *TREM2* is just one example, other genes, such as *SORL1* and *ABCA7*, demonstrate the potential of next-generation sequencing techniques in identifying low-frequency genetic variants. These variants could later be incorporated into commercial microarrays, enhancing the precision of GWAS findings through fine mapping [36]. Despite the technical and computational challenges associated with genotype-by-sequencing techniques, the field is expected to gradually transition to these methods in the coming years, thanks to the increasing availability of low-cost sequencing technologies [26, 37]. Indeed, some studies are already demonstrating the potential of these techniques in elucidating the "*missing heritability*" attributed to rare variants [38], explaining the high phenotypic heterogeneity of the disease, or identifying novel ethnicity-specific genetic variants [39]. Overall, the shift towards genotype-by-sequencing techniques holds promise for unravelling the intricate genetic architecture underlying AD. Figure 1 demonstrates the complex genetic architecture of AD.

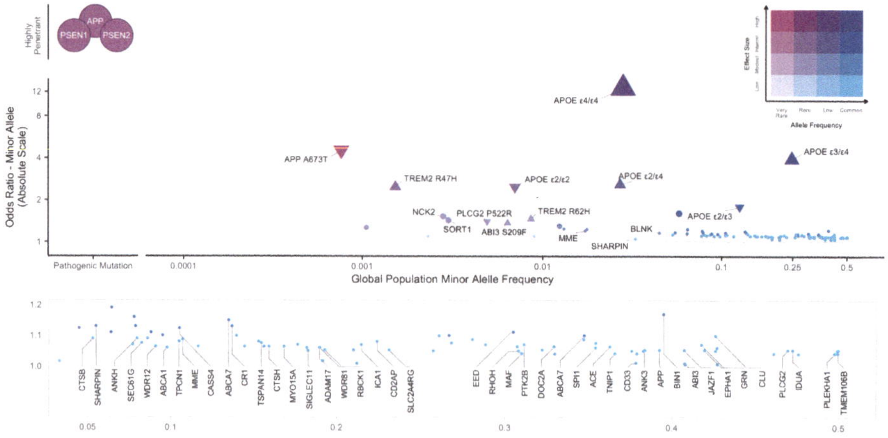

Fig. 1 Complex genetic architecture of AD. This figure highlights mutations associated with autosomal dominant AD (*PSEN1*, *PSEN2*, and *APP* genes) and 81 genome-wide significant loci. GWAS studies identified 101 independent AD-associated single nucleotide polymorphisms across 81 genome-wide significant (p<5e-8) loci. The shaded area represents 80% power to detect genome-wide significant association for variants at a given effect size and population frequency between an effective sample size of 382,472 (top) [8] and one million (bottom) assuming an AD prevalence of 0.18. Odds ratios are reported on the absolute scale with triangles indicating directionality of *APOE4* genotypes and rare variants with moderate to high SnpEff impact annotations. (Used with permission from Andrews et al. 2023 [28]; Copyright of © 2023 The Author(s). Published by Elsevier B.V.; permission obtained through the Copyright Clearance Center, Inc).

2.4 Post-discovery Genetic Analyses

The identification of genetic variants has greatly improved our understanding of the genetics behind AD, identifying many genetic regions linked to the risk of the disease. However, simply identifying these regions is only the first step. Post-discovery genetic analyses are essential for the next phase, as they facilitate the identification of genes within these regions that significantly affect AD risk [40].

Techniques such as fine statistical mapping are critical at this stage, as they allow researchers to narrow the wide-ranging results of GWAS to specific causal variants, which are most likely to directly affect the disease process [41]. Co-localization plays a role in aligning these genetic variants with changes in gene expression, particularly in brain tissue, validating whether the same variants associated with AD risk also correlate with gene activity levels.

Functional inference extends beyond the simple detection of genetic variants. It aims to clarify the regulatory networks these genes are involved in, identifying potential impacts on cellular functions. For instance, a genetic variation might influence a gene that plays a role in microglia functions, altering how these brain cells manage amyloid plaques, which are crucial in AD development. Moreover, by integrating GWAS data with additional biological datasets, such as proteomic [42, 43] and metabolomic [41, 44] profiles, researchers can construct a more holistic view of the disrupted biological pathways in AD [45].

Complementary, Mendelian randomization (MR) studies harness these genetic variants as instrumental variables to explore the biological interplay between genetic predispositions and disease pathology, offering a unique perspective on how genetic and environmental factors converge in AD. For instance, a recent study identified genetically determined increases in high-density lipoprotein cholesterol and systolic blood pressure as factors associated with a heightened risk of AD [46]. Additionally, investigations into potentially modifiable risk factors revealed significant associations between genetically predicted educational attainment and AD risk [47]. These findings underscore the diversity of biological pathways involved in AD pathogenesis and highlight the potential for MR studies to uncover novel disease mechanisms. However, it is essential to approach MR findings cautiously, considering the inherent limitations and the intricate interplay between genetic and environmental factors in AD pathogenesis.

Overall, through post-discovery GWAS strategies, we are moving closer to unravelling the complex molecular mechanisms driving AD, paving the way for precision medicine approaches in AD management.

2.5 Risk Prediction Models in AD

While the *APOE* ε4 allele is the most significant genetic risk factor for AD, it does not work alone. Over time, GWAS has highlighted a more complex picture, underscoring the fact that genetic predisposition to AD is not determined by any single genetic anomaly but by the combined effect of multiple genes, a concept known as polygenicity [48]. Polygenicity refers to the influence of many different genes, each contributing a small effect to the overall risk of a disease. The most recent GWAS of sporadic AD and related dementias have identified more than 70 genetic variants that modify the risk of developing AD, beyond *APOE* [49]. The identification of these genes is crucial because each one is involved in various biological pathways that potentially influence the onset and progression of AD.

Polygenic Risk Scores (PRSs) are an emerging tool in the field of genetics that aggregate the effects of numerous genetic variants across the genome to predict an individual's likelihood of developing a particular disease. Building on a reference GWAS, a PRS for a disease, such as AD, can be computed as the sum of risk alleles that an individual has, weighted by the risk allele effect sizes as estimated by the GWAS. As a measure of an individual's genetic risk for a specific trait, PRSs have increasingly become a tool of interest in both research and clinical frameworks, holding the possibility of increasing the efficiency of disease diagnosis and prognosis [50].

In the clinical context of AD, where cognitive symptoms manifest as the end result of a long silent pathophysiological progression, PRSs represent a powerful predictive instrument for early identification of at-risk individuals, thus facilitating personalized therapeutic interventions. Patient stratification based on an individuals' genetic risk might in fact enable more effective and timely accurate medications, with, for instance, high-risk individuals benefiting from earlier and more aggressive preventive strategies, such as lifestyle changes, medication, or participation in clinical trials for promising new treatments [51].

Beyond clinical applications, PRSs have also proven useful in research settings to stratify participants into different risk categories based on their genetic predisposition for a specific trait. These stratifications are crucial to understand how genetic factors interact with other risk factors, such as environment and lifestyle, along the disease course. In AD research, PRSs have also demonstrated their predictive value for the presence of brain pathologies, such as amyloid [52] and tau [53] deposition in presymptomatic individuals, and have generally been related to AD diagnosis, cognitive trajectories, and brain atrophy [54–56]. These studies provide evidence of how PRSs can help in identifying at-risk individuals in the earliest stages and improving our understanding of the evolution of the disease.

While the majority of the AD risk variants affect Aβ42 production and clearance, other genes involved in lipid metabolism, immune and inflammatory response, and endocytosis have also been shown to contribute to the AD polygenicity [48, 57]. These additional pathways might be responsible for the evident heterogeneity of neuropathological features observed early in the disease [58], and their characterization is crucial for the understanding of AD complexity. In this context, recent advances have proposed the possibility of linking risk variants to genes, and genes to associated biological pathways, allowing the computation of pathway-specific PRSs, which retain information about how the burden of genetic risk varies across biological processes [59]. Pathway PRSs can thus be used as a proxy of the genetic predisposition to certain biological processes, and have demonstrated how distinct biological processes might be underlying the early neuropathological heterogeneity of AD [16, 59].

In a similar way, the expression profile of candidate genes may be relevant for the understanding of the underlying pathophysiology. Many of the AD at-risk variants are highly expressed in microglia and astrocytes, but their functional role in the AD progression has not yet been elucidated. In a recent work [60], cell-type PRSs were created by grouping variants based on their expression profiles. The results of this study demonstrated a differential timescale in which astrocytes and microglia operate in the development of AD pathology, and advocate for distinct cellular phases of AD [61].

In conclusion, PRSs are an important tool for both clinical and research settings and provide a solid instrument for risk assessment and patient stratification, thus contributing to the refining of personalized medicine approaches. Moreover, by taking into consideration the expression profiles and related biological pathways of candidate genes, pathway- and cell-type PRSs constitute a novel approach to further characterize disease heterogeneity in a disease-stage-dependent manner.

3 Transcriptomics of AD

3.1 RNA Sequencing and Relevance of Transcriptomics for AD

Shifting focus from the genetic underpinnings revealed by GWAS, the field of AD research has expanded into transcriptomics to delve deeper into the functional consequences of these genetic variations. Transcriptomics, the comprehensive study of RNA transcripts produced by the genome, offers insights into how genes are expressed differently in AD patients compared to healthy controls. The understanding of this aspect may allow for identification of new disease biomarkers, improved recruitment stratification for clinical trials and early disease diagnosis and provide targets for further research for drug development and treatments. This is highly applicable since there are a limited number of AD biomarkers in use, especially beyond the known pathological hallmarks of amyloid-β and tau. There are also a limited number of treatment targets, where only amyloid-lowering therapies are currently at the forefront of clinical trials [62, 63].

Previously, gene expression studies were performed using low-throughput methods, such as quantitative polymerase chain reaction (qPCR), a limitation of which is measuring single transcripts, or hybridization-based microarrays, a limitation of which is the inability to accurately quantify lowly expressed genes. However, advances in methodology have resulted in the development of high-throughput methods such as RNA sequencing (RNA-Seq). There is a higher coverage with RNA-Seq than the preceding methods, thus providing a superior resolution of the dynamic transcriptome. RNA-Seq is, therefore, a method that can be employed to uncover disease-associated changes and reveal novel therapeutic targets. Lastly, such sequencing studies can be performed on a variety of samples in different ways. For example, bulk RNA-Seq analyses pooled cell populations giving an averaged readout of expression levels of individual genes, and can be performed, for example, using peripheral whole blood or postmortem brain homogenate. On the other hand, single-cell RNA-Seq analyses individual cell types giving an average readout of expression levels of individual genes per cell type, and can be performed, for example, using peripheral blood mononuclear cells. Success of both bulk and single-cell RNA-Seq studies has already been achieved in the field of cancer, providing promise to the AD and neurodegenerative field at large.

Analysing changes in gene expression from RNA-Seq data can take many forms. For example, changes in single gene expression can be investigated using differential gene expression (DGE), or expression changes in networks of co-expressed genes using weighted-gene co-expression network analysis (WGCNA). Top differentially expressed genes (DEGs) from DGE analyses may highlight candidate genes that are dysregulated in a disease and therefore, could be potential disease markers. Similarly, highly interconnected genes, or hub genes, within a particular gene co-expression network may represent a critical gene responsible for driving pathological changes within a particular biological pathway, highlighting a potential disease mechanism that could be a novel therapeutic drug target. Performing functional enrichment and pathway analysis with such genes or networks can provide insight into dysregulated or preserved biological pathways, cellular components, or molecular functions. Thus, transcriptomics can provide a means to disentangle biological mechanisms beyond established risk factors and/or pathological biomarkers to advance our understanding of the underlying pathophysiology of disease.

3.2 Description of Key Findings

Since AD is a brain disease, many transcriptomic studies have been performed on postmortem brain tissue. Postmortem brain studies have provided insight into both up- and down-regulated genes and networks that function within various pathways. Such studies have found that AD brains have many more up- and down-regulated genes compared to control subjects, which are, for example, functionally enriched for mitochondrial activity, intracellular (calcium) signalling, and inflammation, and are consistently found across studies. The advantage of postmortem studies is a known (i.e., neuropathologically confirmed) diagnosis of AD when performing group comparisons. However, given samples are obtained upon death, it is likely that the stage of AD represented is the end stage of disease. As a result, there may be age-related confounders influencing observed differences, given that changes in gene expression as part of the healthy aging process are still largely unknown. Nonetheless, using postmortem brain samples allows for changes in gene expression to be analysed between different brain regions or cell types, thus providing insight into differences using a fine-grained approach.

Early postmortem brain studies using low-throughput methodology enabled the AD field to begin piecing together the pathological changes occurring in AD. An early study in tau tangle-bearing CA1 neurons found that these cells had a significant reduction in mRNAs related to, for example, cytoskeletal and synaptic proteins, and an upregulation in cathepsin D [64]. Similar results were observed in later studies using laser-captured CA1 neurons and microarray analyses [65], RNA-Seq and qPCR [66], and protein–protein interaction and network analyses, which also revealed enrichments such as signal transduction and cellular metabolism [67], and transcription and chemical synaptic transmission [68]. Similarly, a study analysing cholinergic nucleus basalis neurons in AD patients also demonstrated an upregulation of cathepsin D, as well as a downregulation of trkB and trkC in AD patients [69]. Furthermore, the neocortex of AD patients was found to have over 5000 DEGs compared to controls, many of which were enriched for synaptic dysfunction and neuroinflammation [70]. Moreover, transcriptome profiles appeared to differ between the frontal (e.g., regulation of synaptic transmission and synaptic plasticity), and temporal (e.g., neurotransmitter and axonal transport) lobes of AD patients [71], as well as the parietal cortex, which showed an upregulation of lipid metabolism pathways [72]. Thus, these early transcriptional studies suggest gene expression differences between brain regions, which may be reflecting the differences in the spread and progression of AD neuropathology.

More recent postmortem brain studies using the high-throughput RNA-Seq methodology have found largely similar results, building on those earlier studies with the broader coverage technique. Synaptic transmission and/or assembly, transcriptional regulation, mitochondrial, cell death, and inflammatory pathways are just some of the umbrella terms that emphasize the biological processes found to be dysregulated in AD brains, and have been replicated in the most recent AD RNA-Seq studies, validating previously reported findings [73–81]. Importantly, these biological pathway changes have now been shown to be associated with AD endophenotypes, such as dysregulation in relation to tau deposition, amyloid deposition, and cognitive decline [82]. Hence, an AD brain molecular fingerprint has begun to emerge.

In recent years, there has been a shift in the AD field, whereby non-brain tissues are being analysed given their ease of sampling. Furthermore, biofluids such as blood are not only much more easily accessible than brain tissue, but they allow for sampling across the disease spectrum in longitudinal study designs, as well as serial sampling from the same individual. Similar studies are inherently impossible when relying on postmortem tissue for analyses. Additionally, there is a large overlap in the transcriptional profiles of the brain and blood, which further advocates for the use of blood as a surrogate tissue for transcriptional analyses [83, 84]. AD peripheral blood transcriptomic studies have shown that AD patients have a similar dysregulation of genes, networks, and functional enrichment terms as observed in the post-mortem brain studies [39, 81, 85–89]. Consistently found across studies, and replicated in the most recently published AD blood transcriptomic studies, is a dysregulation of protein synthesis, mitochondrial/metabolic pathways, apoptosis, and inflammation [90–93].

Peripheral blood samples for transcriptomic studies can be collected alongside other data modalities such as brain imaging, providing the ability to determine (non-)synergistic relationships between the two during life, and across the course of disease. This allows insight into any relationships between changes in gene expression in the periphery and more specific AD neuropathology or endophenotypes. A network-based AD blood transcriptomic analysis found that a co-expressed gene network was significantly related to an AD diagnosis and enriched for the Fc gamma receptor-mediated phagocytosis pathway, endocytosis, and B cell receptor signalling, amongst others [94]. This network of co-expressed genes was also positively associated with cortical amyloid-β deposition, and negatively associated with executive function and hippocampal volume. Building on this study, the authors later published that levels of *APH1B* (encoding a subunit of the γ-secretase complex) were increased in AD patients compared to controls in DGE analyses, which was associated with decreased entorhinal cortical thickness and increased cortical amyloid-β deposition [95]. Enrichment analysis with other DEGs found in this study revealed significant enrichments for similar pathways as those described above: protein oligomerization, positive regulation of programmed cell death, positive regulation of apoptotic processes, and vesicle-mediated transport. A similar

DGE study found that *CREB5* (involved in synaptic strengthening and memory formation) was significantly upregulated in AD patients compared to controls, which was also significantly associated with a decline in memory and executive function, greater brain atrophy, and higher cortical amyloid-β deposition in these individuals [96]. Similarly, gene co-expression networks and their enrichment terms, which again mirror those described above, have also been significantly associated with the key AD plasma biomarkers currently in use: Aβ42/Aβ40 ratio, total tau, tau phosphorylated at threonine 181 (p-tau181), and neurofilament-light chain (NfL) [92]. These results highlight that both an AD diagnosis and its endophenotypes can be detected in blood via alterations in transcription in those individuals with an AD diagnosis.

With a shift in the use of biofluids in AD research, there has also been a collective shift to further understand the preclinical phase of the disease continuum where clinical symptoms are not yet apparent, given this phase provides a window of opportunity for therapeutic intervention and prevention. Preclinical RNA AD studies are still few and far between. However, those that have been published are again replicating the previous postmortem and blood transcriptomic findings. Common dysregulated pathways include those umbrella terms mentioned above, such as mitochondrial/metabolism, cell signalling, immune pathways, and ribosomal alterations, that are associated with amyloid accumulation [97], Braak stages and cognitive outcomes [98, 99], conversion to mild cognitive impairment (MCI) [100], and an early AD diagnosis (MCI due to AD) [101]. Despite emerging evidence from preclinical AD transcriptomic studies that supports findings from studies with a case-control study design, there is still an absence of evidence in this field, prompting an area in critical need of investigation and validation to further delineate early changes.

To conclude, RNA-Seq has allowed for the detection of transcriptomic dysregulation in AD, which points towards key biological pathways that go awry in disease, such as metabolism, cellular/synaptic signalling, protein dynamics, and inflammation. These biological pathways are both linked to AD diagnosis and its endophenotypes providing evidence for the pathophysiological underpinnings of AD and demonstrating its highly complex, dynamic, and interwoven biological nature. The overlap of expression profiles between different sample types is key for allowing a progression to more accessible tissue types, providing the ability to translate findings into clinical practice or clinical trial design. This may eventually allow for alternative disease treatment targets to be identified, improved disease monitoring, creating risk profiles for stratified trial recruitment, and a move towards personalized medicine.

3.3 Integration of Genomics and Transcriptomics with Other Omics Data in AD

Technological progress has enabled the generation of data spanning various biological domains, such as DNA sequencing, RNA expression profiling, methylation patterns, proteomics, and metabolomics. Multi-omics refer to the integration of these approaches to gain a more comprehensive understanding of disease complexity and deeper insight into the complex genotype–phenotype associations [102, 103].

Given the multifactorial and heterogeneous nature of AD [104, 105], multi-omics approaches can be applied to underlying molecular pathways and mechanisms driving AD heterogeneity, paving the way for more accurate diagnosis, prognosis, and personalized treatment interventions [106, 107].

In a multicohort approach, Swarup et al. integrated proteomics and transcriptomics to identify robust early and late proteomic changes in AD and other neurodegenerative dementias [108]. The authors integrated genetic risk and GWAS to understand the relationship of each multi-omics module with potential causal mechanisms and found that genetic risk is differentially enriched across disorders, with glial-immune modules being enriched only in AD. Additionally, proteomics and transcriptomics reported shared co-expression modules showing consistent molecular alterations at multi-omics levels. In a more recent cerebrospinal fluid (CSF) proteomics study, Tijms et al. [109] reported the identification of five different subtypes of AD. Integrating genetic data on specific AD genetic risk variants, they found variations in the prevalence of AD risk variants among subtypes, suggesting unique underlying molecular mechanisms. This integration marks a significant advance in AD molecular characterization, prompting further investigation into subtype-specific treatment responses and subject selection for clinical trials, thereby advancing personalized medicine for AD.

Large multi-omics datasets, such as proteomics datasets comprising hundreds to thousands of identified proteins, face limitations in statistical power when analysed within single cohorts. However, the establishment of initiatives for data (and sample) sharing, such as the ROS and MAP cohorts, the Alzheimer's Disease Neuroimaging Initiative (ADNI) cohort, as well as large-scale omics consortia like the Global Neuro Proteomics Consortia (GNPC) or the Community using Olink for Research on AD and other neurological diseases (CORAL), holds a big potential for the design of innovative multi-omics data integrative studies.

In a recent study, Avelar-Pereira et al. employed an innovative approach to integrate data from the ADNI cohort. The authors utilized a multi-dimensional network framework, which combined data from various modalities into a unified model. Unlike traditional methods, this framework excels in managing interactions across different data types, revealing connections that might have been overlooked otherwise. Their approach showed promise in capturing the diverse nature of AD across the spectrum and identifying individuals who either have or will develop AD [110, 111].

Furthermore, technological advancements in methodologies, such as single-cell RNA sequencing and epigenomics, have revolutionized the ability to acquire RNA transcripts and chromatin statuses at an unprecedented resolution—down to the level of individual cells or nuclei. These breakthroughs have opened up new avenues for exploring the intricate molecular changes underlying AD through the lens of single-cell omics.

In essence, the convergence of multi-omics integration, collaborative data sharing initiatives, and the adoption of state-of-the-art methodologies signify a transformative phase in AD research. This concerted effort holds immense promise in facilitating the development of refined diagnostic modalities, precision-targeted therapeutic interventions, and, ultimately, enhanced clinical outcomes for individuals at risk of AD.

4 Linking Genetics and Endophenotypes of AD

Over the last decade, GWAS targeting AD has identified an ever-growing number of genetic loci associated with AD (see Sect. 2.3). This discovery surge correlates strongly with increased sample sizes and expansion towards underrepresented ethnicity groups, highlighting a significant hurdle, yet an exciting opportunity, in the field to advance our understanding of the disease. Yet, challenges persist in assessing the reliability of the genetic associations, particularly beyond the *APOE* genes, unravelling the "missing heritability" [110] by harnessing the potential of WGS data and validating the genetic signals with biological interpretation and causal links. These include the need for larger sample sizes, inclusion of cross-ancestry genetic data, and more refined disease definitions—the definition of "AD-by-proxy" [8] may be a noisy approximation for the trade-off of sample sizes, considering the comorbidities and symptom heterogeneity of AD. Changes in brain structure follow the sequence of temporal changes that occur before cognitive impairment. This wide and diverse range of AD-related markers suggest that AD might be understood as a complex multifactorial and heterogeneous disease [111]. Previous studies [112, 113] have found different clusters of cognitively unimpaired (CU) individuals, some of whom behave like MCI and AD dementia patients, based on their CSF and magnetic resonance imaging (MRI) markers. These findings highlight the need to move beyond the clinical outcome to explore what are the biological mechanisms linked to AD-pathological hallmarks, even in CU individuals, who may be at early stages of the disease *continuum* and could be targeted for preventive strategies. Therefore, working with AD-related endophenotypes (i.e., brain imaging derived phenotypes or AD core biomarkers) improves the understanding of the biological mechanisms leading to the disease.

4.1 Genetic Studies of Brain Imaging Derived Phenotypes

Brain imaging genetics is a burgeoning multidisciplinary field that combines techniques from genetics and neuroimaging to study how genetic variations influence the structure and function of the brain [114–116]. It aims to identify the genetic factors that underlie individual differences in brain structure, function, and connectivity. These neuroimaging imaging-derived phenotypes (IDPs) have become increasingly important in understanding the genetic basis of various neurological and psychiatric disorders, such as late-life depression, autism spectrum disorder, and AD, as well as healthy brain aging [117–120]. Prior studies show a high degree of heritability via both twin-based and SNP-based approaches, suggesting that much of their variability is genetically determined. These IDPs are closely tied to gene action, potentially allowing genetic variants to exert more direct effects on brain measures than on broader conditions such as AD itself [121]. Moreover, these neuroimaging IDPs are genetically correlated with brain diseases [122], suggesting that genetic influences on the brain can shed light on those affecting AD, possibly requiring smaller sample sizes for significant discovery. The importance of this approach is underscored by the fact that brain alterations linked to AD often precede the onset of clinical symptoms by several years [123]. This early manifestation provides a critical window for intervention, making brain imaging genetics a valuable field for early detection and possibly preventing the progression of AD.

In brain imaging genetics, various analytical challenges and advancements are being addressed. Among these, one of the significant challenges in brain imaging genetics is the high dimensionality and complexity of the data, referred to as the "course of dimensionality," necessitating sophisticated modelling strategies. Conventional candidate gene approaches focus on the association between candidate genetic variants and specific neuroimaging features, typically concentrating on known factors such as *APOE* and targeting brain features implicated in AD, such as the hippocampus [124, 125]. These methods rely on a priori factors, hindering the generalizability and breadth of AD knowledge.

With the increased sample sizes, researchers conducted GWAS of specific or brain-wide brain IDPs. Recent studies in this area have uncovered new genetic variants influencing brain structures, such as the putamen and caudate nucleus, alongside finding stronger evidence for loci previously associated with the hippocampal and intracranial volumes (ICV) [126, 127]. This work identifies genetic loci related to ICV, some of which were also related to childhood and adult cognitive functions. In this case, despite the brain reserve hypothesis suggesting a premorbid brain size can modify resilience to age-related brain pathology, there was no genetic correlation between ICV and AD. Further, imaging-genetics collaborations also identified specific genetic effects on focal structures rather than global effects across the brain [126]. While the individual SNPs related to subcortical brain volumes were not related to neuropathological traits, the studies noted that cortical messenger RNA expression of many of the genes is associated with neuropathological alterations typically observed in AD such as β-amyloid load, the presence of neuritic plaques, and tau or neurofibrillary tangle density.

The investigation of imaging phenotypes in relation to candidate genotypes related to AD has predominantly focused on *APOE* and/or AD PRS. Results have suggested that carriers of the *APOE* ε4 allele exhibit lower medial temporal lobe volumes, and specifically, that carriers have accelerated hippocampal atrophy [128–130]. This effect is not restricted to those with a clinical diagnosis of AD, and is apparent in non-demented, cognitively normal *APOE* ε4 carriers at midlife or older. To echo this, including more genetic markers within an AD-PRS within cognitively healthy individuals suggests a high AD-PRS results in a thinner cortex and smaller brain structures with the effect increasing with age [131–133]. These relationships were additionally seen to have further significant *APOE* genotype-specific interactions between age and genetic risk.

While most of the research is undertaken within structural MRI, targeted analysis has been increasingly explored within positron emission tomography (PET) imaging, followed by diffusion and functional imaging. Results within PET have reported significant differences in *APOE* ε4 carriers in relation to amyloid-β pathology and cerebral glucose metabolism over the lifespan [129, 134]. Diffusion-weighted imaging has further revealed that white matter tracts, such as the fornix and cingulum, have reduced integrity in those with high AD-PRS and the *APOE* ε4 allele [132, 135, 136]. Lastly, functional MRI (fMRI) has described altered memory-related functional activity in AD-specific regions such as the hippocampus within *APOE* ε4 carriers [137, 138]. Resting state fMRI functional connectivity has also been implicated to be disrupted in *APOE* ε4 carriers in similar areas to those with preclinical AD [139]. Other imaging modalities, such as magnetic resonance spectroscopy, within this field are still in their infancy with conflicting results in regards to the relationships of *APOE* and metabolite concentrations in the AD continuum [140, 141].

More recently, the community has bridged complex multivariate analysis of neuroimaging and genetic data. This modelling segment uses sophisticated methodologies, such as regularized and sparse techniques, machine learning, and deep learning, to address the complexities of the data. This approach aims to generate new intermediate phenotypes that are believed to offer more accurate representations of the disease. For example, Patel and colleagues [142] employed unsupervised deep networks to extract data-driven endophenotypes of brain structure. They conducted large-scale brain IDP GWAS, showcasing how unsupervised deep learning can yield robust, unbiased, heritable, and interpretable brain imaging phenotypes. Integrating machine learning/artificial intelligence (ML/AI)-derived IDPs into modern GWAS has powered new discoveries in AD research. In recent studies, Wen and colleagues [143] derived 12 biological age gaps across nine human organ systems and unveiled the underlying genetic architecture responsible for shaping the heterogeneity observed within these organ systems. Using GWAS and post-GWAS analytic approaches, they dissect the genetic associations to causality in the context of human aging and disease, revealing robust causal links between AD and multiple organ systems. Inspired by the close associations between ML/AI-derived IDPs and genetics, Yang et al. [144] introduced a multi-view weakly supervised clustering model named Gene-SGAN. This approach jointly models imaging and genetic data to simultaneously study disease heterogeneity and enhances the discovery capabilities in modern GWAS of AD. Another aspect to consider in brain imaging genetics related to AD is the validation of the genetic signals identified in large-scale GWAS, particularly given the ongoing increase in sample sizes for brain IDP GWAS. The development of state-of-the-art computational genomics and statistical methodologies has played a significant role in supporting these recent discoveries and the biological validation of modern AD GWAS [145]. For example, LD Score Regression (LDSC) has been widely used in brain imaging genetics to elucidate the genetic overlaps between brain IDPs and other clinical traits, such as cognition and neurodegenerative diseases. The calculation of SNP-based heritability is one of the key genetic parameters in understanding the genetic architecture of AD and the brain IDPs, determined through individual-level genetic data and summary statistics from GWAS [146]. Likewise, deep learning has emerged as a pioneering force in modern computational genomics analysis, which has been the preferred method for numerous genomics modelling endeavours, such as forecasting the effects of genetic variation on gene regulatory processes like DNA accessibility and splicing [147]. With the recent surge in progress in brain imaging and genetics, the AD community is seizing exceptional opportunities to investigate AD by integrating large-scale brain imaging and genetic datasets. This integration holds great promise for unravelling the complex genetic underpinnings of AD and developing more effective prevention and treatment strategies.

4.2 *Genetic Studies of AD-Related Biomarkers*

The complex and multifactorial nature of AD, along with the lack of consensus in the field regarding the continuum from cognitive health to AD, sets the stage for categorizing AD biomarkers at the individual level, independent of clinical diagnosis. The "A/T/N" system, as described by Jack et al. and Olsson et al. [148, 149], classifies core AD biomarkers into three categories based on the pathophysiology they measure: "A" for β-amyloid, "T" for tau, and "N" for neurodegeneration or neuronal injury. These biomarkers are derived from various sources including structural MRI, PET, CSF, and plasma. Within the AD continuum, primary outcomes include changes in protein levels, such as CSF Aβ42, phosphorylated tau (pTau), total tau (tTau), and NfL, as well as plasma tTau. These biomarkers have been identified as effective in differentiating cognitively unimpaired (CU) individuals from AD patients [150].

Analysing the biological mechanisms related to these biomarkers can provide crucial information for targeting new proteins to prevent AD pathophysiology in CU individuals who may be at the preclinical stages of the disease. GWAS have identified specific genetic variants independently associated with CSF Aβ42 and pTau levels. For instance, Jansen et al. [151] found associations between pTau- and Aβ42-specific variants and disease progression, AD risk, and age of onset. This GWAS [151] explored genetic factors across the full spectrum of the disease, identifying significant variants in the *CR1* gene for Aβ42 and in the *BIN1*, *GMNC*, and *C16orf95* genes for pTau, suggesting distinct genetic architectures for these AD hallmarks. Moreover, these studies highlight that the strongest associations for both biomarkers occur in the *APOE* locus.

Further exploration of genome-wide significant variants from clinical-based AD GWAS with CSF AD biomarker levels has led to gene-set enrichment analysis, revealing clusters of genes overrepresented in key biological pathways: amyloid-based (e.g. *BIN1*, *PICALM*, *ABCA7*, *CLU*), immune-based (e.g. *GRN*, *TMEM107B*, *SNX1*, *MAPT*), and migration-based (e.g. *TREM2*, *GRN*) clusters. These findings underscore diverse pathways contributing to AD aetiology, reflecting the heterogeneity among AD individuals. Recent studies have also shown that pathway polygenic risk scores, estimating genetic predisposition through specific biological mechanisms (see Sect. 2.5), are differentially associated with CSF AD biomarkers and cognition [59, 152]. For example, the PRS for amyloid-beta protein precursor and the PRS for cholesterol predict changes in CSF Aβ42 levels earlier than the overall PRS for AD.

However, challenges remain. GWAS typically assesses associations in cross-sectional studies, which may not capture the dynamic nature of biomarker levels that vary over time. Longitudinal studies could provide a more accurate picture of genetic factors linked to the dynamics of AD pathophysiology. Furthermore, common GWAS often overlook biomarker interactions by focusing on univariate outcomes. Recent studies advocate for considering protein-level ratios or linear

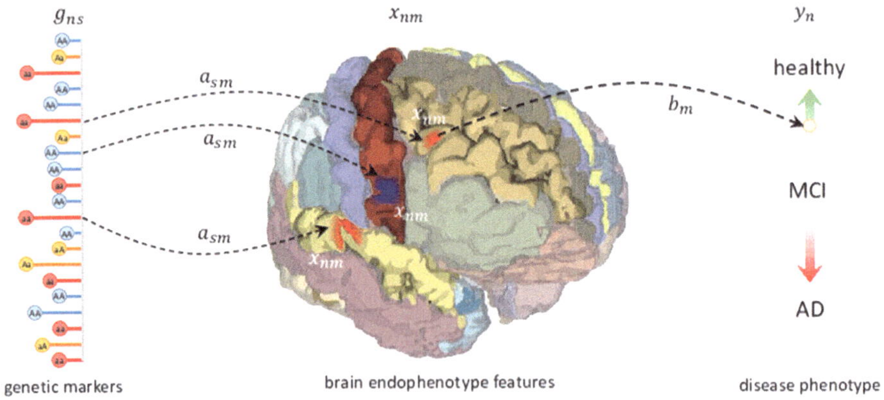

Fig. 2 General Model for the relationship between genetics, imaging, and cognition. In this model, y_n represents the disease phenotype (0 or 1) for participant n in the study ($1 \leq n \leq N$). x_n and g_n are vectors of *M* imaging biomarkers (features) and *S* genetic markers (SNPs) for participant *n*, respectively. The overall process is captured by two coupled regression models: a logistic regression predicts class label y_n from image features x_n; a ridge regression associates genetic variants g_n with image features x_n. AD refers to AD dementia. (Used with permission from Batmanghelich et al. 2013 [155]; Copyright © Springer-Verlag Berlin Heidelberg 2013; permission obtained through the Copyright Clearance Center, Inc)

combinations of multiple AD CSF biomarker levels as outcomes in GWAS to address these interactions [153, 154]. Additionally, traditional GWAS do not usually account for sex-specific effects, even though AD prevalence is higher in women. Recent findings suggest that understanding these sex-specific mechanisms could lead to more effective preventive strategies for both men and women.

Figure 2 illustrates a general model for the relationship between genetics, imaging, and clinical measures.

5 Future Directions

As research in genetics progresses, several future directions are likely to shape our understanding and management of AD. These include delving into gene-environment and gene–gene interactions, broadening the scope of studies to encompass sex-specific analyses and greater genetic diversity, and improving access to and sharing of data, as well as addressing ethical concerns.

5.1 Genetic and Environmental Interactions

Although AD heritability, defined as the proportion of phenotypic variance explained by genotypic variation, is estimated to be between 60% and 80% in twin studies [156], the variability explained by common risk variants targeted in GWAS is often much lower [157]. The gap between these two estimates is known as "missing heritability" and represents one of the major unresolved challenges, even as the field continues to progress into the post-GWAS era.

Many factors are thought to contribute to this knowledge gap, such as the heterogeneity in AD clinical manifestations and the genetic factors contributing to it, the previously ignored contribution of rare variants, or the insufficient efforts put into creating more genetically and socioculturally diverse GWAS cohorts. Additionally, some of the reasons contributing to this gap are based on methodological limitations that make it difficult to eliminate the assumption that all individual genetic variants in our genome exert their risk on disease in a univariate, dose-dependent, and linear manner, following the classical additive model [28, 158].

From a biological perspective, it is unlikely to assume that genetic variants do not interact with each other in intermediate downstream biological pathways. Indeed, the epistasis phenomenon—when genetic variants interact with each other to contribute to phenotypic variation—is thought to be one of the major contributors to missing heritability in AD. Nonetheless, most gene–gene (G × G) interaction studies exploring epistatic mechanisms in AD have focused on previously identified genetic risk loci and/or pairs of genes with a prior hypothetical interaction [159, 160], attempting to decrease the intensive computational burden required to detect epistatic interactions on a genome-wide scale. Still, advanced statistical methodologies are slowly contributing to uncover novel gene–gene interactions now known to contribute to the hidden genetic risk of AD [161].

To further complicate matters, genes are known to interact with environmental exposures in modulating an individual's risk of developing a wide range of diseases [162], including AD [163]. Some environmental factors that increase the risk of AD when interacting with genetic risk factors include well-described modifiable risk factors for dementia development [164], such as air pollution [165], adherence to a healthy lifestyle [166], or cardiometabolic disease [167].

Until recently, gene–environment (G × E) interaction studies have not been a priority in the field, partially due to the complexity of characterizing those interactions. Not all GxE interactions act synergistically, as those described earlier; some necessitate both genotype and environmental risk to co-occur for developing the disease, while others suggest that one factor—genetic or environmental—exacerbates the other, leading to an increased risk of the disease [168, 169]. This complexity, coupled with the heterogeneity in AD genetic determinants, makes these effects challenging to disentangle and translate into population-based and clinical settings. In addition, a need to better estimate the contribution of other worldwide and region-specific relevant environmental risks for AD [170] and their interaction with diverse genetic ancestries will be necessary to improve our understanding of the disease in diverse scenarios, uncover novel drug targets, and enhance the implementation of prevention strategies worldwide [171].

5.2 Sex-Specific Genetic Insights

Research into sex-specific genetic differences in AD highlights the critical role of biological sex in disease susceptibility and progression. Studies like those conducted by Deming et al. [172] and Wang et al. [173] have started identifying sex-specific genetic associations, such as the roles of *SERPINB1* in amyloidosis and *OSTN* and *CLDN16* in tau pathology for females, or the differential role of specific genetic variants associated with sex differences in regional tau aggregation measures by PET. However, these studies face challenges such as the need for large and diverse sample sizes and the complexity of genetic interactions with varying environmental and lifestyle factors across sexes.

Future research must continue to prioritize sex as a key variable, employing advanced methodologies such as single-cell RNA sequencing, metabolomics, and multi-omics analyses [174]. This perspective will deepen our understanding of AD pathophysiology, revealing how sex hormones and chromosomes interact with disease mechanisms like inflammation and metabolism. Emphasizing sex-specific research will lead to more personalized and effective treatments, addressing unique disease progression characteristics between men and women.

5.3 Enhancing Genetic Diversity in AD

Historically, genetic studies have predominantly included participants of European descent [175], which limits the generalizability of findings across different ethnicities. Expanding the diversity in genetic studies for AD is not only about improving scientific precision, but it also tackles major inequalities in healthcare and research. The dominance of European descent in genetic databases has resulted in a biased perception of AD, which might lead to interventions that are less effective or not suitable for non-European populations. To address this issue, a comprehensive approach is necessary.

While the genetic databases are expanding in their diversity and more global sites are beginning to undertake genetic research into AD, there is still a long way to go. One of the discoveries from this expansion reports the main genetic marker of AD, *APOE* ε4 allele, has widely differing prevalences globally [176, 177]. This differing global prevalence is also apparent within AD itself [178]. Research in this area has begun to reveal that this relationship between AD and *APOE* is additionally dependent on genetic ancestry. However, most of the genetic research into diverse ancestral groups is still limited in its regional applicability with the majority of research being undertaken within western countries, such as the UK or the USA. This restricts the ability to understand any gene–environmental interactions with respect to global variability of AD prevalence, as mentioned above. This lack of global representation is due to many barriers including elements pertaining to the social, cultural, and infrastructure aspects of genetic research. Both reducing the costs of genetic processing and improving education surrounding the topic would have a large impact on the field.

A group often further left out of genetic research is those with mixed ancestry, a growing group in today's societies. With arbitrary cut-offs for global ancestry classification, those who do not reach this cut-off for one ancestry often are omitted from research. Recent developments in genetic methodology have allowed better inclusion of these individuals and further revealed the importance of the ancestry of specific genes under investigation (ancestral origin of the distinct chromosomal segment surrounding the gene), beyond the global ancestry of the whole genome [176, 179–181]. This is possibly due to aggregated effects of both common and rare genetic variations and/or due to differences in *APOE* expression [182]. Studies have reported that ancestry of ε4 alleles confers differences in risk for AD and in AD pathology, such as the burden of neuritic plaques [180, 181]. Despite these crucial insights, the examination of how this ancestry changes the effects of *APOE* on a population level remains an underexplored area of research. This is an important development within the current focus on disease-modifying treatments, many of which rely on group classification through genetic markers of which the research foundations lie in predominantly European populations.

5.3.1 Broader Recruitment Strategies and Educational Initiatives

To enhance genetic diversity, researchers must implement broader and more inclusive recruitment strategies. This involves partnerships with communities historically underrepresented in genetic research, including those of African, Hispanic/Latino, Asian, and Indigenous descent. Engaging these communities through local leaders and healthcare providers can help build trust and awareness about the importance of genetic research in these populations. Moreover, educational initiatives tailored to diverse communities can help overcome mistrust and misinformation about genetic research. Clear communication about the goals of AD research, the importance of diversity in genetic studies, and the safeguards in place to protect participants' privacy and rights is crucial. These efforts should be linguistically and culturally appropriate to resonate with different communities.

5.3.2 Infrastructure and Support

Improving the infrastructure to support genetic research in diverse settings is a crucial endeavour that involves several key steps. Establishing research centres in areas with large populations of underrepresented groups is essential. These centres would not only bring research closer to these communities but also allow for studies that are specifically tailored to their unique genetic backgrounds and health needs. Providing local researchers with the necessary resources and training is equally important. This includes access to modern laboratory equipment, digital databases, and knowledge about advanced genetic analysis techniques, which would help in building local expertise and reduce reliance on external laboratories. It is also vital that funding agencies and institutions prioritize grants and projects that focus on genetic diversity. This would encourage more research in this field and help ensure that the findings are relevant to a broader range of populations. Furthermore, there is a need to develop more affordable methods for genetic processing. This could involve partnerships with tech companies and academic institutions aimed at creating cost-effective genetic analysis technologies that can be widely implemented, particularly in regions like South America where samples are currently sent out of the continent for processing. By focusing on these areas, genetic studies can become more inclusive and play a significant role in addressing health disparities across different populations.

5.3.3 Collaborative International Studies

Participating in international studies can also significantly enhance genetic datasets. By incorporating populations from various geographical regions and backgrounds, researchers can acquire a more detailed understanding of the genetic structure of AD. International collaborations are especially valuable as they bring together diverse genetic ancestries and environmental contexts, offering insights into how these elements combine to impact AD risk and progression. Furthermore, these collaborations can standardize methodologies and increase the robustness of genetic findings by comparing results across different populations. This approach not only broadens the scope of research but also improves the reliability and applicability of the conclusions drawn, paving the way for globally relevant therapeutic strategies and interventions.

Moreover, enhancing diversity in AD research is not just about including diverse data from different populations; it is equally crucial to ensure diversity among the researchers themselves. Researchers and clinicians from various cultural and ethnic backgrounds can bring unique insights into the genetic and environmental factors that might influence AD differently across populations. This diversity in the research team can help formulate more culturally sensitive research questions and methodologies, which are essential for addressing specific needs and nuances of different groups and can help mitigate biases that might arise from a more homogeneous group making all the decisions. By incorporating researchers with different life experiences and viewpoints, the research becomes more inclusive, enhancing its relevance and potential impact on public health.

5.4 Advances in Data Access and Sharing

The proliferation of rich data across multiple modalities (e.g. clinical, digital, imaging, -omics, and more), alongside rapid advances in analytics capabilities, creates unprecedented opportunities for research breakthroughs in our understanding of the causes, prevention, and treatment of neurodegenerative diseases. The scope and scale of research questions that can now be addressed have never been larger. Practically, however, the fragmented nature of both the data and the tools to analyse it still presents barriers. The trustworthy sharing and analysis of data is essential to promote the full power of advanced analytics in delivering benefits to patients.

Patient and public involvement (PPIE) in research is vital for ensuring that the research improves patient care and fosters the trust of patients and the public. In best practice research, patients and the public should be partners in design and decision making; the needs of patients and the public should be understood and embedded in the work; and people should have trust and confidence in the use of health data for research and innovation. There is consensus that involving patients and members of the public in research results in an increase in the quality and relevance of studies through informed and inclusive study design [183, 184].

While protecting patient data and privacy is central to health data research, patients and the public may also have an interest in contributing to research advances that improve health through data sharing. The goal is to balance the achievement of the greatest possible scientific impact and societal benefit while still protecting individuals' privacy. Several promising frameworks are helping to advance this conversation. The Five Safes framework [185] (safe people, safe projects, safe settings, safe data, safe outputs), first developed by the UK's Office for National Statistics, provides a template for designing efficient data access that is also safe. Wider lessons on the ethical use of health data to address global health challenges can also be drawn from leading initiatives such as the Global Alliance for Genomics and Health [186], the International Severe Acute Respiratory and Emerging Infection Consortium (ISARIC; [187]), and the International COVID-19 Data Alliance (ICODA; [188]).

Impactful research using health data also involves advanced adherence to FAIR (findable, accessible, interoperable, reusable) [189] principles, which provide guidance for data management to ensure that data are shared in a way that are reusable by both humans and machines. Through the adoption of common, open, and consistent standards, health data can be made "FAIR"—enabling it to be analysed using new and powerful computational approaches. As the scale and breadth of data collected increase, the use of a common data model (e.g. Observational Medical Outcomes Partnership (OMOP)) brings consistency to health data. Adoption of data standards outlined by a common data model means that researchers do not need to repeat time-consuming data mapping to perform their analyses. Harmonization to a common data model can also enable federated analysis approaches—which can simplify information governance challenges, enhance data security, and support international collaboration.

Finally, there are several examples of technological solutions that operate within existing governance frameworks to expand data sharing and facilitate collaborative analyses. For example, trusted research environments (TREs) such as the Alzheimer's Disease Data Initiative's (ADDI) AD Workbench and Dementia's Platform UK's (DPUK) Data Portal are becoming increasingly common. TREs provide approved researchers with a secure environment where they can access datasets, tools for analysis, and collaborate on analyses with permissioned colleagues who may reside in different geographical locations.

The objective is to promote research that protects people's privacy while providing solutions that allow researchers to do work that can provide benefit to all. Achieving this requires significant and sustainable investment in the infrastructure and capacity building required to collect, link, access, and ultimately (re)use data—alongside data that are inclusive and representative of all the people who stand to benefit.

Acknowledgements Natalia Vilor-Tejedor is supported by the Spanish Ministry of Science and Innovation - State Research Agency grant RYC2022-038136-I, cofunded by the European Union FSE+, and grant PID2022-143106OA-I00 cofunded by the European Union FEDER. Additionally, Natalia Vilor-Tejedor is supported by the William H. Gates Sr. Fellowship (cohort I) from the Alzheimer's Disease Data Initiative, grant 23S06083-001 funded by the Ajuntament de Barcelona and "la Caixa" Foundation and InterAct-beurs WE.08-2024-07 funded by the AlzheimerNederland. Federica Anastasi receives funding from the JDC2022-049347-I grant, funded by the MCIU/AEI/10.13039/501100011033 and the European Union NextGenerationEU/PRTR. Tavia E. Evans is supported by the William H. Gates Sr. Fellowship (cohort II) from the Alzheimer's Disease Data Initiative. This work was supported by Health Data Research UK, which is funded by the Medical Research Council (UKRI), the National Institute for Health Research, the British Heart Foundation, Cancer Research UK, the Economic and Social Research Council (UKRI), the Engineering and Physical Sciences Research Council (UKRI), Health and Care Research Wales, Chief Scientist Office of the Scottish Government Health and Social Care Directorates, and Health and Social Care Research and Development Division (Public Health Agency, Northern Ireland).

References

1. DeTure MA, Dickson DW. The neuropathological diagnosis of Alzheimer's disease. Mol Neurodegener. 2019;14:32.
2. Pathan N, Kharod MK, Nawab S, Di Scipio M, Paré G, Chong M. Genetic determinants of vascular dementia. Can J Cardiol. 2024; https://doi.org/10.1016/j.cjca.2024.03.025.
3. Bertram L, Tanzi RE. Genomic mechanisms in Alzheimer's disease. Brain Pathol. 2020;30:966–77.
4. Lambert J-C, Ramirez A, Grenier-Boley B, Bellenguez C. Step by step: towards a better understanding of the genetic architecture of Alzheimer's disease. Mol Psychiatry. 2023;28:2716–27.
5. Lanoiselée H-M, Nicolas G, Wallon D, et al. APP, PSEN1, and PSEN2 mutations in early-onset Alzheimer disease: a genetic screening study of familial and sporadic cases. PLoS Med. 2017;14:e1002270.
6. Jansen IE, Savage JE, Watanabe K, et al. Genome-wide meta-analysis identifies new loci and functional pathways influencing Alzheimer's disease risk. Nat Genet. 2019;51:404–13.
7. Wightman DP, Jansen IE, Savage JE, et al. A genome-wide association study with 1,126,563 individuals identifies new risk loci for Alzheimer's disease. Nat Genet. 2021;53:1276–82.
8. Bellenguez C, Küçükali F, Jansen IE, et al. New insights into the genetic etiology of Alzheimer's disease and related dementias. Nat Genet. 2022;54:412–36.
9. Kunkle BW, Grenier-Boley B, Sims R, et al. Genetic meta-analysis of diagnosed Alzheimer's disease identifies new risk loci and implicates Aβ, tau, immunity and lipid processing. Nat Genet. 2019;51:414–30.
10. Liu C-C, Kanekiyo T, Xu H, Bu G. Apolipoprotein E and Alzheimer disease: risk, mechanisms and therapy. Nat Rev Neurol. 2013;9:106–18.
11. Fahed AC, Philippakis AA, Khera AV. The potential of polygenic scores to improve cost and efficiency of clinical trials. Nat Commun. 2022;13:2922.
12. Natarajan P, Young R, Stitziel NO, et al. Polygenic risk score identifies subgroup with higher burden of atherosclerosis and greater relative benefit from statin therapy in the primary prevention setting. Circulation. 2017;135:2091–101.
13. Arneson D, Zhang Y, Yang X, Narayanan M. Shared mechanisms among neurodegenerative diseases: from genetic factors to gene networks. J Genet. 2018;97:795–806.
14. Ibanez L, Farias FHG, Dube U, Mihindukulasuriya KA, Harari O. Polygenic risk scores in neurodegenerative diseases: a review. Curr Genet Med Rep. 2019;7:22–9.
15. Leonenko G, Baker E, Stevenson-Hoare J, Sierksma A, Fiers M, Williams J, de Strooper B, Escott-Price V. Identifying individuals with high risk of Alzheimer's disease using polygenic risk scores. Nat Commun. 2021;12:4506.
16. Harrison JR, Foley SF, Baker E, Bracher-Smith M, Holmans P, Stergiakouli E, Linden DEJ, Caseras X, Jones DK, Escott-Price V. Pathway-specific polygenic scores for Alzheimer's disease are associated with changes in brain structure in younger and older adults. Brain Commun. 2023;5:fcad229.
17. Vacher M, Doré V, Porter T, et al. Assessment of a polygenic hazard score for the onset of pre-clinical Alzheimer's disease. BMC Genomics. 2022;23:401.
18. Dai M-H, Zheng H, Zeng L-D, Zhang Y. The genes associated with early-onset Alzheimer's disease. Oncotarget. 2018;9:15132–43.
19. Mol MO, van der Lee SJ, Hulsman M, et al. Mapping the genetic landscape of early-onset Alzheimer's disease in a cohort of 36 families. Alzheimers Res Ther. 2022;14:77.
20. Ray NR, Ayodele T, Jean-Francois M, et al. The early-onset Alzheimer's disease whole-genome sequencing project: study design and methodology. Alzheimers Dement. 2023;19:4187–95.
21. Raulin A-C, Doss SV, Trottier ZA, Ikezu TC, Bu G, Liu C-C. ApoE in Alzheimer's disease: pathophysiology and therapeutic strategies. Mol Neurodegener. 2022;17:72.

22. Troutwine BR, Hamid L, Lysaker CR, Strope TA, Wilkins HM. Apolipoprotein E and Alzheimer's disease. Acta Pharm Sin B. 2022;12:496–510.
23. Vance JE, Hayashi H. Formation and function of apolipoprotein E-containing lipoproteins in the nervous system. Biochim Biophys Acta. 2010;1801:806–18.
24. Kim J, Basak JM, Holtzman DM. The role of apolipoprotein E in Alzheimer's disease. Neuron. 2009;63:287–303.
25. Ye S, Roccati E, Wang W, et al. Leading determinants of incident dementia among individuals with and without the apolipoprotein E ε4 genotype: a retrospective cohort study. BMC Neurol. 2024;24:71.
26. Uffelmann E, Huang QQ, Munung NS, de Vries J, Okada Y, Martin AR, Martin HC, Lappalainen T, Posthuma D. Genome-wide association studies. Nat Rev Methods Primers. 2021; https://doi.org/10.1038/s43586-021-00056-9.
27. Visscher PM, Wray NR, Zhang Q, Sklar P, McCarthy MI, Brown MA, Yang J. 10 years of GWAS discovery: biology, function, and translation. Am J Hum Genet. 2017;101:5–22.
28. Andrews SJ, Renton AE, Fulton-Howard B, Podlesny-Drabiniok A, Marcora E, Goate AM. The complex genetic architecture of Alzheimer's disease: novel insights and future directions. EBioMedicine. 2023;90:104511.
29. Lambert JC, Ibrahim-Verbaas CA, Harold D, et al. Meta-analysis of 74,046 individuals identifies 11 new susceptibility loci for Alzheimer's disease. Nat Genet. 2013; https://doi.org/10.1038/ng.2802.
30. Marioni RE, Harris SE, Zhang Q, et al. GWAS on family history of Alzheimer's disease. Transl Psychiatry. 2018;8:99.
31. Belloy ME, Guen YL, Stewart I, et al. The role of X chromosome in Alzheimer's disease genetics. medRxiv. 2024; https://doi.org/10.1101/2024.04.22.24306094.
32. Guerreiro RJ, Gustafson DR, Hardy J. The genetic architecture of Alzheimer's disease: beyond APP, PSENs and APOE. Neurobiol Aging. 2012;33:437–56.
33. Le Ber I, De Septenville A, Guerreiro R, et al. Homozygous TREM2 mutation in a family with atypical frontotemporal dementia. Neurobiol Aging. 2014;35:2419.e23–5.
34. Guerreiro RJ, Lohmann E, Brás JM, et al. Using exome sequencing to reveal mutations in TREM2 presenting as a frontotemporal dementia-like syndrome without bone involvement. JAMA Neurol. 2013;70:78–84.
35. Guerreiro RJ, Wojtas A, Bras J, et al. TREM2 variants in Alzheimer's disease. N Engl J Med. 2013;368:117–27.
36. Khani M, Gibbons E, Bras J, Guerreiro R. Challenge accepted: uncovering the role of rare genetic variants in Alzheimer's disease. Mol Neurodegener. 2022;17:3.
37. Negro SS, Millet EJ, Madur D, Bauland C, Combes V, Welcker C, Tardieu F, Charcosset A, Nicolas SD. Genotyping-by-sequencing and SNP-arrays are complementary for detecting quantitative trait loci by tagging different haplotypes in association studies. BMC Plant Biol. 2019;19:1–22.
38. Sundarrajan S, Venkatesan A, Kumar SU, Gopikrishnan M, Tayubi IA, Aditya M, Siddaiah GB, George Priya Doss C, Zayed H. Exome sequence analysis of rare frequency variants in Late-Onset Alzheimer Disease. Metab Brain Dis. 2023;38:2025–36.
39. Shigemizu D, Asanomi Y, Akiyama S, Mitsumori R, Niida S, Ozaki K. Whole-genome sequencing reveals novel ethnicity-specific rare variants associated with Alzheimer's disease. Mol Psychiatry. 2022;27:2554–62.
40. Cano-Gamez E, Trynka G. From GWAS to function: using functional genomics to identify the mechanisms underlying complex diseases. Front Genet. 2020;11:424.
41. Broekema RV, Bakker OB, Jonkers IH. A practical view of fine-mapping and gene prioritization in the post-genome-wide association era. Open Biol. 2020;10:190221.
42. Ferkingstad E, Sulem P, Atlason BA, et al. Large-scale integration of the plasma proteome with genetics and disease. Nat Genet. 2021;53:1712–21.
43. Madrid L, Labrador SC, González-Pérez A, Sáez ME, The Alzheimer's Disease Neuroimaging Initiative Adni. Integrated genomic, transcriptomic and proteomic analysis for identifying

markers of Alzheimer's disease. Diagnostics (Basel). 2021; https://doi.org/10.3390/diagnostics11122303.
44. Darst BF, Lu Q, Johnson SC, Engelman CD. Integrated analysis of genomics, longitudinal metabolomics, and Alzheimer's risk factors among 1,111 cohort participants. Genet Epidemiol. 2019;43:657–74.
45. Aerqin Q, Wang Z-T, Wu K-M, He X-Y, Dong Q, Yu J-T. Omics-based biomarkers discovery for Alzheimer's disease. Cell Mol Life Sci. 2022;79:585.
46. European Alzheimer's & Dementia Biobank Mendelian Randomization (EADB-MR) Collaboration, Luo J, Thomassen JQ, et al. Genetic associations between modifiable risk factors and Alzheimer disease. JAMA Netw Open. 2023;6:e2313734.
47. Larsson SC, Traylor M, Malik R, Dichgans M, Burgess S, Markus HS, CoSTREAM Consortium, on behalf of the International Genomics of Alzheimer's Project. Modifiable pathways in Alzheimer's disease: Mendelian randomisation analysis. BMJ. 2017;359:j5375.
48. Hardy J, Bogdanovic N, Winblad B, Portelius E, Andreasen N, Cedazo-Minguez A, Zetterberg H. Pathways to Alzheimer's disease. J Intern Med. 2014;275:296–303.
49. Clark K, Leung YY, Lee W-P, Voight B, Wang L-S. Polygenic risk scores in Alzheimer's disease genetics: methodology, applications, inclusion, and diversity. J Alzheimers Dis. 2022;89:1–12.
50. Tan CH, Hyman BT, Tan JJX, et al. Polygenic hazard scores in preclinical Alzheimer disease. Ann Neurol. 2017;82:484–8.
51. Luckett ES, Abakkouy Y, Reinartz M, et al. Association of Alzheimer's disease polygenic risk scores with amyloid accumulation in cognitively intact older adults. Alzheimers Res Ther. 2022;14:138.
52. Kumar A, Janelidze S, Stomrud E, Palmqvist S, Hansson O, Mattsson-Carlgren N, for the Alzheimer's Disease Neuroimaging Initiative. β-Amyloid–dependent and –independent genetic pathways regulating CSF tau biomarkers in Alzheimer disease. Neurology. 2022;99:e476–87.
53. de Rojas I, Moreno-Grau S, Tesi N, et al. Common variants in Alzheimer's disease and risk stratification by polygenic risk scores. Nat Commun. 2021;12:3417.
54. Vilor-Tejedor N, Genius P, Rodríguez-Fernández B, et al. Genetic characterization of the ALFA study: uncovering genetic profiles in the Alzheimer's continuum. Alzheimers Dement. 2024;20:1703–15.
55. Mormino EC, Sperling RA, Holmes AJ, Buckner RL, De Jager PL, Smoller JW, Sabuncu MR, Alzheimer's Disease Neuroimaging Initiative. Polygenic risk of Alzheimer disease is associated with early- and late-life processes. Neurology. 2016;87:481–8.
56. Van Cauwenberghe C, Van Broeckhoven C, Sleegers K. The genetic landscape of Alzheimer disease: clinical implications and perspectives. Genet Med. 2016;18:421–30.
57. Goyal D, Tjandra D, Migrino RQ, Giordani B, Syed Z, Wiens J. Characterizing heterogeneity in the progression of Alzheimer's disease using longitudinal clinical and neuroimaging biomarkers. Alzheimers Dement. 2018;10:629–37.
58. Darst BF, Koscik RL, Racine AM, et al. Pathway-specific polygenic risk scores as predictors of amyloid-β deposition and cognitive function in a sample at increased risk for Alzheimer's disease. J Alzheimers Dis. 2017;55:473–84.
59. Ahmad S, Bannister C, van der Lee SJ, et al. Disentangling the biological pathways involved in early features of Alzheimer's disease in the Rotterdam study. Alzheimers Dement. 2018;14:848–57.
60. Yang H-S, Teng L, Kang D, et al. Cell-type-specific Alzheimer's disease polygenic risk scores are associated with distinct disease processes in Alzheimer's disease. Nat Commun. 2023;14:7659.
61. De Strooper B, Karran E. The cellular phase of Alzheimer's disease. Cell. 2016;164:603–15.
62. Sims JR, Zimmer JA, Evans CD, et al. Donanemab in early symptomatic Alzheimer disease: the TRAILBLAZER-ALZ 2 randomized clinical trial. JAMA. 2023;330:512–27.

63. Cummings J, Apostolova L, Rabinovici GD, et al. Lecanemab: appropriate use recommendations. J Prev Alzheimers Dis. 2023;10:362–77.
64. Ginsberg SD, Hemby SE, Lee VM, Eberwine JH, Trojanowski JQ. Expression profile of transcripts in Alzheimer's disease tangle-bearing CA1 neurons. Ann Neurol. 2000;48:77–87.
65. Blalock EM, Buechel HM, Popovic J, Geddes JW, Landfield PW. Microarray analyses of laser-captured hippocampus reveal distinct gray and white matter signatures associated with incipient Alzheimer's disease. J Chem Neuroanat. 2011;42:118–26.
66. Magistri M, Velmeshev D, Makhmutova M, Faghihi MA. Transcriptomics profiling of Alzheimer's disease reveal neurovascular defects, altered amyloid-β homeostasis, and deregulated expression of long noncoding RNAs. J Alzheimers Dis. 2015;48:647–65.
67. van Rooij JGJ, Meeter LHH, Melhem S, Nijholt DAT, Wong TH, Netherlands Brain Bank, Rozemuller A, Uitterlinden AG, van Meurs JG, van Swieten JC. Hippocampal transcriptome profiling combined with protein-protein interaction analysis elucidates Alzheimer's disease pathways and genes. Neurobiol Aging. 2019;74:225–33.
68. Quan X, Liang H, Chen Y, Qin Q, Wei Y, Liang Z. Related network and differential expression analyses identify nuclear genes and pathways in the hippocampus of Alzheimer disease. Med Sci Monit. 2020;26:e919311.
69. Mufson EJ, Counts SE, Ginsberg SD. Gene expression profiles of cholinergic nucleus basalis neurons in Alzheimer's disease. Neurochem Res. 2002;27:1035–48.
70. Tan MG, Chua W-T, Esiri MM, Smith AD, Vinters HV, Lai MK. Genome wide profiling of altered gene expression in the neocortex of Alzheimer's disease. J Neurosci Res. 2010;88:1157–69.
71. Twine NA, Janitz K, Wilkins MR, Janitz M. Whole transcriptome sequencing reveals gene expression and splicing differences in brain regions affected by Alzheimer's disease. PLoS One. 2011;6:e16266.
72. Mills JD, Nalpathamkalam T, Jacobs HIL, Janitz C, Merico D, Hu P, Janitz M. RNA-Seq analysis of the parietal cortex in Alzheimer's disease reveals alternatively spliced isoforms related to lipid metabolism. Neurosci Lett. 2013;536:90–5.
73. Liu A, Fernandes BS, Citu C, Zhao Z. Unraveling the intercellular communication disruption and key pathways in Alzheimer's disease: an integrative study of single-nucleus transcriptomes and genetic association. Alzheimers Res Ther. 2024;16:1–17.
74. Soreq L, Bird H, Mohamed W, Hardy J. Single-cell RNA sequencing analysis of human Alzheimer's disease brain samples reveals neuronal and glial specific cells differential expression. PLoS One. 2023;18:e0277630.
75. Weber LM, Divecha HR, Tran MN, et al. The gene expression landscape of the human locus coeruleus revealed by single-nucleus and spatially-resolved transcriptomics. elife. 2024; https://doi.org/10.7554/eLife.84628.
76. Adewale Q, Khan AF, Bennett DA, Iturria-Medina Y. Single-nucleus RNA velocity reveals critical synaptic and cell-cycle dysregulations in neuropathologically confirmed Alzheimer's disease. Sci Rep. 2024;14:7269.
77. de Vries LE, Jongejan A, Monteiro Fortes J, Balesar R, Rozemuller AJM, Moerland PD, Huitinga I, Swaab DF, Verhaagen J. Gene-expression profiling of individuals resilient to Alzheimer's disease reveals higher expression of genes related to metallothionein and mitochondrial processes and no changes in the unfolded protein response. Acta Neuropathol Commun. 2024;12:68.
78. Das S, Li Z, Wachter A, et al. Distinct transcriptomic responses to Aβ plaques, neurofibrillary tangles, and APOE in Alzheimer's disease. Alzheimers Dement. 2024;20:74–90.
79. Guennewig B, Lim J, Marshall L, McCorkindale AN, Paasila PJ, Patrick E, Kril JJ, Halliday GM, Cooper AA, Sutherland GT. Defining early changes in Alzheimer's disease from RNA sequencing of brain regions differentially affected by pathology. Sci Rep. 2021;11:4865.
80. Zhang J, Li X, Xiao J, Xiang Y, Ye F. Analysis of gene expression profiles in Alzheimer's disease patients with different lifespan: A bioinformatics study focusing on the disease heterogeneity. Front Aging Neurosci. 2023;15:1072184.

81. Ding Q, Markesbery WR, Chen Q, Li F, Keller JN. Ribosome dysfunction is an early event in Alzheimer's disease. J Neurosci. 2005;25:9171–5.
82. Williams JB, Cao Q, Yan Z. Transcriptomic analysis of human brains with Alzheimer's disease reveals the altered expression of synaptic genes linked to cognitive deficits. Brain Commun. 2021;3:fcab123.
83. Liew C-C, Ma J, Tang H-C, Zheng R, Dempsey AA. The peripheral blood transcriptome dynamically reflects system wide biology: a potential diagnostic tool. J Lab Clin Med. 2006;147:126–32.
84. Sullivan PF, Fan C, Perou CM. Evaluating the comparability of gene expression in blood and brain. Am J Med Genet B Neuropsychiatr Genet. 2006;141B:261–8.
85. Xue W, Li J, Fu K, Teng W. Differential expression of mRNAs in peripheral blood related to Prodrome and progression of Alzheimer's disease. Biomed Res Int. 2020;2020:4505720.
86. Shimohama S. Apoptosis in Alzheimer's disease–an update. Apoptosis. 2000;5:9–16.
87. Bottero V, Potashkin JA. Meta-analysis of gene expression changes in the blood of patients with mild cognitive impairment and Alzheimer's disease dementia. Int J Mol Sci. 2019; https://doi.org/10.3390/ijms20215403.
88. Fehlbaum-Beurdeley P, Jarrige-Le Prado AC, Pallares D, et al. Toward an Alzheimer's disease diagnosis via high-resolution blood gene expression. Alzheimers Dement. 2010;6:25–38.
89. Wang H, Han X, Gao S. Identification of potential biomarkers for pathogenesis of Alzheimer's disease. Hereditas. 2021;158:23.
90. Wang Y, Zhan D, Wang L. Ribosomal proteins are blood biomarkers and associated with CD4+ T cell activation in Alzheimer's disease: a study based on machine learning strategies and scRNA-Seq data validation. Am J Transl Res. 2023;15:2498–514.
91. Cao THM, Le APH, Tran TT, et al. Plasma cell-free RNA profiling of Vietnamese Alzheimer's patients reveals a linkage with chronic inflammation and apoptosis: a pilot study – PubMed. Front Mol Neurosci. 2023;16:1308610.
92. Zhong H, Zhou X, Uhm H, et al. Using blood transcriptome analysis for Alzheimer's disease diagnosis and patient stratification. Alzheimers Dement. 2024;20:2469–84.
93. Mondal T, Noreen Z, Loffredo CA, et al. Transcriptomic analysis of Alzheimer's disease pathways in a Pakistani population. J Alzheimers Dis Rep. 2024;8:479–93.
94. Park YH, Hodges A, Risacher SL, et al. Dysregulated fc gamma receptor-mediated phagocytosis pathway in Alzheimer's disease: network-based gene expression analysis. Neurobiol Aging. 2020;88:24–32.
95. Park YH, Pyun J-M, Hodges A, Jang J-W, Bice PJ, Kim S, Saykin AJ, Nho K. Dysregulated expression levels of APH1B in peripheral blood are associated with brain atrophy and amyloid-β deposition in Alzheimer's disease. Alzheimers Res Ther. 2021;13:1–10.
96. Nho K, Nudelman K, Allen M, et al. Genome-wide transcriptome analysis identifies novel dysregulated genes implicated in Alzheimer's pathology. Alzheimers Dement. 2020;16:1213–23.
97. Luckett ES, Zielonka M, Kordjani A, Schaeverbeke J, Adamczuk K, De Meyer S, Van Laere K, Dupont P, Cleynen I, Vandenberghe R. Longitudinal APOE4- and amyloid-dependent changes in the blood transcriptome in cognitively intact older adults. Alzheimers Res Ther. 2023;15:121.
98. Kelley CM, Maloney B, Beck JS, Ginsberg SD, Liang W, Lahiri DK, Mufson EJ, Counts SE. Micro-RNA profiles of pathology and resilience in posterior cingulate cortex of cognitively intact elders. Brain Commun. 2024;6:fcae082.
99. Kelley CM, Ginsberg SD, Liang WS, Counts SE, Mufson EJ. Posterior cingulate cortex reveals an expression profile of resilience in cognitively intact elders. Brain Commun. 2022;4:fcac162.
100. Song L, Chen J, Lo C-YZ, Guo Q, ZIB Consortium, Feng J, Zhao X-M. Impaired type I interferon signaling activity implicated in the peripheral blood transcriptome of preclinical Alzheimer's disease. EBioMedicine. 2022;82:104175.

101. Peña-Bautista C, Tarazona-Sánchez A, Braza-Boils A, Balaguer A, Ferré-González L, Cañada-Martínez AJ, Baquero M, Cháfer-Pericás C. Plasma microRNAs as potential biomarkers in early Alzheimer disease expression. Sci Rep. 2022;12:15589.
102. Hasin Y, Seldin M, Lusis A. Multi-omics approaches to disease. Genome Biol. 2017; https://doi.org/10.1186/s13059-017-1215-1.
103. Ritchie MD, Holzinger ER, Li R, Pendergrass SA, Kim D. Methods of integrating data to uncover genotype-phenotype interactions. Nat Rev Genet. 2015;16:85–97.
104. Duara R, Barker W. Heterogeneity in Alzheimer's disease diagnosis and progression rates: implications for therapeutic trials. Neurotherapeutics. 2022;19:8–25.
105. Jellinger KA. Recent update on the heterogeneity of the Alzheimer's disease spectrum. J Neural Transm. 2022;129:1–24.
106. Babu M, Snyder M. Multi-omics profiling for health. Mol Cell Proteomics. 2023;22:100561.
107. Rahimzadeh N, Srinivasan SS, Zhang J, Swarup V. Gene networks and systems biology in Alzheimer's disease: insights from multi-omics approaches. Alzheimers Dement. 2024;20:3587–605.
108. Swarup V, Chang TS, Duong DM, Dammer EB, Dai J, Lah JJ, Johnson ECB, Seyfried NT, Levey AI, Geschwind DH. Identification of conserved proteomic networks in neurodegenerative dementia. Cell Rep. 2020;31:107807.
109. Tijms BM, Vromen EM, Mjaavatten O, et al. Cerebrospinal fluid proteomics in patients with Alzheimer's disease reveals five molecular subtypes with distinct genetic risk profiles. Nat Aging. 2024;4:33–47.
110. Young AI. Discovering missing heritability in whole-genome sequencing data. Nat Genet. 2022;54:224–6.
111. Avelar-Pereira B, Belloy ME, O'Hara R, Hosseini SMH, Alzheimer's Disease Neuroimaging Initiative. Decoding the heterogeneity of Alzheimer's disease diagnosis and progression using multilayer networks. Mol Psychiatry. 2023;28:2423–32.
112. Nettiksimmons J, Beckett L, Schwarz C, Carmichael O, Fletcher E, Decarli C. Subgroup of ADNI normal controls characterized by atrophy and cognitive decline associated with vascular damage. Psychol Aging. 2013;28:191–201.
113. Nettiksimmons J, Harvey D, Brewer J, et al. Subtypes based on cerebrospinal fluid and magnetic resonance imaging markers in normal elderly predict cognitive decline. Neurobiol Aging. 2010;31:1419–28.
114. Zhao B, Li T, Yang Y, et al. Common genetic variation influencing human white matter microstructure. Science. 2021; https://doi.org/10.1126/science.abf3736.
115. Wen J, Nasrallah IM, Abdulkadir A, et al. Genomic loci influence patterns of structural covariance in the human brain. Proc Natl Acad Sci USA. 2023;120:e2300842120.
116. Grasby KL, Jahanshad N, Painter JN, et al. The genetic architecture of the human cerebral cortex. Science. 2020; https://doi.org/10.1126/science.aay6690.
117. Skampardoni I, Nasrallah IM, Abdulkadir A, et al. Genetic and clinical correlates of AI-based brain aging patterns in cognitively unimpaired individuals. JAMA Psychiatry. 2024;81:456–67.
118. Leonardsen EH, Vidal-Piñeiro D, Roe JM, et al. Genetic architecture of brain age and its causal relations with brain and mental disorders. Mol Psychiatry. 2023;28:3111–20.
119. Xin Y, Sheng J, Miao M, Wang L, Yang Z, Huang H. A review ofimaging genetics in Alzheimer's disease. J Clin Neurosci. 2022;100:155–63.
120. Kim M, Wu R, Yao X, Saykin AJ, Moore JH, Shen L, Alzheimer's Disease Neuroimaging Initiative. Identifying genetic markers enriched by brain imaging endophenotypes in Alzheimer's disease. BMC Med Genet. 2022;15:168.
121. Kendler KS, Neale MC. Endophenotype: a conceptual analysis. Mol Psychiatry. 2010;15:789–97.
122. Zhao B, Zhang J, Ibrahim JG, et al. Large-scale GWAS reveals genetic architecture of brain white matter microstructure and genetic overlap with cognitive and mental health traits (n = 17,706). Mol Psychiatry. 2019;26:3943–55.

123. Jack CR Jr, Knopman DS, Jagust WJ, Shaw LM, Aisen PS, Weiner MW, Petersen RC, Trojanowski JQ. Hypothetical model of dynamic biomarkers of the Alzheimer's pathological cascade. Lancet Neurol. 2010;9:119–28.
124. Khan W, Giampietro V, Banaschewski T, et al. A multi-cohort study of ApoE ε4 and amyloid-β effects on the hippocampus in Alzheimer's disease. J Alzheimers Dis. 2017;56:1159–74.
125. Brandon JA, Farmer BC, Williams HC, Johnson LA. APOE and Alzheimer's disease: neuroimaging of metabolic and cerebrovascular dysfunction. Front Aging Neurosci. 2018;10:180.
126. Satizabal CL, Adams HHH, Hibar DP, et al. Genetic architecture of subcortical brain structures in 38,851 individuals. Nat Genet. 2019;51:1624–36.
127. Knol MJ, Poot RA, Evans TE, et al. Genetic variants for head size share genes and pathways with cancer. Cell Rep Med. 2024;5:101529.
128. Wolk DA, Dickerson BC, Initiative TADN, et al. Apolipoprotein E (APOE) genotype has dissociable effects on memory and attentional–executive network function in Alzheimer's disease. Proc Natl Acad Sci. 2010;107:10256–61.
129. Mishra S, Blazey TM, Holtzman DM, Cruchaga C, Su Y, Morris JC, Benzinger TLS, Gordon BA. Longitudinal brain imaging in preclinical Alzheimer disease: impact of APOE ε4 genotype. Brain. 2018;141:1828–39.
130. Lumsden AL, Mulugeta A, Zhou A, Hyppönen E. Apolipoprotein E (APOE) genotype-associated disease risks: a phenome-wide, registry-based, case-control study utilising the UK Biobank. EBioMedicine. 2020;59:102954.
131. He X-Y, Wu B-S, Kuo K, et al. Association between polygenic risk for Alzheimer's disease and brain structure in children and adults. Alzheimers Res Ther. 2023;15:109.
132. Foley SF, Tansey KE, Caseras X, Lancaster T, Bracht T, Parker G, Hall J, Williams J, Linden DEJ. Multimodal brain imaging reveals structural differences in Alzheimer's disease polygenic risk carriers: a study in healthy Young adults. Biol Psychiatry. 2017;81:154–61.
133. Sabuncu MR, Buckner RL, Smoller JW, Lee PH, Fischl B, Sperling RA, Alzheimer's Disease Neuroimaging Initiative. The association between a polygenic Alzheimer score and cortical thickness in clinically normal subjects. Cereb Cortex. 2012;22:2653–61.
134. Wolf AB, Caselli RJ, Reiman EM, Valla J. APOE and neuroenergetics: an emerging paradigm in Alzheimer's disease. Neurobiol Aging. 2013;34:1007–17.
135. Lacalle-Aurioles M, Iturria-Medina Y. Fornix degeneration in risk factors of Alzheimer's disease, possible trigger of cognitive decline. Cereb Circ Cogn Behav. 2023;4:100158.
136. Gold BT, Johnson NF, Powell DK, Smith CD. White matter integrity and vulnerability to Alzheimer's disease: preliminary findings and future directions. Biochim Biophys Acta. 2012;1822:416–22.
137. Filbey FM, Chen G, Sunderland T, Cohen RM. Failing compensatory mechanisms during working memory in older apolipoprotein E-epsilon4 healthy adults. Brain Imaging Behav. 2010;4:177–88.
138. Håglin S, Koch E, Schäfer Hackenhaar F, Nyberg L, Kauppi K. APOE ε4, but not polygenic Alzheimer's disease risk, is related to longitudinal decrease in hippocampal brain activity in non-demented individuals. Sci Rep. 2023;13:8433.
139. McKenna F, Koo B-B, Killiany R, Alzheimer's Disease Neuroimaging Initiative. Comparison of ApoE-related brain connectivity differences in early MCI and normal aging populations: an fMRI study. Brain Imaging Behav. 2016;10:970–83.
140. Riese F, Gietl A, Zölch N, et al. Posterior cingulate γ-aminobutyric acid and glutamate/glutamine are reduced in amnestic mild cognitive impairment and are unrelated to amyloid deposition and apolipoprotein E genotype. Neurobiol Aging. 2015;36:53–9.
141. Suri S, Emir U, Stagg CJ, et al. Effect of age and the APOE gene on metabolite concentrations in the posterior cingulate cortex. NeuroImage. 2017;152:509–16.
142. Patel K, Xie Z, Yuan H, et al. Unsupervised deep representation learning enables phenotype discovery for genetic association studies of brain imaging. Commun Biol. 2024;7:414.
143. Wen J, Zhao B, Yang Z, et al. The genetic architecture of multimodal human brain age. Nat Commun. 2024;15:2604.

144. Yang Z, Wen J, Abdulkadir A, et al. Gene-SGAN: discovering disease subtypes with imaging and genetic signatures via multi-view weakly-supervised deep clustering. Nat Commun. 2024;15:354.
145. Yang J, Lee SH, Goddard ME, Visscher PM. GCTA: a tool for genome-wide complex trait analysis. Am J Hum Genet. 2011;88:76–82.
146. Bulik-Sullivan BK, Loh P-R, Finucane HK, Ripke S, Yang J, Schizophrenia Working Group of the Psychiatric Genomics Consortium, Patterson N, Daly MJ, Price AL, Neale BM. LD score regression distinguishes confounding from polygenicity in genome-wide association studies. Nat Genet. 2015;47:291–5.
147. Eraslan G, Avsec Ž, Gagneur J, Theis FJ. Deep learning: new computational modelling techniques for genomics. Nat Rev Genet. 2019;20:389–403.
148. Jack CR, Bennett DA, Blennow K, et al. A/T/N: an unbiased descriptive classification scheme for Alzheimer disease biomarkers. Neurology. 2016;87:539.
149. Olsson B, Lautner R, Andreasson U, et al. CSF and blood biomarkers for the diagnosis of Alzheimer's disease: a systematic review and meta-analysis. Lancet Neurol. 2016;15:673–84.
150. Cruchaga C, Kauwe JSK, Mayo K, et al. SNPs associated with cerebrospinal fluid phospho-tau levels influence rate of decline in Alzheimer's disease. PLoS Genet. 2010;6:e1001101.
151. Jansen IE, van der Lee SJ, Gomez-Fonseca D, et al. Genome-wide meta-analysis for Alzheimer's disease cerebrospinal fluid biomarkers. Acta Neuropathol. 2022;144:821–42.
152. Xu Y, Vasiljevic E, Deming YK, et al. Effect of pathway-specific polygenic risk scores for Alzheimer's disease (AD) on rate of change in cognitive function and AD-related biomarkers among asymptomatic individuals. medRxiv. 2023; https://doi.org/10.1101/2023.01.30.23285142.
153. Suhre K. Genetic associations with ratios between protein levels detect new pQTLs and reveal protein-protein interactions. Cell Genom. 2024;4:100506.
154. Neumann A, Ohlei O, Küçükali F, et al. Multivariate GWAS of Alzheimer's disease CSF biomarker profiles implies GRIN2D in synaptic functioning. Genome Med. 2023;15:79.
155. Batmanghelich NK, Dalca AV, Sabuncu MR, Golland P. Joint modeling of imaging and genetics. In: Gee JC, Joshi S, Pohl KM, Wells WM, Zöllei L, editors. Information processing in medical imaging, IPMI 2013 lecture notes in computer science, vol. 7917. Berlin/Heidelberg: Springer; 2013. https://doi.org/10.1007/978-3-642-38868-2_64.
156. Karlsson IK, Escott-Price V, Gatz M, Hardy J, Pedersen NL, Shoai M, Reynolds CA. Measuring heritable contributions to Alzheimer's disease: polygenic risk score analysis with twins. Brain Commun. 2022;4:fcab308.
157. Baker E, Leonenko G, Schmidt KM, et al. What does heritability of Alzheimer's disease represent? PLoS One. 2023;18:e0281440.
158. Brandes N, Weissbrod O, Linial M. Open problems in human trait genetics. Genome Biol. 2022;23:131.
159. Hohman TJ, Bush WS, Jiang L, et al. Discovery of gene-gene interactions across multiple independent data sets of late onset Alzheimer disease from the Alzheimer Disease Genetics Consortium. Neurobiol Aging. 2016;38:141–50.
160. Louwersheimer E, Cohn-Hokke PE, Pijnenburg YAL, et al. Rare genetic variant in SORL1 may increase penetrance of Alzheimer's disease in a family with several generations of APOE-ε4 homozygosity. J Alzheimers Dis. 2017;56:63–74.
161. Lundberg M, Sng LMF, Szul P, Dunne R, Bayat A, Burnham SC, Bauer DC, Twine NA. Novel Alzheimer's disease genes and epistasis identified using machine learning GWAS platform. Sci Rep. 2023;13:17662.
162. Virolainen SJ, VonHandorf A, Viel KCMF, Weirauch MT, Kottyan LC. Gene-environment interactions and their impact on human health. Genes Immun. 2023;24:1–11.
163. Dunn AR, O'Connell KMS, Kaczorowski CC. Gene-by-environment interactions in Alzheimer's disease and Parkinson's disease. Neurosci Biobehav Rev. 2019;103:73–80.
164. Livingston G, Huntley J, Sommerlad A, et al. Dementia prevention, intervention, and care: 2020 report of the Lancet Commission. Lancet. 2020;396:413–46.

165. Franz CE, Gustavson DE, Elman JA, et al. Associations between ambient air pollution and cognitive abilities from midlife to early old age: modification by APOE genotype. J Alzheimers Dis. 2023;93:193–209.
166. Ward DD, Ranson JM, Wallace LMK, Llewellyn DJ, Rockwood K. Frailty, lifestyle, genetics and dementia risk. J Neurol Neurosurg Psychiatry. 2022;93:343–50.
167. Chen Y, Zhang Y, Li S, Zhou L, Li H, Li D, Wang Y, Yang H. Cardiometabolic diseases, polygenic risk score, APOE genotype, and risk of incident dementia: a population-based prospective cohort study. Arch Gerontol Geriatr. 2023;105:104853.
168. Ottman R. An epidemiologic approach to gene-environment interaction. Genet Epidemiol. 1990;7:177–85.
169. Kauffmann F, Demenais F. Gene-environment interactions in asthma and allergic diseases: challenges and perspectives. J Allergy Clin Immunol. 2012;130:1229–40. quiz 1241–2
170. Elonheimo HM, Andersen HR, Katsonouri A, Tolonen H. Environmental substances associated with Alzheimer's disease-a scoping review. Int J Environ Res Public Health. 2021; https://doi.org/10.3390/ijerph182211839.
171. Stephan BCM, Pakpahan E, Siervo M, et al. Prediction of dementia risk in low-income and middle-income countries (the 10/66 study): an independent external validation of existing models. Lancet Glob Health. 2020;8:e524–35.
172. Deming Y, Dumitrescu L, Barnes LL, et al. Sex-specific genetic predictors of Alzheimer's disease biomarkers. Acta Neuropathol. 2018;136:857–72.
173. Wang X, Broce I, Deters KD, Fan CC, Banks SJ. Identification of sex-specific genetic variants associated with tau PET. Neurol Genet. 2022;8:e200043.
174. Lopez-Lee C, Torres ERS, Carling G, Gan L. Mechanisms of sex differences in Alzheimer's disease. Neuron. 2024;112:1208–21.
175. Mills MC, Rahal C. A scientometric review of genome-wide association studies. Commun Biol. 2019;2:9.
176. Belloy ME, Andrews SJ, Le Guen Y, Cuccaro M, Farrer LA, Napolioni V, Greicius MD. APOE genotype and Alzheimer disease risk across age, sex, and population ancestry. JAMA Neurol. 2023;80:1284–94.
177. Ju D, Hui D, Hammond DA, Wonkam A, Tishkoff SA. Importance of including non-European populations in large human genetic studies to enhance precision medicine. Annu Rev Biomed Data Sci. 2022;5:321–39.
178. GBD 2019 Dementia Forecasting Collaborators. Estimation of the global prevalence of dementia in 2019 and forecasted prevalence in 2050: an analysis for the Global Burden of Disease Study 2019. Lancet Public Health. 2022;7:e105–25.
179. Atkinson EG, Maihofer AX, Kanai M, et al. Tractor uses local ancestry to enable the inclusion of admixed individuals in GWAS and to boost power. Nat Genet. 2021;53:195–204.
180. Marca-Ysabel MV, Rajabli F, Cornejo-Olivas M, et al. Dissecting the role of Amerindian genetic ancestry and the ApoE ε4 allele on Alzheimer disease in an admixed Peruvian population. Neurobiol Aging. 2021;101:298.e11–5.
181. Naslavsky MS, Suemoto CK, Brito LA, et al. Global and local ancestry modulate APOE association with Alzheimer's neuropathology and cognitive outcomes in an admixed sample. Mol Psychiatry. 2022;27:4800–8.
182. Griswold AJ, Celis K, Bussies PL, et al. Increased APOE ε4 expression is associated with the difference in Alzheimer's disease risk from diverse ancestral backgrounds. Alzheimers Dement. 2021;17:1179–88.
183. Vale CL, Thompson LC, Murphy C, Forcat S, Hanley B. Involvement of consumers in studies run by the Medical Research Council Clinical Trials Unit: results of a survey. Trials. 2012;13:9.
184. Price A, Albarqouni L, Kirkpatrick J, Clarke M, Liew SM, Roberts N, Burls A. Patient and public involvement in the design of clinical trials: an overview of systematic reviews. J Eval Clin Pract. 2018;24:240–53.
185. UK Data Service What is the Five Safes framework? — UK Data Service. https://ukdataservice.ac.uk/help/secure-lab/what-is-the-five-safes-framework/. Accessed 24 May 2024.

186. Bernier A, Raven-Adams M, Zaccagnini D, Knoppers BM. Recording the ethical provenance of data and automating data stewardship. Big Data Soc. 2023; https://doi.org/10.2053/9517231163174.
187. Sigfrid L, Maskell K, Bannister PG, et al. Addressing challenges for clinical research responses to emerging epidemics and pandemics: a scoping review. BMC Med. 2020;18:190.
188. Boylan S, Arsenault C, Barreto M, et al. Data challenges for international health emergencies: lessons learned from ten international COVID-19 driver projects. Lancet Digit Health. 2024;6:e354–66.
189. Wilkinson MD, Dumontier M, Aalbersberg IJJ, et al. The FAIR Guiding Principles for scientific data management and stewardship. Sci Data. 2016;3:160018.

Part IV
Biomarkers in Clinical and Translational Alzheimer Disease Research

Biomarkers in Clinical Trials of Alzheimer Disease: An Overview

Fabricio Ferreira de Oliveira, Raphael Machado Castilhos, Artur Martins Coutinho, Alan Cronemberger Andrade, Yun Jin Kim, Fardin Nabizadeh, Fábio Henrique de Gobbi Porto, and Gustavo Alves Andrade dos Santos

Abstract Harmonized biomarker assessments could improve diagnoses and the differential diagnoses of neurodegenerative diseases, enhance prognostication, allow monitoring disease progression, and improve stratification of participants for inclusion into clinical trials. Different contexts of use require diverse biomarkers that can facilitate disease staging, assess the severity of clinical symptoms, select participants for personalized therapeutic approaches, assist with differential diagnoses, monitor longitudinal progression and therapeutic response, and establish prognoses and meaningful outcomes. Biomarkers should always be integrated with cognitive, functional, and behavioral variables and can be used to measure and track downstream disease-modifying effects of investigational treatments. The combination of multiple biomarkers may offer the best strategy to be employed for patient selection and outcome assessments at present. Biomarkers for patient selection and

F. F. de Oliveira (✉)
ELYSIAN® Clínica Médica & Neurologia, São Paulo, SP, Brazil

R. M. Castilhos
Hospital de Clínicas de Porto Alegre, Porto Alegre, RS, Brazil
e-mail: rcastilhos@hcpa.edu.br

A. M. Coutinho · F. H. de Gobbi Porto
Universidade de São Paulo, São Paulo, SP, Brazil

A. Cronemberger Andrade
Universidade Federal de São Paulo, São Paulo, SP, Brazil
e-mail: alan.andrade@unifesp.br

Y. J. Kim
Xiamen University Malaysia, Sepang, Selangor, Malaysia
e-mail: yjkim@xmu.edu.my

F. Nabizadeh
Iran University of Medical Sciences, Tehran, Tehran, Iran

G. A. A. dos Santos
Universidade de São Paulo, Ribeirão Preto, SP, Brazil

R. Tarawneh (ed.), *Biomarkers of Neurodegenerative Disorders*,
https://doi.org/10.1007/978-3-031-80849-4_14

outcome assessments in clinical and translational studies need to be homogenized; however, many challenges including access and cost issues, the presence of multiple simultaneous neuropathological phenomena, and limited access to many resources of biosample processing or advanced imaging may hinder their applicability in different centers. This chapter explores the practicality of using neuroimaging and biofluid biomarkers for patient selection and outcome assessments in clinical and translational research, along with impactful genetic variants. It is expected that the issues to be discussed may leverage standard formats for the uniformity of biomarker use in different research settings.

Keywords Alzheimer disease · Biomarkers · Clinical Trials · Participant Selection

1 Introduction

1.1 Overview

Harmonization of biomarkers for patient selection and outcome assessments in clinical studies is a pressing need. However, defining homogeneous methods to obtain consistent results faces many challenges, including limited access to resources needed for advanced imaging or biosample processing, though it could improve diagnosis of neurodegenerative diseases, reduce trial costs, and achieve reliable standards of care for such diseases.

Biomarkers are expected to facilitate disease staging and assessment of the severity of clinical symptoms, monitor longitudinal progression, and evaluate participant engagement and therapeutic response in clinical studies. Usefulness of biomarkers for analyses of risk must consider their precision and the accurate interpretation of results, given the fact that not all biomarkers reflect the actual neuropathology in the brain, while sensitivity and specificity issues must also be taken into account [1]. Standardized procedures to be adopted in such studies would enable the use of biomarkers in line with their demonstrated informative value and uniformly across different centers [2].

This chapter explores the feasibility of neuroimaging and biofluid biomarkers for patient selection and outcome assessments in clinical and translational research, while also tackling the impact of genetic variants on study findings. It is expected that the issues to be discussed may leverage standard formats for biomarker use in different research settings.

1.2 Need for Biomarkers

Over the last few years, there has been a growing number of studies to identify and validate biomarkers for neurodegenerative diseases [3]. There is a great need for accurate and timely diagnosis of such diseases because neurodegeneration begins decades before clinical symptoms; a period during which individuals are most likely to respond to disease-modifying therapies that delay disease progression prior to significant irreversible neuronal loss. This need is not limited to providing patients with insights into diagnosis and prognosis of neurodegenerative diseases and also includes the optimization of treatment plans, the provision of appropriate support and care, and the preparation of patients to become participants in clinical trials [4]. Incorrect diagnosis can result in inadequate treatment and care, as well as unnecessary medical assessments and expenses due to uncertainty in diagnosis [5]. For example, proper use of biomarkers for diagnosis is even more important in view of the fact that older people could be invariably suffering from Primary Age-Related Tauopathy (PART) or Age-Related Tau Astrogliopathy (ARTAG), two diseases that display prominent tau pathology and may clinically mimic Alzheimer disease (AD) [6]. Even after symptom onset, a definitive diagnosis is only possible by post-mortem neuropathological analyses [7]. Estimates suggest that 15% of cognitively normal adults above the age of 65 and 35–50% of those above the age of 85 have neuropathological evidence of AD [6].

Currently, there are numerous clinical trials underway to evaluate and develop potential disease-modifying therapies targeting amyloid-β, tau, or α-synuclein [8]. Nevertheless, analyses of clinical trial cohorts clinically diagnosed with mild dementia due to AD suggest that approximately 25% do not have detectable amyloid-β in the brain [9, 10]; therefore, biomarker evidence of pathology is now required for inclusion in AD clinical trials. In addition, older people often bear evidence of multiple heterogeneous neuropathological phenomena simultaneously, while determining the primary cause of symptoms is important for each patient. An individual with considerable amyloid-β and α-synuclein pathology may not be a suitable candidate for a trial of anti-amyloid therapy if the latter pathology is the primary cause of most symptoms [5]. Stratification of research participants according to biomarkers may identify subgroups that are more likely to respond to different therapies. Also, it seems that distinct disease-modifying therapies have different optimal time windows during the course of the disease when their efficacy is increased. Most therapies are expected to be effective during the early symptomatic stages, or pre-symptomatic (i.e., preclinical) disease stages prior to the onset of clinical symptoms [11, 12].

Preclinical diagnostic biomarkers are necessary in order to include asymptomatic participants for clinical trials [4]. In addition, surrogate biomarkers that can demonstrate therapeutic effects on downstream events such as neurodegeneration are highly required for such trials, particularly in the preclinical or prodromal stages [13].

1.3 Availability of Biomarkers for Research

Research with biomarkers in neurodegenerative disorders can serve to enhance outcome evaluations by way of better patient selection and outcome predictions. Use of biomarkers covers multiple applications, from differential diagnoses to the assessment of post-intervention clinical outcomes. Table 1 summarizes some of the contexts of use for biomarkers in AD clinical trials.

Access issues may hinder a more widespread use of biomarkers in research settings. Modern molecular imaging and biofluid platforms are currently applicable to most trials [15, 16], but can be limited to well-resourced settings due to technical demands to implement their analyses [17]. Despite the critical role of multimodal and multisite approaches in enriching research cohorts and expanding the use of

Table 1 Biomarker types and applications in research and clinical contexts in AD

Biomarker types (FDA category)	Contexts of use	Typical examples of applicability in research	Clinical use
Diagnostic biomarkers	Use as a diagnostic tool based on differential diagnoses for inclusion or exclusion criteria	Evaluating biomarker profiles to select eligible participants for clinical trials	Confirming the diagnosis of Alzheimer disease through amyloid imaging or low cerebrospinal fluid amyloid-β 42
Prognostic biomarkers	Make prognostic inferences/assumptions concerning disease progression or likelihood of developing a certain condition	Stratification of samples into subgroups of participants to assess potential disease progression	Using tau PET to forecast rapid cognitive decline or *APOE*-ε4 carrier status for ARIA risk in immunotherapy
Predictive biomarkers	To predict the response to a therapeutic intervention	Enriching clinical trial cohorts with patients more likely to respond to anti-tau therapies	Tailoring anti-tau treatments to patients identified through amyloid PET as likely responders
Response biomarkers	To monitor therapeutic response and assess efficacy	Monitoring changes in biomarkers of amyloidosis or cerebrospinal fluid tau during trials to gauge drug efficacy	Assessing treatment effectiveness by observing reductions in amyloid plaques or changes in brain metabolism via FDG-PET
Safety biomarkers	To detect adverse reactions or safety concerns of a therapy	Using MRI in trials to monitor for adverse effects like ARIAs in novel drug candidates	Assessment of meaningful outcomes of biomarkers during clinical observation to detect treatment-related complications

Adapted from Amur et al. 2015 [14]. Note: The table shows different types of biomarkers as classified by the US Food and Drug Administration (available at https://www.fda.gov/media/119271/download). Abbreviations: *ARIA* amyloid-related imaging abnormalities, *APOE-ε4* E4 allele of apolipoprotein E, *FDG* [^{18}F]fluorodeoxyglucose, *MRI* magnetic resonance imaging, *PET* positron emission tomography biomarkers, logistical barriers may restrict their use in some regions, leaving lower-funded research systems at a disadvantage. Underrepresented participants of clinical studies in the USA are being addressed with specific research recruitment strategies to overcome these issues [18].

The distribution and application of biomarkers in research can also be uneven across different global regions, reflecting a broader issue of inequality in research capabilities. Current research cohorts such as the Alzheimer's Disease Neuroimaging Initiative (ADNI) (USA) or BioFINDER (Sweden) are a testbed for a broad range of biomarkers [15, 16], but with added costs and expenses which render them scarce in some regions. In Europe, specialized research and memory clinics have used AD biomarkers since the early 2000s [19]. Experts from Asia (Japan, Korea, Singapore, Taiwan) state that applicability of advanced imaging modalities remains limited in the continent because of prohibitive costs [20]. In Africa, in accordance with a group of experts representing ten different countries (Benin, Botswana, Cameroon, Ethiopia, Ghana, Kenya, Mozambique, Nigeria, Tanzania, Zimbabwe), datasets on imaging and fluid biomarkers are also limited [21]. These disparities may impact the scope and quality of research while also affecting the inclusion and diversity of research participants, further complicating the interpretation and applicability of research outcomes across different populations. Given these availability challenges, it is imperative that the research community addresses access issues to enhance the equitable distribution of biomarker technologies and to foster more inclusive and comprehensive research.

1.4 Clinical Trial Relevance

Biomarkers are useful in several ways, including diagnosing diseases, predicting disease outcomes, monitoring disease progression or response to therapy, and identifying individuals at risk of developing a particular disease, such as AD, Parkinson's disease (PD), and Huntington's disease [22]. Biomarkers are also used in drug development and clinical trials. Biomarkers can be used to identify individuals who are more likely to respond to a particular treatment, which can help optimize treatment plans and reduce the risk of adverse events. They can also be used to measure the effectiveness of a treatment or drug, which can help in drug development and reduce the costs of clinical trials [23].

Clinical trials of neurodegenerative disorders typically involve the use of specific biomarkers to track disease progression and treatment response [24]. These trials may involve the use of imaging techniques, such as Magnetic Resonance Imaging (MRI), positron emission tomography (PET), and single-photon emission computed tomography (SPECT) scans, as well as tests of cerebrospinal fluid (CSF) or blood samples to measure levels of specific proteins or other markers associated with the disease [25]. The goal of these trials is to identify biomarkers that can be used as surrogate endpoints, which are measures that are strongly correlated with clinical outcomes and can be used to predict treatment effectiveness; medical researchers can more quickly and accurately evaluate the efficacy of potential treatments, which can accelerate the drug development process and improve patient outcomes [26].

Considering the low accuracy of clinical diagnosis to predict the underlying brain pathology, as well as the fact that clinical symptoms may overlap among different pathological mechanisms in neurodegenerative diseases, biomarkers are essential to help in differential diagnoses [27]. In order to minimize the risk of including patients without the pathology of interest in clinical trials, biomarkers specific to the pathology being studied are often used for screening, ensuring that only such patients are included in clinical trials. This can improve the accuracy of the trial results and help verify that new treatments are developed for the specific pathology being studied [10]. For example, clinical trials in mild AD dementia show that about 25% of patients fulfilling clinical criteria for AD do not exhibit amyloid-β pathology [12]; therefore, many AD-related clinical trials recently require biomarker evidence of amyloid-β pathology for inclusion. Unfortunately, several different neurodegenerative mechanisms underlying brain pathology may be found in older patients, and it is important to determine the primary cause of the symptoms in each case [28]. Diagnostic biomarkers identifying disease pathology before the onset of overt symptoms are crucial for recruiting suitable individuals for secondary prevention trials. This approach, known as "preclinical detection" can help identify individuals who are at risk of developing a particular disease and allow for early intervention before the onset of clinical symptoms [29]. Disease-modifying therapies, where drug target engagement in the human brain has been firmly established in early phase I–II studies, are more likely to be shown as clinically effective in subsequent large-scale phase III clinical trials, while in phase II studies, the effect of pharmacodynamic measurements should be correlated with clinical outcomes [13, 30].

However, there are also some challenges associated with the use of biomarkers in clinical trials for neurodegenerative disorders. For example, there is often variability in the measurement of biomarkers, which can affect the accuracy of the results. Additionally, some biomarkers may not be specific to a particular disease, which can make it difficult to interpret the results. In conclusion, biomarkers are important tools in the development of new treatments and in clinical observational studies for neurodegenerative disorders, and ongoing research is focused on identifying new and more accurate biomarkers to improve the diagnosis, prognosis, and treatment of these conditions. The identification of accurate biomarkers of different diseases is also an important research direction when establishing relevant risk factors and protective factors, including studies of genetic, epigenetic, developmental, and lifestyle factors [31].

2 Neuroimaging Biomarkers

2.1 *Structural and Functional MRI*

MRI can be used in clinical trials for participant selection, outcome assessments, and safety monitoring in three key areas:

1. employing imaging markers for participant selection and stratification;
2. utilizing imaging markers as potential outcome measures in clinical trials;
3. and leveraging imaging markers for monitoring trial safety [32, 33].

2.1.1 MRI for Participant Selection and Stratification

MRI Findings as Exclusion Criteria for Anti-amyloid Therapies

MRI plays a crucial role in selecting participants for trials involving anti-amyloid therapies in AD. Several MRI findings have been proposed as exclusion criteria for treatment with these drugs [34]:

- Presence of more than four microhemorrhages, defined as lesions measuring 10 mm or less in diameter.
- Identification of a single macrohemorrhage exceeding 10 mm in diameter.
- Observation of superficial siderosis in a specific area.
- Detection of vasogenic edema.
- Identification of more than two lacunar infarcts or a stroke affecting a major vascular territory.
- Presence of severe subcortical hyperintensities consistent with a Fazekas score of 3.
- Evidence suggesting amyloid-β-related angiitis (ABRA).
- Evidence of cerebral amyloid angiopathy-related inflammation (CAA-ri).
- Identification of other significant intracranial abnormalities that may potentially lead to cognitive impairment.
- Identification of other notable brain lesions during screening that suggest a dementia diagnosis other than AD.
- MRI evidence supporting the presence of a non-AD dementia.

The careful analysis of these MRI features is also important in clinical practice when considering the use of anti-amyloid therapies.

MRI as an Enrichment Biomarker for Clinical Trials

The National Institute on Aging, in conjunction with the Alzheimer's Association, proposed a biological framework for defining AD based on key biomarkers: amyloid, tau, neurodegeneration, or AT(N). Using the AT(N) Research Framework for AD [17], MRI can serve as a marker of neurodegeneration, particularly atrophy in regions typically affected by AD [35, 36].

Hippocampal volume can be used as an enrichment biomarker for selecting participants in clinical trials involving individuals with mild cognitive impairment (MCI) [37]. Enrichment biomarkers are employed to identify individuals who are more likely to experience disease progression and thus, more likely to benefit from the investigational therapy, thereby optimizing trial efficiency and reducing participant numbers and associated costs. Specifically, selecting participants with smaller hippocampal volumes, indicative of neurodegenerative changes associated with AD, can enhance the likelihood of detecting treatment effects in clinical trials involving individuals with amnestic MCI. Decreased entorhinal cortex and anteromedial temporal cortex volumes may also predict cognitive decline [32, 33].

Whole-brain voxel-based studies have revealed a regional atrophy pattern associated with AD, characterized by increased atrophy in specific brain regions, representing an "AD-signature" of brain atrophy [32]. These regions include the temporal neocortex, posterior cingulate, orbitofrontal cortex, and medial temporal lobe. Notably, these findings have been observed in cognitively healthy individuals up to 10 years prior to the onset of dementia. The "AD-signature" patterns may also serve as an enrichment strategy in clinical trials.

An underutilized opportunity is using functional MRI (fMRI) to investigate the effects of interventions on brain circuits, providing evidence of target engagement and supporting proof of pharmacology [35, 38]. Cognition and behavior are supported by complex brain circuits, consisting of interconnected nodes [39]. Some circuits are active at rest, distinct from those engaged during specific cognitive tasks, while fMRI measures circuit activity in vivo.

In AD, connectivity between the posterior default mode network (DMN) and posterior hippocampus, as well as cingulo-parietal-occipital regions, is reduced, while connectivity in the anterior salience network (SN) is intensified [40, 41]. Circuit dysfunction begins in the preclinical stage of AD and worsens through the prodromal and dementia stages [42].

If treatment benefits are demonstrated using fMRI, they would provide evidence of circuit-level effects and increase confidence in the potential cognitive benefits of the treatment [38]. Preliminary data from multisite trials and reliability studies support its implementation in phase II trials [43–45].

Despite its potential, fMRI remains primarily a research tool and has had limited value as an outcome measure. This limitation is due to issues with standardization, difficulties in translating group-level data to the individual level, and challenges in defining cutoffs. Additionally, fMRI results are influenced by numerous confounders, complicating their interpretation and reducing their reliability for clinical use.

2.1.2 MRI as a Potential Outcome Measure in Clinical Trials

MRI is an established tool for the assessment of neurodegeneration and can serve as a valuable indicator of underlying disease processes such as neuronal loss, providing essential evidence of potential disease-modifying effects in clinical trials of disease-modifying therapies.

In AD, brain atrophy is linked to gray matter loss, particularly neuronal and synaptic losses, which correlate with neurofibrillary tangles [46]. This suggests that interventions targeting neurodegeneration and gray matter loss should ideally slow atrophy and result in a noticeable difference between drug and placebo groups at the end of clinical trials, favoring the active treatment. However, a review of MRI studies reveals that this expected outcome has seldom been observed [35]. In most trials reporting MRI results, there was no discernible difference between the drug and placebo groups upon trial completion, and in a few cases, the active treatment group exhibited greater levels of atrophy [47]. It remains uncertain whether the apparent increase in shrinkage or "pseudo-atrophy" in the treatment group reflects amyloid removal, reduced inflammation, or if, in certain treatment mechanisms, it indicates neurotoxicity and genuine exacerbation of atrophy.

Studies of resting-state fMRI in AD consistently indicate that treatment with cholinesterase inhibitors and memantine leads to enhanced activation in various brain regions, including the posterior cingulate cortex, parahippocampal gyrus, dorsolateral frontal cortex, middle cingulate, precuneus, anterior hippocampus, middle frontal gyrus, precentral gyrus, insular cortex, and thalamus [44]. Additionally, these treatments result in increased functional connectivity across multiple networks, such as the DMN, dorsal attention network, control network, and SN. These findings are supported by correlations between the brain's functional responses and cognitive measures, highlighting the relevance of the observed changes [44].

2.1.3 MRI for Monitoring Trial Safety

In AD clinical trials, a key challenge is monitoring and managing potential adverse events associated with anti-amyloid treatments [34, 48]. The term "amyloid-related imaging abnormalities" (ARIAs) encompasses a range of MRI findings observed in patients undergoing anti-amyloid immunotherapies for AD [49]. ARIA can be classified into two main categories: ARIA characterized by edema and effusion (ARIA-E) and ARIA characterized by microhemorrhages and superficial siderosis (ARIA-H) [50, 51]. ARIA is primarily based on imaging observations, with its pathophysiology resulting from monoclonal antibodies binding to accumulated amyloid-β in brain tissue and blood vessels. This binding disrupts vessel wall integrity, leading to the leakage of proteinaceous fluid (ARIA-E) or heme products (ARIA-H) [51].

These abnormalities typically occur early in the treatment process and are more common in individuals who carry the ε4 allele of the apolipoprotein E gene (*APOE*). ARIA-E exhibits a dose-dependent relationship with anti-amyloid treatments, with

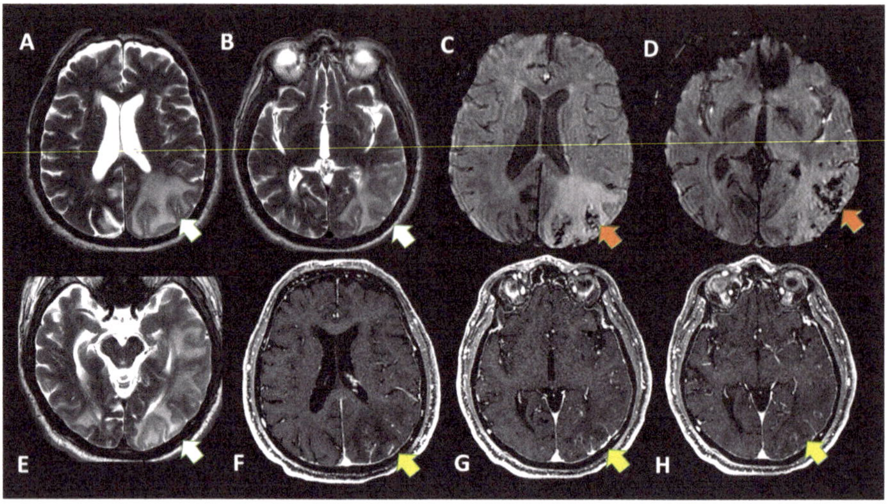

Fig. 1 "Amyloid-Related Imaging Abnormalities" (ARIAs) on MRI: Brain lesions related to treatment with anti-amyloid drugs that appear similar to cerebral amyloid angiopathy-related inflammation (CAA-ri). A large area of cortical-subcortical temporal-parietal-occipital signal alteration is seen on the left, with subcortical predominance, characterized by hyperintensity on T2/FLAIR images, representing edema—ARIA-E (white arrows on A, B, E). Multiple foci of magnetic susceptibility permeating SWI images (black areas indicated by orange arrows on C and D, corresponding to hemosiderin) are signs of ARIA-H. Juxtacortical/leptomeningeal contrast enhancement is also indicated by yellow arrows on post-contrast T1-weighted images in F, G, and H, representing inflammation. The findings jointly produce effacement of adjacent cortical sulci, compression, and partial collapse of the ipsilateral ventricle

a higher incidence in patients receiving higher doses of anti-amyloid immunotherapies [51]. In most cases, ARIA is asymptomatic and resolves spontaneously. However, the identification of ARIA has implications for patient selection and requires careful monitoring during anti-amyloid immunotherapy. In some instances, the development of ARIA may necessitate a temporary pause or discontinuation of treatment. Three fatal cases of ARIA were reported in the donanemab trial [52], and four cases in the open-label extension of the lecanemab study, at the time of this review [53]. In one lecanemab case, the patient had prior signs of cerebral amyloid angiopathy (CAA), highlighting the necessity of neuroimaging screening before using anti-amyloid immunotherapies [54].

With the recent FDA approval of the amyloid-targeting monoclonal antibody lecanemab [55] for AD treatment in 2023 and the subsequent approval of donanemab in 2024 [52], the use of these therapies will extend beyond clinical trials. This will lead to increased encounters with ARIA in clinical practice for radiologists (Fig. 1) [50].

Monoclonal antibodies targeting amyloid-β bind to vascular amyloid-β, further disrupting vascular integrity and increasing permeability. This immune response against vessels, combined with impaired clearance pathways, leads to vascular damage [56]. When therapy with such monoclonal antibodies is initiated, the mobilization of amyloid-β from brain tissue can temporarily overwhelm the perivascular drainage system, potentially leading to vascular deposition similar to what is observed in CAA. Additionally, antibody-mediated inflammation may occur, causing breakdown of the blood–brain barrier and further compromising vascular integrity, akin to ABRA [51]. It is important to note that ABRA can occur spontaneously without immunotherapy administration. Therefore, the MRI findings of ARIA-H (microhemorrhages and superficial siderosis) closely resemble those of CAA, while the MRI findings of ARIA-E closely resemble those of CAA-ri or ABRA. The key distinguishing factor for diagnosing ARIA, compared to these other entities, is a history of monoclonal antibody therapy targeting amyloid-β.

2.2 *Molecular Imaging: SPECT, Amyloid PET, and Tau PET*

2.2.1 General Concepts

Molecular neuroimaging and nuclear medicine methods are represented by conventional scintigraphy, acquired in gamma-cameras in three-dimensional form (SPECT), and by dedicated PET machines. These methods use the injection of radiotracers to detect functional or molecular phenomena in vivo. Radiotracers consist of molecules that mimic a physiological function linked to a radioactive substance (usually ^{99m}Tc, ^{18}F, or ^{11}C), and can be safely administered. In general, SPECT uses ^{99m}Tc- or ^{123}I- marked molecules as tracers, which have a longer half-life than PET radioligands, usually marked with ^{18}F or ^{11}C.

PET and SPECT technologies have continuously evolved and are nowadays performed on multimodal hybrid machines, where such techniques are conjugated with conventional computed tomography (i.e., CT) or, less frequently, MRI (SPECT/CT, PET/CT, and more recently PET/MRI). While most SPECT cameras are still nonhybrid ones, virtually all newly commercialized PET machines have a PET/CT configuration.

These methods are highly sensitive for patient selection and prognostic evaluation/stratification and have been commonly used for outcome assessments in most of the recent AD trials, mainly due to their capacity to detect specific pathologic or physiologic changes. They are, however, of limited use in safety monitoring, due to the higher contrast and spatial resolution of MRI, which has a recognized superiority in the detection of overall conditions such as bleeding, stroke, edema, and ARIA [34, 50].

PET and SPECT tracers also provide information about pathological processes in other diseases. PET scans with amyloid (amyloid PET) and tau (tau PET) ligands detect brain deposition of amyloid-β plaques and intracellular tangles of phosphorylated tau protein, respectively, as well as their removal after potential disease-modifying therapies. PET scans with an assessment of glucose metabolism ([^{18}F]fluorodeoxyglucose, or FDG-PET) and cerebral perfusion SPECT are used as proxies for synaptic activity and can identify signs of neurodegeneration and specific patterns seen in different pathologic processes.

The main limitations of molecular neuroimaging are the unavailability of some tracers (mainly tau PET) in certain environments and the high cost of combining multiple tests.

A recent review addressed and consolidated neuroimaging methods in the AT(N) staging of AD [57], which will be discussed in detail below.

2.2.2 Amyloid PET

Amyloid PET has represented a major technological advance in the evaluation of AD by neuroimaging. This method allows the detection of extracellular brain deposits of amyloid-β in neuritic plaques with high sensitivity and specificity (96% and 100%, respectively) [58]. Therefore, it may be considered an accurate method for the inclusion/exclusion of individuals in AD trials and is suitable for use within the AT(N) Research Framework [59].

Amyloid tracers most commonly used in clinical research include ^{11}C-PiB, ^{18}F-florbetapir, ^{18}F-florbetaben, and ^{18}F-flutemetamol. Tracers labeled with ^{18}F have longer half-lives and can be shipped to locations distant from the production center, making amyloid PET studies possible in centers that do not have their own cyclotron [60].

In the AT(N) Research Framework, amyloid PET scans are generally interpreted as a dichotomous measure being classified as positive or negative (A+ or A−), facilitating the selection of participants for clinical trials. The test becomes positive (A+) often decades before the onset of symptoms, with amyloid load reaching near maximal levels around the time of symptom onset. Most studies suggest that brain amyloid burden is not closely associated with cognition in early symptomatic AD. The absence of brain amyloid on amyloid PET scans (A− tests) renders the diagnosis of AD very unlikely. Incident brain amyloid positivity in amyloid PET in cognitively

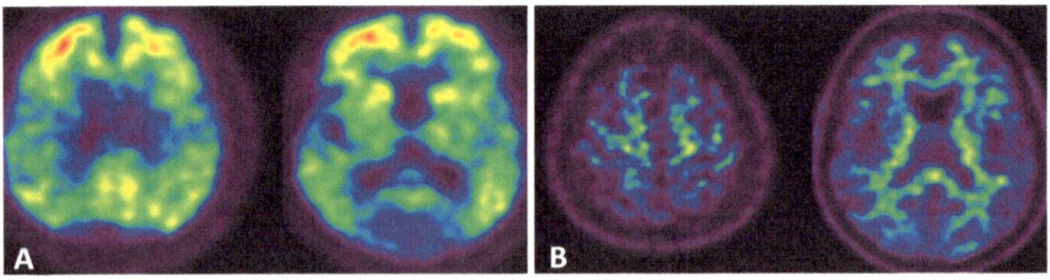

Fig. 2 PET images with the ^{11}C-PiB tracer. On the left (**a**), axial images demonstrate the diffuse concentration of the tracer in the cerebral cortex (A+ staging), corresponding to the deposition of amyloid-β plaques related to Alzheimer disease. On the right (**b**), axial images of a case without cortical deposition of amyloid-β plaques, with tracer concentration restricted to the white matter tracts (normal deposition on myelin sheaths), which renders the diagnosis of Alzheimer disease unlikely (A+ is required, but not sufficient, for the diagnosis of Alzheimer disease)

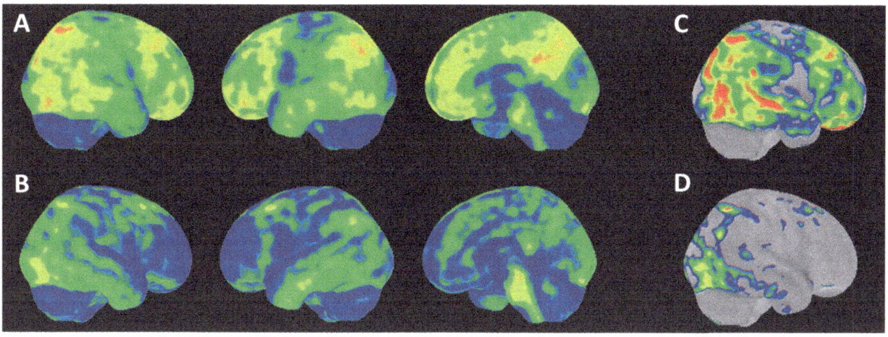

Fig. 3 PET images with the ^{11}C-PiB tracer. Superior row (**a**): 3D-stereotactic surface projection (3D-SSP, software Cortex ID suite, GE Healthcare) showing a diffuse concentration of the tracer in the cerebral cortex (A+ staging), with the uptake normalized by the cerebellar gray matter, which is confirmed in the 3D-SSP Z-score map images (**c**), highlighting the areas with amyloid-β deposition compared to a control group matched by age (Cortex ID Suite, Z-score threshold of 2.0 standard-deviations). Inferior row (**b**): 3D-SSP images showing a reduction in the deposition of amyloid-β plaques in comparison to the images in (**a**) after 18 months of treatment with an anti-amyloid drug (aducanumab), confirmed in the 3D-SSP Z-score map (**d**)

unimpaired people may vary from 2.7% in those aged 50–59 years to 41.3% in those aged 80–89 years [36], generally varying around 10–20% in individuals in their 60s and 70s. Conversely, around 25% of individuals clinically diagnosed with AD have negative (A−) scans in most surveys [61]. The percentage of A+ PET is comparable across different cohorts (Fig. 2) [62].

Amyloid PET will be fundamental in the selection of individuals for treatment with anti-amyloid drugs but also for treatment monitoring. It is recommended that only A+ individuals be treated with these drugs. The removal of amyloid-β aggregates, usually monitored with amyloid PET as a gold standard, was established as an outcome in recent trials involving anti-amyloid therapies and used as a strong argument in favor of the approval of the drugs (Fig. 3) [34, 55, 63, 64].

Normal scans (negative for amyloid deposition) indicate tracer deposition in white matter tracts and differential uptake in these tracts in comparison to cortical uptake. Positive scans demonstrate cortical uptake of the tracer at levels similar to or greater than white matter deposition, usually diffusely across the cortex, sparing only occipital regions. Although not necessary for patient selection, a semi-quantitative assessment of the amyloid load by standardized uptake value ratio (SUVr) quantification using the cortex of the cerebellum or the whole cerebellum as a reference region is welcome in aid of visual interpretation and is critical for therapeutic monitoring and outcome assessments [65]. SUVr monitoring of amyloid and tau PET has been used in major AD trials as an outcome assessment [34, 55, 63, 64].

Considering the technical difficulties of standardizing SUVr data between different institutions, especially for use in trials, several attempts have been made to create a universal amyloid deposition scale. Among these, the most successful is the Centiloid project, which led to the creation of a 100-point scale termed "Centiloid," from an average value of zero in "high certainty" amyloid-negative participants to an average of 100 in "typical" AD patients [66]. Its use is progressively rising in the scientific community in data reporting of amyloid scans.

2.2.3 Tau PET

The development of tau tracers that can reliably detect brain tau pathology has had a major impact on the field. The first generation of tau tracers, epitomized by the radiopharmaceutical [^{18}F]AV-1451 (previously [^{18}F]T807 and commercially named ^{18}F-flortaucipir), had higher affinity for 3R/4R helical tau filaments. Thus, these methods were excellent for the selection of T+ or T- individuals with AD and for staging the disease, as their deposition follows the Braak/Braak staging [67], which could help stratify individuals in the early or late stages of the disease. However, this tracer had significant limitations in identifying 4R tau deposits related to other tauopathies (e.g., frontotemporal lobar degeneration [FTLD], corticobasal degeneration, and progressive supranuclear palsy [PSP]) [68].

Second-generation tau tracers (among the most promising are [^{18}F]MK-6240 and [^{18}F]PI-2620) apparently overcome this limitation in the detection of different variants of tau deposits. Such tracers identify 3R/4R tau deposits (classically seen in AD), as well as 4R tauopathies, with a broader potential use in clinical trial conditions other than AD [69–71].

2.2.4 Molecular Imaging Markers of Neurodegeneration

The nuclear medicine methods capable of investigating status (N) are cerebral perfusion on SPECT and brain metabolism on FDG-PET. Perfusion SPECT with ^{99m}Tc-ECD reflects regional brain perfusion. FDG-PET is a method used to directly assess glycolytic consumption and brain metabolism. Although they provide similar information, PET has notably better performance and higher sensitivity than SPECT [72, 73].

Perfusion SPECT, and mostly FDG-PET, can be used as "enrichment biomarkers" for patient selection, as they show reduced metabolism in brain regions known to be affected by early AD pathology such as the precuneus and the posterior cingulate, and can assist in the differential diagnosis of AD [17]. They can also detect individuals at risk of progression, as was suggested for hippocampal volume in MRI in the previous section [37].

The possibility of quantitative analysis of FDG-PET images, which is possible through easily accessible software, increases the diagnostic accuracy of this method [74] and has significantly increased its reliability, as it has been proven to improve the reading of less-experienced examiners and improve inter-rater reliability. Thus, in the evaluation of FDG-PET images, the gold standard analysis for the (N) status is currently considered to be achieved by visual reading in conjunction with quantification methods, which in practice means an analysis with the 3D-SSP method from various manufacturers against a database of normal individuals matched by age [74].

FDG-PET is today the most sensitive method in clinical practice to assess neurodegeneration, particularly in AD [74–78], and thus to classify individuals as N+/N-. It also has a broad capacity for differential diagnosis by showing patterns of neurodegeneration typical of AD, its presenile variants, and related diseases, such as Lewy body dementias, atypical parkinsonism, FTLD, and LATE (limbic-predominant age-related TDP-43 encephalopathy). It may help in the differential diagnosis of individuals with an initial clinical diagnosis of AD, but with amyloid-negative assessments [62, 79–81].

Typical patterns of neurodegeneration seen on FDG-PET are well defined. In AD and its variants, there is hypometabolism in the temporal-parietal association neocortex, precuneus and posterior cingulate, with a late impairment in prefrontal areas. In Lewy body dementias, there is a classical parieto-occipital involvement including the visual cortex. In FTLD, there is marked anterior, medial, and basal frontal hypometabolism in the behavioral variant, sometimes with insular impairment and asymmetry. In the language variants of frontotemporal dementia (primary progressive semantic and non-fluent aphasias), there is usually left temporal involvement in the polar and more anterior temporal, and left frontal and per-insular, neocortical regions, respectively. Other patterns are also described for most of the neurodegenerative conditions (such as PD dementia, multiple system atrophy, PSP, and corticobasal syndrome [CBS]), and are detailed elsewhere [82, 83].

Among all degenerative processes, FDG-PET can detect with better accuracy the neurodegenerative pattern related to AD and its primary differential diagnoses. A pattern suggestive of AD (posterior temporal-parietal hypometabolism and in the posterior precuneus/posterior cingulate) on FDG-PET can predict amyloid positivity in more than 95% of cases both in amnestic syndromes [62] and in presenile cases such as CBS [84]. So, FDG-PET may also be used as a proxy to the "A" staging if amyloid-β markers are not available.

Sensitivity of FDG-PET and amyloid PET, respectively, in differentiating between AD and diseases on the FTLD spectrum are, respectively, of 96% (89–100%) versus 80% (68–92%), and a specificity of 86% (76–95%) for FDG-PET versus 84% (74–93%) for amyloid PET. The methods can be used in combination, simulating an "A(N) staging," and have significant high accuracy when the results are concordant, with 97% sensitivity (92–100%) and 98% specificity (93–100%) [79]. Importantly, the ability of FDG-PET to generate alternative diagnostic hypotheses, especially in negative cases for amyloid deposition, is noteworthy [58, 62, 79].

FDG-PET is also a superior prognostic marker and an excellent method of assessing the progression of neurodegenerative diseases when compared to amyloid PET. It also has a relatively low cost in comparison to the other methods. In addition to these points, a normal FDG-PET guarantees at least two favorable years for non-conversion to the dementia phase [85] and, in some cases, favors the interpretation of psychiatric cases over neurodegenerative ones, when scans are considered within normal limits and without signs of neurodegeneration [74].

3 Biofluid Biomarkers

3.1 CSF Biomarkers

CSF biomarkers are the most reliable biofluid biomarkers due to their intimate relationship to the brain [86] and reliable changes in pathological states. The fact that concentrations of biomarkers in the CSF are much higher than in other biofluids also contributes to the reliability of their assessments [87]. The main caveat is that collection of CSF is usually more burdensome than other biofluids, an aspect that has led to the search for biomarkers in other biofluids that may be more easily accessible and less invasive.

Current models of disease progression suggest that amyloid-β biomarkers become abnormal decades before detection of cognitive decline, and also before biomarkers of neurodegeneration, which become widespread later and correlate with clinical severity [88]. Such knowledge has been challenged in view of tau pathology apparently preceding amyloid-β deposition at a subthreshold biomarker detection level in time, with later independent amyloid-β deposition rising above the biomarker detection threshold and accelerating tauopathy [89].

When it comes to AD, the most traditional CSF biomarkers to be sought are amyloid-β 42, amyloid-β 40 (and particularly the amyloid-β 42/amyloid-β 40 ratio when abnormal, correcting for potential pre-analytical confounders) [90], tau and phospho-tau Thr_{181} [91]. Early in life, about 90% of amyloid-β in the brain consists of amyloid-β 40; with aging, and particularly at the onset of AD, insoluble amyloid-β 42 rapidly increases leading to the formation of amyloid plaques [92]. CSF amyloid-β 42 is a biomarker of soluble amyloid-β in the brain, while concentrations of amyloid-β 40 reflect the total generation of amyloid-β [93]. As brain amyloid aggregation increases, CSF amyloid-β 42 levels decrease due to a "sink" mechanism. CSF tau mostly reflects axonal degeneration, while concentrations of phospho-tau Thr_{181} are associated with the formation of ubiquitinated neurofibrillary tangles and are more specific to AD [94]. Therefore, the typical CSF biomarker profile of AD is high CSF tau and phospho-tau Thr_{181} and low CSF amyloid-β 42. Emerging isoforms of phospho-tau, such as phospho-tau Thr_{217} and phospho-tau Thr_{231}, seem to be at least equally useful for patient selection [87], though phospho-tau Thr_{217} is more strongly associated with tau tangle load and might distinguish AD from other dementias with higher accuracy [25]. Quality control programs are essential to harmonize procedures and monitor the consistency of the results [91]. In addition, biomarker ratios are usually more useful for outcome assessments than isolated biomarkers [95]. One study validated the CSF tau/amyloid-β 42 ratio level of >0.52 as the most sensitive to predict AD [96]. Conditions associated with severe and rapid neuronal loss such as stroke, traumatic brain injuries, or Creutzfeldt–Jakob disease may also cause very high CSF tau levels (>1000 pg/ml), which are higher than what is typically seen with AD, with close to normal phospho-tau [8].

Carriers of *APOE*-ε4 alleles have an earlier onset of decreased CSF amyloid-β 42 [93]. Such alleles contribute mostly to impaired amyloid-β clearance [97], but also to the dissemination of neurofibrillary tangles, microglial dysregulation, and neuro-inflammation [98]. An important implication of these pathophysiological features is that patient selection for clinical trials based on lower CSF amyloid-β 42 levels may bias the samples towards a higher proportion of *APOE*-ε4 carriers. Nevertheless, lower CSF amyloid-β 42 results from increased cortical amyloid-β independently of *APOE*-ε4 carrier status [99], also leading to faster cognitive and functional decline [100]. It has also been verified that lower CSF amyloid-β 42 implies a higher risk for ARIAs in *APOE*-ε4 non-carriers [101].

In dominantly inherited AD, it has been shown that CSF amyloid-β 42 may decline up to 25 years before the onset of cognitive symptoms reflecting its sensitivity to detect aberrant amyloid-β oligomerization, and even before abnormalities in amyloid PET, while CSF tau may increase as early as 15 years before symptom onset [102]. In Down syndrome, it has been demonstrated that an abnormal CSF amyloid-β 42/amyloid-β 40 ratio may be found as early as the third decade of life, while CSF phospho-tau Thr_{181} and Neurofilament Light (NfL) increase in the fourth decade of life [103]. Nevertheless, both children with pathogenic presenilin-1 (*PSEN1*) mutations and those with Down syndrome may present with higher CSF amyloid-β 42 due to amyloid-β synthesis overcoming clearance at a much higher rate at that age [102, 103].

NfL is a non-specific biomarker that corresponds to large-caliber axonal damage associated with aging or with faster disease progression and higher rates of brain atrophy in most neurodegenerative diseases [25], potentially affected by the *APOE*-ε4 carrier status [104]. Similarly to glial fibrillary acidic protein (GFAP, a marker of astrocyte activity), it could be used for staging and monitoring neurodegenerative diseases, prognostic prediction, and assessing potential response to future disease-modifying therapies [87]. Metabolome profiling of the CSF of patients with AD has revealed several elevated amino acids which are functionally important for energy metabolism and neurotransmission [105], though their clinical use is compromised by low specificity to the disease and divergent results from different studies. Likewise, CSF monoamine metabolites are typically increased in temporal lobe epilepsy, but may be useful only for pathophysiological studies rather than for etiological diagnosis [106].

Synaptic degeneration is a downstream result of amyloid-β oligomer toxicity, tauopathy, and inflammation; specific CSF biomarkers of synaptic degeneration such as neurogranin, synaptosomal-associated protein-25 (SNAP-25), and synaptotagmin-1 may be useful for disease staging and prediction of cognitive decline and brain atrophy, but still require etiological confirmation of AD pathology by amyloid-β and tau biomarkers [107].

CSF amyloid-β peptide profiles are different in dementia with Lewy bodies and AD, particularly considering amyloid-β 42 and the amyloid-β 42/amyloid-β 38 and α-synuclein/amyloid-β 42 ratios [95], and mostly for *APOE*-ε4 non-carriers [104], representing divergent patterns of aberrant amyloid-β metabolism specific to each disease. Nevertheless, insufficient specificity due to high levels of amyloidosis particularly in older patients with dementia with Lewy bodies [95] prevent these biomarkers from being used to differentiate dementia syndromes in clinical trials. The same issue hampers the use of CSF biomarkers for outcome assessments in normal pressure hydrocephalus, given the concomitant AD pathology presented by many patients [86]. It has also been shown in patients with incident PD that CSF amyloid-β 42, amyloid-β 38, and the amyloid-β 42/amyloid-β 40 and amyloid-β 38/amyloid-β 40 ratios are usually reduced only in those with the postural instability and gait disorder phenotype, which bears a close relationship with the subsequent development of dementia [108].

The genetic, phenotypic, and pathologic heterogeneity of FTLD syndromes challenges the use of CSF biomarkers in clinical trials. CSF-soluble amyloid precursor protein fragment (sAPPβ) is usually lower in frontotemporal dementia, while tau-associated FTLD syndromes do not have elevated CSF tau and phospho-tau isoforms, and CSF progranulin is usually lower only in *GRN* mutation carriers; however, insufficient specificity prevents the wide dissemination of these biomarkers for translational research [87].

Concerning vascular cognitive impairment and dementia, CSF-soluble platelet-derived growth factor receptor-β (sPDGFRβ, a marker of brain capillary and blood–brain barrier damage) and placental growth factor seem promising, though validation of such biomarkers has been hampered by issues with standardization of pre-analytical and analytical processes, harmonization, and reliability against clinically relevant outcomes [109]. Likewise, absent specificity to disease processes and the subtlety of changes thwart the use of metabolomics in clinical trials [91]. Nevertheless, blood pressure variability has been associated with evidence of tauopathy, implying the need for selection of anti-hypertensive therapies with neurobiological evidence of benefits [110].

Some studies have already demonstrated associations of CSF biomarkers with behavioral features [111, 112], and cognitive or functional tests [95]. Such biomarkers may also predict future cognitive and functional decline [2]. These are important aspects to be considered in clinical trials.

Definitions of the timing and magnitude of CSF biomarker changes have been validated mostly in cross-sectional studies; however, longitudinal follow-up is essential for clinical trials. Therefore, proper validation for longitudinal preventive and therapeutic studies is required for each biomarker before it can be reliably implemented in clinical trials.

3.2 Plasma Biomarkers

The development of plasma biomarkers for neurodegenerative diseases has the potential to revolutionize the diagnosis of these disorders. In the context of AD, plasma biomarkers of amyloid-β and tau demonstrate high correlation with both neuropathology [113, 114] and in vivo AD pathology through PET [115, 116]. While amyloid and tau biomarkers using PET and CSF effectively predict cognitive decline in individuals with normal cognition and MCI, the accessibility and cost-effectiveness of plasma biomarkers make them highly convenient for widespread

Table 2 Plasma biomarkers of Alzheimer disease frequently used in clinical trials

Plasma biomarker	Clinical trial stage	Clinical trial	Findings
Phospho-tau Thr_{181}	Pre-screening	TRAILBLAZER-ALZ 2 (donanemab)	Higher percentage of amyloid-PET and tau-PET positivity (63% versus 37%)
Phospho-tau Thr_{217}	Sole enrollment criterion	TRAILBLAZER-ALZ 3 (donanemab)	Ongoing
Phospho-tau Thr_{217}	Secondary outcome	TRAILBLAZER-ALZ (donanemab)	Decrease in the treated group (24% decrease versus 6% increase)
Aβ42/Aβ40 ratio	Secondary outcome	Clarity AD (lecanemab)	Increased levels compared with controls
Phospho-tau Thr_{181}	Secondary outcome	Clarity AD (lecanemab)	Reduced levels compared with controls
GFAP	Secondary outcome	Clarity AD (lecanemab)	Reduced levels compared with controls
NfL	Secondary outcome	Clarity AD (lecanemab)	No difference between groups
Phospho-tau Thr_{181}	Secondary outcome	EMERGE (aducanumab*)	13% decrease from baseline in the treated group
Phospho-tau Thr_{181}	Secondary outcome	ENGAGE (aducanumab*)	16% decrease from baseline in the treated group

*off-market drug; Abbrevations: *GFAP* glial fibrillary acidic protein, *NfL* neurofilament light

use in both clinical care and clinical trials [117]. Table 2 lists the plasma AD biomarkers that have been most frequently used in clinical trials.

The discussion of plasma biomarkers will focus primarily on their potential use in clinical trials of neurodegenerative diseases, specifically AD, examining two main aspects: patient selection and outcome assessments.

3.2.1 Plasma Biomarkers for Patient Selection

As plasma biomarkers in AD, particularly phospho-tau Thr_{181} and phospho-tau Thr_{217}, demonstrate strong correlations with, and predictive ability for, amyloid and tau PET [116, 118], their utilization in patient selection for clinical trials can prove beneficial, especially due to cost and optimization of time for participant recruitment. Moreover, considering that these blood-based AD biomarkers begin to rise at the early stages of the disease continuum, their inclusion in preventive clinical trials involving cognitively healthy individuals is highly appealing. Since about 25% of clinically diagnosed AD patients do not have amyloid pathology [25], pre-screening with plasma biomarkers is likely to be a cost-saving strategy. Additionally, in clinical trials focusing on lifestyle interventions, plasma biomarkers may potentially suffice to include individuals with AD pathology, eliminating the need for CSF and PET examinations [117]. A simulation of a theoretical clinical trial employing plasma amyloid-β 42/amyloid-β 40 and *APOE*-ε4 carrier status as pre-screening measures in cognitively unimpaired participants demonstrated a potential reduction of 54% in the number of amyloid PET scans and a 43% decrease in the total recruitment cost [119]. A recent study conducted in the USA showed that the implementation of an AD blood test prior to amyloid PET could reduce the enrollment time by 49% (from 2.8 to 1.4 years) and decrease costs by 25% (from US$2.9 million to US$2.2 million) [120]. Another study, using data from ADNI, estimated that the use of phospho-tau Thr_{181} as a pre-screening strategy would result in savings in both total cost and screening phase time for a potential trial involving 1000 amyloid-β positive asymptomatic individuals [116]. Yet another study estimated the potential reduction in the number of required amyloid PET scans or assessments of CSF phospho-tau Thr_{181} and amyloid-β 42 in a trial using longitudinal tau PET as the outcome measure. The authors calculated a sample reduction of 50% if cognitively unimpaired individuals were recruited, and of 20% if cognitively impaired participants were recruited [121]. The same group analyzed the performance of longitudinal phospho-tau Thr_{181} and NfL in assessing neurodegeneration and clinical progression in cognitively unimpaired and cognitively impaired individuals. While both markers proved to be reliable indicators for predicting and monitoring neurodegeneration and cognitive decline, NfL was found to lack specificity for AD pathology [122]. However, at present, plasma biomarkers are only recommended in clinical trials, provided the AD status is confirmed with PET or CSF testing [123].

Some clinical trials have already incorporated plasma biomarkers as a recruitment strategy. For instance, the TRAILBLAZER-ALZ 2 (Eli Lilly; NCT04437511) donanemab trial for early AD aimed to assess the potential of pre-screening with plasma phospho-tau Thr_{181} to reduce the need for PET imaging. The group with elevated plasma phospho-tau Thr_{181} exhibited a higher percentage of positivity for amyloid PET and tau PET (63% versus 37%) compared to the group with normal values [124]. The same company is currently conducting the groundbreaking study, TRAILBLAZER-ALZ 3 donanemab trial in a large sample of asymptomatic older adults (NCT05026866), using plasma phospho-tau Thr_{217} as the sole enrollment criterion. In this trial, the definitive enrollment decision will be based solely on the result of the plasma biomarker. The outcomes of this trial could potentially pave the way for the use of plasma biomarkers as a recruitment tool for preventive trials.

While the use of plasma biomarkers holds considerable potential for recruitment of participants for clinical trials, there are certain challenges that need to be acknowledged and addressed for their broader application in this context. A study highlighted the contrasting positive predictive value and negative predictive value of plasma phospho-tau Thr_{217} for AD pathology (amyloid PET and tau PET) between cognitively unimpaired and cognitively impaired individuals [125]. For cognitively impaired individuals, the positive predictive value was high (0.94 and 0.93) while the negative predictive value was low (0.66 and 0.65); conversely, for cognitively unimpaired individuals, the positive predictive value was quite low (<0.50) while the negative predictive value was high (>0.94). The authors suggested that phospho-tau Thr_{217} could serve as a pre-screening biomarker for cognitively impaired individuals and a rule-out biomarker for those with normal cognition. This result addresses the challenge of relying solely on plasma biomarkers as a stand-alone screening tool in clinical trials, as confirmation of a positive result with CSF or PET AD biomarkers becomes necessary. Even when plasma biomarkers are limited to pre-screening, the presence of false negative results in cognitively unimpaired individuals may lead to the need to screen a larger number of individuals, as the tests may not be sensitive enough to detect changes at this stage. This caveat could impose limitations for certain studies, not only in terms of costs but also in terms of the difficulty in including patients from underrepresented minority groups [120].

3.2.2 Plasma Biomarkers for Outcome Assessments

In addition to aiding in the recruitment of individuals for clinical trials, plasma biomarkers can also be used as a way to assess target engagement and as surrogate outcomes. In the context of AD, tau-related biomarkers, particularly phospho-tau Thr_{181} and phospho-tau Thr_{217}, seem to predict longitudinal tau PET and cognitive changes, thus offering promising options as surrogate biomarkers in clinical trials [121, 126]. Additionally, the longitudinal stability of plasma phospho-tau Thr_{181} and high intra-individual increases of phospho-tau Thr_{217} could be used to assess therapeutic interventions, including the observation of reductions in plasma levels or a slower increase of these biomarkers [116, 127]. The use of plasma biomarkers provides a less invasive and cost-effective alternative for monitoring drug activity in a clinical trial, both in early development stages and in phase III studies. Furthermore, plasma biomarkers hold potential to inform decisions in clinical trials with adaptive designs [123].

Some recent clinical trials targeting amyloid-β demonstrated the reduction of plasma phospho-tau levels with treatment. The TRAILBLAZER-ALZ study (NCT03367403) evaluated donanemab in a phase II study in patients in the early symptomatic stages of AD. Results showed that over a period of 76 weeks, the concentration of phospho-tau Thr_{217} decreased by 24% in the treated group, while increasing by 6% in the control group. Moreover, the change in phospho-tau Thr_{217} correlated with changes in amyloid plaques and neurofibrillary tangles, as measured by PET [128]. In AD, the number of neurofibrillary tangles correlates better with the stage of dementia than the number of amyloid plaques, while tau accumulation is invariably wild-type and not the result of inherited mutations [129]. Another phase III study with lecanemab (Clarity AD) in early symptomatic AD showed an improvement in several plasma biomarkers in the treatment group, including the amyloid-β 42/amyloid-β 40 ratio, phospho-tau Thr_{181}, and GFAP. However, NfL levels did not decrease with treatment, likely due to its lacking specificity to amyloid pathology [130]. Two previous studies evaluating the effectiveness of aducanumab in early AD (EMERGE and ENGAGE) also showed a reduction in plasma phospho-tau Thr_{181} levels in the treatment group compared with placebo [63], though this drug has since been removed from the market due to lack of proven clinical benefits.

The use of plasma biomarkers as primary outcome measures has not yet been sufficiently studied but shows potential for the future. However, as highlighted in a recent study, careful evaluation of cost estimates is necessary. In this study, using

ADNI data, the authors estimated the reduction of sample size in a clinical trial involving cognitively unimpaired individuals using longitudinal plasma phospho-tau Thr_{181} and NfL as surrogate markers. The findings revealed that a theoretical 24-month trial would require an 85% smaller sample size for phospho-tau Thr_{181} and a 63% smaller sample size for NfL compared to a 12-month trial. Furthermore, the authors demonstrated that population enrichment with individuals who are either plasma amyloid-β positive or who have intermediate levels of amyloid-β in PET scans (Centiloid 20–40) would lead to a 43% and 73% reduction in sample size for phospho-tau Thr_{181} and 16% and 59% for NfL, respectively. However, surprisingly, due to the larger longitudinal variability of plasma phospho-tau Thr_{181} and NfL, a higher number of individuals would be required in this hypothetical trial, resulting in higher costs compared to neuroimaging as an outcome measure [131]. These findings emphasize the performance of plasma biomarkers when estimating costs.

Other potential plasma biomarkers are currently under evaluation for different neurodegenerative diseases. Neurogranin has emerged as a promising post-synaptic biomarker given its increase in early stages of AD, while soluble triggering receptor expressed on myeloid cells 2 (sTREM2) appears to have a bimodal course along the disease continuum [91]. Astrocytic reactivity plays an important role in the pathophysiology of AD and, although most biomarkers in this regard have been developed for CSF, plasma GFAP shows promise in predicting future cognitive decline in MCI and cognitively unimpaired individuals [132, 133].

3.3 Biomarkers of Atypical Biofluids

Typically, the pathophysiological mechanism of a neurodegenerative disease involves a protein, or its precursor, and will result in an aberrant molecule, with unstable structure on an overproduction scale [134]. The question to be answered lies precisely in detecting what these biomarkers are and in which biofluid they can be found. The widely used biofluids are blood and CSF; however, there are others, such as saliva and sebum, which have been extensively researched in the last decade.

A metabolomic study analyzed, qualitatively and quantitatively, the mixture of metabolites present in sebum, a biofluid rich in lipids, and through Liquid Chromatography Mass Spectrometry (LCMS) sought to relate these substances to the prognosis and diagnosis of PD [135]. The use of LCMS facilitates the detection and concentration analysis of a huge variety of molecular species, such as those present in sebum [136, 137] and in saliva, another biofluid much investigated in the last decade [138, 139]. Analytical techniques based on LCMS are routinely employed to characterize the metabolome. A recent study estimated that the amount of different small molecules found in the human body exceeds 19,000. This estimate considered derivatives from food, medication, the microbiome that inhabits the body and the environment, in addition to metabolites directly linked to endogenous enzymatic activities encoded by the human genome [140].

Table 3 Salivary biomarkers in Alzheimer disease

Biomarker	Function	Biomarker type
Aβ40 (amyloid-beta)	Marker of the amyloid cascade	Predictive biomarker
Aβ42 (amyloid-beta)	Marker of the amyloid cascade	Predictive biomarker
t-tau (total tau protein)	Marker of axonal damage	Predictive biomarker
Phospho-tau (phosphorylated tau protein)	Marker of axonal damage	Predictive biomarker
α1-antitrypsin	Immunological marker, modulates microglial-mediated neuroinflammation and protects microglial cells from the toxic effects of amyloid-β	Prognostic biomarker
ADA (adenosine deaminase)	Immunological marker, neuromodulator involved in aging and neurodegenerative diseases	Prognostic biomarker
CC4 (complement C4)	Immunological marker, expression of mRNA for C4 can be up to three times higher in the brains of patients with Alzheimer disease	Prognostic biomarker
ChE (cholinesterase)	Synaptic marker, response to use of cholinesterase inhibitors may indicate therapeutic efficacy	Prognostic biomarker
CRP (C-reactive protein)	Marker of inflammation, directly associated with increased risk of Alzheimer disease in midlife, and inversely associated with such risk in late life	Prognostic biomarker
FRAP (ferric reducing ability of plasma)	Marker of oxidative stress, usually lower in Alzheimer disease compared to healthy adults	Prognostic biomarker
HP (haptoglobin)	Marker of inflammation	Prognostic biomarker
MIP-4 (macrophage inflammatory protein-4)	Marker of inflammation	Prognostic biomarker
PEDF (pigment epithelium-derived factor)	Marker of axonal damage, can be reduced in older people and even more in patients with Alzheimer disease	Prognostic biomarker

Adapted from Tvarijonaviciute et al. [146]

In AD, saliva has been widely studied as a possible fluid where biomarkers related to the evolution of this type of dementia can be found [141–146]. Table 3 lists the main salivary biomarkers of AD.

The use of atypical biofluids, such as sebum and saliva, brings benefits such as non-invasiveness, low cost, easy collection, in addition to being tests capable of identifying diseases such as PD and AD at an early stage.

A case-control study was performed on 60 participants, 30 patients diagnosed with probable AD and 30 healthy participants without AD. Proteins were quantified in saliva by way of enzyme-linked immunosorbent assays (ELISAs). The results pointed to a higher concentration of amyloid-β 42 in the saliva of the group of patients with AD compared to the group of participants without AD. In this same study, the concentrations of total tau in the saliva of groups of participants with and without AD were compared. The result was 134.4 pg/mL in the group without AD, while in the group with AD the total tau concentration was 86.7 pg/mL [139, 144], confirming the hypothesis that the concentration of total tau in the saliva of patients with AD is lower when compared to the control group [145, 147].

4 Genetics

Variants of the moderately penetrant *APOE* gene explain a substantial part of the heritability of late-onset AD, but are not essential for development of the disease [31], possibly due to variable expressivity caused by other genetic factors, such as the *TOMM40* poly-T polymorphisms, which are linked to *APOE* [148]. In fact, it has been shown that genetic heterogeneity causes opposite effects of *APOE*-ε4 alleles in early-onset versus late-onset AD [149]. Nevertheless, the *APOE*-ε4 allele has been implicated in conferring a greater risk for [109], and an earlier onset of, dementia [150], as well as behavioral symptom modulation [151], differences in caregiver burden [152], decreased participation in physical activities [153], pleiotropic effects, and interactions with other genes [109] that may affect risk and clinical manifestations of late-onset AD.

Variants of single genes have been reported to underlie broad phenotypes of neurodegenerative diseases. The *APOE*-ε4 allele also interferes with the incidence and clinical manifestations of Lewy body dementias [154], vascular dementia [109] and even some forms of FTLD [28]. On the other hand, the *APOE*-ε2 allele is a risk factor for hemorrhagic cerebral amyloid angiopathy [101].

Most longitudinal studies show no direct effects of *APOE* alleles on cognitive or functional decline in AD [155]; however, it appears that such effects may happen in a non-linear pattern [156]. Anyway, stratification of samples according to *APOE*-ε4 carrier status has yielded significantly different results concerning protective effects of education and lifetime sanitary conditions [157], and effects of the coronary heart disease risk [158], blood pressure [159], and lipid profile variations [160] on cognitive and functional changes in patients with AD. In addition, *APOE*-ε4 carrier status predisposes participants of trials with anti-amyloid immunotherapies to ARIAs [109], while *APOE*-ε4 non-carriers are more susceptible to the harmful prospective effects of second-generation anti-psychotics and anti-epileptic drugs on cognitive and functional changes in AD [161]. There is enough evidence to sustain the use of the *APOE*-ε4 carrier status as a covariate in all clinical trials involving patients with late-onset AD [98].

Other genes have shown importance for modulation of behavioral features [162] and for longitudinal pharmacogenetic effects in AD, particularly those associated with blood pressure modulation such as the *ACE* gene [163], and those involved with cholesterol metabolism such as *LDLR* [164], *HMGCR* [165], *CETP* and *NR1H2* [166], either in isolation or through stratification according to *APOE*-ε4 carrier status. Genes other than *APOE* could also be important for patient selection and outcome assessments in clinical trials.

Concerning dominantly inherited AD, variants of *APP*, *PSEN1*, and *PSEN2* lead to AD with complete penetrance by interfering with processing of amyloid-β [102], and should be included along with the *APOE*-ε4 carrier status in trials involving large numbers of patients with early-onset AD. This is also relevant for trials with Down syndrome, in which the trisomy of chromosome 21 leads to a triplication of the *APP* gene and lifetime risk of AD surpassing 90% [103].

Though it might be difficult to fully embrace the complex genetics of FTLD in clinical trials, highly penetrant mutations of *GRN* and *MAPT*, as well as the number of GGGGCC expansions in *C9orf72*, are useful to predict the underlying neuropathology despite poor genotype–phenotype correlations [87]. In Lewy body dementias, variants of *SNCA*, *SNCB*, and *SNCG* from the synuclein gene family have been associated with unusual risk [167], while mutations in the *GBA* gene increase the risk for PD and lead to earlier onset and more cognitive changes [168]; such variants could also be assessed in addition to *APOE* alleles in clinical trials.

5 Conclusion

The need to homogenize study participants in clinical and translational research studies, in order to employ focused interventions at different disease stages, has enhanced interest in the identification and validation of biomarkers for patient selection and outcome assessments. In translational research, biomarkers should always be integrated with cognitive, functional, and behavioral variables, and measure not only disease-modifying effects but also symptomatic benefits.

Advancements in highly sensitive quantitative technologies have led to the promising translation of CSF biomarkers to blood. However, blood-based biomarkers are still recommended in clinical trials only when the disease status is confirmed by other objective methods, such as PET or CSF testing in case of patients with AD [123].

Combinations of biomarkers of different modalities are still recommended to increase statistical power to detect differences in decline between the treatment and placebo groups in clinical and translational research. Such strategy might enrich clinical trials with participants likely to have more rapid disease progression, potentially shortening the follow-up time required to detect disease-modifying effects, or alternatively to help warrant that randomized groups are more likely to be balanced concerning rates of disease progression. In conclusion, prognostic panels of combined biomarkers are the best strategy to be employed for patient selection and outcome assessments at present, though cost and access issues may hamper the wide availability of such biomarkers in lower income settings.

Funding Fabricio Ferreira de Oliveira is supported by FAPESP - The State of São Paulo Research Foundation, grant #2015/10109-5. Raphael Machado Castilhos is supported by the Alzheimer's Association, grant #AARGD-21-846545.

References

1. Rosen AC, Arias JJ, Ashford JW, Blacker D, Chhatwal JP, Chine NA, Clark L, Denny SS, Goldman JS, Gleason CE, Grill JD, Heidebrink JL, Henderson VW, Lavacot JA, Lingler JH, Menon M, Nosheny RL, Oliveira FF, Parker MW, Rahman-Filipiak A, Revoori A, Rumbaugh MC, Sanchez DL, Schindler SE, Schwarz CG, Toy L, Tyrone J, Walter S, Wang L, Wijsman EM, Zallen DT, Aggarwal NT. The advisory group on risk evidence education for dementia: multidisciplinary and open to all. J Alzheimers Dis. 2022;90:953–62. https://doi.org/10.3233/JAD-220458.
2. Boccardi M, Monsch A, Ferrari C, Altomare D, Berres M, Bos I, Buchmann A, Cerami C, Didic M, Festari C, Nicolosi V, Sacco L, Aerts L, Albanese E, Annoni JM, Ballhausen N, Chicherio C, Démonet JF, Descloux V, Diener S, Ferreira D, Georges J, Gietl A, Girtler N, Kilimann I, Klöppel S, Kustyniuk N, Mecocci P, Mella N, Pigliautile M, Seeher K, Shirk S, Toraldo A, Brioschi-Guevara A, Chan K, Crane P, Dodich A, Grazia A, Kochan N, Oliveira FF, Nobili F, Kukull W, Peters O, Ramakers I, Sachdev P, Teipel S, Visser PJ, Wagner M, Weintraub S, Westman E, Froelich L, Brodaty H, Dubois B, Cappa S, Salmon D, Winblad B, Frisoni G, Kliegel M. Harmonizing neuropsychological assessment for mild neurocognitive disorders in Europe. Alzheimers Dement. 2022;18:29–42. https://doi.org/10.1002/alz.12365.
3. Raghunathan R, Turajane K, Wong LC. Biomarkers in neurodegenerative diseases: proteomics spotlight on ALS and Parkinson's disease. Int J Mol Sci. 2022;23:9299. https://doi.org/10.3390/ijms23169299.
4. Karran E, Mercken M, De Strooper B. The amyloid cascade hypothesis for Alzheimer's disease: an appraisal for the development of therapeutics. Nat Rev Drug Discov. 2011;10:698–712. https://doi.org/10.1038/nrd3505.
5. Rabbito A, Dulewicz M, Kulczyńska-Przybik A, Mroczko B. Biochemical markers in Alzheimer's disease. Int J Mol Sci. 2020;21:1989. https://doi.org/10.3390/ijms21061989.
6. Silverberg N, Elliott C, Ryan L, Masliah E, Hodes R. NIA commentary on the NIA-AA research framework: towards a biological definition of Alzheimer's disease. Alzheimers Dement. 2018;14:576–8. https://doi.org/10.1016/j.jalz.2018.03.004.
7. Miller DB, O'Callaghan JP. Biomarkers of Parkinson's disease: present and future. Metabolism. 2015;64:S40–6. https://doi.org/10.1016/j.metabol.2014.10.030.
8. Molinuevo JL, Ayton S, Batrla R, Bednar MM, Bittner T, Cummings J, Fagan AM, Hampel H, Mielke MM, Mikulskis A, O'Bryant S, Scheltens P, Sevigny J, Shaw LM, Soares HD, Tong G, Trojanowski JQ, Zetterberg H, Blennow K. Current state of Alzheimer's fluid biomarkers. Acta Neuropathol. 2018;136:821–53. https://doi.org/10.1007/s00401-018-1932-x.
9. Salloway S, Sperling R, Fox NC, Blennow K, Klunk W, Raskind M, Sabbagh M, Honig LS, Porsteinsson AP, Ferris S, Reichert M, Ketter N, Nejadnik B, Guenzler V, Miloslavsky M, Wang D, Lu Y, Lull J, Tudor IC, Liu E, Grundman M, Yuen E, Black R, Brashear HR. Two phase 3 trials of bapineuzumab in mild-to-moderate Alzheimer's disease. N Engl J Med. 2014;370:322–33. https://doi.org/10.1056/NEJMoa1304839.
10. Sevigny J, Suhy J, Chiao P, Chen T, Klein G, Purcell D, Oh J, Verma A, Sampat M, Barakos J. Amyloid PET screening for enrichment of early-stage Alzheimer disease clinical trials: experience in a phase 1b clinical trial. Alzheimer Dis Assoc Disord. 2016;30:1–7. https://doi.org/10.1097/WAD.0000000000000144.
11. Masters CL, Bateman R, Blennow K, Rowe CC, Sperling RA, Cummings JL. Alzheimer's disease. Nat Rev Dis Primers. 2015;1:15056. https://doi.org/10.1038/nrdp.2015.56.
12. Power MC, Mormino E, Soldan A, James BD, Yu L, Armstrong NM, Bangen KJ, Delano-Wood L, Lamar M, Lim YY, Nudelman K, Zahodne L, Gross AL, Mungas D, Widaman KF, Schneider J. Combined neuropathological pathways account for age-related risk of dementia. Ann Neurol. 2018;84:10–22. https://doi.org/10.1002/ana.25246.
13. Bateman RJ, Klunk WE. Measuring target effect of proposed disease-modifying therapies in Alzheimer's disease. Neurotherapeutics. 2008;5:381–90. https://doi.org/10.1016/j.nurt.2008.05.009.

14. Amur S, LaVange L, Zineh I, Buckman-Garner S, Woodcock J. Biomarker qualification: toward a multiple stakeholder framework for biomarker development, regulatory acceptance, and utilization. Clin Pharmacol Ther. 2015;98:34–46. https://doi.org/10.1002/cpt.136.
15. Cullen N, Janelidze S, Palmqvist S, Stomrud E, Mattsson-Carlgren N, Hansson O. Association of CSF Aβ_{38} levels with risk of Alzheimer disease-related decline. Neurology. 2022;98:e958–67. https://doi.org/10.1212/WNL.0000000000013228.
16. Hansson O, Seibyl J, Stomrud E, Zetterberg H, Trojanowski JQ, Bittner T, Lifke V, Corradini V, Eichenlaub U, Batrla R, Buck K, Zink K, Rabe C, Blennow K, Shaw LM. CSF biomarkers of Alzheimer's disease concord with amyloid-β PET and predict clinical progression: a study of fully automated immunoassays in BioFINDER and ADNI cohorts. Alzheimers Dement. 2018;14:1470–81. https://doi.org/10.1016/j.jalz.2018.01.010.
17. Knopman D, Haeberlein S, Carrillo M, Hendrix J, Kerchner G, Margolin R, Maruff P, Miller DS, Tong G, Tome MB, Murray ME, Nelson PT, Sano M, Mattson N, Sultzer DL, Montine TJ, Jack CR Jr, Kolb H, Petersen RC, Vemuri P, Canniere MZ, Schneider JA, Resnick SM, Romano G, van Harten AC, Wolk DA, Bain LJ, Siemers E. The National Institute on Aging and the Alzheimer's Association Research Framework for Alzheimer's disease: perspectives from the research roundtable. Alzheimers Dement. 2018;14:563–75. https://doi.org/10.1016/j.jalz.2018.03.002.
18. Weiner MW, Veitch DP, Miller MJ, Aisen PS, Albala B, Beckett LA, Green RC, Harvey D, Jack CR Jr, Jagust W, Landau SM, Morris JC, Nosheny R, Okonkwo OC, Perrin RJ, Petersen RC, Rivera-Mindt M, Saykin AJ, Shaw LM, Toga AW, Tosun D, Trojanowski JQ. Increasing participant diversity in AD research: plans for digital screening, blood testing, and a community-engaged approach in the Alzheimer's Disease Neuroimaging Initiative 4. Alzheimers Dement. 2023;19:307–17. https://doi.org/10.1002/alz.12797.
19. Zetterberg H, Blennow K. Moving fluid biomarkers for Alzheimer's disease from research tools to routine clinical diagnostics. Mol Neurodegener. 2021;16:10. https://doi.org/10.1186/s13024-021-00430-x.
20. Kandiah N, Choi SH, Hu CJ, Ishii K, Kasuga K, Mok VCT. Current and future trends in biomarkers for the early detection of Alzheimer's disease in Asia: expert opinion. J Alzheimers Dis Rep. 2022;6:699–710. https://doi.org/10.3233/ADR-220059.
21. Akinyemi RO, Yaria J, Ojagbemi A, Guerchet M, Okubadejo N, Njamnshi AK, Sarfo FS, Akpalu A, Ogbole G, Ayantayo T, Adokonou T, Paddick SM, Ndetei D, Bosche J, Ayele B, Damas A, Coker M, Mbakile-Mahlanza L, Ranchod K, Bobrow K, Anazodo U, Damasceno A, Seshadri S, Pericak-Vance M, Lawlor B, Miller BL, Owolabi M, Baiyewu O, Walker R, Gureje O, Kalaria RN, Ogunniyi A. Dementia in Africa: current evidence, knowledge gaps, and future directions. Alzheimers Dement. 2022;18:790–809. https://doi.org/10.1002/alz.12432.
22. Biomarkers Definitions Working Group. Biomarkers and surrogate endpoints: preferred definitions and conceptual framework. Clinical Pharmacol Ther. 2001;69:89–95. https://doi.org/10.1067/mcp.2001.113989.
23. Jeromin A, Bowser R. Biomarkers in neurodegenerative diseases. Adv Neurobiol. 2017;15:491–528. https://doi.org/10.1007/978-3-319-57193-5_20.
24. Simrén J, Ashton NJ, Blennow K, Zetterberg H. An update on fluid biomarkers for neurodegenerative diseases: recent success and challenges ahead. Curr Opin Neurobiol. 2020;61:29–39. https://doi.org/10.1016/j.conb.2019.11.019.
25. Hansson O. Biomarkers for neurodegenerative diseases. Nat Med. 2021;27:954–63. https://doi.org/10.1038/s41591-021-01382-x.
26. Rizzo G, Copetti M, Arcuti S, Martino D, Fontana A, Logroscino G. Accuracy of clinical diagnosis of Parkinson disease: a systematic review and meta-analysis. Neurology. 2016;86:566–76. https://doi.org/10.1212/WNL.0000000000002350.
27. Guzman-Martinez L, Maccioni RB, Farías GA, Fuentes P, Navarrete LP. Biomarkers for Alzheimer's disease. Curr Alzheimer Res. 2019;16:518–28. https://doi.org/10.2174/1567205016666190517121140.

28. Yang HS, Yu L, White CC, Chibnik LB, Chhatwal JP, Sperling RA, Bennett DA, Schneider JA, De Jager PL. Evaluation of TDP-43 proteinopathy and hippocampal sclerosis in relation to *APOE* ε4 haplotype status: a community-based cohort study. Lancet Neurol. 2018;17:773–81. https://doi.org/10.1016/S1474-4422(18)30251-5.
29. Counts SE, Ikonomovic MD, Mercado N, Vega IE, Mufson EJ. Biomarkers for the early detection and progression of Alzheimer's disease. Neurotherapeutics. 2017;14:35–53. https://doi.org/10.1007/s13311-016-0481-z.
30. Suzuki K, Iwata A, Iwatsubo T. The past, present, and future of disease-modifying therapies for Alzheimer's disease. Proc Jpn Acad Ser B Phys Biol Sci. 2017;93:757–71. https://doi.org/10.2183/pjab.93.048.
31. Sierksma A, Escott-Price V, De Strooper B. Translating genetic risk of Alzheimer's disease into mechanistic insight and drug targets. Science. 2020;370:61–6. https://doi.org/10.1126/science.abb8575.
32. Cash DM, Rohrer JD, Ryan NS, Ourselin S, Fox NC. Imaging endpoints for clinical trials in Alzheimer's disease. Alzheimers Res Ther. 2014;6:87. https://doi.org/10.1186/s13195-014-0087-9.
33. Ten Kate M, Ingala S, Schwarz AJ, Fox NC, Chételat G, Van Berckel BNM, Ewers M, Foley C, Gispert JD, Hill D, Irizarry MC, Lammertsma AA, Molinuevo JL, Ritchie C, Scheltens P, Schmidt ME, Visser PJ, Waldman A, Wardlaw J, Haller S, Barkhof F. Secondary prevention of Alzheimer's dementia: neuroimaging contributions. Alzheimers Res Ther. 2018;10:112. https://doi.org/10.1186/s13195-018-0438-z.
34. Rafii MS. Appropriate use recommendations for lecanemab. J Prev Alzheimers Dis. 2023;10. https://doi.org/10.14283/jpad.2023.34.
35. Cummings J. The role of biomarkers in Alzheimer's disease drug development. Adv Exp Med Biol. 2019;1118 https://doi.org/10.1007/978-3-030-05542-4_2.
36. Roberts RO, Aakre JA, Kremers WK, Vassilaki M, Knopman DS, Mielke MM, Alhurani R, Geda YE, Machulda MM, Coloma P, Schauble B, Lowe VJ, Jack CR, Petersen RC. Prevalence and outcomes of amyloid positivity among persons without dementia in a longitudinal, population-based setting. JAMA Neurol. 2018;75:970–9. https://doi.org/10.1001/jamaneurol.2018.0629.
37. Hill DLG, Schwarz AJ, Isaac M, Pani L, Vamvakas S, Hemmings R, Carrillo MC, Yu P, Sun J, Beckett L, Boccardi M, Brewer J, Brumfield M, Cantillon M, Cole PE, Fox N, Frisoni GB, Jack C, Kelleher T, Luo F, Novak G, Maguire P, Meibach R, Patterson P, Bain L, Sampaio C, Raunig D, Soares H, Suhy J, Wang H, Wolz R, Stephenson D. Coalition Against Major Diseases/European Medicines Agency biomarker qualification of hippocampal volume for enrichment of clinical trials in predementia stages of Alzheimer's disease. Alzheimers Dement. 2014;10:421–9. https://doi.org/10.1016/j.jalz.2013.07.003.
38. Cummings J, Zhong K, Cordes D. Drug development in Alzheimer's disease – the role of default mode network assessment in phase II. US Neurol. 2017;13:67. https://doi.org/10.17925/usn.2017.13.02.67.
39. Canter RG, Penney J, Tsai LH. The road to restoring neural circuits for the treatment of Alzheimer's disease. Nature. 2016;539:187–96. https://doi.org/10.1038/nature20412.
40. Sorg C, Riedl V, Mühlau M, Calhoun VD, Eichele T, Läer L, Drzezga A, Förstl H, Kurz A, Zimmer C, Wohlschläger AM. Selective changes of resting-state networks in individuals at risk for Alzheimer's disease. Proc Natl Acad Sci USA. 2007;104:18760–5. https://doi.org/10.1073/pnas.0708803104.
41. Zhou J, Greicius MD, Gennatas ED, Growdon ME, Jang JY, Rabinovici GD, Kramer JH, Weiner M, Miller BL, Seeley WW. Divergent network connectivity changes in behavioural variant frontotemporal dementia and Alzheimer's disease. Brain. 2010;133:1352–67. https://doi.org/10.1093/brain/awq075.
42. McLaren DG, Sreenivasan A, Diamond EL, Mitchell MB, Van Dijk KRA, Deluca AN, O'Brien JL, Rentz DM, Sperling RA, Atri A. Tracking cognitive change over 24 weeks with

longitudinal functional magnetic resonance imaging in alzheimer's disease. Neurodegener Dis. 2012;9:176–86. https://doi.org/10.1159/000335876.
43. Atri A, O'Brien JL, Sreenivasan A, Rastegar S, Salisbury S, DeLuca AN, O'Keefe KM, LaViolette PS, Rentz DM, Locascio JJ, Sperling RA. Test-retest reliability of memory task functional magnetic resonance imaging in alzheimer disease clinical trials. Arch Neurol. 2011;68:599–606. https://doi.org/10.1001/archneurol.2011.94.
44. Guo H, Grajauskas L, Habash B, D'Arcy RCN, Song X. Functional MRI technologies in the study of medication treatment effect on Alzheimer's disease. Aging Med. 2018;1:75–95. https://doi.org/10.1002/agm2.12017.
45. Sutton BP, Goh J, Hebrank A, Welsh RC, Chee MWL, Park DC. Investigation and validation of intersite fMRI studies using the same imaging hardware. J Magn Reson Imaging. 2008;28:21–8. https://doi.org/10.1002/jmri.21419.
46. Whitwell JL, Josephs KA, Murray ME, Kantarci K, Przybelski SA, Weigand SD, Vemuri P, Senjem ML, Parisi JE, Knopman DS, Boeve BF, Petersen RC, Dickson DW, Jack CR. MRI correlates of neurofibrillary tangle pathology at autopsy: a voxel-based morphometry study. Neurology. 2008;71:743–9. https://doi.org/10.1212/01.wnl.0000324924.91351.7d.
47. Ayton S. Brain volume loss due to donanemab. Eur J Neurol. 2021;28:e67–8. https://doi.org/10.1111/ene.15007.
48. Cummings J, Rabinovici GD, Atri A, Aisen P, Apostolova LG, Hendrix S, Sabbagh M, Selkoe D, Weiner M, Salloway S. Aducanumab: appropriate use recommendations update. J Prev Alzheimers Dis. 2022;9:221–30. https://doi.org/10.14283/jpad.2022.34.
49. Sperling RA, Jack CR, Black SE, Frosch MP, Greenberg SM, Hyman BT, Scheltens P, Carrillo MC, Thies W, Bednar MM, Black RS, Brashear HR, Grundman M, Siemers ER, Feldman HH, Schindler RJ. Amyloid-related imaging abnormalities in amyloid-modifying therapeutic trials: recommendations from the Alzheimer's Association Research Roundtable Workgroup. Alzheimers Dement. 2011;7:367–85. https://doi.org/10.1016/j.jalz.2011.05.2351.
50. Filippi M, Cecchetti G, Spinelli EG, Vezzulli P, Falini A, Agosta F. Amyloid-related imaging abnormalities and β-amyloid-targeting antibodies: a systematic review. JAMA Neurol. 2022;79:291–304. https://doi.org/10.1001/jamaneurol.2021.5205.
51. Roytman M, Mashriqi F, Al-Tawil K, Schulz PE, Zaharchuk G, Benzinger TLS, Franceschi AM. Amyloid-related imaging abnormalities: an update. Am J Roentgenol. 2023;220:562–74. https://doi.org/10.2214/ajr.22.28461.
52. Sims JR, Zimmer JA, Evans CD, Lu M, Ardayfio P, Sparks JD, Wessels AM, Shcherbinin S, Wang H, Monkul Nery ES, Collins EC, Solomon P, Salloway S, Apostolova LG, Hansson O, Ritchie C, Brooks DA, Mintun M, Skovronsky DM. Donanemab in early symptomatic Alzheimer disease: the TRAILBLAZER-ALZ 2 randomized clinical trial. JAMA. 2023;330:512–27. https://doi.org/10.1001/jama.2023.13239.
53. Honig LS, Sabbagh MN, van Dyck CH, Sperling RA, Hersch S, Matta A, Giorgi L, Gee M, Kanekiyo M, Li D, Purcell D, Dhadda S, Irizarry M, Kramer L. Updated safety results from phase 3 lecanemab study in early Alzheimer's disease. Alzheimers Res Ther. 2024;16:105. https://doi.org/10.1186/s13195-024-01441-8.
54. Solopova E, Romero-Fernandez W, Harmsen H, Ventura-Antunes L, Wang E, Shostak A, Maldonado J, Donahue MJ, Schultz D, Coyne TM, Charidimou A, Schrag M. Fatal iatrogenic cerebral β-amyloid-related arteritis in a woman treated with lecanemab for Alzheimer's disease. Nat Commun. 2023;14:8220. https://doi.org/10.1038/s41467-023-43933-5.
55. van Dyck CH, Swanson CJ, Aisen P, Bateman RJ, Chen C, Gee M, Kanekiyo M, Li D, Reyderman L, Cohen S, Froelich L, Katayama S, Sabbagh M, Vellas B, Watson D, Dhadda S, Irizarry M, Kramer LD, Iwatsubo T. Lecanemab in early Alzheimer's disease. N Engl J Med. 2023;388:9–21. https://doi.org/10.1056/NEJMoa2212948.
56. Hampel H, Elhage A, Cho M, Apostolova LG, Nicoll JAR, Atri A. Amyloid-related imaging abnormalities (ARIA): radiological, biological and clinical characteristics. Brain. 2023;146:4414–24. https://doi.org/10.1093/brain/awad188.

57. Therriault J, Zimmer ER, Benedet AL, Pascoal TA, Gauthier S, Rosa-Neto P. Staging of Alzheimer's disease: past, present, and future perspectives. Trends Mol Med. 2022;28:726–41. https://doi.org/10.1016/j.molmed.2022.05.008.
58. Chételat G, Arbizu J, Barthel H, Garibotto V, Law I, Morbelli S, van de Giessen E, Agosta F, Barkhof F, Brooks DJ, Carrillo MC, Dubois B, Fjell AM, Frisoni GB, Hansson O, Herholz K, Hutton BF, Jack CR Jr, Lammertsma AA, Landau SM, Minoshima S, Nobili F, Nordberg A, Ossenkoppele R, Oyen WJG, Perani D, Rabinovici GD, Scheltens P, Villemagne VL, Zetterberg H, Drzezga A. Amyloid-PET and 18F-FDG-PET in the diagnostic investigation of Alzheimer's disease and other dementias. Lancet Neurol. 2020;19:951–62. https://doi.org/10.1016/S1474-4422(20)30314-8.
59. Jack CR, Bennett DA, Blennow K, Carrillo MC, Dunn B, Haeberlein SB, Holtzman DM, Jagust W, Jessen F, Karlawish J, Liu E, Molinuevo JL, Montine T, Phelps C, Rankin KP, Rowe CC, Scheltens P, Siemers E, Snyder HM, Sperling R, Elliott C, Masliah E, Ryan L, Silverberg N. NIA-AA research framework: toward a biological definition of Alzheimer's disease. Alzheimers Dement. 2018;14:535–62. https://doi.org/10.1016/j.jalz.2018.02.018.
60. Minoshima S, Drzezga AE, Barthel H, Bohnen N, Djekidel M, Lewis DH, Mathis CA, McConathy J, Nordberg A, Sabri O, Seibyl JP, Stokes MK, Van Laere K. SNMMI procedure standard/EANM practice guideline for amyloid PET imaging of the brain 1.0. J Nucl Med. 2016;57:1316–22. https://doi.org/10.2967/jnumed.116.174615.
61. Rabinovici GD, Gatsonis C, Apgar C, Chaudhary K, Gareen I, Hanna L, Hendrix J, Hillner BE, Olson C, Lesman-Segev OH, Romanoff J, Siegel BA, Whitmer RA, Carrillo MC. Association of amyloid positron emission tomography with subsequent change in clinical management among medicare beneficiaries with mild cognitive impairment or dementia. JAMA. 2019;321:1286–94. https://doi.org/10.1001/jama.2019.2000.
62. Coutinho AM, Busatto GF, de Gobbi Porto FH, de Paula FD, Ono CR, Garcez AT, Squarzoni P, de Souza Duran FL, de Oliveira MO, Tres ES, Brucki SMD, Forlenza OV, Nitrini R, Buchpiguel CA. Brain PET amyloid and neurodegeneration biomarkers in the context of the 2018 NIA-AA research framework: an individual approach exploring clinical-biomarker mismatches and sociodemographic parameters. Eur J Nucl Med Mol Imaging. 2020;47:2666–80. https://doi.org/10.1007/s00259-020-04714-0.
63. Budd Haeberlein S, Aisen PS, Barkhof F, Chalkias S, Chen T, Cohen S, Dent G, Hansson O, Harrison K, von Hehn C, Iwatsubo T, Mallinckrodt C, Mummery CJ, Muralidharan KK, Nestorov I, Nisenbaum L, Rajagovindan R, Skordos L, Tian Y, van Dyck CH, Vellas B, Wu S, Zhu Y, Sandrock A. Two randomized phase 3 studies of aducanumab in early Alzheimer's disease. J Prev Alzheimers Dis. 2022;9:197–210. https://doi.org/10.14283/jpad.2022.30.
64. Cummings J, Aisen P, Apostolova LG, Atri A, Salloway S, Weiner M. Aducanumab: appropriate use recommendations. J Prev Alzheimers Dis. 2021;8:398–410. https://doi.org/10.14283/jpad.2021.41.
65. La Joie R, Ayakta N, Seeley WW, Borys E, Boxer AL, DeCarli C, Doré V, Grinberg LT, Huang E, Hwang JH, Ikonomovic MD, Jack C Jr, Jagust WJ, Jin LW, Klunk WE, Kofler J, Lesman-Segev OH, Lockhart SN, Lowe VJ, Masters CL, Mathis CA, McLean CL, Miller BL, Mungas D, O'Neil JP, Olichney JM, Parisi JE, Petersen RC, Rosen HJ, Rowe CC, Spina S, Vemuri P, Villemagne VL, Murray ME, Rabinovici GD. Multisite study of the relationships between antemortem [11C]PIB-PET Centiloid values and postmortem measures of Alzheimer's disease neuropathology. Alzheimers Dement. 2019;15:205–16. https://doi.org/10.1016/j.jalz.2018.09.001.
66. Klunk WE, Koeppe RA, Price JC, Benzinger TL, Devous MD Sr, Jagust WJ, Johnson KA, Mathis CA, Minhas D, Pontecorvo MJ, Rowe CC, Skovronsky DM, Mintun MA. The Centiloid Project: standardizing quantitative amyloid plaque estimation by PET. Alzheimers Dement. 2015;11:1–15. https://doi.org/10.1016/j.jalz.2014.07.003.
67. Vogel JW, Mattsson N, Iturria-Medina Y, Strandberg OT, Schöll M, Dansereau C, Villeneuve S, van der Flier WM, Scheltens P, Bellec P, Evans AC, Hansson O, Ossenkoppele R. Data-

driven approaches for tau-PET imaging biomarkers in Alzheimer's disease. Hum Brain Mapp. 2019;40:638–51. https://doi.org/10.1002/hbm.24401.
68. Lagarde J, Olivieri P, Caillé F, Gervais P, Baron JC, Bottlaender M, Sarazin M. [18F]-AV-1451 tau PET imaging in Alzheimer's disease and suspected non-AD tauopathies using a late acquisition time window. J Neurol. 2019;266:3087–97. https://doi.org/10.1007/s00415-019-09530-7.
69. Aguero C, Dhaynaut M, Normandin MD, Amaral AC, Guehl NJ, Neelamegam R, Marquie M, Johnson KA, El Fakhri G, Frosch MP, Gomez-Isla T. Autoradiography validation of novel tau PET tracer [F-18]-MK-6240 on human postmortem brain tissue. Acta Neuropathol Commun. 2019;7:37. https://doi.org/10.1186/s40478-019-0686-6.
70. Brendel M, Barthel H, van Eimeren T, Marek K, Beyer L, Song M, Palleis C, Gehmeyr M, Fietzek U, Respondek G, Sauerbeck J, Nitschmann A, Zach C, Hammes J, Barbe MT, Onur O, Jessen F, Saur D, Schroeter ML, Rumpf JJ, Rullmann M, Schildan A, Patt M, Neumaier B, Barret O, Madonia J, Russell DS, Stephens A, Roeber S, Herms J, Bötzel K, Classen J, Bartenstein P, Villemagne V, Levin J, Höglinger GU, Drzezga A, Seibyl J, Sabri O. Assessment of 18F-PI-2620 as a biomarker in progressive supranuclear palsy. JAMA Neurol. 2020;77:1408–19. https://doi.org/10.1001/jamaneurol.2020.2526.
71. Tagai K, Ono M, Kubota M, Kitamura S, Takahata K, Seki C, Takado Y, Shinotoh H, Sano Y, Yamamoto Y, Matsuoka K, Takuwa H, Shimojo M, Takahashi M, Kawamura K, Kikuchi T, Okada M, Akiyama H, Suzuki H, Onaya M, Takeda T, Arai K, Arai N, Araki N, Saito Y, Trojanowski JQ, Lee VMY, Mishra SK, Yamaguchi Y, Kimura Y, Ichise M, Tomita Y, Zhang MR, Suhara T, Shigeta M, Sahara N, Higuchi M, Shimada H. High-contrast in vivo imaging of tau pathologies in Alzheimer's and non-Alzheimer's disease tauopathies. Neuron. 2021;109:42–58. https://doi.org/10.1016/j.neuron.2020.09.042.
72. Davison CM, O'Brien JT. A comparison of FDG-PET and blood flow SPECT in the diagnosis of neurodegenerative dementias: a systematic review. Int J Geriatr Psychiatry. 2014;29:551–61. https://doi.org/10.1002/gps.4036.
73. O'Brien JT, Firbank MJ, Davison C, Barnett N, Bamford C, Donaldson C, Olsen K, Herholz K, Williams D, Lloyd J. 18F-FDG PET and perfusion SPECT in the diagnosis of Alzheimer and Lewy body dementias. J Nucl Med. 2014;55:1959–65. https://doi.org/10.2967/jnumed.114.143347.
74. Nobili F, Arbizu J, Bouwman F, Drzezga A, Agosta F, Nestor P, Walker Z, Boccardi M. European Association of Nuclear Medicine and European Academy of Neurology recommendations for the use of brain 18 F-fluorodeoxyglucose positron emission tomography in neurodegenerative cognitive impairment and dementia: Delphi consensus. Eur J Neurol. 2018;25:1201–17. https://doi.org/10.1111/ene.13728.
75. Burdette JH, Minoshima S, Vander Borght T, Tran DD, Kuhl DE. Alzheimer disease: improved visual interpretation of PET images by using three-dimensional stereotaxic surface projections. Radiology. 1996;198:837–43. https://doi.org/10.1148/radiology.198.3.8628880.
76. Cerami C, Della Rosa AP, Magnani G, Santangelo R, Marcone A, Cappa SF, Perani D. Brain metabolic maps in mild cognitive impairment predict heterogeneity of progression to dementia. Neuroimage Clin. 2015;7:187–94. https://doi.org/10.1016/j.nicl.2014.12.004.
77. Drzezga A, Lautenschlager N, Siebner H, Riemenschneider M, Willoch F, Minoshima S, Schwaiger M, Kurz A. Cerebral metabolic changes accom- panying conversion of mild cognitive impairment into Alzheimer's disease: a PET follow-up study. Eur J Nucl Med Mol Imaging. 2003;30:1104–13. https://doi.org/10.1007/s00259-003-1194-1.
78. Minoshima S, Giordani B, Berent S, Frey K, Foster N, Kuhl D. Metabolic reduction in the posterior cingulate cortex in very early Alzheimer's disease. Ann Neurol. 1997;42:85–94. https://doi.org/10.1002/ana.410420114.
79. Lesman-Segev OH, La Joie R, Iaccarino L, Lobach I, Rosen HJ, Seo SW, Janabi M, Baker SL, Edwards L, Pham J, Olichney J, Boxer A, Huang E, Gorno-Tempini M, DeCarli C, Hepker M, Hwang JL, Miller BL, Spina S, Grinberg LT, Seeley WW, Jagust WJ, Rabinovici GD. Diagnostic accuracy of amyloid versus 18 F-fluorodeoxyglucose positron emission

tomography in autopsy-confirmed dementia. Ann Neurol. 2021;89:389–401. https://doi.org/10.1002/ana.25968.
80. Parmera JB, Oliveira MCB, Rodrigues RD, Coutinho AM. Progressive supranuclear palsy and corticobasal degeneration: novel clinical concepts and advances in biomarkers. Arq Neuropsiquiatr. 2022;80:126–36. https://doi.org/10.1590/0004-282X-ANP-2022-S134.
81. Studart-Neto A, Coutinho AM. From clinical phenotype to proteinopathy: molecular neuroimaging in neurodegenerative dementias. Arq Neuropsiquiatr. 2022;80:24–35. https://doi.org/10.1590/0004-282X-ANP-2022-S138.
82. Meyer PT, Frings L, Rücker G, Hellwig S. ^{18}F-FDG PET in parkinsonism: differential diagnosis and evaluation of cognitive impairment. J Nucl Med. 2017;58:1888–98. https://doi.org/10.2967/jnumed.116.186403.
83. Minoshima S, Mosci K, Cross D, Thientunyakit T. Brain [F-18]FDG PET for clinical dementia workup: differential diagnosis of Alzheimer's disease and other types of dementing disorders. Semin Nucl Med. 2021;51:230–40. https://doi.org/10.1053/j.semnuclmed.2021.01.002.
84. Parmera JB, Coutinho AM, Aranha MR, Studart-Neto A, de Godoi CC, de Almeida IJ, Fontoura Solla DJ, Ono CR, Barbosa ER, Nitrini R, Buchpiguel CA, Brucki SMD. FDG-PET patterns predict amyloid deposition and clinical profile in Corticobasal syndrome. Mov Disord. 2021;36:651–61. https://doi.org/10.1002/mds.28373.
85. Silverman DH, Small GW, Chang CY, Lu CS, Kung de Aburto MA, Chen W, Czernin J, Rapoport SI, Pietrini P, Alexander GE, Schapiro MB, Jagust WJ, Hoffman JM, Welsh-Bohmer KA, Alavi A, Clark CM, Salmon E, de Leon MJ, Mielke R, Cummings JL, Kowell AP, Gambhir SS, Hoh CK, Phelps ME. Positron emission tomography in evaluation of dementia: regional brain metabolism and long-term outcome. JAMA. 2001;286:2120–7. https://doi.org/10.1001/jama.286.17.2120.
86. Scollato A, Terreni A, Caldini A, Salvadori B, Gallina P, Francese S, Mastrobuoni G, Pieraccini G, Moneti G, Bini L, Messeri G, Di Lorenzo N. CSF proteomic analysis in patients with normal pressure hydrocephalus selected for the shunt: CSF biomarkers of response to surgical treatment. Neurol Sci. 2010;31:283–91. https://doi.org/10.1007/s10072-009-0181-0.
87. del Campo M, Zetterberg H, Gandy S, Onyike CU, Oliveira F, Udeh-Momoh C, Lleó A, Teunissen CE, Pijnenburg Y. New developments of biofluid-based biomarkers for routine diagnosis and disease trajectories in frontotemporal dementia. Alzheimers Dement. 2022;18:2292–307. https://doi.org/10.1002/alz.12643.
88. Jack CR Jr, Knopman DS, Jagust WJ, Shaw LM, Aisen PS, Weiner MW, Petersen RC, Trojanowski JQ. Hypothetical model of dynamic biomarkers of the Alzheimer's pathological cascade. Lancet Neurol. 2010;9:119–28. https://doi.org/10.1016/S1474-4422(09)70299-6.
89. Selkoe DJ, Hardy J. The amyloid hypothesis of Alzheimer's disease at 25 years. EMBO Mol Med. 2016;8:595–608. https://doi.org/10.15252/emmm.201606210.
90. De Oliveira FF, Miraldo MC, Castro-Neto EF, Almeida SS, Matas SLA, Bertolucci PHF, Naffah-Mazzacoratti MG. Anthropometric and demographic features affect the interpretation of cerebrospinal fluid biomarkers in patients with different dementia syndromes and cognitively healthy adults. Neuromol Med. 2024;26:43. https://doi.org/10.1007/s12017-024-08810-4.
91. Scheltens P, De Strooper B, Kivipelto M, Holstege H, Chetélat G, Teunissen CE, Cummings J, van der Flier WM. Alzheimer's disease. Lancet. 2021;397:1577–90. https://doi.org/10.1016/S0140-6736(20)32205-4.
92. Wright JW, Harding JW. The brain RAS and Alzheimer's disease. Exp Neurol. 2010;223:326–33. https://doi.org/10.1016/j.expneurol.2009.09.012.
93. Sala A, Nordberg A, Rodriguez-Vieitez E. Longitudinal pathways of cerebrospinal fluid and positron emission tomography biomarkers of amyloid-β positivity. Mol Psychiatry. 2021;26:5864–74. https://doi.org/10.1038/s41380-020-00950-w.
94. Pereira JB, Janelidze S, Ossenkoppele R, Kvartsberg H, Brinkmalm A, Mattsson-Carlgren N, Stomrud E, Smith R, Zetterberg H, Blennow K, Hansson O. Untangling the asso-

ciation of amyloid-β and tau with synaptic and axonal loss in Alzheimer's disease. Brain. 2021;144:310–24. https://doi.org/10.1093/brain/awaa395.
95. De Oliveira FF, Miraldo MC, Castro-Neto EF, Almeida SS, Matas SLA, Bertolucci PHF, Naffah-Mazzacoratti MG. Differential associations of clinical features with cerebrospinal fluid biomarkers in dementia with Lewy bodies and Alzheimer's disease. Aging Clin Exp Res. 2023;35:1741–52. https://doi.org/10.1007/s40520-023-02452-5.
96. Duits FH, Teunissen CE, Bouwman FH, Visser PJ, Mattson N, Zetterberg H, Blennow K, Hansson O, Minthon L, Andreasen N, Marcusson J, Wallin A, Rikkert MO, Tsolaki M, Parnetti L, Herukka SK, Hampel H, De Leon MJ, Schröder J, Aarsland D, Blankenstein MA, Scheltens P, van der Flier WM. The cerebrospinal fluid "Alzheimer profile": easily said, but what does it mean? Alzheimers Dement. 2014;10:713–23. https://doi.org/10.1016/j.jalz.2013.12.023.
97. Leonenko G, Shoai M, Bellou E, Sims R, Williams J, Hardy J, Escott-Price V. Genetic risk for Alzheimer disease is distinct from genetic risk for amyloid deposition. Ann Neurol. 2019;86:427–35. https://doi.org/10.1002/ana.25530.
98. De Oliveira FF. Looking behind the curtain: patient stratification according to genetic or demographic factors may yield unexpected results in studies of neurodegenerative diseases. J Alzheimers Dis. 2023;94:777–80. https://doi.org/10.3233/JAD-230561.
99. Lautner R, Palmqvist S, Mattson N, Andreasson U, Wallin A, Pålsson E, Jakobsson J, Herukka SK, Owenius R, Olsson B, Hampel H, Rujescu D, Ewers M, Landén M, Minthon L, Blennow K, Zetterberg H, Hansson O. Apolipoprotein E genotype and the diagnostic accuracy of cerebrospinal fluid biomarkers for Alzheimer disease. JAMA Psychiatry. 2014;71:1183–91. https://doi.org/10.1001/jamapsychiatry.2014.1060.
100. Okello A, Koivunen J, Edison P, Archer HA, Turkheimer FE, Nagren K, Bullock R, Walker Z, Kennedy A, Fox NC, Rossor MN, Rinne JO, Brooks DJ. Conversion of amyloid positive and negative MCI to AD over 3 years: an ^{11}C-PIB PET study. Neurology. 2009;73:754–60. https://doi.org/10.1212/WNL.0b013e3181b23564.
101. Greenberg SM, Bacskai BJ, Hernandez-Guillamon M, Pruzin J, Sperling R, van Veluw SJ. Cerebral amyloid angiopathy and Alzheimer disease—one peptide, two pathways. Nat Rev Neurol. 2020;16:30–42. https://doi.org/10.1038/s41582-019-0281-2.
102. Bateman RJ, Xiong C, Benzinger TLS, Fagan AM, Goate A, Fox NC, Marcus DS, Cairns NJ, Xie X, Blazey TM, Holtzman DM, Santacruz A, Buckles V, Oliver A, Moulder K, Aisen PS, Ghetti B, Klunk WE, McDade E, Martins RN, Masters CL, Mayeux R, Ringman JM, Rossor MN, Schofield PR, Sperling RA, Salloway S, Morris JC. Clinical and biomarker changes in dominantly inherited Alzheimer's disease. N Engl J Med. 2012;367:795–804. https://doi.org/10.1056/NEJMoa1202753.
103. Fortea J, Vilaplana E, Carmona-Iragui M, Benejam B, Videla L, Barroeta I, Fernández S, Altuna M, Pegueroles J, Montal V, Valldeneu S, Giménez S, González-Ortiz S, Muñoz L, Estellés T, Illán-Gala I, Belbin O, Camacho V, Wilson RL, Annus T, Osorio RS, Videla S, Lehmann S, Holland AJ, Alcolea D, Clarimón J, Zaman SH, Blesa R, Lleó A. Clinical and biomarker changes of Alzheimer's disease in adults with Down syndrome: a cross-sectional study. Lancet. 2020;395:1988–97. https://doi.org/10.1016/S0140-6736(20)30689-9.
104. De Oliveira FF, Miraldo MC, Castro-Neto EF, Almeida SS, Matas SLA, Bertolucci PHF, Naffah-Mazzacoratti MG. Associations of neuropsychiatric features with cerebrospinal fluid biomarkers of amyloidogenesis and neurodegeneration in dementia with lewy bodies compared with Alzheimer's disease and cognitively healthy people. J Alzheimers Dis. 2021;81:1295–309. https://doi.org/10.3233/JAD-210272.
105. Kaiser E, Schoenknecht P, Kassner S, Hildebrandt W, Kinscherf R, Schroeder J. Cerebrospinal fluid concentrations of functionally important amino acids and metabolic compounds in patients with mild cognitive impairment and Alzheimer's disease. Neurodegener Dis. 2010;7:251–9. https://doi.org/10.1159/000287953.

106. Naffah-Mazzacoratti MG, Amado D, Cukiert A, Gronich G, Marino R, Calderazzo L, Cavalheiro EA. Monoamines and their metabolites in cerebrospinal fluid and temporal cortex of epileptic patients. Epilepsy Res. 1996;25:133–7. https://doi.org/10.1016/0920-1211(96)00030-7.
107. Colom-Cadena M, Spires-Jones T, Zetterberg H, Blennow K, Caggiano A, DeKosky ST, Fillit H, Harrison JE, Schneider LS, Scheltens P, de Haan W, Grundman M, van Dyck CH, Izzo NJ, Catalano SM. The clinical promise of biomarkers of synapse damage or loss in Alzheimer's disease. Alzheimers Res Ther. 2020;12:21. https://doi.org/10.1186/s13195-020-00588-4.
108. Alves G, Pedersen KF, Bloem BR, Blennow K, Zetterberg H, Borm GF, Dalaker TO, Beyer MK, Aarsland D, Andreasson U, Lange J, Tysnes OB, Zivadinov R, Larsen JP. Cerebrospinal fluid amyloid-β and phenotypic heterogeneity in de novo Parkinson's disease. J Neurol Neurosurg Psychiatry. 2013;84:537–43. https://doi.org/10.1136/jnnp-2012-303808.
109. Silva NCBS, Bracko O, Nelson A, Oliveira FF, Robison L, Shaaban CE, Hainsworth AH, Price BR. Vascular cognitive impairment and dementia: an early career researcher perspective. Alzheimers Dement. 2022;14:e12310. https://doi.org/10.1002/dad2.12310.
110. De Oliveira FF. Shedding light on the effects of blood pressure on cognitive decline and dementia risk by way of neurobiological evidence. J Alzheimers Dis. 2024;99:485–8. https://doi.org/10.3233/JAD-240032.
111. Agüera-Ortiz L, Babulal GM, Bruneau MA, Creese B, D'Antonio F, Fischer CE, Gatchel JR, Ismail Z, Kumar S, McGeown WJ, Mortby ME, Nuñez NA, Oliveira FF, Pereiro A, Ravona-Springer R, Rouse HJ, Wang H, Lanctôt KL. Psychosis as a treatment target in dementia: a roadmap for designing interventions. J Alzheimers Dis. 2022;88:1203–28. https://doi.org/10.3233/JAD-215483.
112. Bateman DR, Gill S, Hu S, Foster ED, Ruthirakuhan MT, Sellek AF, Mortby ME, Matusková V, Ng KP, Tarawneh RM, Freund-Levi Y, Kumar S, Gauthier S, Rosenberg PB, Oliveira FF, Devanand DP, Ballard C, Ismail Z. Agitation and impulsivity in mid and late life as possible risk markers for incident dementia. Alzheimers Dement. 2020;6:e12016. https://doi.org/10.1002/trc2.12016.
113. Janelidze S, Mattsson N, Palmqvist S, Smith R, Beach TG, Serrano GE, Chai X, Proctor NK, Eichenlaub U, Zetterberg H, Blennow K, Reiman EM, Stomrud E, Dage JL, Hansson O. Plasma P-tau181 in Alzheimer's disease: relationship to other biomarkers, differential diagnosis, neuropathology and longitudinal progression to Alzheimer's dementia. Nat Med. 2020;26:379–86. https://doi.org/10.1038/s41591-020-0755-1.
114. Salvadó G, Ossenkoppele R, Ashton NJ, Beach TG, Serrano GE, Reiman EM, Zetterberg H, Mattsson-Carlgren N, Janelidze S, Blennow K, Hansson O. Specific associations between plasma biomarkers and postmortem amyloid plaque and tau tangle loads. EMBO Mol Med. 2023;15:e17123. https://doi.org/10.15252/emmm.202217123.
115. Gendron TF, Heckman MG, White LJ, Veire AM, Pedraza O, Burch AR, Bozoki AC, Dickerson BC, Domoto-Reilly K, Foroud T, Forsberg LK, Galasko DR, Ghoshal N, Graff-Radford NR, Grossman M, Heuer HW, Huey ED, Hsiung GYR, Irwin DJ, Kaufer DI, Leger GC, Litvan I, Masdeu JC, Mendez MF, Onyike CU, Pascual B, Ritter A, Roberson ED, Rojas JC, Tartaglia MC, Wszolek ZK, Rosen H, Boeve BF, Boxer AL, Petrucelli L. Comprehensive cross-sectional and longitudinal analyses of plasma neurofilament light across FTD spectrum disorders. Cell Rep Med. 2022;3:100607. https://doi.org/10.1016/j.xcrm.2022.100607.
116. Karikari TK, Benedet AL, Ashton NJ, Rodriguez JL, Snellman A, Suárez-Calvet M, Saha-Chaudhuri P, Lussier F, Kvartsberg H, Rial AM, Pascoal TA, Andreasson U, Schöll M, Weiner MW, Rosa-Neto P, Trojanowski JQ, Shaw LM, Blennow K, Zetterberg H. Diagnostic performance and prediction of clinical progression of plasma phospho-tau181 in the Alzheimer's Disease Neuroimaging Initiative. Mol Psychiatry. 2021;26:429–42. https://doi.org/10.1038/s41380-020-00923-z.
117. Hansson O, Blennow K, Zetterberg H, Dage J. Blood biomarkers for Alzheimer's disease in clinical practice and trials. Nat Aging. 2023;3:506–19. https://doi.org/10.1038/s43587-023-00403-3.

118. Palmqvist S, Janelidze S, Quiroz YT, Zetterberg H, Lopera F, Stomrud E, Su Y, Chen Y, Serrano GE, Leuzy A, Mattsson-Carlgren N, Strandberg O, Smith R, Villegas A, Sepulveda-Falla D, Chai X, Proctor NK, Beach TG, Blennow K, Dage JL, Reiman EM, Hansson O. Discriminative accuracy of plasma phospho-tau217 for Alzheimer disease vs other neurodegenerative disorders. JAMA. 2020;324:772–81. https://doi.org/10.1001/jama.2020.12134.
119. Cullen NC, Janelidze S, Stomrud E, Bateman RJ, Palmqvist S, Hansson O, Mattsson-Carlgren N. Plasma amyloid-β42/40 and apolipoprotein E for amyloid PET pre-screening in secondary prevention trials of Alzheimer's disease. Brain Commun. 2023;5:fcad015. https://doi.org/10.1093/braincomms/fcad015.
120. Schindler SE, Li Y, Li M, Despotis A, Park E, Vittert L, Hamilton BH, Womack KB, Saef B, Holtzman DM, Morris JC, Bateman RJ, Gupta MR. Using Alzheimer's disease blood tests to accelerate clinical trial enrollment. Alzheimers Dement. 2023;19:1175–83. https://doi.org/10.1002/alz.12754.
121. Moscoso A, Karikari TK, Grothe MJ, Ashton NJ, Lantero-Rodriguez J, Snellman A, Zetterberg H, Blennow K, Schöll M. CSF biomarkers and plasma p-tau181 as predictors of longitudinal tau accumulation: implications for clinical trial design. Alzheimers Dement. 2022;18:2614–26. https://doi.org/10.1002/alz.12570.
122. Moscoso A, Grothe MJ, Ashton NJ, Karikari TK, Lantero Rodríguez J, Snellman A, Suárez-Calvet M, Blennow K, Zetterberg H, Schöll M. Longitudinal associations of blood phosphorylated Tau181 and neurofilament light chain with neurodegeneration in Alzheimer disease. JAMA Neurol. 2021;78:396–406. https://doi.org/10.1001/jamaneurol.2020.4986.
123. Hansson O, Edelmayer RM, Boxer AL, Carrillo MC, Mielke MM, Rabinovici GD, Salloway S, Sperling R, Zetterberg H, Teunissen CE. The Alzheimer's Association appropriate use recommendations for blood biomarkers in Alzheimer's disease. Alzheimers Dement. 2022;18:2669–86. https://doi.org/10.1002/alz.12756.
124. Gonzalez-Ortiz F, Kac PR, Brum WS, Zetterberg H, Blennow K, Karikari TK. Plasma phospho-tau in Alzheimer's disease: towards diagnostic and therapeutic trial applications. Mol Neurodegener. 2023;18:18. https://doi.org/10.1186/s13024-023-00605-8.
125. Doré V, Doecke JD, Saad ZS, Triana-Baltzer G, Slemmon R, Krishnadas N, Bourgeat P, Huang K, Burnham S, Fowler C, Rainey-Smith SR, Bush AI, Ward L, Robertson J, Martins RN, Masters CL, Villemagne VL, Fripp J, Kolb HC, Rowe CC. Plasma p217+tau versus NAV4694 amyloid and MK6240 tau PET across the Alzheimer's continuum. Alzheimers Dement. 2022;14:e12307. https://doi.org/10.1002/dad2.12307.
126. Leuzy A, Smith R, Cullen NC, Strandberg O, Vogel JW, Binette AP, Borroni E, Janelidze S, Ohlsson T, Jögi J, Ossenkoppele R, Palmqvist S, Mattsson-Carlgren N, Klein G, Stomrud E, Hansson O. Biomarker-based prediction of longitudinal tau positron emission tomography in Alzheimer disease. JAMA Neurol. 2022;79:149–58. https://doi.org/10.1001/jamaneurol.2021.4654.
127. Mattsson-Carlgren N, Janelidze S, Palmqvist S, Cullen N, Svenningsson AL, Strandberg O, Mengel D, Walsh DM, Stomrud E, Dage JL, Hansson O. Longitudinal plasma p-tau217 is increased in early stages of Alzheimer's disease. Brain. 2020;143:3234–41. https://doi.org/10.1093/brain/awaa286.
128. Mintun MA, Lo AC, Duggan Evans C, Wessels AM, Ardayfio PA, Andersen SW, Shcherbinin S, Sparks J, Sims JR, Brys M, Apostolova LG, Salloway SP, Skovronsky DM. Donanemab in early Alzheimer's disease. N Engl J Med. 2021;384:1691–704. https://doi.org/10.1056/NEJMoa2100708.
129. Walsh DM, Selkoe DJ. Deciphering the molecular basis of memory failure in Alzheimer's disease. Neuron. 2004;44:181–93. https://doi.org/10.1016/j.neuron.2004.09.010.
130. van Dyck CH, Sabbagh M, Cohen S. Lecanemab in early Alzheimer's disease. Reply. N Engl J Med. 2023;388:1631–2. https://doi.org/10.1056/NEJMc2301380.
131. Ferreira P, Ferrari-Souza JP, Tissot C, Bellaver B, Leffa D, Lussier F, Povala G, Therriault J, Benedet AL, Ashton NJ, Cohen AD, Lopez OL, Tudorascu D, Klunk WE, Soucy JP, Gauthier S, Villemagne V, Zetterberg H, Blennow K, Rosa-Neto P, Zimmer ER, Karikari

TK, Pascoal T. Potential utility of plasma P-tau and neurofilament light chain as surrogate biomarkers for preventive clinical trials. Neurology. 2023;101:38. https://doi.org/10.1212/WNL.0000000000207115.
132. Cicognola C, Janelidze S, Hertze J, Zetterberg H, Blennow K, Mattsson-Carlgren N, Hansson O. Plasma glial fibrillary acidic protein detects Alzheimer pathology and predicts future conversion to Alzheimer dementia in patients with mild cognitive impairment. Alzheimers Res Ther. 2021;13:68. https://doi.org/10.1186/s13195-021-00804-9.
133. Verberk IMW, Thijssen E, Koelewijn J, Mauroo K, Vanbrabant J, Wilde A, Zwan MD, Verfaillie SCJ, Ossenkoppele R, Barkhof F, van Berckel BNM, Scheltens P, van der Flier W, Stoops E, Vanderstichele HM, Teunissen CE. Combination of plasma amyloid beta and glial fibrillary acidic protein strongly associates with cerebral amyloid pathology. Alzheimers Res Ther. 2020;12:118. https://doi.org/10.1186/s13195-020-00682-7.
134. Basha FH, Waseem M, Srinivasan H. Cellular and molecular mechanism in neurodegeneration: possible role of neuroprotectants. Cell Biochem Funct. 2021;39:613–22. https://doi.org/10.1002/cbf.3630.
135. Sinclair E, Trivedi DK, Sarkar D, Walton-Doyle C, Milne J, Kunath T, Rijs AM, de Bie RMA, Goodacre R, Silverdale M, Barran P. Metabolomics of sebum reveals lipid dysregulation in Parkinson's disease. Nat Commun. 2021;12:1592. https://doi.org/10.1038/s41467-021-21669-4.
136. Havelund JF, Heegaard NHH, Færgeman NJK, Gramsbergen JB. Biomarker research in Parkinson's disease using metabolite profiling. Metabolites. 2017;7:42. https://doi.org/10.3390/metabo7030042.
137. Shao Y, Le W. Recent advances and perspectives of metabolomics-based investigations in Parkinson's disease. Mol Neurodegener. 2019;14:3. https://doi.org/10.1186/s13024-018-0304-2.
138. Santos GA, Paulino N, Pardi PC. [P2–240]: salivary biomarkers for the detection of Alzheimer's disease. Alzheimers Dement. 2017;13:P703. https://doi.org/10.1016/j.jalz.2017.06.892.
139. Santos GAA, Olave E, Pardi PC. Salivary biomarkers in Alzheimer's disease. Int J Morphol. 2020;38:230–4. https://doi.org/10.4067/S0717-95022020000100230.
140. Wishart DS, Jewison T, Guo AC, Wilson M, Knox C, Liu Y, Djoumbou Y, Mandal R, Aziat F, Dong E, Bouatra S, Sinelnikov I, Arndt D, Xia J, Liu P, Yallou F, Bjorndahl T, Perez-Pineiro R, Eisner R, Allen F, Neveu V, Greiner R, Scalbert A. HMDB 3.0--The human metabolome database in 2013. Nucleic Acids Res. 2013;41:D801–7. https://doi.org/10.1093/nar/gks1065.
141. Andrews JL, Fernandez F. Chapter 15 - Salivary biomarkers in Alzheimer's disease. In: Martin CR, Preedy VR, editors. Diagnosis and management in dementia. Academic Press; 2020. p. 239–54., ISBN 9780128158548. https://doi.org/10.1016/B978-0-12-815854-8.00015-X.
142. Ashton NJ, Ide M, Zetterberg H, Blennow K. Salivary biomarkers for Alzheimer's disease and related disorders. Neurol Ther. 2019;8:83–94. https://doi.org/10.1007/s40120-019-00168-1.
143. Pekeles H, Qureshi HY, Paudel HK, Schipper HM, Gornistky M, Chertkow H. Development and validation of a salivary tau biomarker in Alzheimer's disease. Alzheimers Dement. 2018;10:53–60. https://doi.org/10.1016/j.dadm.2018.03.003.
144. Santos GAA, Fazan VPS, Pardi PC. The salivary tau protein as a biomarker for Alzheimer's disease. Alzheimers Dement. 2021;17:e053217. https://doi.org/10.1002/alz.053217.
145. Shi M, Sui YT, Peskind ER, Li G, Hwang H, Devic I, Ginghina C, Edgar JS, Pan C, Goodlett DR, Furay AR, Gonzalez-Cuyar LF, Zhang J. Salivary tau species are potential biomarkers of Alzheimer's disease. J Alzheimers Dis. 2011;27:299–305. https://doi.org/10.3233/JAD-2011-110731.
146. Tvarijonaviciute A, Zamora C, Ceron JJ, Bravo-Cantero AF, Pardo-Marin L, Valverde S, Lopez-Jornet P. Salivary biomarkers in Alzheimer's disease. Clin Oral Invest. 2020;24:3437–44. https://doi.org/10.1007/s00784-020-03214-7.
147. Marksteiner J, Defrancesco M, Humpel C. Saliva tau and phospho-tau-181 measured by Lumipulse in patients with Alzheimer's disease. Front Aging Neurosci. 2022;14:1014305. https://doi.org/10.3389/fnagi.2022.1014305.

148. Roses AD. An inherited variable poly-T repeat genotype in *TOMM40* in Alzheimer disease. Arch Neurol. 2010;67:536–41. https://doi.org/10.1001/archneurol.2010.88.
149. De Luca V, Orfei MD, Gaudenzi S, Caltagirone C, Spalletta G. Inverse effect of the APOE epsilon4 allele in late- and early-onset Alzheimer's disease. Eur Arch Psychiatry Clin Neurosci. 2016;266:599–606. https://doi.org/10.1007/s00406-015-0663-4.
150. De Oliveira FF, Bertolucci PHF, Chen ES, Smith MAC. Risk factors for age at onset of dementia due to Alzheimer's disease in a sample of patients with low mean schooling from São Paulo, Brazil. Int J Geriatr Psychiatry. 2014;29:1033–9. https://doi.org/10.1002/gps.4094.
151. Oliveira FF, Almeida SS, Smith MC, Bertolucci PHF. Behavioural effects of the *ACE* insertion/deletion polymorphism in Alzheimer's disease depend upon stratification according to *APOE*-ε4 carrier status. Cogn Neuropsychiatry. 2021;26:293–305. https://doi.org/10.1080/13546805.2021.1931085.
152. De Oliveira FF, Wajman JR, Bertolucci PHF, Chen ES, Smith MAC. Correlations among cognitive and behavioural assessments in patients with dementia due to Alzheimer's disease. Clin Neurol Neurosurg. 2015a;135:27–33. https://doi.org/10.1016/j.clineuro.2015.05.010.
153. De Oliveira FF, Pivi GAK, Chen ES, Smith MAC, Bertolucci PHF. Risk factors for cognitive and functional change in one year in patients with Alzheimer's disease dementia from São Paulo, Brazil. J Neurol Sci. 2015b;359:127–32. https://doi.org/10.1016/j.jns.2015.10.051.
154. De Oliveira FF, Machado FC, Sampaio G, Marin SMC, Naffah-Mazzacoratti MG, Bertolucci PHF. Neuropsychiatric feature profiles of patients with Lewy body dementia. Clin Neurol Neurosurg. 2020;194:105832. https://doi.org/10.1016/j.clineuro.2020.105832.
155. De Oliveira FF, Pereira FV, Pivi GAK, Smith MAC, Bertolucci PHF. Effects of *APOE* haplotypes and measures of cardiovascular risk over gender-dependent cognitive and functional changes in one year in Alzheimer's disease. Int J Neurosci. 2018;128:472–6. https://doi.org/10.1080/00207454.2017.1396986.
156. Martins CAR, Oulhaj A, de Jager CA, Williams JH. *APOE* alleles predict the rate of cognitive decline in Alzheimer disease—a nonlinear model. Neurology. 2005;65:1888–93. https://doi.org/10.1212/01.wnl.0000188871.74093.12.
157. De Oliveira FF, Almeida SS, Chen ES, Smith MC, Naffah-Mazzacoratti MG, Bertolucci PHF. Lifetime risk factors for functional and cognitive outcomes in patients with Alzheimer's disease. J Alzheimers Dis. 2018;65:1283–99. https://doi.org/10.3233/JAD-180303.
158. Oliveira FF, Chen ES, Smith MAC, Bertolucci PHF. Predictors of cognitive and functional decline in patients with Alzheimer disease dementia from Brazil. Alzheimer Dis Assoc Disord. 2016;30:243–50. https://doi.org/10.1097/WAD.0000000000000117.
159. De Oliveira FF, Chen ES, Smith MAC, Bertolucci PHF. Associations of blood pressure with functional and cognitive changes in patients with Alzheimer's disease. Dement Geriatr Cogn Disord. 2016;41:314–23. https://doi.org/10.1159/000447585.
160. De Oliveira FF, Chen ES, Smith MAC, Bertolucci PHF. Longitudinal lipid profile variations and clinical change in Alzheimer's disease dementia. Neurosci Lett. 2017;646:36–42. https://doi.org/10.1016/j.neulet.2017.03.003.
161. De Oliveira FF, Almeida SS, Chen ES, Smith MC, Bertolucci PHF. APOE ε4 carrier status as mediator of effects of psychotropic drugs on clinical changes in patients with Alzheimer's disease. J Neuropsychiatry Clin Neurosci. 2022;34:351–60. https://doi.org/10.1176/appi.neuropsych.21060160.
162. De Oliveira FF, Chen ES, Smith MAC, Bertolucci PHF. Associations of cerebrovascular metabolism genotypes with neuropsychiatric symptoms and age at onset of Alzheimer's disease dementia. Rev Bras Psiquiatr. 2017;39:95–103. https://doi.org/10.1590/1516-4446-2016-1991.
163. Oliveira FF, Almeida SS, Chen ES, Smith MC, Bertolucci PHF. Pharmacogenetics of angiotensin modulators according to *APOE*-ε4 alleles and the *ACE* insertion/deletion polymorphism in Alzheimer's disease. Acta Neuropsychiatrica. 2023;35:346–61. https://doi.org/10.1017/neu.2023.38.

164. De Oliveira FF, Chen ES, Smith MAC, Bertolucci PHF. Selected *LDLR* and *APOE* polymorphisms affect cognitive and functional response to lipophilic statins in Alzheimer's disease. J Mol Neurosci. 2020;70:1574–88. https://doi.org/10.1007/s12031-020-01588-7.
165. Leduc V, De Beaumont L, Théroux L, Dea D, Aisen P, Petersen RC, Dufour R, Poirier J. *HMGCR* is a genetic modifier for risk, age of onset and MCI conversion to Alzheimer's disease in a three cohorts study. Mol Psychiatry. 2015;20:867–73. https://doi.org/10.1038/mp.2014.81.
166. De Oliveira FF, Bertolucci PHF, Chen ES, Smith MAC. Pharmacogenetic analyses of therapeutic effects of lipophilic statins on cognitive and functional changes in Alzheimer's disease. J Alzheimers Dis. 2022;87:359–72. https://doi.org/10.3233/JAD-215735.
167. Oliveira FF, Machado FC, Sampaio G, Marin SMC, Chen ES, Smith MAC, Bertolucci PHF. Contrasts between patients with Lewy body dementia syndromes and *APOE*-ε3/ε3 patients with late-onset Alzheimer disease dementia. Neurologist. 2015;20:35–41. https://doi.org/10.1097/NRL.0000000000000045.
168. Sidransky E, Lopez G. The link between the *GBA* gene and parkinsonism. Lancet Neurol. 2012;11:986–98. https://doi.org/10.1016/S1474-4422(12)70190-4.

Biomarkers as Outcome Measures in Clinical Trials of Alzheimer Disease

Rawan Tarawneh

Abstract Over the last decade, biomarkers have played an increasingly important role in Alzheimer disease (AD) drug development. Biomarkers can improve patient selection and stratification, increase diagnostic accuracy, provide important prognostic information regarding future risk of clinical or radiological disease progression, and assist in monitoring disease severity or response to treatment. Through these roles, biomarkers can enrich clinical trial populations with study cohorts that have the target pathology and facilitate drug advancement using potentially smaller cohorts and shorter durations of trials. In this chapter, we review the main roles of biomarkers in AD drug development, explain the context of use for different biomarkers, and provide examples of successful biomarker utilization in AD clinical trials. Further, we discuss the different stages of biomarker development and present future directions for how biomarkers can be used to increase the chances of success of clinical trials of emerging AD therapies.

Keywords Biomarkers · Outcome measure · Clinical trials · Drug development · Alzheimer disease

1 Introduction

Alzheimer disease (AD) is the most common cause of dementia in individuals above the age of 65 years, and a major cause of morbidity and mortality in older populations [1]. In the absence of effective ways to prevent or slow disease progression, AD has become a global epidemic and a major economic burden. A new diagnosis of AD is made every 65 s in the US, and it is estimated to be the fifth leading cause of death among US elderly [2]. Age is the most important risk factor for AD as the incidence doubles every 5 years above the age of 65 [1]. With the increased

R. Tarawneh (✉)
Department of Neurology, University of New Mexico, Albuquerque, NM, USA
e-mail: raltarawneh@salud.unm.edu

R. Tarawneh (ed.), *Biomarkers of Neurodegenerative Disorders*,
https://doi.org/10.1007/978-3-031-80849-4_15

life expectancy and aging of populations, the incidence of AD is expected to almost quadruple from an estimated 35 million individuals worldwide in 2020 to more than 130 million by 2050 [3]. AD is also a major economic health burden, costing the US more than $231 billion in 2023 due to nursing home placement and lost wages and productivity [2].

Despite significant research efforts and billions of dollars invested in the AD drug pipeline over the last 20 years, results of clinical trials of disease-modifying therapies in AD have largely been disappointing [4, 5]. No new drugs were approved between 2003 and 2021 and AD drug development programs have faced a failure rate typically exceeding 98% [6]. In many instances, clinical trial failures may be attributed to flaws in trial design, methodological concerns, or inadequate power to detect desired study outcomes [5, 7, 8]. Further, there are inherent challenges in attempting to translate data from animal studies into human therapeutics [7, 8]. However, it has become increasingly clear that past failures of AD drug development cannot be solely attributed to flaws in clinical trial design or study methods, and rather, may reflect significant gaps in our current understanding of AD from a molecular and pathophysiological standpoint [5, 8]. Emerging data from clinical trials, clinicopathological, and animal studies of AD supports the notion that AD is a complex multifactorial disease, which involves various pathologies whose interactions culminate in synaptic loss and neurodegeneration, the latter being the main pathological surrogates of cognitive decline [9–11]. Further, recent studies converge on the notion that there is a significant degree of genetic, molecular, and pathological heterogeneity in AD [12–14], as relative contributions of different pathologies to cognitive outcomes may vary across individuals and different disease stages [14].

Table 1 defines the different roles of biomarkers, and Figure 1 demonstrates examples of different contexts of use for biomarkers, in drug development. In this chapter, we will focus on discussing the use of biomarkers as measures of target engagement and disease modification in clinical trials of AD therapies.

The current proposed model of AD progression provides additional possible explanations for the failure of most clinical trials of AD therapeutics [16]. This model, which is a hypothetical proposition of the temporal sequence of pathological changes in AD brains derived from clinicopathological studies and supported by the use of biomarkers, suggests the presence of a long preclinical phase of disease that precedes symptom onset, and starts 10–15 years prior to the first signs of cognitive impairment. This long preclinical phase of AD is characterized by early amyloid aggregation followed by tau pathology, and other pathologies such as inflammation, calcium-mediated excitotoxicity, mitochondrial damage, and oxidative stress, which eventually culminate in neuronal loss and synaptic dysfunction. Besides amyloid and tau, other pathologies such as α-synuclein and TAR DNA binding protein-43 (TDP-43) aggregation and vascular disease, are also present to variable extents in AD brains and contribute to the neurodegenerative process; however, these are not currently integrated into the AD model [5]. By the time individuals with AD develop the earliest clinically detectable signs of memory loss or cognitive impairment, significant brain pathology and irreversible neuronal and synaptic loss have already occurred in vulnerable brain regions [17]. In other words, early symptomatic AD

Table 1 The US Food and Drug Administration (FDA)—National Institute of Health (NIH)—BEST (*B*iomarkers, *E*ndpoint*s*, and Other *T*ools) Definition of Biomarkers in Drug Development (by the FDA-NIH Biomarker Working Group) [15]

Biomarker	Definition
Diagnosis	A biomarker that detects or confirms the presence of a disease or condition or identifies an individual with a subtype of the disease
Monitoring	A biomarker that is measured repeatedly for assessing the status of a disease or medical condition or for evidence of exposure to (or effect of) a medical product or an environmental agent
Response/ Pharmacodynamic	Response biomarker: a biomarker that is used to show that a biological response, potentially beneficial or harmful, has occurred in an individual who has been exposed to a medical product or an environmental agent. Pharmacodynamic biomarker: A response biomarker that indicates biologic activity of a medical product or environmental agent without necessarily drawing conclusions about efficacy or disease outcome or necessarily linking this activity to an established mechanism of action
Predictive	A biomarker that is used to identify individuals who are more likely than similar individuals without the biomarker to experience a favorable or unfavorable effect from exposure to a medical product or an environmental agent
Prognostic	A biomarker that is used to identify the likelihood of a clinical event, disease recurrence or progression in patients who have the disease or medical condition of interest
Risk/susceptibility	A biomarker that indicates the potential for developing a disease or medical condition in an individual who does not currently have clinically apparent disease or the medical condition
Safety	A biomarker that is measured before or after an exposure to a medical product or an environmental agent to indicate the likelihood, presence, or extent of toxicity as an adverse effect

typically represents an advanced disease stage from a molecular and pathological perspective. Therefore, the preclinical phase of AD represents an important time-window during which disease-modifying therapies that target various pathologies may have the highest chance of success prior to significant synaptic or neuronal loss [18, 19].

2 Biomarkers in Patient Selection

The importance of biomarkers in characterizing the pathobiological processes associated with AD was highlighted by the updated research framework for AD, which was proposed by the National Institute of Aging—Alzheimer's Association (NIA-AA) Work Group in 2018 [20]. This framework emphasizes that AD should be conceptualized as a clinicopathological spectrum of AD pathology, which can be detected using fluid or imaging biomarkers independently of clinical status or the presence of cognitive impairment. Specifically, the AD research framework

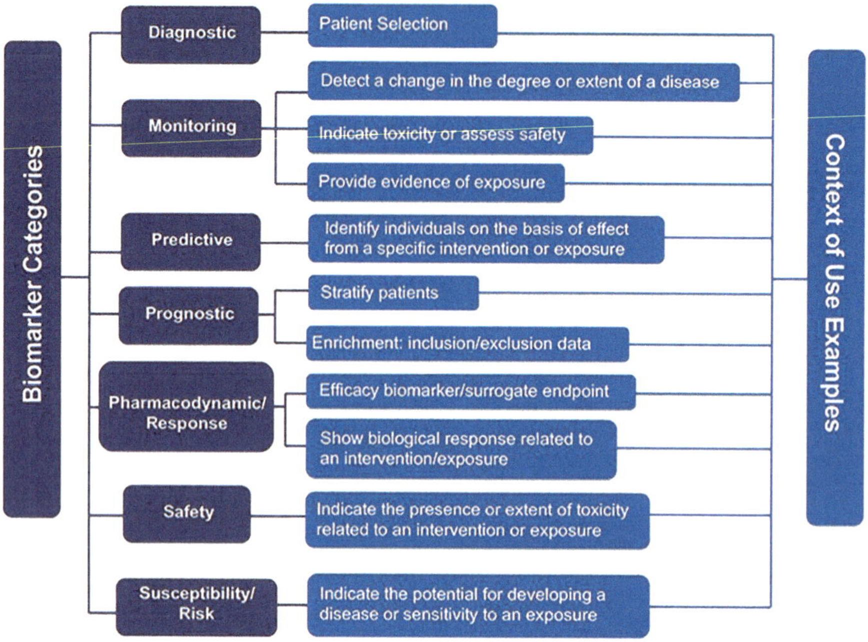

Fig. 1 Biomarker categories and contexts of use as defined by the FDA-NIH Biomarker Working Group [15]. (Adapted from [15])

proposes the use of a classification system for AD which includes fluid or imaging biomarkers of amyloid pathology (A), tau pathology (T), and neurodegeneration (N) [referred to as the "ATN" classification] to indicate the presence or absence of each of these 3 pathologies. Evidence of amyloid pathology "A" in the ATN framework is provided by an abnormal cerebrospinal fluid (CSF) amyloid-β (Aβ)42/Aβ40 ratio or amyloid positron emission tomography (PET) imaging, tau pathology "T" by abnormal CSF p-tau181 levels or tau PET imaging, and neurodegeneration "N" by abnormal CSF total tau levels, brain atrophy on magnetic resonance imaging (MRI), or hypometabolism on 18F-fluoro-deoxy-glucose (18F-FDG)-PET scans [20]. Individuals with only positive "A" markers are considered to have AD-related pathological change, while the presence of positive biomarkers for both "A" and "T", with or without positive "N" markers, is required for a biological diagnosis of AD. The implementation of this framework has shifted the conceptualization of AD from a clinical diagnosis to that of a biological construct which encompasses both preclinical and clinical stages and allows a biomarker-based definition of AD prior to cognitive impairment [21]. Further, the ATN framework closely parallels the new AD staging system proposed by the US Food and Drug Administration (FDA) [22].

Observations from AD clinical trials suggest that a clinical diagnosis of AD alone is insufficient for inclusion into clinical trials, and that biomarker confirmation of the presence of AD pathology is essential for accurate patient selection and stratification [23]. Examination of clinical trial cohorts of disease-modifying therapies in AD, which relied on clinical diagnosis alone, suggests that as many as 15–25% of these cohorts were later found to have alternative diagnoses [24, 25]. The identification of cognitively normal individuals who have pathological evidence of AD (i.e., individuals with preclinical AD) represents another important way by which biomarkers can be implemented in clinical trial design and patient selection. As clinical trials of emerging AD therapies have shifted focus to targeting disease pathology in these preclinical stages prior to symptom onset, biomarkers are expected to play an increasingly important role in identifying at-risk individuals. Therefore, fluid or imaging biomarkers that accurately reflect the core AD pathologies of amyloid and tau can be utilized in patient selection and stratification for enrollment in AD clinical trials, including the identification of individuals with preclinical AD for secondary prevention trials and increasing the diagnostic accuracy for AD in clinical trials that target the early symptomatic stages.

Further, emerging data from AD clinical trials suggests that clinical outcomes associated with anti-amyloid treatments may be influenced by the baseline burden of other AD pathologies such as tau [26], and possibly α-synuclein or TDP-43. Therefore, biomarkers that can reliably measure other key cellular and molecular substrates involved in AD pathogenesis such as tau, α-synuclein, or TDP-43 aggregation, and therefore, provide detailed molecular characterization of individuals with AD, will facilitate participant stratification for clinical trial enrollment and potentially assist in predicting study outcomes, based on the presence or degree of concomitant AD pathologies [26].

3 Biomarkers as Measures of Target Engagement

In drug development, Phase 2a trials are typically needed to establish proof of concept (POC) to support potential drug efficacy, and Phase 2b trials help establish the appropriate drug dosage, before advancing the drug into Phase 3 trials [27]. For symptomatic medications that are designed to improve or stabilize cognitive measures, POC may be demonstrated using clinical assessments alone. Conversely, for drugs designed to modify the underlying disease process and subsequently slow clinical progression, long trial durations in relatively large cohorts, which are often confounded by participants dropping out or being lost to follow-up, are needed to demonstrate POC using clinical outcomes alone. The reliance on the need to demonstrate POC using clinical outcomes has led to a "Phase 2 conundrum", in which drug sponsors may advance investigational drugs to Phase 3 without sufficient evidence of POC in Phase 2, further exacerbating the high rate of AD clinical trial failures in Phase 3 [27, 28].

The implementation of biomarkers to demonstrate adequate engagement of the study drug with its biological target can provide evidence to support POC in Phase 2 studies. Biomarkers can be used to confirm the effective binding of the drug to its key biological targets, and if applicable, adequate neutralization of these targets with treatment, either by receptor occupancy, proof of pharmacology, or both [27].

By using biomarkers as key measures of target engagement in Phase 2 studies, drug-placebo differences can be demonstrated in relatively shorter trials in smaller cohorts, and promising therapies may be advanced earlier into Phase 3 [27, 29]. Demonstrating convincing evidence for an early biomarker response to treatment can complement clinical outcome data, inform outcome assessments at the end of Phase 2 trials, and increase the chances of advancement into Phase 3, thereby, addressing the "Phase 2 conundrum". Therefore, Phase 2 trials should provide evidence of target engagement and establish the drug dose to be advanced to Phase 3. In Phase 3 trials of disease-modifying AD therapies, biomarkers can also be used to demonstrate biological evidence to support the drug's ability to modify the underlying disease pathology and provide a biological explanation for any downstream effects on clinical outcomes such as slowing clinical or functional progression (i.e., evidence of disease modification). Similarly, the absence of a significant biomarker response may suggest that a failed clinical response may have been due to poor target engagement or neutralization by the study drug, rather than other aspects of study design [30].

While symptomatic agents can be advanced into Phase 3 based on clinical responses alone, demonstrating target engagement provides valuable insights into brain penetration, dose optimization, and efficacy [27]. Conversely, disease-modifying therapies should not be advanced to Phase 3 without biomarker evidence of target engagement in Phase 2 [27]. Biomarkers that are suitable as measures of target engagement should be quantifiable markers whose levels are directly and specifically influenced by the investigational drug.

3.1 Fluid and Imaging Markers of Amyloid

A low CSF Aβ42 level, or a low CSF Aβ42/Aβ40 ratio, is a core biomarker feature of AD and is thought to result from the abnormal aggregation of amyloid into the brain parenchyma via a sink mechanism [31]. This is supported by the observation that low CSF Aβ42 and CSF Aβ42/Aβ40 levels correlate with increased amyloid burden on amyloid PET [31].

CSF Aβ42 levels have been used as a pharmacodynamic measure of target engagement for anti-amyloid agents with variable success [27]. For β- or γ-secretase inhibitors whose pharmacological effects include reduced Aβ42 production, demonstration of target engagement can be achieved by stable isotope labeling kinetic (SILK) studies [32]. This method involves labeling an amino acid in peripheral blood and utilizing mass-spectrometry to track its appearance in the amyloid protein, indicating amyloid synthesis, and its disappearance, indicating amyloid clearance. In the trial of the γ-secretase inhibitor, semagacestat (LY450139), participants underwent a 9-hour infusion of labeled leucine following oral administration of the drug, and CSF samples were obtained continuously for 36 h [33]. Using the SILK technique, the investigators demonstrated that this drug was associated with a short-term dose-dependent decrease in Aβ production ranging from 47% to 84%, while no significant effects were observed on Aβ clearance [33]. Despite these promising results, a 14-week study of semagacestat was not clearly associated with significant long-term reductions in Aβ production, as reflected by CSF Aβ42 levels [34]. No significant differences in CSF Aβ42 levels between the drug and placebo arms were observed in a 104-week Phase 2 trial of the γ-secretase inhibitor, avagacestat, in mild cognitive impairment (MCI) [35], while another 24-week Phase 2 study of this drug in mild to moderate AD dementia was associated with significant reductions in baseline CSF Aβ42 levels in the high-dose group [36]. γ-secretase inhibitors reduce CSF Aβ42 production, yet increase the production of shorter fragments, such as Aβ15–16, which may also be used as short-term evidence of target engagement [37]. The β-secretase inhibitor, verubecestat, was associated with both short-term (36 h) and long-term (78 weeks) reductions in CSF Aβ levels [38, 39], although these did not result in improved cognitive outcomes.

Results of studies using CSF Aβ levels as measures of target engagement in trials of anti-Aβ monoclonal antibodies (mAbs) have been mixed [40]. While some anti-amyloid mAbs have been associated with changes in CSF Aβ42 levels at the end of the trial period, results have been less predictable with others. Further, the direction of change is influenced by the drug's mechanism of action and the predominant Aβ species targeted by the drug [40, 41]. Phase 1 and 2 studies of solanezumab demonstrated dose-dependent increases in plasma and CSF total Aβ42 levels as evidence of target engagement [42, 43], and a 12-week Phase 2 study provided evidence for a dose-dependent increase in free CSF Aβ42 levels suggesting the mobilization of Aβ42 from plaques into the CSF [43]. A Phase 3 study of lecanemab was also associated with increases in both CSF Aβ42 and plasma Aβ42/Aβ40 levels in the treatment arm compared to placebo, but no changes in CSF Aβ40 levels, at 12 months and 18 months [44]. A Phase 2 trial of crenezumab showed an increase in CSF Aβ42 levels with treatment with no significant changes in amyloid PET [45, 46]. Consistent dose-dependent increases in total plasma and CSF Aβ42 or Aβ40 levels were observed in Phase 3 trials of crenezumab providing evidence of target engagement [47], while no significant effects were observed on CSF total tau, p-tau181 or Aβ oligomers [47].

The 76-week Phase 2 trial (TRAILBLAZER-ALZ) of donanemab showed no significant changes in plasma Aβ42/Aβ40 levels in the treatment arm compared to placebo [48]; however, a significant reduction of brain amyloid on amyloid PET was observed in the treatment arm and was associated with slower clinical progression in the primary clinical outcome (i.e., the integrated Alzheimer Disease Rating Scale; iADRS) [49]. Significant changes in CSF Aβ42 levels have not been consistently observed with other mAbs such as gantenerumab [50].

When employing CSF or plasma Aβ42 levels as endpoints of treatment efficacy or measures of target engagement in AD clinical trials of anti-amyloid mAbs, it is important to consider that these antibodies vary considerably in the main neurotoxic Aβ species that they target (e.g., protofibrils vs oligomers) and their ability to bind Aβ monomers which may differentially influence their effects on fibrillar Aβ clearance (Table 2). Another consideration is whether brain entry, and the ability of these

Table 2 Characteristics and targets of anti-amyloid monoclonal antibodies

Antibody	Origin	Subclass	Epitope	Monomer	Oligomer	Fibril
Bapineuzumab	Humanized	IgG1	AA 1–5	Yes	Yes	Yes
Solanezumab	Humanized	IgG1	AA 16–26	Yes[a]	No	No
Gantenerumab	Human	IgG1	AA 3–12, 18–27 (N-terminal and central)	Weak	Yes	Yes[a]
Crenezumab	Humanized	IgG4	AA 13–24	Yes	Yes	Yes
Lecanemab	Humanized	IgG1	Protofibrils[a]	No/Very weak	Yes (large)	Yes
Aducanumab	Human	IgG1	AA 3–6 (N-terminal)	No/Very weak	Yes[a]	Yes
Donanemab	Humanized	IgG1	N-terminal pyroglutamate Aβ in plaques[a]	No	No	No

Adapted from [40, 41]

[a]Solanezumab, aducanumab, gantenerumab, lecanemab, and donanemab preferentially bind Aβ monomers, oligomers, fibrils, protofibrils, and plaques, respectively

antibodies to cross the blood-brain barrier (BBB), is needed for their therapeutic effects on brain amyloid [40, 51]. This issue remains unresolved and requires further investigation. A few of these monoclonal antibodies, such as solanezumab, may be able to reduce brain amyloid via a peripheral sink mechanism which depends on the ability of Aβ to cross the BBB from the brain into the periphery and the presence of an equilibrium between central and peripheral Aβ levels [52, 53].

Demonstrating the ability of anti-amyloid agents to reduce fibrillar Aβ aggregates in the brain, measured as cortical amyloid load on amyloid PET imaging using ligands such as Pittsburgh Compound B (PiB) or ^{18}F-AV45 (^{18}F-florbetapir; Amyvid®), may provide more direct evidence of target engagement in clinical trials of anti-amyloid agents than CSF or plasma Aβ, as PET imaging avoids possible fluctuations in Aβ levels associated with circadian rhythms or the presence of other peripheral sources of plasma Aβ [54, 55].

In the Phase 1b PRIME trial of the anti-amyloid mAb, aducanumab, time- and dose-dependent reductions in brain amyloid on PET were observed at 26 and 54 weeks [56], and significant reductions in brain amyloid load were also observed in two Phase 3 trials of this drug (the EMERGE and ENGAGE trials) [57]. The ability of aducanumab to significantly reduce brain amyloid load provided the basis for the accelerated FDA approval of this drug in 2021 [58]. However, amyloid reduction was associated with possible clinical benefits in only one (i.e., the EMERGE trial) of the two Phase 3 trials of aducanumab [57]; further, clinical benefits in that trial were modest and not likely to be clinically meaningful [59] (following its approval, aducanumab has been discontinued by its manufacturer, Biogen, in 2024). Conversely, the Phase 3 study of the mAb, lecanemab, demonstrated significant reductions in brain amyloid load on amyloid PET with ~68% of participants reaching amyloid-negative status at 18-months [44, 60]. In this trial, amyloid reduction was associated with statistically significant favorable effects on several cognitive and functional outcomes, including the Clinical Dementia Rating®-sum of boxes (CDR-SB) and the Alzheimer Disease Cooperative Study—Activities of Daily Living for Mild Cognitive Impairment (ADCS-MCI-ADL); however, there is some controversy whether these benefits were clinically meaningful [61]. In participants with low-medium tau burden at baseline, the anti-amyloid mAb, donanemab, was associated with a significant reduction in amyloid PET burden, reducing this to below-positivity levels in 34% of treated participants at 24 weeks and 80% at 76 weeks compared to 0.2% and 0%, respectively, in the placebo arm [26]. Reductions of brain amyloid on PET have also been reported with other anti-amyloid mAbs, such as gantenerumab [62] and the small molecules, verubecestat [39] and bexarotene [63]. However, other mAbs such as crenezumab and solanezumab were not associated with significant changes on amyloid PET in the treatment compared to placebo arms [47, 64, 65].

It is important to note that even a significant reduction in brain amyloid plaque burden was not reliably or convincingly associated with favorable effects on cognitive or functional outcomes in many AD trials. These observations may be potentially explained by the fact that amyloid deposition is an early event that begins a decade or longer before symptom onset and has nearly plateaued by the time the first signs of cognitive impairment appear [16]. Several studies converge on the notion that amyloid pathology is not closely associated with cognitive outcomes in early symptomatic AD, a stage in which other AD pathologies, such as tau, are more closely associated with cognitive decline [66–68]. It has also been suggested that, given the relatively short durations of clinical trials, longer durations of treatment may be needed to achieve near complete amyloid reduction in most participants [61]. Therefore, the demonstration of target engagement in Phase 2 studies does not reliably predict a clinical response in Phase 3 [27]; this may be particularly true for anti-amyloid agents when examined in early symptomatic AD. Recent studies suggest that amyloid pathology may influence cognitive trajectories in earlier preclinical stages [69, 70]. Therefore, there has been an interest in shifting the focus of amyloid-targeting therapies to earlier preclinical stages of the disease, when they may have a better chance of achieving clinically relevant disease modification.

3.2 *Fluid and Imaging Markers of Tau*

Elevated CSF levels of several phosphorylated tau species, including p-tau181, p-tau231, and p-tau217, are observed in AD and are thought to reflect neurofibrillary tangle pathology [71, 72]. The release of p-tau into the extracellular space, and subsequently the CSF, may occur due to the death of tangle-bearing neurons or leakage of p-tau during its prion-like inter-neuronal spread [73]. High CSF p-tau181, in combination with low CSF Aβ42, levels represent the typical CSF biomarker profile of AD. Recent studies suggest that other p-tau species, such as p-tau217, provide comparable, and potentially higher, sensitivity and specificity for AD compared to p-tau181 [72]. Plasma assays for p-tau181, −217, and −231 have been developed and are currently being validated across different centers [74]. Plasma measures of p-tau species offer less invasive means to measure brain tau pathology compared to CSF.

CSF p-tau181 has been used as a measure of target engagement in several clinical trials of investigational anti-tau therapies. Monoclonal antibodies that target the N-terminal region of the tau protein have been examined in AD and other tauopathies. In the Phase 2 TAURIEL study, treatment with the anti-tau agent, semorinemab, in prodromal to mild AD dementia was associated with a dose-dependent increase in plasma levels of mid-domain tau, considered the target engagement marker in this trial, and decreased CSF levels of p-tau181 and total tau; however, this was not associated with reduced brain tangle pathology as measured by tau PET [75]. Another trial of semorinemab in mild to moderate AD dementia showed 42% slowing in clinical progression on the 11-item cognitive subscale of the Alzheimer Disease Assessment Scale (ADAS-Cog11), although no significant benefits were observed in the co-primary endpoint (Alzheimer Disease Cooperative Study—Activities of Daily Living; ADCS-ADL) or any of the secondary clinical endpoints [76]. Reductions in CSF total tau, p-tau181, and p-tau217 levels, but no significant effect on tau-PET burden, were observed in the treatment arm [76]. The study investigators concluded that changes in levels of soluble CSF tau and p-tau provide evidence to support target engagement by semorinemab; however, additional studies are needed to explain the absence of significant effects on the aggregation or spread of insoluble tau measured by tau-PET.

In the TANGO study, which examined another anti-tau agent, gosuranemab, in early AD, target engagement by the drug was demonstrated by up to 93% reductions in baseline CSF levels of the unbound N-terminal tau fragment in the treatment arms [77]. CSF total tau and p-tau181 levels were also reduced in some of the treatment arms, which examined different drug doses, compared to placebo. Despite this, no beneficial effects on cortical tau burden on PET, and no slowing of clinical progression, were observed with treatment [77].

Similar results have been reported in clinical trials of anti-tau therapies examined in other neurodegenerative disorders, such as progressive supranuclear palsy (PSP), including a lack of clinical benefit despite effective reduction of brain tau pathology. A Phase 2 trial of the anti-tau agent, tilavonemab, in PSP failed to show a beneficial effect on the primary study endpoint of the PSP rating scale (PSPRS), with scores possibly worsening in participants who received the higher doses, although this drug was shown to reduce CSF free tau levels, and increase plasma total tau levels, compared to placebo, as proof of target engagement [78]. A fourth anti-tau agent, zagotenemab, failed to reach its primary clinical endpoint in a Phase 2 study in early AD [79]. $MAPT_{RX}$ is an anti-sense oligonucleotide that targets tau and has been shown to produce >50% reductions in CSF total tau concentrations from baseline in a Phase 1b trial in mild AD dementia [80].

Similarly to what has been observed with anti-amyloid agents, the failure of anti-tau agents to produce clinical benefits despite their ability to reduce CSF tau or p-tau levels further supports the notion that demonstrating target engagement in Phase 2 studies does not necessarily translate into favorable clinical benefits in Phase 3 studies. For anti-tau therapies that target the N-terminal region of the tau peptide, it has been postulated that the lack of a clinical response may be due to loss of the normal physiological functions of the tau protein in spacing microtubules, a function for which the N-terminal region appears to be essential. Therefore, it has been suggested that anti-tau antibodies that target the microtubule binding region, which spans residues 224–369, of the peptide may be more effective in reducing the intercellular propagation of aggregated tau and thus, producing a clinical response [81]. This hypothesis remains to be examined in clinical trials.

An imbalance between tau kinases, such as glycogen synthase kinase-3 (GSK-3), and tau phosphatases, such as protein phosphatase A2, has been implicated in the abnormal tau hyperphosphorylation that occurs in AD. GSK-3 inhibitors such as valproic acid (or divalproex sodium) and lithium are among the anti-tau agents that have been investigated in AD clinical trials [82, 83]. These agents have failed to demonstrate consistent results in reducing CSF p-tau or slowing clinical disease progression [82–84]. Measuring tau kinetics can also be achieved using labeled leucine, although -so far- this appears to be underutilized in clinical trials of anti-tau agents.

Clinical trials of drugs that target other pathologies such as inflammation, hormonal pathways, and oxidative stress also provide examples of the utilization of biomarkers as measures of target engagement. Neflamapimod, a p38 α-kinase inhibitor with anti-inflammatory activity, showed reduced CSF p-tau181 and total tau levels, with no significant effects on CSF Aβ42, Aβ40, or neurofilament-light chain (NFL) levels or cognitive (i.e., episodic memory) outcomes in a Phase 2 study [85]. Rifaximin is an antimicrobial agent which alters gut microbiota and reduces pro-inflammatory mediators and was found to lower serum NFL levels despite no significant effect on inflammatory cytokine levels [86]. Cytochrome oxidase and citrate synthase in platelet-derived mitochondria were measured as target engagement markers in a small trial of the estrogen-receptor β agonist, S-equol, which induces mitochondrial activity, and were increased following 2 weeks of treatment in most treated patients [87].

A systematic review of 272 clinical trials of disease-modifying therapies in AD found that only 26% ($n = 72$) of the trials (26% of Phase 1, 31% of Phase 2, and 15% of Phase 3 trials) employed fluid-based biomarkers of target engagement in the study design [88]. Fluid-based biomarkers of target engagement were most commonly used in trials of drugs targeting inflammation (17 of 40) or amyloid (16 of 54). Conversely, there was a paucity of target engagement markers in trials of drugs targeting neurotransmitter receptors, neurogenesis, vascular mechanisms, epigenetic regulation, proteostasis, and the gut-brain axis. Perhaps the most striking was the paucity of target engagement markers in clinical trials of drugs that target synaptic plasticity and neuroprotection, with biomarkers being used in only 7 of 36 trials.

4 Biomarkers of Disease Modification

4.1 Fluid Biomarkers As Measures of Disease Modification

A major role for biomarkers in current drug development programs is to provide evidence in support of disease modification. Disease modification is defined by treatments or interventions that can change the disease course by influencing its underlying pathophysiology and the cellular and molecular pathways that lead to neurodegeneration, and subsequently, are expected to be associated with improved clinical outcomes [15]. Biomarkers that can reliably measure disease progression can be used to demonstrate evidence of disease modification in response to treatment and have been increasingly included as secondary, tertiary, or exploratory endpoints in clinical trials of emerging AD therapeutics.

Amyloid and tau biomarkers can provide evidence of target engagement of anti-amyloid and anti-tau agents, respectively. Further, changes in biomarkers that reflect tau pathology, or other pathologies potentially downstream to Aβ, may provide evidence of disease modification in clinical trials of anti-amyloid agents. Biomarker analyses in the Phase 3 trial of lecanemab in early symptomatic AD (CLARITY-AD) suggest favorable downstream effects of amyloid reduction on CSF p-tau181 levels at the end of the 18-month trial [44]. The Phase 3 trial of the anti-amyloid agent, bapineuzumab, showed reductions in CSF p-tau181 in apolipoprotein E4 (*APOE4*) carriers [89]. Lower CSF p-tau181 levels were also reported in participants treated with gantenerumab compared to placebo, with possible sex-based differences in p-tau181 responses observed in this trial [62, 90]. In the TRAILBLAZER-ALZ2 Phase 3 trial of the anti-amyloid agent, donanemab, plasma p-tau217 was examined as an exploratory measure of disease-modification associated with amyloid reduction [26]. Plasma p-tau217 in the donanemab arm was significantly reduced from baseline, compared with placebo, in the low-medium tau and combined cohorts at 76 weeks. However, no significant differences in the frontal tau load on tau PET imaging were observed between the treatment and placebo arms. Conversely, crenezumab was not associated with significant changes in CSF total tau or CSF p-tau181 in Phase 3 trials [47].

Fluid or imaging biomarkers that can reliably measure neuronal and synaptic loss, and are closely associated with cognitive outcomes, offer valuable tools to demonstrate disease modification [5]. Neuronal and synaptic loss reflects the cumulative outcome of various pathological substrates in AD and is more closely associated with cognitive outcomes than markers of amyloid or tau [17]. Markers of synaptic, neuronal, and axonal injury offer valuable tools to measure responses to therapeutic interventions which target different pathologies, and monitor their change over time, independently of changes to primary therapeutic targets such as CSF Aβ42 or p-tau [91, 92]. Therefore, these markers may provide useful secondary or tertiary outcome measures in clinical trials, and complement information provided by clinical, functional, or imaging assessments.

Biomarkers that measure "N" in the ATN framework, including CSF total tau, brain atrophy on MRI, or hypometabolism on 18F-FDG-PET have been used as endpoints, and measures of disease-modification, in several clinical trials of AD therapeutics [27]. Other emerging markers such as the post-synaptic injury marker, neurogranin [93], markers of neuronal injury such as visinin-like protein-1 (VILIP-1) [91, 94, 95], markers of axonal injury such as NFL [92], and markers of astrocytic reactivity which are closely associated with neurodegeneration such as glial fibrillary acidic protein (GFAP) [96] have also been implemented as outcome measures [44, 47]. Further supporting their value as measures of disease progression is the strong correlation of these markers with cognitive outcomes and imaging measures of neurodegeneration (e.g., brain atrophy) in studies of early symptomatic AD [91, 93–95].

Neurogranin (Ng) is a calmodulin-binding post-synaptic protein, which is abundantly expressed in the post-synaptic membrane and is involved in synaptic plasticity and long-term potentiation (LTP), including memory functions [93]. CSF Ng is a promising marker of synaptic injury which has demonstrated value as a diagnostic and prognostic marker across different cohorts and using different immunoassays [93, 97–99]. Baseline CSF Ng levels have been shown to predict future cognitive decline in cognitively normal individuals and are closely associated with whole brain and regional atrophy in AD, including atrophy of regions vulnerable to early AD pathology, such as the hippocampus, entorhinal cortex, and parahippocampal gyrus [93]. Further, CSF Ng levels are closely associated with other markers of AD pathology, such as CSF tau and p-tau181, and correlate with higher cortical amyloid burden on amyloid PET scans in cognitively normal individuals [93]. Moreover, CSF Ng levels may offer relative diagnostic specificity for AD as they differentiate AD from most other neurodegenerative disorders [93, 99]. CSF Ng has been used as a secondary or exploratory biomarker outcome measure in clinical trials of emerging AD therapies. In the DIAN-TU-001 trial of gantenerumab in autosomal-dominant AD, treatment was associated with decreased CSF Ng levels [100]. The CLARITY-AD Phase 3 trial of lecanemab showed reduced CSF Ng levels in association with amyloid reduction [44]. In both of these studies, reduced CSF Ng levels provide evidence for downstream disease-modifying effects of reducing brain amyloid. CSF Ng, and other markers such as NFL, are also included in several ongoing trials (e.g., the AHEAD 3–45 study of lecanemab in preclinical AD) [101]. Conversely, studies suggest that plasma Ng levels, measured using enzyme-linked immunosorbent assays (ELISAs), are not significantly elevated in AD compared to controls, and thus, offer limited value as outcome measures in AD trials [102]. Recent studies examining Ng in neuronally-derived blood extracellular vesicles (EVs) report lower levels, reflective of reduced brain expression, in AD compared to controls [103], and in individuals with MCI who progressed to dementia compared to those with MCI who did not progress [104].

VILIP-1 is a widely expressed neuronal calcium-sensor protein which has been implicated in calcium-mediated excitotoxicity in AD and is abundantly expressed in neuronal perikarya [105]. Studies across different cohorts suggest that CSF VILIP-1 is a promising diagnostic and prognostic marker of neuronal injury in AD [91, 94, 95]. In 3 large studies examining a well-characterized cohort of AD and healthy controls [91, 94, 95], CSF VILIP-1 levels were significantly higher in individuals with preclinical and early symptomatic AD compared to healthy controls and other dementias [91]. Further, elevated CSF VILIP-1 levels, reflective of neuronal loss, provided comparable utility to CSF tau, p-tau181, and Aβ42 in predicting the presence of brain amyloid on amyloid PET scans or a clinical diagnosis of AD [91]. CSF VILIP-1 levels were strongly associated with CSF markers of tau pathology and whole brain and regional atrophy of the hippocampus, entorhinal cortex, parahippocampal gyrus, fusiform, and posterior cingulate; regions which are involved in early AD [91]. Higher baseline CSF VILIP-1 levels in cognitively normal individuals strongly predicted a higher risk of future cognitive decline over 7 years of follow-up [94]; high baseline CSF VILIP-1 or the VILIP-1/Aβ42 ratio predicted progression from CDR 0 to CDR > 0 over time to a better extent than baseline CSF tau, p-tau181, or tau/Aβ42 and p-tau181/Aβ42, respectively [91]. In individuals with early symptomatic AD, higher baseline CSF VILIP-1 levels were closely associated with higher rates of cognitive decline in global cognition and measures of episodic memory, working memory, language, and visuospatial functions in longitudinal studies [94]. Further, whole brain and regional atrophy progressed more rapidly in individuals with early symptomatic AD who had higher baseline CSF VILIP-1 levels, compared to those with lower baseline levels, over follow-up [95]. These studies also suggest that CSF VILIP-1 may offer relative specificity for neuronal loss associated with AD compared to other neurodegenerative diseases such as frontotemporal dementia (FTD), Parkinson's disease (PD), and Dementia with Lewy bodies (DLB); however, studies with larger non-AD cohorts are needed [91]. Elevated plasma VILIP-1 levels have also been observed in AD [91]. The Single Molecule Array (SIMOA, Quanterix®) technology has recently been applied to measure VILIP-1 levels in blood samples with higher sensitivity than ELISAs, which will further expand the potential of VILIP-1 as a measure of disease modification in future clinical trials of AD therapeutics [106]. Several clinical trials of emerging AD therapies have included VILIP-1 as a secondary or exploratory outcome measure including Phase 2 and 2/3 studies of the BACE inhibitor, atabecestat (Clinical Trials NCT02569398 and NCT02406027), studies of the anti-tau antisense oligonucleotide, $MAPT_{RX}$ (NCT03186989) in AD, as well as large observational studies of AD such as the China Cognition and Aging Study (COAST) [107] and the Dominantly Inherited Alzheimer's Network (DIAN) [108].

NFL is a soluble and highly expressed component of neuronal axons which is released in the setting of neurodegeneration [109]. CSF NFL levels have been shown to correlate with brain atrophy, cognitive decline, and reduced survival in several cross-sectional and longitudinal studies of AD [110–112]. However, elevated CSF NFL levels have also been reported in other neurodegenerative disorders, traumatic and ischemic brain injury, and demyelinating or inflammatory neurological disorders, in which they provide useful information regarding the severity and progression of neuronal injury. Therefore, despite its value to measure and track disease progression in AD, CSF NFL elevation is not specific for AD-related neurodegeneration [113]. The implementation of ultrasensitive biomarker assays using SIMOA has allowed the reliable quantification of NFL in blood samples [114]. Plasma or serum NFL levels differentiate preclinical and early symptomatic AD from controls in studies of familial and sporadic AD [114, 115], and accurately predict rates of disease progression over time, offering a useful, although relatively nonspecific, peripheral measure of disease severity [115, 116]. CSF or plasma NFL levels have been utilized as secondary or exploratory endpoints in clinical trials of AD therapies. So far, a few trials of anti-amyloid agents have failed to demonstrate significant reductions in CSF NFL levels [44], which has been potentially explained by the fact that changes in NFL levels may require longer durations of treatment than the relatively short durations of trials. However, plasma or CSF NFL levels significantly increased with a few anti-amyloid agents, such as solanezumab, compared to placebo [100]. Further, a trial of the anti-inflammatory agent, neflamapimod, was not associated with significant changes in CSF NFL levels. Therefore, more research is needed to better understand the dynamics of CSF or plasma NFL levels in response to amyloid reduction or targeting inflammation and the extent and direction by which emerging AD therapies that target core AD pathologies may influence neurodegeneration.

In addition to Ng, other emerging markers of synaptic injury which may offer value as potential outcome measures in AD clinical trials include synaptosomal-associated protein-25 (SNAP-25) and synaptotagmin-1 (Syt-1). SNAP-25 is involved in vesicle docking, neurotransmitter release, and neurite outgrowth [117]. Higher CSF levels of soluble SNAP-25 fragments are observed in AD compared to controls, and correlate with brain atrophy and future risk of cognitive decline [118, 119]. However, altered CSF SNAP-25 levels have also been reported in primary psychiatric and other neurodegenerative conditions, suggesting relatively limited diagnostic specificity of SNAP-25 for AD [117]. Therefore, further research is warranted before CSF SNAP-25 can be more widely implemented as a measure of synaptic loss or neurodegeneration in AD clinical trials. Two splicing variants of SNAP-25, referred to as SNAP-25a and SNAP-25b which differ in 9 amino acid residues, have been identified [120]. Whether CSF levels of the 2 isoforms have different diagnostic or prognostic utility in AD remains to be investigated. SNAP-25 and/or Syt-1 have been included as outcome measures in a few studies examining anti-tau agents [121] or other treatment effects on brain synaptic density [122].

CSF levels of growth-associated protein-43 (GAP-43, neuromodulin) may also offer promise as outcome measures of axonal or pre-synaptic injury, although larger studies are needed to validate their value as measures of neurodegeneration in AD. GAP-43 has important roles in axonal branching and synaptic plasticity [123], and facilitates vesicle recycling via its interactions with the pre-synaptic proteins, synaptophysin and SNAP-25 [124]. CSF GAP-43 levels are increased in early symptomatic AD compared to controls and most other neurodegenerative conditions, and correlate with amyloid and tau pathology in AD [125, 126]. Despite these promising results, GAP-43 has not yet been widely adopted as an outcome measure of neurodegeneration in AD clinical trials.

Other markers that measure astrocytic reactivity, such as GFAP, YKL-40 (also known as chitinase-3-like protein-1 [CHI3L1]) and S-100b, which have demonstrated diagnostic value in AD, have also been included as endpoints in several trials of anti-amyloid therapies. GFAP is a type-III intermediate filament that is predominantly expressed by astrocytes near amyloid plaques [127]. Increased expression of both GFAP isoforms, GFAP-α and -β, has been observed in long astrocytic processes in AD brains, and a gradual increase in GFAP+1 isoform expression in a subset of astrocytes is associated with disease progression [96]. Higher CSF and plasma GFAP levels have been reported in AD and other neurodegenerative disorders such as PD, compared to controls, and are thought to reflect reactive astrogliosis [128–130]. Further, higher CSF GFAP levels have been observed in FTD compared to AD or DLB [131]. CSF and plasma GFAP levels have been shown to be associated with cognition in baseline and longitudinal studies of AD and PD [130, 132, 133]. There has recently been a growing interest in implementing plasma or CSF GFAP as a marker of disease modification to provide evidence supporting downstream effects of therapeutic agents that target amyloid or tau [44]. Plasma GFAP appears to outperform CSF GFAP as an indicator of brain amyloid pathology in AD [134]. Amyloid reduction was associated with lower plasma GFAP levels in clinical trials of gantenerumab, donanemab, and lecanemab [44, 48, 100]. In these trials, lower plasma GFAP levels associated with treatment likely signify downstream effects of disease modification; however, further research is warranted to understand the underlying mechanisms for these findings.

YKL-40 is a glycoprotein which is expressed by astrocytes near amyloid plaques with putative roles in astrocytic reactivity, inflammation, and angiogenesis [135]. Several studies have reported elevated CSF YKL-40 levels in early symptomatic AD compared to controls, and suggest that YKL-40 levels are associated with markers of amyloid, tau, and synaptic injury, but not *APOE4*, in AD [136–138]. CSF YKL-40 levels predict progression from MCI to AD dementia [139] and are associated with brain atrophy in *APOE4* carriers [140] or individuals with low CSF Aβ42 [141]. Despite its value as a marker of reactive astrogliosis in AD, CSF YKL-40 is elevated in several neurological disorders, and thus, may not be useful in differentiating AD from other dementias [142, 143]. It has been suggested that CSF YKL-40 levels are influenced by sex, ethnicity, and genetic polymorphisms of the *CHI3L1* gene [144], which may also need to be considered when evaluating the potential value of this marker as an outcome measure in AD clinical trials. Recent studies suggest that CSF YKL-40 levels may not be reflective of YKL-40 brain pathology in AD and rather, may be more closely associated with cerebral amyloid angiopathy [145]; however, further investigation is needed.

Elevated CSF S-100b levels in AD likely reflect increased S-100b expression in astrocytes, and possibly microglia, surrounding amyloid plaques [146], but are not specific for AD [147]. In AD, S-100b appears to be involved in the inflammatory response, and may contribute to tau pathology and increased amyloid deposition through activation of β-secretase [148]. While YKL-40 and S-100b have not been widely adopted as outcome measures in AD clinical trials targeting amyloid, they may offer value in clinical trials of other treatment modalities targeting inflammation. Recent studies suggest that plasma YKL-40 may offer predictive value for the future risk of all-cause dementia although further studies are needed to determine its association with the risk of dementia due to AD [149].

Several clinical trials have successfully implemented AD biomarkers as measures of target engagement or disease modification. The main findings from these studies are summarized in Table 3.

Table 3 Treatment effects on cognitive and biomarker outcomes in Alzheimer disease clinical trials

Monoclonal anti-amyloid antibodies	
Lecanemab [44]	Favorable effects on primary and secondary clinical outcomes including slowing of progression on CDR-SB® (primary), ADAS-Cog14, ADCS-MCI-ADL, and ADCOMS at end of the 18-mo trial compared to placebo Significant reduction in brain amyloid load on PET with treatment Increased CSF Aβ42 but not CSF Aβ40; decrease in CSF tau, p-tau181, and Ng with treatment; No changes in CSF NFL Increased plasma Aβ42/40 ratio, decreases in plasma p-tau181 and GFAP with treatment
Donanemab [26, 48]	Favorable effects of treatment on primary (iADRS) and several secondary endpoints (including CDR-SB®) Significant reduction in brain amyloid load on PET with treatment Plasma p-tau217 and GFAP decreased with treatment
Aducanumab [57, 156]	Possible slowing of cognitive decline (observed in only 1 of the 2 Phase 3 trials) Significant reduction in brain amyloid load on PET with treatment
Gantenerumab [50, 62, 100]	No significant effect on cognition with treatment Significant reduction in brain amyloid load on PET with treatment CSF tau, p-tau181, and Ng decreased, while CSF sTREM2 increased, with treatment No significant change in CSF Aβ42 levels in some studies; increased CSF Aβ42 in others Plasma GFAP decreased with treatment
Crenezumab [47]	No significant effect on cognition with treatment No significant change in amyloid PET Increase in CSF and plasma Aβ42 and Aβ40 with treatment No significant change in CSF tau, p-tau181, or Aβ oligomers with treatment
Solanezumab [42, 43, 100]	No significant effect on cognition with treatment No significant change in amyloid PET CSF and plasma Aβ42 and Aβ40, and CSF NFL increased with treatment
Bapineuzumab [89]	No significant effect on cognition with treatment Significant reduction in brain amyloid load on PET with treatment in *APOE4+* individuals Reduced CSF p-tau181 in *APOE4+* individuals
Monoclonal anti-tau antibodies	
Semorinemab [75, 76]	No significant effect on cognition with treatment CSF tau, p-tau181, and p-tau217 decreased with treatment; no effect on tau PET Increase in plasma mid-domain tau (target engagement)
Gosuranemab [77]	No significant effect on cognition with treatment CSF free N-terminal tau decreased with treatment (target engagement) No effect on tau PET
Tilavonemab [78]	No significant effect on clinical outcomes in PSP; no cognitive benefits in other trials in AD CSF free tau decreased with treatment (target engagement) Increased plasma total tau (target engagement)

Abbreviations: *CDR-SB* Clinical Dementia Rating-sum of boxes, *iADRS* integrated Alzheimer Disease Rating Scale, *ADAS-Cog14* 14-item cognitive subscale of the Alzheimer Disease Assessment Scale, *ADCS-MCI-ADL* Alzheimer Disease Cooperative Study-Activities of Daily Living for Mild Cognitive Impairment, *ADCOMS* Alzheimer Disease Composite Score, *CSF* cerebrospinal fluid, *Aβ42* amyloid peptide 1–42, *Aβ40* amyloid β peptide 1–40, *Ng* neurogranin, *NFL* neurofilament-light chain, *p-tau181* tau phosphorylated at threonine 181, *p-tau217* tau phosphorylated at threonine 217, *GFAP* glial fibrillary acidic protein, *APOE4* apolipoprotein E4, *PSP* progressive supranuclear palsy

4.2 Imaging Measures of Neurodegeneration

Volumetric brain MRI and 18F-FDG-PET are considered markers of neurodegeneration reflected by "N" in the updated research framework proposed by the NIA-AA (referred to as the ATN classification with CSF or imaging markers of amyloid represented by "A" and CSF or imaging markers of tau represented by "T"; CSF markers of "N" include CSF total tau and NFL with the possibility for emerging CSF and plasma markers such as VILIP-1 and GAP-43 to be included).

Brain atrophy on MRI reflects neuronal loss which follows neurodegeneration and typically progresses over time. Theoretically, disease-modifying treatments that slow the progression of AD pathologies contributing to neurodegeneration are expected to slow whole brain atrophy, or atrophy in regions affected by early AD such as the hippocampus and entorhinal cortex, over the trial period leading to measurable drug-placebo differences at the end of the trial with slower progression of brain atrophy in the treatment, compared to placebo, groups. However, this result has rarely been achieved in clinical trials of disease-modifying treatments in AD [27]. Most trials have failed to demonstrate drug-placebo differences in brain atrophy at the end of the trial period [50, 64, 65, 89, 150–152], and in fact, a few studies suggest possible worsening of brain atrophy, demonstrated by increased ventricular volumes, in the treatment arm compared to placebo [153–155]. The exact clinical significance of these findings, and the underlying mechanisms by which they occur, are yet to be determined [61]. It remains unclear whether possible worsening of brain atrophy with treatment may reflect reduced inflammation and edema [155], or whether certain treatments may be associated with more rapid neuronal loss. Observations from other clinical trials of anti-amyloid therapies (e.g., solanezumab) that have found higher levels of CSF NFL [100], another marker of neurodegeneration, associated with treatment may also need to be considered when examining these findings. It will be important for future studies to provide an understanding of the mechanisms by which anti-amyloid agents, or potentially other disease-modifying treatments, may paradoxically worsen neurodegeneration and whether specific participant subgroups are more susceptible than others.

Hypometabolism on 18F-FDG-PET is another imaging marker that reflects "N" in the ATN classification as it measures reduced metabolic activity as a result of neuronal or synaptic degeneration. Only a few clinical trials have included 18F-FDG-PET as an outcome measure likely due to technical and methodological limitations [27]. However, several advances in 18F-FDG-PET techniques and data from studies on neuropathologically-confirmed cohorts and those with longitudinal follow-up suggest that 18F-FDG-PET may be suitable to demonstrate treatment effects in multi-site trials and may allow the detection of treatment effects in relatively small cohorts [27, 157, 158]. As several medications, including cholinesterase inhibitors used to stabilize symptoms in AD, may influence hypometabolism on 18F-FDG-PET [159], this will need to be considered when including this marker as a potential outcome in clinical trials of AD therapies [27].

5 Current Landscape of Biomarkers in AD Clinical Trials

A recent systematic review of the current landscape of fluid biomarker utilization in AD clinical trials examined 272 clinical trials of AD disease-modifying therapies between 2017 and 2024 [88]. Findings from this study suggest that only 44% of the trials (n = 121 trials) employed fluid biomarkers as endpoints, either by measuring target engagement and/or disease-modification. Among trials that used fluid biomarker endpoints, the majority were in Phase 2 (64%; n = 77) and less frequently in Phase 1 (23%; n = 28) or Phase 3 (13%; n = 16). Throughout all study phases, biomarkers were mostly used as exploratory or secondary, rather than primary, outcome measures. Blood and CSF biomarker use was generally similar across all phases. As expected, phase 3 studies, the design of which requires the demonstration of clinical efficacy, were the least likely to use fluid biomarkers as primary endpoints (6–7%) compared to Phase 1 and 2 studies (15%) and relied on cognitive and functional outcomes as primary endpoints. Academia-sponsored trials (71%) were more likely than industry-sponsored trials (36%) to employ fluid biomarkers as endpoints. Of the fluid biomarkers employed by clinical trials, CSF markers of tau pathology, often in combination with markers of amyloid or other pathologies, were most frequently used. The use of fluid biomarkers for target engagement was generally limited.

The trials involving donanemab, lecanemab, and gantenerumab (Table 3) provide valuable insights into the association of changes in AD biomarkers with cognitive outcomes, and thus, the potential value of these markers as biomarker surrogate endpoints in trials of emerging AD therapies. CSF or plasma p-tau181, plasma p-tau217, and plasma GFAP emerge as the most promising candidates for surrogate biomarkers in the amyloid pathway. Conversely, data from these trials suggests that CSF or plasma NFL may not be a robust outcome measure for disease-modifying effects in AD clinical trials, and further research is needed to explain the increase in CSF NFL levels observed in some trials of anti-amyloid agents. Conversely, in other trials of disease-modifying treatments (e.g., anti-tau therapies: semorinemab, gosuranemab, and tilavonemab), biomarker changes were not associated with disease-modification or clinical benefits, and therefore, it is important to note that target engagement alone is not enough to demonstrate disease-modifying effects or reliably predict cognitive outcomes.

5.1 *Biomarkers in Drug Development*

Several factors need to be considered when evaluating the appropriateness of a biomarker as a measure of target engagement or disease modification in clinical trials. The proper choice of biomarker endpoints depends largely on the context of use, including the patient population, trial duration, biomarker dynamics influencing CSF or blood levels, and the therapeutic target. Biomarkers that are intended to measure target engagement should directly relate to the ability of the investigational drug to bind and modulate its target and therefore, should be highly specific to the drug and target class. At this time, there is insufficient evidence to suggest that any single fluid AD biomarker, alone, can reliably predict clinical outcomes and therefore, serve as a primary endpoint; however, fluid biomarkers that are known to be associated with cognition may serve as surrogate endpoints and complement information provided by cognitive or functional outcome assessments [27, 160]. Importantly, as clinical outcomes associated with emerging AD therapies may not be fully apparent by the end of the relatively short clinical trial durations, several clinical trials of AD therapies have recently incorporated fluid or imaging biomarkers as secondary or exploratory outcome measures to demonstrate drug efficacy (Table 3). By implementing dynamic biomarkers that are known to track with disease progression, biomarkers can provide biological evidence of disease modification relatively early in the course of trials and complement information provided by clinical or cognitive outcomes regarding treatment efficacy.

Another important consideration is confidence in the technical performance of the assays used to measure the biomarkers of interest, including their sensitivity, specificity, analytical variability due to internal or external factors (e.g., diurnal variations, sample handling, processing, and storage, tube material, method of collection and freeze-thaw cycles), reproducibility, and potential for standardization. While many proteins are known to be associated with AD pathogenesis, advanced assays with adequate technical performance may not be readily available for many of these, limiting their applicability as clinical trial endpoints. Further, there is a need to validate and standardize assay methods used for biomarker measurements across different cohorts and centers which will facilitate comparisons across different clinical trials that measure similar biomarker endpoints. The European Medicines Agency (EMA) recommends that biomarker qualification consider information regarding the ability of biomarkers to monitor disease progression, intra-individual variability in biomarker measurements, effects of population variability, and longitudinal changes over time in response to disease progression or treatment [88, 161]. Further, it will be important to understand how plasma or CSF biomarker levels may vary by age, sex, race/ethnicity, or the *APOE4* genotype and the influence of concomitant disease such as hypertension, diabetes, or renal insufficiency on biomarker levels.

Biomarker-supported drug development has been widely implemented in other fields such as oncology and cardiology and is thought to facilitate market approval by reducing trial failure rates [28]. In AD, the integration of biomarkers into drug development programs of AD therapeutics has been increasingly utilized and encouraged by regulatory agencies such as the US-FDA and the EMA. While these agencies emphasize the need for cognitive and functional outcomes in evaluating drug efficacy, biomarkers may be used to potentially provide evidence to support disease-modifying effects and complement information provided by clinical assessments [162, 163]. In an updated draft guidance for drug development programs in early-stage AD which are focused on targeting the earliest pre-symptomatic stages, the FDA statement indicates that the agency strongly supports and encourages continued research in understanding the role of biomarkers in AD and stresses the potential importance of biomarkers in the successful development of effective treatments appropriate for use in the earliest stages of AD [164]. The guidance also indicates that direct measures of clinical benefit or validated surrogate endpoints may support a traditional approval, while surrogate endpoints or intermediate clinical endpoints that do not directly measure clinical benefit, but are considered reasonably likely to predict clinical benefit, may support an accelerated approval. However, the guidance also clarifies that the appropriateness of a surrogate endpoint will depend on the disease stage, trial population, therapeutic mechanism of action, and availability of existing treatments so that a biomarker that may be a useful surrogate endpoint in one trial may not necessarily be an appropriate surrogate endpoint in another trial. According to this guidance, the FDA encourages sponsors who are considering the use of a biomarker as the primary measure of effect to discuss their plans with the FDA in the early stages of drug development.

Recently, two anti-amyloid agents, aducanumab and lecanemab, received accelerated FDA approval based on their effects on a surrogate endpoint (i.e., amyloid PET) as these drugs significantly and reliably reduced brain amyloid load on amyloid PET scans in the treatment arm compared to placebo. Despite effective amyloid clearance, aducanumab did not demonstrate clinical benefits in one of the two Phase 3 trials, and clinical benefits were only modest and not likely clinically meaningful in the second trial. Due to several factors including controversial benefits and low sales, aducanumab has been discontinued. Conversely, lecanemab proceeded to full FDA approval based on results of a confirmatory Phase 3 trial which demonstrated statistically significant slowing of progression on several clinical outcome measures including the primary outcome measure, CDR-SB, and several secondary clinical outcomes such as the ADCS-MCI-ADL, ADAS-Cog14, and the ADCOMS. The FDA also granted accelerated approval for tofersen for amyotrophic lateral sclerosis (ALS) due to its ability to lower CSF NFL levels, reflective of neurodegeneration [165]. In all these examples, accelerated approval was based on the ability of these treatments to demonstrate "an effect on a surrogate endpoint (i.e., amyloid PET for AD and CSF NFL for ALS) that is reasonably likely to predict a clinical benefit to patients".

CSF p-tau181/Aβ42 and total tau/Aβ42, measured by Elecsys® assays (Roche Diagnostics), and the Lumipulse® G CSF Aβ42/Aβ40 assay (Fujirebio Diagnostics), have been validated and FDA-approved for use as diagnostic markers, and therefore, can be used to support timely diagnosis and enrich clinical trial cohorts. FDA Breakthrough Device designation has also been granted to the Elecsys® amyloid plasma panel and the Elecsys® plasma p-tau217 assay (Roche Diagnostics). However, to the best of our knowledge, none of the AD fluid biomarkers have received FDA approval for use as clinical trial endpoints at the time of this review. CSF Aβ42 and CSF total tau have been approved by the EMA for the identification of prodromal AD and for participant enrichment in clinical trials of mild to moderate AD dementia.

As AD clinical trials have appropriately begun to shift focus from targeting early symptomatic AD to targeting even earlier pre-symptomatic stages, the role of

Table 4 Five phases of biomarker development

Phase 1	Biomarker discovery and validation (including omics) Preclinical studies in animal models related to the target mechanism with recommendation to develop biomarker and drug simultaneously
Phase 2	Technical: Development and/or validation of assays Evaluating assay parameters including sensitivity, specificity, predictive values, reproducibility and analytical variability Identifying effect of pre-analytical factors on biomarker measurements Clinical: Correlate with gold standard measures and disease features
Phase 3	Technical Establish cut-offs for normal vs abnormal values Clinical: Examine diagnostic accuracy in AD cohorts at various stages
Phase 4	Technical: In Vitro Diagnostic (IVD) assay development and validation Clinical: Establish context of use in clinical trials Examine cost-efficacy Demonstrate predictive validity Conduct prospective diagnostic studies
Phase 5	Regulatory IVD assay certification Clinical: Establish biomarker performance in different clinical settings and across different patient cohorts Integrate biomarker use in clinical guidelines Evaluate biomarker role in clinical decision-making Reimbursement

Adapted from [88, 160]

biomarkers will become increasingly important as outcome measures, as evaluating clinical efficacy in individuals with no or minimal cognitive or functional deficits would require large cohorts to be followed over long periods. Therefore, biomarkers may provide useful surrogate endpoints to evaluate treatment efficacy and facilitate shorter trial durations. Table 4 demonstrates a proposed framework for biomarker development and incorporation into clinical trials of AD therapies [88, 160].

Additional concepts that are relevant to the implementation of biomarkers in AD drug development are "companion" and "complementary" diagnostics. A companion diagnostic is an in vitro diagnostic which is required for the safe and effective use of a corresponding therapeutic, and typically may be used to identify individuals who are more likely to respond to treatment or more likely to have side effects, or to monitor treatment response [166, 167]. Detailed clinical and technical validation is typically needed to approve companion diagnostics, and when approved, their use is mandatory prior to administering the drug [168]. Amyloid PET scans provide an example of a companion diagnostic in clinical trials of anti-amyloid agents; if an anti-amyloid agent is approved based on a clinical trial that defined study eligibility based on abnormal amyloid PET scan results, then amyloid PET imaging would be considered a companion diagnostic that should be obtained and evaluated prior to drug administration in clinical settings [27]. An important aspect of validation of companion diagnostics is the identification of proper thresholds or cut-off values that differentiate "normal" from "abnormal" results. Conversely, a complementary diagnostic is not required for prescribing a therapeutic agent but provides important information to facilitate risk or benefit stratification, as it may identify subsets of participants who may have a higher chance of a therapeutic response or a higher risk for side effects [169]. For example, in most clinical trials of anti-amyloid therapies, the risk of amyloid-related imaging abnormalities (ARIAs) appears to be higher in *APOE4* carriers compared to non-carriers; therefore, testing for the *APOE4* genotype would be considered a complementary diagnostic in these settings (note: the appropriate use recommendations for the use of lecanemab in clinical settings include APOE4 genotyping to better inform risk assessments related to ARIAs).

6 Conclusions and Future Directions

There has been great progress over the last few years in integrating biomarkers into the design of AD clinical trials, particularly regarding the use of amyloid PET for patient selection, and the utilization of fluid markers, in CSF or plasma, as secondary outcome measures in clinical trials of anti-amyloid therapies [26, 44, 170]. As most drug development efforts have focused on amyloid-targeting agents, biomarkers that measure amyloid, tau, and to a lesser extent, neuronal or axonal injury have been included as secondary, tertiary or exploratory outcomes in trials of anti-amyloid agents. However, overall, biomarkers remain underutilized in AD drug development. A recent meta-analysis suggests that only a third of industry-sponsored trials of emerging AD therapies used fluid biomarkers [88]. Until recently, there has been a paucity of investigational treatments in the AD drug pipeline which target other pathological substrates including tau, α-synuclein, TDP-43, oxidative stress, or calcium-mediated excitotoxicity [171, 172]. As newer agents that target these pathologies enter more advanced stages of drug development, there will be a need to identify and validate fluid and imaging markers that can reliably measure AD pathologies beyond amyloid and tau, as these will aid participant selection and stratification by confirming the presence of the target pathology (e.g., the presence of α-synuclein co-pathology in synuclein-targeting therapies) and potentially identifying individuals who may be at higher risk for complications associated with treatments (e.g., vascular or endothelial injury biomarkers may identify participants with baseline endothelial injury who may be at higher risk for ARIAs associated with anti-amyloid agents) [5, 173].

AD is a neuropathologically complex disease that extends beyond amyloid and tau. Studies suggest that non-amyloid co-pathologies, including tau, α-synuclein, TDP-43, and vascular disease, account for >50% of variance in cognitive outcomes

[174, 175]. Given the pathological heterogeneity of AD, and the variability of clinical phenotypes associated with any particular pathology, biomarkers are needed to provide adequate molecular characterization and stratification of individuals with variable degrees of AD co-pathologies [18]. The significance of this has been highlighted by emerging data from clinical trials of anti-amyloid agents (e.g., donanemab) that suggests that cognitive outcomes associated with amyloid reduction are influenced by the baseline burden of tau pathology [26].

Novel biomarker discovery is at the forefront of AD research and will become even more important as AD trials appropriately shift focus to preclinical AD [101]. Based on their context of use, biomarkers can improve patient stratification for clinical trials of AD therapies and enrich study cohorts with individuals who are at high risk for future cognitive decline [176]. This is particularly relevant in clinical trials of preclinical AD in which cognitive symptoms are minimal or absent. In this scenario, biomarkers are valuable in identifying participants who are the most likely to respond to disease-modifying drugs, based on biomarker evidence of disease pathology, and thereby, increase the power of studies to detect drug effects during the relatively short durations of follow-up [18]. This approach has been successfully applied in several clinical trials of investigational anti-amyloid therapies in cognitively normal individuals with significant amyloid deposition on amyloid PET scans [101], and those with genetic risk factors for AD or AD-causing mutations.

Importantly, demonstrating the ability of new biomarkers to track the progression of relevant AD pathologies over time or in response to treatment will support their potential value as outcome measures in clinical trials of disease-modifying therapies. Biomarker studies in large well-characterized cohorts who are followed longitudinally will be important to understand the dynamic changes of these biomarkers naturally with disease progression and how this may be altered with treatment [5]. Biomarkers that demonstrate dynamic changes over the trial period and are closely associated with cognitive outcomes will further facilitate drug development by providing biomarker surrogate co-endpoints, along with cognitive or functional measures, in support of effective disease modification [27]. Therefore, establishing associations of biomarkers with AD pathologies in cross-sectional studies, without demonstrating their ability to change with disease-modification over the average duration of clinical trials (~2 years), may not reliably predict their appropriateness as trial outcome measures. For example, despite a large number of studies demonstrating the ability of CSF or plasma NFL to measure disease severity, cognitive decline, and brain atrophy in AD, surprisingly, this biomarker had limited value as an outcome measure of disease-modification in several trials of anti-amyloid agents, failing to show a reduction with treatment over the trial period despite the ability of these trials to demonstrate beneficial effects of these therapies on clinical progression and their ability to reduce levels of other markers of synaptic loss or neuronal injury (e.g., Ng or tau) [44, 100].

Multi-omics approaches to novel biomarker and drug target discovery are the future of AD biomarker research. Data-driven approaches which implement bioinformatics, mechanism-based modeling, simulation approaches, and unsupervised clustering will provide valuable tools to elucidate novel pathways or mechanisms involved in AD pathogenesis [177–179]. Further, integrating network biology and pathway databases to examine multidirectional interactions and nonlinear associations between different pathologies is more likely to capture the neuropathological heterogeneity of AD than empiric approaches which focus on single biomarker data [180, 181]. Systems biology allows the integration of biomarker data with other domains (i.e., clinical, cognitive, and imaging) and may be useful, not only to reflect the molecular, genetic, and pathological heterogeneity of AD, but also to measure the spatiotemporal evolution of AD pathologies over time [180, 181]. Importantly, computational modeling methods will be important to generate novel predictive models of disease progression or response to mechanism-based treatments among individuals with different clinical phenotypes. Overall, advanced data analytics that extend beyond empiric approaches can identify different signatures of AD pathology, or different molecular subtypes of AD among different individuals, and will pave the way towards personalized approaches to disease treatment and the prevention of cognitive decline.

Funding This work was supported by the National Institute of Health-National Institute of Aging Grant RF1AG083744 (RT).

References

1. Tarawneh R, Holtzman DM. The clinical problem of symptomatic Alzheimer disease and mild cognitive impairment. Cold Spring Harb Perspect Med. 2012;2(5):a006148.
2. Alzheimer's Association. The Alzheimer's Association Report: facts and figures; 2023. Available from: https://www.alz.org/alzheimers-dementia/facts-figures
3. World Health Organization and Alzheimer's disease international. Dementia: A Public Health Priority; 2012. Available from: https://www.who.int/publications/i/item/dementia-a-public-health-priority.
4. Cummings JL, Morstorf T, Zhong K. Alzheimer's disease drug-development pipeline: few candidates, frequent failures. Alzheimers Res Ther. 2014;6(4):37.
5. Tarawneh R. Biomarkers: our path towards a cure for Alzheimer disease. Biomark Insights. 2020;15:1177271920976367.
6. Kim CK, Lee YR, Ong L, Gold M, Kalali A, Sarkar J. Alzheimer's disease: key insights from two decades of clinical trial failures. J Alzheimers Dis. 2022;87(1):83–100.
7. Mehta D, Jackson R, Paul G, Shi J, Sabbagh M. Why do trials for Alzheimer's disease drugs keep failing? A discontinued drug perspective for 2010–2015. Expert Opin Investig Drugs. 2017;26(6):735–9.
8. Aisen PS, Cummings J, Jack CR Jr, Morris JC, Sperling R, Frölich L, et al. On the path to 2025: understanding the Alzheimer's disease continuum. Alzheimers Res Ther. 2017;9(1):60.
9. Iqbal K, Grundke-Iqbal I. Alzheimer's disease, a multifactorial disorder seeking multitherapies. Alzheimers Dement. 2010;6(5):420–4.
10. Carreiras MC, Mendes E, Perry MJ, Francisco AP, Marco-Contelles J. The multifactorial nature of Alzheimer's disease for developing potential therapeutics. Curr Top Med Chem. 2013;13(15):1745–70.
11. Iturria-Medina Y, Carbonell FM, Sotero RC, Chouinard-Decorte F, Evans AC. Multifactorial causal model of brain (dis)organization and therapeutic intervention: application to Alzheimer's disease. NeuroImage. 2017;152:60–77.

12. Ferreira D, Wahlund LO, Westman E. The heterogeneity within Alzheimer's disease. Aging (Albany NY). 2018;10(11):3058–60.
13. Devi G, Scheltens P. Heterogeneity of Alzheimer's disease: consequence for drug trials? Alzheimers Res Ther. 2018;10(1):122.
14. Au R, Piers RJ, Lancashire L. Back to the future: Alzheimer's disease heterogeneity revisited. Alzheimers Dement (Amst). 2015;1(3):368–70.
15. FDA-NIH Biomarker Working Group. BEST (Biomarkers, EndpointS, and other Tools) Resource; 2021. Available from: https://www.ncbi.nlm.nih.gov/books/NBK338449/
16. Jack CR Jr, Knopman DS, Jagust WJ, Petersen RC, Weiner MW, Aisen PS, et al. Tracking pathophysiological processes in Alzheimer's disease: an updated hypothetical model of dynamic biomarkers. Lancet Neurol. 2013;12(2):207–16.
17. Price JL, Ko AI, Wade MJ, Tsou SK, McKeel DW, Morris JC. Neuron number in the entorhinal cortex and CA1 in preclinical Alzheimer disease. Arch Neurol. 2001;58(9):1395–402.
18. Tarawneh R, Holtzman DM. Biomarkers in translational research of Alzheimer's disease. Neuropharmacology. 2010;59(4–5):310–22.
19. Tarawneh R, Holtzman DM. Critical issues for successful immunotherapy in Alzheimer's disease: development of biomarkers and methods for early detection and intervention. CNS Neurol Disord Drug Targets. 2009;8(2):144–59.
20. Jack CR Jr, Bennett DA, Blennow K, Carrillo MC, Dunn B, Haeberlein SB, et al. NIA-AA research framework: toward a biological definition of Alzheimer's disease. Alzheimers Dement. 2018;14(4):535–62.
21. Sperling RA, Karlawish J, Johnson KA. Preclinical Alzheimer disease-the challenges ahead. Nat Rev Neurol. 2013;9(1):54–8.
22. Sabbagh MN, Hendrix S, Harrison JE. FDA position statement "early Alzheimer's disease: developing drugs for treatment, guidance for industry". Alzheimer's Dementia. 2019;5:13–9.
23. Cummings J. Lessons learned from Alzheimer disease: clinical trials with negative outcomes. Clin Transl Sci. 2018;11(2):147–52.
24. Landau SM, Horng A, Fero A, Jagust WJ. Amyloid negativity in patients with clinically diagnosed Alzheimer disease and MCI. Neurology. 2016;86(15):1377–85.
25. Sevigny J, Suhy J, Chiao P, Chen T, Klein G, Purcell D, et al. Amyloid PET screening for enrichment of early-stage Alzheimer disease clinical trials: experience in a phase 1b clinical trial. Alzheimer Dis Assoc Disord. 2016;30(1):1–7.
26. Sims JR, Zimmer JA, Evans CD, Lu M, Ardayfio P, Sparks J, et al. Donanemab in early symptomatic Alzheimer disease: the TRAILBLAZER-ALZ 2 randomized clinical trial. JAMA. 2023;330(6):512–27.
27. Cummings J. The role of biomarkers in Alzheimer's disease drug development. Adv Exp Med Biol. 2019;1118:29–61.
28. Cummings JL. Optimizing phase II of drug development for disease-modifying compounds. Alzheimers Dement. 2008;4(1S1):S15–20.
29. Cook D, Brown D, Alexander R, March R, Morgan P, Satterthwaite G, et al. Lessons learned from the fate of AstraZeneca's drug pipeline: a five-dimensional framework. Nat Rev Drug Discov. 2014;13(6):419–31.
30. Simon GM, Niphakis MJ, Cravatt BF. Determining target engagement in living systems. Nat Chem Biol. 2013;9(4):200–5.
31. Lewczuk P, Matzen A, Blennow K, Parnetti L, Molinuevo JL, Eusebi P, et al. Cerebrospinal fluid Aβ42/40 corresponds better than Aβ42 to amyloid PET in Alzheimer's disease. J Alzheimers Dis. 2017;55(2):813–22.
32. Bateman RJ, Munsell LY, Morris JC, Swarm R, Yarasheski KE, Holtzman DM. Human amyloid-beta synthesis and clearance rates as measured in cerebrospinal fluid in vivo. Nat Med. 2006;12(7):856–61.
33. Bateman RJ, Siemers ER, Mawuenyega KG, Wen G, Browning KR, Sigurdson WC, et al. A gamma-secretase inhibitor decreases amyloid-beta production in the central nervous system. Ann Neurol. 2009;66(1):48–54.

34. Fleisher AS, Raman R, Siemers ER, Becerra L, Clark CM, Dean RA, et al. Phase 2 safety trial targeting amyloid β production with a γ-secretase inhibitor in Alzheimer disease. Arch Neurol. 2008;65(8):1031–8.
35. Coric V, Salloway S, van Dyck CH, Dubois B, Andreasen N, Brody M, et al. Targeting prodromal Alzheimer disease with Avagacestat: a randomized clinical trial. JAMA Neurol. 2015;72(11):1324–33.
36. Coric V, van Dyck CH, Salloway S, Andreasen N, Brody M, Richter RW, et al. Safety and tolerability of the γ-secretase inhibitor Avagacestat in a phase 2 study of mild to moderate Alzheimer disease. Arch Neurol. 2012;69(11):1430–40.
37. Portelius E, Zetterberg H, Dean RA, Marcil A, Bourgeois P, Nutu M, et al. Amyloid-β(1-15/16) as a marker for γ-secretase inhibition in Alzheimer's disease. J Alzheimers Dis. 2012;31(2):335–41.
38. Kennedy ME, Stamford AW, Chen X, Cox K, Cumming JN, Dockendorf MF, et al. The BACE1 inhibitor verubecestat (MK-8931) reduces CNS β-amyloid in animal models and in Alzheimer's disease patients. Sci Transl Med. 2016;8(363):363ra150.
39. Egan MF, Kost J, Tariot PN, Aisen PS, Cummings JL, Vellas B, et al. Randomized trial of Verubecestat for mild-to-moderate Alzheimer's disease. N Engl J Med. 2018;378(18):1691–703.
40. van Dyck CH. Anti-amyloid-β monoclonal antibodies for Alzheimer's disease: pitfalls and promise. Biol Psychiatry. 2018;83(4):311–9.
41. Söderberg L, Johannesson M, Nygren P, Laudon H, Eriksson F, Osswald G, et al. Lecanemab, Aducanumab, and Gantenerumab – binding profiles to different forms of amyloid-Beta might explain efficacy and side effects in clinical trials for Alzheimer's disease. Neurotherapeutics. 2023;20(1):195–206.
42. Siemers ER, Friedrich S, Dean RA, Gonzales CR, Farlow MR, Paul SM, et al. Safety and changes in plasma and cerebrospinal fluid amyloid beta after a single administration of an amyloid beta monoclonal antibody in subjects with Alzheimer disease. Clin Neuropharmacol. 2010;33(2):67–73.
43. Farlow M, Arnold SE, van Dyck CH, Aisen PS, Snider BJ, Porsteinsson AP, et al. Safety and biomarker effects of solanezumab in patients with Alzheimer's disease. Alzheimers Dement. 2012;8(4):261–71.
44. van Dyck CH, Swanson CJ, Aisen P, Bateman RJ, Chen C, Gee M, et al. Lecanemab in early Alzheimer's disease. N Engl J Med. 2023;388(1):9–21.
45. Honigberg L, Clayton D, Cho W, Rabe C, et al., (editor) Biomarker results from the crenezumab anti-Aβ phase 2 biomarker trial. In: Presented at the Clinical trials on Alzheimer's disease (CTAD). Philadelphia, PA; 2014.
46. Salloway S, Honigberg LA, Cho W, Ward M, Friesenhahn M, Brunstein F, et al. Amyloid positron emission tomography and cerebrospinal fluid results from a crenezumab anti-amyloid-beta antibody double-blind, placebo-controlled, randomized phase II study in mild-to-moderate Alzheimer's disease (BLAZE). Alzheimers Res Ther. 2018;10(1):96.
47. Ostrowitzki S, Bittner T, Sink KM, Mackey H, Rabe C, Honig LS, et al. Evaluating the safety and efficacy of Crenezumab vs placebo in adults with early Alzheimer disease: two phase 3 randomized placebo-controlled trials. JAMA Neurol. 2022;79(11):1113–21.
48. Pontecorvo MJ, Lu M, Burnham SC, Schade AE, Dage JL, Shcherbinin S, et al. Association of Donanemab Treatment with exploratory plasma biomarkers in early symptomatic Alzheimer disease: a secondary analysis of the TRAILBLAZER-ALZ randomized clinical trial. JAMA Neurol. 2022;79(12):1250–9.
49. Mintun Mark A, Lo Albert C, Duggan Evans C, Wessels Alette M, Ardayfio Paul A, Andersen Scott W, et al. Donanemab in early Alzheimer's disease. N Engl J Med. 2021;384(18):1691–704.
50. Ostrowitzki S, Lasser RA, Dorflinger E, Scheltens P, Barkhof F, Nikolcheva T, et al. A phase III randomized trial of gantenerumab in prodromal Alzheimer's disease. Alzheimers Res Ther. 2017;9(1):95.

51. Lemere CA. Immunotherapy for Alzheimer's disease: hoops and hurdles. Mol Neurodegener. 2013;8:36.
52. Zhang Y, Lee DHS. Sink hypothesis and therapeutic strategies for attenuating Aβ levels. Neuroscientist. 2011;17(2):163–73.
53. Liu YH, Wang YR, Xiang Y, Zhou HD, Giunta B, Mañucat-Tan NB, et al. Clearance of amyloid-beta in Alzheimer's disease: shifting the action site from center to periphery. Mol Neurobiol. 2015;51(1):1–7.
54. Huang Y, Potter R, Sigurdson W, Kasten T, Connors R, Morris JC, et al. β-amyloid dynamics in human plasma. Arch Neurol. 2012;69(12):1591–7.
55. Shigemori K, Nomura S, Umeda T, Takeda S, Tomiyama T. Peripheral Aβ acts as a negative modulator of insulin secretion. Proc Natl Acad Sci. 2022;119(12):e2117723119.
56. Sevigny J, Chiao P, Bussière T, Weinreb PH, Williams L, Maier M, et al. The antibody aducanumab reduces Aβ plaques in Alzheimer's disease. Nature. 2016;537(7618):50–6.
57. Budd Haeberlein S, Aisen PS, Barkhof F, Chalkias S, Chen T, Cohen S, et al. Two randomized phase 3 studies of Aducanumab in early Alzheimer's disease. J Prev Alzheimers Dis. 2022;9(2):197–210.
58. The US Food and Drug Administration (FDA). https://www.fda.gov/drugs/postmarket-drug-safety-information-patients-and-providers/aducanumab-marketed-aduhelm-information; 2021.
59. Alexander GC, Knopman David S, Emerson Scott S, Ovbiagele B, Kryscio Richard J, Perlmutter Joel S, et al. Revisiting FDA approval of Aducanumab. N Engl J Med. 2021;385(9):769–71.
60. Knopman DS, Hershey L. Implications of the approval of Lecanemab for Alzheimer disease patient care: incremental step or paradigm shift? Neurology. 2023;101(14):610–20.
61. Tarawneh R, Pankratz VS. The search for CLARITY regarding "clinically meaningful outcomes" in Alzheimer disease clinical trials: CLARITY-AD and beyond. Alzheimers Res Ther. 2024;16(1):37.
62. Bateman Randall J, Smith J, Donohue Michael C, Delmar P, Abbas R, Salloway S, et al. Two phase 3 trials of Gantenerumab in early Alzheimer's disease. N Engl J Med. 2023;389(20):1862–76.
63. Cummings JL, Zhong K, Kinney JW, Heaney C, Moll-Tudla J, Joshi A, et al. Double-blind, placebo-controlled, proof-of-concept trial of bexarotene Xin moderate Alzheimer's disease. Alzheimers Res Ther. 2016;8:4.
64. Doody Rachelle S, Thomas Ronald G, Farlow M, Iwatsubo T, Vellas B, Joffe S, et al. Phase 3 trials of solanezumab for mild-to-moderate Alzheimer's disease. N Engl J Med. 2014;370(4):311–21.
65. Honig Lawrence S, Vellas B, Woodward M, Boada M, Bullock R, Borrie M, et al. Trial of Solanezumab for mild dementia due to Alzheimer's disease. N Engl J Med. 2018;378(4):321–30.
66. Biel D, Brendel M, Rubinski A, Buerger K, Janowitz D, Dichgans M, et al. Tau-PET and in vivo Braak-staging as prognostic markers of future cognitive decline in cognitively normal to demented individuals. Alzheimers Res Ther. 2021;13(1):137.
67. Pontecorvo MJ, Devous MD, Kennedy I, Navitsky M, Lu M, Galante N, et al. A multicentre longitudinal study of flortaucipir (18F) in normal ageing, mild cognitive impairment and Alzheimer's disease dementia. Brain. 2019;142(6):1723–35.
68. St-Onge F, Chapleau M, Breitner JC, Villeneuve S, Binette AP. Tau accumulation and its spatial progression across the Alzheimer's disease spectrum. medRxiv. 2023; https://doi.org/10.1101/2023.06.02.23290880.
69. Insel PS, Weiner M, Mackin RS, Mormino E, Lim YY, Stomrud E, et al. Determining clinically meaningful decline in preclinical Alzheimer disease. Neurology. 2019;93(4):e322–e33.
70. Donohue MC, Sperling RA, Petersen R, Sun CK, Weiner MW, Aisen PS. Association between elevated brain amyloid and subsequent cognitive decline among cognitively Normal persons. JAMA. 2017;317(22):2305–16.

71. Hampel H, Buerger K, Zinkowski R, Teipel SJ, Goernitz A, Andreasen N, et al. Measurement of phosphorylated tau epitopes in the differential diagnosis of Alzheimer disease: a comparative cerebrospinal fluid study. Arch Gen Psychiatry. 2004;61(1):95–102.
72. Janelidze S, Stomrud E, Smith R, Palmqvist S, Mattsson N, Airey DC, et al. Cerebrospinal fluid p-tau217 performs better than p-tau181 as a biomarker of Alzheimer's disease. Nat Commun. 2020;11(1):1683.
73. Coughlin D, Irwin DJ. Emerging diagnostic and therapeutic strategies for Tauopathies. Curr Neurol Neurosci Rep. 2017;17(9):72.
74. Milà-Alomà M, Ashton NJ, Shekari M, Salvadó G, Ortiz-Romero P, Montoliu-Gaya L, et al. Plasma p-tau231 and p-tau217 as state markers of amyloid-β pathology in preclinical Alzheimer's disease. Nat Med. 2022;28(9):1797–801.
75. Teng E, Manser PT, Pickthorn K, Brunstein F, Blendstrup M, Sanabria Bohorquez S, et al. Safety and efficacy of Semorinemab in individuals with prodromal to mild Alzheimer disease: a randomized clinical trial. JAMA Neurol. 2022;79(8):758–67.
76. Monteiro C, Toth B, Brunstein F, Bobbala A, Datta S, Ceniceros R, et al. Randomized phase II study of the safety and efficacy of Semorinemab in participants with mild-to-moderate Alzheimer disease: Lauriet. Neurology. 2023;101(14):e1391–e401.
77. Shulman M, Kong J, O'Gorman J, Ratti E, Rajagovindan R, Viollet L, et al. TANGO: a placebo-controlled randomized phase 2 study of efficacy and safety of the anti-tau monoclonal antibody gosuranemab in early Alzheimer's disease. Nat Aging. 2023;3(12):1591–601.
78. Höglinger GU, Litvan I, Mendonca N, Wang D, Zheng H, Rendenbach-Mueller B, et al. Safety and efficacy of tilavonemab in progressive supranuclear palsy: a phase 2, randomised, placebo-controlled trial. Lancet Neurol. 2021;20(3):182–92.
79. Fleisher AS, Munsie LM, Perahia DGS, Andersen SW, Higgins IA, Hauck PM, et al. Assessment of efficacy and safety of Zagotenemab. Neurology. 2024;102(5):e208061.
80. Mummery CJ, Börjesson-Hanson A, Blackburn DJ, Vijverberg EGB, De Deyn PP, Ducharme S, et al. Tau-targeting antisense oligonucleotide MAPTRx in mild Alzheimer's disease: a phase 1b, randomized, placebo-controlled trial. Nat Med. 2023;29(6):1437–47.
81. Imbimbo BP, Balducci C, Ippati S, Watling M. Initial failures of anti-tau antibodies in Alzheimer's disease are reminiscent of the amyloid-β story. Neural Regen Res. 2023;18(1):117–8.
82. Profenno LA, Jakimovich L, Holt CJ, Porsteinsson A, Tariot PN. A randomized, double-blind, placebo-controlled pilot trial of safety and tolerability of two doses of divalproex sodium in outpatients with probable Alzheimer's disease. Curr Alzheimer Res. 2005;2(5):553–8.
83. Hampel H, Ewers M, Bürger K, Annas P, Mörtberg A, Bogstedt A, et al. Lithium trial in Alzheimer's disease: a randomized, single-blind, placebo-controlled, multicenter 10-week study. J Clin Psychiatry. 2009;70(6):922–31.
84. Kang JH, Ryoo NY, Shin DW, Trojanowski JQ, Shaw LM. Role of cerebrospinal fluid biomarkers in clinical trials for Alzheimer's disease modifying therapies. Korean J Physiol Pharmacol. 2014;18(6):447–56.
85. Prins ND, Harrison JE, Chu HM, Blackburn K, Alam JJ, Scheltens P. A phase 2 double-blind placebo-controlled 24-week treatment clinical study of the p38 alpha kinase inhibitor neflamapimod in mild Alzheimer's disease. Alzheimers Res Ther. 2021;13(1):106.
86. Suhocki PV, Ronald JS, Diehl AME, Murdoch DM, Doraiswamy PM. Probing gut-brain links in Alzheimer's disease with rifaximin. Alzheimers Dement (N Y). 2022;8(1):e12225.
87. Wilkins HM, Mahnken JD, Welch P, Bothwell R, Koppel S, Jackson RL, et al. A mitochondrial biomarker-based study of S-Equol in Alzheimer's disease subjects: results of a single-arm, Pilot Trial. J Alzheimers Dis. 2017;59(1):291–300.
88. Oosthoek M, Vermunt L, de Wilde A, Bongers B, Antwi-Berko D, Scheltens P, et al. Utilization of fluid-based biomarkers as endpoints in disease-modifying clinical trials for Alzheimer's disease: a systematic review. Alzheimers Res Ther. 2024;16(1):93.
89. Salloway S, Sperling R, Fox NC, Blennow K, Klunk W, Raskind M, et al. Two phase 3 trials of bapineuzumab in mild-to-moderate Alzheimer's disease. N Engl J Med. 2014;370(4):322–33.

90. The Alzheimer's Association AlzForum. Biomarkers and efficacy: not (Yet?) a perfect union; 2023. Available from: https://www.alzforum.org/news/conference-coverage/biomarkers-and-efficacy-not-yet-perfect-union.
91. Tarawneh R, D'Angelo G, Macy E, Xiong C, Carter D, Cairns NJ, et al. Visinin-like protein-1: diagnostic and prognostic biomarker in Alzheimer disease. Ann Neurol. 2011;70(2):274–85.
92. Zetterberg H, Skillbäck T, Mattsson N, Trojanowski JQ, Portelius E, Shaw LM, et al. Association of cerebrospinal fluid neurofilament light concentration with Alzheimer disease progression. JAMA Neurol. 2016;73(1):60–7.
93. Tarawneh R, D'Angelo G, Crimmins D, Herries E, Griest T, Fagan AM, et al. Diagnostic and prognostic utility of the synaptic marker Neurogranin in Alzheimer disease. JAMA Neurol. 2016;73(5):561–71.
94. Tarawneh R, Lee JM, Ladenson JH, Morris JC, Holtzman DM. CSF VILIP-1 predicts rates of cognitive decline in early Alzheimer disease. Neurology. 2012;78(10):709–19.
95. Tarawneh R, Head D, Allison S, Buckles V, Fagan AM, Ladenson JH, et al. Cerebrospinal fluid markers of neurodegeneration and rates of brain atrophy in early Alzheimer disease. JAMA Neurol. 2015;72(6):656–65.
96. Kamphuis W, Middeldorp J, Kooijman L, Sluijs JA, Kooi EJ, Moeton M, et al. Glial fibrillary acidic protein isoform expression in plaque related astrogliosis in Alzheimer's disease. Neurobiol Aging. 2014;35(3):492–510.
97. Mattsson N, Insel PS, Palmqvist S, Portelius E, Zetterberg H, Weiner M, et al. Cerebrospinal fluid tau, neurogranin, and neurofilament light in Alzheimer's disease. EMBO Mol Med. 2016;8(10):1184–96.
98. Thorsell A, Bjerke M, Gobom J, Brunhage E, Vanmechelen E, Andreasen N, et al. Neurogranin in cerebrospinal fluid as a marker of synaptic degeneration in Alzheimer's disease. Brain Res. 2010;1362:13–22.
99. Wellington H, Paterson RW, Portelius E, Törnqvist U, Magdalinou N, Fox NC, et al. Increased CSF neurogranin concentration is specific to Alzheimer disease. Neurology. 2016;86(9):829–35.
100. Wagemann O, Liu H, Wang G, Shi X, Bittner T, Scelsi MA, et al. Downstream biomarker effects of Gantenerumab or Solanezumab in dominantly inherited Alzheimer disease: the DIAN-TU-001 randomized clinical trial. JAMA Neurol. 2024;81(6):582–93.
101. Rafii MS, Sperling RA, Donohue MC, Zhou J, Roberts C, Irizarry MC, et al. The AHEAD 3-45 study: design of a prevention trial for Alzheimer's disease. Alzheimers Dement. 2023;19(4):1227–33.
102. De Vos A, Jacobs D, Struyfs H, Fransen E, Andersson K, Portelius E, et al. C-terminal neurogranin is increased in cerebrospinal fluid but unchanged in plasma in Alzheimer's disease. Alzheimers Dement. 2015;11(12):1461–9.
103. Goetzl EJ, Kapogiannis D, Schwartz JB, Lobach IV, Goetzl L, Abner EL, et al. Decreased synaptic proteins in neuronal exosomes of frontotemporal dementia and Alzheimer's disease. FASEB J. 2016;30(12):4141–8.
104. Winston CN, Goetzl EJ, Akers JC, Carter BS, Rockenstein EM, Galasko D, et al. Prediction of conversion from mild cognitive impairment to dementia with neuronally derived blood exosome protein profile. Alzheimer's Dement. 2016;3:63–72.
105. Schnurra I, Bernstein HG, Riederer P, Braunewell KH. The neuronal calcium sensor protein VILIP-1 is associated with amyloid plaques and extracellular tangles in Alzheimer's disease and promotes cell death and tau phosphorylation in vitro: a link between calcium sensors and Alzheimer's disease? Neurobiol Dis. 2001;8(5):900–9.
106. Halbgebauer S, Steinacker P, Riedel D, Oeckl P, Anderl-Straub S, Lombardi J, et al. Visinin-like protein 1 levels in blood and CSF as emerging markers for Alzheimer's and other neurodegenerative diseases. Alzheimers Res Ther. 2022;14(1):175.
107. ClinicalTrials.Gov. China Cognition and Aging Study; 2020. Available from: https://cdn.clinicaltrials.gov/large-docs/56/NCT03653156/Prot_000.pdf.

108. Schindler SE, Li Y, Todd KW, Herries EM, Henson RL, Gray JD, et al. Emerging cerebrospinal fluid biomarkers in autosomal dominant Alzheimer's disease. Alzheimers Dement. 2019;15(5):655–65.
109. Khalil M, Teunissen CE, Otto M, Piehl F, Sormani MP, Gattringer T, et al. Neurofilaments as biomarkers in neurological disorders. Nat Rev Neurol. 2018;14(10):577–89.
110. Dhiman K, Gupta VB, Villemagne VL, Eratne D, Graham PL, Fowler C, et al. Cerebrospinal fluid neurofilament light concentration predicts brain atrophy and cognition in Alzheimer's disease. Alzheimers Dement (Amst). 2020;12(1):e12005.
111. Dhiman K, Villemagne VL, Fowler C, Bourgeat P, Li QX, Collins S, et al. Cerebrospinal fluid Neurofilament light predicts risk of dementia onset in cognitively healthy individuals and rate of cognitive decline in mild cognitive impairment: a prospective longitudinal study. Biomedicines. 2022;10(5):1045.
112. Rolstad S, Berg AI, Eckerström C, Johansson B, Wallin A. Differential impact of Neurofilament light subunit on cognition and functional outcome in memory clinic patients with and without vascular burden. J Alzheimers Dis. 2015;45(3):873–81.
113. Uphaus T, Steffen F, Muthuraman M, Ripfel N, Fleischer V, Groppa S, et al. NfL predicts relapse-free progression in a longitudinal multiple sclerosis cohort study. EBioMedicine. 2021;72:103590.
114. Mattsson N, Andreasson U, Zetterberg H, Blennow K. Association of plasma neurofilament light with neurodegeneration in patients with Alzheimer disease. JAMA Neurol. 2017;74(5):557–66.
115. Preische O, Schultz SA, Apel A, Kuhle J, Kaeser SA, Barro C, et al. Serum neurofilament dynamics predicts neurodegeneration and clinical progression in presymptomatic Alzheimer's disease. Nat Med. 2019;25(2):277–83.
116. Cantó E, Barro C, Zhao C, Caillier SJ, Michalak Z, Bove R, et al. Association between serum neurofilament light chain levels and long-term disease course among patients with multiple sclerosis followed up for 12 years. JAMA Neurol. 2019;76(11):1359–66.
117. Antonucci F, Corradini I, Fossati G, Tomasoni R, Menna E, Matteoli M. SNAP-25, a known presynaptic protein with emerging postsynaptic functions. Front Synaptic Neurosci. 2016;8:7.
118. Zhang H, Therriault J, Kang MS, Ng KP, Pascoal TA, Rosa-Neto P, et al. Cerebrospinal fluid synaptosomal-associated protein 25 is a key player in synaptic degeneration in mild cognitive impairment and Alzheimer's disease. Alzheimers Res Ther. 2018;10(1):80.
119. Brinkmalm A, Brinkmalm G, Honer WG, Frölich L, Hausner L, Minthon L, et al. SNAP-25 is a promising novel cerebrospinal fluid biomarker for synapse degeneration in Alzheimer's disease. Mol Neurodegener. 2014;9:53.
120. Bark IC, Wilson MC. Human cDNA clones encoding two different isoforms of the nerve terminal protein SNAP-25. Gene. 1994;139(2):291–2.
121. ClinicalTrials.Gov. A Randomized, Double-Blind, Placebo-Controlled Study, Followed by an Open-Label Extension, to Evaluate the Safety, Tolerability, Pharmacokinetics and Pharmacodynamics of Multiple Ascending Doses of Intrathecally Administered ISIS 814907 in Patients With Mild Alzheimer's Disease. NCT03186989; 2022. Available from: https://cdn.clinicaltrials.gov/large-docs/89/NCT03186989/Prot_SAP_000.pdf.
122. ClinicalTrials.Gov. Effect of CT1812 treatment on brain synaptic density (NCT03493282); 2023. Available from: https://clinicaltrials.gov/study/NCT03493282?tab=history&a=8.
123. Denny JB. Molecular mechanisms, biological actions, and neuropharmacology of the growth-associated protein GAP-43. Curr Neuropharmacol. 2006;4(4):293–304.
124. Holahan MR. A shift from a pivotal to supporting role for the growth-associated protein (GAP-43) in the coordination of axonal structural and functional plasticity. Front Cell Neurosci. 2017;11:266.
125. Sjögren M, Davidsson P, Gottfries J, Vanderstichele H, Edman A, Vanmechelen E, et al. The cerebrospinal fluid levels of tau, growth-associated protein-43 and soluble amyloid precursor protein correlate in Alzheimer's disease, reflecting a common pathophysiological process. Dement Geriatr Cogn Disord. 2001;12(4):257–64.

126. Sandelius Å, Portelius E, Källén Å, Zetterberg H, Rot U, Olsson B, et al. Elevated CSF GAP-43 is Alzheimer's disease specific and associated with tau and amyloid pathology. Alzheimers Dement. 2019;15(1):55–64.
127. Kim KY, Shin KY, Chang KA. GFAP as a potential biomarker for Alzheimer's disease: a systematic review and meta-analysis. Cells. 2023;12(9):1309.
128. Abdelhak A, Foschi M, Abu-Rumeileh S, Yue JK, D'Anna L, Huss A, et al. Blood GFAP as an emerging biomarker in brain and spinal cord disorders. Nat Rev Neurol. 2022;18(3):158–72.
129. Heimfarth L, Passos FRS, Monteiro BS, Araújo AAS, Quintans Júnior LJ, Quintans JSS. Serum glial fibrillary acidic protein is a body fluid biomarker: a valuable prognostic for neurological disease – a systematic review. Int Immunopharmacol. 2022;107:108624.
130. Pereira JB, Janelidze S, Smith R, Mattsson-Carlgren N, Palmqvist S, Teunissen CE, et al. Plasma GFAP is an early marker of amyloid-β but not tau pathology in Alzheimer's disease. Brain. 2021;144(11):3505–16.
131. Ishiki A, Kamada M, Kawamura Y, Terao C, Shimoda F, Tomita N, et al. Glial fibrillar acidic protein in the cerebrospinal fluid of Alzheimer's disease, dementia with Lewy bodies, and frontotemporal lobar degeneration. J Neurochem. 2016;136(2):258–61.
132. Bartl M, Dakna M, Galasko D, Hutten SJ, Foroud T, Quan M, et al. Biomarkers of neurodegeneration and glial activation validated in Alzheimer's disease assessed in longitudinal cerebrospinal fluid samples of Parkinson's disease. PLoS One. 2021;16(10):e0257372.
133. Tang Y, Han L, Li S, Hu T, Xu Z, Fan Y, et al. Plasma GFAP in Parkinson's disease with cognitive impairment and its potential to predict conversion to dementia. npj Parkinson's Dis. 2023;9(1):23.
134. Benedet AL, Milà-Alomà M, Vrillon A, Ashton NJ, Pascoal TA, Lussier F, et al. Differences between plasma and cerebrospinal fluid glial fibrillary acidic protein levels across the Alzheimer disease continuum. JAMA Neurol. 2021;78(12):1471–83.
135. Libreros S, Iragavarapu-Charyulu V. YKL-40/CHI3L1 drives inflammation on the road of tumor progression. J Leukoc Biol. 2015;98(6):931–6.
136. Baldacci F, Toschi N, Lista S, Zetterberg H, Blennow K, Kilimann I, et al. Two-level diagnostic classification using cerebrospinal fluid YKL-40 in Alzheimer's disease. Alzheimers Dement. 2017;13(9):993–1003.
137. Baldacci F, Lista S, Cavedo E, Bonuccelli U, Hampel H. Diagnostic function of the neuroinflammatory biomarker YKL-40 in Alzheimer's disease and other neurodegenerative diseases. Expert Rev Proteomics. 2017;14(4):285–99.
138. Olsson B, Lautner R, Andreasson U, Öhrfelt A, Portelius E, Bjerke M, et al. CSF and blood biomarkers for the diagnosis of Alzheimer's disease: a systematic review and meta-analysis. Lancet Neurol. 2016;15(7):673–84.
139. Kester MI, Teunissen CE, Sutphen C, Herries EM, Ladenson JH, Xiong C, et al. Cerebrospinal fluid VILIP-1 and YKL-40, candidate biomarkers to diagnose, predict and monitor Alzheimer's disease in a memory clinic cohort. Alzheimers Res Ther. 2015;7(1):59.
140. Gispert JD, Monté GC, Suárez-Calvet M, Falcon C, Tucholka A, Rojas S, et al. The APOE ε4 genotype modulates CSF YKL-40 levels and their structural brain correlates in the continuum of Alzheimer's disease but not those of sTREM2. Alzheimer's Dement. 2017;6(1):50–9.
141. Alcolea D, Vilaplana E, Pegueroles J, Montal V, Sánchez-Juan P, González-Suárez A, et al. Relationship between cortical thickness and cerebrospinal fluid YKL-40 in predementia stages of Alzheimer's disease. Neurobiol Aging. 2015;36(6):2018–23.
142. Mattsson N, Tabatabaei S, Johansson P, Hansson O, Andreasson U, Månsson JE, et al. Cerebrospinal fluid microglial markers in Alzheimer's disease: elevated chitotriosidase activity but lack of diagnostic utility. NeuroMolecular Med. 2011;13(2):151–9.
143. Craig-Schapiro R, Perrin RJ, Roe CM, Xiong C, Carter D, Cairns NJ, et al. YKL-40: a novel prognostic fluid biomarker for preclinical Alzheimer's disease. Biol Psychiatry. 2010;68(10):903–12.
144. Baldacci F, Lista S, Palermo G, Giorgi FS, Vergallo A, Hampel H. The neuroinflammatory biomarker YKL-40 for neurodegenerative diseases: advances in development. Expert Rev Proteomics. 2019;16(7):593–600.

145. Hok AHYS, Hoozemans JJM, Hu WT, Wouters D, Howell JC, Rábano A, et al. YKL-40 changes are not detected in post-mortem brain of patients with Alzheimer's disease and frontotemporal lobar degeneration. Alzheimers Res Ther. 2022;14(1):100.
146. Cristóvão JS, Gomes CM. S100 Proteins in Alzheimer's Disease. Front Neurosci. 2019;13:463.
147. Steiner J, Bogerts B, Schroeter ML, Bernstein HG. S100B protein in neurodegenerative disorders. Clin Chem Lab Med. 2011;49(3):409–24.
148. Mori T, Asano T, Town T. Targeting S100B in Cerebral Ischemia and in Alzheimer's Disease. Cardiovasc Psychiatry Neurol. 2010;2010:687067. https://doi.org/10.1155/2010/687067. Epub 2010 Sep 2. PMID: 20862385; PMCID: PMC2939387.
149. Tarawneh R. YKL-40 as a new plasma biomarker for dementia risk. Neurology. 2024;102(4):e209145.
150. Alves F, Kalinowski P, Ayton S. Accelerated Brain Volume Loss Caused by Anti-β-Amyloid Drugs: A Systematic Review and Meta-analysis. Neurology. 2023 May 16;100(20):e2114–e2124. https://doi.org/10.1212/WNL.0000000000207156. Epub 2023 Mar 27. PMID: 36973044; PMCID: PMC10186239.
151. Doody RS, Raman R, Farlow M, Iwatsubo T, Vellas B, Joffe S, et al. A phase 3 trial of semagacestat for treatment of Alzheimer's disease. N Engl J Med. 2013;369(4):341–50.
152. Barkhof F, Knopman DS. Brain Shrinkage in Anti-β-Amyloid Alzheimer Trials: Neurodegeneration or Pseudoatrophy? Neurology. 2023 May 16;100(20):941–942. https://doi.org/10.1212/WNL.0000000000207268. Epub 2023 Mar 27. PMID: 36973045.
153. Swanson CJ, Zhang Y, Dhadda S, Wang J, Kaplow J, Lai RYK, et al. A randomized, double-blind, phase 2b proof-of-concept clinical trial in early Alzheimer's disease with lecanemab, an anti-Aβ protofibril antibody. Alzheimers Res Ther. 2021;13(1):80.
154. Knopman DS. Lecanemab reduces brain amyloid-β and delays cognitive worsening. Cell Rep Med. 2023;4(3):100982.
155. Ayton S. Brain volume loss due to donanemab. Eur J Neurol. 2021;28(9):e67–e8.
156. Knopman DS, Jones D, Greicius MD. Failure to demonstrate efficacy of aducanumab: An analysis of the EMERGE and ENGAGE trials as reported by Biogen, December 2019. Alzheimer's Dement. 2021;17:696–701. https://doi.org/10.1002/alz.12213
157. Alexander GE, Chen K, Pietrini P, Rapoport SI, Reiman EM. Longitudinal PET evaluation of cerebral metabolic decline in dementia: a potential outcome measure in Alzheimer's disease treatment studies. Am J Psychiatry. 2002;159(5):738–45.
158. Bohnen NI, Djang DSW, Herholz K, Anzai Y, Minoshima S. Effectiveness and safety of ^{18}F-FDG PET in the evaluation of dementia: a review of the recent literature. J Nucl Med. 2012;53(1):59–71.
159. Mega MS, Dinov ID, Porter V, Chow G, Reback E, Davoodi P, et al. Metabolic patterns associated with the clinical response to galantamine therapy: a fludeoxyglucose F 18 positron emission tomographic study. Arch Neurol. 2005;62(5):721–8.
160. Cummings J, Kinney J. Biomarkers for Alzheimer's disease: context of use, qualification, and roadmap for clinical implementation. Medicina (Kaunas). 2022;58(7) https://doi.org/10.3390/medicina58070952.
161. European Medicines Agency (EMA). Letter of support of neurofilament light in childhood neurological diseases; 2022. Available from: https://www.ema.europa.eu/en/documents/other/letter-support-neurofilament-light-childhood-neurological-diseases_en.pdf.
162. US Food and Drug Administration (FDA). Early Alzheimer's disease: developing drugs for treatment, guidance for industry: Draft Guidance; 2018. Available from https://www.fda.gov/files/drugs/published/Alzheimer's-Disease---Developing-Drugs-for-Treatment-Guidance-for-Industy.pdf
163. European Medicines Agency (EMA). Guideline on the clinical investigation of medicines for the treatment of Alzheimer's disease; 2018. Available from: https://www.ema.europa.eu/en/documents/scientific-guideline/guideline-clinical-investigation-medicines-treatment-alzheimers-disease-revision-2_en.pdf.

164. US Food and Drug Admininstration (FDA). Early Alzheimer's disease: developing drugs for treatment; 2024.
165. US Food and Drug Admininstration (FDA). FDA approves treatment of amyotrophic lateral sclerosis associated with a mutation in the SOD1 gene; 2023. Available from: https://www.fda.gov/drugs/news-events-human-drugs/fda-approves-treatment-amyotrophic-lateral-sclerosis-associated-mutation-sod1-gene.
166. Jørgensen JT, Hersom M. Clinical and regulatory aspects of companion diagnostic development in oncology. Clin Pharmacol Ther. 2018;103(6):999–1008.
167. The US Food and Drug Administration (FDA). Qualification Process for Drug Development Tools Guidance for Industry and FDA Staff; 2020. Available from: https://www.fda.gov/regulatory-information/search-fda-guidance-documents/qualification-process-drug-development-tools-guidance-industry-and-fda-staff.
168. Kraus VB. Biomarkers as drug development tools: discovery, validation, qualification and use. Nat Rev Rheumatol. 2018;14(6):354–62.
169. Scheerens H, Malong A, Bassett K, Boyd Z, Gupta V, Harris J, et al. Current status of companion and complementary diagnostics: strategic considerations for development and launch. Clin Transl Sci. 2017;10(2):84–92.
170. Thal LJ, Kantarci K, Reiman EM, Klunk WE, Weiner MW, Zetterberg H, Galasko D, Praticò D, Griffin S, Schenk D, Siemers E. The role of biomarkers in clinical trials for Alzheimer disease. Alzheimer Dis Assoc Disord. 2006 Jan-Mar;20(1):6–15. https://doi.org/10.1097/01.wad.0000191420.61260.a8. PMID: 16493230; PMCID: PMC1820855.
171. Cummings J, Zhou Y, Lee G, Zhong K, Fonseca J, Cheng F. Alzheimer's disease drug development pipeline: 2023. Alzheimer's Dement. 2023;9(2):e12385.
172. Cummings J, Lee G, Ritter A, Sabbagh M, Zhong K. Alzheimer's disease drug development pipeline: 2020. Alzheimer's Dement. 2020;6(1):e12050.
173. Tarawneh R, Kasper RS, Sanford J, Phuah C-L, Hassenstab J, Cruchaga C. Vascular endothelial-cadherin as a marker of endothelial injury in preclinical Alzheimer disease. Ann Clin Transl Neurol. 2022;9(12):1926–40.
174. Kapasi A, DeCarli C, Schneider JA. Impact of multiple pathologies on the threshold for clinically overt dementia. Acta Neuropathol. 2017;134(2):171–86.
175. Boyle PA, Yu L, Leurgans SE, Wilson RS, Brookmeyer R, Schneider JA, et al. Attributable risk of Alzheimer's dementia attributed to age-related neuropathologies. Ann Neurol. 2019;85(1):114–24.
176. Mattsson N, Carrillo MC, Dean RA, Devous MD, Nikolcheva T, Pesini P, et al. Revolutionizing Alzheimer's disease and clinical trials through biomarkers. Alzheimer's Dement. 2015;1(4):412–9.
177. Toschi N, Lista S, Baldacci F, Cavedo E, Zetterberg H, Blennow K, et al. Biomarker-guided clustering of Alzheimer's disease clinical syndromes. Neurobiol Aging. 2019;83:42–53.
178. Geerts H, Dacks PA, Devanarayan V, Haas M, Khachaturian ZS, Gordon MF, et al. Big data to smart data in Alzheimer's disease: the brain health modeling initiative to foster actionable knowledge. Alzheimers Dement. 2016;12(9):1014–21.
179. Haas M, Stephenson D, Romero K, Gordon MF, Zach N, Geerts H. Big data to smart data in Alzheimer's disease: real-world examples of advanced modeling and simulation. Alzheimers Dement. 2016;12(9):1022–30.
180. Hampel H, Toschi N, Babiloni C, Baldacci F, Black KL, Bokde ALW, et al. Revolution of Alzheimer precision neurology. Passageway of systems biology and neurophysiology. J Alzheimers Dis. 2018;64(s1):S47–s105.
181. Hampel H, Vergallo A, Aguilar LF, Benda N, Broich K, Cuello AC, et al. Precision pharmacology for Alzheimer's disease. Pharmacol Res. 2018;130:331–65.

Impact of Pre-analytical Factors on Fluid Biomarker Measurements in Alzheimer Disease

Javier Torres-Torronteras, Mariam Gouda, Charlotte E. Teunissen, Inge M. W. Verberk, and Marta del Campo

Abstract The field of cerebrospinal fluid (CSF) and plasma-based biomarkers in Alzheimer disease (AD) is advancing quickly. These biomarkers are being included in current guidelines and recommendations to biologically diagnose AD. However, reliable and reproducible measurements within and across different centers worldwide are necessary for the implementation of these markers in clinical routines. The reliability of such measurements is influenced, to some extent, by pre-analytical confounding factors (i.e., factors associated with sample collection, processing, and storage). These factors have been thoroughly evaluated for the measurements of the core AD CSF biomarkers and are currently being evaluated for their plasma counterparts. Results from such scientific efforts are compiled in CSF-based or plasma-based standardized operating procedures (SOPs) that are advised to be followed by centers assessing the biomarkers, thereby minimizing the variability of biomarker

---

Javier Torres-Torronteras and Mariam Gouda contributed equally with all other contributors.

J. Torres-Torronteras (✉)
Barcelonaβeta Brain Research Center (BBRC), Pasqual Maragall Foundation, Barcelona, Spain

Hospital del Mar Research Institute (IMIM), Barcelona, Spain
e-mail: jtorres@barcelonabeta.org

M. Gouda · C. E. Teunissen · I. M. W. Verberk
Neurochemistry Laboratory, Department of Laboratory Medicine, Amsterdam Neuroscience, VU University Medical Center, Vrije Universiteit Amsterdam, Amsterdam, The Netherlands
e-mail: m.m.t.e.e.gouda@amsterdamumc.nl; c.teunissen@amsterdamumc.nl; i.verberk@amsterdamumc.nl

M. del Campo (✉)
Barcelonaβeta Brain Research Center (BBRC), Pasqual Maragall Foundation, Barcelona, Spain

Hospital del Mar Research Institute (IMIM), Barcelona, Spain

Departamento de Ciencias Farmacéuticas y de la Salud, Facultad de Farmacia, Universidad San Pablo-CEU, CEU Universities, Alcorcón, Spain
e-mail: mcampo@barcelonabeta.org

R. Tarawneh (ed.), *Biomarkers of Neurodegenerative Disorders*,
https://doi.org/10.1007/978-3-031-80849-4_16

measurements. We here jointly review the available literature on the effect of different pre-analytical factors and the most updated recommendations for the measurements of the core biomarkers for AD (amyloid beta peptides, total and phosphorylated forms of Tau) and related processes i.e., neurofilament light as a marker associated with neurodegenerative processes, or glial fibrillary acidic protein as a marker associated with astrogliosis, in CSF and blood. We also review the impact of some pre-analytical factors on the CSF and plasma proteome using technologies employed for discovery of novel potential biofluid biomarkers. Lastly, we describe ongoing international initiatives aiming to standardize CSF- and blood-based neurodegenerative biomarker measurements. Understanding the impact of the different pre-analytical factors on biomarker measurements and compliance with consensus guidelines will facilitate the implementation of AD biomarkers in the clinical diagnostic workup.

Keywords Alzheimer Disease · Cerebrospinal Fluid Biomarkers · Blood-based Biomarkers · Pre-analytical Factors

1 General Introduction

The field of biofluid-based biomarkers for Alzheimer disease (AD) and associated pathophysiological processes, such as amyloid and tau pathology, astrogliosis, and axonal degeneration, is advancing rapidly. These fluid biomarkers can be reliably measured in cerebrospinal fluid (CSF) and, more recently, in blood [1]. Therefore, these biomarkers are likely to become an integral part of the clinical diagnostic workflow, thanks in part to the recent advancements in fully automated assays. Additionally, these markers will also play an essential role in clinical trials for pre-screening purposes, selection criteria, assessing target engagement, or as surrogate measures [1].

The most reliable and studied fluid biomarkers related to AD pathology to date are amyloid beta (Aβ) peptides 42 (Aβ42) and 40 (Aβ40), different forms of phosphorylated Tau (pTau) (e.g., pTau181, pTau217, pTau231, pTau235), total Tau (tTau), glial fibrillary acidic protein (GFAP), and neurofilament light (NfL) [1]. The analysis of amyloid peptides and their ratio (Aβ42/40) in both CSF and plasma can be used to reliably detect amyloid pathology along the full AD continuum, even before the appearance of cognitive symptoms [1–3]. Different isoforms of pTau can also be measured in both CSF and plasma [3, 4]. CSF tTau can be measured as a proxy of the AD neurodegenerative process [5], but its blood counterpart has shown variable results [6–8]. Recently, a brain-derived form of tau has been identified in both CSF and plasma, which reflects AD-type neurodegeneration [9]. Concentrations of CSF and plasma GFAP, a marker reflecting reactive astrogliosis, are increased in AD but also in other neurodegenerative dementias or after acute neuronal injury (e.g., due to stroke or traumatic brain injury) [1, 10]. Similarly, NfL is a marker reflecting general axonal degeneration, and its concentrations are increased in both CSF and blood in different neurodegenerative disorders, including AD [1]. This marker could be especially useful to indicate a neurodegenerative condition rather than a psychiatric disorder [11] or as a surrogate marker in clinical trials, at least in the contexts of specific disorders (e.g., Multiple Sclerosis or Amyotrophic Lateral Sclerosis) [12].

CSF and blood biomarkers are widely used or ready for use in clinical settings and in trials to support the biological diagnosis of AD. However, variations in fluid biomarker sampling and sample handling before the biomarker measurement, i.e., the pre-analytical phase, can result in analytical variability that affects the reliability of biomarker measurements and the comparability of biomarker results across centers. The development and use of standard operating procedures (SOPs) allows to minimize such effects and is an essential step towards the implementation of biofluid biomarkers in the clinical routine. Consensus reports for CSF have already been developed and are continuously being updated [13–16], with the latest international guidelines from the Alzheimer's Association (CSF AA guidelines) published in 2021 [17]. The learnings from the CSF field have allowed us to efficiently and swiftly evaluate many of these factors in blood, resulting in the first version of an SOP for blood analysis [18]. This chapter summarizes the findings of the effects of pre-analytical factors in the measurement of fluid biomarkers associated with AD and neurodegenerative disorders in both CSF and plasma. Moreover, we not only discuss how these confounding factors influence established and emerging biomarkers but also highlight the minimal pre-analytical factors that should be evaluated when validating a novel potential fluid biomarker across different matrices. Lastly, we outline current international initiatives focused on the standardization and validation of biomarker tests for their use in clinical routine worldwide.

2 Impact of Different Pre-analytical Factors in Cerebrospinal Fluid (CSF)

2.1 CSF Sampling Factors

2.1.1 Fasting Conditions

The requirement of fasting before CSF collection varies between centers, with some centers requiring 6 hrs of fasting before the lumbar puncture (LP), while other centers do not [19]. Only one study has analyzed the impact of fasting, which reported no effect of fasting on CSF Aβ42 [20]. The CSF AA guidelines thus recommend that fasting is not necessary when collecting CSF for the core AD biomarkers analysis [17].

2.1.2 Effect of Circadian Rhythms and Diurnal Variation

CSF Aβ42 and Aβ40 concentrations may fluctuate along with circadian rhythms [21–23]. However, the reported level of fluctuation varies significantly across different studies. Some studies showed around 50% variability in CSF Aβ42 and Aβ40 concentrations over 6 hrs, while others show fluctuations of only around 10% over 6 hrs. Some authors have hypothesized that these differences could, however, be attributed to differences in the sampling procedure used in different studies, the age of the donors, and/or the presence of amyloidosis [23, 24]. Lumbar punctures for AD diagnostics are routinely performed during working hours. However, due to the evidence of remarkable Aβ variations, it may be recommended to obtain CSF samples in the same time period (i.e., from 8 a.m. to 12 p.m.) [19, 24].

2.1.3 CSF Collection and Blood Contamination

For the main AD-related biomarkers, it is recommended to collect CSF via LP, between the L3/L4 or L4/L5 intervertebral space, using an atraumatic needle to prevent post-puncture headaches and other possible complications [13, 25]. During LP, CSF might be collected using a syringe or by gravity-drip directly from the needle. The different approaches did not affect the concentrations of Aβ peptides, tTau, pTau, α-synuclein, neurogranin, and beta-site amyloid precursor protein cleaving enzyme 1 [26, 27]. Still, current guidelines recommend using the gravity drip collection of CSF directly from the needle without any aspiration methods for diagnostic purposes [17], as the sample volume needed is small (2 ml) and the use of an aspiration method could cause Aβ or other proteins to attach to the plastic walls of the syringe, affecting the accuracy of the results. For research or biobanking purposes that require higher sample volumes, gravity drip should be avoided because it is time-consuming and can be uncomfortable for the patient. The volume of extraction might also be a variable to consider since the concentration of the biomarker could be influenced by the rostrocaudal concentration gradient. That is, brain-derived proteins are more concentrated in the ventricular than in lumbar CSF, and blood-derived proteins have a lower ventricular than lumbar CSF concentration [28]. Several studies have shown that there are no significant variations in Aβ42 concentrations in sample volume extraction ranging from 20 to 40 ml [21, 29–31], while others have shown increasing Aβ42 concentrations in sample aliquots obtained after the first 5 ml [20].

Another factor to be considered during the CSF collection is blood contamination, usually occurring at the beginning of the extraction. Of note, 80% of proteins in CSF are derived from the blood stream [32] and the total protein concentration in CSF is 50 to 200 times lower compared to blood serum [33, 34]. Small blood contamination during CSF sample extraction can thus cause inaccurate quantification of both brain- and plasma-derived proteins in CSF [35]. In addition, interactions with other plasma proteins or degradation through blood-specific proteases can alter the quantification of CSF proteins [36]. Blood contamination was shown to reduce

CSF Aβ42 concentrations, which was explained by Aβ42 interactions with plasma proteins [13, 21]. In plasma-contaminated CSF samples (5% plasma, corresponding to a CSF/serum albumin ratio of 55), significant reductions of up to 50% in Aβ42 concentrations have been reported, but low levels of blood contamination (up to 0.1% CSF volume) did not affect Aβ42 concentrations [21]. The confounding effect of blood contamination has been demonstrated in other studies in which a time- and dose-dependent decrease in Aβ42 and Aβ40 concentrations was observed in CSF samples contaminated with 0.04 to 10% blood. The effect of blood contamination was reduced when samples were centrifuged or stored at 4 °C and −20 °C [21, 31, 37]. However, when samples are stored for a long term (−80 °C for 4 years), small blood contamination (0.25%) results in reduced CSF Aβ42 levels, suggesting potential degradation of amyloid peptides likely during long-term storage or after thawing due to proteases released from blood cells during the freeze-thaw process [20]. It should be noted that in some studies showing a reduction in Aβ42 and Aβ40, the Aβ42/40 ratio was not affected by blood contamination [31, 37]. Centrifugation after sample collection (see Sect. 2.2.1 below) would be a potential strategy to reduce blood contamination interference, but for diagnosis purposes, the safest approach is to discard blood-contaminated samples to avoid blood contamination interference. Considering all the above, SOPs recommend discarding the first 1–2 ml of CSF and using a sample obtained from the first 20 ml extracted from the patient for the AD biomarkers diagnostics [17].

2.1.4 Collection Tube

The composition of the tube material is a very relevant aspect that has been extensively studied since it can modulate the adhesion of CSF proteins to the tube walls. The use of polystyrene and glass tubes results in decreased concentrations of pTau181, tTau, and especially Aβ40 and Aβ42 due to their sticky nature [21, 38]. Polypropylene (PP) tubes are therefore the preferred tube composition [21], though it should be noted that there is still a wide heterogeneity of polymer composition and surface treatment in PP tubes. The use of different commercially available PP tubes can lead to high variability in concentrations of amyloid peptides reaching up to >50% in the case of Aβ42 [39]. The concentration variation observed for tTau and pTau was lower than for Aβ42 (below ±10%), probably because they are more hydrophilic proteins [39]. Tube sterilization was shown to not affect Aβ42 concentrations [31]. The greatest recoveries of Aβ42 were obtained when PP low-binding tubes were used, therefore they are the recommended tubes in standard guidelines [31, 40]. Although these PP low-binding tubes are available from different manufacturers, some specific assays recommend a particular reference tube, and thus, the use of any other type of tube must be compared with the recommended one before implementing it in the clinical routine for CSF diagnostics [17]. An additional factor that might cause variability is CSF transfer across tubes, as proteins can then adhere to the new tubes [41]. It has been reported that 5–10% of the Aβ42 concentration is lost with every tube transfer due to plastic adsorption not only in the new tube but also in the plastic tip [42]. Clinically, such effects were, however, mitigated by the analysis of the Aβ42/40 ratio [42]. The latest guidelines, however, recommend avoiding tube transfers [17].

2.1.5 Filling Volume

The ratio of the surface tube area to the sample volume is inversely proportional to CSF Aβ42 concentrations. Higher surface/volume ratios are thus associated with lower protein recovery, suggesting that tube filling may influence the accuracy of the results [43]. The same tendency has been observed in several studies, although it is partially mitigated depending on the use of specific PP tubes [31, 44]. Based on such findings, the general recommendation is to minimize the empty portion of the tube and to fill the aliquot tubes to a maximum of 75% of their capacity [15]. However, it is important to follow the tube specifications, if available, for each specific assay according to the manufacturer's guidelines or validate procedures when new volumes or tube types are used [17].

2.2 *Sample Processing and Storage Factors*

2.2.1 Centrifugation

It has been proposed by several authors that when blood contamination is visible in CSF samples (it becomes visually evident from 0.1% blood contamination) [37], a centrifugation step should be included in the protocol within 2 hrs after sample extraction to prevent blood interference at standardized protocol conditions (2000 *g* for 10 min at room temperature (RT)) [13, 15, 16, 45]. CSF centrifugation has also been recommended to remove other types of cells that are typically present. However, there is no consensus on the need for centrifugation in nonhemorrhagic samples due to the inconsistency observed between studies on the impact of CSF centrifugation on Aβ peptide measurements [21, 30, 37, 46–49]. Considering that CSF samples with visible blood contamination should be excluded from the AD core biomarkers analysis in CSF, current guidelines for the analysis of these markers in clinical routine recommend not centrifuging fresh CSF [17].

2.2.2 Sample Mixing

Mixing samples by vortexing, roller shaking, or inversion is a common practice before analysis to obtain homogeneous samples. However, mixing may contribute to the adhesion of amyloid peptides to the tube surface, thus reducing their levels in the analyzed sample. Some studies have reported no effect of mixing with a vortex or roller on CSF Aβ42 and Aβ40 concentrations [37, 42]. A recent study has, however, observed that different mixing procedures in fresh CSF reduced the concentrations of CSF Aβ42, with the strongest effect observed after vortexing (mean reduction of Aβ42 concentration of 13%). The impact was dependent on the filling volumes of the tube, with a stronger effect in tubes with smaller CSF content [31]. In conclusion, there is not a clear benefit of mixing samples for biomarker analysis in fresh CSF. However, in frozen samples, mixing could be more critical due to the concentration gradients induced by the freeze/thaw procedure. It has been observed that analysis of non-mixed samples after the freeze/thaw procedure results in high variability and unreliable Aβ40 results [37]. Based on that, current guidelines recommend avoiding mixing for fresh sample analysis or before freezing, but gentle mixing is recommended after thawing frozen samples to reduce variability [17].

2.2.3 Processing Delays and Transportation Before Sample Analysis or Storage

Different laboratories and studies follow different time and temperature protocols before analyzing or storing samples [19, 24]. Some studies have investigated the stability of Aβ42, Aβ40, tTau, and pTau in CSF samples directly maintained at RT or in cooled conditions for different lengths of time [24]. The concentration of CSF Aβ40, tTau, and pTau remains stable at RT for short periods beyond 24 hrs and can even remain stable for up to 14 days in these conditions [50]. Concentrations of CSF Aβ42 appear stable at RT for 24 hrs [24]; however, different results have been obtained with periods longer than 24 hrs at RT. While some authors have reported changes in CSF Aβ42 concentration in samples kept at RT for longer than 24 hrs [50–52], others have shown stable Aβ42 concentrations in samples maintained at RT for 5 days [51], 5–7 days [52], or even up to 14 days [50]. The interaction between time and temperature over the stability of the Aβ peptides has also been studied. CSF Aβ42 concentrations showed a 10% decrease after 2 weeks stored at RT, but did not change when stored at 2–8 °C. Because Aβ40 was stable at both RT and cooled conditions, the CSF Aβ42/40 ratio showed reductions at both temperatures [37]. Following the main findings of the different studies, current guidelines now recommend keeping samples at RT for periods no longer than 2 days before analysis and at 4 °C for up to 14 days. Samples can also be directly stored at −20 °C or −80 °C before analysis [17].

Depending on the place of sample extraction, samples might be sent to specialized laboratories for their analysis. Sample storage conditions during transportation from the extraction point to the reference laboratory or biobank might also be a potential source of variability in protein measurements. A recent study investigated the impact of time, temperature, tube position, and filling volume during transportation on the CSF concentration of Aβ40, Aβ42, tTau, and pTau [31]. It was observed that Aβ42 levels varied when samples were transported in a horizontal position, but this could be prevented by refrigeration at 4 °C or filling the tube to the maximum volume. Thus, the recommendation is that samples must always be transported with the predetermined fill volume for specific tubes (see above) in cold blocks for a maximum period of 14 days [17].

2.2.4 Freeze-Thaw Cycles and Aliquoting

Freezing of CSF samples allows longer storage to prevent degradation before analysis. However, freezing is a stressing procedure that can induce protein degradation during the different stages. Hence, exposing the CSF sample to several freeze-thaw cycles and using different freezing and thawing protocols can be another source of variability. Of note, the different impact of freeze/thaw cycles on the stability of the different AD core biomarkers has shown different results across studies. Some of these discrepancies might be explained, however, by different experimental designs that do or do not account for other confounding factors such as the kind of the plastic tube [53], the sample volume [54], the freeze/thaw protocol [21], or the use of different cryoprotection compounds [53]. Still, most studies indicate that the least stable biomarker is CSF Aβ42 with freeze/thaw cycles ranging between 1 cycle [21, 55] and 5 cycles [47] across studies. The most frequent observation is that more than 2 freeze/thaw cycles result in a significant Aβ42 concentration drop, which agrees with previously reported recommendations [13, 16]. To prevent exceeding the number of freeze/thaw cycles, it is important to aliquot the CSF sample before freezing, considering the factors previously discussed (e.g., low-binding PP tubes, or fill the tubes up to 75% or following manufacturer's recommendations) [15, 17, 24, 43].

2.2.5 Long-Term Storage

The stability of the CSF proteins might be affected by the duration of storage. This could be especially relevant when performing longitudinal studies using biobanked cohorts collected over extended periods. One potential factor that can influence the CSF biomarker concentration during long-term storage at −80 °C is sample evaporation/sublimating, but its significance has been ruled out, at least during a period of 2 years [13, 56]. In addition, protein degradation during long-term storage at −80 °C can contribute to the variability observed when comparing samples collected at different time points. The use of accelerated stability test mathematical models has predicted that Tau and Aβ42 are stable in CSF stored at −80 °C for very long periods [46]. Subsequent studies based on real measurements have shown that CSF Aβ42, pTau, and total Tau are stable in samples stored at −80 °C up to 12 years [21, 44, 57].

3 Impact of Different Pre-analytical Factors on AD-Relevant Biomarkers in Blood

3.1 Blood Sampling Factors

3.1.1 Fasting Conditions

As mentioned earlier, fasting is not a necessity before the collection of CSF for the measurement of core AD biomarkers. However, this may not be the case for blood collection. For instance, there are significant alterations in the levels of hematological analytes (hemoglobin, red and white blood cells, triglycerides, albumin, and C-reactive protein) after food consumption compared to those in fasting conditions [58, 59]. There is, however, a paucity of studies confirming the influence of food intake on the levels of blood-based AD biomarkers. One recent study investigated the influence of acute food intake on a wide panel of plasma AD-related biomarkers (pTau217, pTau181, pTau231, Aβ42, Aβ40, GFAP, and NfL) measured with the high-sensitivity Simoa platform and showed that all measured biomarkers were changed after food intake compared to the fasted condition [60]. Noteworthy, the cohort investigated in this study may not represent the real-world scenario of AD blood biomarker testing, as the effect of fasting was evaluated only in obese individuals and after consuming a high-protein drink that does not resemble a standard diet [60]. This calls for additional studies to confirm such findings. Indeed, a recent study found no difference in levels of plasma Aβ42 and NfL in a plasma sample collected in the morning in a fasted state compared to a sample collected in the evening in a non-fasted state [61], though the sample size was limited. Considering the few studies conducted so far and their intrinsic limitations, additional studies are needed to establish the effect of fasting conditions on AD plasma biomarker levels.

3.1.2 Effect of Circadian Rhythms and Diurnal Variation

Only one study has investigated the effect of circadian rhythm on plasma Aβ levels in samples collected at three time points over a day (8 am, 12 pm, and 3 pm) and found that plasma Aβ42 and Aβ40 levels increased by 5–9% on average from morning until 3 pm [62]. This variability is not trivial considering that the difference in plasma amyloid levels between controls and AD patients is less than 20%. The Aβ42/40 ratio was, however, stable over the tested time points [62]. It is still unknown how circadian rhythms affect other AD-related plasma biomarkers.

3.1.3 Blood Collection

Most errors of the pre-analytical phase are attributed to improper blood collection [63, 64], caused by, for example, needle size, location of sampling, variation in the types of blood collection tubes used (tube material and additives), tube filling volume, or the order of blood tube type withdrawal [18, 65]. Venipuncture is performed with different types of needles and methods, which can differentially impact blood sample hemolysis and thus biomarker measurements [66]. The strongest predictor of hemolysis identified so far is the device used for collecting blood. For example, the use of butterfly needles resulted in significantly lower hemolysis rates (2.7%) in comparison to IV catheters (14.6%) [67]. The location of blood withdrawal or the size of the needle can also impact sample hemolysis. For instance, hemolysis rates significantly increased in blood samples withdrawn using an IV catheter from locations other than the antecubital fossa, or when it was withdrawn using a small-gauge catheter or with prolonged tourniquet time of more than 1 minute [67]. This was confirmed by another study, which found that using a small gauge catheter to draw blood from veins outside the antecubital area significantly increased the risk of hemolysis [68]. To date, there is no evidence of the influence of the type of needle used, or hemolysis, on the levels of AD blood biomarkers.

3.1.4 Collection Tube and Filling Volume

Whole blood is collected in different types of tubes with different clot activators or anticoagulant types recognizable through the cap color depending on the ultimate matrix that needs to be obtained: serum or plasma [65]. Serum is the fluid remaining after whole blood is allowed to clot and is collected in tubes with or without a clot activator. Plasma is collected from tubes containing different types of anticoagulants, and it is the fluid remaining after centrifuging out whole blood constituents, whereas clot-forming proteins are preserved. The most used anticoagulant tubes in clinical chemistry practice are ethylenediaminetetraacetic acid (EDTA), sodium citrate, and lithium heparin [63]. Most AD biobanks have collected serum and EDTA plasma samples, while in clinical chemistry and immunochemistry, EDTA and heparin tubes are most often used [63]. Several studies have assessed the influence of using different types of tubes on the levels of the core AD blood-based biomarkers [18, 62, 69–73]. Most studies found that the samples collected in heparin tubes had higher levels of Aβ42, Aβ40, tTau, pTau181, GFAP, and NfL, than samples collected in EDTA, citrate, or serum tubes [18, 62, 70]. Other studies, however, found no differences in the levels of Aβ42, Aβ40, pTau217, GFAP, and NfL between samples collected in heparin and EDTA tubes [62, 72, 73]. The discrepancies observed might be partly explained by the type of technology used. For example, heparin can interact with tau protein, promoting its aggregation in vivo [72, 74]. Such structural changes may interfere with antigen-antibody binding and thus might be less suitable for immune-based methods [63]. Different studies have also compared the levels of AD biomarkers between plasma EDTA and serum samples. Most of them have shown that levels of Aβ42, Aβ40, GFAP, and tTau were lower in serum than in plasma EDTA [69–71, 75]. For NfL, most studies showed higher levels in serum compared to plasma EDTA samples [70, 71, 76], though higher levels in plasma EDTA have also been detected [75]. These data overall suggest that biomarker values coming from different matrices or plasma tubes might not be directly comparable and that this information should always be annotated. Still, the fact that some biomarkers showed higher levels with some tube types can be especially relevant for those markers that are measured close to the lower limit of detection of the assays. It is important to note that the diagnostic performance of the markers Aβ42, Aβ40, GFAP, tTau, and NfL seems not to be affected by the collection tube type (serum, EDTA, or heparin plasma) [71, 77]. Considering that the use of serum results in lower biomarker levels and heparin is more prone to interact with other proteins in the samples, for now, it may be recommended to consider the use of plasma EDTA tubes for the measurements of AD blood biomarkers in clinical routine. Of note, the main SOP on blood sample handling for AD biomarkers published so far is developed for plasma EDTA samples [18].

3.1.5 Order of Blood Collection

The Clinical & Laboratory Standards Institute (CLSI) has established guidelines for the sequence of blood tube collection. This guideline was established to minimize the potential cross-contamination of tube additives over blood tubes, which could impact the accuracy of analytical measurements [69]. The recommended order for blood drawing is as follows: (1) blood culture tubes, (2) sodium citrate tubes, (3) serum tubes with or without clot activator and/or serum gel separator, (4) heparin tubes, and (5) EDTA tubes. Currently, no studies have been conducted to examine the impact of altering the standardized blood tube collection sequence on the measurement of AD-related blood-based biomarkers. An inventory study among AD biobanks has revealed that those centers consistently adhere to the established standardized sequence, and thus, the blood tube collection sequence is not a pre-analytical factor that varies across different biobanks and therefore, has not been deemed relevant to study experimentally [20].

3.1.6 Filling Volume

It has been reported that 14% of pre-analytical errors could be attributed to an incorrect blood/anticoagulant ratio [78]. It is advised to ensure the complete filling of the primary collection tubes according to the tube mark labeling. Incomplete filling of tubes (e.g., <80% of the tube mark) changes the ratio of plasma to anticoagulant, consequently changing the dilution factor of plasma proteins in the tube [78]. For AD blood biomarkers, only one study has investigated the impact of incomplete tube filling on the levels of Aβ42, Aβ40, and tTau. Half-filled EDTA plasma tubes reduced tTau levels by 12–14%, but there was no influence noted on Aβ42 and Aβ40 levels [62]. Additional studies are needed to understand the effect of underfilled tubes on the levels of other AD blood-based biomarkers, though a survey conducted among AD biobanks has shown that all centers follow the practice of complete tube filling, and thus this factor may not strongly impact AD plasma biomarker studies [18].

3.2 Sample Processing and Storage Factors

3.2.1 Centrifugation

Centrifugation is the first processing step after blood withdrawal, through which serum or plasma is separated from the rest of the blood constituents (red blood cells and platelets). Careful, complete, and timely separation of blood constituents from the plasma or serum might be important to prevent the interaction of cellular blood components with the serum or plasma and prevent interference of these components with the biomarker levels. The World Health Organization (WHO) broadly recommends centrifugation of samples for 15 minutes at 2000–3000 *g* for plasma and 10 minutes at 1500 *g* for serum [79], which are the standardized practices applied in most biobanks and clinical routine labs. Few studies have investigated the effect of different centrifugation speeds, force, duration, and temperature on AD blood-based biomarker levels. Only one study analyzed the influence of centrifugation temperature on the levels of Aβ42, Aβ40, GFAP, NfL, pTau181, and tTau, where only tTau showed different levels when centrifuging at 4 °C compared to RT [18]. Several studies have, however, investigated the influence of delayed centrifugation of EDTA plasma collection tubes on AD blood-based biomarkers [18, 62, 73, 75]. Plasma Aβ42 and Aβ40 were most sensitive to delayed centrifugation among the AD blood biomarkers. Their levels declined after 3 and 6 hrs [18, 73] of centrifugation delay when tubes were kept at RT. Keeping samples in the fridge prolonged Aβ42 and Aβ40 stability, where their levels remained stable for up to 6 and 24 hrs [18, 62, 73]. tTau showed instability after centrifugation delays of 3 hrs at RT and 1 hr at 4 °C [18]. The other investigated plasma AD biomarkers (pTau181, GFAP, NfL, and pTau217) were found to be stable in unprocessed EDTA and serum blood tubes both at RT and in the fridge for up to 24 hrs [18, 62, 73, 75]. One study examined up to 72 hrs of blood tube processing delay at 4 °C and observed a significant, yet modest, decrease in EDTA plasma levels of NfL and GFAP and an increase in the EDTA plasma levels of pTau181 [75].

In summary, the studies on delayed centrifugation results indicate the critical importance of promptly processing blood tubes after collection for accurate AD biomarker measurements, especially for Aβ42 and Aβ40. If direct processing is not possible, it may be advisable to keep the blood tubes refrigerated for not more than 24 hrs to ensure accurate AD plasma biomarker measurements. Knowing the limits of sample processing delays provides the needed flexibility, especially for large, geographically scattered populations where there is a need for long-distance sample shipment.

As for centrifugation settings, it seems not to influence the majority of blood-based biomarkers, yet there was a slight effect on tTau for the centrifugation temperature [18].

3.2.2 Sample Mixing and Tube Transfer

It is crucial to thoroughly mix whole blood in the collection tube to guarantee the even distribution of the clot activator or anticoagulant throughout the blood samples. According to a survey conducted among AD biobanks, it was shown that all centers invert blood collection tubes up to 10 inversions, immediately after blood withdrawal. Thus, it is assumed that this factor may not impact AD plasma biomarkers studies [18]. As for sample transfer from one tube to another, only a few studies have investigated such influence on the levels of AD-related plasma biomarkers [62, 73]. Both studies only detected a slight decrease in the levels of Aβ42 and Aβ40 proteins (<5%) after 5 tube transfers. The levels of plasma GFAP, NfL, and pTau217 remained stable over multiple tube transfers [73]. Accordingly, it would not be recommended to perform more than four transfers before AD blood biomarker measurements.

3.2.3 Processing Delays and Transportation Before Sample Analysis or Storage

As blood-based biomarkers are advantageous in that they are more convenient to use, they are often assumed to have clinical applications in primary care settings. However, primary care clinics typically lack the advanced technology required for precise blood biomarker measurements as well as −80 °C freezers for proper longer-term storage. As a result, processed blood samples (i.e., plasma/serum aliquots) may need to be temporarily kept at RT, in the fridge, or in −20 °C freezers before being sent to central laboratories for −80 °C storage and/or AD blood biomarker measurements. Several studies have investigated variations in storage conditions and their influence on AD blood-based biomarker levels [18, 53, 73, 80]. All studies revealed similar trends: Aβ42 and Aβ40 were the most sensitive markers to delayed storage with their levels declining after 6 hrs when kept at RT. [18, 53, 73] These markers, however, remained stable when plasma was kept for up to 24 and 48 hrs in the fridge [18, 53, 73, 80], or for up to 2 weeks when stored at −20 °C [81]. Applying the Aβ42/40 ratio only partly mitigated the observed reduction in Aβ42 and Aβ40 levels during delayed storage [53, 73]. Similarly, leaving samples at RT or in the fridge for long durations after freezing, but prior to biomarker measurements, resulted in decreased Aβ42 and Aβ40 values, with earlier and more pronounced effects when samples were kept at RT compared to samples that were kept in the fridge [53]. Other AD-related plasma biomarkers (i.e., GFAP, NfL, pTau181, and tTau [18, 53, 73]) showed good stability against delayed storage conditions. Their levels remained stable for up to 24 hrs, both when plasma was kept at RT or in the fridge [18, 53, 73]. One study showed that pTau181 levels were lower after 2 days of storage in the fridge [80], while the levels of tTau decreased after 2 weeks of storage in the fridge [18]. However, all these aforementioned markers (GFAP, NFL, pTau181, tTau, Aβ42, and Aβ40) remained stable under −20° storage for up to 9 days or even two weeks [18, 80]. Of note, it has been observed that pTau217 levels increased when plasma was kept at RT for 24 hrs, but remained stable at 4 °C [73]. Overall, storage delays particularly affect Aβ42 and Aβ40 stability, but keeping blood samples cold can prolong their stability until 24 hrs. Knowing the maximum delay before freezing samples at −80 °C after being processed is important, as it provides flexibility in sample handling, which is essential to employ the AD blood-based biomarkers in diverse and/or geographically distant settings.

3.2.4 Freeze-Thaw Cycles

The effect of sample freeze-thawing on AD blood biomarkers has been investigated in numerous studies [18, 53, 70, 73, 75, 76]. Generally, plasma levels of Aβ42 and Aβ40 were more stable than their serum levels during multiple freeze-thaw (FT) cycles. In independent studies, it was observed that serum Aβ42 levels decreased significantly after already 1 FT cycle, and levels decreased by 20–26% after 3 and 4 FT cycles [18, 53, 70]. A similar trend was observed for serum Aβ40, as it was shown to be significantly decreased even after 1 FT cycle [70, 75]. On the other hand, both plasma Aβ42 and Aβ40 levels remained stable across all tested FT cycles (up to 3 FT [64] and 4 FT [20, 72]). Interestingly, both serum and plasma Aβ42/40 ratios remained stable for up to 4 FT cycles [18, 62, 70, 75]. Similarly, serum and plasma levels of pTau181 and plasma levels of pTau217 were tested to be stable for up to 4 FT cycles [18, 70, 73, 75]. Controversial results were obtained for NfL and GFAP in serum. In one study, both markers increased after 1 FT cycle, while levels returned to normal in the next FT cycle followed by a significant reduction in levels at the third FT cycle [75]. Other studies found that serum NfL and GFAP levels remained stable for up to 4 FT cycles [70, 76]. In plasma, NfL remained stable for up to 4 FT cycles, while a slight increase in plasma GFAP was observed at the fourth FT cycle [18, 70, 73]. In conclusion, one might be careful to apply multiple FT cycles for clinical AD blood biomarker measurements, especially for serum samples, though up to 4 FT cycles of plasma samples seem acceptable.

3.2.5 Long-Term Storage

Few studies have investigated the impact of long-term −80 °C storage (5, 16, and 20 years) on AD blood biomarkers in serum and plasma [81, 82]. Generally, the effect of prolonged −80 °C storage on blood biomarkers such as Aβ42, Aβ40, tTau [82], and NfL [81] appeared minimal on diagnostic performance [82] or measurement reliability [83]. Overall, the impact of extended storage at −80 °C on the different AD-related plasma biomarkers needs to be extensively investigated and researchers working with biobanked samples may need to consider adjusting for storage duration in their statistical analyses.

4 The Discovery of Novel Fluid Biomarkers and the Effect of Pre-analytical Factors

Novel promising biomarker candidates are constantly being identified for different disorders, especially since mass-spectrometry discovery platforms and innovative aptamer- and antibody-based protein arrays allow high-throughput proteomics screening of samples with large protein coverage. Understanding how pre-analytical cofounding factors may affect proteome stability might be useful for the selection of optimal samples already in the discovery phase of the biomarker development workflow. Proteomic studies have shown that CSF samples left at RT for 30 minutes led to changes in the abundance of specific proteins [15, 83]. Hok-A-Hin et al. previously investigated the effect of delayed storage and freeze/thaw conditions on a total of 1603 CSF proteins measured by two different large protein arrays widely used for the identification of fluid biomarkers [15] and observed that up to 25% of the 1129 proteins measured through aptamer-based arrays were affected when CSF processing was delayed 24 h, whether the samples were kept at 4 °C or RT. Interestingly, more than 85% of the proteins measured with both platforms remained stable when storage of the samples was delayed for 1 week at 4 °C. Similarly, most of the CSF proteins (>80%) remained stable after 4 freeze/thaw cycles. Such a study forms a large database of CSF proteins affected by different pre-analytical factors, which may serve as a catalog for studies assessing the potential value of novel biomarker candidates for clinical practice. Overall, the data suggest that the CSF proteome is – in general – stable, even when undergoing unusual processing conditions. This information might be helpful as precious human biobanked samples can then be used for different purposes (i.e., this usually includes several freeze/thaw cycles). Other large-scale proteomic studies on plasma protein stability have been performed. Pre-centrifugation delay, centrifugal force, or biobank storage time were the factors that most influenced protein abundance [84–86]. Freeze/thaw cycles, however, did not strongly impact plasma protein measurements [84]. Pre-analytical confounding factors are not often an issue when using samples coming from the same biobank for biomarker discovery studies, as SOPs are often followed. Still, researchers may consider performing some preliminary tests with specific candidates especially on the effects of freeze/thaw, so further analysis and validation in other cohorts following similar procedures can be initiated. The different pre-analytical factors should, however, be tested before implementation of any biomarker candidate following the international recommendations on both CSF [17] and plasma [18].

5 Initiatives to Standardize Pre-analytical Procedures for Fluid Biomarkers Analysis

International collaborations are imperative to facilitate and accelerate the implementation of biomarkers in clinical diagnostics routines and facilitate their global feasibility and interpretation. As a result, several global, collaborative initiatives have been established together, aiming to standardize CSF-based and blood-based neurodegenerative biomarker measurements. The Alzheimer's Association established the Global Biomarker Standardization Consortium (GBSC) in 2010 in which the expertise of academic researchers, clinicians, industry, and regulatory representatives is combined [87, 88]. Today's programs executed under the guidance of the GBSC are focused on analytical biomarker measurement variations (e.g., the Quality Control Program, the Reference Material Program, and the Reference Methods Program) and pre-analytical biomarker measurement variations (the CSF Pre-Analytics Consortium and the Standardization of Alzheimer's Blood Biomarkers (SABB) Program). The analytically focused programs aim to tackle between-laboratory variations, e.g., by developing universal reference materials to which mass spectrometry and immunoassays can be calibrated so that biomarker values are globally harmonized. The pre-analytically focused programs aim to develop consensus, evidence-based SOPs to mitigate any pre-analytical sample handling effects on the CSF or blood-based biomarker measurements. Hallmark sample handling SOPs have been published as a result of these GBSC efforts [17, 18] and are continuously being refined. Additional large collaborative efforts have also been established under the umbrella of the Innovative Medicines Initiatives (IMI), which has the overall objective to improve human health by accelerating the development of, and access to, new therapeutics. IMI-funded efforts focus on having solid research infrastructures for the development of novel biomarkers for neurodegenerative diseases (e.g., the European Platform for Neurodegenerative Disease [EPND] and 3TR), for example, by establishing best practices for biospecimens biobanking but also focusing on data storage and sharing [89–91]. Another example of an ongoing initiative for biomarker measurement standardization is the European Union's Horizon 2020 research NeuroMET, now continued as NeuroBioStand, which aims to develop reference measurement procedures, clinical protocols, and measurement infrastructures to establish confidence in new technology and diagnostic tools resulting in their clinical uptake [92].

6 Conclusions

The implementation of CSF and plasma biomarkers in clinical practice and trials requires harmonized protocols that minimize the effect of pre-analytical confounding factors. This will provide robust and reproducible measurements across centers and projects. International evidence-based consensus guidelines for processing and biobanking of CSF and, more recently, also blood, have been developed for the optimal analysis of core AD and related biomarkers [17, 18], which have been summarized in the current chapter. Additional studies are still needed to determine the impact of different factors in plasma AD biomarker analysis (e.g., fasting conditions, long-term storage) (Fig. 1). Considering that the field of fluid biomarkers is continuously evolving, we also reviewed the studies that investigate the impact of some of the pre-analytical factors on the global CSF and plasma proteome, which can be useful to understand which type of samples can be used in the discovery phase. In addition, the different international initiatives that are also described in

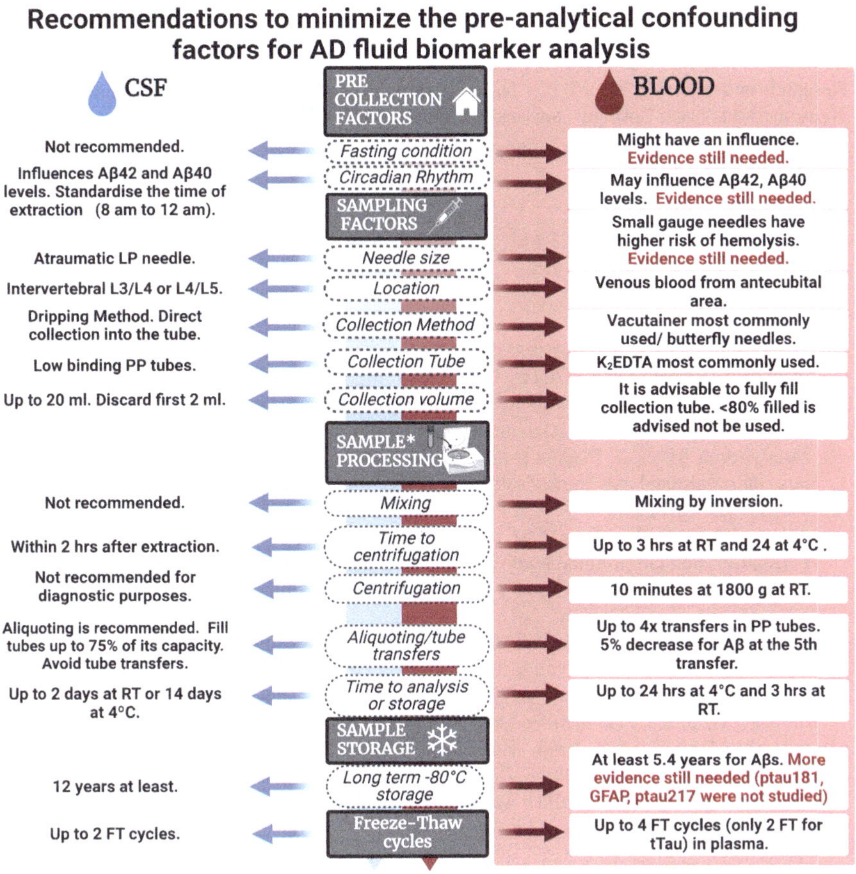

Fig. 1 Workflow for minimizing pre-analytical factors in CSF and plasma samples. The recommendations provided are based on existing literature. However, further studies are required, particularly regarding the handling of blood samples. *The recommendations for sample processing pertain to fresh samples. Abbreviations: Aβ Amyloid beta, LP lumbar puncture, L3/4 lumbar vertebrae 3 or 4, K_2EDTA dipotassium ethylenediaminetetraacetic acid, RT room temperature, hrs hours, PP polypropylene, FT Freeze-thaw cycles. (Created with BioRender.com)

this chapter will facilitate the development of consensus and validated SOPs that can be implemented by the different labs worldwide, thereby facilitating bench-to-bedside translation of biomarker findings and the implementation of AD biomarkers in the clinical diagnostic workup.

Funding MC research is supported by "PROYECTOS I+D+I – 2020"- Retos de investigación from the Ministerio Español de Ciencia e innovación (PID2020-115613RA-I00).

Competing Interests Statement MC has been an invited speaker at Eisai, is an associate editor at Alzheimer's Research & Therapy, and has been an invited writer for Springer Healthcare. CET has a collaboration contract with ADx Neurosciences, Quanterix and Eli Lilly, performed contract research or received grants from AC-Immune, Axon Neurosciences, Bioconnect, Bioorchestra, Brainstorm Therapeutics, Celgene, EIP Pharma, Eisai, Grifols, Novo Nordisk, PeopleBio, Roche, Toyama, Vivoryon. She serves on editorial boards of Medidact Neurologie/Springer, Alzheimer Research and Therapy, Neurology: Neuroimmunology & Neuroinflammation, and is editor of a Neuromethods book Springer. She had speaker contracts for Roche, Grifols, Novo Nordisk. The rest of the authors declare no competing interest.

References

1. Teunissen CE, et al. Blood-based biomarkers for Alzheimer's disease: towards clinical implementation. Lancet Neurol. 2021;21:66. https://doi.org/10.1016/S1474-4422(21)00361-6.
2. Milà-Alomà M, et al. Amyloid beta, tau, synaptic, neurodegeneration, and glial biomarkers in the preclinical stage of the Alzheimer's continuum. Alzheimers Dement. 2020;16:1358–71.
3. Milà-Alomà M, et al. Plasma p-tau231 and p-tau217 as state markers of amyloid-β pathology in preclinical Alzheimer's disease. Nat Med. 2022;28:1797. https://doi.org/10.1038/s41591-022-01925-w.
4. Ashton NJ, et al. Plasma and CSF biomarkers in a memory clinic: head-to-head comparison of phosphorylated tau immunoassays. Alzheimers Dement. 2022;5:1913.
5. Jack CR, et al. NIA-AA research framework: toward a biological definition of Alzheimer's disease. Alzheimers Dement. 2018;14:535–62.
6. Mattsson N, et al. Plasma tau in Alzheimer disease. Neurology. 2016;87:1827–35.
7. Zetterberg H, et al. Plasma tau levels in Alzheimer's disease. Alzheimers Res Ther. 2013;5:9.
8. Simrén J, et al. The diagnostic and prognostic capabilities of plasma biomarkers in Alzheimer's disease. Alzheimers Dement. 2021;17:1145–56.
9. Gonzalez-Ortiz F, et al. Brain-derived tau: a novel blood-based biomarker for Alzheimer's disease-type neurodegeneration. Brain. 2023;146:1152–65.
10. Abdelhak A, et al. Blood GFAP as an emerging biomarker in brain and spinal cord disorders. Nat Rev Neurol. 2022;3(18):158–72.
11. Vijverberg EGB, et al. Cerebrospinal fluid biomarker examination as a tool to discriminate behavioral variant frontotemporal dementia from primary psychiatric disorders. Alzheimers Dement. 2017;7
12. Miller TM, et al. Trial of antisense oligonucleotide tofersen for SOD1 ALS. N Engl J Med. 2022;387:1099–110.
13. Del Campo M, et al. Recommendations to standardize preanalytical confounding factors in Alzheimer's and Parkinson's disease cerebrospinal fluid biomarkers: an update. Biomark Med. 2012;6:419–30.

14. Willemse EAJ, Teunissen CE. Importance of pre-analytical stability for CSF biomarker testing. In: Cerebrospinal fluid in clinical neurology; 2015. p. 59–77. https://doi.org/10.1007/978-3-319-01225-4.
15. Hok-A-Hin YS, Willemse EAJ, Teunissen CE, Del Campo M. Guidelines for CSF processing and biobanking: impact on the identification and development of optimal CSF protein biomarkers. In: Methods in molecular biology, vol. 2044. Humana Press Inc.; 2019. p. 27–50.
16. Vanderstichele H, et al. Standardization of preanalytical aspects of cerebrospinal fluid biomarker testing for Alzheimer's disease diagnosis: a consensus paper from the Alzheimer's Biomarkers Standardization Initiative. Alzheimers Dement. 2012;8:65–73.
17. Hansson O, et al. The Alzheimer's Association international guidelines for handling of cerebrospinal fluid for routine clinical measurements of amyloid β and tau. Alzheimers Dement. 2021:1575. https://doi.org/10.1002/alz.12316.
18. Verberk IMW, et al. Characterization of pre-analytical sample handling effects on a panel of Alzheimer's disease–related blood-based biomarkers: results from the Standardization of Alzheimer's Blood Biomarkers (SABB) working group. Alzheimers Dement. 2021;18:1484. https://doi.org/10.1002/ALZ.12510.
19. Park SA, et al. A consensus in Korea regarding a protocol to reduce preanalytical sources of variability in the measurement of the cerebrospinal fluid biomarkers of Alzheimer's disease. J Clin Neurol. 2015;11:132–41.
20. Darrow JA, et al. Effect of patient-specific preanalytic variables on CSF Aβ1-42 concentrations measured on an automated chemiluminescent platform. J Appl Lab Med. 2021;6:397–408.
21. Bjerke M, et al. Confounding factors influencing amyloid Beta concentration in cerebrospinal fluid. Int J Alzheimers Dis. 2010;2010:1–11.
22. Bateman RJ, Wen G, Morris JC, Holtzman DM. Fluctuations of CSF amyloid-beta levels: implications for a diagnostic and therapeutic biomarker. Neurology. 2007;68:666–9.
23. Dobrowolska JA, et al. Diurnal patterns of soluble amyloid precursor protein metabolites in the human central nervous system. PLoS One. 2014;9:e89998.
24. Hansson O, et al. The impact of preanalytical variables on measuring cerebrospinal fluid biomarkers for Alzheimer's disease diagnosis: a review. Alzheimers Dement. 2018;14:1313–33.
25. Engelborghs S, et al. Consensus guidelines for lumbar puncture in patients with neurological diseases. Alzheimers Dement (Amst). 2017;8:111–26.
26. Rembach A, et al. Alzheimer's disease cerebrospinal fluid biomarkers are not influenced by gravity drip or aspiration extraction methodology. Alzheimers Res Ther. 2015;7:71.
27. Doecke JD, et al. Core Alzheimer's disease cerebrospinal fluid biomarker assays are not affected by aspiration or gravity drip extraction methods. Alzheimers Res Ther. 2021;13:79.
28. Reiber H. Proteins in cerebrospinal fluid and blood: barriers, CSF flow rate and source-related dynamics. Restor Neurol Neurosci. 2003;21:79–96.
29. Brandner S, et al. Ventricular and lumbar cerebrospinal fluid concentrations of Alzheimer's disease biomarkers in patients with normal pressure hydrocephalus and posttraumatic hydrocephalus. J Alzheimers Dis. 2014;41:1057–62.
30. Le Bastard N, De Deyn PP, Engelborghs S. Importance and impact of preanalytical variables on Alzheimer disease biomarker concentrations in cerebrospinal fluid. Clin Chem. 2015;61:734–43.
31. Hansson O, et al. Pre-analytical protocol for measuring Alzheimer's disease biomarkers in fresh CSF. Alzheimers Dement (Amst). 2020;12
32. Reiber H. Dynamics of brain-derived proteins in cerebrospinal fluid. Clin Chim Acta. 2001;310:173–86.
33. Schilde LM, et al. Protein variability in cerebrospinal fluid and its possible implications for neurological protein biomarker research. PLoS One. 2018;13:e0206478.
34. Jankovska E, Svitek M, Holada K, Petrak J. Affinity depletion versus relative protein enrichment: a side-by-side comparison of two major strategies for increasing human cerebrospinal fluid proteome coverage. Clin Proteomics. 2019;16:9.

35. Aasebø E, et al. Effects of blood contamination and the rostro-caudal gradient on the human cerebrospinal fluid proteome. PLoS One. 2014;9:e90429.
36. You J-S, et al. The impact of blood contamination on the proteome of cerebrospinal fluid. Proteomics. 2005;5:290–6.
37. Janelidze S, Stomrud E, Brix B, Hansson O. Towards a unified protocol for handling of CSF before β-amyloid measurements. Alzheimers Res Ther. 2019;11:63.
38. Perret-Liaudet A, et al. Risk of Alzheimer's disease biological misdiagnosis linked to cerebrospinal collection tubes. J Alzheimers Dis. 2012;30:1–8.
39. Pica-Mendez AM, Tanen M, Dallob A, Tanaka W, Laterza OF. Nonspecific binding of Aβ42 to polypropylene tubes and the effect of Tween-20. Clin Chim Acta. 2010;411:1833.
40. Vanderstichele HMJ, et al. Optimized standard operating procedures for the analysis of cerebrospinal fluid Aβ42 and the ratios of Aβ isoforms using low protein binding tubes. J Alzheimers Dis. 2016;53:1121–32.
41. Toombs J, Paterson RW, Schott JM, Zetterberg H. Amyloid-beta 42 adsorption following serial tube transfer. Alzheimers Res Ther. 2014;6:5.
42. Willemse E, et al. How to handle adsorption of cerebrospinal fluid amyloid β (1–42) in laboratory practice? Identifying problematic handlings and resolving the issue by use of the Aβ42/Aβ40 ratio. Alzheimers Dement. 2017;13:885–92.
43. Toombs J, et al. Identification of an important potential confound in CSF AD studies: aliquot volume. Clin Chem Lab Med. 2013;51:2311–7.
44. Willemse EAJ, van Uffelen KWJ, van der Flier WM, Teunissen CE. Effect of long-term storage in biobanks on cerebrospinal fluid biomarker Aβ1-42, T-tau, and P-tau values. Alzheimers Dement (Amst). 2017;8:45–50.
45. Fourier A, et al. Pre-analytical and analytical factors influencing Alzheimer's disease cerebrospinal fluid biomarker variability. Clin Chim Acta. 2015;449:9–15.
46. Schoonenboom NSM, et al. Effects of processing and storage conditions on amyloid beta (1-42) and tau concentrations in cerebrospinal fluid: implications for use in clinical practice. Clin Chem. 2005;51:189–95.
47. Leitão MJ, et al. Chasing the effects of pre-analytical confounders – a multicenter study on CSF-AD biomarkers. Front Neurol. 2015;6:1–8.
48. Hu WT, et al. CSF beta-amyloid 1-42 - what are we measuring in Alzheimer's disease? Ann Clin Transl Neurol. 2015;2:131–9.
49. Berge G, et al. Effect of tween-20 on core biomarkers measured in cerebrospinal fluid from patients with Alzheimer's disease, mild cognitive impairment, or healthy control individuals. J Alzheimers Dis. 2016;49:493–502.
50. Simonsen AH, et al. Pre-analytical factors influencing the stability of cerebrospinal fluid proteins. J Neurosci Methods. 2013;215:234–40.
51. Zimmermann R, et al. Preanalytical sample handling and sample stability testing for the neurochemical dementia diagnostics. J Alzheimers Dis. 2011;25:739–45.
52. Lelental N, et al. Comparison of different matrices as potential quality control samples for neurochemical dementia diagnostics. J Alzheimers Dis. 2016;52:51–64.
53. Rozga M, Bittner T, Höglund K, Blennow K. Accuracy of cerebrospinal fluid Aβ1-42 measurements: evaluation of pre-analytical factors using a novel Elecsys immunosassay. Clin Chem Lab Med. 2017;55:1545–54.
54. Gervaise-Henry C, et al. Cerebrospinal fluid Aβ42/Aβ40 as a means to limiting tube- and storage-dependent pre-analytical variability in clinical setting. J Alzheimers Dis. 2017;57:437–45.
55. Sjögren M, et al. Tau and Abeta42 in cerebrospinal fluid from healthy adults 21-93 years of age: establishment of reference values. Clin Chem. 2001;47:1776–81.
56. Willemse EAJ, Koel-Simmelink MJA, Durieux-Lu S, van der Flier WM, Teunissen CE. Standard biobanking conditions prevent evaporation of body fluid samples. Clin Chim Acta. 2015;442:141–5.

57. Schipke CG, et al. Long-term stability of Alzheimer's disease biomarker proteins in cerebrospinal fluid. J Alzheimers Dis. 2011;26:255–62.
58. Lima-Oliveira G, et al. Influence of a regular, standardized meal on clinical chemistry analytes. Ann Lab Med. 2012;32:250–6.
59. Lippi G, et al. Influence of a light meal on routine haematological tests. Blood Transfus. 2010;8:94–9.
60. Huber H, et al. Levels of Alzheimer's disease blood biomarkers are altered after food intake-A pilot intervention study in healthy adults. Alzheimers Dement. 2023;19:5531.
61. Mengel D, et al. Dynamics of plasma biomarkers in Down syndrome: the relative levels of Aβ42 decrease with age, whereas NT1 tau and NfL increase. Alzheimers Res Ther. 2020;12:27.
62. Rózga M, Bittner T, Batrla R, Karl J. Preanalytical sample handling recommendations for Alzheimer's disease plasma biomarkers. Alzheimers Dement. 2019;11:291–300.
63. Revuelta-López E, et al. Pre-analytical considerations in biomarker research: focus on cardiovascular disease. Clin Chem Lab Med. 2021;59:1747–60.
64. Carraro P, Plebani M. Errors in a stat laboratory: types and frequencies 10 years later. Clin Chem. 2007;53:1338–42.
65. Lippi G, Simundic AM. The EFLM strategy for harmonization of the preanalytical phase. Clin Chem Lab Med. 2017;56:1660–6.
66. Howanitz PJ, Lehman CM, Jones BA, Meier FA, Horowitz GL. Clinical laboratory quality practices when hemolysis occurs. Arch Pathol Lab Med. 2015;139:901–6.
67. Wollowitz A, Bijur PE, Esses D, Gallagher EJ. Use of butterfly needles to draw blood is independently associated with marked reduction in hemolysis compared to intravenous catheter. Acad Emerg Med. 2013;20:1151–5.
68. Tanabe P, Kyriacou DN, Garland F. Factors affecting the risk of blood bank specimen hemolysis. Acad Emerg Med. 2003;10:897–900.
69. Lachno DR, et al. The influence of matrix type, diurnal rhythm and sample collection and processing on the measurement of plasma beta-amyloid isoforms using the INNO-BIA plasma Abeta forms multiplex assay. J Nutr Health Aging. 2009;13:220–5.
70. Ashton NJ, et al. Effects of pre-analytical procedures on blood biomarkers for Alzheimer's pathophysiology, glial activation, and neurodegeneration. Alzheimers Dement. 2021;13:e12168.
71. Jonaitis EM, et al. Crosswalk study on blood collection-tube types for Alzheimer's disease biomarkers. Alzheimers Dement (Amst). 2022;14
72. Jiang L, et al. Head-to-head comparison of different blood collecting tubes for quantification of Alzheimer's disease biomarkers in plasma. Biomol Ther. 2022;12
73. Kurz C, et al. Impact of pre-analytical sample handling factors on plasma biomarkers of Alzheimer's disease. J Neurochem. 2023;165:95–105.
74. Goedert M, et al. Assembly of microtubule-associated protein tau into Alzheimer-like filaments induced by sulphated glycosaminoglycans. Nature. 1996;383:550–3.
75. Panikkar D, et al. Pre-analytical variables influencing stability of blood-based biomarkers of neuropathology. J Alzheimers Dis. 2023;95:735–48.
76. Van Lierop ZYGJ, et al. Pre-analytical stability of serum biomarkers for neurological disease: neurofilament-light, glial fibrillary acidic protein and contactin-1. Clin Chem Lab Med. 2022;60:842–50.
77. Kac PR, et al. Diagnostic value of serum versus plasma phospho-tau for Alzheimer's disease. Alzheimers Res Ther. 2022;14:65.
78. Kitchen S, et al. International Council for Standardisation in Haematology (ICSH) recommendations for collection of blood samples for coagulation testing. Int J Lab Hematol. 2021;43:571–80.
79. Tantisaranon P, Dumkengkhachornwong K, Hnoonual A. Influence of reduced centrifugation time on clinical chemistry analytes and literature review. Turk J Biochem. 2023;48:376–87.
80. Mansilla A, et al. Effects of storage conditions on the stability of blood-based markers for the diagnosis of Alzheimer's disease. Clin Chem Lab Med. 2023;61:1580–9.

81. Schubert CR, Paulsen AJ, Pinto AA, Merten N, Cruickshanks KJ. Effect of long-term storage on the reliability of blood biomarkers for Alzheimer's disease and neurodegeneration. J Alzheimers Dis. 2022;85:1021.
82. Chiu MJ, et al. Long-term storage effects on stability of Aβ1-40, Aβ1-42, and total Tau proteins in human plasma samples measured with immunomagnetic reduction assays. Dement Geriatr Cogn Dis Extra. 2019;9:77–86.
83. Rosenling T, et al. The effect of preanalytical factors on stability of the proteome and selected metabolites in cerebrospinal fluid (CSF). J Proteome Res. 2009;8:5511–22.
84. Shen Q, et al. Strong impact on plasma protein profiles by precentrifugation delay but not by repeated freeze-thaw cycles, as analyzed using multiplex proximity extension assays. Clin Chem Lab Med (CCLM). 2018;56:582–94.
85. Hassis ME, et al. Evaluating the effects of preanalytical variables on the stability of the human plasma proteome. Anal Biochem. 2015;478:14–22.
86. Enroth S, Hallmans G, Grankvist K, Gyllensten U. Effects of long-term storage time and original sampling month on biobank plasma protein concentrations. EBioMedicine. 2016;12:309–14.
87. Carrillo MC, et al. Global standardization measurement of cerebral spinal fluid for Alzheimer's disease: an update from the Alzheimer's Association Global Biomarkers Consortium. Alzheimers Dement. 2013;9:137–40.
88. Global Biomarker Standardization Consortium (GBSC) | Alzheimer's Association. https://www.alz.org/research/for_researchers/partnerships/gbsc.
89. Home | 3TR. https://3tr-imi.eu/.
90. Bose N, Brookes AJ, Scordis P, Visser PJ. Data and sample sharing as an enabler for large-scale biomarker research and development: the EPND perspective. Front Neurol. 2022;13
91. Creating a New Future for Neurodegenerative Research | EPND. https://epnd.org/.
92. The NeuroMET Project - LGC Group. https://www.lgcgroup.com/our-programmes/empir-neuromet/neuromet-landing-page/.

Mild Behavioral Impairment and Alzheimer Disease Biomarkers: Sample Enrichment for Clinical Trial Screening and Recruitment

Maryam Ghahremani, Rebeca Leon, Daniella Vellone, Dylan Guan, Zahra Hosseinpour, Jasper Crockford, Sabika Azhar, Sergio Sharif, Ibadat Warring, and Zahinoor Ismail

Abstract Recent breakthroughs in Alzheimer disease (AD) therapeutics show promise, particularly in early-stage disease. Yet, identifying individuals in these early stages is challenging. Relying solely on cognitive markers to detect preclinical AD is limited, since cognition at this stage is normal by definition. Research is shifting toward noncognitive markers of disease, such as late-onset neuropsychiatric symptoms (NPS), which are common in dementia but can also emerge before

M. Ghahremani · Z. Hosseinpour
Psychiatry, Cumming School of Medicine, University of Calgary, Calgary, AB, Canada

Hotchkiss Brain Institute, Cumming School of Medicine, University of Calgary, Calgary, AB, Canada

R. Leon
Clinical Neurosciences, Cumming School of Medicine, University of Calgary, Calgary, AB, Canada

D. Vellone
Psychiatry, Cumming School of Medicine, University of Calgary, Calgary, AB, Canada

D. Guan
Hotchkiss Brain Institute, Cumming School of Medicine, University of Calgary, Calgary, AB, Canada

J. Crockford
Psychiatry, Cumming School of Medicine, University of Calgary, Calgary, AB, Canada

Hotchkiss Brain Institute, Cumming School of Medicine, University of Calgary, Calgary, AB, Canada

Clinical Neurosciences, Cumming School of Medicine, University of Calgary, Calgary, AB, Canada

S. Azhar · S. Sharif · I. Warring
Department of Biological Sciences, University of Calgary, Calgary, AB, Canada

R. Tarawneh (ed.), *Biomarkers of Neurodegenerative Disorders*,
https://doi.org/10.1007/978-3-031-80849-4_17

dementia diagnosis. Mild Behavioral Impairment (MBI) is a framework for identifying high-risk individuals for incident dementia based on later-life persistent new-onset NPS. The inclusion of MBI in the National Institute on Aging-Alzheimer's Association (NIA-AA) Clinical Staging Framework underscores its importance in early disease detection, necessitating further research into reliability, predictive value, and associations with AD pathology. Longitudinal studies consistently demonstrate MBI associations with incident cognitive decline and dementia, outperforming traditional NPS models. Neuroimaging findings link MBI with early AD-related changes in brain structure, function, and pathology. Biomarker studies show MBI associations with elevated amyloid-beta and tau deposition, supporting MBI as an early clinical manifestation of AD. Leveraging MBI in conjunction with cognitive status could offer an accessible and effective strategy for identifying individuals with early-stage AD, enhancing recruitment strategies for AD clinical trials, and potentially accelerating development of disease-modifying therapies to improve patient and caregiver outcomes.

Keywords Alzheimer disease · Preclinical stage · Biomarkers · Neuropsychiatric symptoms · Mild behavioral impairment

1 Treatment of Alzheimer Disease

Alzheimer disease (AD) is a progressive neurodegenerative disease accounting for 60–70% of all dementia cases (World Health Organization). It poses a significant global public health challenge, carrying substantial economic and social consequences. Currently, more than 55 million people worldwide are living with dementia, and this number is projected to increase to 152 million by 2050 (Alzheimer's Disease International). Considering the anticipated rise in dementia cases in the coming decades, the associated costs such as medical care, long-term care facilities, lost

Z. Ismail (✉)
Psychiatry, Cumming School of Medicine, University of Calgary, Calgary, AB, Canada

Hotchkiss Brain Institute, Cumming School of Medicine, University of Calgary, Calgary, AB, Canada

Clinical Neurosciences, Cumming School of Medicine, University of Calgary, Calgary, AB, Canada

Community Health Sciences, Cumming School of Medicine, University of Calgary, Calgary, AB, Canada

O'Brien Institute for Public Health, Cumming School of Medicine, University of Calgary, Calgary, AB, Canada

Clinical and Biomedical Sciences, Faculty of Health and Life Sciences, University of Exeter, Exeter, UK
e-mail: ismailz@ucalgary.ca; z.ismail@exeter.ac.uk

productivity due to caregiving responsibilities, and decreased workforce participation pose a significant concern. In 2019, the total annual economic burden of dementia globally was estimated to be $1.3 trillion USD (World Health Organization). Within the societal realm, dementia introduces an additional layer of complexity, manifesting in heightened caregiver burden [1] and a noticeable decline in quality of life [2]. The prospect of earlier prevention, detection, and treatment holds promise in alleviating the burdens of dementia. Thus, there is a critical need to improve early detection to facilitate prompt intervention and improve outcomes for patients and caregivers.

Most current therapies for AD dementia are symptomatic in nature, and until recently, no new drugs had been approved for almost 20 years [3]. Recent therapeutics aimed at targeting amyloid-beta (Aβ) accumulation, such as aducanumab, donanemab, and lecanemab, have demonstrated promising outcomes in terms of Aβ reduction and treatment of either Mild Cognitive Impairment (MCI) due to AD or mild AD dementia [4]. It is important to note that these drugs have demonstrated effectiveness primarily in earlier stages of the disease when cognitive impairment and other disease symptoms are just beginning to emerge. Thus, a significant challenge in the development of disease-modifying therapies (DMTs) has been the identification of individuals in the early stages of the disease [3]. Identifying this therapeutic window requires biomarker verification, which is expensive and not scalable, highlighting the need for better approaches to achieve greater sensitivity and specificity in early case detection [5, 6]. Inefficient recruitment strategies, primarily based on cognitive measures, have resulted in high screen-failure rates for Aβ-positivity and inflated trial costs [3]. Indeed, the development of these drugs incurs substantial expenses. While the estimated cost of developing cancer treatment is well under $1 billion USD per agent and takes approximately 7 years for approval, the cost for AD amounts to $5.6 billion USD, with a significantly longer process of 13 years for approval [7, 8].

2 The Cognocentric Paradigm of Alzheimer Disease

Clinically, AD lies on a continuum, progressing from cognitively unimpaired (CU), characterized by intact cognition and function, to MCI, where cognition is impaired and function remains largely intact, and ultimately to dementia, where both cognition and function are impaired [9]. Despite using this cognitive and functional continuum, clinical diagnosis can be imprecise. Evidence suggests that 25% of those diagnosed with AD lack Aβ pathology [10]. Historically, these individuals may have enrolled in AD clinical trials without having the pathology specifically targeted by anti-amyloid mechanisms [10]. Neuropathologically, AD is characterized by Aβ plaques and tau tangles, which gradually affect brain structure and function, but only result in cognitive and functional impairments at a late stage of the disease. In fact, AD pathology can begin accumulating 15 years before cognitive and functional symptoms emerge [11]. Thus, the current diagnostic paradigm, contingent on changes in cognition and function, becomes relevant only later in the disease process.

In pursuit of earlier and more accurate detection, the field has shifted toward identifying *prodromal* (biomarker-positive MCI) and *preclinical* (biomarker-positive CU) disease stages. However, as detection moves earlier along the cognitive continuum, the specificity tends to diminish. Research has shown that 50% of individuals diagnosed with MCI lack Aβ pathology [10]. Moreover, employing a cognocentric paradigm to identify preclinical AD poses significant challenges, given that cognition is normal by definition. Additionally, screening all older CU individuals for biomarkers is impractical. One approach to enriching samples for those at risk has been leveraging the risk associated with subjective amnestic symptoms (i.e., Subjective Cognitive Decline [SCD]). However, samples as large as 1000 participants may be required to detect important between-group differences on neuropsychological tests [12], with tenuous linkages to AD biomarkers [13]. For instance, in the Mayo Clinic Study of Aging, while 74% of older adults who performed normally on standard cognitive tests reported subjective cognitive impairment, only 14% (i.e., 10% of the sample) represented a group at higher risk of incident MCI [14]. Furthermore, medical issues, stressors, and even medications can lead to subtle cognitive impairment in the absence of preclinical dementia markers, falsely triggering an SCD diagnosis. Therefore, while SCD does indeed represent an at-risk group, too few people with SCD exhibit neurodegenerative disease to justify biomarker screening for therapeutics.

3 Clarifying the Disease State

To address issues around diagnostic imprecision in AD, the National Institute on Aging and Alzheimer's Association (NIA-AA) proposed a biological definition based on a biomarker continuum [13]. The goal was to define AD as a unique neurodegenerative disease and foster research on relationships between "pathologic processes (biomarkers) and cognitive symptoms" [13]. In this ATN framework, AD is defined by the presence of **A**myloid ($A\beta_{42}$), hyperphosphorylated **T**au (p-tau), and **N**eurodegeneration (total tau). Thus, A + T-N- is considered Alzheimer pathologic change, with A + T + N ± considered AD. Staging along the AD continuum then incorporates clinical symptoms, primarily cognitive. Preclinical AD encompasses Stage 1 (asymptomatic, normal objective testing, i.e., CU) and Stage 2 (subjective cognitive symptoms, normal objective testing, i.e., SCD). Prodromal AD encompasses Stage 3 (cognitive symptoms, abnormal testing, i.e., MCI) [9].

4 Assessing Alzheimer Disease Pathology In Vivo: Scaling Up Biomarker Use

Positron emission tomography (PET) and cerebrospinal fluid (CSF) biomarker assessment techniques are the standard techniques for identifying AD and are required for clinical trial case enrolment. However, high costs, logistical barriers to scaling up, and risk to patients render these methods unavailable to many and unsuitable for screening [15]. Ultra-low concentrations of biomarkers in blood have posed challenges to accurate measurement, limiting the sensitivity and specificity of traditional techniques. However, the field has been revolutionized by the development of ultrasensitive immunoassays, positioning blood-based biomarkers as a promising approach to screening.

Over the last decade, significant progress has been made in the quest for blood-based biomarkers for AD and other dementia pathologies [16–20]. Assays measuring plasma Aβ fragments ($A\beta_{42}/A\beta_{40}$ ratio) reflect brain amyloidosis [21, 22]. Plasma tau, phosphorylated at threonine 181 (p-tau_{181}), 217 (p-tau_{217}), and 231 (p-tau_{231}), serves as a highly specific pathological marker of AD [23], which starts to increase around 15 years prior to clinical disease onset in familial AD [24] and predicts disease neuropathology at least 8 years prior to autopsy in sporadic disease [25]. Elevated glial fibrillary acidic protein (GFAP) indicates astrocyte dysfunction and neuronal damage [26], for which levels correlate with cognitive impairment and AD [27, 28]. Neurofilament light (NfL), an axonal cytoskeleton protein, is a marker of general neuroaxonal injury, which is increased in AD and other neurodegenerative disorders [18, 29–31]. NfL also serves as a biomarker to differentiate early behavioral symptoms that are the onset of a neurodegenerative disorder (i.e., Frontotemporal dementia (FTD), with higher NfL levels) and primary psychiatric disorders (with unchanged NfL levels) [32–34]. Thus, using blood-based biomarkers may facilitate early disease detection and identify those who may benefit from further PET or CSF investigations. Conceivably, blood-based biomarkers could supplant these conventional investigations for many cases. Nonetheless, even blood-based biomarker assays could benefit from targeted use to save costs and effort. For instance, in the US, the plasma p-tau_{217} assay is now commercially available but may cost between \$200 and \$500 (ALZpath, Inc.). Therefore, implementing sample enrichment strategies, even at the blood test screening level, could result in significant cost savings.

Due to the imprecise selection of at-risk older adults for biomarker confirmation and subsequent enrolment into trials, the screen failure rate for Aβ biomarkers remains persistently high [3]. Current trials targeting preclinical AD often need to screen numerous additional participants to fulfill recruitment targets for individuals who test positive for Aβ. This imprecision in case detection not only contributes to the profound expense of trials, but also slows down the pace of recruitment considerably. Clinical care may mirror this imprecision, highlighting the need for more efficient screening methods. The absence of a scalable cognitive marker of early disease that is specific, reliable, fast, and cost-effective has been a major limiting factor. There is a pressing need for a reliable and low-cost clinical marker capable of screening large samples to identify individuals most likely to exhibit biomarker positivity. Investigation of a putative noncognitive marker of early disease, such as depression, may provide insights on how to improve yield.

5 Depression as an Example of a Potential Prodromal Symptom

Depression has long been associated with dementia and is often incorporated into dementia risk models. A seminal meta-analysis found that individuals with a history of lifetime depression were twice as likely to develop dementia compared to those without depression, establishing depression as an independent risk factor for dementia [35]. Indeed, the Lancet Commission acknowledges depression as a modifiable risk factor, suggesting that treatment could reduce rates of incident dementia [36]. However, conducting long-term treatment trials to investigate whether managing midlife depression can prevent dementia is impractical. Consequently, the epidemiological literature has played a crucial role in informing the field. Observational cohorts did not demonstrate a significant effect of depression treatment on dementia risk. In fact, several studies reported an increased risk of dementia associated with pharmacotherapy for depression [37, 38]. Recent literature has begun to incorporate the natural history of depression as a germane factor to improve our understanding of the nature of depressive symptoms in older adults. The Whitehall study is an example, demonstrating that the link between depression and dementia is only evident when depression precedes the diagnosis of dementia by less than 10 years, but not when depression is diagnosed earlier [39]. This finding raises the possibility of reverse causation, suggesting that depression observed shortly before dementia diagnosis could be a prodromal symptom and a consequence of disease, rather than a causal factor. Indeed, the authors suggest that later-life onset depressive symptoms might represent preclinical or prodromal disease symptoms rather than independent risk factors for dementia [39]. However, focusing solely on depression, as observed in these studies, still lacks sufficient specificity to meaningfully capture risk, necessitating exploration of broader psychiatric symptomatology. This notion is supported by other studies establishing linkages within shorter timeframes, such as those involving psychosis [40]. Notably, a nationwide Finnish study of almost 60,000 participants since 1972, spanning multiple pharmacological and non-pharmacological treatments, determined that later-life mental health conditions may be associated with incident AD, but only in the 5-year window between the emergence of symptoms and the development of dementia. Thus, the authors proposed that *any mental health condition emerging in later life* might represent prodromal symptoms of dementia [41].

6 Neuropsychiatric Symptoms as Markers of Dementia Risk

Neuropsychiatric symptoms (NPS), such as depression, anxiety, psychosis, apathy, agitation, and sleep/appetite disturbances, are well known to occur in dementia, but are also observed in later life in advance of dementia. Previous studies have indicated that NPS are present in 35–85% of MCI patients [42] and are associated with a 1.3- to 1.8-fold greater risk of progression to dementia [43]. Similarly, NPS in CU older adults confer a higher likelihood of progression to MCI or dementia, with psychotic symptoms demonstrating greater risk than affective symptoms [44]. Unfortunately, it is still unclear whether NPS, as included in the modeling, represent preclinical and prodromal AD, biologically.

Both FTD and dementia with Lewy bodies (DLB) can have behavioral prodromes, but findings for AD are inconsistent. In a systematic review of NPS and AD biomarkers [45], few meaningful links were found, with depression having the least consistent results. The lack of consistent findings was attributed to measurement differences, diagnostic imprecision, and a lack of appreciation for the natural history of symptoms [45, 46].

Cross-sectional assessment of mental health symptomatology can identify important symptoms and capture distress and risk to self or others. Essential to psychiatric care, both the Diagnostic and Statistical Manual of Mental Disorders (DSM) and International Classification of Diseases (ICD) use this approach, organizing symptoms over a short duration into syndromic clusters. Major depression, for example, requires symptom presence for 2 weeks. Accordingly, the Beck Depression Inventory [47] has a 2-week reference range for symptom presence, while the Geriatric Depression Scale (GDS) [48] assesses symptoms over just one week. The Neuropsychiatric Inventory (NPI) [49], developed to measure NPS in dementia, measures a meaningfully longer period of 4 weeks. Nonetheless, in dementia-free older persons, when it comes to NPS representing behavioral sequelae of underlying dementia proteinopathies, i.e., serving as *markers* of dementia, this approach may require reevaluation. Rating scales and clinical criteria with short reference ranges can capture transient symptoms that present due to life events, psychosocial stressors, or medical conditions, especially in older individuals who are particularly vulnerable. Thus, scales like the GDS have a low signal-to-noise ratio and limited specificity for capturing behavioral manifestations of early-phase AD.

Understanding the natural history of NPS is fundamental to identifying behavioral manifestations of preclinical and prodromal AD. The DSM and ICD are essentially silent on natural history. Two weeks of depression in adolescence, adulthood, or later life could meet criteria. However, as per the Whitehall study and others [39, 40], later-life onset of depression, psychosis, or other NPS may be more likely to reflect Aβ or tau pathology than symptoms with earlier onset. Substantial evidence supports this assertion. In the National Alzheimer Coordinating Center (NACC) data repository, over 59% of participants develop NPS before the diagnosis of a cognitive disorder, including 30% of those who develop AD [50]. Considering the high prevalence of AD relative to other dementias, and the overall prevalence of dementia, 30% of AD cases is a large number. To put this into perspective, in the US alone, nearly 500,000 new cases of AD will be diagnosed this year (Alzheimer's San Diego)—30% of this figure translates to approximately 150,000 AD cases in which NPS will be the presenting symptoms. These findings align with the Finnish study of mental health presentations to hospital services harboring persons with prodromal dementia and studies demonstrating the AD prodrome hiding in geriatric psychiatry clinics as well [41, 51]. Studies in neurology or memory clinics have found that patients are often initially provided a psychiatric diagnosis, and neurodegenerative disease and referral to a specialty clinic are not considered until after failed treatment with psychiatric medications or when substantial cognitive issues become apparent [52, 53]. These results emphasize how late-onset psychiatric symptoms like depression and anxiety might be erroneously diagnosed as psychiatric conditions instead of being recognized as early signs of neurodegenerative disease. Overall, these findings stress the importance of eliciting the natural history of symptoms, as opposed to just cross-sectional symptom measurement, in determining dementia risk.

7 Mild Behavioral Impairment Construct

To address nosological issues in categorization of later-life neuropsychiatric symptomatology, and to operationalize the assessment of NPS as potential dementia risk markers, an Alzheimer's Association consensus group (International Society to Advance Alzheimer's Research and Treatment [ISTAART] Neuropsychiatric Syndromes Professional Interest Area) developed research diagnostic criteria for Mild Behavioral Impairment (MBI) [54]. MBI, since validated, is a neurobehavioral syndrome that is characterized by the de novo onset of NPS in later life, which persist for at least 6 months and represent a change from longstanding behavior or personality. MBI identifies a high-risk group for cognitive decline and incident dementia by leveraging the risk associated with later-life emergent and persistent NPS. For some, MBI represents preclinical or prodromal symptoms of AD (Fig. 1). Importantly, MBI is not a competing construct to MCI; rather it serves as a complementary behavioral analogue to the cognitive changes described by MCI. Changes in cognition and behavior are not mutually exclusive and must be considered together. When emerging after dementia, behavioral changes are termed behavioral and psychological symptoms of dementia (BPSD) [55]. When emerging before dementia, these behavioral changes are MBI. Thus, reporting MBI status in conjunction with cognitive status (i.e., CU, SCD, MCI) leverages risk associated with both groups and can significantly enhance risk assessment [56].

MBI symptoms are clustered into 5 domains: (1) decreased drive/motivation (apathy/indifference); (2) affective dysregulation (depression/dysphoria, anxiety, and elation/euphoria); (3) impulse dyscontrol (agitation/aggression, irritability/lability, and aberrant motor behavior); (4) social inappropriateness (impaired social cognition); and (5) psychotic symptoms (hallucinations and delusions) [54]. The incorporation of this broad spectrum of domains allows for more sensitive detection of the later-life NPS that can precede dementia [57]. During early stages, these behavioral domains may not be fully formed or discrete, sometimes presenting as broad-spectrum or subthreshold behavioral changes, akin to the early stages of psychiatric disorders in younger people [58]. Thus, exploring the global NPS construct of MBI is a logical next step toward more precise and efficient detection of early-stage AD, representing a ripe opportunity to exploit for enrichment of clinical trials, improvement of cohort-based recruitment, and identification of persons with AD.

ISTAART Research Diagnostic Criteria for MBI

1. Changes in behavior or personality observed by patient, informant, or clinician, starting later in life (age ≥50 years) and persisting at least intermittently for ≥6 months. These represent clear change from the person's usual behavior or personality as evidenced by at least one of the following:
   a. Decreased motivation (e.g., apathy, aspontaneity, indifference)
   b. Affective dysregulation (e.g., anxiety, dysphoria, changeability, euphoria, irritability)
   c. Impulse dyscontrol (e.g., agitation, disinhibition, gambling, obsessiveness, behavioral perseveration, stimulus bind)
   d. Social inappropriateness (e.g., lack of empathy, loss of insight, loss of social graces or tact, rigidity, exaggeration of previous personality traits)
   e. Abnormal perception or thought content (e.g., delusions, hallucinations)
2. Behaviors are of sufficient severity to produce at least minimal impairment in at least one of the following areas:
   a. Interpersonal relationships
   b. Other aspects of social functioning
   c. Ability to perform in the workplace
   d. The patient should generally maintain his/her independence of function in daily life, with minimal aids or assistance.
3. Although comorbid conditions may be present, the behavioral or personality changes are not attributable to another current psychiatric disorder (e.g., generalized anxiety disorder, major depression, manic or psychotic disorders), traumatic or general medical causes, or the physiological effects of a substance or medication.
4. The patient does not meet criteria for a dementia syndrome (e.g., Alzheimer's disease, frontotemporal dementia, dementia with Lewy bodies, vascular dementia, other dementia). MCI can be concurrently diagnosed with MBI.

Abbreviations: ISTAART, International Society to Advance Alzheimer's Research and Treatment; MBI, mild behavioral impairment; MCI, mild cognitive impairment.

Fig. 1 ISTAART research diagnostic criteria for Mild Behavioral Impairment (MBI)

Importantly, NPS have been included in the NIA-AA Clinical Staging Framework. The ATN framework states: "while cognitive symptoms represent the core clinical feature of AD, in some individuals, the initial presentation may be neurobehavioral (e.g., depression, anxiety, and apathy) rather than cognitive" [9]. These NPS have been included as early as Stage 2 with a qualification: "neurobehavioral symptoms should have a clearly defined recent onset, which persist" [9]. This qualification invokes the criteria for MBI, i.e., the later-life emergence and persistence of NPS [54]. With the inclusion of MBI in the NIA-AA ATN clinical staging and given that 30% of individuals with AD develop NPS in advance of a cognitive diagnosis [50], it is imperative to quantify the relationship between MBI, incident dementia, and AD proteinopathies. For MBI to be considered a marker for disease in a subset of individuals with late-life NPS, several fundamental questions must be answered: (1) Can MBI be reliably measured?; (2) Does MBI provide better predictive value for dementia compared to current nosological frameworks for mental health disorders and NPS?; and (3) Is there biomarker evidence indicating that MBI predicts pathology more effectively and accurately than current approaches? The subsequent sections aim to provide comprehensive answers to these questions, shedding light on the potential of MBI as a robust marker of dementia.

8 Measurement of Mild Behavioral Impairment

In addressing the first fundamental question regarding the reliability of MBI measurement, it is important to note that previously, no rating scale existed to capture later-life onset of persistent NPS in accordance with the MBI criteria. Early studies utilized NPI data, which may have generated inflated estimates of prevalence due to the inability to confirm symptom persistence for at least 6 months [1]. Subsequent studies improved specificity by using measures over 2 timepoints for the MBI symptom persistence criterion [56]. While improving prognostication, not all MBI domains were sufficiently identified using this approach. Moreover, the logistics of requiring 2 assessments spaced at least 6 months apart rendered this approach sub-optimal—determining MBI status at a single visit is a simpler, more practical, and more efficient approach. Thus, the MBI checklist (MBI-C) was developed by an ISTAART working group as the case ascertainment instrument for MBI [59]. The MBI-C stipulates explicitly that symptoms emerge in later life, persist for at least 6 months, and represent change from longstanding patterns of behavior or personality [59]. Since its development, the MBI-C has been validated for in-person, telephone, and online use, and can be completed by the individual, informant, and/or clinician [60–63].

9 Mild Behavioral Impairment Epidemiology

Various methodologies have been used to determine the prevalence of MBI, which accounts for the high heterogeneity in estimates from meta-analyses that pool studies with different case ascertainment approaches [64]. MBI case definitions have predominantly relied on operationalizations using the NPI or NPI Questionnaire (NPI-Q) [65], but with the establishment of the MBI-C, studies are now starting to report prevalence using this scale, which was developed specifically to measure MBI.

Studies employing the NPI or its variants have either determined status from a single timepoint or from two or more timepoints, the latter used to confirm symptom persistence. Studies using a single timepoint tend to report higher prevalence estimates likely due to the inclusion of transient or impersistent symptoms, captured over the shorter reference range of the scale. For instance, in an Australian community-based study of older adults, MBI, as measured by the NPI at a single timepoint, had an overall prevalence of 34.1%, with 27.6% in the CU group, 43.1% in the SCD group, and 48.9% in the MCI group [66]. In a sample of dementia-free older adults from the Florida AD Research Centre, MBI had a prevalence of 38.6% in the CU group and 67.9% in the MCI group [67]. In a Canadian cognitive neurology clinic, the prevalence of MBI was even higher at 76.5% in patients with SCD and 85.3% in those with MCI [1]. While NPS prevalence is higher in clinical vs. community samples [68], these single timepoint estimates are still inflated. More recent studies have defined MBI based on NPI data across 2 consecutive visits, to operationalize the MBI symptom persistence criterion. For example, in a prospective clinical cohort of dementia-free Asians, MBI was operationalized using NPI data from both baseline and year-one visits. The overall prevalence of MBI was 14.5%, with 7.1% in participants with normal cognition and 12.9% in those with cognitive impairment not meeting criteria for dementia (CIND) [69]. A NACC study using the NPI-Q across 2 consecutive annual visits demonstrated an MBI prevalence of 32.6% in the CU group and 48.2% in SCD [56]. Very recently, the utility of various MBI operational definitions was published to foster the use of legacy datasets to determine MBI case status. Appreciating these definitions helps in understanding the heterogeneity observed in published prevalence rates [70].

As the MBI-C becomes established, an increasing number of studies are employing this instrument to estimate the prevalence of MBI. By directly accounting for the MBI core criteria of later-life symptom emergence and persistence for at least 6 months, the MBI-C has yielded lower, yet more specific and accurate, prevalence estimates of MBI. In a study from the UK involving a large cohort of CU older adults, MBI captured online with the self-rated MBI-C had a prevalence of 10% [71]. In a Spanish primary care population study, using conservative inclusion criteria, the prevalence of MBI captured using the MBI-C was 5.8% in individuals with SCD [72] and 14.2% in those with MCI [60]. In a Chinese primary care clinic study of dementia-free older adults, 18.2% of the sample exhibited MBI [73]. In a Singaporean outpatient neurology clinic study of participants with normal cognition and MCI, 17.9% of the sample had MBI, with 7.5% in the normal cognition group and 28% in the MCI group [74]. In an Indian memory clinic study of older adults with SCD or MCI, 41.1% of the combined sample met the diagnostic criteria for MBI [75]. These results were echoed in another memory clinic study from Canada, among individuals with SCD and MCI, where 36.62% and 53.98% had MBI, respectively [62]. Similarly, an Iranian memory clinic study of patients with MCI reported that 50% of the sample were diagnosed with MBI [76]. Overall, MBI prevalence increases with declining cognitive status from CU to SCD to MCI; prevalence is also higher in clinical vs. community samples and in memory clinics vs. primary care. While the MBI-C provides a more accurate characterization of MBI, understanding the sample (e.g., cognitive category and setting) is required to fully appreciate the prevalence rates.

Irrespective of the scale used, and mostly irrespective of setting, affective dysregulation, impulse dyscontrol, and decreased motivation have been reported as the most common MBI domains [1, 51, 60, 61, 72–76]. A recent meta-analysis on the prevalence of MBI domains on 12,067 participants across the cognitive continuum from CU to SCD to MCI further confirms these observations [77]. Among the studies analyzed, five utilized the NPI-Q at a single time point, one used the NPI-Q at 2 timepoints, one study adopted the ISTAART criteria, and three studies utilized the MBI-C to operationalize MBI. The MBI affective dysregulation and impulse dyscontrol domains were the most prevalent with frequencies of 32.84% and 26.67%, respectively, followed by the decreased motivation domain with a frequency of 12.58%. The prevalence for other MBI domains in decreasing order was 6.05% for social inappropriateness and 2.81% for abnormal perception/thought content [77].

Overall, these findings indicate that MBI is prevalent across all stages preceding dementia and its prevalence increases as one progresses along the cognitive continuum, with the highest prevalence in MCI. Findings also highlight the importance of the MBI criteria of symptom emergence and persistence for at least 6 months, to enhance the specificity in capturing behavioral sequelae of neurodegenerative disease. The lower prevalence of MBI when captured with the MBI-C, as opposed to NPI, may in fact be a reflection of a greater diagnostic specificity, capturing a high-risk group for further assessments or interventions.

10 Enhancing Dementia Prediction: Longitudinal Studies on Mild Behavioral Impairment

To address the second fundamental question on whether MBI can predict dementia more effectively than current nosological frameworks, several studies have validated the MBI criteria of later-life emergence and persistence of symptoms for early detection and prognostication of AD. Cross-sectionally, MBI criteria identify groups with worse cognition. In a study of 499 Brain Health Registry participants, individuals with MBI performed worse on measures of memory and executive function

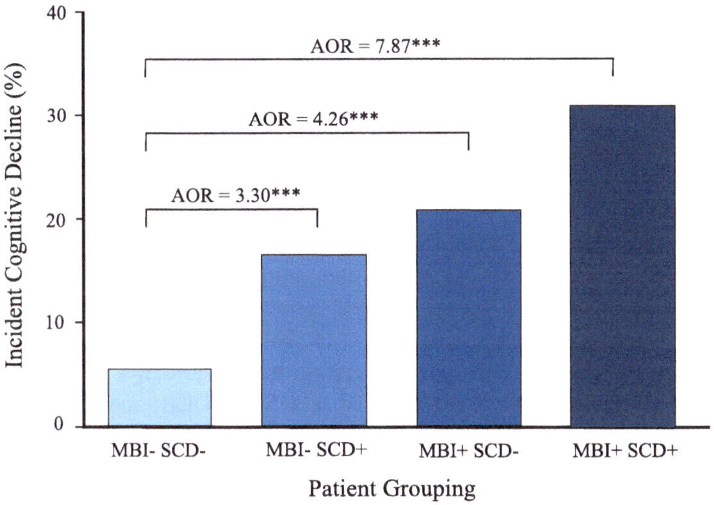

Fig. 2 Adjusted odds ratios (AOR) of incident cognitive decline (Clinical Dementia Rating > 0) after 3 years, across MBI/SCD categories. Abbreviations: MBI mild behavioral impairment, SCD subjective cognitive decline

compared to those without MBI [63]. These findings are consistent with studies in NACC [67] and UK PROTECT cohorts [71]. Longitudinally in CU, SCD, and MCI, the presence of MBI consistently confers a greater risk for incident cognitive decline and dementia compared to those without MBI [69, 78]. In CU participants with and without SCD, the subgroups with concurrent MBI demonstrated a greater 3-year progression rate to MCI or dementia compared to participants without MBI [56] (Fig. 2). In a 5-year NACC study of CU participants, MBI was associated with both clinically diagnosed and neuropathologically confirmed AD [79]. In an online sample of CU community-dwelling older adults from the UK PROTECT study, MBI was associated with cognitive decline over 1 year in the domains of attention and working memory [71]. Subsequent analyses at 3.5 years showed that decline in cognition was faster in MBI compared to those without MBI, which was evident in both males and females via sex-stratified analyses [80]. An Alzheimer's Disease Neuroimaging Initiative (ADNI) study of participants with normal cognition or MCI demonstrated that MBI was longitudinally associated with decline in episodic memory and executive function and 3.92-fold greater risk for AD dementia. In comparison, such associations were not present in the non-MBI NPS group [81]. In MCI participants, emergent and persistent NPS meeting MBI criteria were associated with a greater 3-year progression rate to dementia and a lower reversion rate to normal cognition [82]. These studies confirmed that adding behavioral risk, operationalized as MBI, to cognitive risk resulted in improved prognostic model performance.

In addition to the prognostic utility of global MBI status, studies demonstrate the prognostic utility of MBI domains. In a UK PROTECT study of CU older adults, individuals with MBI-psychosis, assessed with the MBI-C, had a 3.6-fold greater risk of incident cognitive impairment over approximately 4 years, relative to those with no psychosis [83]. Similarly, a NACC study of CU participants reported that the MBI-psychosis had the greatest hazard ratio (HR) for clinically diagnosed AD [79]. Another NACC study of participants with normal cognition or MCI demonstrated that compared to no NPS, individuals with MBI-psychosis had a 3.76-fold greater dementia incidence rate, while those with non-MBI psychosis had a 1.92-fold greater incidence rate [84]. In a sample of older adults with normal cognition or MCI, MBI-apathy had the greatest dementia incidence rate compared to no NPS or non-MBI apathy [85]. A recent study in dementia-free older adults found a HR for 10-year dementia incidence of 1.76 for the MBI-affective dysregulation domain [86]. Notably, of the 25.3% of participants with MBI-affective dysregulation who progressed to dementia, 85.5% developed AD, 3.0% DLB, and 2.5% FTD. These findings suggest that the utility of MBI criteria to predict incident dementia generalizes to the affective dysregulation domain, with progression primarily to AD dementia. Further analyses showed the importance of including cognitive status in conjunction with MBI due to effect modification. Dementia incidence was greater in MCI than CU, but when MBI-affective dysregulation emerged in CU older adults, the HR was 2.59 relative to no NPS, whereas in MCI the HR was 1.50 [86]. The moderating effect of cognitive status on MBI-associated risk has been replicated in MBI-apathy [85] and MBI-psychosis [84]. These findings provided further evidence that stratifying the cognitive risk group by MBI status improves the prognostic model performance and allows earlier identification of at-risk individuals for incident dementia.

11 Mild Behavioral Impairment Versus Neuropsychiatric Symptoms Versus Psychiatric Disorders

To determine whether MBI offers any advantage over conventional approaches, several studies have also assessed NPS not meeting MBI criteria, referred to as non-MBI NPS, as an additional comparator to test the specificity of MBI for predicting AD. This group represents conventional approaches to the incorporation of psychiatric symptomatology in dementia research (e.g., using the GDS or NPI to determine behavioral risk). One such study in MCI demonstrated that adding MBI to MCI identified a subgroup with a greater progression rate to dementia and a lower reversion rate to normal cognition. The non-MBI NPS group in this study included NPS that were still later-life in onset, but not persistent for more than 6 months. Thus, the study only tested differences based on the MBI persistence criterion. Compared to no NPS, the non-MBI NPS group was more likely to progress to dementia but was no less likely to revert to NC [82]. Indeed, MBI outperforming conventional NPS approaches has been shown in multiple studies testing the MBI persistence and/or later-life emergence criteria [81, 84, 85].

Longitudinal data have also validated MBI as separate from psychiatric conditions persisting into late life, for which MBI can be mistaken. In a dementia-clinic study, individuals with MBI had a higher progression rate to dementia than a psychiatric comparator group consisting of late-life psychiatric conditions [87]. In a geriatric psychiatric clinic study of 2853 patients, MBI had an 8-fold higher incidence rate of dementia compared to other DSM geriatric psychiatric disorders [51]. Thus, the MBI core criteria of symptom emergence in later life and symptom persistence for at least 6 months identify a higher-risk subgroup of individuals experiencing NPS and distinguish MBI from other mental health conditions, providing context for studies such as the Whitehall study [39] or the Finnish hospital data study [41].

12 Mild Behavioral Impairment and Neuroimaging

An emerging body of evidence has begun to address the third fundamental question regarding whether there is biomarker evidence supporting the utility of MBI to predict pathology more accurately than current approaches. Some of this evidence comes from neuroimaging studies demonstrating that MBI may be associated with neurodegeneration in regions linked to early-stage AD [88]. In a structural magnetic resonance imaging (MRI) study of older adults with SCD or MCI, the MBI-C total score and impulse dyscontrol and decreased motivation domain scores were associated with atrophy in the entorhinal cortex and hippocampus [89], regions known to be affected in early AD [90]. Similar findings were observed in a NACC study of dementia-free participants, where lower grey matter volume in the entorhinal cortex and hippocampus was associated with MBI [91]. Furthermore, in dementia-free older adults, a machine learning study using structural MRI data found the MBI total score to be as powerful as hippocampal volume in predicting dementia 40 months later [92]. One diffusion tensor imaging study in a sample of mostly dementia-free participants identified an association between MBI-impulse dyscontrol and lower white matter integrity in the fornix, superior fronto-occipital fasciculus, and cingulum, and lower cortical thickness in the parahippocampal gyrus [93]. MRI studies have also revealed an association between MBI and white matter hyperintensities (WMH), which reflect small vessel cerebrovascular disease [94] and are associated with cognitive decline [95] and incident AD [96]. In a memory clinic sample of patients with MCI, a higher WMH volume was observed in those with MBI compared to those without [97].

Functional imaging studies have also identified associations between MBI and disruption in specific resting-state networks implicated in AD. In a resting-state functional MRI (fMRI) study of participants with normal cognition or amnestic MCI, a link was found between MBI and disruptions within the fronto-parietal control network, a resting-state network that plays an important role in executive function and working memory [98, 99]. Notably, among participants with MBI, particularly those with affective dysregulation symptoms, an association was found with decreased functional connectivity between the left posterior parietal cortex and the right middle frontal gyrus, as well as the left frontal pole and superior frontal gyrus [99]. In another resting-state fMRI study of dementia-free participants, MBI was associated with lower functional connectivity between the posterior cingulate cortex and the medial prefrontal cortex within the default mode network, regions that are implicated in memory [100] and social and affective behaviors [101]. Lower functional connectivity was also reported between the anterior cingulate cortex and the anterior insula within the salience network, regions implicated in neuropsychiatric symptoms [102–104].

PET studies have also provided evidence on the association of MBI with elevated Aβ and tau deposition in brain regions implicated in early stages of AD. One study in a sample of CU elderly individuals demonstrated that higher MBI-C scores were associated with higher Aβ deposits in the left frontal lobe, posterior cingulate cortex, caudate nucleus, and thalamus, regions known to exhibit amyloidosis in the first phases of hierarchical amyloidosis in AD [105]. More recently, an ADNI study in older adults with normal cognition or MCI reported an association between MBI and tau uptake in Braak I and Braak III regions [90] among Aβ-positive individuals. This association was not present in the Aβ-negative group, indicating the high specificity of MBI in capturing AD pathology [106]. Similar findings were reported in another PET study of CU Aβ-positive participants, where notable associations were observed between MBI and early tau pathology within Braak I-II regions [107]. Across the MBI-C domains, only affective dysregulation and impulse dyscontrol domains demonstrated associations with tau pathology, although the sample was small [107]. In another study of CU older adults with moderate to severe depressive symptoms, the decreased motivation domain of the MBI-C was associated with elevated tau-PET uptake in the amygdala and hippocampus [108].

These neuroimaging findings indicate that MBI is associated with early structural, functional, and pathological changes in the brain and add to the evidence base that MBI can represent an early clinical manifestation of AD.

13 Mild Behavioral Impairment and Alzheimer Disease Fluid Biomarkers

Building on the growing body of evidence supporting the prognostic value of MBI, there is mounting evidence indicating that MBI may be associated with AD proteinopathies. This emerging evidence indicates that MBI screening could serve as an effective enrichment strategy to identify individuals with early-stage AD. MBI is associated with axonal loss in dementia-free older adults, showing a significant increase in plasma NfL over 2 years, suggesting that MBI may represent a group with accelerated neurodegeneration [109]. In a mixed MCI/CU sample of 139 participants, lower plasma $A\beta_{42}/A\beta_{40}$ ratio was associated with higher MBI total score and MBI affective dysregulation domain score [57]. Furthermore, in 571 dementia-free participants without psychiatric history, MBI (but not non-MBI NPS) had

Table 1 Longitudinal association between neuropsychiatric symptoms (NPS) and cerebrospinal fluid biomarkers in a span of 4 years, using data from ADNI and MEMENTO cohorts

Outcome	Predictor	Standardized β	95% CI		p-value
ADNI					
$A\beta_{42}$	MBI	−2.21%	−3.33%	−1.09%	**<0.001**
	Non-MBI NPS	−0.003%	−0.50%	0.50%	0.990
Aβ40	MBI	0.02%	−0.74%	0.79%	0.957
	Non-MBI NPS	−0.10%	−0.48%	0.30%	0.633
p-tau	MBI	2.65%	1.43%	3.89%	**<0.001**
	Non-MBI NPS	0.38%	−0.007%	0.77%	0.055
t-tau	MBI	2.28%	1.20%	3.37%	**<0.001**
	Non-MBI NPS	0.44%	0.01%	0.86%	**0.045**
$A\beta_{42}/A\beta_{40}$	MBI	−2.27%	−3.23%	−1.29%	**<0.001**
	Non-MBI NPS	−0.0004%	−0.30%	0.30%	0.998
p-tau/$A\beta_{42}$	MBI	4.77%	2.83%	6.76%	**<0.001**
	Non-MBI NPS	0.06%	−0.37%	0.48%	0.796
t-tau/$A\beta_{42}$	MBI	4.37%	2.61%	6.15%	**<0.001**
	Non-MBI NPS	0.05%	−0.38%	0.47%	0.819
MEMENTO					
$A\beta_{42}$	MBI	−0.79%	−1.86%	0.30%	0.163
	Non-MBI NPS	0.02%	−0.84%	0.89%	0.971
Aβ40	MBI	−0.45%	−1.33%	0.43%	0.325
	Non-MBI NPS	−0.74%	−1.54%	0.03%	0.067
p-tau	MBI	1.47%	0.25%	2.72%	**0.022**
	Non-MBI NPS	−0.26%	−0.74%	0.21%	0.293
t-tau	MBI	1.69%	−0.08%	3.49%	0.067
	Non-MBI NPS	−0.41%	−1.19%	0.35%	0.298
$A\beta_{42}/A\beta_{40}$	MBI	−0.31%	−1.57%	0.96%	0.636
	Non-MBI NPS	0.75%	−0.33%	1.87%	0.185
p-tau/$A\beta_{42}$	MBI	2.17%	0.23%	4.16%	**0.033**
	Non-MBI NPS	−0.47%	−1.53%	0.58%	0.390
t-tau/$A\beta_{42}$	MBI	2.61%	0.18%	5.12%	**0.040**
	Non-MBI NPS	−0.40%	−1.60%	0.78%	0.516

The reference group for all models is the no-NPS group. Models adjusted for age, sex, education, and the source of NPS data. Abbreviations: *CI* confidence interval, *MBI* mild behavioral impairment, *NPS* neuropsychiatric symptoms

higher levels of plasma p-tau$_{181}$ at baseline compared to no NPS; the group with MBI at baseline also exhibited greater declines in memory and executive function over 4 years and a 3.42 times greater rate of incident dementia over 10 years compared to no NPS [81]. A very recent study in MCI participants in ADNI and MEMENTO assessed CSF biomarkers in association with MBI [110]. Both cross-sectionally and longitudinally over 4 years, MBI was associated with a greater AD biomarker burden, including lower Aβ$_{42}$ and higher p-tau$_{181}$ levels. In contrast, non-MBI NPS did not differ from the no-NPS group (Table 1) [110]. In a small sample

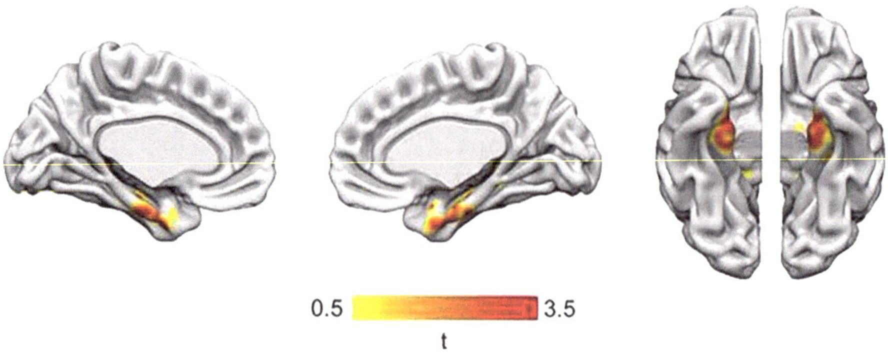

Fig. 3 Whole brain voxel-based analysis between Mild Behavioral Impairment checklist (MBI-C) scores and tau-PET in cognitively unimpaired Aβ-positive individuals from the Swedish BioFINDER−2 study. (Figure was taken from Johansson, Stomrud et al., 2021 [107] through an Open Access license http://creativecommons.org/licenses/by/4.0/)

of Aβ-positive participants in the Swedish BioFinder2 study, while higher MBI-C scores were associated with tau PET uptake in the entorhinal cortex and hippocampus (Fig. 3), fluid biomarker findings were consistent, demonstrating an association between MBI and higher CSF p-tau$_{181}$ [107]. While p-tau was predicted by MBI-C score, it was not predicted by the Alzheimer Disease Assessment Scale (ADAS) delayed recall score, suggesting that MBI is associated with tau independent of memory effects in this preclinical AD sample [107]. These findings provide strong evidence further supporting MBI as a sequela of neurodegeneration that could be leveraged to improve screening efficiency for AD.

14 Implications for Screening, Case Finding, and Recruitment into Clinical Trials

In summary, MBI can be reliably measured and distinguished from psychiatric conditions and transient or reactive NPS that exhibit similar overt clinical presentations but have non-neurodegenerative etiologies. The MBI-C, being the only assessment tool designed to address the MBI core criteria of symptom emergence and persistence, is the ideal instrument for measuring MBI symptoms in dementia-free older persons. Nonetheless, in situations where MBI-C data are unavailable, other NPS assessment tools can serve as valid alternatives using different operational definitions depending on source data [70]. A robust body of evidence suggests that older individuals with MBI are at a higher risk of incident dementia, compared to traditional measurements of NPS and psychiatric conditions [51, 82, 84–87]. Likewise, several biomarker studies have demonstrated that MBI offers enhanced value for detecting prevalent AD and predicting incident AD pathology in comparison to current approaches. Studies have found associations between MBI and elevated Aβ and tau deposition as measured by PET imaging [105–107], lower levels of $A\beta_{42}$ [110], higher levels of p-tau_{181} [81], and elevated plasma NfL levels [109], the latter of which indicates axonal loss. These associations have been observed in both dementia-free older adults and those with MCI, providing further evidence that, like MCI, MBI can be an early manifestation of AD and leveraging MBI in conjunction with cognitive status could serve as an effective enrichment strategy for identifying individuals with early-stage AD and those at risk of developing AD dementia.

As the prevalence of AD continues to rise worldwide, there is a growing need for enhanced screening methods for preclinical and prodromal AD. These methods should not only demonstrate efficacy in disease prognostication, but also be applicable to large populations and diverse settings. The construct of MBI can serve as a simple, accessible, and easy-to-administer clinical marker that enhances specificity of the at-risk group for AD. For instance, the MBI-C is free to use (www.mbitest.org), is available in 30 languages, has been validated in community and clinical samples, and takes approximately 7 minutes to administer in-person, by telephone, or online by participants, informants, or clinicians [59–63, 72, 111]. As such, the MBI-C can serve as a screening tool in primary care or other community settings to identify individuals within the population who are more likely to have preclinical or prodromal AD. These individuals may then be referred to speciality clinics for additional clinical examination and biomarker testing, as has been described in some dementia guidelines [112, 113]. Although further research is still needed, MBI may also facilitate the interpretation of biomarker findings. It is well-recognized that not all individuals with AD pathology will necessarily exhibit or even develop symptoms of dementia [114]. Therefore, those who test positive for biomarkers and exhibit MBI symptoms may represent the group of individuals most likely to experience cognitive, behavioral, and functional decline as a result of the disease (i.e., those who may be less resilient). Disease-modifying therapies and secondary prevention programs may yield the greatest therapeutic benefit for these individuals; when administered at an early stage, such as in those marked by MBI, these interventions may prove more efficacious due to the minimal cognitive, behavioral, and functional impairments present in this population.

MBI may also play an important role in AD clinical trials. Historically, AD clinical trials have been plagued by several obstacles, including a requirement for large sample sizes and a relatively long study duration of approximately 18 months per participant (i.e., long efficacy readouts), resulting in slow progress and considerable expenses. These challenges stem from a variety of sources, including noisy outcome measures and uncertainty about whether recruited participants in the placebo arm will necessarily decline sufficiently over the study period to show separation from the treatment group. Notably, rates of progression from MCI to dementia remain highly variable, even in clinical trial settings [115]. Enriching clinical trial samples through the use of MBI at enrollment, in combination with other prodromal constructs such as MCI, may increase the likelihood of including participants with underlying AD who experience more rapid decline, thereby shortening the time to an efficacy readout. Given that NPS are core features of dementia that are increasingly recognized as direct consequences of the underlying neurodegenerative process, as opposed to being secondary to cognitive impairment, MBI may also hold utility as a valuable outcome measure in AD clinical trials [116]. As with cognitive and functional outcome measures, therapies that successfully target AD should, in theory, also modify the trajectory of behavioral symptoms. Beyond dementia risk, MBI has been linked to several other adverse health outcomes, including poorer gait [117], hearing loss [118, 119], diabetes [120], frailty [73, 121], loneliness [122], poorer quality of life [123], and greater caregiver burden [1]. Interventions that mitigate symptoms of MBI will play an important role in addressing the overall burden of the disease for both patients and their families.

Future research should prioritize refining screening and case-finding strategies. This includes improving the sensitivity, specificity, and cost-effectiveness of blood-based biomarkers, as well as further validating and refining MBI criteria. Collaboration between researchers, clinicians, and industry partners is essential to develop and implement effective screening and recruitment strategies, ultimately advancing the development of disease-modifying therapies for AD.

Funding ZI is supported by the Canadian Institutes of Health Research (BCA2633), the National Institute for Health and Care Research Exeter Biomedical Research Centre, and Gordie Howe CARES. MG is funded by the Mathison Centre for Mental Health, Research, and Education at the University of Calgary. DXG is supported by the Hotchkiss Brain Institute, the Killam Trust, the Alzheimer Society of Canada, the Canadian Institutes of Health Research, and the Vascular Training Platform. DV is funded by the William H. Davies Medical Research Scholarship and the Alberta Graduate Excellence Scholarships (AGES) for Doctoral Research.

References

1. Sheikh F, Ismail Z, Mortby ME, et al. Prevalence of mild behavioral impairment in mild cognitive impairment and subjective cognitive decline, and its association with caregiver burden. Int Psychogeriatr. 2018;30(2):233–44. https://doi.org/10.1017/S104161021700151X.
2. Shearer J, Green C, Ritchie CW, Zajicek JP. Health state values for use in the economic evaluation of treatments for Alzheimer's disease. Drugs Aging. 2012;29(1):31–43. https://doi.org/10.2165/11597380-000000000-00000.
3. Gauthier S, Albert M, Fox N, et al. Why has therapy development for dementia failed in the last two decades? Alzheimers Dement. 2016;12(1):60–4. https://doi.org/10.1016/j.jalz.2015.12.003.
4. Rabinovici GD, La Joie R. Amyloid-targeting monoclonal antibodies for Alzheimer disease. JAMA. 2023;330(6):507–9. https://doi.org/10.1001/jama.2023.11703.
5. Frank A, Ismail Z, Wilson M, et al. Health system change for Alzheimer's disease-modifying therapies in Canada: beginning the discussion. Can J Neurol Sci. 2023:1–9. https://doi.org/10.1017/cjn.2023.322.
6. Gauthier S, Ismail Z, Goodarzi Z, Ng KP, Rosa-Neto P. Viewpoint: Clinicians' perspectives on how disease modifying drugs for Alzheimer's disease impact specialty care. J Prev Alzheimers Dis. 2023;10(3):339–41. https://doi.org/10.14283/jpad.2023.72.
7. Marsden G, Mestre-Ferrandiz J. Dementia: the R&D landscape. Office of Health Economics: Research; 2015.
8. Cummings J, Reiber C, Kumar P. The price of progress: funding and financing Alzheimer's disease drug development. Alzheimers Dement (N Y). 2018;4:330–43. https://doi.org/10.1016/j.trci.2018.04.008.

9. Jack CR Jr, Bennett DA, Blennow K, et al. NIA-AA research framework: toward a biological definition of Alzheimer's disease. Alzheimers Dement. 2018;14(4):535–62. https://doi.org/10.1016/j.jalz.2018.02.018.
10. Cummings J, Ritter A, Zhong K. Clinical trials for disease-modifying therapies in Alzheimer's disease: a primer, lessons learned, and a blueprint for the future. J Alzheimers Dis. 2018;64(s1):S3–S22. https://doi.org/10.3233/JAD-179901.
11. Younes L, Albert M, Moghekar A, Soldan A, Pettigrew C, Miller MI. Identifying changepoints in biomarkers during the preclinical phase of Alzheimer's disease. Front Aging Neurosci. 2019;11:74. https://doi.org/10.3389/fnagi.2019.00074.
12. Donohue MC, Sperling RA, Salmon DP, et al. The preclinical Alzheimer cognitive composite: measuring amyloid-related decline. JAMA Neurol. 2014;71(8):961–70. https://doi.org/10.1001/jamaneurol.2014.803.
13. Elman JA, Panizzon MS, Gustavson DE, et al. Amyloid-beta positivity predicts cognitive decline but cognition predicts progression to amyloid-beta positivity. Biol Psychiatry. 2020;87(9):819–28. https://doi.org/10.1016/j.biopsych.2019.12.021.
14. van Harten AC, Mielke MM, Swenson-Dravis DM, et al. Subjective cognitive decline and risk of MCI: the Mayo Clinic Study of Aging. Neurology. 2018;91(4):e300–12. https://doi.org/10.1212/WNL.0000000000005863.
15. Blennow K, Zetterberg H. Biomarkers for Alzheimer's disease: current status and prospects for the future. J Intern Med. 2018;284(6):643–63. https://doi.org/10.1111/joim.12816.
16. Hye A, Riddoch-Contreras J, Baird AL, et al. Plasma proteins predict conversion to dementia from prodromal disease. Alzheimers Dement. 2014;10(6):799–807 e2. https://doi.org/10.1016/j.jalz.2014.05.1749.
17. Ashton NJ, Hye A, Rajkumar AP, et al. An update on blood-based biomarkers for non-Alzheimer neurodegenerative disorders. Nat Rev Neurol. 2020;16(5):265–84. https://doi.org/10.1038/s41582-020-0348-0.
18. Pilotto A, Imarisio A, Carrarini C, et al. Plasma neurofilament light chain predicts cognitive progression in prodromal and clinical dementia with Lewy bodies. J Alzheimers Dis. 2021;82(3):913–9. https://doi.org/10.3233/JAD-210342.
19. Teunissen CE, Verberk IMW, Thijssen EH, et al. Blood-based biomarkers for Alzheimer's disease: towards clinical implementation. Lancet Neurol. 2022;21(1):66–77. https://doi.org/10.1016/S1474-4422(21)00361-6.
20. Hansson O, Blennow K, Zetterberg H, Dage J. Blood biomarkers for Alzheimer's disease in clinical practice and trials. Nat Aging. 2023;3(5):506–19. https://doi.org/10.1038/s43587-023-00403-3.
21. Janelidze S, Stomrud E, Palmqvist S, et al. Plasma beta-amyloid in Alzheimer's disease and vascular disease. Sci Rep. 2016;6:26801. https://doi.org/10.1038/srep26801.
22. Nakamura A, Kaneko N, Villemagne VL, et al. High performance plasma amyloid-beta biomarkers for Alzheimer's disease. Nature. 2018;554(7691):249–54. https://doi.org/10.1038/nature25456.
23. Karikari TK, Ashton NJ, Brinkmalm G, et al. Blood phospho-tau in Alzheimer disease: analysis, interpretation, and clinical utility. Nat Rev Neurol. 2022;18(7):400–18. https://doi.org/10.1038/s41582-022-00665-2.
24. O'Connor A, Karikari TK, Poole T, et al. Plasma phospho-tau181 in presymptomatic and symptomatic familial Alzheimer's disease: a longitudinal cohort study. Mol Psychiatry. 2021;26(10):5967–76. https://doi.org/10.1038/s41380-020-0838-x.
25. Lantero Rodriguez J, Karikari TK, Suarez-Calvet M, et al. Plasma p-tau181 accurately predicts Alzheimer's disease pathology at least 8 years prior to post-mortem and improves the clinical characterisation of cognitive decline. Acta Neuropathol. 2020;140(3):267–78. https://doi.org/10.1007/s00401-020-02195-x.
26. Hol EM, Pekny M. Glial fibrillary acidic protein (GFAP) and the astrocyte intermediate filament system in diseases of the central nervous system. Curr Opin Cell Biol. 2015;32:121–30. https://doi.org/10.1016/j.ceb.2015.02.004.

27. Oeckl P, Halbgebauer S, Anderl-Straub S, et al. Glial fibrillary acidic protein in serum is increased in Alzheimer's disease and correlates with cognitive impairment. J Alzheimers Dis. 2019;67(2):481–8. https://doi.org/10.3233/JAD-180325.
28. Elahi FM, Casaletto KB, La Joie R, et al. Plasma biomarkers of astrocytic and neuronal dysfunction in early- and late-onset Alzheimer's disease. Alzheimers Dement. 2020;16(4):681–95. https://doi.org/10.1016/j.jalz.2019.09.004.
29. Mattsson N, Andreasson U, Zetterberg H, Blennow K, Neuroimaging A's D, I. Association of plasma neurofilament light with neurodegeneration in patients with Alzheimer disease. JAMA Neurol. 2017;74(5):557–66. https://doi.org/10.1001/jamaneurol.2016.6117.
30. Lewczuk P, Ermann N, Andreasson U, et al. Plasma neurofilament light as a potential biomarker of neurodegeneration in Alzheimer's disease. Alzheimers Res Ther. 2018;10(1):71. https://doi.org/10.1186/s13195-018-0404-9.
31. Mattsson N, Cullen NC, Andreasson U, Zetterberg H, Blennow K. Association between longitudinal plasma neurofilament light and neurodegeneration in patients with Alzheimer disease. JAMA Neurol. 2019;76(7):791–9. https://doi.org/10.1001/jamaneurol.2019.0765.
32. Al Shweiki MR, Steinacker P, Oeckl P, et al. Neurofilament light chain as a blood biomarker to differentiate psychiatric disorders from behavioural variant frontotemporal dementia. J Psychiatr Res. 2019;113:137–40. https://doi.org/10.1016/j.jpsychires.2019.03.019.
33. Ducharme S, Dols A, Laforce R, et al. Recommendations to distinguish behavioural variant frontotemporal dementia from psychiatric disorders. Brain. 2020;143(6):1632–50. https://doi.org/10.1093/brain/awaa018.
34. Katisko K, Cajanus A, Jaaskelainen O, et al. Serum neurofilament light chain is a discriminative biomarker between frontotemporal lobar degeneration and primary psychiatric disorders. J Neurol. 2020;267(1):162–7. https://doi.org/10.1007/s00415-019-09567-8.
35. Ownby RL, Crocco E, Acevedo A, John V, Loewenstein D. Depression and risk for Alzheimer disease: systematic review, meta-analysis, and metaregression analysis. Arch Gen Psychiatry. 2006;63(5):530–8. https://doi.org/10.1001/archpsyc.63.5.530.
36. Livingston G, Huntley J, Sommerlad A, et al. Dementia prevention, intervention, and care: 2020 report of the lancet commission. Lancet. 2020;396(10248):413–46. https://doi.org/10.1016/S0140-6736(20)30367-6.
37. Wang YC, Tai PA, Poly TN, et al. Increased risk of dementia in patients with antidepressants: a meta-analysis of observational studies. Behav Neurol. 2018;2018:5315098. https://doi.org/10.1155/2018/5315098.
38. Kodesh A, Sandin S, Reichenberg A, et al. Exposure to antidepressant medication and the risk of incident dementia. Am J Geriatr Psychiatry. 2019;27(11):1177–88. https://doi.org/10.1016/j.jagp.2019.05.019.
39. Singh-Manoux A, Dugravot A, Fournier A, et al. Trajectories of depressive symptoms before diagnosis of dementia: a 28-year follow-up study. JAMA Psychiatry. 2017;74(7):712–8. https://doi.org/10.1001/jamapsychiatry.2017.0660.
40. Almeida OP, Ford AH, Hankey GJ, Yeap BB, Golledge J, Flicker L. Risk of dementia associated with psychotic disorders in later life: the health in men study (HIMS). Psychol Med. 2019;49(2):232–42. https://doi.org/10.1017/S003329171800065X.
41. Tapiainen V, Hartikainen S, Taipale H, Tiihonen J, Tolppanen AM. Hospital-treated mental and behavioral disorders and risk of Alzheimer's disease: a nationwide nested case-control study. Eur Psychiatry. 2017;43:92–8. https://doi.org/10.1016/j.eurpsy.2017.02.486.
42. Martin E, Velayudhan L. Neuropsychiatric symptoms in mild cognitive impairment: a literature review. Dement Geriatr Cogn Disord. 2020;49(2):146–55. https://doi.org/10.1159/000507078.
43. Liew TM. Symptom clusters of neuropsychiatric symptoms in mild cognitive impairment and their comparative risks of dementia: a cohort study of 8530 older persons. J Am Med Dir Assoc. 2019;20(8):1054 e1-1054 e9. https://doi.org/10.1016/j.jamda.2019.02.012.
44. Liew TM. Neuropsychiatric symptoms in cognitively normal older persons, and the association with Alzheimer's and non-Alzheimer's dementia. Alzheimers Res Ther. 2020;12(1):35. https://doi.org/10.1186/s13195-020-00604-7.

45. Showraki A, Murari G, Ismail Z, et al. Cerebrospinal fluid correlates of neuropsychiatric symptoms in patients with Alzheimer's disease/mild cognitive impairment: a systematic review. J Alzheimers Dis. 2019;71(2):477–501. https://doi.org/10.3233/JAD-190365.
46. Creese B, Ismail Z. Mild behavioral impairment: measurement and clinical correlates of a novel marker of preclinical Alzheimer's disease. Alzheimers Res Ther. 2022;14(1):2. https://doi.org/10.1186/s13195-021-00949-7.
47. Beck AT, Ward CH, Mendelson M, Mock J, Erbaugh J. An inventory for measuring depression. Arch Gen Psychiatry. 1961;4:561–71. https://doi.org/10.1001/archpsyc.1961.01710120031004.
48. Yesavage JA, Brink TL, Rose TL, et al. Development and validation of a geriatric depression screening scale: a preliminary report. J Psychiatr Res. 1982;17(1):37–49. https://doi.org/10.1016/0022-3956(82)90033-4.
49. Cummings JL, Mega M, Gray K, Rosenberg-Thompson S, Carusi DA, Gornbein J. The neuropsychiatric inventory: comprehensive assessment of psychopathology in dementia. Neurology. 1994;44(12):2308–14. https://doi.org/10.1212/wnl.44.12.2308.
50. Wise EA, Rosenberg PB, Lyketsos CG, Leoutsakos JM. Time course of neuropsychiatric symptoms and cognitive diagnosis in National Alzheimer's coordinating centers volunteers. Alzheimers Dement (Amst). 2019;11:333–9. https://doi.org/10.1016/j.dadm.2019.02.006.
51. Matsuoka T, Ismail Z, Narumoto J. Prevalence of mild behavioral impairment and risk of dementia in a psychiatric outpatient clinic. J Alzheimers Dis. 2019;70(2):505–13. https://doi.org/10.3233/JAD-190278.
52. Cieslak A, Smith EE, Lysack J, Ismail Z. Case series of mild behavioral impairment: toward an understanding of the early stages of neurodegenerative diseases affecting behavior and cognition. Int Psychogeriatr. 2018;30(2):273–80. https://doi.org/10.1017/S1041610217001855.
53. Woolley JD, Khan BK, Murthy NK, Miller BL, Rankin KP. The diagnostic challenge of psychiatric symptoms in neurodegenerative disease: rates of and risk factors for prior psychiatric diagnosis in patients with early neurodegenerative disease. Research Support, N.I.H., Extramural Research Support, Non-U.S. Gov't. J Clin Psychiatry. 2011;72(2):126–33. https://doi.org/10.4088/JCP.10m06382oli.
54. Ismail Z, Smith EE, Geda Y, et al. Neuropsychiatric symptoms as early manifestations of emergent dementia: provisional diagnostic criteria for mild behavioral impairment. Alzheimers Dement. 2016;12(2):195–202. https://doi.org/10.1016/j.jalz.2015.05.017.
55. Finkel SI, Costa e Silva J, Cohen G, Miller S, Sartorius N. Behavioral and psychological signs and symptoms of dementia: a consensus statement on current knowledge and implications for research and treatment. Int Psychogeriatr. 1996;8(Suppl 3):497–500. https://doi.org/10.1017/s1041610297003943.
56. Ismail Z, McGirr A, Gill S, Hu S, Forkert ND, Smith EE. Mild behavioral impairment and subjective cognitive decline predict cognitive and functional decline. J Alzheimers Dis. 2021;80(1):459–69. https://doi.org/10.3233/JAD-201184.
57. Miao R, Chen HY, Gill S, et al. Plasma beta-amyloid in mild behavioural impairment - neuropsychiatric symptoms on the Alzheimer's continuum. J Geriatr Psychiatry Neurol. 2022;35(3):434–41. https://doi.org/10.1177/08919887211016068.
58. Nogovitsyn N, Souza R, Muller M, et al. Aberrant limbic brain structures in young individuals at risk for mental illness. Psychiatry Clin Neurosci. 2020;74(5):294–302. https://doi.org/10.1111/pcn.12985.
59. Ismail Z, Aguera-Ortiz L, Brodaty H, et al. The mild behavioral impairment checklist (MBI-C): a rating scale for neuropsychiatric symptoms in pre-dementia populations. J Alzheimers Dis. 2017;56(3):929–38. https://doi.org/10.3233/JAD-160979.
60. Mallo SC, Ismail Z, Pereiro AX, et al. Assessing mild behavioral impairment with the mild behavioral impairment-checklist in people with mild cognitive impairment. J Alzheimers Dis. 2018;66(1):83–95. https://doi.org/10.3233/JAD-180131.
61. Creese B, Griffiths A, Brooker H, et al. Profile of mild behavioral impairment and factor structure of the mild behavioral impairment checklist in cognitively normal older adults. Int Psychogeriatr. 2020;32(6):705–17. https://doi.org/10.1017/S1041610219001200.

62. Hu S, Patten S, Charlton A, et al. Validating the mild behavioral impairment checklist in a cognitive clinic: comparisons with the neuropsychiatric inventory questionnaire. J Geriatr Psychiatry Neurol. 2023;36(2):107–20. https://doi.org/10.1177/08919887221093353.
63. Kassam F, Chen H, Nosheny RL, et al. Cognitive profile of people with mild behavioral impairment in Brain Health Registry participants. Int Psychogeriatr. 2023;8:1–10. https://doi.org/10.1017/S1041610221002878.
64. Pan Y, Shea YF, Li S, et al. Prevalence of mild behavioural impairment: a systematic review and meta-analysis. Psychogeriatrics. 2021;21(1):100–11. https://doi.org/10.1111/psyg.12636.
65. Kaufer DI, Cummings JL, Ketchel P, et al. Validation of the NPI-Q, a brief clinical form of the neuropsychiatric inventory. J Neuropsychiatry Clin Neurosci. 2000;12(2):233–9. https://doi.org/10.1176/jnp.12.2.233.
66. Mortby ME, Ismail Z, Anstey KJ. Prevalence estimates of mild behavioral impairment in a population-based sample of pre-dementia states and cognitively healthy older adults. Int Psychogeriatr. 2018;30(2):221–32. https://doi.org/10.1017/S1041610217001909.
67. Rouse HJ, Small BJ, Schinka JA, Loewenstein DA, Duara R, Potter H. Mild behavioral impairment as a predictor of cognitive functioning in older adults. Int Psychogeriatr. 2021;33(3):285–93. https://doi.org/10.1017/S1041610220000678.
68. Ismail Z, Elbayoumi H, Fischer CE, et al. Prevalence of depression in patients with mild cognitive impairment: a systematic review and meta-analysis. JAMA Psychiatry. 2017;74(1):58–67. https://doi.org/10.1001/jamapsychiatry.2016.3162.
69. Kan CN, Cano J, Zhao X, Ismail Z, Chen CL, Xu X. Prevalence, clinical correlates, cognitive trajectories, and dementia risk associated With mild behavioral impairment in Asians. J Clin Psychiatry. 2022;83(3) https://doi.org/10.4088/JCP.21m14105.
70. Guan DX, Smith EE, Pike GB, Ismail Z. Persistence of neuropsychiatric symptoms and dementia prognostication: a comparison of three operational case definitions of mild behavioral impairment. Alzheimers Dement (Amst). 2023;15(4):e12483. https://doi.org/10.1002/dad2.12483.
71. Creese B, Brooker H, Ismail Z, et al. Mild behavioral impairment as a marker of cognitive decline in cognitively normal older adults. Am J Geriatr Psychiatry. 2019;27(8):823–34. https://doi.org/10.1016/j.jagp.2019.01.215.
72. Mallo SC, Ismail Z, Pereiro AX, et al. Assessing mild behavioral impairment with the mild behavioral impairment checklist in people with subjective cognitive decline. Int Psychogeriatr. 2019;31(2):231–9. https://doi.org/10.1017/S1041610218000698.
73. Fan S, Liang X, Yun T, et al. Mild behavioral impairment is related to frailty in non-dementia older adults: a cross-sectional study. BMC Geriatr. 2020;20(1):510. https://doi.org/10.1186/s12877-020-01903-2.
74. Wong F, Kok Pin N, Chathuri Y, Audrey L, Zahinoor I, Nagaendran K. Characterising mild behavioural impairment in Asian mild cognitive impairment and cognitively normal individuals. Alzheimers Dement. 2020;16(S6) https://doi.org/10.1002/alz.045059.
75. Rao AR, Chatterjee P, Thakral M, Dwivedi SN, Dey AB. Behavioural issues in late life may be the precursor of dementia- a cross sectional evidence from memory clinic of AIIMS, India. PLoS One. 2020;15(6):e0234514. https://doi.org/10.1371/journal.pone.0234514.
76. Kianimehr G, Fatehi F, Noroozian M. Prevalence of mild behavioral impairment in patients with mild cognitive impairment. Acta Neurol Belg. 2022;122(6):1493–7. https://doi.org/10.1007/s13760-021-01724-z.
77. Pan Y, Shea YF, Ismail Z, et al. Prevalence of mild behavioural impairment domains: a meta-analysis. Psychogeriatrics. 2022;22(1):84–98. https://doi.org/10.1111/psyg.12782.
78. Rouse HJ, Ismail Z, Andel R, Molinari VA, Schinka JA, Small BJ. Impact of mild behavioral impairment on longitudinal changes in cognition. J Gerontol A Biol Sci Med Sci. 2024;79(1) https://doi.org/10.1093/gerona/glad098.
79. Ruthirakuhan M, Ismail Z, Herrmann N, Gallagher D, Lanctot KL. Mild behavioral impairment is associated with progression to Alzheimer's disease: a clinicopathological study. Alzheimers Dement. 2022;18(11):2199–208. https://doi.org/10.1002/alz.12519.

80. Wolfova K, Creese B, Aarsland D, et al. Gender/sex differences in the association of mild behavioral impairment with cognitive aging. J Alzheimers Dis. 2022;88(1):345–55. https://doi.org/10.3233/JAD-220040.
81. Ghahremani M, Wang M, Chen HY, et al. Plasma phosphorylated tau at threonine 181 and neuropsychiatric symptoms in preclinical and prodromal Alzheimer disease. Neurology. 2023;100(7):e683–93. https://doi.org/10.1212/WNL.0000000000201517.
82. McGirr A, Nathan S, Ghahremani M, Gill S, Smith EE, Ismail Z. Progression to dementia or reversion to normal cognition in mild cognitive impairment as a function of late-onset neuropsychiatric symptoms. Neurology. 2022;98(21):e2132–9. https://doi.org/10.1212/WNL.0000000000200256.
83. Creese B, Arathimos R, Aarsland D, et al. Late-life onset psychotic symptoms and incident cognitive impairment in people without dementia: Modification by genetic risk for Alzheimer's disease. Alzheimers Dement (N Y). 2023;9(2):e12386. https://doi.org/10.1002/trc2.12386.
84. Ismail Z, Ghahremani M, Munir MA, Fischer CE, Smith EE, Creese B. A longitudinal study of late-life psychosis and incident dementia and the potential effects of race and cognition. Nat Mental Health. 2023;1(4):273–83. https://doi.org/10.1038/s44220-023-00043-x.
85. Vellone D, Ghahremani M, Goodarzi Z, Forkert ND, Smith EE, Ismail Z. Apathy and APOE in mild behavioral impairment, and risk for incident dementia. Alzheimers Dement (N Y). 2022;8(1):e12370. https://doi.org/10.1002/trc2.12370.
86. Ebrahim IM, Ghahremani M, Camicioli R, Smith EE, Ismail Z. Effects of race, baseline cognition, and APOE on the association of affective dysregulation with incident dementia: a longitudinal study of dementia-free older adults. J Affect Disord. 2023;332:9–18. https://doi.org/10.1016/j.jad.2023.03.074.
87. Taragano FE, Allegri RF, Heisecke SL, et al. Risk of conversion to dementia in a mild behavioral impairment group compared to a psychiatric group and to a mild cognitive impairment group. J Alzheimers Dis. 2018;62(1):227–38. https://doi.org/10.3233/JAD-170632.
88. Matsuoka T, Imai A, Narumoto J. Neuroimaging of mild behavioral impairment: a systematic review. Psychiatry Clin Neurosci Rep. 2023;2(1):e81.
89. Matuskova V, Ismail Z, Nikolai T, et al. Mild behavioral impairment is associated with atrophy of entorhinal cortex and hippocampus in a memory clinic cohort. Front Aging Neurosci. 2021;13:643271. https://doi.org/10.3389/fnagi.2021.643271.
90. Braak H, Alafuzoff I, Arzberger T, Kretzschmar H, Del Tredici K. Staging of Alzheimer disease-associated neurofibrillary pathology using paraffin sections and immunocytochemistry. Acta Neuropathol. 2006;112(4):389–404. https://doi.org/10.1007/s00401-006-0127-z.
91. Guan DX, Rehman T, Nathan S, Durrani R, Smith EE, Ismail Z. Mild behavioral impairment as an indicator of early dementia neurodegeneration: evidence from structural neuroimaging. Alzheimers Dement. 2023;19:e062692.
92. Gill S, Mouches P, Hu S, et al. Using machine learning to predict dementia from neuropsychiatric symptom and neuroimaging data. J Alzheimers Dis. 2020;75(1):277–88. https://doi.org/10.3233/JAD-191169.
93. Gill S, Wang M, Mouches P, et al. Neural correlates of the impulse dyscontrol domain of mild behavioral impairment. Int J Geriatr Psychiatry. 2021;36(9):1398–406. https://doi.org/10.1002/gps.5540.
94. Smith EE, Egorova S, Blacker D, et al. Magnetic resonance imaging white matter hyperintensities and brain volume in the prediction of mild cognitive impairment and dementia. Arch Neurol. 2008;65(1):94–100. https://doi.org/10.1001/archneurol.2007.23.
95. Brickman AM, Honig LS, Scarmeas N, et al. Measuring cerebral atrophy and white matter hyperintensity burden to predict the rate of cognitive decline in Alzheimer disease. Arch Neurol. 2008;65(9):1202–8. https://doi.org/10.1001/archneur.65.9.1202.
96. Bos D, Wolters FJ, Darweesh SKL, et al. Cerebral small vessel disease and the risk of dementia: a systematic review and meta-analysis of population-based evidence. Alzheimers Dement. 2018;14(11):1482–92. https://doi.org/10.1016/j.jalz.2018.04.007.

97. Miao R, Chen HY, Robert P, Smith EE, Ismail Z, Group MS. White matter hyperintensities and mild behavioral impairment: findings from the MEMENTO cohort study. Cereb Circ Cogn Behav. 2021;2:100028. https://doi.org/10.1016/j.cccb.2021.100028.
98. Seeley WW, Menon V, Schatzberg AF, et al. Dissociable intrinsic connectivity networks for salience processing and executive control. J Neurosci. 2007;27(9):2349–56. https://doi.org/10.1523/JNEUROSCI.5587-06.2007.
99. Matsuoka T, Ueno D, Ismail Z, et al. Neural correlates of mild behavioral impairment: a functional brain connectivity study using resting-state functional magnetic resonance imaging. J Alzheimers Dis. 2021;83(3):1221–31. https://doi.org/10.3233/JAD-210628.
100. Raichle ME. The brain's default mode network. Annu Rev Neurosci. 2015;38:433–47. https://doi.org/10.1146/annurev-neuro-071013-014030.
101. Li W, Mai X, Liu C. The default mode network and social understanding of others: what do brain connectivity studies tell us. Front Hum Neurosci. 2014;8:74. https://doi.org/10.3389/fnhum.2014.00074.
102. Rosenberg PB, Nowrangi MA, Lyketsos CG. Neuropsychiatric symptoms in Alzheimer's disease: what might be associated brain circuits? Mol Aspects Med. 2015;43–44:25–37. https://doi.org/10.1016/j.mam.2015.05.005.
103. Chen Y, Dang M, Zhang Z. Brain mechanisms underlying neuropsychiatric symptoms in Alzheimer's disease: a systematic review of symptom-general and -specific lesion patterns. Mol Neurodegener. 2021;16(1):38. https://doi.org/10.1186/s13024-021-00456-1.
104. Ghahremani M, Nathan S, Smith EE, McGirr A, Goodyear B, Ismail Z. Functional connectivity and mild behavioral impairment in dementia-free elderly. Alzheimers Dement (N Y). 2023;9(1):e12371. https://doi.org/10.1002/trc2.12371.
105. Lussier FZ, Pascoal TA, Chamoun M, et al. Mild behavioral impairment is associated with beta-amyloid but not tau or neurodegeneration in cognitively intact elderly individuals. Alzheimers Dement. 2020;16(1):192–9. https://doi.org/10.1002/alz.12007.
106. Naude J, Wang M, Leon R, Smith E, Ismail Z. Tau-PET in early cortical Alzheimer brain regions in relation to mild behavioral impairment in older adults with either normal cognition or mild cognitive impairment. Neurobiol Aging. 2024;138:19–27. https://doi.org/10.1016/j.neurobiolaging.2024.02.006.
107. Johansson M, Stomrud E, Insel PS, et al. Mild behavioral impairment and its relation to tau pathology in preclinical Alzheimer's disease. *Transl.* Psychiatry. 2021;11(1):76. https://doi.org/10.1038/s41398-021-01206-z.
108. Mayblyum DV, Quiroz YT, Langella S, et al. Associations between mild behavioral impairment (MBI-C) affective domains and Alzheimer's disease neuroimaging biomarkers in older adults. Alzheimers Dement. 2022;18:e068302.
109. Naude JP, Gill S, Hu S, et al. Plasma neurofilament light: a marker of neurodegeneration in mild behavioral impairment. J Alzheimers Dis. 2020;76(3):1017–27. https://doi.org/10.3233/JAD-200011.
110. Ismail Z, Leon R, Creese B, Ballard C, Robert P, Smith EE. Optimizing detection of Alzheimer's disease in mild cognitive impairment: a 4-year biomarker study of mild behavioral impairment in ADNI and MEMENTO. Mol Neurodegener. 2023;18(1):50. https://doi.org/10.1186/s13024-023-00631-6.
111. Cui Y, Dai S, Miao Z, et al. Reliability and validity of the Chinese version of the mild behavioral impairment checklist for screening for Alzheimer's disease. J Alzheimers Dis. 2019;70(3):747–56. https://doi.org/10.3233/JAD-190113.
112. Tang-Wai DF, Smith EE, Bruneau MA, et al. CCCDTD5 recommendations on early and timely assessment of neurocognitive disorders using cognitive, behavioral, and functional scales. Alzheimers Dement (N Y). 2020;6(1):e12057. https://doi.org/10.1002/trc2.12057.
113. Ismail Z, Black SE, Camicioli R, et al. Recommendations of the 5th Canadian consensus conference on the diagnosis and treatment of dementia. Alzheimers Dement. 2020;16(8):1182–95. https://doi.org/10.1002/alz.12105.
114. Gomez-Isla T, Frosch MP. Lesions without symptoms: understanding resilience to Alzheimer disease neuropathological changes. Nat Rev Neurol. 2022;18(6):323–32. https://doi.org/10.1038/s41582-022-00642-9.

115. Petersen RC, Thomas RG, Aisen PS, et al. Randomized controlled trials in mild cognitive impairment: sources of variability. Neurology. 2017;88(18):1751–8. https://doi.org/10.1212/WNL.0000000000003907.
116. Soto M, Rosenberg P, Ballard C, et al. CTAD task force paper: neuropsychiatric symptoms in AD: clinical trials targeting mild behavioral impairment: a report from the International CTAD Task Force. J Prev Alzheimers Dis. 2024;11(1):56–64. https://doi.org/10.14283/jpad.2023.125.
117. Guan DX, Chen HY, Camicioli R, Montero-Odasso M, Smith EE, Ismail Z. Dual-task gait and mild behavioral impairment: The interface between non-cognitive dementia markers. Exp Gerontol. 2022;162:111743. https://doi.org/10.1016/j.exger.2022.111743.
118. Gosselin P, Guan DX, Chen HY, et al. The relationship between hearing and mild behavioral impairment and the influence of sex: a study of older adults without dementia from the COMPASS-ND study. J Alzheimers Dis Rep. 2022;6(1):57–66. https://doi.org/10.3233/ADR-210045.
119. Gosselin P, Guan DX, Smith EE, Ismail Z. Temporal associations between treated and untreated hearing loss and mild behavioral impairment in older adults without dementia. Alzheimers Dement (N Y). 2023;9(4):e12424. https://doi.org/10.1002/trc2.12424.
120. Soo SA, Ng KP, Wong F, et al. The association between diabetes mellitus and mild behavioral impairment among mild cognitive impairment: findings from Singapore. J Alzheimers Dis. 2021;82(1):411–20. https://doi.org/10.3233/JAD-210037.
121. Guan DX, Rockwood K, Smith EE, Ismail Z. Sex moderates the association between frailty and mild behavioral impairment. J Prev Alzheimers Dis. 2022;9(4):692–700. https://doi.org/10.14283/jpad.2022.61.
122. Matsuoka T, Ismail Z, Imai A, et al. Relationship between loneliness and mild behavioral impairment: validation of the Japanese version of the MBI checklist and a cross-sectional study. J Alzheimers Dis. 2024;97(4):1951–60. https://doi.org/10.3233/JAD-230923.
123. Tsai CF, Huang MH, Lin YS, Chen CY, Fuh JL. Health-related quality of life and mild behavioral impairment in older adults without dementia. Int J Geriatr Psychiatry. 2023;38(9):e6003. https://doi.org/10.1002/gps.6003.

Part V
Future Directions in Biomarker Research

Extracellular Vesicles as Biomarkers of Alzheimer Disease and Related Dementias

Zainab Mianoor, Jolene Phelps, and AmanPreet Badhwar

Abstract Extracellular vesicles (EVs) have emerged as promising biomarkers for neurodegenerative disorders, offering unique insights into disease pathogenesis and progression. These vesicles, including exosomes, microvesicles, and apoptotic bodies, are small, membrane-bound particles released by cells that carry diverse molecular cargoes, including proteins, lipids, and nucleic acids. EVs are involved in intercellular communication and may facilitate spread of pathogenic proteins across the nervous system, potentially contributing to the progression of neurodegenerative disorders. Emerging evidence supports the ability of EVs isolated from biofluids such as blood, cerebrospinal fluid (CSF), and urine to reliably reflect the molecular signature of affected neural and other cell types, highlighting their importance as tools in noninvasive biomarker discovery. This chapter reviews the value of EVs as biomarkers in neurodegenerative disorders, explores their molecular composition, and discusses the latest advancements in EV isolation and characterization techniques. Furthermore, we examine the challenges and opportunities of EV-based diagnostics, emphasizing the potential of EVs as biomarkers for early detection and monitoring of Alzheimer disease (AD) and related dementias (ADRDs).

Keywords Alzheimer disease and related dementias · Biofluid-based biomarkers · Brain-derived extracellular vesicles · Brain cell-specificity · Cargo content

Z. Mianoor · A. Badhwar (✉)
Multiomics Investigation of Neurodegenerative Diseases (MIND) Laboratory, Montréal, QC, Canada

Département de pharmacologie et physiologie, Institut de Génie Biomédical, Université de Montréal, Montréal, QC, Canada

Centre de Recherche de l'Institut Universitaire de Gériatrie (CRIUGM), Montréal, QC, Canada
e-mail: amanpreet.badhwar@umontreal.ca

J. Phelps
Department of Clinical Neurosciences, University of Calgary, Calgary, AB, Canada

R. Tarawneh (ed.), *Biomarkers of Neurodegenerative Disorders*,
https://doi.org/10.1007/978-3-031-80849-4_18

1 Introduction: Extracellular Vesicles as Biomarkers

Extracellular vesicles (EVs) are universally secreted by cells across all biological kingdoms, from bacteria to humans, and play critical roles in health and disease [1]. These heterogeneous, nanosized, membrane-bound structures are secreted into the extracellular milieu and are essential for the intercellular transfer of bioactive molecules, such as nucleic acids, proteins, lipids, and metabolites [1, 2]. Beyond this, EVs are implicated in various other functions such as immunity [3], cell signaling [4], and tissue repair [5]. In general, EVs contain surface- and cargo-proteins specific to their tissue (and cell) of origin, offering insights into the health status of their parent cells [6]. This unique attribute positions EVs as a rich biomarker source for a wide range of diseases, including Alzheimer disease (AD) and related dementias (ADRDs) [6]. Emerging ADRD literature has associated EVs with the propagation and/or accumulation of several hallmark misfolded proteins in the brain, as well as their clearance into peripheral biofluids [7]. Two prominent examples include (a) the involvement of EVs in the spread of β-amyloid and phosphorylated tau in the brain, two established AD biomarkers, and (b) significantly elevated levels of the Parkinson's disease (PD) biomarker, α-synuclein, in plasma EVs of diagnosed individuals [8–11].

This chapter will introduce EVs (Sects. 1.1, 1.2, and 1.3), how to isolate them from biofluids and characterize them (Sect. 2), and discuss their implications as biomarkers in ADRDs (Sects. 3 and 4).

1.1 EV Subtypes and Biogenesis

EVs are broadly categorized into three main subtypes: exosomes, ectosomes or shedding microvesicles, and apoptotic bodies. Emerging literature also indicates the presence of additional subtypes, such as oncosomes, migrasomes, exomeres, supermeres, to name a few [1, 12]. Below, we describe the three main EV subtypes.

Exosomes, which range between 30 and150 nm in diameter, originate from intraluminal vesicles (ILVs) formed by the inward budding of the endosomal membrane in late endosomes, also known as multivesicular bodies (MVBs) [13]. The creation of ILVs inside MVBs during the maturation process depends on the endosomal sorting complex required for transport (ESCRT), which consists of four protein subunit complexes (ESCRT-0 to III) [14]. The ESCRT machinery, along with proteins Alix and ARRDC1, helps assemble specific molecules into the ILVs. If an MVB merges with the cell surface to release its contents, the ILVs released are then termed exosomes [13, 15].

Ectosomes are larger vesicles ranging from 100 to 1000 nm in diameter. They form through outward budding and shedding of the plasma membrane when triggered by stimuli such as cell activation or injury [12, 16]. The formation of ectosomes entails actin cytoskeleton reorganization, which facilitates membrane budding, followed by the separation and release of vesicles [12]. Ectosome shedding is believed to occur in most (if not all) cells, with frequencies varying among cell-types [12, 17]. Resting or nondividing cells typically show minimal shedding rates [12, 16, 17].

Apoptotic bodies constitute the largest class of vesicles, ranging from 50 to 5000 nm. They are released during the late stages of apoptosis, a type of programmed cell death marked by nuclear condensation and fragmentation. Although apoptotic bodies are typically engulfed by neighboring phagocytes to aid in tissue homeostasis, they can also be released into body fluids [18, 19].

The well-recognized International Society for Extracellular Vesicles (ISEV) recommends the term EVs for all lipid bilayer-delimited particles incapable of replication that are naturally released from cells [20]. In addition, ISEV advises that terms linked to EV biogenesis pathways, such as "exosome" and "ectosome", should be used only with solid evidence of specific biogenesis pathways [20]. This is because despite existing literature on biogenesis of the main EV subtypes, there remains a lack of consensus regarding specific markers that can distinguish these subtypes post-cellular release [21]. Current separation techniques, therefore, do not effectively and/or consistently isolate EV subtypes, thereby making the use of biogenesis-based terms problematic [20–22]. ISEV further stresses the need for caution when distinguishing EV subtypes based on size, density, molecular composition, or cellular origin. For example, while size cut-off descriptors, such as small (<200 nm) or large (>200 nm) EVs are used to categorize EVs sorted by methods such as filtration or differential ultracentrifugation, caution should be exercised due to overlapping size profiles and the absence of agreed-upon size cut-offs [20–22].

1.2 Molecular Composition of EVs

Generally, EVs are composed of a lipid bilayer, internal cargo, and surface-associated molecules [23]. More recently, a corona composed of proteins that are adsorbed onto the surface when exposed to various biofluids has been described [24]. We provide a brief summary of each component below.

1.2.1 Lipid Bilayer

Each EV consists of a phospholipid bilayer, which forms its outer membrane and provides structural integrity and protection, allowing it to travel through bodily fluids [21]. The phospholipid bilayer contains various biomolecules, such as lipids, proteins, glycans, and DNA, that are intrinsically and/or peripherally associated with the EV membrane. Several of these molecules reflect the EVs' cell of origin and/or are crucial for their stability, biodistribution, and cellular interactions [25, 26].

Lipids form the essential structural framework of the EV phospholipid bilayer [27, 28] and include ceramides, cholesterol, sphingolipids, glycerolipids, and others [29]. These lipids in the bilayer contribute to EV stability and diverse bodily functions [29] and can potentially enhance our understanding of certain ADRDs [26]. For instance, Wang et al. reported ceramides as key in triggering amyloid-induced astrocyte death in AD by facilitating the release of exosomes that signal apoptosis [28]. Similarly, unique lipid compositions in brain-derived EVs were reported [29], with higher levels of glycerophospholipids, and lower levels of sphingolipids, which may impact their role in ADRDs.

Proteins, both integral and peripheral to the phospholipid bilayer membrane, have been reported [26]. Integral proteins contain α-helical transmembrane domains and include tetraspanins (e.g., CD9, CD63, CD81), transport/fusion proteins, antigen presentation proteins (MHC class I and II), integrins, and receptors. In contrast, peripheral proteins interact with integral membrane proteins and phospholipid head groups via non-covalent bonds or electrostatic interactions [30]. Examples of peripheral proteins include cell adhesion proteins, chaperonins, heparin-binding proteins, enzymes, coagulation factors, nucleic acid-binding proteins, and immunoglobulin superfamily proteins [30, 31].

Glycans are another important molecule peripherally associated with the phospholipid bilayer through covalent attachment to protein amino-acid side chains. EVs are enriched with transmembrane glycosaminoglycans and proteoglycans, which facilitate EV internalization into recipient cells and influence recipient cell activity [26, 32] EVs released by different cell-types, and under different pathophysiological states, bear unique glycan signatures, showing promise as predictive and diagnostic biomarkers [26, 32].

Finally, different types of *DNA* fragments, encompassing both chromosomal and mitochondrial origins, have also been identified on the surfaces of EVs [30, 33, 34]. This DNA presence influences several EV characteristics, such as imparting negative charge and promoting aggregation [35]. Over time, these DNA fragments can also be transferred to recipient cells, potentially facilitating processes like horizontal gene transfer and interaction with components of the extracellular matrix, such as fibronectin [33]. The mechanism by which DNA is structurally associated with EVs, and when this association occurs, is not currently well understood [36].

1.2.2 Internal Cargo

There has been considerable interest in characterizing internal EV cargo, found within the phospholipid bilayer, to leverage them as potential disease biomarkers [25]. Nucleic acids, micro-RNAs (miRs), proteins, lipids, and metabolites represent various types of cargo that can be encapsulated and transported within EVs [37]. The molecular cargo of EVs is not randomly selected, but rather involves selective sorting mechanisms that target specific components into EVs, facilitating the transfer of chemical signals across cells and tissues [38].

EVs contain cargo specific to their biogenesis pathways. The main cytosolic proteins incorporated in exosomes, regardless of their cell of origin, include proteins associated with exosome biogenesis (the ESCRT complex, ALIX, and TSG101), proteins involved in membrane transport/fusion (Rab GTPases and annexins), heat shock proteins (HSP70 and HSP90), and cytoskeletal proteins [39]. The molecular mechanisms involved in ectosome biogenesis are less well understood, with many of the incorporated proteins overlapping with the aforementioned proteins that are incorporated into exosomes (e.g., ESCRT subunits, Alix, TSG101, GTPases) [40]. Apoptotic bodies contain DNA fragments, histones, chromatin remnants, cytosolic fractions, and degraded proteins [22, 39]. Annexin V, a marker for apoptosis, is also enriched in apoptotic bodies [41]. Specific cargo components may enable identification of EV subtypes, although no specific markers are universally accepted at this time [20]. For example, Lamp1 has been identified as a marker specific to exosomes [42], while Annexin A1 [43], SLC3A2, and BSG [44] are enriched in ectosomes.

The composition of molecular cargo within EVs can vary depending on the cell type and physiological state of the originating cell. For instance, in AD, neurons affected by pathological changes release EVs containing misfolded proteins such as β-amyloid (Aβ) and hyperphosphorylated tau [45]. These EVs can be taken up by neighboring neurons and glial cells, facilitating the spread of pathological proteins throughout the brain. This process likely contributes to the progressive nature of AD, as the spread of pathological proteins leads to neuronal dysfunction, synaptic loss, and ultimately neurodegeneration [46].

1.2.3 Protein Corona

Recent findings show that upon exposure to biological fluids, EVs acquire a corona composed of extracellular proteins and other molecules that affect their interactions with recipient cells [47, 48]. For example, Toth et al. reported nine shared EV corona proteins (ApoA1, ApoB, ApoC3, ApoE, complement factors 3 and 4B, fibrinogen α-chain, immunoglobulin heavy constant γ2 and γ4 chains), which appear to be common among EVs in blood plasma [47]. These proteins adsorb onto the EV surface upon exposure, thereby altering the surface properties of EVs and impacting their biodistribution, cellular interactions, and biological activities [47, 48]. For example, specific proteins in the EV corona can modulate their recognition by immune cells and may enhance or suppress the uptake of EVs by macrophages or dendritic cells [47, 48]. While the role of the EV protein corona is still emerging and is not fully understood, the protein corona attaching to lipid-based nanoparticles has been explored as a mode for biomarker discovery in other fields (e.g., cancer) [49]. While limited investigations have been conducted in the ADRD field, a recent study showed that Aβ proteins aggregate on the EV surface [50], suggesting that Aβ might be both surface-associated and present in internal cargo.

1.3 EV Trafficking and Communication

EVs play crucial roles in intercellular communication by transferring biomolecules between cells. Moreover, the ability of brain cell-derived EVs (BEVs) to cross the blood-brain barrier (BBB) enables their enrichment/isolation from blood for biomarker investigations in the ADRDs and other brain diseases [23, 38, 51]. The BBB tightly controls substance exchange between the bloodstream and the brain, maintaining homeostasis and protecting against harmful agents [52]. EVs often carry specific surface markers or ligands that facilitate their interaction with BBB endothelial cells and promote their transport across the barrier [52, 53]. Mechanisms used by EVs to cross the BBB include receptor-mediated and adsorption-mediated transcytosis, direct fusion, or endocytosis by endothelial cells [53]. In addition, BBB dysfunction and resulting increased permeability during inflammatory conditions in the brain, such as those experienced during ADRD, may increase circulation of BEVs in blood [54].

Once in blood circulation, EVs have varied circulation times (from minutes to hours), influenced by factors such as their size, surface charge, and cargo composition [55, 56]. Larger EVs typically circulate longer due to decreased clearance by the reticuloendothelial system and extended retention in the bloodstream [57]. EVs may have enhanced or reduced clearance due to the presence of specific surface markers. For example, EVs expressing CD47 may evade phagocytosis and clearance by immune cells, thereby extending their circulation time [58]. In contrast, EVs with high levels of opsonins, such as complement proteins or immunoglobulins, may experience accelerated clearance due to enhanced recognition by phagocytic cells [59]. Extended circulation enhances the detection of EVs in biomarker studies, since the window for detection increases, enhancing the likelihood of reliable disease diagnosis.

Overall, the ability of BEVs to cross the BBB and retain stability in circulation allows them to be detected and isolated from remote biofluids for biomarker work in the ADRD field. Methods to isolate and characterize these EV populations are described in the following section.

2 Isolating and Analyzing EVs from Biofluids

The isolation and analysis of EVs is highly challenging as there is no current isolation method that can generate a pure EV population, nor is there a universally recommended standard for EV characterization [20]. Methods chosen for EV isolation and characterization, as well as the biofluid and conditions in which they are handled, can have a profound impact on ADRD biomarker study results. Therefore, it is important to understand the advantages and limitations of available approaches.

ISEV published a set of guidelines called the Minimal Information for Studies of Extracellular Vesicles (MISEV) in 2014 [60]. Subsequent updates in 2018

(MISEV2018) and 2023 (MISEV2023) offer insights into the evolving understanding of EVs in this rapidly growing field [20, 22]. MISEV presents advantages and limitations of available approaches for the production/collection, separation, and characterization of EV populations from multiple sources (i.e., body fluids, cell culture, solid tissues) and lists major reporting recommendations to support the reproducibility and comparison of studies. For EV analyses in biofluids, numerous pre-analytical processing factors including method and time of collection, fasting status, the type of collection tubes used, anticoagulants, centrifugation protocols, storage, and handling can all impact downstream findings [20]. In addition to MISEV, ISEV has published guidelines specific to EVs circulating in the blood [61], cerebrospinal fluid (CSF) [62], and urine [63].

The upcoming Sects. 2.1, 2.3, and 2.4 will outline the advantages and limitations of current methods and highlight the most commonly used methods and recommendations specific to the ADRD biomarker field.

2.1 Collection and Preprocessing of Biofluids for EV Analysis

While EVs can be collected from various biofluids, the majority of biomarker studies in the ADRD field have focused on EVs isolated from CSF and blood (both plasma and serum) [9]. More recently, saliva and urine EVs have been evaluated as biomarkers for ADRDs, which may have high potential due to their ease of accessibility. Considerations for collecting and preprocessing each of these biofluids are listed below.

EVs in *CSF* exhibit a higher specificity to the brain [64], suggesting a higher diagnostic and research value for ADRDs. However, sampling of CSF is invasive, which limits the sample volume available. CSF also has a lower concentration of EVs as compared to blood [64]. Common contaminants in CSF include both cells and blood, requiring care during collection and preprocessing to prevent introduction of cell debris.

Blood is relatively accessible and routinely sampled. Isolation of pure EVs from blood can be challenging due to the abundance of EVs from multiple sources, non-EV components such as lipoproteins and platelets that overlap in size/density, and free soluble proteins that contribute to the protein corona [20]. Platelets are an abundant source of EVs that can impact and complicate downstream analyses and should be removed as efficiently as possible. Plasma is often preferred over serum due to the higher presence of contaminating EVs released by platelets during clot formation and processing in the serum fraction [24, 38]. Platelet removal via filtration (0.8 μm) prior to EV isolation from plasma has been shown to improve reproducibility and is endorsed by ISEV [65]. In addition to the removal of platelets, preprocessing steps to remove fibrinogen and thrombin are commonly used to enhance sample quality.

Saliva is an easily accessible biofluid being explored for ADRD biomarker studies [66]. To note, donor variables (e.g., sex, age, and medical conditions) are associated with differences in viscosity, pH, protein concentrations, and flow rate, which may impact EV populations [20]. Further, the composition depends on the relative contributions of major salivary glands, determined by the method of collection (i.e., depending on where the swab is done, the composition may change due to the location of those salivary glands, and caution must be taken to ensure consistent swab techniques across participants). Common co-isolates include eukaryotic cells, proteins, electrolytes, food debris, and bacterial cells and their EVs.

Urine collection, similar to saliva, is less invasive than CSF and blood and can be obtained in higher volumes. EVs from urine have recently been implicated as biomarkers in AD [67, 68]. Considerations when isolating EVs from urine include (a) normalization, which is challenged by varying concentrations due to differing levels of hydration and other factors, and (b) co-isolated urine proteins [20]. Tamm-Horsfall protein (THP) is one specific abundant protein in urine that co-isolates but also polymerizes into lattice-like networks that can trap EVs and reduce their yield [20]. THP can be depolymerized by reducing urine ionic strength or pH, or by treating with reducing agents [69].

2.2 Storage of Biofluids and EVs

Following collection and preprocessing of biofluid(s), storage is often required prior to isolation and analysis of EVs. The impact of various storage conditions on EV integrity has been extensively studied, and ISEV provides recommendations on storage-associated variables such as temperature, time in storage, and freeze-thaw cycles [20, 61]. Storage of samples at −80 °C, using protease inhibitors, and limiting freeze-thaw cycles are generally recommended to preserve the integrity and facilitate the complete recovery of EVs [70, 71]. The use of cryoprotectants such as albumin or trehalose may reduce effects of freeze-thaw; however, single freeze-thaw cycles (i.e., only thawed once, without cryoprotectants) do not appear to significantly affect the size or membrane integrity of EVs [72]. Promisingly, a recent study compared fresh versus frozen plasma (storage conditions not reported) and found no significant difference in protein content of EVs, indicating that bio-banked specimens can be used reproducibly across sites [73].

2.3 Isolation of EVs from Biofluids

A variety of techniques can be used to isolate EV populations, leveraging biophysical properties such as size and density (ultracentrifugation, ultrafiltration, size exclusion chromatography), membrane receptors (antibody capture), solubility (precipitation), or electrical properties (anion exchange chromatography). Different techniques may be more suited towards isolation of EVs from different types of biofluids, due to the varying composition and contaminants in each biofluid (as listed in Sect. 2.1).

As previously stated, no current isolation method exists to isolate a completely pure population of EVs, and preparations often contain protein aggregates, lipoproteins, and nucleic acids which co-isolate with EVs due to similarities in size/density and/or due to molecular interactions at the EV surface [22]. Depending on the method used for EV isolation, the isolated fraction may be enriched in different subsets of EVs (i.e., if the isolation method targets a specific size range, density, charge, or surface marker, it may be more enriched in exosomes, ectosomes, or apoptotic bodies). In addition, the choice of isolation method for extracting EVs can influence the EV protein corona and other surface-associated molecules and lead to false results due to association or dissociation of molecules from the surface during the isolation process [74].

Table 1 provides an overview of common EV isolation techniques used in biomarker analysis, highlighting their respective advantages and disadvantages.

Critical considerations in the context of biomarker studies and clinical trials are the volume of the sample and the availability of specialized equipment. Large sample volumes are frequently unavailable and while specialized equipment like ultracentrifuges may offer higher precision, they are less accessible in clinical settings compared to commercially available kits. Polymer-based precipitation (ExoQuick ULTRA) has been reported to isolate the highest quantity of EVs (measured by overall particles/ml, as well as binding to CD9 and CD63 beads) from low plasma volumes, as compared to other commercially available kits including membrane affinity (exoEasy Maxi Kit), size-exclusion chromatography (SEC) (qEVoriginal), iodixanol gradient (OptiPrep), and phosphatidylserine affinity (MagCapture) [79]. ExoQuick is easy to perform, fast, and of relatively low cost; however, its separation of EVs from proteins and non-vesicular extracellular particles such as lipoproteins is limited [79]. The more recent ExoQuick-LP kit improves EV separation by using a lipoprotein preclearing reagent [79]. Due to the lack of EV specificity from such kits, further processing steps are required. A combination of techniques is commonly used to concentrate and purify samples in biomarker research, with the majority of current studies using polymer-based precipitation for EV enrichment followed by immunocapture [80].

Immunocapture is required to specifically isolate and purify BEVs, as the EVs cannot be separated from other systemic-derived EVs by physical properties alone (e.g., size, density, charge). A recent review performed by the *ISTAART BBB Extracellular Vesicle working group* found that most studies evaluating BEVs as biomarkers in the ADRD field targeted neuronal EVs using anti-L1CAM. While L1CAM (CD171) is a transmembrane protein widely expressed in neurons, it is also expressed in other cell types (e.g., renal collecting duct cells and melanocytes) [38, 81]. Therefore, in addition to L1CAM, the specificity of alternative neuronal proteins (e.g., SYT1, SYP, NRXN2, and GRIA4) for enrichment/isolation of neuronal EVs from biofluids needs to be determined [80]. Targets for other brain cell types that have been utilized in the ADRD field include GLAST for astrocyte-derived EVs [82–85], TMEM119 and TMEM120 for microglia-derived EVs [84, 86], and MOG and PDGFRα for oligodendrocyte-derived EVs [84, 87].

Table 1 Advantages and disadvantages of EV isolation methods for biomarker analysis

Method; time for isolation	*Isolation basis;* affected by	Advantages	Disadvantages	Ref.
Differential ultracentrifugation; 140–600 mins	*Sedimentation coefficient (size, density)*; time, speed, rotor type, tube type, sample volume, and concentration	Low cost, high concentration, no additional chemicals	Equipment, complexity, low purity, low yield, damage and aggregation due to high forces	[75]
Density gradient ultra centrifugation; 250 mins to 2 days	*Density*; buffer composition, sample volume and concentration, preparation	High purity	Equipment, complexity, low yield, time, requires further concentration	[75]
Size exclusion chromatography; 1 mL/min + column washing	*Size*; pore size, flow rate, packing method, volume, column dimensions	Preserves EV integrity, high yield, reproducible	Low purity, high cost, sample dilution, limited to one sample per procedure	[75]
Polymer-based precipitation; 45–65 mins	*Solubility*; polymer concentration, pH, sample volume	Low cost, simple procedure, preserves EV integrity, appropriate for small sample volume, high yield	Contamination and retention of polymer, low purity, minimal specificity, high ion concentration	[75]
Immunocapture/immuno precipitation; 240 mins	*Affinity*; specificity of antibody used, sample concentration	High purity, high specificity, selects for specific EV populations (i.e., brain-derived)	Cost, low yield, difficult to detach EVs from bound material	[75]
Ultrafiltration/filter concentration; 130 mins	*Size*; filter cut-off, time, speed	Simple, high concentration	Low purity, deformation of EVs, low yield, filter plugging	[75]
Tangential flow filtration; 60 mins	*Size*; filter cut-off, flow rate, time	High yield, scalable, gentle, concentrates sample	Low purity	[76]
Anion exchange chromatography; 2–10 mL/min	*Zeta potential*; ion concentration, pH, flow rate	High purity, high efficiency, preserves EV integrity	Specialized equipment, availability	[77]
Asymmetric-flow field-flow fractionation (AF4); < 60 mins	*Diffusion coefficient, size;* flow rate, membrane, channel height, sample concentration	Gentle, fast, reproducible, scalable	Sample dilution, co-isolation of non-EV components	[78]

Higher purity workflows might consider an alternate method to polymer-based precipitation to increase specificity to EVs from co-isolating proteins and lipids. SEC separates components based on size, can effectively separate EVs from free soluble proteins, and can be optimized by altering the column dimensions, flow rate, and matrix material to increase EV yield and purity [88]. Several commercial SEC columns are available that can increase reproducibility, such as Sephacryl, Sepharose, Superdex, qEV, miniPURE-EV, and ExoSpin columns. SEC can further be combined with affinity-based methods by modifying the matrix of the column. For example, CaptoCore resins enable SEC and anion exchange (AEX) chromatography separation simultaneously and offer higher or similar EV recovery to Sepharose columns with higher purity by removing larger portions of free plasma proteins [89]. However, the dilutive nature of SEC necessitates a concentration step (e.g., filtration, ultracentrifugation, or polymer precipitation) before immunocapture.

Emerging microfluidic techniques such as the ExoDisc or the Total Exosome Isolation Chip [90, 91] may offer high clinical applicability due to their simplicity, small size, reproducibility, and limited sample volume needed. Microfluidic devices work off of the same principles for isolation as listed in Table 1, but on a smaller scale, and therefore are subject to the same limitations. As the field develops, we can expect further understanding of EV properties, allowing for increased EV purity and yield on smaller scale devices. Similarly, with further understanding of the pathogenesis of ADRDs, we will be able to identify more specific markers to isolate EV subtypes associated with the disease state.

EVs isolated from different biofluids have unique considerations, as these samples are associated with differences in concentration, viscosity, and different types of contaminants. Regardless of the isolation process chosen, it is important to verify the utility of the process through thorough characterization of the isolated EV fraction, as discussed next.

2.4 *Characterization and Analysis of EVs*

Characterization of EV populations is needed to understand their quality, quantity, and composition, and to validate their presence in isolated fractions. Throughout EV characterization, it is important to consider contributions of non-EV components which may confound analyses. Several methods exist to examine molecular components (proteins, RNAs, lipids, and metabolites) within or associated with EVs, which are briefly listed in the following sections. EV characterization is challenged by the lack of EV-specificity of available techniques and the fact that no single method exists that can satisfy all EV characterization requirements [20]. The use of orthogonal methods that utilize different measurement techniques is recommended [20].

2.4.1 EV Validation

To validate the presence of EVs in an isolated preparation, the following analyses are recommended by ISEV [20]:

1. Approximate the quantity of EVs by particle number, protein, lipid, and/or RNA content.
2. Conduct single EV analyses to determine size distribution and validate EV morphology.
3. Test for the presence of components that are associated with EVs and/or their subtypes.
4. Determine the degree to which co-isolated components are present.

Points 1 and 2 are addressed in the following Sects. 2.4.2 and 2.4.3. For points 3 and 4 listed above, ISEV recommends the "five-component framework" introduced in MISEV2018 and summarized in Table 2 [22]. In this framework, five categories of proteins are described to assess the presence of EVs (categories 1 and 2), the purity of EVs from common contaminants (category 3), EV subtypes and/or pathological/atypical states (category 4), and secreted proteins isolated with the EVs (category 5).

2.4.2 EV Quantification

When measuring differences in EV biomarkers between samples, samples can be compared on a basis of EV quantity (i.e., particle number), or protein, lipid, or RNA content. The majority of studies in the ADRD field measure EV quantity based on particle number. Each of these approximations hosts its own limitations and none can currently provide an accurate quantification of EV number [20]. However, quantification measures are important to provide a basis of comparison across studies and assays and to validate isolation techniques. It is also essential to report the quantity of EV source, such as the volume of biofluid used, to enable study reproducibility.

Particle number Methods such as nanoparticle tracking analysis (NTA), flow cytometry (FC), single-particle interferometric reflectance imaging sensing (SP-IRIS), and resistive pulse sensing (RPS) can provide an estimate of particle concentration, a metric widely used for standardizing assay inputs and outputs [20]. However, these techniques either lack specificity for EVs or sensitivity to detect all EV types and can therefore over- or underestimate total EV concentration. Because of this, EV concentrations from blood plasma have been reported to span six orders of magnitude depending on the method used for quantification [92]. To enhance accuracy in distinguishing EVs from other particles like protein aggregates and lipoproteins, the use of fluorescent dyes or antibody tags has aimed to mitigate the overestimation of EVs. Despite these efforts, no dye currently exists that can universally stain all EV types [20]. Given these challenges, it is advisable to use the term

Table 2 Protein-based EV characterization requirements as per MISEV2018 [22]

Category 1: proteins associated with lipid bilayer (at least 1)	Category 2: cargo proteins (at least 1)	Category 3: non-EV co-isolated structures (at least 1)	Category 4: specific to EV subtypes and/or pathological state (required if claiming analysis of EV subtype or state)	Category 5: secreted proteins recovered with EVs (required for documenting functional activity)
1a: Nontissue specific: e.g., CD63, CD81, CD47, HLA-A/B/C, LAMP1/2 1b: Tissue-specific: e.g., PECAM1 (endothelial), AChE-S (neurons)	2a: Lipid/membrane protein binding: e.g., TSG101, FLOT1/2, syntenin 2b: Promiscuous incorporation: e.g., HSP70, GAPDH, tubulin	3a: Lipoproteins, albumin, apolipoproteins 3b: Protein and protein/nucleic acid aggregates, THP, ribosomal proteins	4a: Nucleus: e.g., histones, Lamin A/C 4b: Mitochondria: e.g., IMMT, cytochrome C, TOMM20 4c: Secretory pathway: e.g., calnexin, Grp94, GM130 4d: Others	5a: Cytokines and growth factors: e.g., FGF, VEGF, interleukins 5b: Adhesion and ECM proteins: e.g., fibronectin, collagen

"particle number" rather than "EV number" and to carefully consider the limit of detection for each technique [20].

Protein content Overall protein content in EV samples is estimated through colorimetric or fluorometric assays, global protein stains on SDS-PAGE, or absorbance readings. The particle (as measured from the above aforementioned techniques) to protein ratio is often used as a measure of EV purity (higher ratio = higher purity), as protein quantity is often overestimated by the abundance of co-isolating proteins in EV preparations. Notably, protein quantity is heavily dependent on whether intact or disrupted EVs are measured, thus, any physical or chemical disruptions should be reported [20].

Lipid content Lipid quantification can be done through several techniques including colorimetric assays [93], chromatography [94], fluorescence of membrane intercalating dyes, or total reflection Fourier-transform infrared spectroscopy (FTIR), though the latter two may be inappropriate for the small quantities of EVs isolated for biomarker studies [20]. Total lipid quantity may be overestimated due to co-isolation of lipoproteins.

RNA content RNA is typically quantified through spectrometry-based methods. More sensitive techniques such as the Agilent Bioanalyzer pico chip or Quant-iT RiboGreen RNA Assay are recommended over less sensitive methods such as NanoDrop [95]. For small RNAs, specialized kits such as the Qubit microRNA Assay kit are recommended [20]. Isolation kits are often not specific to RNA over DNA and pretreatment with RNase-free DNase is required for more accurate RNA quantification [96]. Quantifying RNA in EVs is less common for overall EV quantification, but can serve as a measure of quality control or a normalization method in RNA profiling studies.

2.4.3 Single EV Analysis

ISEV [22] advocates for the integration of two complementary techniques when characterizing single EVs. This approach should pair a technology that can estimate biophysical properties of EVs such as size (i.e., dynamic light scattering, NTA, SP-IRIS, RPS, FC) and a high-resolution microscopy technique such as transmission electron microscopy (TEM), scanning electron microscopy (SEM), or atomic force microscopy (AFM) to verify size and morphology. NTA and TEM are the most commonly used methods for single EV analysis [97]. Size distribution measurements should be compared to those obtained from microscopic techniques to confirm their size, as EVs have a tendency to aggregate, and non-EV components are often present [22]. Of note, microscopy techniques such as TEM measure the core diameter of EVs, while techniques that rely on the mobility of particles such as

NTA measure the larger hydrodynamic diameter in solution. EVs can be identified in TEM images by their cup-shape morphology, which is a result of the drying process required to prepare the spherical EVs for TEM imaging [98].

2.4.4 Biomolecular Composition

The characterization of RNAs, proteins, and lipids in isolated EV fractions is critical for the validation and discovery of biomarkers in ADRDs. Recent advancements in RNA sequencing (RNAseq) and mass-spectrometry (MS)-based methods, particularly on samples enriched for BEVs, have revealed subsets of RNAs, proteins, and lipids that are differentially expressed in ADRDs, shedding light on potential biomarkers and pathways [80, 99–101].

RNAs are commonly recognized for their biomarker potential due to their capacity to elicit changes in recipient cell function via translation or alteration of gene expression. Indeed, several groups have found alterations in miR profiles in the blood of AD patients [102–106]. Specific RNAs can be isolated and quantified using quantitative PCR (qPCR) analysis, while more extensive RNAseq can be used to identify biological pathways affected in ADRDs. Protein-based biomarkers are attractive due to their relatively high stability and abundance in comparison to RNA-based biomarkers. The composition of targeted proteins in EVs can be measured using (a) singular detection methods such as enzyme linked immunosorbent assays (ELISA) or immunoblotting, (b) multiplex methods such as SIngle MOlecule Array (SIMOA) or Luminex, or (c) high-throughput mass spectrometry (MS)-based methods. MS can be applied for nontargeted proteomic approaches that can identify all detectable ions to give a holistic view and comprehensive understanding of protein composition, ideal for the discovery of novel protein biomarkers. Lipids and metabolites can also be identified and quantified by MS-based approaches. Although lipidomic studies are less prevalent than proteomic analyses, the variation in lipid species and their biochemical properties (e.g., head group charge, fatty acid tail length and saturation) have been noted to shift under ADRDs [107, 108].

3 EV Biomarkers in ADRDs

EVs, specifically BEVs circulating in accessible biofluids (e.g., blood), have emerged as a promising minimally invasive and widely accessible biomarker source for the ADRDs [38]. Increasing evidence indicates that changes in the pathophysiology of ADRDs, such as amyloidosis, tauopathy, synaptic dysfunction, insulin dysregulation, differential expression of lysosomal proteins, and transcription factor alterations, may manifest in the quantity and makeup of BEVs [38, 102, 109]. Studies have also identified alterations in EV-associated miRs and other noncoding RNAs in ADRD patients [38, 110]. In Sects. 3.1, 3.2, and 3.3 and Table 3, we summarize the literature on EVs as biomarkers of ADRDs, especially for EVs

Table 3 Validated ADRD EV-associated biomarkers

Biomarker	Biofluid	Source	Conditions	References
Amyloid				
Aβ42 ↑	Whole Blood	Neurons	AD, MCI	[111]
	Plasma		AD, DS, FTD, MCI	[83, 112–120]
	Serum		AD	[121]
	Plasma	Endothelial	MCI + SCeVD	[122]
	Urine	n/a	AD	[123]
Aβ42 ↓	Plasma	Astrocytes	AD	[83]
		Neurons	DS + AD	[114]
Aβ40 ↑	Plasma	Neurons	AD, MCI	[115]
	Plasma	Endothelial	MCI + SCeVD	[122]
Aβ42/40 ↑	Plasma	Neurons	AD, MCI	[112]
Amyloid precursor protein (APP)-pathway proteins				
BACE-1 ↑, s-APPα ↑ s-APPβ ↑, Septin-8 ↑	Plasma	Astrocytes	AD	[83]
Tau				
T-tau ↑	Whole Blood	Neurons	AD, MCI	[111]
	Plasma		AD, FTD, MCI, PD	[112, 113, 118, 124]
	Serum		AD	[125]
P-T181-tau ↑	Whole Blood	Neurons	AD, MCI	[111]
	Plasma		AD, DS, FTD, MCI, PD	[112–114, 116, 118–120, 126]
	Plasma	Endothelial	MCI + SCeVD	[122]
P-S396-tau ↑	Plasma	Neurons	AD, DS	[113, 114, 116, 118]
	Urine	n/a	AD	[123]
pTau217 EVs ↓	Plasma	Neurons	AD	[127]
p-tau (S202) ↑	Serum	Neurons	AD	[125]

Synaptic proteins				
Neurogranin ↓	Plasma	Neurons	AD, FTD, MCI	[45, 116–118, 128–130]
Neurogranin ↑	Plasma		AD	[120]
GAP43 ↓	Plasma		AD, MCI	[120, 128–130]
Synaptotagmin ↓	Plasma		AD, FTD, MCI	[117, 128–130]
Synaptopodin ↓, Synaptophysin ↓	Plasma		AD, MCI, FTD	[117, 129]
SNAP-25 ↓	Plasma		AD, MCI	[128, 130]
	Serum		AD	[131]
Synapsin 1 ↓	Plasma		AD	[120, 129]
Synapsin 1 ↑	Plasma		FTD	[129]
proBDNF ↓, GluR2 ↓, PSD95 ↓	Plasma		AD	[120]
Syntaxin-1 ↓	Plasma		AD	[120]
AMPA4 ↓, NPTX2 ↓, NLGN1 ↓, NRXN2 ↓	Plasma		AD	[82]
α-Synuclein ↑	CSF	n/a	PD, MSA	[132]
	Plasma	Neurons	PD, MSA	[87, 133–136]
		Oligodendrocytes	PD, MSA	[87]
		Astrocytes	PD	[85]
	Serum	Neurons	PD, DLB	[10, 87, 136–139]
		Oligodendrocytes	PD, MSA	[87]
		n/a	PD	[140]
α-Synuclein ↓	Plasma	Neurons	PD	[126]
	Serum			[141]
α-Synuclein EVs ↑	Plasma	n/a	PD	[142]

(continued)

Table 3 (continued)

Biomarker	Biofluid	Source	Conditions	References
Proteins associated with insulin dysregulation				
IRS-1 ↓, p-panTyr-IRS-1 ↓	Plasma	Neurons	AD	[143]
pSer312-IRS-1 ↑			AD, FTD, PD	[143, 144]
NDUFS3 ↓, SDHB ↓			AD, MCI	[145]
pS616-IRS-1 ↑, pY612-IRS-1 ↑, Erk ↑			PD	[144]
pY-IRS-1 ↓			PD	[126]
Lysosomal proteins and transcription factors				
cathepsin B ↑	CSF	n/a	AD	[146]
cathepsin D ↑	Plasma	Neurons	AD, FTD	[147]
		n/a	FTLD	[148]
LAMP1 ↑, ubiquitinated proteins ↑		Neurons	AD	[147]
HSP70 ↓			AD, FTD	[147]
LAMP-2 ↑			AD, FTD	[45]
Let-7e ↑			AD	[149]
LRP6 ↓, HSF1 ↓			AD	[150]
REST ↓			AD	[116, 118, 150]
REST ↑			FTD	[150]
C1q ↑, C4b ↑, C3d ↑, factor B ↑, factor D ↑, Bb ↑, C3b ↑, and C5b-C9 terminal complement complex ↑, CD59 ↓, CD46 ↓, DAF ↓, CR1 ↓		Astrocytes	AD	[151]
C1QC ↑, CO9 ↑, CFAH ↑, KVD30 ↑, GP1BB ↑, RSU1 ↑, A0A0G2JRQ6 ↑, A2MG ↓, ADA10 ↓, A1AG2 ↓, IGHA1 ↓, HV428 ↓		n/a	AD	[152]
C1q ↑, C4 ↑, and C3 ↑			FTLD	[148]

Neurotrophic factors				
HGF ↓, FGF-2 ↓, FGF-13 ↓, IGF1 ↓	Plasma	CSPG4E	AD	[153]
GDNF ↓		Astrocytes	AD	[83]
Proteins associated with neuronal function (e.g., *neuronal cytoskeleton, neurotransmitter release)*				
TDP-43 ↑	Plasma	Neurons	AD	[154]
GFAP ↑, Glusyn ↓, NF-Lch ↑, NS-enolase ↑				[83]
Nfl ↑			AD, MCI, FTD	[45, 112]
GFAP ↓		Astrocytes	AD	[83]
STX-1A ↓, VAMP-2 ↓	Serum		PD	[10]
Immune response and inflammation				
CCL2, CXCL12 ↑	Plasma	Neurons	PD	[135]
IL-6 ↑, TNF-α ↑, IL-1β ↑		Astrocytes	AD	[151]
Cellular structure and trafficking				
HSPA1A ↑, NPEPPS ↑, PTGFRN ↑	CSF	n/a	AD	[155]
MFN-2 ↓, golgin A4 ↑	Plasma	Neurons	AD, FTD	[45]

Abbreviations: *AD* Alzheimer disease, *DLB* Dementia with Lewy bodies, *DS* Down syndrome, *FTD* Frontotemporal dementia, *FTLD* frontotemporal lobar degeneration, *MCI* mild cognitive impairment, *MSA* multiple system atrophy, *SCD* subjective cognitive decline, *SCeVD* small vessel cerebro vascular disease, *PD* Parkinson's disease circulating in biofluids such as CSF, blood, saliva, and urine. BEVs are known to originate from various components of the brain such as neurons, glia, and cerebral blood vessels, to name a few [38].

3.1 CSF

Despite the widespread use of CSF for the detection of established ADRD biomarkers (e.g., amyloid, tau), the study of EVs or BEVs in CSF has been limited. This is in part due to the invasive nature of CSF collection, limited volumes that can be attained, and the lower EV concentrations in CSF as compared to blood, as discussed earlier [62, 156, 157]. Regardless, a handful of CSF studies have attempted to evaluate EVs, and very few BEVs (often identified via indirect means using protein profiling), as biomarkers of ADRDs. In total CSF EVs, Yuyama et al. demonstrated an increase in cathepsin B levels during the transition from amyloid-negative to amyloid-positive status [146]. Using label-free proteomics on EVs isolated from CSF, another study found that 16.1% of detected proteins were enriched in neurons and 83.8% in glia [155]. Moreover, of the detected proteins, the authors found that levels of heat shock 70 kDa protein 1A (HSPA1A), puromycin-sensitive aminopeptidase (NPEPPS), and prostaglandin F2 receptor negative regulator (PTGFRN) were significantly elevated in EVs from individuals with AD dementia compared to those with mild cognitive impairment (MCI). The study further reported that the combination of HSPA1A and NPEPPS in CSF EVs could serve in monitoring the conversion of MCI to AD dementia [155]. In general, a greater number of studies assessed miR levels in CSF EVs than proteins. For example, in AD, miR-125b-5p levels were observed to be increased in EVs from CSF, while miR-451a, miR-193b, and miR-605-5p levels were decreased [106, 158]. In PD, EVs from CSF contain elevated levels of α-synuclein, miR-153, miR-409-3p, miR-10a-5p, and let-7 g-3p and reduced levels of miR-1 and miR-19b-3p [132, 159].

3.2 Blood

3.2.1 Neuron-Derived EVs

Amyloid and Tau

Levels of several Aβ- and tau-species, detected using positron emission tomography (PET) imaging in the brain and biochemical assays in CSF and blood, are considered core biomarkers of AD [156, 157]. Specifically, amyloid PET demonstrates an increased Aβ load in the AD brain, while CSF- and blood plasma-assays demonstrate decreases in Aβ42. On the other hand, tau PET demonstrates increased brain tau load, with CSF- and plasma-assays also demonstrating increases in phosphorylated (p)-tau and total (t)-tau levels [156, 157]. Aβ- and tau-species can also be detected in circulating EVs in biofluids, with most ADRD studies investigating L1CAM-isolated neuron-derived EVs in blood (whole blood, plasma, or serum) [38]. We provide a summary of these studies next.

Aβ and other amyloid precursor protein(APP)-pathway proteins in EVs To date, ADRD studies investigating Aβ42, Aβ40, or the Aβ42/Aβ40 ratio in neuron-derived plasma EVs report increased levels in AD continuum individuals (i.e., individuals with MCI or AD dementia) compared to healthy controls (HC) (Table 3) [83, 111–113, 116–119]. Interestingly, a decreased Aβ42 level was found in neuron-derived EVs from individuals with Down syndrome who had dementia compared to those without dementia [114]. With regard to other proteins in the APP-pathway, one study investigating soluble amyloid precursor protein α and β (sAPPβ, sAPPα), β-site amyloid precursor protein-cleaving enzyme 1 (BACE-1), and γ-secretase in neuron-derived plasma EVs reported elevated levels of soluble amyloid precursor protein α and β in AD continuum individuals compared to HC [83]. Similar to the findings in AD, Aβ42 was reported to be increased in one study investigating neuron-derived plasma EVs in frontotemporal dementia (FTD) [113].

Tau in EVs It is widely accepted that level alterations in phosphorylated tau species could reflect the stage of disease in AD or offer a differential perspective on the neurodegenerative process [127]. Multiple studies in AD continuum individuals have reported increased levels of t-tau and p-tau (P-T181, P-S396, p-tau S202) in neuron-derived EVs from whole blood, plasma, and serum [83, 111–114, 116–120, 124–126]. On the other hand, neuron-derived plasma EVs isolated using novel neuronal markers, gamma-aminobutyric acid type A receptor subunit delta or G protein-coupled receptor 162 (GABRD or GPR162), demonstrated decreased levels of pTau217 in AD individuals compared to HC [127]. Roche and Eli Lilly's new blood test for pTau217 recently received the Breakthrough Device designation by the FDA [160]. Similar to AD, p-tau isoforms, such as P-T181, were reported to be increased in neuron-derived plasma EVs in FTD [113] and PD [126].

Synaptic Proteins

Early synaptic dysfunction is a hallmark of AD [161, 162], prompting investigation into synaptic proteins as potential biomarkers. Several studies examined synaptic protein cargo in plasma- or serum-isolated neuron-derived EVs, all utilizing L1CAM for enrichment. Decreased levels of growth-associated protein 43, synaptotagmin, synaptopodin, synaptophysin, synapsin, as well as synaptosomal-associated protein-25 (SNAP-25), precursor brain-derived neurotrophic factor (proBDNF), glutamate receptor 2 (GluR2), postsynaptic density protein (PSD95), and syntaxin-1 have been observed in the AD continuum (Table 3) [45, 82, 116–118, 120, 128–131]. For neurogranin, while most AD continuum studies reported a decrease [45, 116–118, 128–130], one study reported an increase in early AD individuals [120]. Moreover, in neuron-derived plasma EVs from AD individuals, both pre-synaptic (neuronal pentraxin 2, NPTX2; neurexin 2α, NRXN2) and post-synaptic (GluA4-containing glutamate receptor; neuroligin 1, NLGN1) protein interactions were reduced in a longitudinal cohort, suggesting their promise as biomarkers for AD progression [82]. The main PD biomarker, α-synuclein, is reported to be increased in neuron-derived plasma and serum EVs [10, 87, 133–139, 163]. Two studies reported a decrease of α-synuclein in both plasma and serum neuron-derived EVs in PD [126, 141]. Interestingly, one study identified an increase in plasma α-synuclein EVs [142]. In FTD, similar to the findings in AD, synaptotagmin, synaptopodin, synaptophysin, and neurogranin were reported to be decreased in neuron-derived plasma EVs [45, 129]. Interestingly, the synaptic protein synapsin exhibited an increase in FTD and a decrease in AD, suggesting a potential distinction in the synaptic alterations between these two conditions [129].

Insulin Dysregulation

Accumulating evidence suggests that disrupted brain insulin signaling is characteristic of AD, with insulin resistance proposed as a critical link between Aβ and tau pathologies [164]. Given that the phosphorylation patterns of insulin-receptor-substrate-1 (IRS-1) play a pivotal role in both normal insulin signaling and insulin resistance [164, 165], it was investigated in neuron-derived plasma EVs. In neuron-derived plasma EVs of individuals with AD, (a) increased levels of phosphorylated IRS-1 at serine residue 312 (pSer312-IRS-1) [143] and (b) decreased levels of p-pan-tyrosine-IRS-1 (p-panTyr-IRS-1) [143] – both indicative of ineffective insulin signaling, were observed (Table 3). It was further demonstrated that levels of pSer312-IRS-1 and p-panTyr-IRS-1 in EVs from individuals with AD correlated with gray matter volume – pSer312-IRS-1 was negatively associated, while p-panTyr-IRS-1 was positively associated, with brain volume [165]. Additionally, neuron-derived EVs with IRS-1 may potentially come from brain regions with high IRS-1 expression. Thus, the phosphorylation pattern seen in these EVs may indicate its status in brain regions affected by early AD and insulin resistance [166]. Similar to the findings in AD, increased pSer312-IRS-1 was also observed in the neuron-derived plasma EVs of individuals with FTD and PD [143, 144] with lower levels of IRS-1 phosphorylated tyrosine (pY-IRS-1), a facilitator of forward insulin signal propagation, also detected in PD [126].

Moreover, evidence increasingly links mitochondrial dysfunction with diabetes in AD progression [145]. Mitochondrial proteins, NADH ubiquinone oxidoreductase core subunit S3 (NDUFS3) and succinate dehydrogenase complex subunit B (SDHB), in neuronal-derived plasma EVs mirror brain neuronal levels. These levels were notably lower in the AD continuum compared to HC. Additionally, their lower levels correlated with increased brain and hippocampal atrophy over time [145].

Differential Expression of Lysosomal Proteins

Lysosomes are cell organelles that break down waste materials, with lysosomal proteins serving as enzymes and membrane components in this process [167]. A cross-sectional study investigated lysosomal protein levels in neuron-derived EVs in plasma and found that increases in these proteins can differentiate individuals with AD from HC individuals [147]. This study reported significant increases in cathepsin D, type-1 lysosome-associated membrane protein (LAMP1), and ubiquitinated proteins in AD, while levels of heat-shock protein-70 (HSP70) were reduced. Longitudinal investigation indicated that these protein levels differed from those of HC individuals up to 10 years before clinical onset [147]. In FTD, neuron-derived plasma EV levels of the three lysosomal proteins elevated in AD were notably lower [147]. While further studies are necessary, current evidence suggests that lysosomal protein levels in neuron-derived plasma EVs hold potential for distinguishing AD from other age-related dementias.

Transcription Factor Alterations

Neuron-derived plasma EVs contain transcription factors that promote cellular resilience against stressors like the accumulation of Aβ and tau [116]. Several studies have analyzed levels of repressor element 1-silencing transcription factor (REST), a crucial stress response regulator in the aging brain, in neuron-derived EVs isolated from plasma. Lower levels of REST have been observed in AD [116, 118, 150], including those diagnosed with mild to moderate stages of the disease [118]. Meanwhile, higher levels of REST have been associated with FTD [150]. Changes in REST levels provide a promising basis for biomarkers that could distinguish AD and FTD.

3.2.2 Neuronal Precursor Cell-Derived EVs

Chondroitin sulfate proteoglycan 4 (CSPG4) cells, a subtype of neural precursor cells, are widely distributed throughout the brain and exhibit remarkable developmental plasticity. These cells release essential neurotrophic factors that support neuronal growth, repair, and survival [153, 168]. Protein levels of four key neurotrophic factors (hepatocyte growth factor, HGF; fibroblast growth factors 2 and 13, FGF-2 and FGF-13; and type 1 insulin-like growth factor, IGF1) were found to be significantly elevated in CSPG4-derived compared to neuron- and astrocyte-derived EVs in plasma. Interestingly, reductions in all four neurotrophic factor levels within CSPG4-derived EVs were observed during the preclinical stage of AD, with no further significant decreases observed in later stages [153].

3.2.3 Glia-Derived EVs

Astrocyte-derived EVs extracted from plasma exhibit similar size characteristics to those derived from neurons, but are notably less abundant [83]. Like their neuronal counterparts, astrocyte-derived EVs have been found to carry hallmark AD proteins, namely tau (P-T181-tau and P-S396-tau) and Aβ42 (soluble), along with proteins associated with the APP-pathway such as sAPPβ, sAPPα, BACE-1, and γ-secretase. In both AD and HC individuals, astrocyte-derived EVs exhibit significantly higher levels of these proteins compared to neuron-derived EVs [83]. In astrocyte-derived EVs from AD individuals, levels of complement or immune system proteins that enhance pathogen clearance and cell destruction are increased, while glial-derived neurotrophic factor (GDNF) and complement regulatory proteins (CD59, CD46, DAF, and CR1) are decreased suggesting a potential imbalance in complement system activity that may exacerbate AD pathology [151]. Additionally, a study investigating frontotemporal lobar degeneration also reported an increase in plasma levels of complement proteins. Specifically, the complement proteins C1q, C4, and C3 were found to be increased in astrocyte-derived plasma EVs compared to HC, though the EVs' cellular origin was not determined [148]. This dysregulation highlights complement proteins as biomarkers and suggests a shared mechanism in the immune-related aspects of neurodegenerative diseases, emphasizing the complement system's important role in neuroinflammation [169].

3.2.4 Cerebrovascular-Derived EVs

Cerebrovascular dysfunction is an early occurrence in AD [170–172]. Endothelial EVs in plasma, containing surface markers- and cargo-proteins enriched in brain vascular endothelial cells, were recently identified [173]. In cognitively normal adults, as well as in preclinical and prodromal AD with white matter hyperintensities (WMHs), an indicator of vascular injury in the brain, the isolation of cerebrovascular-derived EVs from blood using sequential immunoprecipitations and vascular endothelial-specific antibodies (CD3 and CD146) reported significantly higher concentrations of four cargo proteins: GLUT1, large neutral amino acid transporter (LAT1), P-glycoprotein 1 (P-GP), and nitric oxide synthase trafficker compared to those without WMHs [122, 174]. Elevated levels of these proteins were linked to lower cognitive function, regardless of WMH burden. Additionally, LAT1 and P-GP levels were inversely associated with magnetic resonance imaging (MRI)-detected global gray matter volumes, while concentrations of GLUT1, LAT-1, and P-GP showed significant associations with systemic interleukin (IL)-6 levels [174].

3.2.5 AD–Associated miRs Cargo EVs

miRs are small RNAs that regulate gene expression by binding to messenger RNAs [175]. Among the EV-associated miRs, many are predicted to bind to one or more mRNAs of genes associated with AD, including *APP*, *PSEN1*, *BACE1*, and other top-ranking genes in the AlzGene database (http://www.alzgene.org/) (e.g., *APOE, BIN1, CLU, ABCA7, CR1, PICALM, MS4A6A, CD33, MS4A4E, CD2AP*). However, in the AD continuum [84, 86, 102, 103, 105, 106, 112, 158, 177] and/or in PD [176], only a few miRs were consistently reported by at least two studies in CSF, plasma and/or serum, namely: decrease in miR-193b [102, 103, 106, 158] and miR-342-3p [102, 105] levels, increase in miR-106b-5p 1 [84, 86, 105], miR-223-3p [86, 176, 177], and miR-384 levels [103, 112] (Table 4). The reduction in miR-193b expression was more pronounced in AD dementia compared to MCI. Given that miR-193b is known to target *APP* and *PSEN1* mRNAs, its reduced expression may exacerbate amyloidosis [106]. miR-342-3p, abundantly expressed in the human brain [178], is predicted to modulate tau phosphorylation and aggregation by binding to the 3'-UTR of protein phosphatase 2 (PP2A) and phosphatase and tensin homolog (PTEN) [179]. In serum and in plasma, miR-223-3p is increased in AD dementia [86, 176], while miR-384 and several others are increased in the AD continuum [103, 112]. Yang et al. (2018) found that combining three miRs (193b and 384 more consistently reported, and 135a) was more effective in diagnosing early AD than using them individually [103]. miR-384 distinguished AD from vascular dementia (VD) and PD dementias, while miR-223 levels varied between AD and VD [104]. miR-223, linked to inflammation in dementia, correlated with clinical parameters such as the Mini-Mental State Examination (MMSE) and Clinical Dementia Rating® scores [104]. These miRNAs, therefore, show promise as biomarkers for ADRDs.

3.3 Emerging Biofluids: Saliva and Urine

In general, saliva is much easier to access than blood, and thus of interest as a biomarker source. In the context of ADRDs, the investigation of salivary biomarkers is emerging due to the relationship between declines in salivary gland secretion and AD progression [191]. Given the association between the deterioration of salivary gland function and AD, it is possible that AD-specific proteins and RNA are expressed in saliva [192]. For instance, salivary epithelial cells express amyloid precursor protein (APP) and Aβ, potentially mirroring changes seen in the CSF [193]. In the context of EVs, one study reported significantly higher levels of neuronal-derived Aβ42 and p-tau in saliva in individuals with AD compared to HC [66]. Another study observed that miRNA-485-3p concentration in salivary EVs is notably elevated in individuals with AD compared to HC, particularly in Aβ-PET-positive participants compared to Aβ-PET-negative participants [190].

EVs from urine offer another easily accessible biomarker source [67], and the proteome of urinary EVs has been recently investigated in the ADRDs [68, 123].

Table 4 Validated EV-associated miRs in ADRDs

Biomarker	Biofluid	Source	Conditions	References
miR-6798-3p ↑	CSF	Adipocytes		[180]
miR-125b-5p ↑				[158]
miR-451a ↓, miR-193b ↓, miR-605-5p ↓		n/a	AD	[158]
miR-193b ↓				[106]
miR-153 ↑, miR-409-3p ↑, miR-10a-5p ↑, let-7 g-3p ↑			PD	[159]
miR-1 ↓, miR-19b-3p ↓				[159]
miR-212 ↓, miR-132 ↓	Plasma	Neurons	AD	[181]
miR-23a-3p ↑, miR-223-3p ↑, miR-190a-5p ↑, miR-100-3p ↑				[177]
miRNA-122 ↓, miRNA-3591 ↓				[182]
miR-106b-5p ↑, miR-9-5p ↑, miR-29a-5p ↓, miR-132-5p ↑				[84]
miR-210-3p ↑			MCI	[84]
miRNA-181c ↓			FTD	[182]
miR-384 ↑			AD, MCI, SCD	[112]
miR-4639-5p ↑			PD	[183]
miR-29a-5p, miR-125b-5p, miR-132-5p ↑		Astrocytes	AD	[84]
miR-107 ↑		Astrocytes	AD, MCI	[84]
miR-132-5p ↑, miR-106b-5p ↑, miR-29a-5p ↓		Microglia	AD, MCI	[84]
miR-29a-5p ↑, miR-125b-5p ↑		Oligodendrocytes	AD	[84]
miR-9-5p ↓			AD, MCI	[84]
miR-23b-3p ↓, miR-24-3p ↓, miR-29b-3p ↓, miR-125b-5p ↓, miR-138-5p ↑, miR-548at-5p ↑, miR-659-5p ↑, miR-5001-3p ↑, miR-139-5p ↓, miR-141-3p ↓, miR-150-5p ↓, miR-152-3p ↓, miR-185-5p ↓, miR-338-3p ↓, miR-342-3p ↓, miR-342-5p ↓, miR-3065-5p ↓, miR-3613-3p ↓, miR-3916 ↓, miR-4772-3p ↓, miR-193b ↓		n/a	AD	[102]
miR-106a-5p ↑, miR-16-5p ↑, miR-223-3p ↑, miR-25-3p ↑, miR-30b-5p ↑, miR-92a-3p ↑, miR-451a ↑				[86]
miR-483-5p ↑, miR-502-5p ↑, miR-132-3p ↑			MCI	[184]
hsa-miR-30c-2-3p ↑, hsa-miR-15b-5p ↓, hsa-miR-138-5p ↓, hsa-miR-338-3p ↓, hsa-miR-106b-3p ↓, hsa-miR-431-5p ↓			PD	[185]

Biomarker	Biofluid	Source	Conditions	References
miR-204-5p ↓, miR-373-3p ↓	Serum	Neurons	AD	[186]
miR-6760-3p ↑		Adipocytes		[180]
miR-361-5p ↑, miR-30e-5p ↑, miR-93-5p ↑, miR-15a-5p ↑, miR-143-3p ↑, miR-335-5p ↑, miR-106b-5p ↑, miR-101-3p ↑, miR-424-5p ↑, miR-106a-5p ↑, miR-18b-5p ↑, miR-3065-5p ↑, miR-20a-5p ↑, miR-582-5p ↑, miR-1306-5p ↓, miR-342-3p ↓, miR-15b-3p ↓		n/a	AD	[105]
miR-125b ↑, miR-15a ↑, and miR-361 ↑				[187]
miR-223 ↓			AD, VD	[104]
miR-135a ↑, miR-384 ↑, miR-193b ↓			AD, MCI	[103]
miR-125b ↑, miR-135a ↑, miR-15a ↑, and miR-20a ↑				[188]
miR-501-3p ↑			MCI	[189]
miR-223-3p ↑			PD	[176]
miR-485-3p ↑	Saliva	n/a	AD	[190]

Although urinary EV biomarkers are still emerging, a study reported increased levels of Aβ42 and P-S396-tau in the urinary exosomes of individuals with AD compared to matched HC [123]. It is important to note that these exosomes were not enriched, so their brain specificity could not be determined [123]. Bioinformatics analyses from Wang et al. (2019) unveiled a notable enrichment of endolysosomal proteins associated with AD, PD, and Huntington's diseases. Tissue and biofluid enrichment assessments were performed in this study to check if it is conceivable that urinary EVs derive from organs across the body, including the brain. Notably, among proteins linked to neurological disorders, SNAP23 and calbindin displayed the most significant elevation in PD cases. These proteins enabled successful prediction of PD diagnosis, achieving 86% accuracy in the discovery cohort and 76% accuracy in the replication cohort [68].

4 Future Directions and Opportunities for EV Biomarkers in ADRDs

EVs represent promise as biomarkers for ADRDs, as demonstrated by the increasing number of studies showing correlations to disease pathology. Given that EVs carry markers of their host cells, they are of particular importance as it is possible to isolate EVs from accessible biofluids-specific subsets that originate from the brain, and therefore, EVs have potential to demonstrate high specificity for neurodegenerative conditions [9]. This not only enables researchers to find specific biomarkers that can identify individuals that have or are at risk for ADRDs, but can also help pinpoint targets for therapeutic intervention. As much as EVs hold excitement in the biomarker field due to their ability to cross the BBB and maintain stability in circulation long enough to isolate them from peripheral biofluids, these abilities also offer them high potential as therapeutics, which can travel to the BBB and elicit positive responses [130].

National and international recommendations for the diagnosis of ADRD currently recognize CSF biomarkers, including low CSF Aβ42 and elevated CSF tau (total tau, p-tau) [194], and revised guidelines are taking into account plasma-based biomarkers [157]. As the EV field is still relatively in its infancy, and EV analysis requires more thorough isolation prior to biochemical analysis, few large-scale studies have evaluated EVs as biomarkers compared to a more general protein or RNA fraction. However, the increased specificity that EVs may offer enables them to be isolated from more peripheral and less invasive biofluids such as urine and saliva. Increased understanding of EVs, and the development of more robust methods for isolating and characterizing EVs, will increase their feasibility and applicability in the clinic.

Acknowledgments We would like to thank Flavie E. Detcheverry for reading and providing feedback on the book chapter.

References

1. Kumar MA, Baba SK, Sadida HQ, Marzooqi SA, Jerobin J, Altemani FH, et al. Extracellular vesicles as tools and targets in therapy for diseases. Signal Transduct Target Ther. 2024;9(1):27. https://doi.org/10.1038/s41392-024-01735-1.
2. Yáñez-Mó M, Siljander PRM, Andreu Z, Zavec AB, Borràs FE, Buzas EI, et al. Biological properties of extracellular vesicles and their physiological functions. J Extracell Vesicles. 2015;4:27066. https://doi.org/10.3402/jev.v4.27066.
3. Buzas EI. The roles of extracellular vesicles in the immune system. Nat Rev Immunol. 2023;23(4):236–50. https://doi.org/10.1038/s41577-022-00763-8.
4. Salomon C, Das S, Erdbrügger U, Kalluri R, Kiang Lim S, Olefsky JM, et al. Extracellular vesicles and their emerging roles as cellular messengers in endocrinology: an Endocrine Society scientific statement. Endocr Rev. 2022;43(3):441–68. https://doi.org/10.1210/endrev/bnac009.
5. Ma Z, Wang Y, Li H. Applications of extracellular vesicles in tissue regeneration. Biomicrofluidics. 2020;14(1):011501. https://doi.org/10.1063/1.5127077.
6. Berumen Sánchez G, Bunn KE, Pua HH, Rafat M. Extracellular vesicles: mediators of intercellular communication in tissue injury and disease. Cell Commun Signal. 2021;19(1):104. https://doi.org/10.1186/s12964-021-00787-y.
7. Kalluri R, LeBleu VS. The biology, function, and biomedical applications of exosomes. Science. 2020;367(6478) https://doi.org/10.1126/science.aau6977.
8. Coughlan C, Lindenberger J, Jacot JG, Johnson NR, Anton P, Bevers S, et al. Specific binding of Alzheimer's Aβ peptides to Extracellular vesicles. Int J Mol Sci. 2024 [cited 2024 Apr 3];25(7):3703. Available from: https://www.mdpi.com/1422-0067/25/7/3703
9. Gomes P, Tzouanou F, Skolariki K, Vamvaka-Iakovou A, Noguera-Ortiz C, Tsirtsaki K, et al. Extracellular vesicles and Alzheimer's disease in the novel era of Precision Medicine: implications for disease progression, diagnosis and treatment. Exp Neurol. 2022;358:114183. https://doi.org/10.1016/j.expneurol.2022.114183.
10. Agliardi C, Guerini FR, Meloni M, Clerici M. Alpha-synuclein as a biomarker in Parkinson's disease: focus on neural derived extracelluar vesicles. Neural Regeneration Res. 2022;17(7):1503–4. https://doi.org/10.4103/1673-5374.330604.
11. Alberro A, Iparraguirre L, Fernandes A, Otaegui D. Extracellular vesicles in blood: sources, effects, and applications. Int J Mol Sci. 2021;22(15) https://doi.org/10.3390/ijms22158163.
12. Kalra H, Drummen GPC, Mathivanan S. Focus on Extracellular vesicles: introducing the next small big thing. Int J Mol Sci. 2016;17(2):170. https://doi.org/10.3390/ijms17020170.
13. van Niel G, D'Angelo G, Raposo G. Shedding light on the cell biology of extracellular vesicles. Nat Rev Mol Cell Biol. 2018;19(4):213–28. https://doi.org/10.1038/nrm.2017.125.
14. Colombo M, Moita C, van Niel G, Kowal J, Vigneron J, Benaroch P, et al. Analysis of ESCRT functions in exosome biogenesis, composition and secretion highlights the heterogeneity of extracellular vesicles. J Cell Sci. 2013;126(Pt 24):5553–65. https://doi.org/10.1242/jcs.128868.
15. Meldolesi J. Exosomes and Ectosomes in intercellular communication. Curr Biol. 2018;28(8):R435–44. https://doi.org/10.1016/j.cub.2018.01.059.
16. Colombo M, Raposo G, Théry C. Biogenesis, secretion, and intercellular interactions of exosomes and other extracellular vesicles. Annu Rev Cell Dev Biol. 2014;30:255–89. https://doi.org/10.1146/annurev-cellbio-101512-122326.
17. Jeppesen DK, Zhang Q, Franklin JL, Coffey RJ. Extracellular vesicles and nanoparticles: emerging complexities. Trends Cell Biol. 2023;33(8):667–81. https://doi.org/10.1016/j.tcb.2023.01.002.
18. Yu L, Zhu G, Zhang Z, Yu Y, Zeng L, Xu Z, et al. Apoptotic bodies: bioactive treasure left behind by the dying cells with robust diagnostic and therapeutic application potentials. J Nanobiotechnol. 2023;21(1):218. https://doi.org/10.1186/s12951-023-01969-1.

19. Elmore S. Apoptosis: a review of programmed cell death. Toxicol Pathol. 2007;35(4):495–516. https://doi.org/10.1080/01926230701320337.
20. Welsh JA, Goberdhan DCI, O'Driscoll L, Buzas EI, Blenkiron C, Bussolati B, et al. Minimal information for studies of extracellular vesicles (MISEV2023): from basic to advanced approaches. J Extracell Vesicles. 2024;13(2):e12404. https://doi.org/10.1002/jev2.12404.
21. Di Bella MA. Overview and update on Extracellular vesicles: considerations on exosomes and their application in modern medicine. Biology. 2022;11(6) https://doi.org/10.3390/biology11060804.
22. Théry C, Witwer KW, Aikawa E, Alcaraz MJ, Anderson JD, Andriantsitohaina R, et al. Minimal information for studies of extracellular vesicles 2018 (MISEV2018): a position statement of the International Society for Extracellular Vesicles and update of the MISEV2014 guidelines. J Extracell Vesicles. 2018;7(1):1535750. https://doi.org/10.1080/20013078.2018.1535750.
23. Cano A, Ettcheto M, Bernuz M, Puerta R, Esteban de Antonio E, Sánchez-López E, et al. Extracellular vesicles, the emerging mirrors of brain physiopathology. Int J Biol Sci. 2023;19(3):721–43. https://doi.org/10.7150/ijbs.79063.
24. Palviainen M, Saraswat M, Varga Z, Kitka D, Neuvonen M, Puhka M, et al. Extracellular vesicles from human plasma and serum are carriers of extravesicular cargo—implications for biomarker discovery. PLoS One. 2020 [cited 2024 Apr 30];15(8):e0236439. Available from: https://journals.plos.org/plosone/article/file?id=10.1371/journal.pone.0236439&type=printable
25. Anand S, Samuel M, Kumar S, Mathivanan S. Ticket to a bubble ride: cargo sorting into exosomes and extracellular vesicles. Biochim Biophys Acta, Proteins Proteomics. 2019;1867(12):140203. https://doi.org/10.1016/j.bbapap.2019.02.005.
26. Hallal S, Tűzesi Á, Grau GE, Buckland ME, Alexander KL. Understanding the extracellular vesicle surface for clinical molecular biology. J Extracell Vesicles. 2022;11(10):e12260. https://doi.org/10.1002/jev2.12260.
27. Brzozowski JS, Jankowski H, Bond DR, McCague SB, Munro BR, Predebon MJ, et al. Lipidomic profiling of extracellular vesicles derived from prostate and prostate cancer cell lines. Lipids Health Dis. 2018;17(1):211. https://doi.org/10.1186/s12944-018-0854-x.
28. Wang G, Dinkins M, He Q, Zhu G, Poirier C, Campbell A, et al. Astrocytes secrete exosomes enriched with proapoptotic ceramide and prostate apoptosis response 4 (PAR-4): potential mechanism of apoptosis induction in Alzheimer disease (AD). J Biol Chem. 2012;287(25):21384–95. Available from: https://www.jbc.org/article/S0021-9258(20)49830-8/abstract
29. Ghadami S, Dellinger K. The lipid composition of extracellular vesicles: applications in diagnostics and therapeutic delivery. Front Mol Biosci. 2023;10:1198044. https://doi.org/10.3389/fmolb.2023.1198044.
30. Xu R, Greening DW, Chen M, Rai A, Ji H, Takahashi N, et al. Surfaceome of exosomes secreted from the colorectal cancer cell line SW480: peripheral and integral membrane proteins analyzed by proteolysis and TX114. Proteomics. 2019;19(8):e1700453. https://doi.org/10.1002/pmic.201700453.
31. Cvjetkovic A, Jang SC, Konečná B, Höög JL, Sihlbom C, Lässer C, et al. Detailed analysis of protein topology of extracellular vesicles–evidence of unconventional membrane protein orientation. Sci Rep. 2016 [cited 2024 May 20];6(1):1–12. Available from: https://www.nature.com/articles/srep36338
32. Nyalwidhe JO, Betesh LR, Powers TW, Jones EE, White KY, Burch TC, et al. Increased bisecting N-acetylglucosamine and decreased branched chain glycans of N-linked glycoproteins in expressed prostatic secretions associated with prostate cancer progression. Proteomics Clin Appl. 2013;7(9–10):677–89. Available from: https://onlinelibrary.wiley.com/doi/10.1002/prca.201200134
33. Fischer S, Cornils K, Speiseder T, Badbaran A, Reimer R, Indenbirken D, et al. Indication of horizontal DNA gene transfer by Extracellular vesicles. PLoS One. 2016;11(9):e0163665. https://doi.org/10.1371/journal.pone.0163665.

34. Németh A, Orgovan N, Sódar BW, Osteikoetxea X, Pálóczi K, Szabó-Taylor KÉ, et al. Antibiotic-induced release of small extracellular vesicles (exosomes) with surface-associated DNA. Sci Rep. 2017;7(1):8202. https://doi.org/10.1038/s41598-017-08392-1.
35. Shelke GV, Jang SC, Yin Y, Lässer C, Lötvall J. Human mast cells release extracellular vesicle-associated DNA. Matters (Zürich). 2016.; Available from: https://www.researchgate.net/profile/Ganesh-Shelke-2/publication/301278249_Human_mast_cells_release_extracellular_vesicle-associated_DNA/links/570fc90a08ae68dc79096a4f/Human-mast-cells-release-extracellular-vesicle-associated-DNA.pdf
36. Malkin EZ, Bratman SV. Bioactive DNA from extracellular vesicles and particles. Cell Death Dis. 2020;11(7):584. https://doi.org/10.1038/s41419-020-02803-4.
37. Zhang Y, Liang F, Zhang D, Qi S, Liu Y. Metabolites as extracellular vesicle cargo in health, cancer, pleural effusion, and cardiovascular diseases: an emerging field of study to diagnostic and therapeutic purposes. Biomed Pharmacother. 2023;157:114046. https://doi.org/10.1016/j.biopha.2022.114046.
38. Badhwar A, Haqqani AS. Biomarker potential of brain-secreted extracellular vesicles in blood in Alzheimer's disease. Alzheimers Dement. 2020;12(1):e12001. Available from: https://alz-journals.onlinelibrary.wiley.com/doi/abs/10.1002/dad2.12001
39. Lee YJ, Shin KJ, Chae YC. Regulation of cargo selection in exosome biogenesis and its biomedical applications in cancer. Exp Mol Med. 2024;56:877. https://doi.org/10.1038/s12276-024-01209-y.
40. Xie S, Zhang Q, Jiang L. Current knowledge on exosome biogenesis, cargo-sorting mechanism and therapeutic implications. Membranes. 2022;12(5) https://doi.org/10.3390/membranes12050498.
41. Mir B, Goettsch C. Extracellular vesicles as delivery vehicles of specific cellular cargo. Cells. 2020;9(7) https://doi.org/10.3390/cells9071601.
42. Mathieu M, Névo N, Jouve M, Valenzuela JI, Maurin M, Verweij FJ, et al. Specificities of exosome versus small ectosome secretion revealed by live intracellular tracking of CD63 and CD9. Nat Commun. 2021;12(1):4389. https://doi.org/10.1038/s41467-021-24384-2.
43. Jeppesen DK, Fenix AM, Franklin JL, Higginbotham JN, Zhang Q, Zimmerman LJ, et al. Reassessment of exosome composition. Cell. 2019;177(2):428–45.e18. https://doi.org/10.1016/j.cell.2019.02.029.
44. Mathieu M, Martin-Jaular L, Lavieu G, Théry C. Specificities of secretion and uptake of exosomes and other extracellular vesicles for cell-to-cell communication. Nat Cell Biol. 2019;21(1):9–17. https://doi.org/10.1038/s41556-018-0250-9.
45. Krishna G, Santhoshkumar R, Sivakumar PT, Alladi S, Mahadevan A, Dahale AB, et al. Pathological (dis)similarities in neuronal exosome-derived synaptic and organellar marker levels between Alzheimer's disease and frontotemporal dementia. J Alzheimers Dis. 2023;94(s1):S387–97. https://doi.org/10.3233/JAD-220829.
46. Gabrielli M, Tozzi F, Verderio C, Origlia N. Emerging roles of extracellular vesicles in Alzheimer's disease: Focus on synaptic dysfunction and vesicle–neuron interaction. Cells. 2022 [cited 2024 Mar 25];12(1):63. Available from: https://www.mdpi.com/2073-4409/12/1/63
47. Tóth EÁ, Turiák L, Visnovitz T, Cserép C, Mázló A, Sódar BW, et al. Formation of a protein corona on the surface of extracellular vesicles in blood plasma. J Extracell Vesicles. 2021;10(11):e12140. https://doi.org/10.1002/jev2.12140.
48. White R, Sotillo J, Ancarola ME, Borup A, Boysen AT, Brindley PJ, et al. Special considerations for studies of extracellular vesicles from parasitic helminths: A community-led roadmap to increase rigour and reproducibility. J Extracell Vesicles. 2023;12(1):e12298. Available from: https://onlinelibrary.wiley.com/doi/10.1002/jev2.12298
49. Hadjidemetriou M, Al-Ahmady Z, Buggio M, Swift J, Kostarelos K. A novel scavenging tool for cancer biomarker discovery based on the blood-circulating nanoparticle protein corona. Biomaterials. 2019;188:118–29. https://doi.org/10.1016/j.biomaterials.2018.10.011.

50. Pérez-González R, Kim Y, Miller C, Pacheco-Quinto J, Eckman EA, Levy E. Extracellular vesicles: where the amyloid precursor protein carboxyl-terminal fragments accumulate and amyloid-β oligomerizes. FASEB J. 2020;34(9):12922–31. https://doi.org/10.1096/fj.202000823R.
51. Pauwels MJ, Vandendriessche C, Vandenbroucke RE. Special delEVery: Extracellular vesicles as promising delivery platform to the brain. Biomedicines. 2021;9(11) https://doi.org/10.3390/biomedicines9111734.
52. Abbott NJ, Rönnbäck L, Hansson E. Astrocyte–endothelial interactions at the blood–brain barrier. Nat Rev Neurosci. 2006 [cited 2024 Mar 25];7(1):41–53. Available from: https://www.nature.com/articles/nrn1824
53. Lai RC, Yeo RWY, Tan KH, Lim SK. Exosomes for drug delivery — a novel application for the mesenchymal stem cell. Biotechnol Adv. 2013;31(5):543–51. Available from: https://www.sciencedirect.com/science/article/pii/S0734975012001474
54. Ramos-Zaldívar HM, Polakovicova I, Salas-Huenuleo E, Corvalán AH, Kogan MJ, Yefi CP, et al. Extracellular vesicles through the blood–brain barrier: a review. Fluids Barriers CNS. 2022;19(1):60. https://doi.org/10.1186/s12987-022-00359-3.
55. Yamamoto T, Kosaka N, Ochiya T. Latest advances in extracellular vesicles: from bench to bedside. Sci Technol Adv Mater. 2019;20(1):746–57. https://doi.org/10.1080/14686996.2019.1629835.
56. Auber M, Svenningsen P. An estimate of extracellular vesicle secretion rates of human blood cells. J Extracell Biol. 2022;1(6). Available from: https://onlinelibrary.wiley.com/doi/10.1002/jex2.46
57. György B, Szabó TG, Pásztói M, Pál Z, Misják P, Aradi B, et al. Membrane vesicles, current state-of-the-art: emerging role of extracellular vesicles. Cell Mol Life Sci. 2011;68(16):2667–88. https://doi.org/10.1007/s00018-011-0689-3.
58. Kooijmans SAA, Vader P, van Dommelen SM, van Solinge WW, Schiffelers RM. Exosome mimetics: a novel class of drug delivery systems. Int J Nanomedicine. 2012;7:1525–41. https://doi.org/10.2147/IJN.S29661.
59. Mulcahy LA, Pink RC, Carter DRF. Routes and mechanisms of extracellular vesicle uptake. J Extracell Vesicles. 2014;3 https://doi.org/10.3402/jev.v3.24641.
60. Lötvall J, Hill AF, Hochberg F, Buzás EI, Di Vizio D, Gardiner C, et al. Minimal experimental requirements for definition of extracellular vesicles and their functions: a position statement from the International Society for Extracellular Vesicles. J Extracell Vesicles. 2014;3:26913. https://doi.org/10.3402/jev.v3.26913.
61. Lucien F, Gustafson D, Lenassi M, Li B, Teske JJ, Boilard E, et al. MIBlood-EV: minimal information to enhance the quality and reproducibility of blood extracellular vesicle research. J Extracell Vesicles. 2023;12(12):e12385. Available from: https://isevjournals.onlinelibrary.wiley.com/doi/10.1002/jev2.12385
62. Sandau US, Magaña SM, Costa J, Nolan JP, Ikezu T, Vella LJ, et al. Recommendations for reproducibility of cerebrospinal fluid extracellular vesicle studies. J Extracell Vesicles. 2024;13(1):e12397. https://doi.org/10.1002/jev2.12397.
63. Erdbrügger U, Blijdorp CJ, Bijnsdorp IV, Borràs FE, Burger D, Bussolati B, et al. Urinary extracellular vesicles: A position paper by the Urine Task Force of the International Society for Extracellular Vesicles. J Extracell Vesicles. 2021;10(7):e12093. https://doi.org/10.1002/jev2.12093.
64. Vacchi E, Burrello J, Burrello A, Bolis S, Monticone S, Barile L, et al. Profiling inflammatory extracellular vesicles in plasma and cerebrospinal fluid: an optimized diagnostic model for Parkinson's disease. Biomedicines. 2021;9(3) https://doi.org/10.3390/biomedicines9030230.
65. Bracht JWP, Los M, van Eijndhoven MAJ, Bettin B, van der Pol E, Pegtel DM, et al. Platelet removal from human blood plasma improves detection of extracellular vesicle-associated miRNA. J Extracell Vesicles. 2023;12(2):e12302. Available from: https://onlinelibrary.wiley.com/doi/10.1002/jev2.12302

66. Rani K, Rastogi S, Vishwakarma P, Bharti PS, Sharma V, Renu K, et al. A novel approach to correlate the salivary exosomes and their protein cargo in the progression of cognitive impairment into Alzheimer's disease. J Neurosci Methods. 2021;347:108980. https://doi.org/10.1016/j.jneumeth.2020.108980.
67. Sun IO, Lerman LO. Urinary Extracellular vesicles as biomarkers of kidney disease: from diagnostics to therapeutics. Diagnostics (Basel). 2020;10(5) https://doi.org/10.3390/diagnostics10050311.
68. Wang S, Kojima K, Mobley JA, West AB. Proteomic analysis of urinary extracellular vesicles reveal biomarkers for neurologic disease. EBioMedicine. 2019;45:351–61. https://doi.org/10.1016/j.ebiom.2019.06.021.
69. Correll VL, Otto JJ, Risi CM, Main BP, Boutros PC, Kislinger T, et al. Optimization of small extracellular vesicle isolation from expressed prostatic secretions in urine for in-depth proteomic analysis. J Extracell Vesicles. 2022;11(2):e12184. Available from: https://onlinelibrary.wiley.com/doi/10.1002/jev2.12184
70. Zhou H, Yuen PST, Pisitkun T, Gonzales PA, Yasuda H, Dear JW, et al. Collection, storage, preservation, and normalization of human urinary exosomes for biomarker discovery. Kidney Int. 2006;69(8):1471–6. https://doi.org/10.1038/sj.ki.5000273.
71. Lőrincz ÁM, Timár CI, Marosvári KA, Veres DS, Otrokocsi L, Kittel Á, et al. Effect of storage on physical and functional properties of extracellular vesicles derived from neutrophilic granulocytes. J Extracell Vesicles. 2014;3:25465. https://doi.org/10.3402/jev.v3.25465.
72. Wu JY, Li YJ, Hu XB, Huang S, Xiang DX. Preservation of small extracellular vesicles for functional analysis and therapeutic applications: a comparative evaluation of storage conditions. Drug Deliv. 2021;28(1):162–70. https://doi.org/10.1080/10717544.2020.1869866.
73. Tsamchoe M, Petrillo S, Lazaris A, Metrakos P. Isolation of extracellular vesicles from human plasma samples: the importance of controls. Biotechnol J. 2023;18(6):e2200575. https://doi.org/10.1002/biot.202200575.
74. Böhmert L, Voß L, Stock V, Braeuning A, Lampen A, Sieg H. Isolation methods for particle protein corona complexes from protein-rich matrices. Nanoscale Adv. 2020;2(2):563–82. https://doi.org/10.1039/c9na00537d.
75. Konoshenko MY, Lekchnov EA, Vlassov AV, Laktionov PP. Isolation of extracellular vesicles: general methodologies and latest trends. Biomed Res Int. 2018;2018:8545347. https://doi.org/10.1155/2018/8545347.
76. Busatto S, Vilanilam G, Ticer T, Lin WL, Dickson DW, Shapiro S, et al. Tangential flow filtration for highly efficient concentration of extracellular vesicles from large volumes of fluid. Cells. 2018;7(12) https://doi.org/10.3390/cells7120273.
77. Heath N, Grant L, De Oliveira TM, Rowlinson R, Osteikoetxea X, Dekker N, et al. Rapid isolation and enrichment of extracellular vesicle preparations using anion exchange chromatography. Sci Rep. 2018;8(1):5730. https://doi.org/10.1038/s41598-018-24163-y.
78. Zhang H, Lyden D. Asymmetric-flow field-flow fractionation technology for exomere and small extracellular vesicle separation and characterization. Nat Protoc. 2019;14(4):1027–53. https://doi.org/10.1038/s41596-019-0126-x.
79. Veerman RE, Teeuwen L, Czarnewski P, Güclüler Akpinar G, Sandberg A, Cao X, et al. Molecular evaluation of five different isolation methods for extracellular vesicles reveals different clinical applicability and subcellular origin. J Extracell Vesicles. 2021;10(9):e12128. https://doi.org/10.1002/jev2.12128.
80. Badhwar A, Hirschberg Y, Tamayo NV, Iulita MF, Udeh-Momoh CT, Matton A, et al. Assessment of brain-derived extracellular vesicle enrichment for blood biomarker analysis in age-related neurodegenerative diseases: An international overview. bioRxiv. 2023; https://doi.org/10.1101/2023.10.02.560210.
81. Karlsson M, Zhang C, Méar L, Zhong W, Digre A, Katona B, et al. A single–cell type transcriptomics map of human tissues Sci Adv. 2021;7(31):eabh2169. Available from: https://www.science.org/doi/abs/10.1126/sciadv.abh2169

82. Goetzl EJ, Abner EL, Jicha GA, Kapogiannis D, Schwartz JB. Declining levels of functionally specialized synaptic proteins in plasma neuronal exosomes with progression of Alzheimer's disease. FASEB J. 2018;32(2):888–93. https://doi.org/10.1096/fj.201700731R.
83. Goetzl EJ, Mustapic M, Kapogiannis D, Eitan E, Lobach IV, Goetzl L, et al. Cargo proteins of plasma astrocyte-derived exosomes in Alzheimer's disease. FASEB J. 2016;30(11):3853–9. https://doi.org/10.1096/fj.201600756R.
84. Kumar A, Su Y, Sharma M, Singh S, Kim S, Peavey JJ, et al. MicroRNA expression in extracellular vesicles as a novel blood-based biomarker for Alzheimer's disease. Alzheimers Dement. 2023;19(11):4952–66. https://doi.org/10.1002/alz.13055.
85. Wang P, Lan G, Xu B, Yu Z, Tian C, Lei X, et al. α-Synuclein-carrying astrocytic extracellular vesicles in Parkinson pathogenesis and diagnosis. Transl Neurodegener. 2023;12(1):40. https://doi.org/10.1186/s40035-023-00372-y.
86. Visconte C, Golia MT, Fenoglio C, Serpente M, Gabrielli M, Arcaro M, et al. Plasma microglial-derived extracellular vesicles are increased in frail patients with mild cognitive impairment and exert a neurotoxic effect. Geroscience. 2023;45(3):1557–71. https://doi.org/10.1007/s11357-023-00746-0.
87. Dutta S, Hornung S, Kruayatidee A, Maina KN, Del Rosario I, Paul KC, et al. Correction to: α-Synuclein in blood exosomes immunoprecipitated using neuronal and oligodendroglial markers distinguishes Parkinson's disease from multiple system atrophy. Acta Neuropathol. 2021;142(3):513. https://doi.org/10.1007/s00401-021-02332-0.
88. Ter-Ovanesyan D, Gilboa T, Budnik B, Nikitina A, Whiteman S, Lazarovits R, et al. Improved isolation of extracellular vesicles by removal of both free proteins and lipoproteins. elife. 2023;12 https://doi.org/10.7554/eLife.86394.
89. Zimmerman AJ, de Oliveira GP, Jr SX, Wood J, Fu Z, Pinckney B, et al. Multimode chromatography-based techniques for high purity isolation of extracellular vesicles from human blood plasma. J Extracell Biol. 2024;3(3). Available from: https://isevjournals.onlinelibrary.wiley.com/doi/10.1002/jex2.147
90. Woo HK, Sunkara V, Park J, Kim TH, Han JR, Kim CJ, et al. Exodisc for rapid, size-selective, and efficient isolation and analysis of nanoscale Extracellular vesicles from biological samples. ACS Nano. 2017;11(2):1360–70. https://doi.org/10.1021/acsnano.6b06131.
91. Liu F, Vermesh O, Mani V, Ge TJ, Madsen SJ, Sabour A, et al. The exosome total isolation Chip. ACS Nano. 2017;11(11):10712–23. https://doi.org/10.1021/acsnano.7b04878.
92. Johnsen KB, Gudbergsson JM, Andresen TL, Simonsen JB. What is the blood concentration of extracellular vesicles? Implications for the use of extracellular vesicles as blood-borne biomarkers of cancer. Biochim Biophys Acta Rev Cancer. 2019;1871(1):109–16. https://doi.org/10.1016/j.bbcan.2018.11.006.
93. Visnovitz T, Osteikoetxea X, Sódar BW, Mihály J, Lőrincz P, Vukman KV, et al. An improved 96 well plate format lipid quantification assay for standardisation of experiments with extracellular vesicles. J Extracell Vesicles. 2019;8(1):1565263. https://doi.org/10.1080/20013078.2019.1565263.
94. Mihály J, Deák R, Szigyártó IC, Bóta A, Beke-Somfai T, Varga Z. Characterization of extracellular vesicles by IR spectroscopy: fast and simple classification based on amide and CH stretching vibrations. Biochim Biophys Acta Biomembr. 2017;1859(3):459–66. https://doi.org/10.1016/j.bbamem.2016.12.005.
95. Hill AF, Pegtel DM, Lambertz U, Leonardi T, O'Driscoll L, Pluchino S, et al. ISEV position paper: extracellular vesicle RNA analysis and bioinformatics. J Extracell Vesicles. 2013;2 https://doi.org/10.3402/jev.v2i0.22859.
96. Verwilt J, Trypsteen W, Van Paemel R, De Preter K, Giraldez MD, Mestdagh P, et al. When DNA gets in the way: a cautionary note for DNA contamination in extracellular RNA-seq studies. Proc Natl Acad Sci USA. 2020;117(32):18934–6. https://doi.org/10.1073/pnas.2001675117.

97. Hartjes TA, Mytnyk S, Jenster GW, van Steijn V, van Royen ME. Extracellular vesicle quantification and characterization: common methods and emerging approaches. Bioengineering (Basel). 2019;6(1) https://doi.org/10.3390/bioengineering6010007.
98. Jung MK, Mun JY. Sample preparation and imaging of exosomes by transmission electron microscopy. J Vis Exp. 2018;131 https://doi.org/10.3791/56482.
99. Nielsen JE, Honoré B, Vestergård K, Maltesen RG, Christiansen G, Bøge AU, et al. Shotgun-based proteomics of extracellular vesicles in Alzheimer's disease reveals biomarkers involved in immunological and coagulation pathways. Sci Rep. 2021;11(1):18518. https://doi.org/10.1038/s41598-021-97969-y.
100. Rajkumar AP, Hye A, Lange J, Manesh YR, Ballard C, Fladby T, et al. Next-generation RNA-sequencing of serum small extracellular vesicles discovers potential diagnostic biomarkers for dementia with Lewy bodies. Am J Geriatr Psychiatry. 2021;29(6):573–84. https://doi.org/10.1016/j.jagp.2020.10.012.
101. Martins TS, Magalhães S, Rosa IM, da Cruz e Silva O, Nunes A, Henriques AG. Lipid profiles of extracellular vesicles are distinct for Alzheimer's disease cases. Alzheimers Dement. 2020;16(S4). Available from: https://alz-journals.onlinelibrary.wiley.com/doi/10.1002/alz.043980
102. Lugli G, Cohen AM, Bennett DA, Shah RC, Fields CJ, Hernandez AG, et al. Plasma Exosomal miRNAs in persons with and without Alzheimer disease: altered expression and prospects for biomarkers. PLoS One. 2015;10(10):e0139233. https://doi.org/10.1371/journal.pone.0139233.
103. Yang TT, Liu CG, Gao SC, Zhang Y, Wang PC. The serum exosome derived microRNA-135a, -193b, and -384 were potential Alzheimer's disease biomarkers. Biomed Environ Sci. 2018;31(2):87–96. https://doi.org/10.3967/bes2018.011.
104. Wei H, Xu Y, Xu W, Zhou Q, Chen Q, Yang M, et al. Serum exosomal miR-223 serves as a potential diagnostic and prognostic biomarker for dementia. Neuroscience. 2018;379:167–76. https://doi.org/10.1016/j.neuroscience.2018.03.016.
105. Cheng L, Doecke JD, Sharples RA, Villemagne VL, Fowler CJ, Rembach A, et al. Prognostic serum miRNA biomarkers associated with Alzheimer's disease shows concordance with neuropsychological and neuroimaging assessment. Mol Psychiatry. 2015;20(10):1188–96. https://doi.org/10.1038/mp.2014.127.
106. Liu CG, Song J, Zhang YQ, Wang PC. MicroRNA-193b is a regulator of amyloid precursor protein in the blood and cerebrospinal fluid derived exosomal microRNA-193b is a biomarker of Alzheimer's disease. Mol Med Rep. 2014;10(5):2395–400. https://doi.org/10.3892/mmr.2014.2484.
107. Wood PL. Lipidomics of Alzheimer's disease: current status. Alzheimers Res Ther. 2012;4(1):5. https://doi.org/10.1186/alzrt103.
108. Haraszti RA, Didiot MC, Sapp E, Leszyk J, Shaffer SA, Rockwell HE, et al. High-resolution proteomic and lipidomic analysis of exosomes and microvesicles from different cell sources. J Extracell Vesicles. 2016;5:32570. https://doi.org/10.3402/jev.v5.32570.
109. Musunuri S, Khoonsari PE, Mikus M, Wetterhall M, Häggmark-Mänberg A, Lannfelt L, et al. Increased levels of Extracellular microvesicle markers and decreased levels of endocytic/Exocytic proteins in the Alzheimer's disease brain. J Alzheimers Dis. 2016;54(4):1671–86. https://doi.org/10.3233/JAD-160271.
110. Dong X, Chen K, Cuevas-Diaz Duran R, You Y, Sloan SA, Zhang Y, et al. Comprehensive identification of long non-coding RNAs in purified cell types from the brain reveals functional LncRNA in OPC fate determination. PLoS Genet. 2015;11(12):e1005669. https://doi.org/10.1371/journal.pgen.1005669.
111. Jia L, Qiu Q, Zhang H, Chu L, Du Y, Zhang J, et al. Concordance between the assessment of Aβ42, T-tau, and P-T181-tau in peripheral blood neuronal-derived exosomes and cerebrospinal fluid. Alzheimers Dement. 2019;15(8):1071–80. https://doi.org/10.1016/j.jalz.2019.05.002.

112. Li Y, Meng S, Di W, Xia M, Dong L, Zhao Y, et al. Amyloid-β protein and microRNA-384 in NCAM-labeled exosomes from peripheral blood are potential diagnostic markers for Alzheimer's disease. CNS Neurosci Ther. 2022;28(7):1093–107. https://doi.org/10.1111/cns.13846.
113. Fiandaca MS, Kapogiannis D, Mapstone M, Boxer A, Eitan E, Schwartz JB, et al. Identification of preclinical Alzheimer's disease by a profile of pathogenic proteins in neurally derived blood exosomes: A case-control study. Alzheimers Dement. 2015;11(6):600–7.e1. https://doi.org/10.1016/j.jalz.2014.06.008.
114. Hamlett ED, Goetzl EJ, Ledreux A, Vasilevko V, Boger HA, LaRosa A, et al. Neuronal exosomes reveal Alzheimer's disease biomarkers in Down syndrome. Alzheimers Dement. 2017;13(5):541–9. https://doi.org/10.1016/j.jalz.2016.08.012.
115. Li TR, Yao YX, Jiang XY, Dong QY, Yu XF, Wang T, et al. β-Amyloid in blood neuronal-derived extracellular vesicles is elevated in cognitively normal adults at risk of Alzheimer's disease and predicts cerebral amyloidosis. Alzheimers Res Ther. 2022;14(1):66. https://doi.org/10.1186/s13195-022-01010-x.
116. Winston CN, Goetzl EJ, Akers JC, Carter BS, Rockenstein EM, Galasko D, et al. Prediction of conversion from mild cognitive impairment to dementia with neuronally derived blood exosome protein profile. Alzheimers Dement. 2016;3:63–72. https://doi.org/10.1016/j.dadm.2016.04.001.
117. Winston CN, Goetzl EJ, Baker LD, Vitiello MV, Rissman RA. Growth hormone-releasing hormone modulation of neuronal exosome biomarkers in mild cognitive impairment. J Alzheimers Dis. 2018;66(3):971–81. https://doi.org/10.3233/JAD-180302.
118. Alvarez XA, Winston CN, Barlow JW, Sarsoza FM, Alvarez I, Aleixandre M, et al. Modulation of amyloid-β and tau in Alzheimer's disease plasma neuronal-derived Extracellular vesicles by Cerebrolysin® and donepezil. J Alzheimers Dis. 2022;90(2):705–17. https://doi.org/10.3233/JAD-220575.
119. Gu D, Liu F, Meng M, Zhang L, Gordon ML, Wang Y, et al. Elevated matrix metalloproteinase-9 levels in neuronal extracellular vesicles in Alzheimer's disease. Ann Clin Transl Neurol. 2020;7(9):1681–91. https://doi.org/10.1002/acn3.51155.
120. Eitan E, Thornton-Wells T, Elgart K, Erden E, Gershun E, Levine A, et al. Synaptic proteins in neuron-derived extracellular vesicles as biomarkers for Alzheimer's disease: novel methodology and clinical proof of concept. Extracell Vesicles Circ Nucl Acids. 2023;4(1):133–50. https://doi.org/10.20517/evcna.2023.13.
121. Li J, Ni W, Jin D, Yu Y, Xiao MM, Zhang ZY, et al. Nanosensor-driven detection of neuron-derived exosomal Aβ42 with graphene electrolyte-gated transistor for Alzheimer's disease diagnosis. Anal Chem. 2023;95(13):5719–28. https://doi.org/10.1021/acs.analchem.2c05751.
122. Abner EL, Elahi FM, Jicha GA, Mustapic M, Al-Janabi O, Kramer JH, et al. Endothelial-derived plasma exosome proteins in Alzheimer's disease angiopathy. FASEB J. 2020;34(4):5967–74. https://doi.org/10.1096/fj.202000034R.
123. Sun R, Wang H, Shi Y, Gao D, Sun Z, Chen Z, et al. A pilot study of urinary exosomes in Alzheimer's disease. Neurodegener Dis. 2019;19(5–6):184–91. https://doi.org/10.1159/000505851.
124. Shi M, Kovac A, Korff A, Cook TJ, Ginghina C, Bullock KM, et al. CNS tau efflux via exosomes is likely increased in Parkinson's disease but not in Alzheimer's disease. Alzheimers Dement. 2016;12(11):1125–31. https://doi.org/10.1016/j.jalz.2016.04.003.
125. Nam E, Lee YB, Moon C, Chang KA. Serum tau proteins as potential biomarkers for the assessment of Alzheimer's disease progression. Int J Mol Sci. 2020;21(14) https://doi.org/10.3390/ijms21145007.
126. Blommer J, Pitcher T, Mustapic M, Eren E, Yao PJ, Vreones MP, et al. Extracellular vesicle biomarkers for cognitive impairment in Parkinson's disease. Brain. 2023;146(1):195–208. https://doi.org/10.1093/brain/awac258.

127. Guo Z, Tian C, Shi Y, Song XR, Yin W, Tao QQ, et al. Blood-based CNS regionally and neuronally enriched extracellular vesicles carrying pTau217 for Alzheimer's disease diagnosis and differential diagnosis. Acta Neuropathol Commun. 2024;12(1):38. https://doi.org/10.1186/s40478-024-01727-w.
128. Jia L, Zhu M, Kong C, Pang Y, Zhang H, Qiu Q, et al. Blood neuro-exosomal synaptic proteins predict Alzheimer's disease at the asymptomatic stage. Alzheimers Dement. 2021;17(1):49–60. Available from: https://alz-journals.onlinelibrary.wiley.com/doi/10.1002/alz.12166
129. Goetzl EJ, Kapogiannis D, Schwartz JB, Lobach IV, Goetzl L, Abner EL, et al. Decreased synaptic proteins in neuronal exosomes of frontotemporal dementia and Alzheimer's disease. FASEB J. 2016;30(12):4141–8. https://doi.org/10.1096/fj.201600816R.
130. Cai H, Pang Y, Ren Z, Fu X, Jia L. Delivering synaptic protein mRNAs via extracellular vesicles ameliorates cognitive impairment in a mouse model of Alzheimer's disease. BMC Med. 2024;22(1):138. https://doi.org/10.1186/s12916-024-03359-2.
131. Agliardi C, Guerini FR, Zanzottera M, Bianchi A, Nemni R, Clerici M. SNAP-25 in serum is carried by exosomes of neuronal origin and is a potential biomarker of Alzheimer's disease. Mol Neurobiol. 2019;56(8):5792–8. https://doi.org/10.1007/s12035-019-1501-x.
132. Guo M, Wang J, Zhao Y, Feng Y, Han S, Dong Q, et al. Microglial exosomes facilitate α-synuclein transmission in Parkinson's disease. Brain. 2020;143(5):1476–97. https://doi.org/10.1093/brain/awaa090.
133. Shi M, Liu C, Cook TJ, Bullock KM, Zhao Y, Ginghina C, et al. Plasma exosomal α-synuclein is likely CNS-derived and increased in Parkinson's disease. Acta Neuropathol. 2014;128(5):639–50. https://doi.org/10.1007/s00401-014-1314-y.
134. Kluge A, Bunk J, Schaeffer E, Drobny A, Xiang W, Knacke H, et al. Detection of neuron-derived pathological α-synuclein in blood. Brain. 2022;145(9):3058–71. https://doi.org/10.1093/brain/awac115.
135. Jiao Y, Zhu X, Zhou X, Li Y, Zhou L, Zhao A, et al. Collaborative plasma biomarkers for Parkinson disease development and progression: a cross-sectional and longitudinal study. Eur J Neurol. 2023;30(10):3090–7. Available from: https://onlinelibrary.wiley.com/doi/10.1111/ene.15964
136. Yan S, Jiang C, Janzen A, Barber TR, Seger A, Sommerauer M, et al. Neuronally derived Extracellular vesicle α-Synuclein as a serum biomarker for individuals at risk of developing Parkinson disease. JAMA Neurol. 2024;81(1):59–68. https://doi.org/10.1001/jamaneurol.2023.4398.
137. Jiang C, Hopfner F, Berg D, Hu MT, Pilotto A, Borroni B, et al. Validation of α-synuclein in L1CAM-immunocaptured exosomes as a biomarker for the stratification of parkinsonian syndromes. Mov Disord. 2021;36(11):2663–9. Available from: https://movementdisorders.onlinelibrary.wiley.com/doi/10.1002/mds.28591
138. Meloni M, Agliardi C, Guerini FR, Zanzottera M, Bolognesi E, Picciolini S, et al. Oligomeric α-synuclein and tau aggregates in NDEVs differentiate Parkinson's disease from atypical parkinsonisms. Neurobiol Dis. 2023;176:105947. https://doi.org/10.1016/j.nbd.2022.105947.
139. Jiang C, Hopfner F, Katsikoudi A, Hein R, Catli C, Evetts S, et al. Serum neuronal exosomes predict and differentiate Parkinson's disease from atypical parkinsonism. J Neurol Neurosurg Psychiatry. 2020;91(7):720–9. https://doi.org/10.1136/jnnp-2019-322588.
140. Ishiguro Y, Tsunnemi T, Shimada T, Yoroisaka A, Ueno SI, Takeshige-Amano H, et al. Extracellular vesicles contain filamentous alpha-synuclein and facilitate the propagation of Parkinson's pathology. Biochem Biophys Res Commun. 2024;703:149620. https://doi.org/10.1016/j.bbrc.2024.149620.
141. Si X, Tian J, Chen Y, Yan Y, Pu J, Zhang B. Central nervous system-derived Exosomal alpha-Synuclein in serum may be a biomarker in Parkinson's disease. Neuroscience. 2019;413:308–16. https://doi.org/10.1016/j.neuroscience.2019.05.015.
142. Lucien F, Benarroch EE, Mullan A, Ali F, Boeve BF, Mielke MM, et al. Poly (ADP-ribose) and α-synuclein extracellular vesicles in patients with Parkinson disease: a possible bio-

marker of disease severity. PLoS One. 2022;17(4):e0264446. https://doi.org/10.1371/journal.pone.0264446.
143. Kapogiannis D, Boxer A, Schwartz JB, Abner EL, Biragyn A, Masharani U, et al. Dysfunctionally phosphorylated type 1 insulin receptor substrate in neural-derived blood exosomes of preclinical Alzheimer's disease. FASEB J. 2015;29(2):589–96. https://doi.org/10.1096/fj.14-262048.
144. Chou SY, Chan L, Chung CC, Chiu JY, Hsieh YC, Hong CT. Altered insulin receptor substrate 1 phosphorylation in blood neuron-derived Extracellular vesicles from patients with Parkinson's disease. Front Cell Dev Biol. 2020;8:564641. https://doi.org/10.3389/fcell.2020.564641.
145. Chi H, Yao R, Sun C, Leng B, Shen T, Wang T, et al. Blood Neuroexosomal mitochondrial proteins predict Alzheimer disease in diabetes. Diabetes. 2022;71(6):1313–23. https://doi.org/10.2337/db21-0969.
146. Yuyama K, Sun H, Fujii R, Hemmi I, Ueda K, Igeta Y. Extracellular vesicle proteome unveils cathepsin B connection to Alzheimer's disease pathogenesis. Brain. 2024;147(2):627–36. https://doi.org/10.1093/brain/awad361.
147. Goetzl EJ, Boxer A, Schwartz JB, Abner EL, Petersen RC, Miller BL, et al. Altered lysosomal proteins in neural-derived plasma exosomes in preclinical Alzheimer disease. Neurology. 2015;85(1):40–7. https://doi.org/10.1212/WNL.0000000000001702.
148. Bellini S, Saraceno C, Benussi L, Squitti R, Cimini S, Ricci M, et al. Plasma small Extracellular vesicles with complement alterations in GRN/C9orf72 and sporadic frontotemporal lobar degeneration. Cells. 2022;11(3) https://doi.org/10.3390/cells11030488.
149. Durur DY, Tastan B, Ugur Tufekci K, Olcum M, Uzuner H, Karakülah G, et al. Alteration of miRNAs in small neuron-derived Extracellular vesicles of Alzheimer's disease patients and the effect of Extracellular vesicles on microglial immune responses. J Mol Neurosci. 2022;72(6):1182–94. https://doi.org/10.1007/s12031-022-02012-y.
150. Goetzl EJ, Boxer A, Schwartz JB, Abner EL, Petersen RC, Miller BL, et al. Low neural exosomal levels of cellular survival factors in Alzheimer's disease. Ann Clin Transl Neurol. 2015;2(7):769–73. https://doi.org/10.1002/acn3.211.
151. Goetzl EJ, Schwartz JB, Abner EL, Jicha GA, Kapogiannis D. High complement levels in astrocyte-derived exosomes of Alzheimer disease. Ann Neurol. 2018;83(3):544–52. https://doi.org/10.1002/ana.25172.
152. Cai H, Pang Y, Wang Q, Qin W, Wei C, Li Y, et al. Proteomic profiling of circulating plasma exosomes reveals novel biomarkers of Alzheimer's disease. Alzheimers Res Ther. 2022;14(1):181. https://doi.org/10.1186/s13195-022-01133-1.
153. Goetzl EJ, Nogueras-Ortiz C, Mustapic M, Mullins RJ, Abner EL, Schwartz JB, et al. Deficient neurotrophic factors of CSPG4-type neural cell exosomes in Alzheimer disease. FASEB J. 2019;33(1):231–8. https://doi.org/10.1096/fj.201801001.
154. Zhang N, Gu D, Meng M, Gordon ML. TDP-43 is elevated in plasma neuronal-derived exosomes of patients with Alzheimer's disease. Front Aging Neurosci. 2020;12:166. https://doi.org/10.3389/fnagi.2020.00166.
155. Muraoka S, Jedrychowski MP, Yanamandra K, Ikezu S, Gygi SP, Ikezu T. Proteomic profiling of Extracellular vesicles derived from cerebrospinal fluid of Alzheimer's disease patients: A pilot study. Cells. 2020;9(9) https://doi.org/10.3390/cells9091959.
156. Jack CR Jr, Bennett DA, Blennow K, Carrillo MC, Dunn B, Haeberlein SB, et al. NIA-AA research framework: toward a biological definition of Alzheimer's disease. Alzheimers Dement. 2018;14(4):535–62. https://doi.org/10.1016/j.jalz.2018.02.018.
157. Jack CR. Revised criteria for diagnosis and staging of Alzheimer's disease: Alzheimer's Association workgroup. Alzheimer's Association, manuscript in progress 2023;38. Available from: https://www.google.com/url?q=https://alz.org/media/Documents/scientific-conferences/Clinical-Criteria-for-Staging-and-Diagnosis-for-Public-Comment-Draft-2.pdf&sa=D&source=docs&ust=1716216405495541&usg=AOvVaw2GCmdu6u5LCnfPGoBGUGgZ

158. McKeever PM, Schneider R, Taghdiri F, Weichert A, Multani N, Brown RA, et al. MicroRNA expression levels are altered in the cerebrospinal fluid of patients with young-onset Alzheimer's disease. Mol Neurobiol. 2018;55(12):8826–41. https://doi.org/10.1007/s12035-018-1032-x.
159. Gui Y, Liu H, Zhang L, Lv W, Hu X. Altered microRNA profiles in cerebrospinal fluid exosome in Parkinson disease and Alzheimer disease. Oncotarget. 2015;6(35):37043–53. https://doi.org/10.18632/oncotarget.6158.
160. Hale C. Alzheimer's blood test from Roche, Eli Lilly nabs FDA breakthrough tag. FIERCE Biotech; 2024. Available from: https://www.fiercebiotech.com/medtech/alzheimers-blood-test-roche-eli-lilly-nabs-fda-breakthrough-tag.
161. Badhwar A, Tam A, Dansereau C, Orban P, Hoffstaedter F, Bellec P. Resting-state network dysfunction in Alzheimer's disease: A systematic review and meta-analysis. Alzheimers Dement. 2017;8:73–85. https://doi.org/10.1016/j.dadm.2017.03.007.
162. Sperling RA, Aisen PS, Beckett LA, Bennett DA, Craft S, Fagan AM, et al. Toward defining the preclinical stages of Alzheimer's disease: recommendations from the National Institute on Aging-Alzheimer's Association workgroups on diagnostic guidelines for Alzheimer's disease. Alzheimers Dement. 2011;7(3):280–92. Available from: https://www.sciencedirect.com/science/article/pii/S1552526011000999
163. Yan YQ, Pu JL, Zheng R, Fang Y, Gu LY, Guo T, et al. Different patterns of exosomal α-synuclein between Parkinson's disease and probable rapid eye movement sleep behavior disorder. Eur J Neurol. 2022;29(12):3590–9. Available from: https://onlinelibrary.wiley.com/doi/10.1111/ene.15537
164. Mullins RJ, Diehl TC, Chia CW, Kapogiannis D. Insulin resistance as a link between amyloid-Beta and tau pathologies in Alzheimer's disease. Front Aging Neurosci. 2017;9:118. https://doi.org/10.3389/fnagi.2017.00118.
165. Mullins RJ, Mustapic M, Goetzl EJ, Kapogiannis D. Exosomal biomarkers of brain insulin resistance associated with regional atrophy in Alzheimer's disease. Hum Brain Mapp. 2017;38(4):1933–40. https://doi.org/10.1002/hbm.23494.
166. Mustapic M, Tran J, Craft S, Kapogiannis D. Extracellular vesicle biomarkers track cognitive changes following intranasal insulin in Alzheimer's disease. J Alzheimers Dis. 2019;69(2):489–98. https://doi.org/10.3233/JAD-180578.
167. Cooper GM, Adams K. The cell: A molecular approach. Oxford University Press; 2022. 816 p. Available from: https://play.google.com/store/books/details?id=GNebEAAAQBAJ
168. Lin JZ, Duan MR, Lin N, Zhao WJ. The emerging role of the chondroitin sulfate proteoglycan family in neurodegenerative diseases. Rev Neurosci. 2021;32(7):737–50. https://doi.org/10.1515/revneuro-2020-0146.
169. Chen Y, Chu JMT, Chang RCC, Wong GTC. The complement system in the central nervous system: from neurodevelopment to neurodegeneration. Biomol Ther. 2022;12(2) https://doi.org/10.3390/biom12020337.
170. Iturria-Medina Y, Sotero RC, Toussaint PJ, Mateos-Pérez JM, Evans AC. Alzheimer's Disease Neuroimaging Initiative. Early role of vascular dysregulation on late-onset Alzheimer's disease based on multifactorial data-driven analysis. Nat Commun. 2016;7:11934. https://doi.org/10.1038/ncomms11934.
171. Haqqani AS, Mianoor Z, Star AT, Detcheverry FE, Delaney CE, Stanimirovic DB, et al. Proteome profiling of brain vessels in a mouse model of cerebrovascular pathology. Biology. 2023;12(12) https://doi.org/10.3390/biology12121500.
172. Sweeney MD, Montagne A, Sagare AP, Nation DA, Schneider LS, Chui HC, et al. Vascular dysfunction-the disregarded partner of Alzheimer's disease. Alzheimers Dement. 2019;15(1):158–67. https://doi.org/10.1016/j.jalz.2018.07.222.
173. Goetzl EJ, Schwartz JB, Mustapic M, Lobach IV, Daneman R, Abner EL, et al. Altered cargo proteins of human plasma endothelial cell-derived exosomes in atherosclerotic cerebrovascular disease. FASEB J. 2017;31(8):3689–94. https://doi.org/10.1096/fj.201700149.

174. Elahi FM, Casaletto KB, Altendahl M, Staffaroni AM, Fletcher E, Filshtein TJ, et al. "Liquid biopsy" of white matter hyperintensity in functionally normal elders. Front Aging Neurosci. 2018;10:343. https://doi.org/10.3389/fnagi.2018.00343.
175. Bartel DP. MicroRNAs: genomics, biogenesis, mechanism, and function. Cell. 2004;116(2):281–97. https://doi.org/10.1016/s0092-8674(04)00045-5.
176. Citterio LA, Mancuso R, Agostini S, Meloni M, Clerici M. Serum and Exosomal miR-7-1-5p and miR-223-3p as possible biomarkers for Parkinson's disease. Biomol Ther. 2023;13(5) https://doi.org/10.3390/biom13050865.
177. Serpente M, Fenoglio C, D'Anca M, Arcaro M, Sorrentino F, Visconte C, et al. MiRNA profiling in plasma neural-derived small Extracellular vesicles from patients with Alzheimer's disease. Cells. 2020;9(6) https://doi.org/10.3390/cells9061443.
178. Lukiw WJ, Andreeva TV, Grigorenko AP, Rogaev EI. Studying micro RNA function and dysfunction in Alzheimer's disease. Front Genet. 2012;3:327. https://doi.org/10.3389/fgene.2012.00327.
179. Wang LL, Min L, Guo QD, Zhang JX, Jiang HL, Shao S, et al. Profiling microRNA from brain by microarray in a transgenic mouse model of Alzheimer's disease. Biomed Res Int. 2017;2017:8030369. https://doi.org/10.1155/2017/8030369.
180. Batabyal RA, Bansal A, Cechinel LR, Authelet K, Goldberg M, Nadler E, et al. Adipocyte-derived small extracellular vesicles from patients with Alzheimer disease carry miRNAs predicted to target the CREB signaling pathway in neurons. Int J Mol Sci. 2023;24(18) https://doi.org/10.3390/ijms241814024.
181. Cha DJ, Mengel D, Mustapic M, Liu W, Selkoe DJ, Kapogiannis D, et al. miR-212 and miR-132 are downregulated in neurally derived plasma exosomes of Alzheimer's patients. Front Neurosci. 2019;13:1208. https://doi.org/10.3389/fnins.2019.01208.
182. Pounders J, Hill EJ, Hooper D, Zhang X, Biesiada J, Kuhnell D, et al. MicroRNA expression within neuronal-derived small extracellular vesicles in frontotemporal degeneration. Medicine. 2022;101(40):e30854. https://doi.org/10.1097/MD.0000000000030854.
183. He L, Chen Y, Lin S, Shen R, Pan H, Zhou Y, et al. Regulation of Hsa-miR-4639-5p expression and its potential role in the pathogenesis of Parkinson's disease. Aging Cell. 2023;22(6):e13840. https://doi.org/10.1111/acel.13840.
184. Liu CG, Wang JL, Li L, Xue LX, Zhang YQ, Wang PC. MicroRNA-135a and -200b, potential biomarkers for Alzheimer′s disease, regulate β secretase and amyloid precursor protein. Brain Res. 2014;1583:55–64. Available from: https://www.sciencedirect.com/science/article/pii/S0006899314005435
185. Xie S, Niu W, Xu F, Wang Y, Hu S, Niu C. Differential expression and significance of miRNAs in plasma extracellular vesicles of patients with Parkinson's disease. Int J Neurosci. 2022;132(7):673–88. https://doi.org/10.1080/00207454.2020.1835899.
186. Taşdelen E, Kizil ETÖ, Aydemir ST, Yalap ÖE, Bingöl AP, Kutlay N. Determination of miR-373 and miR-204 levels in neuronal exosomes in Alzheimer? S disease. Turk J Med Sci. 2022;52(5):1458–67. Available from: https://journals.tubitak.gov.tr/medical/vol52/iss5/4/
187. Song S, Lee JU, Jeon MJ, Kim S, Sim SJ. Detection of multiplex exosomal miRNAs for clinically accurate diagnosis of Alzheimer's disease using label-free plasmonic biosensor based on DNA-assembled advanced plasmonic architecture. Biosens Bioelectron. 2022;199:113864. Available from: https://www.sciencedirect.com/science/article/pii/S0956566321009015?casa_token=k6z6Jm9-OkUAAAAA:0XVtUpRh0QVCiwuTkcsh42TneCCXoQhcVD2370hMKjrCfBZDpdCtcdpxWH11P-n2i_goAvSugqc
188. Song S, Lee JU, Jeon MJ, Kim S, Lee CN, Sim SJ. Precise profiling of exosomal biomarkers via programmable curved plasmonic nanoarchitecture-based biosensor for clinical diagnosis of Alzheimer's disease. Biosens Bioelectron. 2023;230:115269. https://doi.org/10.1016/j.bios.2023.115269.
189. Toyama K, Spin JM, Tsao PS, Maruyama K, Osawa H, Mogi M, et al. Serum microRNA-501-3p is a potential diagnostic tool for detecting mild cognitive impairment: Ehime genome study. J Neurochem. 2023;166(6):960–71. Available from: https://onlinelibrary.wiley.com/doi/10.1111/jnc.15911

190. Ryu IS, Kim DH, Ro JY, Park BG, Kim SH, Im JY, et al. The microRNA-485-3p concentration in salivary exosome-enriched extracellular vesicles is related to amyloid β deposition in the brain of patients with Alzheimer's disease. Clin Biochem. 2023;118:110603. https://doi.org/10.1016/j.clinbiochem.2023.110603.
191. Zürcher C, Humpel C. Saliva: a challenging human fluid to diagnose brain disorders with a focus on Alzheimer's disease. Neural Regener Res. 2023;18(12):2606–10. https://doi.org/10.4103/1673-5374.373675.
192. Ship JA, Puckett SA. Longitudinal study on oral health in subjects with Alzheimer's disease. J Am Geriatr Soc. 1994;42(1):57–63. https://doi.org/10.1111/j.1532-5415.1994.tb06074.x.
193. Ausó E, Gómez-Vicente V, Esquiva G. Biomarkers for Alzheimer's disease early diagnosis J Pers Med. 2020 [cited 2024 Mar 12];10(3):114. Available from: https://www.mdpi.com/2075-4426/10/3/114
194. McKhann GM, Knopman DS, Chertkow H, Hyman BT, Jack CR Jr, Kawas CH, et al. The diagnosis of dementia due to Alzheimer's disease: recommendations from the National Institute on Aging-Alzheimer's Association workgroups on diagnostic guidelines for Alzheimer's disease. Alzheimers Dement. 2011;7(3):263–9. https://doi.org/10.1016/j.jalz.2011.03.005.

Sex Differences in Alzheimer Disease-Related Biomarkers and Dementia Risk

Considerations for Alzheimer Disease Dementia Prediction and Prevention

Sima Toopchiani, Anne Nyambura Njogu, Cynthia Smith, Shireen Sindi, Onah Okoye, Tamlyn Watermeyer, and Chinedu Udeh-Mohmoh

Abstract Alzheimer disease (AD) and related dementias represent intricate, multifaceted neurodegenerative conditions, with AD prevailing as the most common form. The escalating incidence and prevalence observed across global populations could be attributed to a deficiency in primary and/or secondary prevention strategies. Despite affecting both sexes, AD exhibits a higher prevalence among females.

S. Toopchiani
Ageing Epidemiology (AGE) Research Unit, School of Public Health, Imperial College London, London, UK

A. N. Njogu · C. Smith
Brain and Mind Institute, Aga Khan University, Nairobi, Kenya

S. Sindi
Ageing Epidemiology (AGE) Research Unit, School of Public Health, Imperial College London, London, UK

Division of Clinical Geriatrics, Center for Alzheimer Research, Karolinska Institutet, Stockholm, Sweden

O. Okoye
Ageing Epidemiology (AGE) Research Unit, School of Public Health, Imperial College London, London, UK

NHS Healthcare Trust, London, UK

T. Watermeyer
Faculty of Health and Life Sciences, Department of Psychology, Northumbria University, Newcastle, UK

Edinburgh Dementia Prevention, Centre for Clinical Brain Sciences, University of Edinburgh, Edinburgh, UK

R. Tarawneh (ed.), *Biomarkers of Neurodegenerative Disorders*,
https://doi.org/10.1007/978-3-031-80849-4_19

While the extended life expectancy of women might contribute to this discrepancy, studies have revealed sex disparities in incidence, progression, pathology, and risk profiles for AD; the underlying mechanisms of which largely remain unclear.

Comprehensively grasping sex differences in dementia risk is imperative. Although these risk factors present a diverse and intricate landscape, discerning the impact of sex and gender influences on AD risk could foster the development of precise, tailored treatment and prevention approaches. Reported sex differences in risk profiles often exhibit varying magnitudes and inconsistencies across studies, potentially stemming from divergent study designs, heterogeneous definitions, disparate data collection methodologies for risk factors, and the impact of residual confounding.

Simultaneously, encouraging outcomes from interventional studies on modifying risk factors for cognitive health underscore the necessity of targeting specific subgroups and crafting 'personalized' interventions, considering the documented sex and gender disparities. Clarifying sex differences and their implications in the relationship between established and unidentified risk factors and dementia risk or prevalence, as well as diagnostic and predictive biomarkers of the neurodegenerative process, will prove pivotal for devising personalized interventions.

Enhancing our comprehension of sex-based variances in dementia risk holds potential for advancing care and treatment for individuals of all genders. This understanding stands to bolster endeavors in both dementia prevention and treatment. In this chapter, we summarize existing evidence concerning sex-related disparities in AD, as well as risk factors and biomarkers associated with dementia and cognitive decline, providing insights and perspectives around their ramifications for dementia prevention. Furthermore, we offer recommendations for integrating sex-specific strategies into the realm of dementia prevention research and healthcare applications, including from a global health standpoint.

Keywords Sex differences · Risk profiles · Preventative factors · Dementia prevention

C. Udeh-Mohmoh (✉)
Brain and Mind Institute, Aga Khan University, Nairobi, Kenya

Division of Clinical Geriatrics, Center for Alzheimer Research, Karolinska Institutet, Stockholm, Sweden

Division of Public Health Sciences, Department of Epidemiology and Prevention, Wake Forest University School of Medicine, Winston-Salem, NC, USA

Sheffield Institute of Translational Neuroscience, Sheffield University, Sheffield, UK
e-mail: cmomoh@wakehealth.edu

1 Introduction

Alzheimer Disease (AD) and related dementias (ADRD) are expanding medical phenomena imposing significant financial and emotional burden on the health sector and those affected. AD is the most prevalent form of dementia accounting for ≥60% of dementia cases worldwide [1, 2]. The pathological hallmarks of AD include amyloid-β (Aβ) plaques and neurofibrillary tau tangles [3]. In most regions of the world, the occurrence of AD and other dementias disproportionately affects women [1–3]. Differences in longevity have been proposed as the primary reason for the higher dementia rates among women. However, this difference does not explain disparities of age-specific dementia incidence in men and women, and thus, highlights the need to identify and explore sex-specific drivers.

Although a higher prevalence of ADRD among women has been consistently reported, especially among older age groups [4, 5], the literature on AD incidence rates among the sexes has shown conflicting results. Earlier reports from a large European meta-analysis conducted by Niu and colleagues found that women were more likely to experience dementia compared to men [6]. However, results have been inconsistent as some studies suggest no differences [7], while others report higher rates among women [8] or men [9]. The reasons for these discrepancies remain poorly understood, but it is likely that sex and gender differences in the incidence of ADRD may depend on the time period and geographical region [10]. That said, it remains unclear whether the origins of any observed sex differences are sex- or gender-based.

The routine consideration of sex and gender in ADRD research has been inconsistent and challenging. Sex differences refer to the biological, physiological, and genetic differences between men and women, while gender refers to differences in environmental, psychosocial, and cultural influences. Challenges relating to precise measurement, description, determination, and reporting of sex and gender are still being identified in the wider literature. A crucial issue relates to the definition of the construct and further the recognition of the distinct components attributable to gender. Further concerns relate to possible inconsistencies around collection of information on sex versus gender, its description in reported literature, as well as analytical methods to investigate specific differences. Data on biological sex are relatively easy to collect but gender may be more complex [11]. Finally, some risk factors may be influenced separately or differentially by biological sex or gender variables [12]. Although the era of disease modifying drugs is within reach, the implementation of global testing and treatment platforms for ADRD is only nascent. Therefore, there remains great emphasis on developing effective primary and secondary interventions to prevent or postpone AD onset, particularly tailored strategies that consider the entire AD pre/post-clinical spectrum.

Several lines of evidence confirm that sex differences in ADRD risk exist and thus, there are several potential rationales to consider. Sex and gender may play a significant role in the pathopsychological changes associated with disease risk and progression. While some reports on sex differences in certain dementia risk factors are apparent, such as depression, the E4 allele of the apolipoprotein E gene (*APOE-ℰ4*), sleep, and cardiovascular risk factors; others report no significant, or negligible, differences (i.e., of small magnitude). While these findings could potentially explain sex differences in cognitive outcomes, the extent to which these noted disparities in sex and gender determinants of AD arise from biological and/or sociological/behavioral and cultural components of sex and gender remains to be fully elucidated.

Given that AD etiology is multifactorial, it is highly unlikely that one drug or intervention will halt or delay disease onset. Evidence from The Lancet Commission on dementia prevention and care (2020) [13] reported that 40% of dementia cases can be prevented through changes to modifiable risk factors. To date, several large multidomain randomized controlled trials (RCTs) targeting modifiable risk factors have been completed [14]. These include: The Finish Geriatric Intervention Study to Prevent Cognitive Impairment and Dementia (FINGER) [15], the French Multidomain Alzheimer Preventive Trial (MAPT) [16], and the Dutch Prevention of Dementia by Intensive Vascular Care (PreDIVA) [17]. Pooled post-hoc analysis of these three RCTs reported greater disease-risk reduction in women who completed a multidomain intervention compared to men [18]. However, another study reported no sex differences in intervention benefits but noted variations in risk profiles [19]. This emphasizes the importance of understanding and investigating probable sex differences as this could aid in the development of 'personalized' and sex-specific diagnostic and therapeutic strategies for AD.

In this chapter, we will describe substantial research findings from both past and present evidence regarding the disparities in AD prevention research attributable to sex and gender. Additionally, we will delve into the considerations pertinent to future research endeavors and initiatives aimed at risk reduction, encompassing a global health perspective. Furthermore, we will explore the implications for policy development concerning AD.

2 Sex Differences in AD Pathological Hallmarks, Biomarkers, and Biological Risk Factors

As proposed in the AT(N) framework developed by the National Institute on Aging and Alzheimer's Association (NIA-AA) to categorize AD, biomarkers associated with beta amyloid (Aβ) deposition (A), tau pathology (T), and neurodegeneration (N) constitute pivotal hallmark pathologies within the biological spectrum of the disease [20]. Pathogenesis of AD is posited to begin with amyloid deposition, initiating tau phosphorylation, deposition, and dissemination, culminating in neurodegeneration. Supporting evidence corroborating this sequence of events has been derived empirically by employing the AT(N) framework in clinical research. Such studies have revealed median ages at which the prevalence of biomarkers peaks, progressively increasing from A+ T- N- (indicating amyloid positivity, tau negativity, and absence of neurodegeneration), to A+T+ N-, and subsequently to A+ T+ N+. Nevertheless, age-related biomarker abnormality increases were reported even among cognitively unimpaired individuals [21]. Indeed, these hallmark features of AD are discernible even during the preclinical phase in asymptomatic individuals [22] and more than 30% of cognitively normal individuals aged over 70 exhibit heightened brain amyloid burden [23].

Despite the etiology of AD being complex, these abnormal protein deposits uniquely define AD and therefore, still serve as biomarkers of the disease [20]. AD dementia disproportionately affects women, and there is evidence of more severe disease trajectories in women. While most studies observed higher rates of atrophy among men [21], other studies reported faster progression of hippocampal atrophy [24] and disease progression among women with AD pathology compared to age-matched men [25]. These sex differences are also detected in biomarker loads and concentrations in AD. It is important for these differences to be taken into account in current biomarker research, as they may influence the development of reference intervals for future prognostic and diagnostic use.

Aβ is a well-known biomarker for AD, and the reduced concentrations of Aβ-42 in CSF samples are said to reflect Aβ sequestration in senile plaques in the brain. Lower Aβ-42 concentrations are also seen in preclinical AD and mild cognitive impairment (MCI) [26]. Though Aβ burdens were found not to differ by sex, Buckley and colleagues found that females had faster rates of cognitive decline compared to males with similar levels of elevated Aβ [27]. Similarly, in an 11-year follow-up study, decreased plasma Aβ-42 concentrations were more predictive of future risk of memory decline in women than men, and women were 41% more likely to be diagnosed with clinical AD dementia than men [28]. Evidence also shows that men are able to accumulate more Aβ load than women before manifestation of the disease. This suggests that men may be more resilient to AD pathophysiological processes, and that sex may play a modifying role on the risk of AD [29]. Indeed, it has been found that for a given CSF Aβ-42 concentration, women have greater hippocampal atrophy, declines in memory performance, and increased CSF phosphorylated tau (p-tau) [30].

Abnormally phosphorylated tau proteins are biomarkers of AD pathology and are major components of neurofibrillary tangles (NFT) within the brain. Increased t-tau (total tau) levels in AD CSF samples are correlated with more severe neurodegeneration. Increased p-tau concentrations have also been found in AD CSF samples, but these are said to correlate poorly with NFT pathology in AD brains [26]. There is some conflict in measured biomarker levels of tau among men and women. Higher concentrations of total tau and phosphorylated tau have been found among women [31–33]. However, more recent studies reported no sex differences in plasma tau pathology levels [34, 35]. Sex differences in cognitive trajectories, however, have been found in relation to tau. One study found that, for a given plasma p-tau181 (phosphorylated tau at threonine 181) concentration, women have a greater risk of cognitive decline and AD dementia than men. Additionally, for every unit increase in baseline plasma p-tau181, there was a 7% associated increase in the rate of conversion to dementia for Aβ-positive women, compared to a 3% increase in

Aβ-positive men [36]. These findings show that different biomarker concentrations may communicate different disease trajectories in the sexes, and therefore, highlight the need for sex-specific cut-off points when studying and validating biomarkers of AD. The use of a single cut-off point for p-tau181 concentrations, for example, for studies such as clinical trials, might lead to the inclusion of women with more advanced AD neuropathology who may be less likely to benefit from preventative interventions. Furthermore, despite similar concentrations of plasma p-tau181, women had increased CSF p-tau 181, suggesting sex-specific differences as well, in the type of biofluid the biomarker is being collected from. These concentration differences were proposed to be because of a 'leakier' blood brain barrier (BBB) in men, allowing for p-tau to cross into the blood [36]. A leakier BBB in men has been supported with findings of higher Qalb (a liver-derived protein) concentrations in the CSF of men compared to women, from adolescence and throughout adulthood [37].

Sex-specific differences in brain structure can also affect biomarker levels. Sex differences have been found in structural magnetic resonance imaging (MRI) outcomes [38], and men on average have a 10% larger head size and cerebral brain volume, and a higher percentage of white matter than women [30]. Therefore, when concentrations of the biomarker for subcortical axonal degeneration (N), neurofilament light (NfL), are found to be higher in men, it is called into question whether this reflects pathology, or an increased rate of axonal turnover from increased white matter volumes [30]. Furthermore, this sex difference in NfL concentrations has also been found to start in adolescence and continue throughout life [23, 37]. Special attention, therefore, needs to be paid to physiological differences in the sexes that may influence biomarker levels. Nevertheless, one study found that female carriers of the autosomal dominant *PSEN1* gene mutation, who were predisposed to develop early onset dementia, exhibited higher plasma NfL levels compared to non-carriers, particularly in female carriers who were experiencing cognitive impairment. They also showed significantly higher rates of plasma NfL increase than male carriers, suggesting faster neurodegeneration in women as the disease progresses [39]. Results from genetic studies, however, do not always fully inform mechanisms in sporadic forms of the disease. The use of NfL as a biomarker of AD has been contested with studies suggesting that it is more a general measure of neurodegeneration rather than a biomarker for AD. Furthermore, it is not specific to AD and is also elevated in other dementias such as frontotemporal dementia and vascular dementia [26].

Similarly, the use of Neurogranin (Ng)—a dendritic protein involved in the long-term potentiation of synapses—as a biomarker for neurodegeneration in AD is also disputed. Willemse and colleagues suggested Ng was more a biomarker for pathological processes common to all neurodegenerative diseases rather than for AD and did not find it useful for the differential diagnosis of AD [40]. Contrary to this, Portelius and colleagues found significantly increased CSF Ng in AD dementia, compared to other neurodegenerative diseases. Increased Ng was also positively associated with a greater burden of neuritic plaques, tau tangles, as well as Aβ load in the hippocampus [41]. Lashley and colleagues described it as the best-established biomarker for synaptic loss and dysfunction associated with AD [26]. Ng concentrations may reach a plateau in later stages of the disease, possibly reflecting on the findings of Willemse and colleagues [40]. Certainly, CSF Ng levels were seen to be increased in women in the MCI group compared to men; however, levels did not differ by gender among the AD dementia and control groups [42]. Longitudinal studies, therefore, comparing biomarker levels at different disease stages, can be useful in the continued study and development of AD biomarkers. Other evidence also supports higher Ng levels in women than men, with some association between higher CSF Ng and worse cognition for amyloid-positive individuals [23].

Risk factors for AD may also influence biomarker levels. Female *APOE-ℰ4* carriers are reported and observed to have a stronger association with amyloidosis and a greater susceptibility to develop AD [32]. Carriage of the *APOE-ℰ4* allele confers increased susceptibility and may, indeed, account for 50% of the attributable risk for sporadic late-onset AD [43, 44]. This indicates the presence of other risk factors that may be contributing or interacting with AD risk, onset, and progression. Studies have yet to report sex differences in the prevalence of the *APOE-ℰ4* allele, as it is approximated that only 20% of the population are heterozygotes [45] and <5% are homozygotes. However, evidence suggests that women who carry even one copy of the allele have greater disease prevalence, disease progression, and more detrimental cognitive symptoms [32, 45–50]. This stronger association is also apparent in greater manifestation of AD pathology (CSF-tau & Aβ accumulation) in women [38, 51, 52] and a more recent study by Yan et al. (2021) reported a dose effect on brain region-specific tau-deposition in men and not in women [53]. These sex/gender differences of the effect of the *APOE-ℰ4* allele have also been shown to affect biofluid biomarker levels in a sex-specific manner with female carriers having higher CSF p-tau and t-tau levels in earlier disease stages [54].

Current biomarker research is also focusing on developing less invasive techniques for screening, including sampling from blood, saliva, or urine and examining the retina. Circulating microRNAs in the blood are being studied as potential novel biomarkers for AD and recent research has shown that microRNAs are differentially expressed in brain cell-derived small extracellular vesicles in MCI, MCI-AD, and AD dementia, with clear distinction on dementia status [55]. Salivary biomarkers have shown promise with findings of increased Aβ-42 in the saliva of AD patients, as well as unique metabolite profiles [56, 57]. In the retina, decreased retinal neural fiber layer thickness (RNFL), ganglion cell degeneration, and choroidal thinning have been found in AD [58]. There is, however, a noticeable dearth of data on sex-specific differences in newly researched biomarkers, despite considerable evidence on sex-differences found in canonical AD pathology biomarkers (see Fig. 1).

AD is a complex multifactorial disease, and this is reflected in biomarkers of the disease. Sex-specific differences in AD pathologies and trajectories can affect biomarker levels in men and women. Many factors can contribute to sex-differences in AD biomarkers, from basic physiological differences between men and women, to the interaction of sex on other comorbidities that can affect biomarker levels [30]. Sex differences, rather than merely being adjusted for, should be investigated independently [11]. These disparities offer an opportunity to establish sex-specific thresholds within reference intervals for biomarker levels in AD, thereby enhancing

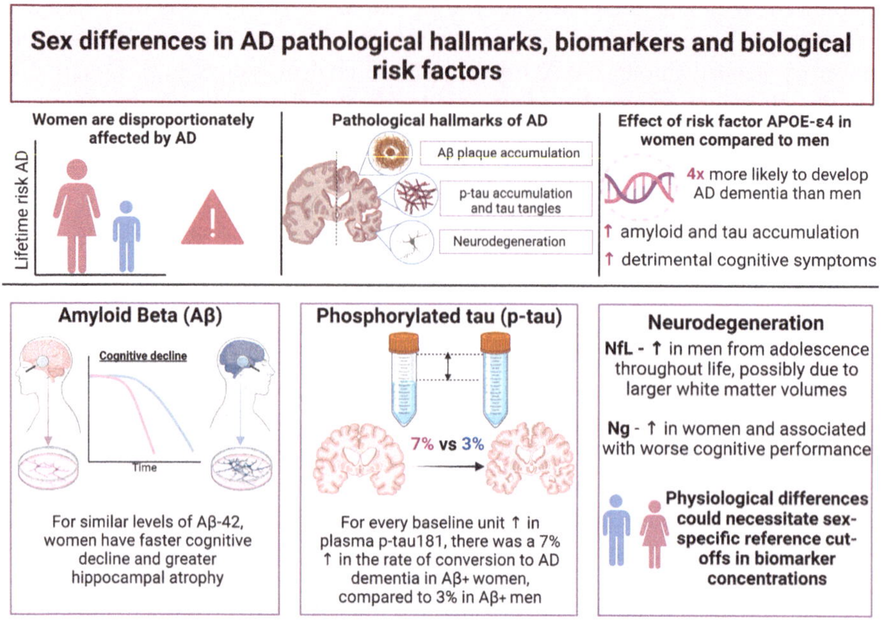

Fig. 1 Conceptual depiction of existing research on sex differences in biological factors associated with Alzheimer disease. Created with Biorender.com

the specificity and validity of the developed biomarkers. Additionally, given the disproportionate impact of AD on women, factors such as menopausal status, hormone therapy usage, and environmental influences must be considered in biomarker research. The paucity of data regarding sex differences across various facets of AD research, including biomarkers and the accumulation of pathology burden, particularly in preclinical AD, underscores a pressing knowledge gap that requires urgent attention. While reports of sex differences in the manifestation of AD pathologies and biomarkers exist, the broader socio-cultural dynamics associated with gender remain to be firmly established.

3 Sex-Specific Risk Factors and Implications in AD Onset and Progression

Accumulating evidence suggests that biological differences between the sexes may contribute to, and explain, sex differences in dementia and AD risk. Historically, in the study of sex differences for AD risk, hormones and fertility proxies, mostly in women, have been prioritized above other biological and social determinants. In both sexes, age-related loss of sex hormones has been shown to increase AD-dementia risk [59, 60]. Estrogen has been suggested to elicit neuroprotective effects in the brain [48, 61, 62] and indeed, even in the male brain, testosterone can be converted into estrogen via the aromatase enzyme [63].

Significant associations between reproductive events and later-life cognition and dementia risk have been indicated, including a study which found that women who spent more months of their life pregnant showed a dose-dependent reduction in risk for developing dementia [64]. Conversely, previous studies found associations in the reverse direction such that higher number of pregnancies tripled AD risk [65] or were associated with earlier AD onset [66]. These inconsistencies might reflect discrepancies in constructs for fertility proxies across studies, where cumulative number of first trimesters [64], rather than number of deliveries [65, 66], may confer protective effects, possibly through pregnancy-induced immunologic effects associated with early gestation.

It may be advantageous to explore associations between the number of successful pregnancies (no complications) compared with the number of pregnancies in which women experience pregnancy-related complications, in relation to late-life cognition. Evidence suggests that women who experience complicated pregnancies such as preeclampsia [67], incomplete pregnancies [68], and pregnancy loss [67] (e.g., miscarriage, still birth) [69] may be at increased risk of developing dementia in later life. As mentioned in these studies, complications during pregnancy contribute to other identified risk factors of dementia including diabetes and atherosclerotic disease. However, what has not been explored is the emotional and psychological changes throughout pregnancy like stress, anxiety, and sleep disturbances. Recently, a causal association between postpartum depression and cognitive impairment was indicated through Mendelian randomization analysis [70], but its importance as a later life risk for AD is yet to be explored.

Regarding genetic and biological factors, one of the main considerations is the loss of endogenous estrogen, such as 17β-estradiol (one of the main estrogens during reproductive years), because estrogen plays a crucial role in memory, cognitive, and executive functions [71]. Women who experience early or premature menopause, whether due to natural causes or interventions such as chemotherapy or surgical procedures like bilateral oophorectomy (ovariectomy), before the age of 40, exhibit an elevated risk of developing MCI, AD dementia, and neurodegeneration in the medial temporal lobe [72]. This risk is more pronounced in women who do not use hormone replacement therapy until their late 50s. Additionally, a 3-year longitudinal study showed that both pre- and postmenopausal women exhibited declines in estrogen-dependent memory tests compared to men when adjusted for age, *APOE-ε4*, and cardiovascular risk factors. Moreover, postmenopausal women showed higher rates of hippocampal volume loss and a higher rate of amyloid beta accumulation. Notably, brain atrophy and excess amyloid beta were observed in both pre- and postmenopausal women compared to men [73].

A few studies have investigated the association of reproductive years on cognitive decline and dementia risk, but findings are inconsistent. While some have found that a longer reproductive period and later menopause is related with reduced risk of dementia and cognitive impairment [74–76], other studies report a higher dementia risk, and an earlier age of onset of AD dementia [77, 78] or no effects [79]. These inconsistencies could be due to confounding factors and selection biases, for instance, inconsistent findings could be because of different stages of menopause;

however, to date, limited research has compared these transitional phases of perimenopausal, menopause, and postmenopausal. At each phase, there are differences in hormone levels and related symptoms. Given that postmenopausal women experience a profound decline in estrogen levels due to the menopausal transition while men either maintain their lifelong testosterone levels or experience a gradual decline [80], such dynamics in the female endocrinological response may account for the increased susceptibility of AD dementia risk in women. In the Framingham Study, postmenopausal women showed significantly higher tauopathy compared to age-matched men, while no differences in tauopathy were found between premenopausal women and men [81]. Interestingly, menopausal status was not associated with amyloid deposition or age. Further, amyloid status and *APOE-ε4* carrier status did not influence sex or menopause associations with tauopathy, suggesting tauopathy might be independent of AD-related pathways [81]. In contrast, data from the Wisconsin Registry for Alzheimer's Prevention cohort suggested that women with elevated amyloid deposition showed higher tau compared with age-matched men. Moreover, earlier age at menopause and late initiation of hormone replacement therapy (HRT) were associated with increased tau vulnerability in women, especially in the presence of elevated neocortical amyloid. In fact, women with more than 5 years delay of HRT initiation following menopause showed higher levels of tau, relative to those women who initiated HRT in close proximity to their menopause [82]. Other work from the European Prevention of Alzheimer's Disease cohort also indicated that earlier introduction of HRT was associated with improved delayed memory and larger entorhinal and amygdala volumes, but only in *APOE-ε4* carriers [83]. Taken together, these studies highlight the interplay of menopausal, and related treatment, timing on disease processes in postmenopausal women, which might further contribute to sex differences in AD risk. Further, they might explain previous contradictory reports of HRT use and dementia risk [84–88]. Presently, the evidence to suggest that HRT use will be beneficial to cognition and dementia risk is still controversial. Given the complexity of possible contributing factors, it would be advantageous to consider HRT efficacy in the context of genetic status, reproductive history, timing, duration, and lifetime exposure to synthetic hormones.

Although less well-studied, increasing age has been accompanied by depleting levels of testosterone in men. More recently, a number of studies have reported an association between age-related reduction in testosterone as a result of andropause and increased risk of cognitive decline and AD-dementia [89–92]. Despite these findings reported, studies were underpowered, and the evidence is mixed and inconclusive. It seems highly likely that a reduction in sex hormones is a potential risk factor for cognitive impairment. Results from studies investigating the correlation between testosterone supplementation therapy and cognitive performance have been inconsistent as they have been when investigating HRT [93]. Although small sample size is the biggest factor to consider, other biological factors, such as reproductive history, hormone treatment timing, and lifetime exposure, could be also relevant.

4 Sex Differences in ADRD Risk Factors

Although ADRD can affect both men and women, women have a 1.6-fold higher chance of being affected in their lifetime than their male counterparts [94]. While there is no known reason for this disparity, empirical evidence indicates that various factors may contribute to and explain these differences in ADRD risk factors between sexes. These factors include genetic and biological factors, cardiometabolic risk factors, mental health conditions, as well as socioeconomic and lifestyle factors [71, 94].

Cardiometabolic risk factors such as metabolic conditions, midlife obesity, midlife diabetes, insulin resistance, midlife hypertension, high cholesterol in midlife, stroke, and physical inactivity have been associated with increased ADRD risk for both men and women [95]. For example, cardiovascular diseases are associated with more than a 50% increase in ADRD risk for both men and women [96]. Although women have a higher incidence of midlife hypertension during menopause than men in their age group, there is still no strong association between blood pressure and ADRD in women [71]. However, women experience higher cardiometabolic risk factors than men during the menopausal transition and menopause due to hormonal changes, which affect energy expenditure, energy intake, and body composition, such as increased fat mass, particularly increased abdominal fat deposition [97]. Moreover, a systematic review conducted in six African countries (Egypt, Central Africa Republic, Democratic Republic of Congo, South Africa, Nigeria, and Benin) indicated that age, female sex, and cardiovascular diseases were risk factors for the prevalence of AD [98]. These results were also replicated in a cross-sectional study conducted in China, which showed that the prevalence of dementia and MCI increased with age, female sex, parental history, and other modifiable factors such as education level and rural residency [99]. However, these two studies did not account for the interaction between sex and cardiovascular factors. This emphasizes the necessity for longitudinal studies in both high-income and low-to middle-income countries to unravel potential sex-specific pathways linking cardiometabolic risk factors to ADRD [71]. It is worth noting that gestation hypertensive syndromes, such as chronic hypertension, preeclampsia, and eclampsia, have been associated with brain lesions, atrophy, and cognitive decline, occurring as early as 5 years and up to 15 years after the index pregnancy. Additionally, these conditions have been linked to an increased risk of cardiovascular diseases and various forms of dementia. This suggests that a maternal history of hypertensive disorders significantly increases the likelihood of developing vascular diseases and dementia [100].

Other mental health conditions that increase an individuals' likelihood of being diagnosed with ADRD include depression, anxiety, and chronic stress, which have detrimental effects on the brain such as inflammation and glucose intolerance, thereby increasing the risk for developing ADRD [95]. A meta-analysis study found that late-life depression is associated with a 1.85-fold increase in all-cause dementia, a 1.65-fold increase in AD dementia, and a 2.52-fold increase in vascular dementia [101]. Studies indicate that women are approximately 1.7 to 2.7 times more likely than men to report major lifetime depression [102]. This vulnerability typically emerges in early adolescence and persists through middle age, lasting into their 50s [103]. Hormonal fluctuations, particularly in estrogen and progesterone levels (which is usually more pronounced during puberty and pregnancy), and heightened sensitivity to neurochemicals involved in regulating mood in the brain may contribute to women's increased susceptibility to depression during this period [104]. This vulnerability also extends into the perimenopausal period, with high levels of depression reported during the menopausal transition. However, there is insufficient consensus supporting a direct association between depression and reproductive status during menopausal transition [103, 104].

Biological and cardiometabolic risk factors, along with other medical conditions, may partially explain sex differences in AD risk. However, non-biological factors such as socioeconomic and lifestyle factors, including education, cultural values, language, diet, sleep, physical activity, immigration history, traumatic brain injury (TBI), social isolation, and environmental factors, have been found to be associated with dementia [94]. It is well established that education plays a crucial role in cognitive functioning and dementia risk later in life, as duration of formal education is associated with both cognitive abilities and the risk of developing dementia in later adulthood [105]. Socially constructed gender norms (household work responsibilities and caregiving), poverty, immigration, armed conflict/violence, and violence against girls and children have led to gender inequalities in access to education and high-demand job roles for women in both low- and middle-income (LMIC), as well as high-income (HIC), countries [71]. However, findings from a recent study by Bloomberg et al. (2021) found that reduced disparities in education attainment and socioeconomic status could lessen and explain the sex differences in disease risk and cognitive decline [106]. In the past decade, there has been progressive improvement in educational attainment and socioeconomic status for women. This will be reflected in current and more recent birth cohorts. However, the effect of residual confounding on the association between education and dementia remains as education is incompletely measured. Most datasets measure educational attainment via years or level of education. Data regarding the subject or content are scarcely collected, which would result in a large variation in cognitive stimulation associated with the same education attainment. This may also be the case when investigating the association between employment and dementia [107]. We do not measure the tasks or hours the employment entails, thus limiting our understanding of sex-differences in lifetime cognitive stimulation associated with employment and dementia risk.

In regard to TBI, few studies have examined how the relationship between TBI and ADRD differs by sex, and the results have been inconsistent [71]. Although women are less predisposed to TBI than men in sports and occupation, which are the two biggest risks for TBI and affect men disproportionately more than women, women, particularly from LMICs, are more likely to experience TBI from head injuries due to domestic violence than their counterparts in HICs [108]. According to the World Health Organization, one in three women worldwide has experienced physical or sexual violence from an intimate partner or non-partner, with women from LMICs being disproportionately affected compared to their counterparts from HICs, where treatment access may be limited [109].

Sex differences exist in sleep, diet, and physical activity due to various physiological, biological, and behavioral norms. While women are less likely to have obstructive sleep apnea (OSA) compared to men, they experience higher rates of anxiety and depression [12, 110], which can impact sleep patterns [111]. Additionally, hormonal changes during puberty, pregnancy, breastfeeding, and the menopausal transition can contribute to sleep disturbances and sleep disorders [112]. Bixler et al.'s study indicates that postmenopausal women have three times the prevalence of OSA compared to premenopausal women [111]. Social and cultural beliefs also influence dietary habits in men and women [71, 113]. While there have been limited population-based studies examining the relationship between diet and ADRD by sex, numerous studies have explored the effects of various dietary patterns such as the Mediterranean, DASH (Dietary Approaches to Stop Hypertension), MIND (Mediterranean-DASH Intervention for Neurodegenerative Delay), ketogenic, and modified Mediterranean-ketogenic diets [114]. These diets have been positively correlated with the prevention and reduction of ADRD, both directly and indirectly, by reducing risk factors such as diabetes, obesity, and cardiovascular disease [115].

Physical activity in midlife is associated with reduced risk for ADRD. On average, women exercise less than men across their lifetime [71]. This is attributed to gender roles such as caregiving and lack of encouragement to engage in physical activities. At the same time, evidence shows that physical inactivity is associated with obesity and diabetes, which are risks for ADRD and particularly for women in midlife [115, 116]. Physical activities such as yoga and aerobic exercises have shown greater cognitive gains for women than men and reduced chances for ADRD in general [117, 118].

5 Discussion

Evidence from current literature suggests that several factors may contribute to higher incidence of dementia among women. It remains unclear to what extent each category of risk factors contributes and explains the sex differences observed. The reported sex differences in risk profiles are, in most cases, small in magnitude and inconsistent across studies, for several reasons. As mentioned above, one of the factors that has – as yet – not been fully considered is age at the time of the life event or risk factor in relation to dementia risk. For example, when considering depression, it is not clear whether earlier onset and longer depression duration exacerbate the risk of dementia. This is extremely important when considering sex-specific risk factors and conditions that may have a detrimental effect on general health and well-being.

Studies investigating the association between sex-specific risk factors and dementia have been sparse. Understanding these biological associations could explain the sex differences in dementia risk on a biological level. To date, no studies have investigated the correlation between lifetime exposure to estrogen and disease risk. Current studies have only looked at reproductive years and pregnancy, but results from these findings cannot be used to draw inferences regarding exposure to sex hormones. Another significant limitation is that – to the best of our knowledge – no study has considered all reproductive events/variables in relation to dementia risk. One of the less studied variables is the use of testosterone therapy and dementia risk. Studies have had small sample sizes partly because the data have not been available. This could be because we have placed more focus on, the more commonly used, HRT. Understanding the role of testosterone could improve our understanding of the biological basis of sex differences in dementia risk.

One of the most significant risk factors to consider when investigating sex differences is the association between education and socioeconomic factors and cognition. Both education and cognitive reserve (i.e., the efficiency, capacity, and flexibility of cognitive processes [119]) heavily influence cognition and disease risk, but the datasets currently available are from a birth cohort of children born during World War II (WWII), a time where women's access to education, employment, and higher socioeconomic status was typically disadvantaged compared to age-matched men [120, 121]. Future datasets from birth cohorts in modern eras might explore whether equitable socioeconomic access between genders/sexes reduces observed sex differences in dementia incidence.

Similarly, ethnicity representation within these datasets predominantly features individuals of White Caucasian descent. Indeed, the bulk of dementia research stems from European nations, Australia, or America, with scant data or studies hailing from Asia, Africa, or the Middle East. Consequently, the applicability of their findings is restricted. Addressing this research gap necessitates intensified focus on sex differences in dementia risk across diverse economic landscapes, encompassing both high and low-income countries.

In response to this imperative, the Brain and Mind Institute at Aga Khan University in Nairobi, Kenya, has launched the Female Brain Health and Endocrine Research in Africa Study (FemBER-AFRICA), in collaboration with international partners. This initiative, sponsored by the Alzheimer's Association, aims to investigate female brain health and endocrine concerns in the African context. Enrolling 250 participants aged 35 and above, the study comprehensively explores the clinical spectrum of AD. The research methodology entails establishing a cohort with thorough phenotyping and integrating diverse biomarker analyses, including genetics, cerebrospinal fluid, plasma, saliva, activity patterns, as well as brain and retinal imaging. Moreover, culturally tailored neuropsychological assessments, encompassing traditional and digital approaches, are incorporated, along with the evaluation of lifestyle factors such as diet and exercise. The study further delves into comprehensive clinical evaluations and gender-specific health metrics, including reproductive and endocrine health. We will additionally examine social and environmental determinants of health, encompassing socio-demographic factors, socioeconomic status, social networks, support systems, lifetime stress, and resilience levels.

This project signifies a pivotal advancement in ADRD research within Africa, establishing a culturally sensitive framework, laying the groundwork for future endeavors focused on risk reduction and preventive measures, tailored to the unique environmental and cultural contexts prevalent across African settings, extending beyond the borders of Kenya.

Notably, the discrepancies observed in studies exploring sex differences in risk factors associated with AD could partly be explained by other factors that are not always considered. One of these factors could be the differences in the AD diagnostic criteria used, different cognitive assessments, differences in other comorbidities, and small statistical power. It is also crucial that studies define sex and gender when exploring sex differences in AD and risk factors associated with disease onset and progression. From an analytical standpoint, independent evaluation of risk factors via sex-stratified analysis or using sex-based norms could be optimal for informing on the direct effect of sex.

Finally, it is important to contemplate if it is a specific risk factor that increases the risk of developing dementia, or could it be the biological and mental impact it has on the individual's cognitive health in the long term. Additionally, for targeting dementia, prevention and intervention studies should consider both cognition and AD pathology as outcomes to target those at the preclinical stages of AD.

6 Concluding Remarks

In summary, there is mounting evidence demonstrating sex differences in disease progression, incidence, symptoms, and risk profiles for AD and dementia. However, the mechanisms by which women are at higher risk of developing cognitive impairment and dementia remain unclear. Sex differences in AD pathological pathways may exist. However, with limited curative treatments, unravelling how sex influences brain pathology, cognitive function, and other established and lesser studied risk factors may aid in the development of more personalized interventions, improve risk reduction guidelines, and facilitate precision medicine strategies for drug development. Moreover, men and women may respond differently to disease-modifying drugs and interventions. Evidence on sex differences continues to emerge and gaps in our understanding pinpoint the importance of considering sex and gender in dementia prevention research. Such work should include a life course approach since sex differences may exist across the life course: from the fetal development stage until death. Finally, evidence on studies investigating the association between gender-based risk factors and dementia is limited but essential.

References

1. 2020 Alzheimer's disease facts and figures. Alzheimers Dement, 2020.
2. Ferri CP, et al. Global prevalence of dementia: a Delphi consensus study. Lancet. 2005;366(9503):2112–7.
3. Jack CR Jr, et al. Age, sex, and APOE epsilon4 effects on memory, brain structure, and beta-amyloid across the adult life span. JAMA Neurol. 2015;72(5):511–9.

4. Cao Q, et al. The prevalence of dementia: a systematic review and meta-analysis. J Alzheimers Dis. 2020;73(3):1157–66.
5. Mazure CM, Swendsen J. Sex differences in Alzheimer's disease and other dementias. Lancet Neurol. 2016;15(5):451–2.
6. Niu H, et al. Prevalence and incidence of Alzheimer's disease in Europe: a meta-analysis. Neurologia. 2017;32(8):523–32.
7. Huque H, et al. Could country-level factors explain sex differences in dementia incidence and prevalence? A systematic review and meta-analysis. J Alzheimers Dis. 2023;91(4):1231–41.
8. Mielke MM, Vemuri P, Rocca WA. Clinical epidemiology of Alzheimer's disease: assessing sex and gender differences. Clin Epidemiol. 2014;6:37–48.
9. Matthews FE, et al. A two decade dementia incidence comparison from the cognitive function and ageing studies I and II. Nat Commun. 2016;7:11398.
10. Mielke MM. Sex and gender differences in Alzheimer's disease dementia. Psychiatr Times. 2018;35(11):14–7.
11. Vila-Castelar C, et al. Sex and gender considerations in dementia: a call for global research. Nat Aging. 2023;3(5):463–5.
12. Udeh-Momoh C, Watermeyer T. Female specific risk factors for the development of Alzheimer's disease neuropathology and cognitive impairment: call for a precision medicine approach. Ageing Res Rev. 2021;71:101459.
13. Livingston G, et al. Dementia prevention, intervention, and care: 2020 report of the Lancet Commission. Lancet. 2020;396(10248):413–46.
14. Zülke AE, et al. Gender-specific design and effectiveness of non-pharmacological interventions against cognitive decline—systematic review and meta-analysis of randomized controlled trials. J Prev Alzheimers Dis. 2023;10(1):69–82.
15. Ngandu T, et al. A 2 year multidomain intervention of diet, exercise, cognitive training, and vascular risk monitoring versus control to prevent cognitive decline in at-risk elderly people (FINGER): a randomised controlled trial. Lancet. 2015;385(9984):2255–63.
16. Yassine HN, Schneider LS. Lessons from the multidomain alzheimer preventive trial. Lancet Neurol. 2017;16(8):585–6.
17. Hoevenaar-Blom MP, et al. Targeting vascular risk factors to reduce dementia incidence in old age: extended follow-up of the prevention of dementia by intensive vascular care (preDIVA) randomized clinical trial. JAMA Neurol. 2021;78(12):1527–8.
18. Kivipelto M, et al. World-wide FINGERS network: a global approach to risk reduction and prevention of dementia. Alzheimers Dement. 2020;16(7):1078–94.
19. Shireen Sindi IK, Ngandue T, Rosenbergf A, Kulmalaa J, Johanssong L, Wetterbergg H, Skoogh J, Sjöberg L, Wang H-X, Fratiglionic L, Skoogg I, Kivipelto M. Sex differences in dementia and response to a lifestyle intervention: evidence from Nordic population-based studies and a prevention trial. K. Institutet, Editor; 2020.
20. Jack CR Jr, et al. NIA-AA research framework: toward a biological definition of Alzheimer's disease. Alzheimers Dement. 2018;14(4):535–62.
21. Jack CR Jr, et al. Age-specific and sex-specific prevalence of cerebral beta-amyloidosis, tauopathy, and neurodegeneration in cognitively unimpaired individuals aged 50-95 years: a cross-sectional study. Lancet Neurol. 2017;16(6):435–44.
22. Sperling RA, et al. Toward defining the preclinical stages of Alzheimer's disease: recommendations from the National Institute on Aging-Alzheimer's Association workgroups on diagnostic guidelines for Alzheimer's disease. Alzheimers Dement. 2011;7(3):280–92.
23. Mielke MM, et al. Comparison of variables associated with cerebrospinal fluid neurofilament, total-tau, and neurogranin. Alzheimers Dement. 2019;15(11):1437–47.
24. Ardekani BA, Convit A, Bachman AH. Analysis of the MIRIAD data shows sex differences in hippocampal atrophy progression. J Alzheimers Dis. 2016;50(3):847–57.
25. Ruitenberg A, et al. Incidence of dementia: does gender make a difference? Neurobiol Aging. 2001;22(4):575–80.
26. Lashley T, et al. Molecular biomarkers of Alzheimer's disease: progress and prospects. Dis Model Mech. 2018;11(5):dmm031781.

27. Buckley RF, et al. Sex, amyloid, and APOE epsilon4 and risk of cognitive decline in preclinical Alzheimer's disease: findings from three well-characterized cohorts. Alzheimers Dement. 2018;14(9):1193–203.
28. Liu C, et al. Sex-specific biomarkers in Alzheimer's disease progression: Framingham heart study. Alzheimers Dement (Amst). 2022;14(1):e12369.
29. Cavedo E, et al. Sex differences in functional and molecular neuroimaging biomarkers of Alzheimer's disease in cognitively normal older adults with subjective memory complaints. Alzheimers Dement. 2018;14(9):1204–15.
30. Mielke MM. Consideration of sex differences in the measurement and interpretation of Alzheimer disease-related biofluid-based biomarkers. J Appl Lab Med. 2020;5(1):158–69.
31. Hohman TJ, et al. Sex-specific association of apolipoprotein E with cerebrospinal fluid levels of tau. JAMA Neurol. 2018;75(8):989–98.
32. Altmann A, et al. Sex modifies the APOE-related risk of developing Alzheimer disease. Ann Neurol. 2014;75(4):563–73.
33. Oveisgharan S, et al. Sex differences in Alzheimer's disease and common neuropathologies of aging. Acta Neuropathol. 2018;136(6):887–900.
34. Buckley RF, et al. Sex differences in the association of global amyloid and regional tau deposition measured by positron emission tomography in clinically normal older adults. JAMA Neurol. 2019;76(5):542–51.
35. Jack CR Jr, et al. Longitudinal tau PET in ageing and Alzheimer's disease. Brain. 2018;141(5):1517–28.
36. Tsiknia AA, et al. Sex differences in plasma p-tau181 associations with Alzheimer's disease biomarkers, cognitive decline, and clinical progression. Mol Psychiatry. 2022;27(10):4314–22.
37. Skillback T, et al. Sex differences in CSF biomarkers for neurodegeneration and blood-brain barrier integrity. Alzheimers Dement (Amst). 2021;13(1):e12141.
38. Ferretti MT, et al. Sex differences in Alzheimer disease – the gateway to precision medicine. Nat Rev Neurol. 2018;14(8):457–69.
39. Vila-Castelar C, et al. Sex differences in blood biomarkers and cognitive performance in individuals with autosomal dominant Alzheimer's disease. Alzheimers Dement. 2023;19(9):4127–38.
40. Willemse EAJ, et al. Neurogranin as biomarker in CSF is non-specific to Alzheimer's disease dementia. Neurobiol Aging. 2021;108:99–109.
41. Portelius E, et al. Cerebrospinal fluid neurogranin concentration in neurodegeneration: relation to clinical phenotypes and neuropathology. Acta Neuropathol. 2018;136(3):363–76.
42. Wang L. I. Alzheimer's Disease Neuroimaging, Association of cerebrospinal fluid Neurogranin with Alzheimer's disease. Aging Clin Exp Res. 2019;31(2):185–91.
43. van der Flier WM, et al. The effect of APOE genotype on clinical phenotype in Alzheimer disease. Neurology. 2006;67(3):526–7.
44. Zuo L, et al. Variation at APOE and STH loci and Alzheimer's disease. Behav Brain Funct. 2006;2:13.
45. Hsu M, et al. Sex differences in gene expression patterns associated with the APOE4 allele. F1000Res. 2019;8:387.
46. Neu SC, et al. Apolipoprotein E genotype and sex risk factors for Alzheimer disease: a meta-analysis. JAMA Neurol. 2017;74(10):1178–89.
47. Nebel RA, et al. Understanding the impact of sex and gender in Alzheimer's disease: a call to action. Alzheimers Dement. 2018;14(9):1171–83.
48. Craig MC, Murphy DG. Estrogen: effects on normal brain function and neuropsychiatric disorders. Climacteric. 2007;10(Suppl 2):97–104.
49. Damoiseaux JS, et al. Gender modulates the APOE epsilon4 effect in healthy older adults: convergent evidence from functional brain connectivity and spinal fluid tau levels. J Neurosci. 2012;32(24):8254–62.
50. Sampedro F, et al. APOE-by-sex interactions on brain structure and metabolism in healthy elderly controls. Oncotarget. 2015;6(29):26663–74.

51. Morgan K, et al. Incidence of dementia in a representative British sample. Br J Psychiatry. 1993;163:467–70.
52. Bachman DL, et al. Incidence of dementia and probable Alzheimer's disease in a general population: the Framingham study. Neurology. 1993;43(3 Pt 1):515–9.
53. Yan S, Zhou Y, Lu J. Sex modifies the APOE e4 dose dependent effect on brain region-specific tau deposition measured by 18F-flortaucipir PET in individuals with cognitive impairment. J Nucl Med. 2021;62(supplement 1):1052.
54. Babapour Mofrad R, et al. Sex differences in CSF biomarkers vary by Alzheimer disease stage and APOE epsilon4 genotype. Neurology. 2020;95(17):e2378–88.
55. Kumar A, et al. MicroRNA expression in extracellular vesicles as a novel blood-based biomarker for Alzheimer's disease. Alzheimers Dement. 2023;19(11):4952–66.
56. Nazir S. Salivary biomarkers: the early diagnosis of Alzheimer's disease. Aging Medicine. 2024;7(2):202–13.
57. Rani P, Vivek S, Ram M. A systematic review on urinary biomarkers for early diagnosis of Alzheimer's disease (AD). Int J Nutr Pharmacol Neurol Dis. 2020;10(3):91–8.
58. Yuan A, Lee CS. Retinal biomarkers for Alzheimer disease: the facts and the future. Asia Pac J Ophthalmol (Phila). 2022;11(2):140–8.
59. Vina J, Lloret A. Why women have more Alzheimer's disease than men: gender and mitochondrial toxicity of amyloid-beta peptide. J Alzheimers Dis. 2010;20(Suppl 2):S527–33.
60. Barron AM, Pike CJ. Sex hormones, aging, and Alzheimer's disease. Front Biosci (Elite Ed). 2012;4:976–97.
61. Craig MC, et al. Physiological variation in estradiol and brain function: a functional magnetic resonance imaging study of verbal memory across the follicular phase of the menstrual cycle. Horm Behav. 2008;53(4):503–8.
62. Brann DW, et al. Neurotrophic and neuroprotective actions of estrogen: basic mechanisms and clinical implications. Steroids. 2007;72(5):381–405.
63. Wu MV, et al. Estrogen masculinizes neural pathways and sex-specific behaviors. Cell. 2009;139(1):61–72.
64. Fox M, et al. Women's pregnancy life history and Alzheimer's risk: can immunoregulation explain the link? Am J Alzheimers Dis Other Dement. 2018;33(8):516–26.
65. Colucci M, et al. The number of pregnancies is a risk factor for Alzheimer's disease. Eur J Neurol. 2006;13(12):1374–7.
66. Sobow T, Kutter E, Kloszewska I. Hormonal decline indicator in women (age at menopause) modifies age of onset in sporadic Alzheimer's disease. Alzheimers Rep. 1999;2(1):27–9.
67. Basit S, Wohlfahrt J, Boyd HA. Pre-eclampsia and risk of dementia later in life: nationwide cohort study. BMJ. 2018;363:k4109.
68. Jang H, et al. Differential effects of completed and incomplete pregnancies on the risk of Alzheimer disease. Neurology. 2018;91(7):e643–51.
69. Gilsanz P, et al. Women's reproductive history and dementia risk. 2018, AAIC: Alzheimier's International Conference (AAIC).
70. Li J, et al. Investigating the causal association of postpartum depression with cerebrovascular diseases and cognitive impairment: a Mendelian randomization study. Front Psych. 2023;14:1196055.
71. Mielke MM, et al. Consideration of sex and gender in Alzheimer's disease and related disorders from a global perspective. Alzheimers Dement. 2022;18(12):2707–24.
72. Sochocka M, et al. Cognitive decline in early and premature menopause. Int J Mol Sci. 2023;24(7):6566.
73. Mosconi L, et al. Increased Alzheimer's risk during the menopause transition: a 3-year longitudinal brain imaging study. PLoS One. 2018;13(12):e0207885.
74. Fox M, Berzuini C, Knapp LA. Cumulative estrogen exposure, number of menstrual cycles, and Alzheimer's risk in a cohort of British women. Psychoneuroendocrinology. 2013;38(12):2973–82.

75. Georgakis MK, et al. Age at menopause and duration of reproductive period in association with dementia and cognitive function: a systematic review and meta-analysis. Psychoneuroendocrinology. 2016;73:224–43.
76. Gilsanz P, et al. Reproductive period and risk of dementia in a diverse cohort of health care members. Neurology. 2019;92(17):e2005–14.
77. Najar J, et al. Reproductive period and dementia: a 44-year longitudinal population study of Swedish women. Alzheimers Dement. 2020;16(8):1153–63.
78. Najar J, et al. Reproductive period and preclinical cerebrospinal fluid markers for Alzheimer disease: a 25-year study. Menopause. 2021;28(10):1099–107.
79. Prince MJ, et al. Reproductive period, endogenous estrogen exposure and dementia incidence among women in Latin America and China; a 10/66 population-based cohort study. PLoS One. 2018;13(2):e0192889.
80. Matsumoto AM. Andropause: clinical implications of the decline in serum testosterone levels with aging in men. J Gerontol A Biol Sci Med Sci. 2002;57(2):M76–99.
81. Buckley RF, et al. Menopause status moderates sex differences in Tau Burden: A Framingham PET Study. Ann Neurol. 2022;92(1):11–22.
82. Coughlan GT, et al. Association of age at menopause and hormone therapy use with tau and beta-amyloid positron emission tomography. JAMA Neurol. 2023;80(5):462–73.
83. Saleh RNM, et al. Hormone replacement therapy is associated with improved cognition and larger brain volumes in at-risk APOE4 women: results from the European prevention of Alzheimer's disease (EPAD) cohort. Alzheimers Res Ther. 2023;15(1):10.
84. Zandi PP, et al. Hormone replacement therapy and incidence of Alzheimer disease in older women: the Cache County study. JAMA. 2002;288(17):2123–9.
85. Nelson HD, et al. Postmenopausal hormone replacement therapy: scientific review. JAMA. 2002;288(7):872–81.
86. Grady D, et al. Effect of postmenopausal hormone therapy on cognitive function: the heart and estrogen/progestin replacement study. Am J Med. 2002;113(7):543–8.
87. Paganini-Hill A, Corrada MM, Kawas CH. Increased longevity in older users of postmenopausal estrogen therapy: the leisure world cohort study. Menopause. 2018;25(11):1256–61.
88. Shumaker SA, et al. Conjugated equine estrogens and incidence of probable dementia and mild cognitive impairment in postmenopausal women: Women's Health Initiative memory study. JAMA. 2004;291(24):2947–58.
89. Hogervorst E, et al. Low free testosterone is an independent risk factor for Alzheimer's disease. Exp Gerontol. 2004;39(11–12):1633–9.
90. Wahjoepramono EJ, et al. The effects of testosterone supplementation on cognitive functioning in older men. CNS Neurol Disord Drug Targets. 2016;15(3):337–43.
91. Lv W, et al. Low testosterone level and risk of Alzheimer's disease in the elderly men: a systematic review and meta-analysis. Mol Neurobiol. 2016;53(4):2679–84.
92. Moffat SD, et al. Free testosterone and risk for Alzheimer disease in older men. Neurology. 2004;62(2):188–93.
93. Hua JT, Hildreth KL, Pelak VS. Effects of testosterone therapy on cognitive function in aging: a systematic review. Cogn Behav Neurol. 2016;29(3):122–38.
94. Sehar U, Kopel J, Reddy PH. Alzheimer's disease and its related dementias in US native Americans: a major public health concern. Ageing Res Rev. 2023;90:102027.
95. Greene AN, Solomon MB, Privette Vinnedge LM. Novel molecular mechanisms in Alzheimer's disease: the potential role of DEK in disease pathogenesis. Front Aging Neurosci. 2022;14:1018180.
96. Titcomb TJ, et al. Association of type 2 diabetes mellitus with dementia-related and non-dementia-related mortality among postmenopausal women: a secondary competing risks analysis of the women's health initiative. Alzheimers Dement. 2024;20(1):234–42.
97. Marlatt KL, et al. Body composition and cardiometabolic health across the menopause transition. Obesity (Silver Spring). 2022;30(1):14–27.

98. George-Carey R, et al. An estimate of the prevalence of dementia in Africa: a systematic analysis. J Glob Health. 2012;2(2):020401.
99. Fu J, et al. Age-and sex-specific prevalence and modifiable risk factors of mild cognitive impairment among older adults in China: a population-based observational study. Front Aging Neurosci. 2020;12:578742.
100. Schliep KC, et al. Association between hypertensive disorders of pregnancy and dementia: a systematic review and meta-analysis. Hypertension. 2023;80(2):257–67.
101. Diniz BS, et al. Late-life depression and risk of vascular dementia and Alzheimer's disease: systematic review and meta-analysis of community-based cohort studies. Br J Psychiatry. 2013;202(5):329–35.
102. Burt VK, Stein K. Epidemiology of depression throughout the female life cycle. J Clin Psychiatry. 2002;63:9–15.
103. Kessler RC, et al. Sex and depression in the National Comorbidity Survey. I: lifetime prevalence, chronicity and recurrence. J Affect Disord. 1993;29(2–3):85–96.
104. Sassarini DJ. Depression in midlife women. Maturitas. 2016;94:149–54.
105. Lovden M, et al. Education and cognitive functioning across the life span. Psychol Sci Public Interest. 2020;21(1):6–41.
106. Mikaela Bloomberg M, et al. Sex differences and the role of education in cognitive ageing: analysis of two UK-based prospective cohort studies. Lancet Public Health. 2021;6(2):e106–15.
107. Subramaniapillai S, et al. Sex and gender differences in cognitive and brain reserve: implications for Alzheimer's disease in women. Front Neuroendocrinol. 2021;60:100879.
108. Adamson MM, et al. Brain injury and dementia in Pakistan: current perspectives. Front Neurol. 2020;11:299.
109. Organization, W.H. Devastatingly pervasive: 1 in 3 women globally experience violence. 2021.
110. Zeng L-N, et al. Gender difference in the prevalence of insomnia: a meta-analysis of observational studies. Front Psych. 2020;11:577429.
111. Bixler EO, et al. Prevalence of sleep-disordered breathing in women: effects of gender. Am J Respir Crit Care Med. 2001;163(3 Pt 1):608–13.
112. Lord C, Sekerovic Z, Carrier J. Sleep regulation and sex hormones exposure in men and women across adulthood. Pathol Biol (Paris). 2014;62(5):302–10.
113. Grzymisławska M, et al. Do nutritional behaviors depend on biological sex and cultural gender? Adv Clin Exp Med. 2020;29(1):165–72.
114. Nucci D, et al. Association between Mediterranean diet and dementia and Alzheimer disease: a systematic review with meta-analysis. Aging Clin Exp Res. 2024;36(1):1–21.
115. Ellouze I, et al. Dietary patterns and Alzheimer's disease: an updated review linking nutrition to neuroscience. Nutrients. 2023;15(14):3204.
116. Hillmeister P, et al. Exercise and cardiovascular diseases. Acta Physiol (Oxf). 2020;229(2):e13476.
117. Kilpatrick LA, et al. Impact of yoga versus memory enhancement training on hippocampal connectivity in older women at risk for Alzheimer's disease. J Alzheimers Dis. 2023;95(1):149–59.
118. Nocera JR, et al. Effects of combined aerobic exercise and cognitive training on verbal fluency in older adults. Gerontol Geriatr Med. 2020;6:2333721419896884.
119. Stern Y, et al. Whitepaper: defining and investigating cognitive reserve, brain reserve, and brain maintenance. Alzheimers Dement. 2020;16(9):1305–11.
120. Skoog I, et al. Decreasing prevalence of dementia in 85-year olds examined 22 years apart: the influence of education and stroke. Sci Rep. 2017;7(1):6136.
121. Wetterberg H, et al. Decreasing incidence and prevalence of dementia among octogenarians: a population-based study on 3 cohorts born 30 years apart. J Gerontol A. 2023;78(6):1069–77.

Imaging Markers of Amyloid and Tau in Alzheimer Disease

Rawan Tarawneh

Abstract The advent of molecular imaging for in vivo visualization of amyloid and tau pathologies has revolutionized the field of Alzheimer disease (AD) research and offered valuable clinical and research tools to facilitate the early detection of AD-related pathology, increase diagnostic accuracy, and provide useful prognostic information regarding disease stage and future risk of cognitive decline in cognitively normal and impaired individuals. Several ligands that can reliably detect neuritic amyloid plaques, and one ligand that can detect neurofibrillary tangles constituted of hyperphosphorylated tau, have been FDA-approved for clinical use, while several other ligands are available for use in research settings.

Amyloid positron emission tomography (PET) scans are useful in detecting early pathology in cognitively normal or impaired individuals, while tau PET scans provide information regarding disease stage and the spatial distribution of tau pathology, which corresponds to regional brain atrophy and impairment in specific cognitive domains. However, a few limitations should be considered when interpreting amyloid and tau PET scans. Amyloid PET scans may be abnormal in a significant portion of older adults, especially those who have one or more copies of the apolipoprotein E4 (APOE4) allele (referred to as *APOE4* carriers) who may not progress to AD or experience symptoms during their lifetimes. Positive tau PET scans are most reliable in detecting advanced stages of tau pathology, typically seen in cognitively impaired individuals with mild to moderate dementia, and may be normal or show minimal changes in earlier disease stages. Amyloid and tau PET scans may assist in the differential diagnosis of AD, and tau PET scans may also assist in disease staging using the updated research criteria for AD. When interpreting amyloid and tau PET imaging, it is also important to consider differences in pharmacokinetics, binding affinities, and off-target binding across different tracers. In clinical settings, the interpretation of amyloid and tau PET scans relies mostly on visual assessments to determine whether a scan is "positive" or "negative"; however,

R. Tarawneh (✉)
Department of Neurology, University of New Mexico, Albuquerque, NM, USA
e-mail: raltarawneh@salud.unm.edu

R. Tarawneh (ed.), *Biomarkers of Neurodegenerative Disorders*,
https://doi.org/10.1007/978-3-031-80849-4_20

several quantitative methods are available to estimate the regional burden of amyloid or tau pathology in research settings.

We here describe the various ligands used in amyloid and tau PET imaging and summarize the current literature regarding their clinical, neuropathological and imaging correlates. We also review the published appropriate use criteria for amyloid PET imaging and discuss future directions in the use of amyloid and tau PET in the diagnosis or staging of AD.

Keywords Positron emission tomography · Imaging · Amyloid · Tau · SUVR

1 Introduction

The extracellular accumulation of aggregated amyloid-β (Aβ) peptides in the form of amyloid plaques and the intracellular aggregation of hyperphosphorylated tau in the form of neurofibrillary tangles are the two main pathological hallmarks of Alzheimer disease (AD) [1, 2]. In AD, increased production of Aβ42 peptides as a result of abnormal processing of the amyloid precursor protein (*APP*), by β- and γ-secretases, in what is referred to as the "amyloidogenic" pathway, and/or reduced Aβ42 clearance results in increased brain Aβ peptides which are prone to aggregation into soluble oligomers, and subsequently, fibrillization into protofibrils, fibrils, and eventually insoluble amyloid plaques [3]. Amyloid aggregates can be in the form of "diffuse" plaques which consist mostly of aggregated Aβ42, or "neuritic" plaques which consist of a core of amyloid proteins surrounded by dystrophic neurites containing hyperphosphorylated tau [4].

The updated research framework for AD, proposed by the National Institute of Aging (NIA)-Alzheimer's Association (AA) in 2018 [5] and recently revised by the AA Workgroup in 2024 [6], defines AD as a biological construct based on biomarker evidence of amyloid and tau pathology regardless of the presence or absence of cognitive impairment, and thus, allows a biomarker-based definition of AD to be made prior to symptom onset, a stage referred to as "preclinical" or "presymptomatic" AD. Ligand-based positron emission tomography (PET) is an established molecular imaging method that can detect, quantify, and map the regional distribution of molecular targets such as aggregated amyloid and tau proteins in vivo using radioactive positron emitting imaging agents (i.e., radiopharmaceuticals or "tracers") that have high sensitivity and specificity to the molecular target [7]. PET imaging using amyloid- and tau-binding ligands has emerged as an important tool to detect the presence of brain amyloid and tau aggregation, respectively, quantify whole brain (i.e., global) and regional estimates of amyloid and tau, and provide in vivo visual assessments of the regional distribution of these pathologies [8–11]. Therefore, amyloid and tau PET scans provide useful noninvasive markers for AD that can improve diagnostic accuracy, assist in differentiating AD from other neurodegenerative disorders, and provide important information regarding disease stage or severity. This chapter reviews different amyloid and tau PET tracers and their applications in various clinical and research settings.

2 Amyloid PET

2.1 Amyloid Ligands

The first successful attempt to image brain amyloid deposits was reported by Klunk et al. in 2004 using the amyloid ligand ^{11}C-labeled Pittsburgh compound B (^{11}C-PiB) [11]. ^{11}C-PiB is an analog of thioflavin-T, a benzothiazole dye used to stain amyloid deposits in histological preparations [12]. ^{11}C-PiB is radiolabeled with carbon-11, a positron-emitting isotope, which, following injection at nanomolar concentrations, crosses the blood-brain barrier and binds with high affinity to fibrillar Aβ deposits in the brain, mostly within neuritic plaques, and to vascular amyloid deposits associated with cerebral amyloid angiopathy (CAA) [12–14]. The positrons emitted by the carbon-11 decay are detected by the PET scanner, allowing the regional distribution of Aβ plaques to be visualized. ^{11}C-PiB is an established amyloid ligand that is widely used in research centers and can be produced on-site. Despite several advantages of ^{11}C-PiB including its high binding affinity to amyloid, rapid crossing of the blood-brain barrier, and quick washout from nontarget regions, a major limitation of ^{11}C-PiB is the short 20-min half-life of the tracer which limits its use to research PET centers equipped with a cyclotron and limits the time window for imaging after its administration [10, 12].

The development of ^{11}C-PiB laid the foundation for the generation of newer fluorinated tracers, labeled with ^{18}F, that have a high affinity to fibrillar amyloid and longer half-lives than ^{11}C-PiB (110-min) [15]. These include ^{18}F-florbetapir (Amyvid; Eli Lilly and Company), ^{18}F-florbetaben (Neuraceq; Life Molecular Imaging), ^{18}F-flutemetamol (Vizamyl; GE Healthcare), and ^{18}F-flutafuranol (formerly known as NAV4694) [15]. ^{11}C-PiB and ^{18}F-flutemetamol belong to the chemical class of benzothiazoles, while ^{18}F-florbetaben and ^{18}F-florbetapir are derived from stilbene and ^{18}F-flutafuranol from benzofuran [10].

^{18}F-florbetapir (Amyvid) was the first amyloid tracer to receive the United States Food and Drug Administration (FDA) approval in 2012 for in vivo estimation of Aβ plaques [16]. Three fluorine-18 (^{18}F-) ligands, ^{18}F-florbetapir (Amyvid), ^{18}F-florbetaben (Neuraceq), and ^{18}F-flutemetamol (Vizamyl), are currently approved for clinical use by the US FDA, the European Medicines Agency [17], and other regulatory agencies around the world "to estimate amyloid neuritic plaque density in adult patients with cognitive impairment who are being evaluated for AD and other causes of cognitive decline". A fourth ^{18}F- amyloid ligand, ^{18}F-flutafuranol (formerly NAV4694), is currently under development and has not yet been approved in the United States or Europe. Table 1 shows the chemical structure and characteristics of different amyloid tracers [10].

Table 1 Characteristics of amyloid ligands used in clinical and research settings

Amyloid ligand	Dose and acquisition protocol	Visualization	Interpretation criteria for positive scan
FDA-approved for clinical use			
^{18}F-Florbetapir (Amyvid) 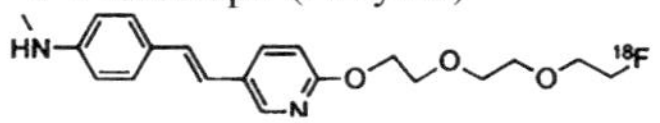	~370 MBq (10 mCi); 10- to 20-min acquisition beginning at 30–50 min p.i. (package insert guidelines) for clinical use or 50–70 min p.i. (optimized kinetics for quantification) for research use	Gray or inverse gray scale Window images to optimize GM/WM contrast in cerebellum	Loss of GM/WM contrast due to increased cortical binding in 2 or more brain regions (each larger than a single gyrus) OR 1 or more regions with intense signal where GM > WM. Regions: temporal, parietal (including precuneus), frontal, and occipital
^{18}F-Florbetaben (Neuraceq) 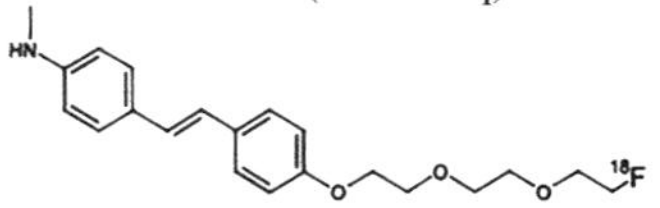	~300 MBq (8.1 mCi); 15- to 20-min acquisition beginning at 45–130 min p.i. (20-min acquisition beginning at 90–110 min p.i. for research use)	Gray or inverse grayscale Window images to optimize GM/WM contrast in cerebellum	Increased GM uptake extending to cortical margin involving most slices in at least 1 of 4 target cortical regions: temporal (lateral), parietal, posterior cingulate/ precuneus, and frontal
^{18}F-Flutemetamol (Vizamyl) 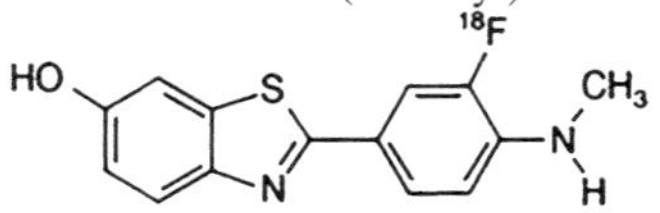	~185 MBq (5 mCi); 10- to 20-min acquisition at 60–120 min p.i. (20-min acquisition at 90–110 min p.i. for research use)	Color scale (NIH); normalize so that pons is at 90% of maximum intensity	Increased GM uptake (>50–60% peak intensity) or loss of GM/WM contrast in at least 1 of 4 cortical regions and 1 subcortical region (striatum) Regions: cortical (frontal, inferolateral parietal, precuneus/posterior cingulate, lateral temporal) and subcortical (striatum)
Research use only			
^{11}C-PiB 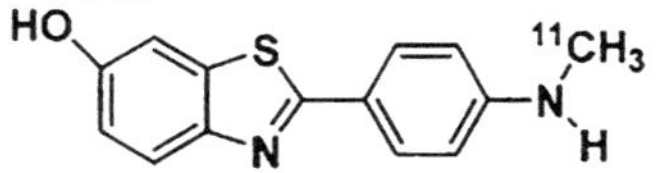	~555 MBq; dynamic 60- to 90-min acquisition (distribution volume ratio) or 20-min acquisition at 50–70 min p.i. (SUV ratio)	Color scale (NIH); window images to optimize GM/WM contrast in cerebellum	No formal guidelines (research use)
^{18}F-flutafuranol 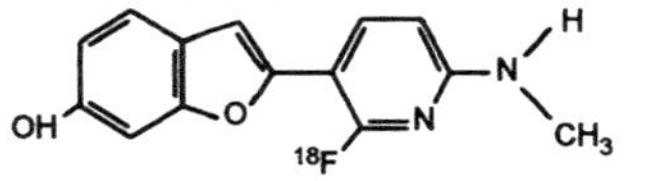	~185 MBq; 20- to 30-min acquisition beginning at 40–50 min p.i.	Color scale (NIH); window images to optimize GM/WM contrast in cerebellum	No formal guidelines (research use)

MBq 1 megabecquerel, *mCi* millicurie, *NIH* National Institute of Health, *p.i.* post-injection, *GM* Gray matter, *WM* white matter, *SUV* standardized uptake value. Adapted with permission and modified from Chapleau et al. The Role of Amyloid PET in Imaging Neurodegenerative Disorders: A Review. J Nucl Med, 2022 [10].

2.2 Neuropathological Targets of Amyloid PET

In neuropathological studies, amyloid plaques can be visualized using thioflavin fluorescent dyes, silver impregnation, Congo red, or antibody-based immunostaining [18–20]. In contrast to diffuse amyloid plaques, neuritic plaques, which are characterized by dense amyloid cores surrounded by tau-positive dystrophic neurites, are considered the pathognomonic plaque type in AD [4, 21]. The level of AD neuropathological change (ADNC) is estimated based on the topographical distribution of amyloid plaques and neurofibrillary tangles as reflected by the ABC score in the NIA-AA neuropathological guidelines [4, 21]. The ABC score includes "A": Amyloid component which is derived from the topographical distribution of any amyloid plaque type using Thal amyloid staging [22], "B": Braak staging of neurofibrillary tau tangle (NFT) pathology [2, 23], and "C": the Consortium to Establish a Registry for AD (CERAD) score which reflects an additional amyloid component to measure neuritic plaques [24]. The ABC score is used to classify individuals as having "no", "low", "intermediate", or "high" ADNC. Intermediate or high ADNC is generally considered to be clinically relevant and sufficient to explain cognitive impairment.

Several staging systems for the spatial progression of amyloid pathology have been proposed [2, 22]. In general, neocortical regions demonstrate early accumulation of amyloid-β in phase 1 including the association cortices, followed by the allocortical regions in phase 2, diencephalic nuclei, the striatum, and the cholinergic nuclei of the basal forebrain in phase 3, additional brainstem nuclei in phase 4, and finally, the cerebellum in phase 5. This pattern of spatial progression of amyloid pathology has been observed across several neuroimaging and neuropathological studies [25–27] and appears to be similar across different clinical phenotypes associated with AD (e.g., memory, language, dysexecutive, and visuospatial-predominant presentations) [28–30].

2.3 Interpretation of Amyloid PET Scans

In AD, cortical amyloid aggregation on PET is typically found in a pattern that closely resembles the distribution of amyloid aggregates on autopsy [31] and is mostly seen across large areas of the neocortex and striatum with relative sparing of the medial temporal lobes and the primary motor or sensory cortices [10]. Amyloid aggregation in the posterior cingulate/precuneus and the medial prefrontal regions is considered one of the earliest findings on amyloid PET with these areas also often showing peak amyloid tracer uptake [27, 32]. In individuals with no/minimal brain amyloid, only nonspecific binding of the tracer in the white matter is observed, whereas individuals with moderate to high amyloid pathology will have increased tracer uptake in the neocortex, which is usually diffuse by the time these individuals become symptomatic. Nonspecific white matter retention is observed with all Aβ tracers and occurs regardless of the presence of amyloid pathology. Amyloid tracers bind more strongly to neuritic than diffuse amyloid plaques [10, 14]. Further, individuals with CAA may demonstrate predominant Aβ tracer uptake in the occipital lobe, a region often affected by CAA and subsequent amyloid deposition in the walls of cerebral blood vessels [33].

Visual interpretation of amyloid PET scans relies on the comparison of the gray matter to white matter binding of the tracer and is the predominant method by which amyloid scans are interpreted in clinical practice. When binding is restricted to the white matter, and the gray matter to white matter contrast is preserved, a scan is read as negative (indicative of non-elevated brain amyloid). Conversely, when cortical gray matter binding is equal to, or more than, white matter binding, a scan is read as positive (indicative of elevated brain amyloid) [10]. Table 1 describes different guidelines for visual interpretation of amyloid scans using different tracers.

Quantitative assessments of cortical amyloid burden can also be performed, generating continuous measures of global (whole brain) and regional amyloid pathology, which are often used in research settings including clinical trials of AD therapies. The most common method involves measuring the ratio of amyloid binding in a target brain region known to accumulate brain amyloid (e.g., neocortical regions in AD) to a nonspecific reference region that is relatively spared of amyloid pathology (e.g., the cerebellum or brainstem), resulting in what is referred to as standardized uptake value ratios (SUVR). These are measured at specific time intervals post-injection when these ratios are known to be stable and differ by tracer [34, 35]. While this semiquantitative method allows for rapid interpretation of scans, it appears to be limited by several factors including susceptibility to effects of blood flow on tracer uptake and possible overestimation of the degree of pathology [36].

Another quantitative method for amyloid scan interpretation is the centiloid method which generates standardized units reflecting cortical amyloid on a scale [35, 37–39]. A centiloid unit value of 0 reflects the mean amyloid uptake in healthy young adults with no brain amyloid pathology, while a value of 100 reflects the mean amyloid uptake in cognitively impaired individuals with mild to moderate AD dementia. On this scale, centiloid unit values between 12 and 25 reflect the threshold for "amyloid positivity", although there is variation in accepted thresholds across studies (range of proposed values, 10–40 centiloid units) [39]. The centiloid approach offers a standardized method to compare amyloid scans using different tracers and across different centers that use different imaging processing methods (Fig. 1).

SUVR and centiloid-based methods of amyloid PET quantification can be performed using several available commercial software packages. Quantitative methods have been added as an adjunct to visual assessments of amyloid PET scans in clinical settings in Europe, but are not currently included in the US FDA labels for any of the approved tracers.

2.4 Clinicopathological and Imaging Correlates

Several clinicopathological studies have validated the ability of amyloid PET tracers to accurately measure brain amyloid by comparing the severity and distribution of amyloid pathology on autopsy to amyloid uptake on PET in individuals who were followed during life [31, 38, 40–42]. Data supporting the FDA approval of amyloid ligands were mostly derived from studies with short intervals between amyloid PET and autopsy (i.e., studies conducted near the end of life) [40–42]. Visual reads of amyloid scans using the 3 FDA-approved ^{18}F-labeled tracers provide 88–98% sensitivity and 80–95% specificity in distinguishing older adults with moderate to frequent neuritic plaque density (according to the CERAD scale) from those with absent to sparse plaques [40–42]. Antemortem amyloid PET scan positivity typically corresponds to Thal Phase 2–3 of amyloid pathology [43].

Marked retention of cortical ^{11}C-PiB correlates with significant cortical deposition of fibrillar Aβ aggregates in individuals with AD compared to controls [11]. Studies using distribution volume ratios have demonstrated increased cortical ^{11}C-PiB, reflective of increased cortical amyloid, in individuals with AD and Lewy body dementia (LBD), but not frontotemporal dementia (FTD) [44]. Other studies suggest that cortical regions with high ^{11}C-PiB binding correspond to regions that show reduced hypometabolism on 18F-fluorodeoxy-glucose (FDG)-PET, especially in the parietal lobes [11]. Quantitative estimates of ^{11}C-PiB scans can distinguish patients in Thal phases 3–5 from those in phases 0–2 on postmortem assessments [38]. In individuals with dominantly inherited forms of AD (e.g., due to *APP* or presenilin [*PSEN*] mutations), the highest degree of amyloid deposition is often seen in the striatum [45]. Conversely, no specific regional predilection for amyloid deposition appears to be associated with apolipoprotein E4 (*APOE4*) carriers, although amyloid deposition appears to occur at an earlier age, and may have a more rapid progression, compared to *APOE4* noncarriers [45].

The prevalence of positive amyloid scans has been examined in various clinical cohorts, including those with or without cognitive impairment. Amyloid PET scans are negative in 70–90% of cognitively normal older adults; however, this is influenced by age and the *APOE* genotype [46, 47]. The prevalence of amyloid PET positivity in cognitively normal older adults increases proportionally with age: amyloid PET positivity is reported in approximately 10% at age 50 years, 15% at age 60 years, 20% at age 70 years, 30% at age 80 years, and 40% at age 90 years [10]. Individuals who have at least 1 *APOE4* allele (i.e., *APOE4* carriers) have a two to threefold higher prevalence of amyloid pathology in any given age group compared to noncarriers [46]. In one study, the annual incidence of cognitively normal older adults over the age of 70 developing a positive amyloid scan was estimated to be 13% per year [48]. Further, uncontrolled hypertension appeared to potentially confer a higher risk of amyloid PET scan positivity in one study [49]. As amyloid PET positivity is typically seen with Thal Phases 2–3, a negative scan does not exclude the presence of amyloid pathology and does not equate to "no amyloid pathology"; rather, it suggests that amyloid pathology, if present, is below the threshold of detection on PET and not likely to be a significant contributor to cognitive impairment.

A significant portion of cognitively normal older adults have high brain amyloid on PET scans [46, 47]. As a group, these individuals carry a higher risk of developing cognitive impairment (i.e., mild cognitive impairment [MCI] or dementia) over time [50]. According to the NIA-AA biological definition of AD, the presence of brain amyloid, defined by abnormal cerebrospinal fluid (CSF) Aβ42 or Aβ42/Aβ40 levels or amyloid PET imaging, indicates the individual is on the "AD continuum"; however, brain amyloid positivity alone is insufficient to make a biomarker-based diagnosis of preclinical AD in cognitively normal adults as the latter requires the presence of both abnormal brain amyloid and tau [5]. In this context, the presence of brain amyloid is required, but alone insufficient, for a diagnosis of AD. The International Working Group recommendations stratify asymptomatic individuals into different risk levels depending on their genetic and biomarker profiles [51]. Importantly, at the individual level, the lifetime risk of developing AD following a positive brain amyloid scan is variable but overall considered to be relatively low [52]. Therefore, it is important to consider that a positive amyloid scan in a cognitively normal older adult may be influenced by age and *APOE4* status and does not necessarily indicate that the individual has AD, or that an AD process is the underlying cause of any future cognitive symptoms. A negative amyloid PET result is useful for excluding AD at any age; however, the positive predictive value of amyloid PET decreases with age, as a positive amyloid scan may be more likely incidental with older age, and other causes of cognitive impairment may need to be considered.

The prevalence of positive brain amyloid scans is higher in cognitively impaired individuals, ranging from 27% to 71% in those with MCI and 70% to 90% in those with AD dementia, depending on the specific criteria used and their *APOE4* status [46, 47, 53, 54]. In MCI, a positive amyloid scan suggests that the cognitive impairment may be due to underlying AD pathology and increases the risk of progression to AD dementia by three to fourfold over 3–5 years after adjusting for other factors such as age and *APOE4* [55, 56]. Amyloid PET positivity is included in the research diagnostic criteria for MCI due to AD [57] and prodromal AD [51]. However, there is considerable variation in individual risk among those with MCI, so risk prediction using amyloid PET data at the individual level may be challenging [58]. The presence of abnormal markers of tau pathology (e.g., elevated CSF p-tau181 or a positive tau PET scan) or markers of neurodegeneration (e.g., atrophy on brain magnetic resonance imaging [MRI] and hypometabolism on FDG-PET) further increases the risk of progression to AD dementia in individuals with MCI [59–61].

The majority of individuals with dementia due to AD (i.e., AD dementia) have abnormal amyloid PET scans [53, 54]. However, a positive amyloid scan alone, in a cognitively impaired individual, does not equate to a biomarker-based definition of AD as abnormal brain amyloid deposition is not specific to AD and can be seen in individuals who have other neurodegenerative disorders that involve amyloid deposition, such as LBD and CAA [62], but do not meet the neuropathological criteria for intermediate-high ADNC. Although a positive amyloid PET scan is not useful in distinguishing AD from these disorders, it can help differentiate AD or other disorders that involve significant brain amyloid deposition from those that do not increase brain amyloid such as FTD [63, 64].

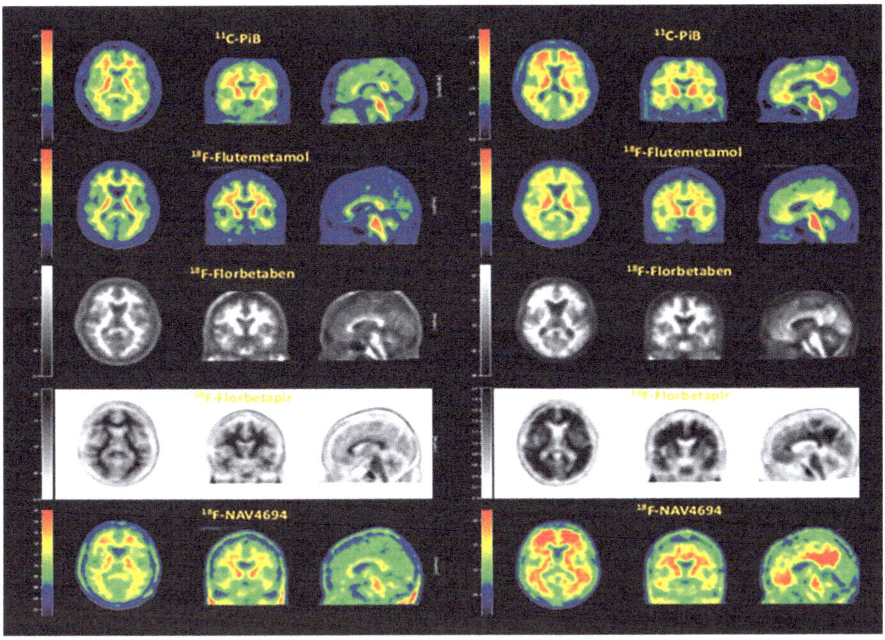

Fig. 1 Examples of positive amyloid brain scans using different amyloid ligands in AD. Negative amyloid scans (~0 centiloid units) are shown in the left panel and positive amyloid scans (all ~50 centiloid units) are shown on the right panel. (Used with permission; Open access through a Creative Commons license http://creativecommons.org/licenses/by/4.0/ from Pemberton et al., 2022 [35])

An abnormal amyloid PET scan is considered to have higher sensitivity in detecting clinically meaningful AD pathology than FDG-PET in cognitively impaired individuals with dementia [64]. Further, a positive amyloid scan predicts a higher risk of brain tau pathology and overall intermediate to high ADNC as amyloid and tau co-pathologies are closely associated in AD [64]. While the approved amyloid tracers cannot reliably detect sparse neuritic plaques, they can reliably detect moderate/frequent neuritic plaque density on the CERAD scale, and most of these individuals will also have Braak III to VI NFT scores and thus, will meet criteria for intermediate/high ADNC [4, 40–42, 65]. Therefore, an abnormal amyloid PET study in a cognitively impaired individual is often considered a strong indicator of intermediate/high ADNC. Interestingly, the prevalence of positive brain amyloid scans in cognitively impaired individuals clinically diagnosed with probable AD decreases with age likely due to the increase in the prevalence of concomitant non-amyloid pathologies associated with amnestic syndromes with older age such as limbic-predominant age-related TDP-43 encephalopathy (i.e., LATE) and primary age-related tauopathy (PART) [54].

Several studies have examined the clinical utility of amyloid PET scans such as the "Imaging Dementia–Evidence for Amyloid Scanning (IDEAS)" study, the

"Amyloid Imaging to Prevent Alzheimer's Disease-Diagnostic and Patient Management (AMYPAD-DPMS)" study, and the "Alzheimer's Biomarkers in Daily Practice (ABIDE)" study [53, 66–69]. The IDEAS study, which was conducted in collaboration with the US Centers for Medicare and Medicaid Services under Coverage with Evidence Development, evaluated the impact of amyloid PET on patient diagnosis, management, and health outcomes in over 16,000 participants with MCI or dementia across nearly 600 specialty clinics in the US who were followed longitudinally [10, 53]. Reported outcomes included the association of amyloid PET with implemented changes in patient management in approximately 60% of patients with MCI and 64% of patients with dementia at 90 days, with the most frequently reported change involving the use of approved symptomatic medications (i.e., cholinesterase inhibitors or memantine) for AD (44% in MCI and 45% in dementia). Following a brain amyloid PET scan, the diagnosis changed in 35% of patients, with 25% of patients originally diagnosed with AD later being diagnosed as non-AD. Conversely, 10% of those originally being diagnosed with non-AD were later diagnosed as AD following the amyloid PET scan. Despite changes in diagnosis and management, overall, the impact of amyloid PET on health outcomes, including rates of 12-month hospitalizations, was modest. Results of another multicenter study were consistent with these findings, suggesting that amyloid PET scan results influence patient diagnosis and management, especially regarding the use of approved symptomatic therapies [66]. Similarly, the ABIDE study conducted in the Netherlands found that 25% of patients had a different diagnosis following amyloid PET scan results with this being more common after a negative, than a positive, scan and that therapeutic plans subsequently changed when amyloid PET data were incorporated into clinical decision-making [69].

Monomeric Aβ peptides, Aβ42 and Aβ40, can be measured in CSF and blood. The CSF Aβ42/Aβ40 and CSF p-tau181/Aβ42 ratios are highly concordant with brain amyloid positivity on PET scans and perform better than individual markers (e.g., CSF Aβ42 or p-tau181) [70]. It is thought that CSF Aβ42, or the Aβ42/Aβ40 ratio, levels decrease shortly before the first evidence of amyloid PET positivity due to a sink mechanism by which levels of soluble monomeric Aβ42 are reduced with the aggregation of Aβ42 into the brain parenchyma (in the form of amyloid plaques detected by amyloid PET) [71]. The CSF Aβ42/Aβ40 or p-tau181/Aβ42 ratios, when measured using high-precision assays, predict the visual classification of amyloid PET scans to a comparable extent to quantitative SUVR measurements [72]. Plasma Aβ42 or Aβ42/Aβ40 levels, measured using mass-spectrometry or single-molecule arrays, also have high concordance with abnormal CSF and PET amyloid markers, although they have not yet been optimized for clinical use [73, 74]. Therefore, amyloid PET scans and/or CSF testing can be used to examine the presence of brain amyloid as these provide similar diagnostic accuracy, and generally, there is no added benefit in combining the 2 modalities. As PET is relatively noninvasive, it is recommended when CSF testing may be contraindicated (e.g., use of anticoagulation) or when CSF results are equivocal [75]. It is also important to note that amyloid PET scans visualize Aβ aggregates, in the form of amyloid plaques, and do not measure soluble forms of Aβ such as the soluble, and highly toxic, Aβ oligomers that are thought to contribute to Aβ-mediated synaptic toxicity and cognitive decline [76].

2.5 *Appropriate Use Criteria (AUC)*

AUCs that describe appropriate and inappropriate clinical use of amyloid PET were developed by a task force from the Alzheimer's Association and the Society of Nuclear Medicine and Molecular Imaging (SNMMI) in 2013 [77]. According to the 2013 AUCs, amyloid PET scans in clinical settings may be considered in individuals who have objective evidence of cognitive impairment (e.g., MCI or dementia) in whom the underlying cause of cognitive impairment remains uncertain following a comprehensive evaluation by a dementia specialist including structural brain imaging (MRI or computed tomography [CT]) and laboratory tests and in whom the amyloid PET scan results are likely to increase diagnostic certainty or result in a change in the treatment plan. Patients with persistent and/or progressive MCI, those with probable AD who have an atypical course or whose presentation suggests the presence of multiple/mixed etiologies, and those with young-onset dementia (before the age of 65 years), are also appropriate candidates for amyloid PET scans. Conversely, other situations such as requesting amyloid PET scans in cognitively normal individuals, to assess dementia severity, or for the purpose of insurance coverage, legal issues, or employment screening are considered inappropriate use. In general, the primary use of amyloid PET scans in clinical settings is to provide evidence regarding the presence or absence of amyloid pathology when there is uncertainty about the diagnosis after a comprehensive evaluation by a dementia expert in patients presenting with cognitive complaints, when AD is a diagnostic consideration, and when knowledge of the amyloid PET results is likely to help determine the underlying etiology or influence treatment decisions (e.g., evaluate eligibility for approved anti-amyloid agents). According to these AUCs, amyloid PET scans should not be used as a substitute for a thorough medical evaluation, including a detailed history and neurological exam, laboratory tests, and cognitive or neurobehavioral assessments, that are typically done to rule out reversible causes of cognitive impairment.

Updated AUCs are currently under development in 2024 for tau PET scans and will also likely address appropriate scenarios for the combined use of both amyloid and tau PET scans. Updated AUCs will also likely address the recent approval of anti-amyloid therapies based on data from clinical trials in which amyloid PET scans were used to confirm the presence of brain amyloid pathology, and thus treatment eligibility, prior to randomization. The anti-amyloid agent, aducanumab, received accelerated FDA approval in 2021 based on its ability to effectively reduce brain amyloid in 2 Phase 3 clinical trials; however, slowing of cognitive decline was only observed in one of the 2 studies and was modest and not likely clinically significant [78]. This drug has since been withdrawn from the market by the manufacturer. Reduction of brain amyloid on amyloid PET scans was also examined as a key outcome measure in Phase 3 clinical trials of 2 other anti-amyloid agents, lecanemab and donanemab, which have received full FDA approval to treat early symptomatic AD in 2023 and 2024, respectively, based on their ability to effectively reduce brain amyloid on PET and to slow cognitive decline in treated patients compared to placebo [79, 80]. In the Phase 3 study of donanemab, amyloid PET was also used to titrate the drug dose during the trial period as participants who had near complete amyloid removal (i.e., "negative" amyloid scans) were switched to placebo [80].

3 Tau PET

Intracellular aggregates of hyperphosphorylated tau, in the form of NFTs, are a key pathological feature of AD and play an important role in driving neurodegeneration and cognitive decline [81]. Studies suggest that accelerated tau aggregation is more proximally associated with the onset of cognitive impairment than amyloid [82]. Tau is encoded by the microtubule-associated protein tau (*MAPT*) gene on chromosome 17, with alternative splicing yielding six different isoforms of the protein [83, 84]. Tau isoforms differ in the number of microtubule-binding regions: three isoforms have three microtubule-binding regions (referred to as 3-repeat [3R] isoforms), while the rest of the transcripts have four (referred to as 4-repeat [4R] isoforms). Similar numbers of 3R and 4R isoforms are seen under normal physiological conditions and with some diseases such as AD, PART, and chronic traumatic encephalopathy (CTE). Conversely, 3R isoforms are predominant in Pick's disease, while 4R isoforms predominate in progressive supranuclear palsy (PSP), corticobasal degeneration (CBD), and argyrophilic grain disease (AGD) [85].

3.1 Tau Ligands

Tau aggregates in AD are formed by the aggregation of tau fibrils, in the form of paired helical filaments (PHF), and include a mixture of 3R/4R tau isoforms. The ability to visualize and measure tau deposits in AD and other neurodegenerative disorders using PET scans has been a significant advancement in the field of AD and other tauopathies.

First-generation tau PET radiotracers include arylquinoline derivatives (^{18}F-THK-5317 and ^{18}F-THK-5351), a pyridoindole derivative (^{18}F-flortaucipir, also known as Tauvid, ^{18}F-AV-1451 or ^{18}F-T807), and a phenyl/pyridinyl-butadienyl-benzothiazone/benzothiazolium (PBB) derivative (^{11}C-PBB3) [86]. In 2020, ^{18}F-flortaucipir (Tauvid) became the first tau ligand to receive FDA approval to estimate the density and regional distribution of aggregated tau NFTs in adult patients with cognitive impairment who were being evaluated for AD [87]. ^{18}F-flortaucipir has high affinity for tau aggregates, in the form of PHF in AD, with minimal to no binding to amyloid-β or other protein aggregates such as TDP-43 or α-synuclein [88, 89]. ^{18}F-flortaucipir has favorable pharmacokinetics including good permeability across the blood–brain barrier due to its lipophilic nature and a favorable metabolic profile [88]. Several studies have shown that ^{18}F-flortaucipir has higher uptake in AD compared to controls and demonstrates regional distributions of tau uptake that closely resemble the neuropathological distribution of NFTs on autopsy, using the Braak and Braak staging [2, 90–92]. In vitro experiments which examined the binding of ^{18}F-flortaucipir to brain specimens with non-AD tauopathy found either no [91] or low-to-moderate binding [93, 94] in non-AD tauopathies, suggesting a relative specificity for AD tau aggregates.

Second-generation tau-selective tracers have recently been developed including ^{18}F-RO-948, ^{18}F-MK-6240, ^{18}F-PI-2620, ^{18}F-JNJ-311, ^{18}F-GTP1, and ^{18}F-PM-PBB3 (also known as ^{18}F-APN-1607) [86, 95]. All of these tau tracers were developed based on their ability to bind AD-related NFTs, and most show absent or weak binding to non-AD tau aggregates such as those in PSP, CBD, CTE, and FTD-tau [96]. Second-generation tau tracers all have minimal to no binding to monoamine oxidase (MAO) and amyloid [97–99]. However, a few of these tau tracers such as ^{18}F-PI-2620 and ^{18}F-PM-PBB3 may be able to bind non-AD tau, and therefore, may have lower specificity to AD-related tau aggregates; these agents may potentially serve as broader spectrum tau imaging agents [100]. ^{18}F-PI-2620 has recently received a Fast Track designation by the FDA for tau PET imaging in AD, PSP, and CBD [101].

Several studies using different tau PET tracers have shown increased tau uptake in disease-relevant brain regions in AD compared to controls and a positive correlation between binding and cognitive decline, supporting the value of tau PET as an emerging imaging marker in AD [86, 99, 102–104]. However, all tau tracers have some degree of off-target binding (i.e., binding to non-tau targets), typically in the basal ganglia, meninges, choroid plexus, and midbrain (red nucleus and substantia nigra), although the degrees and patterns vary across tracers [86]. Off-target binding of ^{18}F-flortaucipir has been reported in pigment-containing structures, including neuromelanin in the substantia nigra, lipofuscin-containing neurons in the lateral geniculate nucleus, and calcified structures such as the choroid plexus [91, 92]. Off-target binding of tau tracers to the choroid plexus, meninges, iron-associated regions, and neuromelanin- and melanin-containing cells may interfere with the accurate quantification of regional tau pathology [93, 105], particularly in key structures such as the hippocampus, or the cerebellum which is often used as a reference region for SUVR measurements [106]. Off-target binding to the choroid plexus is particularly relevant given its close proximity to the hippocampus. It has been suggested that some ^{18}F-flortaucipir binding to the choroid plexus may be on-target as one study reported the presence of tau tangles in the epithelial cells of the choroid plexus which are immunoreactive to tau-specific antibodies [107]. Alternatively, off-target binding to the choroid plexus may be due to leptomeningeal melanocytes or choroid plexus calcifications [93, 108]. ^{18}F-flortaucipir has also been found to bind MAO in vitro, although the significance of this remains unclear. While some studies propose that off-target ^{18}F-flortaucipir binding may occur in brain regions that highly express MAO A/B [109], other studies suggest that in vivo ^{18}F-flortaucipir binding to this enzyme may not be significant and that MAO-B is not a significant off-target binding site in the basal ganglia [110, 111].

3.2 Neuropathological Targets of Tau PET

The Braak and Braak staging of NFT pathology in AD describes six stages of NFT pathology which start in the trans-entorhinal (Stage I) and entorhinal cortices (Stage II), followed by limbic and paralimbic structures including the hippocampus, amygdala, and adjacent neocortical areas in Stage III. Stage IV describes the initial extension of tau pathology into the neocortical association areas including the insula and basal frontal regions, followed by involvement of the association cortices in temporal, parietal, and occipital regions (Stage V) and finally reaching the primary sensorimotor, auditory, and visual cortices in Stage VI (Stages V-VI are referred to as the "isocortical" stage) [2]. For practical purposes, these stages may be simplified into 3 stages: trans-entorhinal (I–II), limbic (III–IV), and isocortical (V–VI) [112]. The Braak and Braak staging system has been integrated into the AD neuropathologic diagnostic criteria [4, 21].

Of the tau tracers, ^{18}F-flortaucipir has the most extensive evidence supporting neuropathologic correlations from a large end-of-life study [92] and case series [113, 114]. ^{18}F-flortaucipir accurately detects AD-like tau pathology, composed of mixed 3R and 4R tau, especially in individuals with advanced Braak stages (i.e., Braak > IV; diagnostic accuracy: 87.5% [95% confidence interval (CI), 77.2–93.5%] [92]). The performance of ^{18}F-flortaucipir may be less accurate in earlier tau stages and a percentage of cases with clinically relevant tau pathology in lower stages may be missed [115]. Further, there are strong correlations between tau PET levels and the quantitative assessments of tau burden in corresponding brain regions in individuals with AD [114, 116] as well as individuals with *MAPT* mutations (who also have mixed 3R/4R tau pathology) [117, 118]. However, the evidence is less conclusive for other tauopathies. For example, a few autopsy-confirmed cases of CBD have been published, with some, but not all, showing moderate to high correlations of tau PET with pathological tau burden [119–121].

The binding of ^{18}F-flortaucipir and ^{18}F-MK-6240 to non-AD tau pathology is limited, which suggests relative specificity of these tracers to AD-related tau. Conversely, ^{18}F-PI-2620 and ^{18}F-PM-PBB3 (a fluorinated version of ^{11}C-PBB3) appear to bind 4R tau inclusions observed in PSP [97, 122]. While, as a group, individuals clinically diagnosed with PSP have higher ^{18}F-flortaucipir or ^{18}F-PI-2620 tau PET burden than controls [97, 118, 123], this is complicated by the observation that tau uptake in PSP occurs mostly in the basal ganglia and substantia nigra, areas that also show off-target binding with several tau tracers leading to overlap in binding patterns. There is a paucity of autopsy-confirmed studies in 4R tauopathies, with only a few studies, which did not show a strong correlation between cortical ^{18}F-flortaucipir PET signal and neuropathologic 4R tau burden and suggested minimal binding outside the off-target regions [121, 124]. Newer tau tracers, such as ^{18}F-PI-2620, which have low off-target binding in the basal ganglia, may offer promise in 4R tauopathies such as PSP.

In summary, while available tau tracers have a relatively high affinity for AD-related tau isoforms, particularly in more advanced disease stages (i.e., Braak stages >IV), tau tracer uptake in non-AD tauopathies appears to be weaker and typically overlaps with areas known to have off-target binding. Newer tau tracers may improve PET detection of non-AD tau.

3.3 Interpretation of Tau PET Scans

Similarly to amyloid PET scans, visual interpretation of tau PET scans is the typical method implemented in clinical practice. A negative scan is one that shows no tracer uptake in the neocortex or when uptake is limited to the medial temporal, anterolateral temporal, or frontal cortex. A positive scan is reported when tracer uptake extends to involve at least the posterolateral temporal or occipital cortex. Further extension into the parietal, posterior cingulate/precuneus, and frontal cortex is seen in more advanced stages [92]. SUVR values can also be used to quantify regional uptake of tau in relation to a reference region. Typically, the earliest changes are noted in the entorhinal cortex and medial temporal lobe followed by involvement of the inferior temporal region, a pattern that is consistent with the Braak stages of tau pathology. Standardized units for tau PET quantification, similar to the centiloid units of amyloid PET scans, are currently being developed [125].

The visual interpretation of antemortem ^{18}F -flortaucipir PET scans as positive provides high sensitivity, and moderate-to-high specificity, in predicting Braak stages V and VI. In the neuropathological validation study of ^{18}F-flortaucipir, positive antemortem tau scans provided 92% sensitivity and 80% specificity for Braak V-VI NFT pathology [92]. Further, as this degree of NFT pathology is almost always associated with amyloid PET positivity, positive tau PET scans also predict the presence of Aβ plaque pathology that is sufficient to meet the criteria for high ADNC [92]. Based on these findings, an abnormal tau PET scan may provide strong evidence to rule in AD pathology as contributing to cognitive impairment. Notably, visually interpreted tau PET scans in this validation study were less reliable in detecting earlier stages of tau pathology (i.e., lower Braak stages) as a positive tau scan predicted tau pathology in only 20% of those with Braak stage III-IV, which is the most frequent Braak stage reported in autopsy-confirmed individuals with MCI. Therefore, visually interpreted tau scans, which is the common practice in clinical settings, have limited utility in detecting early stages of tau pathologies such as those in the preclinical or early symptomatic stages (e.g., MCI). To address this limitation, it has been suggested that quantifying tau PET in the medial temporal lobe may increase diagnostic accuracy for lower Braak stages [126]; however, this is not yet being implemented routinely in clinical practice.

3.4 Clinicopathological and Imaging Correlates

Tau PET is generally considered a useful test that can support a diagnosis of AD, especially in cognitively impaired individuals who are beyond the MCI stage. Most patients with typical amnestic symptoms due to AD follow the Braak and Braak staging of spread of tau pathology with pathology starting in the entorhinal and limbic regions before progressing to the neocortex. However, less commonly, a "hippocampal-sparing" variant is observed in which the hippocampus and limbic structures are relatively preserved and most of the pathology occurs in the neocortex [127, 128]. Hippocampal-sparing variants are more likely to present with atypical nonamnestic symptoms. Further, different clinical variants of AD (e.g., language, visuospatial, and dysexecutive) vary in the distribution of tau pathology which tends to predominate in the regions that correspond to the clinical symptoms [129]. Tau PET scans are able to reliably detect different patterns of NFT distribution that correspond to these clinical variants [130]. For example, studies have shown that ^{18}F-flortaucipir binding was most prominent in the occipital regions in posterior cortical atrophy (PCA; also referred to as the "visuospatial variant of AD") and in the language-dominant left hemisphere in the logopenic variant of primary progressive aphasia (PPA), both of which are considered AD variants (as they are most commonly associated with AD pathology) [9, 131]. These studies have also shown that regional patterns of tau binding correlated with regional atrophy on MRI and hypometabolism on FDG-PET [9, 131]. Conversely, the regional distribution of amyloid does not seem to be strongly correlated with clinical phenotypes in AD [132, 133].

While amyloid PET scans can be defined by global measures of amyloid uptake given its widespread distribution in even the early stages, tau pathology in the neocortex manifests closer to symptom onset, and therefore, there is high importance for a more refined approach to estimate regional tau burden. Studies attempting to predict a spreading pattern of tau uptake on PET, on the basis that areas with more frequent and higher ^{18}F-flortaucipir uptake were most likely to be affected first, predicted a spreading pattern that closely resembled the Braak stages of tau pathology [134] (Fig. 2). Tracer uptake in the temporal cortex, by both ^{18}F-flortaucipir and the second-generation tracer ^{18}F-RO-948, predicted future cognitive decline in cognitively normal individuals, including those with brain amyloid pathology who are likely on the AD continuum, to a better extent than brain MRI and amyloid PET [135]. In another study, models that included tau tracer uptake in the entorhinal cortex, inferior temporal cortex, lateral occipital cortex, and amygdala were found to be predictive of future cognitive decline [136]. Quantitative tau PET has been shown to provide higher accuracy in predicting preclinical and prodromal AD than amyloid PET or brain MRI [135] and can predict short-term progression to MCI in cognitively normal individuals. In one study that followed cognitively normal individuals for 2 years, positive amyloid PET and tau PET scans both predicted future cognitive decline, although the latter appeared to be more strongly associated with a positive tau PET scan [135].

Elevated CSF levels of phosphorylated tau species, such as p-tau181, p-tau217, and p-tau231, offer relative specificity to AD as they reflect amyloid-related changes in tau secretion and phosphorylation [137]. CSF p-tau217, and to a lesser extent

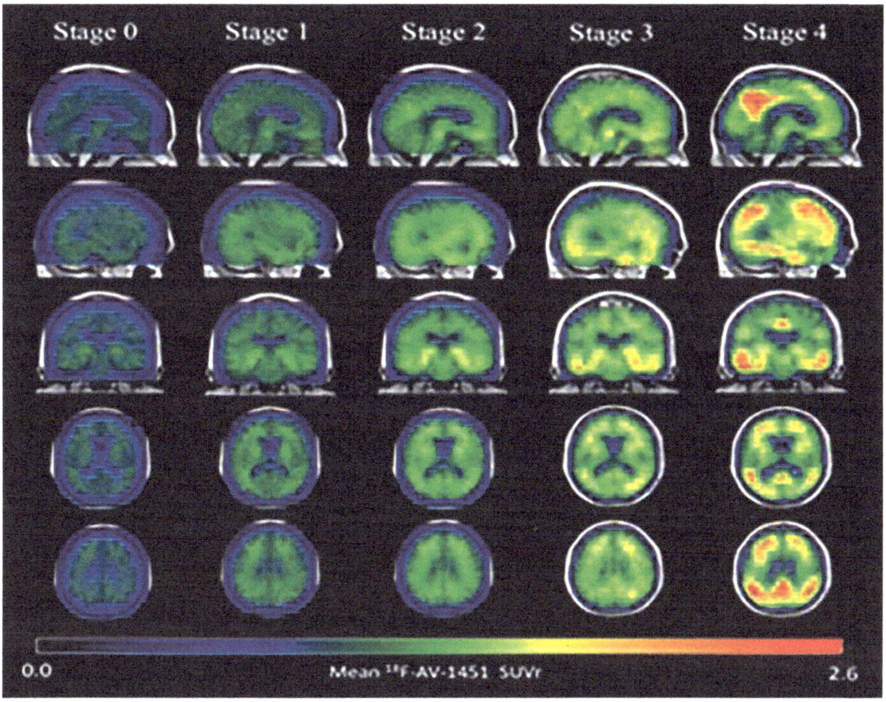

Fig. 2 ^{18}F- flortaucipir Tau PET scans across Braak and Braak stages. Regional SUVRs of tau uptake on ^{18}F- flortaucipir PET scans show gradual progression that corresponds to the Braak and Braak stages of tau pathology. Stage 0 shows no/low tau levels similar to healthy young adults. Tau uptake in the medial temporal regions (Braak I/II stages) is seen in stage 1, extra-medial temporal region in stage 2, inferior and lateral temporal regions (Braak III/IV) in stage 3, and neocortex in stage 4. (Used with permission from Chen et al., 2021 [150]; Open access through a Creative Commons license http://creativecommons.org/licenses/by/4.0/)

p-tau181, levels are closely associated with amyloid and tau burden on PET scans [138, 139]. Studies have shown that the diagnostic utility of tau PET was comparable to CSF p-tau in early prodromal AD, but provided superior diagnostic value to CSF p-tau in differentiating mild to moderate AD dementia [140]. Other CSF p-tau species used in research settings such as the microtubule-binding region of tau at residue 243 (MTBR-tau243) and p-tau205 may be more strongly associated with neuropathological and tau PET burden [141]; however, further research is warranted. Plasma p-tau217, and to a lesser extent p-tau181 and p-tau231, can reliably differentiate cognitively impaired individuals with AD from those with other causes of cognitive impairment [142, 143]. Plasma p-tau217 can also predict future dementia in cognitively normal and impaired individuals and is highly concordant with amyloid PET [144–147]. Measurement of the p-tau217 ratio to other p-tau species (p-tau217/non-p-tau217; % p-tau217) in plasma has been shown to predict amyloid and tau PET positivity with an accuracy of >95% (areas under the curve of 0.95–0.97 and 0.95–0.98, respectively) [148]. It is important to note that CSF or plasma p-tau species reflect various soluble forms of p-tau, whereas tau PET uptake mostly reflects aggregated tau forms, and therefore, CSF or plasma p-tau levels may change before tau PET changes can be detected [149].

3.4.1 Differential Diagnosis of AD

Overall, established tau PET tracers (i.e., ^{18}F-flortaucipir, ^{18}F-MK-6240, and ^{18}F-RO-948) can distinguish AD dementia from non-AD neurodegenerative disorders (including Parkinson's disease [PD], LBD, PSP, CBD, behavioral-variant of FTD, multiple system atrophy [MSA], and semantic and nonfluent variants of PPA) with a relatively high sensitivity and specificity above 90% at a group level [151–153]. However, the utility of tau PET in the differential diagnosis of AD at the individual level remains uncertain, and further studies are needed. LBD, which includes PD, PD-dementia, and Dementia with Lewy Bodies, has minimal tau pathology, and thus, no/minimal uptake on tau PET; therefore, tau PET is considered helpful in differentiating AD from these disorders at a group level.

Substantial overlap in the pattern of tau PET binding has been reported between PSP, AD, and controls. Conversely, tau PET may be more helpful in differentiating PSP, which often presents with a parkinsonian syndrome, from PD which may be a more clinically relevant distinction [154]. Tau PET may help differentiate AD from LBD but is less useful in differentiating AD from CTE and Down syndrome [100, 155, 156]. In terms of differentiating the frontal variant of AD from FTD, one study found heterogeneous frontal-binding patterns in a cohort of patients with the frontal variant of AD [157], which suggests that tau PET may help differentiate this variant from FTD, although this remains to be more extensively investigated. In a few cases with FTD, tau PET binding has been reported in other brain regions such as the basal ganglia, anterior cingulate, and insular white matter, corresponding to areas of neurodegeneration [86]. Interestingly, elevated tau PET signal has been observed in some cases of the semantic variant of PPA which typically involves TDP-43, rather than tau, pathology [158, 159]; tau uptake was mostly noted in areas of asymmetric atrophy. While the exact cause for this is unclear, it has been suggested that this likely reflects off-target binding to non-tau molecules in areas with high neurodegeneration [160]. In summary, tau PET may be useful in differentiating AD from several other neurodegenerative disorders, albeit with limitations. It is also important to note that the accuracy of tau PET is highest in advanced disease stages of AD and is less reliable in earlier preclinical or prodromal stages [9, 161].

3.5 Clinical Use of Tau PET

Appropriate use criteria for tau PET scans, similar to those recommended for amyloid PET, are currently being developed.

Regarding the combined use of amyloid and tau PET, amyloid and tau pathologies vary in their spatial and temporal progression in AD, and therefore, their relation to the onset of clinical symptoms. In general, amyloid PET scans are more sensitive for early AD changes as they are often abnormal in preclinical and early symptomatic stages, such as individuals with MCI or subjective cognitive decline (SCD), when tau PET scans are either normal or show mild changes. In cognitively impaired (i.e., symptomatic) individuals, an abnormal amyloid PET scan alone, with a normal tau PET scan, is not enough for a diagnosis of AD or an attribution of cognitive impairment to an AD process. Conversely, the spatial distribution of tau PET changes closely parallels impaired cognitive domains, and thus, a significantly abnormal tau PET scan in a symptomatic individual provides strong evidence to support that AD is the underlying cause of their cognitive impairment. There are a few scenarios where both amyloid and tau PET scans may be warranted for an accurate diagnosis. One study which randomized participants to receive both amyloid and tau PET scans in a different order (regarding which PET scan was performed first) found that participants who had both positive amyloid and positive tau scans were more likely to progress from normal cognition to cognitive impairment, including MCI or dementia, over follow-up compared to those whose scans were negative for both amyloid and tau or positive for only one type of scan [162]. In this study, the first PET modality changed diagnosis in about 28% of participants, and increased diagnostic confidence by an additional 18–19%. Further, a negative amyloid scan may have a stronger impact on diagnosis than a negative tau scan.

According to the 2018 NIA-AA research framework for AD, amyloid PET imaging or CSF Aβ42 (or Aβ42/Aβ40) provide evidence for the presence of brain amyloid "A", and tau PET or CSF p-tau181 for the presence of tau pathology "T" [5]. The presence of A positivity alone indicates the presence of pathology in the AD continuum, while the presence of both A and T positivity establishes a biological definition of AD. The revised 2024 AA criteria, however, characterize an abnormal amyloid PET scan, validated CSF markers (Aβ42/Aβ40, p-tau181/Aβ42, or total tau/Aβ42), and/or plasma p-tau217 as 'Core 1 Biomarkers' that establish a diagnosis of AD as these correspond to the presence of ADNC (both plaques and tangles), whereas an abnormal tau PET scan is considered a 'Core 2 Biomarker' that can provide useful information regarding disease staging and prognosis in individuals who already have a positive amyloid scan and therefore, provides additional support for an AD diagnosis (i.e., increases diagnostic certainty) [6]. According to this framework, Stage A is defined by a positive amyloid and a negative tau PET scan. Stages B, C, and D all involve a positive amyloid scan with a positive tau scan showing uptake limited to the medial temporal lobe in Stage B, moderate neocortical tau uptake in Stage C, and high neocortical tau uptake in Stage D (Table 2).

These criteria are not intended to provide clinical practice guidelines and remain mostly in the context of research criteria that reflect recent advancements in the AD biomarker field and the emphasis on defining AD as a biological construct rather than a clinical syndrome. Several steps will be needed before these can be incorporated into clinical practice including the need for validated and standardized Core 1 plasma biomarkers and standardization of quantitative tau PET staging as the interpretation of tau PET scans in current clinical settings relies on visual assessment of a neocortical tau signal leading to a dichotomous read of a "positive" vs a "negative" scan.

Table 2 Operationalization of biological staging by positron emission tomography (PET)

Stage	Amyloid PET	Tau PET medial temporal	Tau PET moderate neocortical uptake	Tau PET high neocortical uptake
A	+	–	–	–
B	+	+	–	–
C	+	+	+	–
D	+	+	+	+

Adapted from Jack et al., 2024 [6]

4 Conclusion

In vivo visualization and quantification of amyloid and tau pathologies has been a major advancement in the field of AD biomarkers. Amyloid PET scans are sensitive to early stages of pathology, including preclinical stages, while tau PET scans may not detect early disease stages and are more sensitive in symptomatic patients with moderate to severe dementia. However, tau PET scans are useful in staging disease severity. Regional tau, but not amyloid, uptake closely parallels regional atrophy on MRI and impaired cognitive domains. Interpretation of positive amyloid scans in cognitively normal older adults depends on their age group and *APOE4* genotype. Clinical interpretation of amyloid and tau PET scans relies on visual assessments while quantification techniques are applied in research settings. Appropriate use criteria have been proposed for amyloid scans and are being developed for tau scans and the combined use of both imaging modalities. Both amyloid and tau scans are helpful in differentiating AD from several other neurodegenerative disorders. Emerging tau tracers may be able to accurately visualize non-AD tau pathology.

Funding This work was supported by the National Institute of Health – National Institute of Aging Grant RF1AG083744 (RT).

References

1. Price JL, Morris JC. Tangles and plaques in nondemented aging and "preclinical" Alzheimer's disease. Ann Neurol. 1999;45(3):358–68. https://doi.org/10.1002/1531-8249(199903)45:3<358::aid-ana12>3.0.co;2-x.
2. Braak H, Braak E. Neuropathological stageing of Alzheimer-related changes. Acta Neuropathol. 1991;82(4):239–59. https://doi.org/10.1007/bf00308809.
3. Hampel H, Hardy J, Blennow K, Chen C, Perry G, Kim SH, et al. The Amyloid-β Pathway in Alzheimer's Disease. Mol Psychiatry. 2021;26(10):5481–503. https://doi.org/10.1038/s41380-021-01249-0.
4. Hyman BT, Phelps CH, Beach TG, Bigio EH, Cairns NJ, Carrillo MC, et al. National Institute on Aging-Alzheimer's Association guidelines for the neuropathologic assessment of Alzheimer's disease. Alzheimers Dement. 2012;8(1):1–13. https://doi.org/10.1016/j.jalz.2011.10.007.

5. Jack CR Jr, Bennett DA, Blennow K, Carrillo MC, Dunn B, Haeberlein SB, et al. NIA-AA Research Framework: toward a biological definition of Alzheimer's disease. Alzheimers Dement. 2018;14(4):535–62. https://doi.org/10.1016/j.jalz.2018.02.018.
6. Jack CR Jr, Andrews JS, Beach TG, Buracchio T, Dunn B, Graf A, et al. Revised criteria for diagnosis and staging of Alzheimer's disease: Alzheimer's Association Workgroup. Alzheimers Dement. 2024;20(8):5143–69. https://doi.org/10.1002/alz.13859.
7. Chassé M, Vasdev N. Emerging targets for positron emission tomography imaging in proteinopathies. npj Imaging. 2024;2(1):30. https://doi.org/10.1038/s44303-024-00032-4.
8. Schöll M, Lockhart SN, Schonhaut DR, O'Neil JP, Janabi M, Ossenkoppele R, et al. PET imaging of tau deposition in the aging human brain. Neuron. 2016;89(5):971–82. https://doi.org/10.1016/j.neuron.2016.01.028.
9. Ossenkoppele R, Schonhaut DR, Schöll M, Lockhart SN, Ayakta N, Baker SL, et al. Tau PET patterns mirror clinical and neuroanatomical variability in Alzheimer's disease. Brain. 2016;139(Pt 5):1551–67. https://doi.org/10.1093/brain/aww027.
10. Chapleau M, Iaccarino L, Soleimani-Meigooni D, Rabinovici GD. The role of amyloid PET in imaging neurodegenerative disorders: a review. J Nucl Med. 2022;63(Suppl 1):13s–9s. https://doi.org/10.2967/jnumed.121.263195.
11. Klunk WE, Engler H, Nordberg A, Wang Y, Blomqvist G, Holt DP, et al. Imaging brain amyloid in Alzheimer's disease with Pittsburgh Compound-B. Ann Neurol. 2004;55(3):306–19. https://doi.org/10.1002/ana.20009.
12. Klunk WE, Wang Y, Huang GF, Debnath ML, Holt DP, Mathis CA. Uncharged thioflavin-T derivatives bind to amyloid-beta protein with high affinity and readily enter the brain. Life Sci. 2001;69(13):1471–84. https://doi.org/10.1016/s0024-3205(01)01232-2.
13. Mathis CA, Bacskai BJ, Kajdasz ST, McLellan ME, Frosch MP, Hyman BT, et al. A lipophilic thioflavin-T derivative for positron emission tomography (PET) imaging of amyloid in brain. Bioorg Med Chem Lett. 2002;12(3):295–8. https://doi.org/10.1016/s0960-894x(01)00734-x.
14. Seo SW, Ayakta N, Grinberg LT, Villeneuve S, Lehmann M, Reed B, et al. Regional correlations between [11C]PIB PET and post-mortem burden of amyloid-beta pathology in a diverse neuropathological cohort. Neuroimage Clin. 2017;13:130–7. https://doi.org/10.1016/j.nicl.2016.11.008.
15. Villemagne VL, Doré V, Burnham SC, Masters CL, Rowe CC. Imaging tau and amyloid-β proteinopathies in Alzheimer disease and other conditions. Nat Rev Neurol. 2018;14(4):225–36. https://doi.org/10.1038/nrneurol.2018.9.
16. US Food and Drug Admininstration (FDA). Approval Package for Application Number 202008Orig1s000 (Florbetapir F 18 Injection; Amyvid). https://www.accessdata.fda.gov/drugsatfda_docs/nda/2012/202008Orig1s000Approv.pdf (2012). Accessed.
17. Heurling K, Leuzy A, Zimmer ER, Lubberink M, Nordberg A. Imaging β-amyloid using [(18)F]flutemetamol positron emission tomography: from dosimetry to clinical diagnosis. Eur J Nucl Med Mol Imaging. 2016;43(2):362–73. https://doi.org/10.1007/s00259-015-3208-1.
18. Navarro A, del Valle E, Mártínez E, Ordóñez C, Pérez C, Tolivia J. Highly selective and fast diagnosis of Alzheimer's disease hallmark lesions using Congo Red in isopropyl alcoholic solution. J Alzheimers Dis. 2013;35(3):589–97. https://doi.org/10.3233/jad-122386.
19. Mavrogiorgou P, Gertz HJ, Ferszt R, Wolf R, Bär KJ, Juckel G. Are routine methods good enough to stain senile plaques and neurofibrillary tangles in different brain regions of demented patients? Psychiatr Danub. 2011;23(4):334–9.
20. Wisniewski HM, Wen GY, Kim KS. Comparison of four staining methods on the detection of neuritic plaques. Acta Neuropathol. 1989;78(1):22–7. https://doi.org/10.1007/bf00687398.
21. Montine TJ, Phelps CH, Beach TG, Bigio EH, Cairns NJ, Dickson DW, et al. National Institute on Aging-Alzheimer's Association guidelines for the neuropathologic assessment of Alzheimer's disease: a practical approach. Acta Neuropathol. 2012;123(1):1–11. https://doi.org/10.1007/s00401-011-0910-3.
22. Thal DR, Rüb U, Orantes M, Braak H. Phases of A beta-deposition in the human brain and its relevance for the development of AD. Neurology. 2002;58(12):1791–800. https://doi.org/10.1212/wnl.58.12.1791.

23. Braak H, Alafuzoff I, Arzberger T, Kretzschmar H, Del Tredici K. Staging of Alzheimer disease-associated neurofibrillary pathology using paraffin sections and immunocytochemistry. Acta Neuropathol. 2006;112(4):389–404. https://doi.org/10.1007/s00401-006-0127-z.
24. Mirra SS, Heyman A, McKeel D, Sumi SM, Crain BJ, Brownlee LM, et al. The Consortium to Establish a Registry for Alzheimer's Disease (CERAD). Part II. Standardization of the neuropathologic assessment of Alzheimer's disease. Neurology. 1991;41(4):479–86. https://doi.org/10.1212/wnl.41.4.479.
25. Grothe MJ, Barthel H, Sepulcre J, Dyrba M, Sabri O, Teipel SJ. In vivo staging of regional amyloid deposition. Neurology. 2017;89(20):2031–8. https://doi.org/10.1212/wnl.0000000000004643.
26. Teipel SJ, Temp AGM, Levin F, Dyrba M, Grothe MJ. Association of PET-based stages of amyloid deposition with neuropathological markers of Aβ pathology. Ann Clin Transl Neurol. 2021;8(1):29–42. https://doi.org/10.1002/acn3.51238.
27. Murray ME, Lowe VJ, Graff-Radford NR, Liesinger AM, Cannon A, Przybelski SA, et al. Clinicopathologic and 11C-Pittsburgh compound B implications of Thal amyloid phase across the Alzheimer's disease spectrum. Brain. 2015;138(Pt 5):1370–81. https://doi.org/10.1093/brain/awv050.
28. Johnson JK, Head E, Kim R, Starr A, Cotman CW. Clinical and pathological evidence for a frontal variant of Alzheimer disease. Arch Neurol. 1999;56(10):1233–9. https://doi.org/10.1001/archneur.56.10.1233.
29. Tang-Wai DF, Graff-Radford NR, Boeve BF, Dickson DW, Parisi JE, Crook R, et al. Clinical, genetic, and neuropathologic characteristics of posterior cortical atrophy. Neurology. 2004;63(7):1168–74. https://doi.org/10.1212/01.wnl.0000140289.18472.15.
30. de Souza LC, Corlier F, Habert M-O, Uspenskaya O, Maroy R, Lamari F, et al. Similar amyloid-β burden in posterior cortical atrophy and Alzheimer's disease. Brain. 2011;134(7):2036–43. https://doi.org/10.1093/brain/awr130.
31. Jack CR Jr, Barrio JR, Kepe V. Cerebral amyloid PET imaging in Alzheimer's disease. Acta Neuropathol. 2013;126(5):643–57. https://doi.org/10.1007/s00401-013-1185-7.
32. Villeneuve S, Rabinovici GD, Cohn-Sheehy BI, Madison C, Ayakta N, Ghosh PM, et al. Existing Pittsburgh Compound-B positron emission tomography thresholds are too high: statistical and pathological evaluation. Brain. 2015;138(Pt 7):2020–33. https://doi.org/10.1093/brain/awv112.
33. Johnson KA, Gregas M, Becker JA, Kinnecom C, Salat DH, Moran EK, et al. Imaging of amyloid burden and distribution in cerebral amyloid angiopathy. Ann Neurol. 2007;62(3):229–34. https://doi.org/10.1002/ana.21164.
34. Lopresti BJ, Klunk WE, Mathis CA, Hoge JA, Ziolko SK, Lu X, et al. Simplified quantification of Pittsburgh Compound B amyloid imaging PET studies: a comparative analysis. J Nucl Med. 2005;46(12):1959–72.
35. Pemberton HG, Collij LE, Heeman F, Bollack A, Shekari M, Salvadó G, et al. Quantification of amyloid PET for future clinical use: a state-of-the-art review. Eur J Nucl Med Mol Imaging. 2022;49(10):3508–28. https://doi.org/10.1007/s00259-022-05784-y.
36. Heeman F, Yaqub M, Hendriks J, van Berckel BNM, Collij LE, Gray KR, et al. Impact of cerebral blood flow and amyloid load on SUVR bias. EJNMMI Res. 2022;12(1):29. https://doi.org/10.1186/s13550-022-00898-8.
37. Royse SK, Minhas DS, Lopresti BJ, Murphy A, Ward T, Koeppe RA, et al. Validation of amyloid PET positivity thresholds in centiloids: a multisite PET study approach. Alzheimers Res Ther. 2021;13(1):99. https://doi.org/10.1186/s13195-021-00836-1.
38. La Joie R, Ayakta N, Seeley WW, Borys E, Boxer AL, DeCarli C, et al. Multisite study of the relationships between antemortem [(11)C]PIB-PET Centiloid values and postmortem measures of Alzheimer's disease neuropathology. Alzheimers Dement. 2019;15(2):205–16. https://doi.org/10.1016/j.jalz.2018.09.001.

39. Klunk WE, Koeppe RA, Price JC, Benzinger TL, Devous MD Sr, Jagust WJ, et al. The Centiloid Project: standardizing quantitative amyloid plaque estimation by PET. Alzheimers Dement. 2015;11(1):1–15.e1–4. https://doi.org/10.1016/j.jalz.2014.07.003.
40. Clark CM, Pontecorvo MJ, Beach TG, Bedell BJ, Coleman RE, Doraiswamy PM, et al. Cerebral PET with florbetapir compared with neuropathology at autopsy for detection of neuritic amyloid-β plaques: a prospective cohort study. Lancet Neurol. 2012;11(8):669–78. https://doi.org/10.1016/s1474-4422(12)70142-4.
41. Curtis C, Gamez JE, Singh U, Sadowsky CH, Villena T, Sabbagh MN, et al. Phase 3 trial of flutemetamol labeled with radioactive fluorine 18 imaging and neuritic plaque density. JAMA Neurol. 2015;72(3):287–94. https://doi.org/10.1001/jamaneurol.2014.4144.
42. Sabri O, Sabbagh MN, Seibyl J, Barthel H, Akatsu H, Ouchi Y, et al. Florbetaben PET imaging to detect amyloid beta plaques in Alzheimer's disease: phase 3 study. Alzheimers Dement. 2015;11(8):964–74. https://doi.org/10.1016/j.jalz.2015.02.004.
43. Ikonomovic MD, Buckley CJ, Heurling K, Sherwin P, Jones PA, Zanette M, et al. Post-mortem histopathology underlying β-amyloid PET imaging following flutemetamol F 18 injection. Acta Neuropathol Commun. 2016;4(1):130. https://doi.org/10.1186/s40478-016-0399-z.
44. Rowe CC, Ng S, Ackermann U, Gong SJ, Pike K, Savage G, et al. Imaging beta-amyloid burden in aging and dementia. Neurology. 2007;68(20):1718–25. https://doi.org/10.1212/01.wnl.0000261919.22630.ea.
45. Schmidt ME, Chiao P, Klein G, Matthews D, Thurfjell L, Cole PE, et al. The influence of biological and technical factors on quantitative analysis of amyloid PET: Points to consider and recommendations for controlling variability in longitudinal data. Alzheimers Dement. 2015;11(9):1050–68. https://doi.org/10.1016/j.jalz.2014.09.004.
46. Jansen WJ, Ossenkoppele R, Knol DL, Tijms BM, Scheltens P, Verhey FR, et al. Prevalence of cerebral amyloid pathology in persons without dementia: a meta-analysis. JAMA. 2015;313(19):1924–38. https://doi.org/10.1001/jama.2015.4668.
47. Jansen WJ, Janssen O, Tijms BM, Vos SJB, Ossenkoppele R, Visser PJ, et al. Prevalence estimates of amyloid abnormality across the Alzheimer disease clinical spectrum. JAMA Neurol. 2022;79(3):228–43. https://doi.org/10.1001/jamaneurol.2021.5216.
48. Jack CR Jr, Wiste HJ, Weigand SD, Knopman DS, Lowe V, Vemuri P, et al. Amyloid-first and neurodegeneration-first profiles characterize incident amyloid PET positivity. Neurology. 2013;81(20):1732–40. https://doi.org/10.1212/01.wnl.0000435556.21319.e4.
49. Rodrigue KM, Rieck JR, Kennedy KM, Devous MD Sr, Diaz-Arrastia R, Park DC. Risk factors for β-amyloid deposition in healthy aging: vascular and genetic effects. JAMA Neurol. 2013;70(5):600–6. https://doi.org/10.1001/jamaneurol.2013.1342.
50. Donohue MC, Sperling RA, Petersen R, Sun CK, Weiner MW, Aisen PS. Association between elevated brain amyloid and subsequent cognitive decline among cognitively normal persons. JAMA. 2017;317(22):2305–16. https://doi.org/10.1001/jama.2017.6669.
51. Dubois B, Villain N, Frisoni GB, Rabinovici GD, Sabbagh M, Cappa S, et al. Clinical diagnosis of Alzheimer's disease: recommendations of the International Working Group. Lancet Neurol. 2021;20(6):484–96. https://doi.org/10.1016/s1474-4422(21)00066-1.
52. Brookmeyer R, Abdalla N. Estimation of lifetime risks of Alzheimer's disease dementia using biomarkers for preclinical disease. Alzheimers Dement. 2018;14(8):981–8. https://doi.org/10.1016/j.jalz.2018.03.005.
53. Rabinovici GD, Gatsonis C, Apgar C, Chaudhary K, Gareen I, Hanna L, et al. Association of amyloid positron emission tomography with subsequent change in clinical management among medicare beneficiaries with mild cognitive impairment or dementia. JAMA. 2019;321(13):1286–94. https://doi.org/10.1001/jama.2019.2000.
54. Ossenkoppele R, Jansen WJ, Rabinovici GD, Knol DL, van der Flier WM, van Berckel BN, et al. Prevalence of amyloid PET positivity in dementia syndromes: a meta-analysis. JAMA. 2015;313(19):1939–49. https://doi.org/10.1001/jama.2015.4669.

55. Okello A, Koivunen J, Edison P, Archer HA, Turkheimer FE, Någren K, et al. Conversion of amyloid positive and negative MCI to AD over 3 years: an 11C-PIB PET study. Neurology. 2009;73(10):754–60. https://doi.org/10.1212/WNL.0b013e3181b23564.
56. Jang H, Park J, Woo S, Kim S, Kim HJ, Na DL, et al. Prediction of fast decline in amyloid positive mild cognitive impairment patients using multimodal biomarkers. Neuroimage Clin. 2019;24:101941. https://doi.org/10.1016/j.nicl.2019.101941.
57. Albert MS, DeKosky ST, Dickson D, Dubois B, Feldman HH, Fox NC, et al. The diagnosis of mild cognitive impairment due to Alzheimer's disease: recommendations from the National Institute on Aging-Alzheimer's Association workgroups on diagnostic guidelines for Alzheimer's disease. Alzheimers Dement. 2011;7(3):270–9. https://doi.org/10.1016/j.jalz.2011.03.008.
58. Jutten RJ, Sikkes SAM, Amariglio RE, Buckley RF, Properzi MJ, Marshall GA, et al. Identifying sensitive measures of cognitive decline at different clinical stages of Alzheimer's disease. J Int Neuropsychol Soc. 2021;27(5):426–38. https://doi.org/10.1017/s1355617720000934.
59. Wolk DA, Sadowsky C, Safirstein B, Rinne JO, Duara R, Perry R, et al. Use of flutemetamol F 18-labeled positron emission tomography and other biomarkers to assess risk of clinical progression in patients with amnestic mild cognitive impairment. JAMA Neurol. 2018;75(9):1114–23. https://doi.org/10.1001/jamaneurol.2018.0894.
60. Mormino EC, Betensky RA, Hedden T, Schultz AP, Ward A, Huijbers W, et al. Amyloid and APOE ε4 interact to influence short-term decline in preclinical Alzheimer disease. Neurology. 2014;82(20):1760–7. https://doi.org/10.1212/wnl.0000000000000431.
61. van der Kall LM, Truong T, Burnham SC, Doré V, Mulligan RS, Bozinovski S, et al. Association of β-amyloid level, clinical progression, and longitudinal cognitive change in normal older individuals. Neurology. 2021;96(5):e662–e70. https://doi.org/10.1212/wnl.0000000000011222.
62. Marcus C, Mena E, Subramaniam RM. Brain PET in the diagnosis of Alzheimer's disease. Clin Nucl Med. 2014;39(10):e413–22; quiz e23-6;. https://doi.org/10.1097/rlu.0000000000000547.
63. Rabinovici GD, Rosen HJ, Alkalay A, Kornak J, Furst AJ, Agarwal N, et al. Amyloid vs FDG-PET in the differential diagnosis of AD and FTLD. Neurology. 2011;77(23):2034–42. https://doi.org/10.1212/WNL.0b013e31823b9c5e.
64. Lesman-Segev OH, La Joie R, Iaccarino L, Lobach I, Rosen HJ, Seo SW, et al. Diagnostic accuracy of amyloid versus (18) F-fluorodeoxyglucose positron emission tomography in autopsy-confirmed dementia. Ann Neurol. 2021;89(2):389–401. https://doi.org/10.1002/ana.25968.
65. Lowe VJ, Lundt ES, Albertson SM, Przybelski SA, Senjem ML, Parisi JE, et al. Neuroimaging correlates with neuropathologic schemes in neurodegenerative disease. Alzheimers Dement. 2019;15(7):927–39. https://doi.org/10.1016/j.jalz.2019.03.016.
66. Pontecorvo MJ, Siderowf A, Dubois B, Doraiswamy PM, Frisoni GB, Grundman M, et al. Effectiveness of florbetapir PET imaging in changing patient management. Dement Geriatr Cogn Disord. 2017;44(3–4):129–43. https://doi.org/10.1159/000478007.
67. Cotta Ramusino M, Perini G, Altomare D, Barbarino P, Weidner W, Salvini Porro G, et al. Outcomes of clinical utility in amyloid-PET studies: state of art and future perspectives. Eur J Nucl Med Mol Imaging. 2021;48(7):2157–68. https://doi.org/10.1007/s00259-020-05187-x.
68. Altomare D, Barkhof F, Caprioglio C, Collij LE, Scheltens P, Lopes Alves I, et al. Clinical effect of early vs late amyloid positron emission tomography in memory clinic patients: the AMYPAD-DPMS randomized clinical trial. JAMA Neurol. 2023;80(6):548–57. https://doi.org/10.1001/jamaneurol.2023.0997.
69. de Wilde A, van der Flier WM, Pelkmans W, Bouwman F, Verwer J, Groot C, et al. Association of amyloid positron emission tomography with changes in diagnosis and patient treatment in an unselected memory clinic cohort: the ABIDE project. JAMA Neurol. 2018;75(9):1062–70. https://doi.org/10.1001/jamaneurol.2018.1346.

70. Campbell MR, Ashrafzadeh-Kian S, Petersen RC, Mielke MM, Syrjanen JA, van Harten AC, et al. P-tau/Aβ42 and Aβ42/40 ratios in CSF are equally predictive of amyloid PET status. Alzheimers Dement (Amst). 2021;13(1):e12190. https://doi.org/10.1002/dad2.12190.
71. Palmqvist S, Mattsson N, Hansson O. Cerebrospinal fluid analysis detects cerebral amyloid-β accumulation earlier than positron emission tomography. Brain. 2016;139(Pt 4):1226–36. https://doi.org/10.1093/brain/aww015.
72. Hansson O, Seibyl J, Stomrud E, Zetterberg H, Trojanowski JQ, Bittner T, et al. CSF biomarkers of Alzheimer's disease concord with amyloid-β PET and predict clinical progression: a study of fully automated immunoassays in BioFINDER and ADNI cohorts. Alzheimers Dement. 2018;14(11):1470–81. https://doi.org/10.1016/j.jalz.2018.01.010.
73. Jang H, Kim JS, Lee HJ, Kim CH, Na DL, Kim HJ, et al. Performance of the plasma Aβ42/Aβ40 ratio, measured with a novel HPLC-MS/MS method, as a biomarker of amyloid PET status in a DPUK-KOREAN cohort. Alzheimers Res Ther. 2021;13(1):179. https://doi.org/10.1186/s13195-021-00911-7.
74. Nakamura A, Kaneko N, Villemagne VL, Kato T, Doecke J, Doré V, et al. High performance plasma amyloid-β biomarkers for Alzheimer's disease. Nature. 2018;554(7691):249–254. https://doi.org/10.1038/nature25456. Epub 2018 Jan 31. PMID: 29420472.
75. Engelborghs S, Niemantsverdriet E, Struyfs H, Blennow K, Brouns R, Comabella M, et al. Consensus guidelines for lumbar puncture in patients with neurological diseases. Alzheimers Dement (Amst). 2017;8:111–26. https://doi.org/10.1016/j.dadm.2017.04.007.
76. Sakono M, Zako T. Amyloid oligomers: formation and toxicity of Aβ oligomers. FEBS J. 2010;277(6):1348–58. https://doi.org/10.1111/j.1742-4658.2010.07568.x.
77. Johnson KA, Minoshima S, Bohnen NI, Donohoe KJ, Foster NL, Herscovitch P, et al. Appropriate use criteria for amyloid PET: a report of the Amyloid Imaging Task Force, the Society of Nuclear Medicine and Molecular Imaging, and the Alzheimer's Association. Alzheimers Dement. 2013;9(1):E1–E16. https://doi.org/10.1016/j.jalz.2013.01.002.
78. Dickson SP, Hennessey S, Nicodemus Johnson J, Knowlton N, Hendrix SB. Avoiding future controversies in the Alzheimer's disease space through understanding the aducanumab data and FDA review. Alzheimers Res Ther. 2023;15(1):98. https://doi.org/10.1186/s13195-023-01238-1.
79. van Dyck CH, Swanson CJ, Aisen P, Bateman RJ, Chen C, Gee M, et al. Lecanemab in Early Alzheimer's Disease. N Engl J Med. 2022;388(1):9–21. https://doi.org/10.1056/NEJMoa2212948.
80. Sims JR, Zimmer JA, Evans CD, Lu M, Ardayfio P, Sparks J, et al. Donanemab in early symptomatic Alzheimer disease: the TRAILBLAZER-ALZ 2 randomized clinical trial. JAMA. 2023;330(6):512–27. https://doi.org/10.1001/jama.2023.13239.
81. Hanseeuw BJ, Betensky RA, Jacobs HIL, Schultz AP, Sepulcre J, Becker JA, et al. Association of amyloid and tau with cognition in preclinical Alzheimer disease: a longitudinal study. JAMA Neurol. 2019;76(8):915–24. https://doi.org/10.1001/jamaneurol.2019.1424.
82. Nelson PT, Alafuzoff I, Bigio EH, Bouras C, Braak H, Cairns NJ, et al. Correlation of Alzheimer disease neuropathologic changes with cognitive status: a review of the literature. J Neuropathol Exp Neurol. 2012;71(5):362–81. https://doi.org/10.1097/NEN.0b013e31825018f7.
83. Avila J, Jiménez JS, Sayas CL, Bolós M, Zabala JC, Rivas G, et al. Tau Structures. Front Aging Neurosci. 2016;8:262. https://doi.org/10.3389/fnagi.2016.00262.
84. Neve RL, Harris P, Kosik KS, Kurnit DM, Donlon TA. Identification of cDNA clones for the human microtubule-associated protein tau and chromosomal localization of the genes for tau and microtubule-associated protein 2. Brain Res. 1986;387(3):271–80. https://doi.org/10.1016/0169-328x(86)90033-1.
85. Shi Y, Zhang W, Yang Y, Murzin AG, Falcon B, Kotecha A, et al. Structure-based classification of tauopathies. Nature. 2021;598(7880):359–63. https://doi.org/10.1038/s41586-021-03911-7.

86. Cassinelli Petersen G, Roytman M, Chiang GC, Li Y, Gordon ML, Franceschi AM. Overview of tau PET molecular imaging. Curr Opin Neurol. 2022;35(2):230–9. https://doi.org/10.1097/wco.0000000000001035.
87. Jie C, Treyer V, Schibli R, Mu L. Tauvid™: The first FDA-approved PET tracer for imaging tau pathology in Alzheimer's disease. Pharmaceuticals (Basel). 2021;14(2). https://doi.org/10.3390/ph14020110.
88. Xia CF, Arteaga J, Chen G, Gangadharmath U, Gomez LF, Kasi D, et al. [(18)F]T807, a novel tau positron emission tomography imaging agent for Alzheimer's disease. Alzheimers Dement. 2013;9(6):666–76. https://doi.org/10.1016/j.jalz.2012.11.008.
89. Burnham SC, Iaccarino L, Pontecorvo MJ, Fleisher AS, Lu M, Collins EC, et al. A review of the flortaucipir literature for positron emission tomography imaging of tau neurofibrillary tangles. Brain Commun. 2024;6(1):fcad305. https://doi.org/10.1093/braincomms/fcad305.
90. Chien DT, Bahri S, Szardenings AK, Walsh JC, Mu F, Su MY, et al. Early clinical PET imaging results with the novel PHF-tau radioligand [F-18]-T807. J Alzheimers Dis. 2013;34(2):457–68. https://doi.org/10.3233/jad-122059.
91. Marquié M, Siao Tick Chong M, Antón-Fernández A, Verwer EE, Sáez-Calveras N, Meltzer AC, et al. [F-18]-AV-1451 binding correlates with postmortem neurofibrillary tangle Braak staging. Acta Neuropathol. 2017;134(4):619–28. https://doi.org/10.1007/s00401-017-1740-8.
92. Fleisher AS, Pontecorvo MJ, Devous MD Sr, Lu M, Arora AK, Truocchio SP, et al. Positron emission tomography imaging with [18F]flortaucipir and postmortem assessment of Alzheimer disease neuropathologic changes. JAMA Neurol. 2020;77(7):829–39. https://doi.org/10.1001/jamaneurol.2020.0528.
93. Lowe VJ, Curran G, Fang P, Liesinger AM, Josephs KA, Parisi JE, et al. An autoradiographic evaluation of AV-1451 Tau PET in dementia. Acta Neuropathol Commun. 2016;4 1:58. https://doi.org/10.1186/s40478-016-0315-6.
94. Sander K, Lashley T, Gami P, Gendron T, Lythgoe MF, Rohrer JD, et al. Characterization of tau positron emission tomography tracer [(18)F]AV-1451 binding to postmortem tissue in Alzheimer's disease, primary tauopathies, and other dementias. Alzheimers Dement. 2016;12(11):1116–24. https://doi.org/10.1016/j.jalz.2016.01.003.
95. Bischof GN, Dodich A, Boccardi M, van Eimeren T, Festari C, Barthel H, et al. Clinical validity of second-generation tau PET tracers as biomarkers for Alzheimer's disease in the context of a structured 5-phase development framework. Eur J Nucl Med Mol Imaging. 2021;48(7):2110–20. https://doi.org/10.1007/s00259-020-05156-4.
96. Aguero C, Dhaynaut M, Amaral AC, Moon SH, Neelamegam R, Scapellato M, et al. Head-to-head comparison of [(18)F]-Flortaucipir, [(18)F]-MK-6240 and [(18)F]-PI-2620 postmortem binding across the spectrum of neurodegenerative diseases. Acta Neuropathol. 2024;147(1):25. https://doi.org/10.1007/s00401-023-02672-z.
97. Brendel M, Barthel H, van Eimeren T, Marek K, Beyer L, Song M, et al. Assessment of 18F-PI-2620 as a biomarker in progressive supranuclear palsy. JAMA Neurol. 2020;77(11):1408–19. https://doi.org/10.1001/jamaneurol.2020.2526.
98. Kroth H, Oden F, Molette J, Schieferstein H, Capotosti F, Mueller A, et al. Discovery and preclinical characterization of [(18)F]PI-2620, a next-generation tau PET tracer for the assessment of tau pathology in Alzheimer's disease and other tauopathies. Eur J Nucl Med Mol Imaging. 2019;46(10):2178–89. https://doi.org/10.1007/s00259-019-04397-2.
99. Yap SY, Frias B, Wren MC, Schöll M, Fox NC, Årstad E, et al. Discriminatory ability of next-generation tau PET tracers for Alzheimer's disease. Brain. 2021;144(8):2284–90. https://doi.org/10.1093/brain/awab120.
100. Leuzy A, Chiotis K, Lemoine L, Gillberg PG, Almkvist O, Rodriguez-Vieitez E, et al. Tau PET imaging in neurodegenerative tauopathies-still a challenge. Mol Psychiatry. 2019;24(8):1112–34. https://doi.org/10.1038/s41380-018-0342-8.
101. Life Molecular Imaging Press Release. Life molecular imaging receives FDA fast track designation for [18F]PI-2620 in tau PET imaging across three neurodegenerative condi-

tions. https://life-mi.com/life-molecular-imaging-receives-fda-fast-track-designation-for-18fpi-2620-in-tau-pet-imaging-across-three-neurodegenerative-conditions/ (2024). Accessed 2024.
102. Mormino EC, Toueg TN, Azevedo C, Castillo JB, Guo W, Nadiadwala A, et al. Tau PET imaging with (18)F-PI-2620 in aging and neurodegenerative diseases. Eur J Nucl Med Mol Imaging. 2021;48(7):2233–44. https://doi.org/10.1007/s00259-020-04923-7.
103. Mueller A, Bullich S, Barret O, Madonia J, Berndt M, Papin C, et al. Tau PET imaging with (18)F-PI-2620 in patients with Alzheimer disease and healthy controls: a first-in-humans study. J Nucl Med. 2020;61(6):911–9. https://doi.org/10.2967/jnumed.119.236224.
104. Tezuka T, Takahata K, Seki M, Tabuchi H, Momota Y, Shiraiwa M, et al. Evaluation of [(18) F]PI-2620, a second-generation selective tau tracer, for assessing four-repeat tauopathies. Brain Commun. 2021;3 4:fcab190. https://doi.org/10.1093/braincomms/fcab190.
105. Marquié M, Normandin MD, Vanderburg CR, Costantino IM, Bien EA, Rycyna LG, et al. Validating novel tau positron emission tomography tracer [F-18]-AV-1451 (T807) on postmortem brain tissue. Ann Neurol. 2015;78(5):787–800. https://doi.org/10.1002/ana.24517.
106. Chen J, Li Y, Pirraglia E, Okamura N, Rusinek H, de Leon MJ. Quantitative evaluation of tau PET tracers (18)F-THK5351 and (18)F-AV-1451 in Alzheimer's disease with standardized uptake value peak-alignment (SUVP) normalization. Eur J Nucl Med Mol Imaging. 2018;45(9):1596–604. https://doi.org/10.1007/s00259-018-4040-1.
107. Ikonomovic MD, Abrahamson EE, Price JC, Mathis CA, Klunk WE. [F-18]AV-1451 positron emission tomography retention in choroid plexus: more than "off-target" binding. Ann Neurol. 2016;80(2):307–8. https://doi.org/10.1002/ana.24706.
108. Marquié M, Verwer EE, Meltzer AC, Kim SJW, Agüero C, Gonzalez J, et al. Lessons learned about [F-18]-AV-1451 off-target binding from an autopsy-confirmed Parkinson's case. Acta Neuropathol Commun. 2017;5(1):75. https://doi.org/10.1186/s40478-017-0482-0.
109. Vermeiren C, Motte P, Viot D, Mairet-Coello G, Courade JP, Citron M, et al. The tau positron-emission tomography tracer AV-1451 binds with similar affinities to tau fibrils and monoamine oxidases. Mov Disord. 2018;33(2):273–81. https://doi.org/10.1002/mds.27271.
110. Hansen AK, Brooks DJ, Borghammer P. MAO-B Inhibitors Do Not Block In Vivo Flortaucipir([18F]-AV-1451) Binding. Mol Imaging Biol. 2018;20(3):356–60. https://doi.org/10.1007/s11307-017-1143-1.
111. Smith R, Schöll M, Londos E, Ohlsson T, Hansson O. (18)F-AV-1451 in Parkinson's Disease with and without dementia and in Dementia with Lewy Bodies. Sci Rep. 2018;8(1):4717. https://doi.org/10.1038/s41598-018-23041-x.
112. Nagy Z, Yilmazer-Hanke DM, Braak H, Braak E, Schultz C, Hanke J. Assessment of the pathological stages of Alzheimer's disease in thin paraffin sections: a comparative study. Dement Geriatr Cogn Disord. 1998;9(3):140–4. https://doi.org/10.1159/000017038.
113. Lowe VJ, Lundt ES, Albertson SM, Min HK, Fang P, Przybelski SA, et al. Tau-positron emission tomography correlates with neuropathology findings. Alzheimers Dement. 2020;16(3):561–71. https://doi.org/10.1016/j.jalz.2019.09.079.
114. Pontecorvo MJ, Keene CD, Beach TG, Montine TJ, Arora AK, Devous MD Sr, et al. Comparison of regional flortaucipir PET with quantitative tau immunohistochemistry in three subjects with Alzheimer's disease pathology: a clinicopathological study. EJNMMI Res. 2020;10(1):65. https://doi.org/10.1186/s13550-020-00653-x.
115. Mattay VS, Fotenos AF, Ganley CJ, Marzella L. Brain Tau imaging: food and drug administration approval of (18)F-flortaucipir injection. J Nucl Med. 2020;61(10):1411–2. https://doi.org/10.2967/jnumed.120.252254.
116. Smith R, Wibom M, Pawlik D, Englund E, Hansson O. Correlation of in vivo [18F] flortaucipir with postmortem Alzheimer disease tau pathology. JAMA Neurol. 2019;76(3):310–7. https://doi.org/10.1001/jamaneurol.2018.3692.
117. Smith R, Puschmann A, Schöll M, Ohlsson T, van Swieten J, Honer M, et al. 18F-AV-1451 tau PET imaging correlates strongly with tau neuropathology in MAPT mutation carriers. Brain. 2016;139(Pt 9):2372–9. https://doi.org/10.1093/brain/aww163.

118. Tsai RM, Bejanin A, Lesman-Segev O, LaJoie R, Visani A, Bourakova V, et al. (18) F-flortaucipir (AV-1451) tau PET in frontotemporal dementia syndromes. Alzheimers Res Ther. 2019;11(1):13. https://doi.org/10.1186/s13195-019-0470-7.
119. Josephs KA, Whitwell JL, Tacik P, Duffy JR, Senjem ML, Tosakulwong N, et al. [18F]AV-1451 tau-PET uptake does correlate with quantitatively measured 4R-tau burden in autopsy-confirmed corticobasal degeneration. Acta Neuropathol. 2016;132(6):931–3. https://doi.org/10.1007/s00401-016-1618-1.
120. McMillan CT, Irwin DJ, Nasrallah I, Phillips JS, Spindler M, Rascovsky K, et al. Multimodal evaluation demonstrates in vivo (18)F-AV-1451 uptake in autopsy-confirmed corticobasal degeneration. Acta Neuropathol. 2016;132(6):935–7. https://doi.org/10.1007/s00401-016-1640-3.
121. Soleimani-Meigooni DN, Iaccarino L, La Joie R, Baker S, Bourakova V, Boxer AL, et al. 18F-flortaucipir PET to autopsy comparisons in Alzheimer's disease and other neurodegenerative diseases. Brain. 2020;143(11):3477–94. https://doi.org/10.1093/brain/awaa276.
122. Tagai K, Ono M, Kubota M, Kitamura S, Takahata K, Seki C, et al. High-contrast in vivo imaging of tau pathologies in Alzheimer's and Non-Alzheimer's disease tauopathies. Neuron. 2021;109(1):42–58.e8. https://doi.org/10.1016/j.neuron.2020.09.042.
123. Cho H, Choi JY, Hwang MS, Lee SH, Ryu YH, Lee MS, et al. Subcortical (18) F-AV-1451 binding patterns in progressive supranuclear palsy. Mov Disord. 2017;32(1):134–40. https://doi.org/10.1002/mds.26844.
124. Smith R, Pawlik D, Nilsson CF, Englund E, Hansson O. [18F]Flortaucipir distinguishes Alzheimer's disease from progressive supranuclear palsy pathology in a mixed-pathology case. Acta Neuropathol. 2020;139(2):411–3. https://doi.org/10.1007/s00401-019-02121-w.
125. Villemagne VL, Leuzy A, Bohorquez SS, Bullich S, Shimada H, Rowe CC, et al. CenTauR: toward a universal scale and masks for standardizing tau imaging studies. Alzheimers Dement (Amst). 2023;15(3):e12454. https://doi.org/10.1002/dad2.12454.
126. Kotari V, Southekal S, Navitsky M, Kennedy IA, Lu M, Morris A, et al. Early tau detection in flortaucipir images: validation in autopsy-confirmed data and implications for disease progression. Alzheimers Res Ther. 2023;15(1):41. https://doi.org/10.1186/s13195-023-01160-6.
127. Graff-Radford J, Yong KXX, Apostolova LG, Bouwman FH, Carrillo M, Dickerson BC, et al. New insights into atypical Alzheimer's disease in the era of biomarkers. Lancet Neurol. 2021;20(3):222–34. https://doi.org/10.1016/s1474-4422(20)30440-3.
128. Murray ME, Graff-Radford NR, Ross OA, Petersen RC, Duara R, Dickson DW. Neuropathologically defined subtypes of Alzheimer's disease with distinct clinical characteristics: a retrospective study. Lancet Neurol. 2011;10(9):785–96. https://doi.org/10.1016/s1474-4422(11)70156-9.
129. Petersen C, Nolan AL, de Paula França Resende E, Miller Z, Ehrenberg AJ, Gorno-Tempini ML, et al. Alzheimer's disease clinical variants show distinct regional patterns of neurofibrillary tangle accumulation. Acta Neuropathol. 2019;138(4):597–612. https://doi.org/10.1007/s00401-019-02036-6.
130. Charil A, Shcherbinin S, Southekal S, Devous MD, Mintun M, Murray ME, et al. Tau subtypes of Alzheimer's disease determined in vivo using flortaucipir PET imaging. J Alzheimers Dis. 2019;71(3):1037–48. https://doi.org/10.3233/jad-190264.
131. Xia C, Makaretz SJ, Caso C, McGinnis S, Gomperts SN, Sepulcre J, et al. Association of in vivo [18F]AV-1451 Tau PET imaging results with cortical atrophy and symptoms in typical and atypical Alzheimer disease. JAMA Neurol. 2017;74(4):427 36. https://doi.org/10.1001/jamaneurol.2016.5755.
132. Arriagada PV, Marzloff K, Hyman BT. Distribution of Alzheimer-type pathologic changes in nondemented elderly individuals matches the pattern in Alzheimer's disease. Neurology. 1992;42(9):1681–8. https://doi.org/10.1212/wnl.42.9.1681.
133. Bierer LM, Hof PR, Purohit DP, Carlin L, Schmeidler J, Davis KL, et al. Neocortical neurofibrillary tangles correlate with dementia severity in Alzheimer's disease. Arch Neurol. 1995;52(1):81–8. https://doi.org/10.1001/archneur.1995.00540250089017.

134. Cho H, Choi JY, Hwang MS, Kim YJ, Lee HM, Lee HS, et al. In vivo cortical spreading pattern of tau and amyloid in the Alzheimer disease spectrum. Ann Neurol. 2016;80(2):247–58. https://doi.org/10.1002/ana.24711.
135. Ossenkoppele R, Smith R, Mattsson-Carlgren N, Groot C, Leuzy A, Strandberg O, et al. Accuracy of Tau positron emission tomography as a prognostic marker in preclinical and prodromal Alzheimer disease: a head-to-head comparison against amyloid positron emission tomography and magnetic resonance imaging. JAMA Neurol. 2021;78(8):961–71. https://doi.org/10.1001/jamaneurol.2021.1858.
136. Mishra S, Gordon BA, Su Y, Christensen J, Friedrichsen K, Jackson K, et al. AV-1451 PET imaging of tau pathology in preclinical Alzheimer disease: defining a summary measure. NeuroImage. 2017;161:171–8. https://doi.org/10.1016/j.neuroimage.2017.07.050.
137. Ashton NJ, Brum WS, Di Molfetta G, Benedet AL, Arslan B, Jonatis E, et al. Diagnostic accuracy of the plasma ALZpath pTau217 immunoassay to identify Alzheimer's disease pathology. medRxiv. 2023; https://doi.org/10.1101/2023.07.11.23292493.
138. Janelidze S, Stomrud E, Smith R, Palmqvist S, Mattsson N, Airey DC, et al. Cerebrospinal fluid p-tau217 performs better than p-tau181 as a biomarker of Alzheimer's disease. Nat Commun. 2020;11(1):1683. https://doi.org/10.1038/s41467-020-15436-0.
139. Leuzy A, Janelidze S, Mattsson-Carlgren N, Palmqvist S, Jacobs D, Cicognola C, et al. Comparing the clinical utility and diagnostic performance of CSF P-Tau181, P-Tau217, and P-Tau231 assays. Neurology. 2021;97(17):e1681–e94. https://doi.org/10.1212/wnl.0000000000012727.
140. Mattsson N, Smith R, Strandberg O, Palmqvist S, Schöll M, Insel PS, et al. Comparing (18)F-AV-1451 with CSF t-tau and p-tau for diagnosis of Alzheimer disease. Neurology. 2018;90(5):e388–e95. https://doi.org/10.1212/wnl.0000000000004887.
141. Horie K, Salvadó G, Barthélemy NR, Janelidze S, Li Y, He Y, et al. CSF MTBR-tau243 is a specific biomarker of tau tangle pathology in Alzheimer's disease. Nat Med. 2023;29(8):1954–63. https://doi.org/10.1038/s41591-023-02443-z.
142. Janelidze S, Bali D, Ashton NJ, Barthélemy NR, Vanbrabant J, Stoops E, et al. Head-to-head comparison of 10 plasma phospho-tau assays in prodromal Alzheimer's disease. Brain. 2023;146(4):1592–601. https://doi.org/10.1093/brain/awac333.
143. Janelidze S, Mattsson N, Palmqvist S, Smith R, Beach TG, Serrano GE, et al. Plasma P-tau181 in Alzheimer's disease: relationship to other biomarkers, differential diagnosis, neuropathology and longitudinal progression to Alzheimer's dementia. Nat Med. 2020;26(3):379–86. https://doi.org/10.1038/s41591-020-0755-1.
144. Cullen NC, Leuzy A, Janelidze S, Palmqvist S, Svenningsson AL, Stomrud E, et al. Plasma biomarkers of Alzheimer's disease improve prediction of cognitive decline in cognitively unimpaired elderly populations. Nat Commun. 2021;12(1):3555. https://doi.org/10.1038/s41467-021-23746-0.
145. Mattsson-Carlgren N, Salvadó G, Ashton NJ, Tideman P, Stomrud E, Zetterberg H, et al. Prediction of longitudinal cognitive decline in preclinical Alzheimer disease using plasma biomarkers. JAMA Neurol. 2023;80(4):360–9. https://doi.org/10.1001/jamaneurol.2022.5272.
146. Thijssen EH, La Joie R, Strom A, Fonseca C, Iaccarino L, Wolf A, et al. Plasma phosphorylated tau 217 and phosphorylated tau 181 as biomarkers in Alzheimer's disease and frontotemporal lobar degeneration: a retrospective diagnostic performance study. Lancet Neurol. 2021;20(9):739–52. https://doi.org/10.1016/s1474-4422(21)00214-3.
147. Mattsson-Carlgren N, Collij LE, Stomrud E, Pichet Binette A, Ossenkoppele R, Smith R, et al. Plasma biomarker strategy for selecting patients with Alzheimer disease for anti-amyloid immunotherapies. JAMA Neurol. 2024;81(1):69–78. https://doi.org/10.1001/jamaneurol.2023.4596.
148. Barthélemy NR, Salvadó G, Schindler SE, He Y, Janelidze S, Collij LE, Saef B, Henson RL, Chen CD, Gordon BA, Li Y, La Joie R, Benzinger TLS, Morris JC, Mattsson-Carlgren N, Palmqvist S, Ossenkoppele R, Rabinovici GD, Stomrud E, Bateman RJ, Hansson O. Highly accurate blood test for Alzheimer's disease is similar or superior to clinical cerebrospinal fluid

tests. Nat Med. 2024 Apr;30(4):1085-1095. doi: 10.1038/s41591-024-02869-z. Epub 2024 Feb 21. PMID: 38382645; PMCID: PMC11031399.

149. Brum WS, Cullen NC, Janelidze S, Ashton NJ, Zimmer ER, Therriault J, et al. A two-step workflow based on plasma p-tau217 to screen for amyloid β positivity with further confirmatory testing only in uncertain cases. Nat Aging. 2023;3(9):1079–90. https://doi.org/10.1038/s43587-023-00471-5.
150. Chen SD, Lu JY, Li HQ, Yang YX, Jiang JH, Cui M, et al. Staging tau pathology with tau PET in Alzheimer's disease: a longitudinal study. Transl Psychiatry. 2021;11(1):483. https://doi.org/10.1038/s41398-021-01602-5.
151. Ossenkoppele R, Rabinovici GD, Smith R, Cho H, Schöll M, Strandberg O, et al. Discriminative accuracy of [18F]flortaucipir positron emission tomography for Alzheimer disease vs other neurodegenerative disorders. JAMA. 2018;320(11):1151–62. https://doi.org/10.1001/jama.2018.12917.
152. Jack CR, Wiste HJ, Botha H, Weigand SD, Therneau TM, Knopman DS, et al. The bivariate distribution of amyloid-β and tau: relationship with established neurocognitive clinical syndromes. Brain. 2019;142(10):3230–42. https://doi.org/10.1093/brain/awz268.
153. Leuzy A, Pascoal TA, Strandberg O, Insel P, Smith R, Mattsson-Carlgren N, et al. A multicenter comparison of [(18)F]flortaucipir, [(18)F]RO948, and [(18)F]MK6240 tau PET tracers to detect a common target ROI for differential diagnosis. Eur J Nucl Med Mol Imaging. 2021;48(7):2295–305. https://doi.org/10.1007/s00259-021-05401-4.
154. Schonhaut DR, McMillan CT, Spina S, Dickerson BC, Siderowf A, Devous MD Sr, et al. (18) F-flortaucipir tau positron emission tomography distinguishes established progressive supranuclear palsy from controls and Parkinson disease: a multicenter study. Ann Neurol. 2017;82(4):622–34. https://doi.org/10.1002/ana.25060.
155. Tudorascu DL, Laymon CM, Zammit M, Minhas DS, Anderson SJ, Ellison PA, et al. Relationship of amyloid beta and neurofibrillary tau deposition in Neurodegeneration in Aging Down Syndrome (NiAD) study at baseline. Alzheimers Dement (N Y). 2020;6(1):e12096. https://doi.org/10.1002/trc2.12096.
156. Ferreira D, Przybelski SA, Lesnick TG, Lemstra AW, Londos E, Blanc F, et al. β-Amyloid and tau biomarkers and clinical phenotype in dementia with Lewy bodies. Neurology. 2020;95(24):e3257–e68. https://doi.org/10.1212/wnl.0000000000010943.
157. Singleton E, Hansson O, Pijnenburg YAL, La Joie R, Mantyh WG, Tideman P, et al. Heterogeneous distribution of tau pathology in the behavioural variant of Alzheimer's disease. J Neurol Neurosurg Psychiatry. 2021;92(8):872–80. https://doi.org/10.1136/jnnp-2020-325497.
158. Smith R, Santillo AF, Waldö ML, Strandberg O, Berron D, Vestberg S, et al. 18F-Flortaucipir in TDP-43 associated frontotemporal dementia. Sci Rep. 2019;9(1):6082. https://doi.org/10.1038/s41598-019-42625-9.
159. Pascoal TA, Therriault J, Benedet AL, Savard M, Lussier FZ, Chamoun M, et al. 18F-MK-6240 PET for early and late detection of neurofibrillary tangles. Brain. 2020;143(9):2818–30. https://doi.org/10.1093/brain/awaa180.
160. Makaretz SJ, Quimby M, Collins J, Makris N, McGinnis S, Schultz A, et al. Flortaucipir tau PET imaging in semantic variant primary progressive aphasia. J Neurol Neurosurg Psychiatry. 2018;89(10):1024–31. https://doi.org/10.1136/jnnp-2017-316409.
161. Leuzy A, Smith R, Ossenkoppele R, Santillo A, Borroni E, Klein G, et al. Diagnostic performance of RO948 F 18 tau positron emission tomography in the differentiation of Alzheimer disease from other neurodegenerative disorders. JAMA Neurol. 2020;77(8):955–65. https://doi.org/10.1001/jamaneurol.2020.0989.
162. Altomare D, Caprioglio C, Assal F, Allali G, Mendes A, Ribaldi F, et al. Diagnostic value of amyloid-PET and tau-PET: a head-to-head comparison. Eur J Nucl Med Mol Imaging. 2021;48(7):2200–11. https://doi.org/10.1007/s00259-021-05246-x.

Fluid Markers of Brain Endothelial Injury

Rawan Tarawneh

Abstract Brain endothelial injury is a common pathology in Alzheimer disease (AD) brains. Structural changes consistent with brain endothelial injury are observed in almost all AD brains at autopsy and occur independently of other forms of vascular pathology. Emerging evidence supports an important role for the brain endothelium in mediating amyloid and tau aggregation and contributing to impaired neurogenesis and altered synaptic plasticity. Despite its importance, our understanding of the role of endothelial dysfunction in AD pathogenesis, and the mechanisms by which endothelial alterations contribute to neurodegeneration and cognitive impairment in AD, has been limited by the absence of reliable fluid markers for brain endothelial injury. Vascular endothelial-cadherin (VE-cadherin) is an abundantly expressed endothelium-specific protein that plays an important role in maintaining the blood-brain barrier. Recent studies using proteomics and transcriptomics suggest VE-cadherin may be a useful cerebrospinal fluid (CSF) marker for brain endothelial injury in AD, including the early preclinical stages, and has strong potential for use as a diagnostic, prognostic, and outcome measure in clinical studies of AD. Further research to identify and validate other biomarkers of brain endothelial injury will be important to elucidate endothelial contributions to the onset and progression of AD pathology and the onset of cognitive impairment.

Keywords Endothelium · Vascular endothelial-cadherin · Blood-brain barrier · Alzheimer disease · Biomarker

R. Tarawneh (✉)
Department of Neurology, University of New Mexico, Albuquerque, NM, USA
e-mail: raltarawneh@salud.unm.edu

R. Tarawneh (ed.), *Biomarkers of Neurodegenerative Disorders*,
https://doi.org/10.1007/978-3-031-80849-4_21

1 Introduction

The brain endothelium is an important component of the blood-brain barrier and represents a highly active interface that regulates blood flow and neuronal metabolism [1, 2]. Brain endothelial functions are mediated via several proteins including extracellular matrix proteins (e.g., proteoglycans, laminin, and fibronectin), growth factors (e.g., endothelium-derived growth factor [EDGF], platelet-derived growth factor [PDGF], and fibroblast growth factors [FGF]), proteases (e.g., thrombin and matrix metalloproteinases [MMPs]), vasoconstrictors (e.g., endothelin-1, leukotrienes, and thromboxane A2) and vasodilators (e.g., nitrous oxide [NO], prostacyclin/prostaglandin E2, and endothelium-derived hyperpolarizing factor [EDHF]), antithrombotic (e.g., antithrombin III) and prothrombotic factors (e.g., thromboplastin and platelet-activating factor), and various inflammatory mediators (e.g., interleukin (IL)-1, IL-6, IL-8, monocyte chemoattractant proteins (MCP) 1–2, leukotrienes, and cell adhesion molecules) [1, 2].

Structural changes in the brain endothelium are observed in almost all Alzheimer disease (AD) brains, including areas that are devoid of, or distant from, other AD pathologies [3]. These changes include increased pinocytic vesicles, laminin deposition in the basement membrane, collagen accumulation, and focal necrotic changes. Endothelial cell loss in AD is demonstrated by reduced staining for endothelial markers, such as CD31 and CD34 [4]. Further, these changes appear to occur independently of other forms of vascular pathology involving smooth muscle or connective tissue components of brain vasculature, such as those observed with arteriolosclerosis, atherosclerosis, or cerebral amyloid angiopathy [3, 5–9]. While endothelial and other vascular changes in AD were previously considered to be secondary pathologies that occur as the result of amyloid or tau aggregation, recent evidence suggests that endothelial dysfunction is an early AD pathology that may precede amyloid and tau accumulation and contribute directly to neuronal and synaptic loss [3, 10, 11]. The term "endothelial dysfunction" is used to refer to structural or functional alterations at the intracellular or cell membrane level that interfere with endothelial cell integrity or function [4–9, 12–14]. Reports from animal studies that examined the three-dimensional (3D) arrangement of brain vessels have found that significant morphological changes in cerebral vessels and endothelial dysfunction precede amyloid deposition and cognitive deficits in AD transgenic [11] or insulin-resistant mice [15]. Further, human studies using RNA-sequencing (RNA-seq) analyses suggest that endothelial pathways are among the most differentially expressed in human AD brains [16]. Consistent with these reports, 30 out of the top 45 AD risk genes are expressed in vascular cell types, and several of these have their highest expression levels in endothelial structures [17]. Interestingly, the current proposed model of AD progression, which is based on a hypothetical ordering of biomarker changes in AD, has recently been challenged by data-driven models derived from longitudinal AD cohorts which suggest that vascular changes are the earliest pathological changes in AD and likely precede amyloid or tau aggregation [18].

Brain endothelial injury is associated with reduced expression of pre-synaptic and post-synaptic proteins in human AD brains [19–21] and displays regional and laminar patterns that parallel those of neuronal loss [14]. Further, brain endothelial injury is preferentially localized in the vicinity of dystrophic neurites [1, 12–14]. While the mechanisms that underly endothelial contributions to neurodegeneration and cognitive impairment remain to be fully elucidated, emerging evidence suggests that the brain endothelium exerts direct effects on synaptic and neuronal function which are independent of other AD pathologies [1]. Proposed mechanisms include impaired release of neurotrophic growth factors (brain-derived neurotrophic factor [BDNF], vascular endothelial growth factor [VEGF], and FGF) and reduced hippocampal expression of NO, which plays an important role in long-term potentiation and memory functions [1, 3]. Further, recent studies have identified several novel pathways by which endothelial dysfunction influences axonal repair and synaptic plasticity, including interactions of the endothelial marker vascular endothelial-cadherin (VE-cadherin, also known as cadherin-5 [CDH5]) with semaphorin-3A (SEMA3A), an important mediator of memory and learning functions.

2 Fluid Markers of Endothelial Injury

Despite its prevalence, the role of endothelial dysfunction in AD pathogenesis and its relationship to other AD pathologies has not been adequately investigated mainly due to the absence of reliable fluid markers of brain endothelial injury. This knowledge gap has also led to this important pathological substrate being excluded from current biomarker-based definitions of AD or proposed disease models of AD onset and progression [22, 23]. Therefore, this area warrants further investigation. The identification of fluid or imaging markers that can reliably measure and track brain endothelial injury will have important implications in improving our understanding of endothelial contributions to early AD pathogenesis.

VE-cadherin (cadherin-5) is an abundant endothelium-specific adhesion molecule which is a key component of intercellular endothelial junctions and regulates endothelial barrier permeability [24]. Using RNA-seq studies from brains with neuropathologically confirmed AD and controls (n = 96), we have found that brain VE-cadherin expression is significantly reduced in AD compared to controls (log twofold change, −0.41; $p = 0.02$) reflecting endothelial cell and capillary loss in AD [25]. Other studies have also shown reduced brain VE-cadherin levels in AD mouse models [26]. A recent study that measured VE-cadherin levels in cerebrospinal fluid (CSF) using proteomics in a large well-characterized cohort of individuals with preclinical AD (i.e., cognitively normal individuals with biomarker evidence of brain amyloid and tau pathologies) and cognitively normal individuals with AD-related pathological change (i.e., those with biomarker evidence of brain amyloid only) identified CSF VE-cadherin as a potential marker of brain endothelial injury in AD [25]. In this study, CSF VE-cadherin levels were found to be increased in even the earliest preclinical stages of AD compared to healthy controls and increased with the progression of AD pathology across the ATN staging system [25]. CSF VE-cadherin complemented the ability of each of CSF Aβ42/Aβ40, p-tau181, and total tau to predict brain amyloid status using positron emission tomography (PET) [25]. CSF VE-cadherin levels correlated with CSF or imaging markers of amyloid and tau pathologies. Further, in this study, CSF VE-cadherin levels were closely associated with cognitive outcomes, including the Preclinical Alzheimer's Cognitive Composite (PACC), episodic memory (i.e., Free and Cued Selective Reminding Test), language (i.e., animal fluency), and executive function (i.e., digit-symbol substitution) scores independently of amyloid, tau, and imaging measures of small vessel disease, and partially mediated the relationship between amyloid or tau and cognitive outcomes [25].

Interestingly, functional pathway analyses have identified significant interactions between VE-cadherin and the E4 allele of apolipoprotein-E (*APOE4*). VE-cadherin interacts with the vascular endothelial growth factor receptor-2 (VEGFR2; KDR) which is a cell surface VEGF-receptor and has important roles in endothelial cell growth, migration, and differentiation [27]. VEGFA forms a complex with another endothelial protein, neuropilin-1 (NRP-1), on the endothelial cell surface [28]. It has been reported that *APOE4* is associated with increased NRP-1 expression and increased VEGFR signaling which results in VE-cadherin phosphorylation and the subsequent loosening of endothelial tight junctions [29] (Fig. 1).

3 Conclusion

Together, these findings highlight the importance of further investigating the role of brain endothelial dysfunction in AD pathogenesis with an emphasis on endothelial cell contributions to neuronal or synaptic loss and impaired synaptic plasticity. While the role of vascular pathology in large, medium, and small arteries (e.g., atherosclerosis and arteriolosclerosis) in AD or vascular dementia has been an active area of research for decades, little attention has been given to the specific role of brain endothelial changes in mediating AD onset or progression. Accumulating evidence suggests that brain endothelial cells have important roles in mediating amyloid and tau accumulation, inflammation, and immune dysregulation and may have direct effects on neuronal and synaptic function [33]. Therefore, further examination of brain endothelial contributions to AD and related dementias (ADRD) and the identification and validation of additional brain endothelial injury markers will be an important area for future research. Brain endothelial injury markers offer potential useful diagnostic and prognostic biomarkers that can complement information provided by other markers of AD pathology. Brain endothelial injury markers may also be used as secondary or tertiary outcome measures to examine downstream effects of disease-modifying therapies on blood-brain barrier integrity [34]. Importantly, as the newly approved anti-amyloid monoclonal therapies for AD,

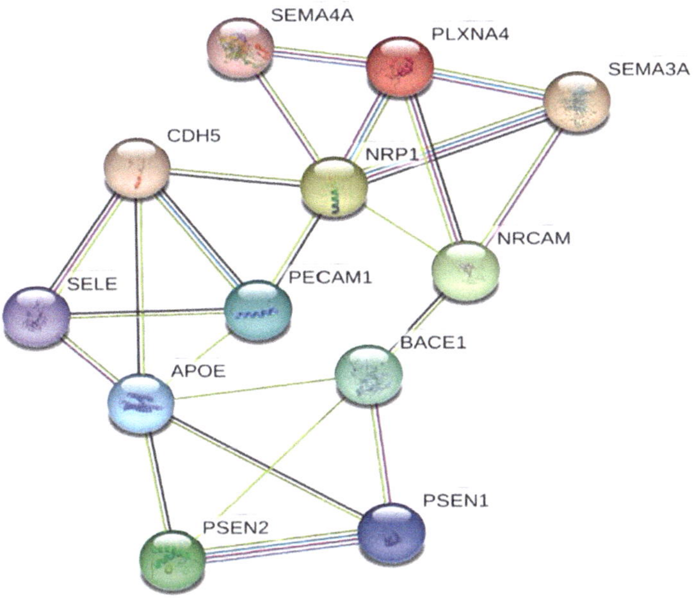

Fig. 1 Endothelial proteins interact with axonal and synaptic pathways. Functional pathway analyses of endothelial and neuronal proteins in AD using STRING are shown. These functional pathway analyses suggest significant interactions between endothelial proteins and proteins involved in axonal repair and synaptic plasticity. Network nodes represent proteins; splice isoforms or post-translational modifications are collapsed so that each node represents all the proteins produced by a single protein-coding gene locus. Edges represent protein–protein associations that are meant to be specific and meaningful. Color coding for the interactions is as follows: known interactions from curated databases (teal); known experimentally determined interactions (purple); predicted interactions from gene neighborhood (green), gene fusions (red), or gene co-occurrence (blue); predicted interactions from text-mining (yellow), co-expression (black), and protein homology (lavender). Abbreviations: APOE apolipoprotein E, BACE β-secretase, CDH5 cadherin-5 (also known as vascular endothelial-cadherin; VE-cadherin), NRCAM neuronal cell adhesion molecule, NRP-1 neuropilin-1, PECAM1 platelet endothelial cell adhesion molecule-1 (also known as CD31), PSEN1 presenilin-1 (aka PS1), PSEN2 presenilin-2 (aka PS2), PLXNA4 Plexin A4, SELE E-Selectin, SEMA3A semaphorin-3A, SEMA4A semaphorin-4A. (Figure created with STRINGv11.5 for functional protein association networks (https://string-db.org) [30–32]. Used with permission from Tarawneh, Biomolecules 2023 [33] (https://creativecommons.org/licenses/by/4.0/)

lecanemab and donanemab, are associated with adverse effects that result from targeting vascular amyloid, referred to as amyloid-related imaging abnormalities (ARIAs), brain endothelial injury markers offer strong potential as fluid surrogates for ARIAs which may detect or predict their onset. In this context, these markers may offer useful risk stratification or safety monitoring tools in clinical and research settings. Further research in this area is warranted.

Funding This work was supported by the National Institute of Health- National Institute of Aging Grant RF1AG083744 (RT).

References

1. Grammas P. Neurovascular dysfunction, inflammation and endothelial activation: implications for the pathogenesis of Alzheimer's disease. J Neuroinflammation. 2011;8:26. https://doi.org/10.1186/1742-2094-8-26.
2. Sumpio BE, Riley JT, Dardik A. Cells in focus: endothelial cell. Int J Biochem Cell Biol. 2002;34(12):1508–12. https://doi.org/10.1016/s1357-2725(02)00075-4.
3. Kelleher RJ, Soiza RL. Evidence of endothelial dysfunction in the development of Alzheimer's disease: is Alzheimer's a vascular disorder? Am J Cardiovasc Dis. 2013;3(4):197–226.
4. Kalaria RN, Hedera P. Differential degeneration of the cerebral microvasculature in Alzheimer's disease. Neuroreport. 1995;6(3):477–80. https://doi.org/10.1097/00001756-199502000-00018.
5. Christov A, Ottman J, Hamdheydari L, Grammas P. Structural changes in Alzheimer's disease brain microvessels. Curr Alzheimer Res. 2008;5(4):392–5. https://doi.org/10.2174/156720508785132334.
6. de la Torre JC. Cerebromicrovascular pathology in Alzheimer's disease compared to normal aging. Gerontology. 1997;43(1–2):26–43. https://doi.org/10.1159/000213834.
7. Claudio L. Ultrastructural features of the blood-brain barrier in biopsy tissue from Alzheimer's disease patients. Acta Neuropathol. 1996;91(1):6–14. https://doi.org/10.1007/s004010050386.
8. Kalaria RN. Cerebral vessels in ageing and Alzheimer's disease. Pharmacol Ther. 1996;72(3):193–214. https://doi.org/10.1016/s0163-7258(96)00116-7.
9. Farkas E, Luiten PG. Cerebral microvascular pathology in aging and Alzheimer's disease. Prog Neurobiol. 2001;64(6):575–611. https://doi.org/10.1016/s0301-0082(00)00068-x.
10. de la Torre JC. Is Alzheimer's disease a neurodegenerative or a vascular disorder? Data, dogma, and dialectics. Lancet Neurol. 2004;3(3):184–90. https://doi.org/10.1016/s1474-4422(04)00683-0.
11. Meyer EP, Ulmann-Schuler A, Staufenbiel M, Krucker T. Altered morphology and 3D architecture of brain vasculature in a mouse model for Alzheimer's disease. Proc Natl Acad Sci USA. 2008;105(9):3587–92. https://doi.org/10.1073/pnas.0709788105.
12. Miyakawa T, Shimoji A, Kuramoto R, Higuchi Y. The relationship between senile plaques and cerebral blood vessels in Alzheimer's disease and senile dementia. Morphological mechanism of senile plaque production. Virchows Arch B Cell Pathol Incl Mol Pathol. 1982;40(2):121–9. https://doi.org/10.1007/bf02932857.
13. Araki K, Miyakawa T, Katsuragi S. Ultrastructure of senile plaque using thick sections in the brain with Alzheimer's disease. Jpn J Psychiatry Neurol. 1991;45(1):85–9. https://doi.org/10.1111/j.1440-1819.1991.tb00510.x.
14. Buée L, Hof PR, Bouras C, Delacourte A, Perl DP, Morrison JH, et al. Pathological alterations of the cerebral microvasculature in Alzheimer's disease and related dementing disorders. Acta Neuropathol. 1994;87(5):469–80. https://doi.org/10.1007/bf00294173.
15. Takechi R, Lam V, Brook E, Giles C, Fimognari N, Mooranian A, et al. Blood-brain barrier dysfunction precedes cognitive decline and neurodegeneration in diabetic insulin resistant mouse model: an implication for causal link. Front Aging Neurosci. 2017;9:399. https://doi.org/10.3389/fnagi.2017.00399.
16. Lau SF, Cao H, Fu AKY, Ip NY. Single-nucleus transcriptome analysis reveals dysregulation of angiogenic endothelial cells and neuroprotective glia in Alzheimer's disease. Proc Natl Acad Sci USA. 2020;117(41):25800–9. https://doi.org/10.1073/pnas.2008762117.
17. Yang AC, Vest RT, Kern F, Lee DP, Agam M, Maat CA, et al. A human brain vascular atlas reveals diverse mediators of Alzheimer's risk. Nature. 2022;603(7903):885–92. https://doi.org/10.1038/s41586-021-04369-3.
18. Iturria-Medina Y, Sotero RC, Toussaint PJ, Mateos-Pérez JM, Evans AC, Weiner MW, et al. Early role of vascular dysregulation on late-onset Alzheimer's disease based on multifactorial data-driven analysis. Nat Commun. 2016;7 1:11934. https://doi.org/10.1038/ncomms11934.

19. VanGuilder HD, Farley JA, Yan H, Van Kirk CA, Mitschelen M, Sonntag WE, et al. Hippocampal dysregulation of synaptic plasticity-associated proteins with age-related cognitive decline. Neurobiol Dis. 2011;43(1):201–12. https://doi.org/10.1016/j.nbd.2011.03.012.
20. Yang J, Yao Y, Wang L, Yang C, Wang F, Guo J, et al. Gastrin-releasing peptide facilitates glutamatergic transmission in the hippocampus and effectively prevents vascular dementia induced cognitive and synaptic plasticity deficits. Exp Neurol. 2017;287(Pt 1):75–83. https://doi.org/10.1016/j.expneurol.2016.08.008.
21. Wang F, Cao Y, Ma L, Pei H, Rausch WD, Li H. Dysfunction of cerebrovascular endothelial cells: prelude to vascular dementia. Front Aging Neurosci. 2018;10:376. https://doi.org/10.3389/fnagi.2018.00376.
22. Jack CR Jr, Bennett DA, Blennow K, Carrillo MC, Dunn B, Haeberlein SB, et al. NIA-AA research framework: toward a biological definition of Alzheimer's disease. Alzheimers Dement. 2018;14(4):535–62. https://doi.org/10.1016/j.jalz.2018.02.018.
23. Jack CR Jr, Knopman DS, Jagust WJ, Petersen RC, Weiner MW, Aisen PS, et al. Tracking pathophysiological processes in Alzheimer's disease: an updated hypothetical model of dynamic biomarkers. Lancet Neurol. 2013;12(2):207–16. https://doi.org/10.1016/s1474-4422(12)70291-0.
24. Li W, Chen Z, Chin I, Chen Z, Dai H. The role of VE-cadherin in blood-brain barrier integrity under central nervous system pathological conditions. Curr Neuropharmacol. 2018;16(9):1375–84. https://doi.org/10.2174/1570159X16666180222164809.
25. Tarawneh R, Kasper RS, Sanford J, Phuah CL, Hassenstab J, Cruchaga C. Vascular endothelial-cadherin as a marker of endothelial injury in preclinical Alzheimer disease. Ann Clin Transl Neurol. 2022;9(12):1926–40. https://doi.org/10.1002/acn3.51685.
26. Lee D, Cho S-J, Lim HJ, Seok J, Jo C, Jo SA, et al. Alteration of vascular endothelial Cadherin in Alzheimer's disease patient and mouse model. bioRxiv. 2018:430140. https://doi.org/10.1101/430140.
27. Ruiz de Almodovar C, Lambrechts D, Mazzone M, Carmeliet P. Role and therapeutic potential of VEGF in the nervous system. Physiol Rev. 2009;89(2):607–48. https://doi.org/10.1152/physrev.00031.2008.
28. Soker S, Takashima S, Miao HQ, Neufeld G, Klagsbrun M. Neuropilin-1 is expressed by endothelial and tumor cells as an isoform-specific receptor for vascular endothelial growth factor. Cell. 1998;92(6):735–45. https://doi.org/10.1016/s0092-8674(00)81402-6.
29. Moore AM, Mahoney E, Dumitrescu L, De Jager PL, Koran MEI, Petyuk VA, et al. APOE ε4-specific associations of VEGF gene family expression with cognitive aging and Alzheimer's disease. Neurobiol Aging. 2020;87:18–25. https://doi.org/10.1016/j.neurobiolaging.2019.10.021.
30. Szklarczyk D, Gable AL, Nastou KC, Lyon D, Kirsch R, Pyysalo S, et al. The STRING database in 2021: customizable protein-protein networks, and functional characterization of user-uploaded gene/measurement sets. Nucleic Acids Res. 2021;49(D1):D605–d12. https://doi.org/10.1093/nar/gkaa1074.
31. Szklarczyk D, Gable AL, Lyon D, Junge A, Wyder S, Huerta-Cepas J, et al. STRING v11: protein-protein association networks with increased coverage, supporting functional discovery in genome-wide experimental datasets. Nucleic Acids Res. 2019;47(D1):D607–d13. https://doi.org/10.1093/nar/gky1131.
32. Szklarczyk D, Morris JH, Cook H, Kuhn M, Wyder S, Simonovic M, et al. The STRING database in 2017: quality-controlled protein-protein association networks, made broadly accessible. Nucleic Acids Res. 2017;45(D1):D362–d8. https://doi.org/10.1093/nar/gkw937.
33. Tarawneh R. Microvascular contributions to Alzheimer disease pathogenesis: is Alzheimer disease primarily an endotheliopathy? Biomolecules. 2023;13(5):830. https://doi.org/10.3390/biom13050830.
34. Oosthoek M, Vermunt L, de Wilde A, Bongers B, Antwi-Berko D, Scheltens P, et al. Utilization of fluid-based biomarkers as endpoints in disease-modifying clinical trials for Alzheimer's disease: a systematic review. Alzheimers Res Ther. 2024;16 1:93. https://doi.org/10.1186/s13195-024-01456-1.

Epilogue: Current Gaps and Future Directions in ADRD Biomarker Research

Rawan Tarawneh

Introduction

Significant advances in Alzheimer disease and related dementias (ADRD) biomarker research have been made over the last decade, culminating in the integration of fluid and imaging biomarkers into updated diagnostic research criteria and staging systems for AD [1-3]. This framework represents a major step that has shifted our conceptualization of AD from a clinical syndrome to that of a biological construct, the diagnosis of which relies on the presence of biomarker evidence of core AD pathologies [1, 3]. Importantly, this lays the foundation for a similar approach to be applied in other neurodegenerative disorders. Further, the use of biomarkers as in vivo surrogates of brain pathology has revolutionized our understanding of the molecular underpinnings of neurodegenerative disorders and highlighted the significant degree of pathological heterogeneity associated with these disorders and the limitations of relying on clinical phenotypes alone to predict underlying brain pathology or response to potential disease-modifying treatments [2]. The implementation of biomarkers in diagnosis, disease staging, and molecular characterization of individuals has significant implications in clinical trials of new ADRD therapies by improving patient selection and stratification and providing useful tools to measure target engagement and downstream disease-modifying effects. Together, the integration of biomarkers into different phases of ADRD drug development will be a major step towards more effective treatments and improved patient outcomes.

Despite these significant advances, there are several limitations in current ADRD biomarker research which represent major knowledge gaps as well as important areas for future research. These include the need for standardization of both

R. Tarawneh
Department of Neurology, University of New Mexico, Albuquerque, NM, USA
e-mail: raltarawneh@salud.unm.edu

R. Tarawneh (ed.), *Biomarkers of Neurodegenerative Disorders*,
https://doi.org/10.1007/978-3-031-80849-4

analytical and pre-analytical factors in fluid biomarker measurements (discussed in Chapter Impact of Pre-analytical Factors on Fluid Biomarker Measurements in Alzheimer Disease) and understanding sex-based differences in ADRD risk and biomarker levels (discussed in Chapter Sex Differences in Alzheimer Disease-Related Biomarkers and Dementia Risk). Additional gaps that need to be addressed in future studies include (a) the lack of reliable fluid or imaging biomarkers to measure 43-kDa transactive response DNA-binding protein (TDP-43) pathology, (b) the need for a better understanding of racial and ethnic differences in biomarker levels or dementia risk, and (c) exploring the potential influence of systemic factors on peripheral ADRD biomarker levels.

Fluid and Imaging Biomarkers of TDP-43 Pathology

TDP-43 is a neuronal protein that is involved in several cellular processes, including the regulation of RNA metabolism (e.g., cryptic RNA splicing and RNA transport), microRNA biogenesis, vesicular and axonal transport, mitochondrial autophagy, and regulation of synaptic protein expression [4, 5]. Neuronal cytoplasmic aggregation of pathological insoluble and ubiquitinated forms of TDP-43, often associated with neuronal loss and gliosis, is an important pathological substrate of various neuropathological disorders, collectively referred to as "TDP-43 proteinopathies". TDP-43 pathology is considered the core pathological feature of amyotrophic lateral sclerosis (ALS; being present in 97% of all cases) and is present in 50% of individuals with frontotemporal lobar degeneration (FTLD), particularly those presenting with the frontal/behavioral variant of frontotemporal dementia (FTD) or the semantic variant of primary progressive aphasia [6–9]. Emerging evidence suggests an increasingly important role for TDP-43 pathology in other neurodegenerative disorders, as it is present in 30% of AD brains [10, 11], and has also been observed in chronic traumatic encephalopathy (CTE) [12], dementia with Lewy bodies (DLB) [13], and Huntington's disease (HD) [14]. TDP-43 pathology, with or without hippocampal sclerosis, underlies the condition referred to as limbic-predominant age-related TDP-43 encephalopathy (LATE), which often occurs in individuals above the age of 75 years and clinically mimics AD [15]. Other less common disorders associated with TDP-43 pathology include Perry syndrome (a syndrome characterized by neuropsychiatric features, parkinsonism, and weight loss often due to *DCTN1* gene mutations), in which TDP-43 aggregates are found in the cortex and brainstem [16], and sporadic inclusion body myopathy in which TDP-43 pathology is observed in skeletal muscle [17]. Further, TDP-43 pathology, mostly in the amygdala, is reported in 24% of cognitively normal aging brains [18], particularly in the "oldest old" population (those over 90 years of age) [19].

There are currently no fluid or imaging markers that can reliably detect pathological forms of TDP-43, representing an important opportunity for future research. Several antibodies have been used to measure TDP-43, including N-terminal antibodies which detect full-length TDP-43 aggregates, C-terminal antibodies which can also detect cleaved cytotoxic fragments [20], and other antibodies that can detect TDP-43 phosphorylated at serine residues 409/410 [21]. However, the accurate detection and measurement of TDP-43 using immunoassays in human biospecimens, including cerebrospinal fluid (CSF) or blood, has been challenging due to several factors. These include the inability of TDP-43 immunoassays to differentiate between normal and pathological forms of this protein [22], the relatively low concentrations of pathological TDP-43 in CSF and plasma, and the tendency of soluble TDP-43 to aggregate which potentially influences the levels of soluble TDP-43 in examined biospecimens. Further, as TDP-43 is ubiquitously expressed in the body, its blood levels may be influenced by several peripheral sources. Therefore, reports have been conflicting regarding the diagnostic value of CSF and plasma TDP-43 levels in various clinically and neuropathologically defined cohorts.

Improved sensitivity of TDP-43 measurements in biofluids may be achieved using single molecule arrays (SIMOA), although the specificity of these measures for pathological TDP-43 remains limited [23, 24]. Real-time quaking-induced conversion (RT-QuIC) assays that can measure aggregated TDP-43 forms are currently being developed and offer promise, akin to the use of this method to measure aggregated α-synuclein and prion proteins [25]. Several research groups are currently working to improve TDP-43 quantification in blood, CSF, and extracellular vesicles. Further, the detection of cryptic exon neoepitopes as a surrogate of loss of neuronal TDP-43 function, due to impaired splicing, has emerged as a potentially new approach to measure TDP-43 pathology. For example, elevated levels of TDP-43-dependent cryptic epitopes have been recently reported in the CSF of symptomatic and asymptomatic individuals who harbor the *C9ORF72* hexanucleotide repeat expansions (HRE) compared to controls [26]. Another approach involves the quantification of TDP-43 in blood cells using peripheral blood mononuclear cell (PBMC)-derived-induced pluripotent stem cells (iPSCs). There are currently no TDP-43 positron emission tomography (PET) ligands that are widely available for research use.

Therefore, there is an urgent need to identify and validate fluid and imaging markers that can detect, measure, and visualize the regional distribution of brain TDP-43 pathology. This will be an important area for future research.

Racial and Ethnic Differences in ADRD Fluid Biomarker Levels

Results of studies regarding racial and ethnic differences in ADRD biomarker levels, dementia risk, and degree of AD neuropathological change have been conflicting [27]. Studies have consistently reported lower CSF total tau and tau phosphorylated at threonine 181 (p-tau181) levels in African American, compared to White, individuals and these differences appear to be influenced by the *APOE4* genotype [27, 28]. Other studies have found that African American individuals have lower Aβ42 and Aβ40 levels and that these differences appear to be linked to self-reported race rather than genetic ancestry [29]. Conversely, the Study of Race to

Understand Alzheimer's Biomarkers (SORTOUT-AB) found that Black participants had higher average plasma Aβ42/Aβ40 levels at baseline compared to White participants, although longitudinal changes in plasma Aβ42/Aβ40 levels were similar between the two groups [30]. African American participants may also have lower CSF concentrations of soluble triggering receptor expressed on myeloid cells 2 (sTREM2) [31]. In other studies, Black participants were found to have a greater than twofold increased chance of a positive amyloid status [32], while other studies suggest no significant racial differences in CSF Aβ42 levels between Black and White participants [27].

As race is a psychosocial rather than a biological construct, understanding racial and ethnic differences in dementia risk or brain pathology is complex and needs to consider several factors related to disease expression including socioeconomic disparities (e.g., educational attainment), psychosocial factors, and comorbid diseases such as vascular risk factors and cardiovascular disease [27]. It is also important to address current limitations related to the underrepresentation of non-White populations in ADRD research as data from most large clinical trials or databases of ADRD are mostly derived from White participants [27]. Increasingly higher numbers of individuals from underrepresented groups have been enrolled in more recent AD clinical trials, although these numbers remain relatively low. The National Institute of Aging has initiated the Health Disparities Research Framework with the aim to contextualize findings from studies identifying racial differences in biomarker levels [33]. Further research into racial differences in the prevalence of ADRD pathology and dementia risk, and the influence of race on ADRD biomarker levels, is warranted.

Influence of Systemic Disease on Peripheral Fluid Biomarker Levels

The field of blood-based AD biomarkers has witnessed significant growth in the last few years with the hope that plasma biomarkers will offer less invasive and more widely implemented tools for screening and diagnosis. However, recent evidence suggests that blood levels of AD biomarkers may be influenced by systemic factors, including chronic kidney disease (CKD), obesity, and cardiac disease. Higher blood levels of Aβ, p-tau, neurofilament-light (NFL), and glial fibrillary acidic protein (GFAP) have all been reported in individuals with low estimated glomerular filtration rate (GFR), or high creatinine, and have been attributed to reduced renal clearance of these proteins [34–37]. Therefore, this needs to be considered when interpreting results of blood biomarker levels in individuals with CKD, as the presence of CKD may lead to false positive, or false negative, results for blood-based p-tau and Aβ measurements, respectively. Reports suggest that abnormal blood AD biomarker levels in the presence of CKD do not necessarily suggest a higher dementia risk [38]. Conversely, other studies suggest that CKD may differentially influence CSF AD biomarker levels as low GFR in these studies was found to be associated with lower CSF Aβ42 and higher p-tau181 levels in individuals with mild cognitive impairment (MCI), and to a lesser extent, those with normal cognition [39]. Obesity and acute food intake in individuals with obesity have both been associated with altered levels of blood-based AD biomarkers [40]. The underlying mechanisms for these observations remain to be fully elucidated. Importantly, it has been proposed that the use of biomarker ratios may provide more accurate assessments of blood-based biomarker levels in the presence of CKD or obesity [35, 37, 41]. Cardiac or vascular amyloid deposits may influence peripheral measures of Aβ, and cardiovascular disease burden can influence plasma tau and NFL measures [42–44]. Therefore, it will be important to consider these factors when interpreting the results of blood-based biomarkers in clinical or research settings.

Further research is warranted to understand the mechanisms by which systemic disease influences biomarker levels and to identify methods by which more accurate estimates of brain pathology may be achieved using blood samples in the presence of systemic disease.

In conclusion, ADRD biomarker research is an exciting and expanding field of research that holds great promise. Over the last decade, biomarkers have been an important tool in ADRD drug development, taking us one step closer to more effective treatments and prevention strategies for AD and other neurodegenerative disorders.

References

1. Jack CR Jr, Bennett DA, Blennow K, Carrillo MC, Dunn B, Haeberlein SB, et al. NIA-AA research framework: toward a biological definition of Alzheimer's disease. Alzheimers Dement. 2018;14(4):535–62. https://doi.org/10.1016/j.jalz.2018.02.018.
2. Tarawneh R. Biomarkers: our path towards a cure for Alzheimer disease. Biomark Insights. 2020;15:1177271920976367. https://doi.org/10.1177/1177271920976367.
3. Jack CR Jr, Andrews JS, Beach TG, Buracchio T, Dunn B, Graf A, et al. Revised criteria for diagnosis and staging of Alzheimer's disease: Alzheimer's Association Workgroup. Alzheimers Dement. 2024;20(8):5143–69. https://doi.org/10.1002/alz.13859.
4. Liao YZ, Ma J, Dou JZ. The role of TDP-43 in neurodegenerative disease. Mol Neurobiol. 2022;59(7):4223–41. https://doi.org/10.1007/s12035-022-02847-x.
5. Heyburn L, Moussa CE. TDP-43 in the spectrum of MND-FTLD pathologies. Mol Cell Neurosci. 2017;83:46–54. https://doi.org/10.1016/j.mcn.2017.07.001.
6. Neumann M, Sampathu DM, Kwong LK, Truax AC, Micsenyi MC, Chou TT, et al. Ubiquitinated TDP-43 in frontotemporal lobar degeneration and amyotrophic lateral sclerosis. Science. 2006;314(5796):130–3. https://doi.org/10.1126/science.1134108.
7. Arai T, Hasegawa M, Akiyama H, Ikeda K, Nonaka T, Mori H, et al. TDP-43 is a component of ubiquitin-positive tau-negative inclusions in frontotemporal lobar degeneration and amyotrophic lateral sclerosis. Biochem Biophys Res Commun. 2006;351(3):602–11. https://doi.org/10.1016/j.bbrc.2006.10.093.
8. Mackenzie IR, Neumann M. Molecular neuropathology of frontotemporal dementia: insights into disease mechanisms from postmortem studies. J Neurochem. 2016;138(Suppl 1):54–70. https://doi.org/10.1111/jnc.13588.

9. Mackenzie IR, Neumann M, Bigio EH, Cairns NJ, Alafuzoff I, Kril J, et al. Nomenclature for neuropathologic subtypes of frontotemporal lobar degeneration: consensus recommendations. Acta Neuropathol. 2009;117(1):15–8. https://doi.org/10.1007/s00401-008-0460-5.
10. Amador-Ortiz C, Lin WL, Ahmed Z, Personett D, Davies P, Duara R, et al. TDP-43 immunoreactivity in hippocampal sclerosis and Alzheimer's disease. Ann Neurol. 2007;61(5):435–45. https://doi.org/10.1002/ana.21154.
11. Josephs KA, Whitwell JL, Knopman DS, Hu WT, Stroh DA, Baker M, et al. Abnormal TDP-43 immunoreactivity in AD modifies clinicopathologic and radiologic phenotype. Neurology. 2008;70(19 Pt 2):1850–7. https://doi.org/10.1212/01.wnl.0000304041.09418.b1.
12. McKee AC, Stein TD, Kiernan PT, Alvarez VE. The neuropathology of chronic traumatic encephalopathy. Brain Pathol. 2015;25(3):350–64. https://doi.org/10.1111/bpa.12248.
13. Toledo JB, Abdelnour C, Weil RS, Ferreira D, Rodriguez-Porcel F, Pilotto A, et al. Dementia with Lewy bodies: impact of co-pathologies and implications for clinical trial design. Alzheimers Dement. 2023;19(1):318–32. https://doi.org/10.1002/alz.12814.
14. Toyoshima Y, Takahashi H. TDP-43 pathology in polyglutamine diseases: with reference to amyotrphic lateral sclerosis. Neuropathology. 2014;34(1):77–82. https://doi.org/10.1111/neup.12053.
15. Nelson PT, Dickson DW, Trojanowski JQ, Jack CR, Boyle PA, Arfanakis K, et al. Limbic-predominant age-related TDP-43 encephalopathy (LATE): consensus working group report. Brain. 2019;142(6):1503–27. https://doi.org/10.1093/brain/awz099.
16. Kim DD, Alghefari H, Jenkins M, Ang LC, Pasternak SH. Neuropathology of Perry Syndrome: evidence of medullary and hypothalamic involvement. Mov Disord Clin Pract. 2021;8(5):713–6. https://doi.org/10.1002/mdc3.13235.
17. Weihl CC, Temiz P, Miller SE, Watts G, Smith C, Forman M, et al. TDP-43 accumulation in inclusion body myopathy muscle suggests a common pathogenic mechanism with frontotemporal dementia. J Neurol Neurosurg Psychiatry. 2008;79(10):1186–9. https://doi.org/10.1136/jnnp.2007.131334.
18. Nascimento C, Di Lorenzo Alho AT, Bazan Conceição Amaral C, Leite REP, Nitrini R, Jacob-Filho W, et al. Prevalence of transactive response DNA-binding protein 43 (TDP-43) proteinopathy in cognitively normal older adults: systematic review and meta-analysis. Neuropathol Appl Neurobiol. 2018;44(3):286–97. https://doi.org/10.1111/nan.12430.
19. Carlos AF, Sekiya H, Koga S, Gatto RG, Casey MC, Pham NTT, et al. Clinicopathologic features of a novel star-shaped transactive response DNA-binding protein 43 (TDP-43) pathology in the oldest old. J Neuropathol Exp Neurol. 2023;83(1):36–52. https://doi.org/10.1093/jnen/nlad105.
20. Kwong LK, Irwin DJ, Walker AK, Xu Y, Riddle DM, Trojanowski JQ, et al. Novel monoclonal antibodies to normal and pathologically altered human TDP-43 proteins. Acta Neuropathol Commun. 2014;2:33. https://doi.org/10.1186/2051-5960-2-33.
21. Inukai Y, Nonaka T, Arai T, Yoshida M, Hashizume Y, Beach TG, et al. Abnormal phosphorylation of Ser409/410 of TDP-43 in FTLD-U and ALS. FEBS Lett. 2008;582(19):2899–904. https://doi.org/10.1016/j.febslet.2008.07.027.
22. Cordts I, Wachinger A, Scialo C, Lingor P, Polymenidou M, Buratti E, et al. TDP-43 proteinopathy specific biomarker development. Cells. 2023;12(4):597. https://doi.org/10.3390/cells12040597.
23. Kasai T, Kojima Y, Ohmichi T, Tatebe H, Tsuji Y, Noto YI, et al. Combined use of CSF NfL and CSF TDP-43 improves diagnostic performance in ALS. Ann Clin Transl Neurol. 2019;6(12):2489–502. https://doi.org/10.1002/acn3.50943.
24. Kojima Y, Kasai T, Noto YI, Ohmichi T, Tatebe H, Kitaoji T, et al. Amyotrophic lateral sclerosis: correlations between fluid biomarkers of NfL, TDP-43, and tau, and clinical characteristics. PLoS One. 2021;16(11):e0260323. https://doi.org/10.1371/journal.pone.0260323.
25. Scialò C, Tran TH, Salzano G, Novi G, Caponnetto C, Chiò A, et al. TDP-43 real-time quaking induced conversion reaction optimization and detection of seeding activity in CSF of amyotrophic lateral sclerosis and frontotemporal dementia patients. Brain Commun. 2020;2(2):fcaa142. https://doi.org/10.1093/braincomms/fcaa142.

26. Irwin KE, Jasin P, Braunstein KE, Sinha IR, Garret MA, Bowden KD, et al. A fluid biomarker reveals loss of TDP-43 splicing repression in presymptomatic ALS-FTD. Nat Med. 2024;30(2):382–93. https://doi.org/10.1038/s41591-023-02788-5.
27. Morris JC, Schindler SE, McCue LM, Moulder KL, Benzinger TLS, Cruchaga C, et al. Assessment of racial disparities in biomarkers for Alzheimer disease. JAMA Neurol. 2019;76(3):264–73. https://doi.org/10.1001/jamaneurol.2018.4249.
28. Garrett SL, McDaniel D, Obideen M, Trammell AR, Shaw LM, Goldstein FC, et al. Racial disparity in cerebrospinal fluid amyloid and tau biomarkers and associated cutoffs for mild cognitive impairment. JAMA Netw Open. 2019;2(12):e1917363. https://doi.org/10.1001/jamanetworkopen.2019.17363.
29. Hajjar I, Yang Z, Okafor M, Liu C, Waligorska T, Goldstein FC, et al. Association of plasma and cerebrospinal fluid Alzheimer disease biomarkers with race and the role of genetic ancestry, vascular comorbidities, and neighborhood factors. JAMA Netw Open. 2022;5(10):e2235068. https://doi.org/10.1001/jamanetworkopen.2022.35068.
30. Xiong C, Luo J, Wolk DA, Shaw LM, Roberson ED, Murchison CF, et al. Baseline levels and longitudinal changes in plasma Aβ42/40 among Black and white individuals. Nat Commun. 2024;15(1):5539. https://doi.org/10.1038/s41467-024-49859-w.
31. Schindler SE, Cruchaga C, Joseph A, McCue L, Farias FHG, Wilkins CH, et al. African Americans have differences in CSF soluble TREM2 and associated genetic variants. Neurol Genet. 2021;7(2):e571. https://doi.org/10.1212/nxg.0000000000000571.
32. Gleason CE, Zuelsdorff M, Gooding DC, Kind AJH, Johnson AL, James TT, et al. Alzheimer's disease biomarkers in Black and non-Hispanic White cohorts: a contextualized review of the evidence. Alzheimers Dement. 2022;18(8):1545–64. https://doi.org/10.1002/alz.12511.
33. National Institute of Aging. Health Disparities Framework. https://www.nia.nih.gov/research/osp/framework (2023). Accessed 2024.
34. Mielke MM, Dage JL, Frank RD, Algeciras-Schimnich A, Knopman DS, Lowe VJ, et al. Performance of plasma phosphorylated tau 181 and 217 in the community. Nat Med. 2022;28(7):1398–405. https://doi.org/10.1038/s41591-022-01822-2.
35. Pichet Binette A, Janelidze S, Cullen N, Dage JL, Bateman RJ, Zetterberg H, et al. Confounding factors of Alzheimer's disease plasma biomarkers and their impact on clinical performance. Alzheimers Dement. 2023;19(4):1403–14. https://doi.org/10.1002/alz.12787.
36. Lehmann S, Schraen-Maschke S, Vidal J-S, Delaby C, Blanc F, Paquet C, et al. Plasma phosphorylated tau 181 predicts amyloid status and conversion to dementia stage dependent on renal function. J Neurol Neurosurg Psychiatry. 2023;94(6):411. https://doi.org/10.1136/jnnp-2022-330540.
37. Syrjanen JA, Campbell MR, Algeciras-Schimnich A, Vemuri P, Graff-Radford J, Machulda MM, et al. Associations of amyloid and neurodegeneration plasma biomarkers with comorbidities. Alzheimers Dement. 2022;18(6):1128–40. https://doi.org/10.1002/alz.12466.
38. Stocker H, Beyer L, Trares K, Perna L, Rujescu D, Holleczek B, et al. Association of kidney function with development of Alzheimer disease and other dementias and dementia-related blood biomarkers. JAMA Netw Open. 2023;6(1):e2252387-e. https://doi.org/10.1001/jamanetworkopen.2022.52387.
39. Hajjar I, Neal R, Yang Z, Lah JJ. Alzheimer's disease cerebrospinal fluid biomarkers and kidney function in normal and cognitively impaired older adults. Alzheimers Dement (Amst). 2024;16(2):e12581. https://doi.org/10.1002/dad2.12581.
40. Huber H, Ashton NJ, Schieren A, Montoliu-Gaya L, Molfetta GD, Brum WS, et al. Levels of Alzheimer's disease blood biomarkers are altered after food intake—a pilot intervention study in healthy adults. Alzheimers Dement. 2023;19(12):5531–40. https://doi.org/10.1002/alz.13163.
41. Janelidze S, Barthélemy NR, He Y, Bateman RJ, Hansson O. Mitigating the associations of kidney dysfunction with blood biomarkers of Alzheimer disease by using phosphorylated tau to total tau ratios. JAMA Neurol. 2023;80(5):516–22. https://doi.org/10.1001/jamaneurol.2023.0199.

42. Kim KY, Kim E, Lee JY. Impact of amyloid and cardiometabolic risk factors on prognostic capacity of plasma neurofilament light chain for neurodegeneration. Alzheimers Res Ther. 2024;16(1):202. https://doi.org/10.1186/s13195-024-01564-y.
43. Jiang X, O'Bryant SE, Johnson LA, Rissman RA, Yaffe K. Association of cardiovascular risk factors and blood biomarkers with cognition: the HABS-HD study. Alzheimers Dement (Amst). 2023;15(1):e12394. https://doi.org/10.1002/dad2.12394.
44. Elia A, Parodi-Rullan R, Vazquez-Torres R, Carey A, Javadov S, Fossati S. Amyloid β induces cardiac dysfunction and neuro-signaling impairment in the heart of an Alzheimer's disease model. bioRxiv. 2023; https://doi.org/10.1101/2023.07.11.548558.

Index

R. Tarawneh (ed.), *Biomarkers of Neurodegenerative Disorders*,
https://doi.org/10.1007/978-3-031-80849-4

GPSR Compliance

The European Union's (EU) General Product Safety Regulation (GPSR) is a set of rules that requires consumer products to be safe and our obligations to ensure this.

If you have any concerns about our products, you can contact us on ProductSafety@springernature.com

In case Publisher is established outside the EU, the EU authorized representative is:

Springer Nature Customer Service Center GmbH
Europaplatz 3
69115 Heidelberg, Germany

Batch number: 08644011

Printed by Printforce, the Netherlands